Inferences about two measures of central tendency based on paired samples

Procedure	Assumptions	Advantages	Limitations	Section
Base inferences on the differences between observations within a pair. With these differences, use a procedure for making inferences about a single measure of central tendency (for the distribution of differences).	Random sample of pairs of observations; other assumptions depending on the procedure used (whether based on a *t* distribution, Wilcoxon signed rank distribution, or a binomial distribution)	Advantages of the procedure we select are the same as for the one-sample case.	Limitations of the procedure we select are the same as for the one-sample case.	11-6

Comparing several means in a single-factor experiment

Procedure	Assumptions	Advantages	Limitations	Section
One-way analysis of variance	Independent random samples from Gaussian distributions with the same variance	Procedure tends to be robust to deviations from the Gaussian assumption, and to small deviations from the equal-variance assumption.	Exact probabilities for inferences depend on the Gaussian assumption and the equal-variance assumption.	12-2
Kruskal–Wallis test	Independent random samples from distributions that have the same shape and variation	We do not need to assume Gaussian observations.	This procedure is not as powerful as one-way analysis of variance when that procedure is valid.	12-3

Comparing several means (or treatments) in a randomized block experiment

Procedure	Assumptions	Advantages	Limitations	Section
Parametric analysis, using analysis of variance	Observations are independent, from Gaussian distributions with equal variances; relative treatment effects are the same for each block	Inferences tend to be robust to deviations from the Gaussian assumption, and to small deviations from the equal-variance assumption.	Exact probabilities for inferences depend on all of the assumptions being satisfied.	12-4
Friedman's test	Observations are independent, from distributions with similar shape and variation; relative treatment effects are the same for each block	We do not need to assume Gaussian observations.	Procedure is not as powerful as the parametric analysis when the assumptions for the parametric analysis are satisfied.	12-5

Assessing the effects of two factors upon a response variable

Procedure	Assumptions	Advantages	Limitations	Section
Two-factor analysis of variance	Independent observations from Gaussian distributions with equal variances	Procedure tends to be robust to deviations from the Gaussian assumption, and to small deviations from the equal-variance assumption.	Exact probabilities for inferences depend on all of the assumptions being satisfied.	13-1

AN INTRODUCTION TO STATISTICS WITH DATA ANALYSIS

The Brooks/Cole Series in Statistics:

An Introduction to Statistics with Data Analysis, by Shelley Rasmussen

Modern Data Analysis: A First Course in Applied Statistics, by Lawrence Hamilton

Statistics with STATA, by Lawrence Hamilton

Statistics for Business and Economics, Oltman & Lackritz

Probability and Statistics for Engineering and the Sciences, Third Edition, Jay L. Devore

Introduction to Statistics for the Social and Behavioral Sciences, Second Edition, Christensen & Stoup

Basic Statistics: Tales of Distributions, Fourth Edition, Spatz & Johnson

Applied Statistics for Public Administration, Meier & Brudney

Regression with Graphics: A Second Course in Applied Statistics, by Lawrence Hamilton

Also available from Brooks/Cole, Wadsworth & Brooks/Cole Advanced Books & Software:

Statistics: A Guide to the Unknown, Third Edition, by Tanur, Mosteller, Kruskal, Lehmann, Link, Pieters, & Rising

Counterexamples in Probability and Statistics, by Romano & Siegel

Elements of Graphing Data, William S. Cleveland

Graphical Methods for Data Analysis, Chambers, Cleveland, Kleiner, & Tukey

Dynamic Graphics for Statistics, Cleveland & McGill

Applied Regression Analysis: A Research Tool, John O. Rawlings

The Analysis of Linear Models, Ronald R. Hocking

Matrices with Applications in Statistics, Second Edition, Franklin A. Graybill

Theory and Application of the Linear Model, Franklin A. Graybill

Statistics: Theory and Methods, Berry & Lindgren

Statistical Inference, Casella & Berger

Mathematical Statistics and Data Analysis, John Rice

Probability: Theory and Examples, Richard Durrett

Real Analysis and Probability, R. M. Dudley

Measure Theory and Probability, Adams & Guillemin

Quality Control, Robust Design, and the Taguchi Method, Khosrow Dehnad

Enhanced Simulation and Statistics Package, Lewis, Orav, & Uribe

Simulation Methodology for Statisticians, Operations Analysts, and Engineers, Lewis & Orav

AN INTRODUCTION TO STATISTICS WITH DATA ANALYSIS

Shelley Rasmussen

Brooks/Cole Publishing Company
Pacific Grove, California

Brooks/Cole Publishing Company
A Division of Wadsworth, Inc.

Printed in the United States of America

10 9 8 7 6 5 4 3 2 1

Library of Congress Cataloging-in-Publication Data
Rasmussen, Shelley, [date]
An introduction to statistics with data analysis / by Shelley Rasmussen.
p. cm.
Includes bibliographical references and index.
ISBN 0-534-13578-1 :
1. Mathematical statistics. I. Title.
QA276.R375 1991
519.5—dc20 91-9035
CIP

International Student Edition ISBN: 0-534-98585-8

Sponsoring Editor: *Michael Sugarman*
Marketing Representative: *John Moroney*
Editorial Assistant: *Lainie Giuliano*
Production Editor: *Ben Greensfelder*
Manuscript Editor: *Susan Reiland*
Permissions Editor: *Carline Haga*
Interior Design: *Vernon T. Boes*
Cover Design: *Michael A. Rogondino*
Cover Photo: *Lee Hocker*
Art Coordinator: *Cloyce Wall*
Typesetting: *G&S Typesetters, Inc.*
Cover Printing: *Lehigh Press Lithographers/Autoscreen*
Printing and Binding: *R. R. Donnelley & Sons Company*

To my family

About the Author

Shelley Rasmussen received a Ph.D. in Statistics from the University of Michigan. She has taught statistics at the Massachusetts Institute of Technology and universities within the state systems of Texas, New Hampshire, and Massachusetts. As a practicing statistician, she has worked in the pharmaceutical industry and for a cancer research center. She currently teaches and consults in statistics, quality control, and experimental design for industries involved in engineering, high technology, and new product development.

This book is intended for a one- or two-semester introduction to statistics. The discussion is not calculus-based; the only prerequisite is high school algebra.

The emphasis is on the art of statistical thinking. I believe that a course emphasizing statistical thinking about applied problems ought to be anyone's introduction to statistics, no matter what major or year in college. Everyone should understand the usefulness of statistics in addressing real-world problems. Such an understanding would enrich the lives of all students and motivate some to further study in the theory and application of statistics.

Almost all of the examples and exercises in this book are based on real data sets. In a few cases I felt forced to invent a data set to illustrate an idea, because I did not have a real example at hand. Even then I based the example on a realistic application. As a student and a teacher, I have always appreciated real examples in references. I believe that students will be more motivated to study statistics if its usefulness is immediately apparent. This will be most obvious in a book if examples and exercises illustrate the use of statistics in real investigations.

Data analysis is introduced at the beginning of the book, in Part I, and used throughout. Data analysis involves the use of simple graphical and tabular techniques to gain an understanding of the information in a data set. Regrettably, techniques of data analysis are not familiar to many college graduates, not to mention high school graduates. At a recent multidisciplinary workshop for a select group of exceptional high school teachers (funded by the National Science Foundation and run by the Tsongas Industrial History Center in Lowell, Massachusetts), one social studies teacher did not understand why we would ever want to graph data and several others said they always skipped graphs in textbooks. My response was that they were missing the opportunity to help their students to understand the many graphical presentations of data, some good and some bad, that appear daily in the media.

In data analysis and formal statistical analysis, the more carefully a data set is collected, the more useful information can be derived from it. When we use the ideas of experimental design, we plan a study in order to address the questions of interest as efficiently as possible. A well-designed study often needs very little formal statistical analysis. A poorly designed study may yield little useful information no matter how much we massage the data. The importance of data collection and experimental design is emphasized throughout the book.

In formal statistical analysis, we use a sample of data to make inferences about a larger population. These inferences take the form of probability statements about the population, based on what we see in the sample. (We have to make certain assumptions about the sample in order for these probability statements to make sense; a good experimental design helps to assure the validity of some of these assumptions.) Since probability statements form the basis of formal statistical inference, we have to discuss some probability. I have kept this discussion to a minimum. Part II contains the essential concepts in probability that we need for statistical inference. Two optional sections, Sections 6-4 and 6-6, contain interesting applications of probability that are not used again later. The reader who does not want to cover the median test (Section 11-5) or Fisher's exact test (Section 16-5) can skip the discussion of the hypergeometric probability distributions in Section 7-3.

A number of topics and techniques of formal statistical inference are presented in Part III. Classical analysis that depends on the assumption of Gaussian (normal) data is discussed for each appropriate application. In addition, for many applications I have included one and sometimes two alternatives to the classical analysis. Section 10-4, for instance, discusses nonparametric inferences about a population mean or median, based on ranks; Section 10-5 covers inferences about a population median based on signs; Section 14-4 discusses robust inferences about two or more variances. I think it is important for students to realize that not all data sets follow a Gaussian distribution and that there are straightforward alternatives to the classical analysis for many applications. Readers who want to consider only classical analyses, however, may skip the sections on alternative approaches without loss of continuity.

Many students and friends helped me by providing data sets, reviews, suggestions, and encouragement during the writing of this book. Among them are Paul Catalano, Dennie Clarke-Hundley, Paul Gavelis, Janet LaBonte, Nicole LaVallee, Mary Lundquist, Alex Olsen, Michele Walsh, and Penny Angus Yepez. Miin-Show Chao helped with a number of computer runs. Lee Panas contributed data sets, reviewed chapters, provided useful advice, and solved all the exercises for the solutions manual.

I appreciate the contributions of the many reviewers who patiently read the various versions of the manuscript, each version better than the previous one in large part because of their comments and advice. These reviewers include: Dr. Richard Alo, University of Houston; Professor David Banks, Carnegie-Mellon University; Dr. Lynne Billard, University of Georgia; Dr. Bill Korin, The American University; Professor Robert Lacher, South Dakota State University; Professor Ed Landauer, Clackamas Community College; Ms. Mary Parker, Austin Community College; Professor Robert Schaefer, Miami University; Professor Paul Speckman, University of Missouri; Professor Jeff Spielman, Roanoke College; Professor George Terrell, Virginia Polytechnic Institute; and Dr. Cindy van Es, Cornell University. I am grateful to the developmental editor, John Bergez, whose detailed criticisms of my organization and writing style, though punishing, helped me with revisions that made the book much easier to read. I am also grateful to copy editor Susan Reiland for her careful reading of the manuscript and many helpful suggestions.

My thanks to all the investigators cited in the references, from whose work I benefited. My apologies to anyone who should have been cited and was not. My thanks also to all of the teachers, colleagues, and students from whom I have learned probability and statistics, especially Michael Woodroofe of the University of Michigan.

I am grateful most of all to my family for all of their love and support. My mother, Jackie Guernsey; my sister, Pam Grant; my brothers, John Rasmussen and Bill Guernsey; and my mother-in-law, Mary Olsen, provided much encouragement. My children, James, Emily, and Vin were extremely patient and supportive. My husband, Dick Olsen, deserves the most gratitude, for encouraging me to start this project, for providing advice and inspiration, and for giving me the time (the children were all preschoolers when I started) to do it. Thank you very much!

Shelley Rasmussen

CONTENTS

PART ONE DATA ANALYSIS

CHAPTER 1 Introduction 1

CHAPTER 2 Studying One Variable at a Time: Lists, Tables, and Plots 27

PART TWO PROBABILITY

PART THREE
STATISTICAL INFERENCE

CHAPTER 1

Introduction

IN THIS CHAPTER

Statistics

Data analysis

Case, variable, data value

Quantitative and qualitative variables

Unit of measurement

Missing value

Statistics are numbers. Statisticians use numbers (or statistics) to expand our knowledge of the universe, if only a very small part of the universe. We are all statisticians when we use numbers in this way. This book is about such use of numbers. It is not intended as a comprehensive manual of statistical techniques, but rather as an introduction to the art of statistical thinking.

By a **statistic** we mean either a number—a numerical piece of information or datum—or a number calculated from a set of data values.

When practicing the *art of statistics,* we use numerical information to increase our knowledge in some way. Used in this sense, statistics refers to the branch of mathematics dealing with theory and techniques of collecting, organizing, and interpreting numerical information.

By **statistics** we mean either a collection of numerical information, or the branch of mathematics dealing with theory and techniques of collecting, organizing, and interpreting numerical information.

We may use information from a market analysis to select cities for introducing a new product. Or, we might study racing forms to decide how to place a bet in the next horse race. Perhaps we want to examine individual or team performance in major-league baseball. In each of these cases, we study a collection of information, called a *data set.*

A **data set** is a collection of information.

When we try to make sense of a data set, we are engaging in *data analysis.*

By **data analysis** we mean making sense of a data set.

Baseball is extremely conducive to data analysis, since baseball statistics are readily available by player and by team. A baseball fan might study individual variables such as batting average: What is a typical batting average for a player in the major leagues? What is an exceptionally good (or poor) batting average? The fan might also examine relationships between variables: What is the relationship between team batting average and winning percentage? Is this relationship different for the American League than for the National League?

Data analysis involves studying variables and relationships between variables in a collection of information. Often we want to do more. We may want to use a sample of information to learn about a larger population. For instance, we might want to use a sample of the thousands of parts produced in a day to decide whether too much gold is being electroplated onto components used in personal computer hardware. Or, we may want to conduct a taste test of two products in a sample of consumers to make decisions regarding product preference in a larger group of consumers. We might want to compare a new treatment with a standard treatment in patients with a particular form of cancer. In each of these cases, it is impractical to study the entire population (parts electroplated in a day, consumers in a product market, or cancer patients). Instead, we look at a sample or subset of the population. We use the information from the sample to learn about the population. This is *statistical inference.*

By **statistical inference** we mean drawing conclusions about a population based on a sample from that population.

The **population** is the group or collection of interest to us.

A **sample** is a subset of the population. We use the observations in the sample to learn about the population.

Data analysis can aid in statistical inference. Medical researchers routinely study characteristics of patients with a particular form of cancer. They look for relationships among such variables as age, sex, stage of illness, response to treatment, and survival.

Estimation is a part of statistical inference. Investigators might use average survival time for patients in a sample to estimate average survival time for all patients in the population. They might then calculate a range of reasonable values for this average survival time. Interpreting such a range of reasonable values, called a confidence interval, depends on ideas in probability.

Statistical inference also involves hypothesis testing. In testing hypotheses, we compare two statements about the state of nature, such as:

Average survival with the new treatment is the same as for the standard treatment.
Average survival with the new treatment is longer than for the standard treatment.

Which of these two statements does the sample support? To decide, we use ideas in probability.

Both estimation and hypothesis testing use probability. We make probability statements about the population based on what we see in the sample. For these statements to make sense, the sample must be similar to the population, a *representative sample.* Suppose the cancer patients in a sample all have very advanced disease. Then researchers cannot make inferences about a larger population that includes patients with less advanced disease. This leads to the idea of experimental design. We want to collect a sample, or carry out an experiment, so that statistical inferences make sense.

1-1 An Overview of the Book

Our study of statistics begins with data analysis. Though the techniques are fairly simple, they can provide a lot of insight into a collection of information.

Some data analysis tools are tabular. A table can summarize certain types of information in a data set. For instance, we might use a table, called a frequency table, to display the number of baseball players with 1991 salaries in each of several intervals (say, less than $500,000, $500,000 to $1,000,000, and so on). Other tools of data analysis are graphical. A histogram, sometimes called a bar graph, is a graphical tool for displaying the information in a frequency table. We could use such a graph to display the information on numbers of players per salary range, instead of listing these numbers in a table.

Chapter 2 discusses tabular and graphical techniques for studying one variable at a time.

Certain information about a variable in a data set can be summarized with a *descriptive statistic.*

> A **descriptive statistic** is a number used to summarize information in a set of data values.

For instance, the average is a summary measure that helps to describe the location or center of a set of values. A baseball player's batting average is a measure of his average performance at bat over one or more seasons. The range, or difference between the largest and smallest values, is a summary measure of the variation or spread in a set of values. Saying that baseball player salaries span a range of $5,000 conveys a very different idea of variation than saying that their salaries span a range of $5,000,000! Chapter 3 covers descriptive statistics as measures of location and measures of variation for a single variable.

We can use tabular and graphical tools to examine relationships between two variables, as we will see in Chapter 4. A scatterplot, for instance, is a graphical tool that could be used to display the relationship between player salary and batting average for the 1991 season. In Chapter 5, we extend these ideas to studying more than two variables at a time. Perhaps we want to look at the relationship between baseball player salary and batting average by playing position, or separately for the American League and the National League.

Data analysis techniques can be applied in a wide variety of situations, including most problems involving statistical inference. We make extensive use throughout the text of the data analysis tools we discuss here in Part I.

After data analysis comes probability, in Part II. Statistical inference involves making some assumptions about the sample observations in order to build a probability model. We then use this probability model to make probability statements about the population.

Chapter 6 provides some background information on probability that allows us to build the probability models we need for statistical inference. This chapter also includes some topics in probability of interest for their own sake: odds ratios, used in public health and gambling; and conditional probabilities, used in assessing the usefulness of medical screening procedures.

Some important probability models for statistical inference are based on counting techniques. Chapter 7 discusses some of these counting techniques and introduces two sets of probability models derived from these techniques: the binomial and hypergeometric models. Gaussian models, the basis for most classical statistical inference, are the subject of Chapter 8.

After probability, we move to statistical inference in Part III. The reasoning involved in statistical inference is more formal than in data analysis. Through definitions and examples, Chapter 9 introduces the traditional concepts of statistical inference, including estimation, confidence intervals, and hypothesis testing. This chapter also emphasizes ideas in experimental design. A well-designed experiment yields much valuable information. A poorly designed experiment can result in no information of value.

Often we are interested in making inferences about one or more averages. Does the average gold thickness on electroplated components exceed the target value? Do athletes run faster on average when competing against themselves or against a rival? Which of several treatments is associated with longest survival time, on average? Chapters 10–13 deal with inferences about averages (also called means or measures of central tendency).

Sometimes questions about the variation in a population are at least as interesting as questions about averages. We may know, for instance, that each of several production processes puts the same amount of dog food in 25-pound bags, *on average*. That is, over many bags filled, the average fill weight is very close to 25 pounds per bag. Our concern is with the variation in the processes. A process that results in a weight close to 25 pounds for each bag is preferable to a process with great variation in bag weights, even though both processes have the same *average* fill weight. We say the first process has less variation (a smaller variance) than the second. Chapter 14 discusses inferences about variances.

Often we want to study relationships between variables. What is the relationship between years in the major leagues and baseball player salary? Is this relationship roughly linear? If so, can we assess the extent of the linear association? If not, can we describe the relationship between the two variables in some other way? Such questions are addressed in Chapter 15.

Sometimes variables have categories rather than numerical values (team, league, and playing position in baseball are examples). Inferences involving such categorical variables are the subject of Chapter 16.

1–2

Data Analysis and the World Bank Data Set

Some fairly rigid requirements must be met for formal procedures in statistical inference to be used correctly. Many data sets do not meet these requirements, but they may nevertheless provide valuable information for learning or decision-making. We can use the techniques of data analysis to study such a data set. These techniques are tools, not ends in themselves. We will emphasize flexibility by showing different uses and formats for a number of techniques. The goal is not to define every possible technique, but to convey the spirit of data analysis. Since no two collections of data are the same, there is no rigid procedure for analysis. Rather, we approach a data set, armed with some techniques and a desire to explore.

It is easy to come out of an introductory statistics course without any idea of what to do with a real data set, because data analysis techniques are often defined as separate entities and illustrated with different examples. They are not presented as a package, each technique with its use in examining a single collection of data. We will try to avoid this dilemma by using a single example, based on a World Bank data set, throughout the next four chapters. For variety, the exercises cover problems from many fields of application.

Our World Bank data set is a collection of indicators of social and economic development for 128 countries with populations of 1 million or more

(World Bank, 1987). We will study these indicators to address such questions as: What is the range of economic development over these World Bank countries? What is the relationship between level of economic development and indicators of quality of life such as calorie supply, percentage of school-age children attending school, and life expectancy? How do birth rates and fertility rates relate to level of education among females and extent of contraception use?

The countries in our World Bank data set are shown on the map in Figure 1-1. The indicators are defined in the first appendix at the end of this chapter. The 128 countries are listed by economic category in Table 1-1. Economic designations, political boundaries, and country names change over time. We are using designations defined by the World Bank in 1987.

Thirty-seven countries are classified as low-income developing nations. Middle-income developing nations include 36 lower-middle-income countries and 23 upper-middle-income countries. Four nations are listed as high-income oil exporters and 19 as industrial market countries. There are 9 countries classified as nonmember nations because of lack of reliable economic information.

The indicators compiled by the World Bank constitute a *data set,* a collection of information. A *case* is an individual sampling unit, the subject of measurement.

A **case** is the individual sampling unit in a data set; it is the basic unit sampled and subject of measurement.

In the World Bank data set, each country is a case because a country is the basic unit for which information is recorded. In a medical study comparing two treatments for lung cancer, case refers to an individual patient participating in the study. If baseball statistics are compiled by team, a team is a case. If statistics are compiled by player, a player is a case. In a quality control setting, a case is an individual object subjected to testing.

A *variable* is a particular piece of information recorded for a case. Many variables may be available for each case. Variables in the World Bank data set include the name of the country, economic category, gross national product per capita, life expectancy, and birth rate. A *data value* is the value a variable takes on for a particular case. For example, one data value for life expectancy is 76 years, the number recorded for the United States.

Variable refers to a particular piece of information recorded for a case.

A **data value** is the value a variable takes on for a particular case.

A *quantitative variable* has numerical values that are measurements or counts. Examples of quantitative variables are life expectancy, population size, number of cities with over 500,000 people, calorie supply per capita, primary school enrollment, birth rate, and percentage of contraception use.

A **quantitative variable** has numerical values that are measurements or counts.

A *qualitative variable,* or *categorical variable,* has values with no intrinsic meaning as numbers. The name of a country is a qualitative variable, as is region of the world. A country is in North America, Africa, Western Europe, and so on. These values are categories, with no numerical meaning.

> A **qualitative,** or **categorical, variable** has values that cannot be interpreted as numbers.

The World Bank classifies countries by whether they are oil exporters, exporters of manufactured goods, or highly indebted. Each of these variables

TABLE 1-1 Listing of 128 countries by economic category. Listings are alphabetical within categories.

Low-income countries

Afghanistan, Bangladesh, Benin, Bhutan, Burkina Faso, Burma, Burundi, Central African Republic, Chad, China, Ethiopia, Ghana, Guinea, Haiti, India, Kampuchea, Kenya, Lao PDR, Madagascar, Malawi, Mali, Mozambique, Nepal, Niger, Pakistan, Rwanda, Senegal, Sierra Leone, Somalia, Sri Lanka, Sudan, Tanzania, Togo, Uganda, Viet Nam, Zaire, Zambia

Lower-middle-income countries

Bolivia, Botswana, Cameroon, Chile, Colombia, Congo, Costa Rica, Côte d'Ivoire, Dominican Republic, Ecuador, Egypt, El Salvador, Guatemala, Honduras, Indonesia, Jamaica, Jordan, Lebanon, Lesotho, Liberia, Mauritania, Mauritius, Morocco, Nicaragua, Nigeria, Papua New Guinea, Paraguay, Peru, Philippines, Syria, Thailand, Tunisia, Turkey, Yemen, Arab Republic, Yemen, PDR, Zimbabwe

Upper-middle-income countries

Algeria, Argentina, Brazil, Greece, Hong Kong, Hungary, Iran, Iraq, Israel, Korea, South, Malaysia, Mexico, Oman, Panama, Poland, Portugal, Romania, Singapore, South Africa, Trinidad and Tobago, Uruguay, Venezuela, Yugoslavia

High-income oil exporters

Kuwait, Libya, Saudi Arabia, United Arab Emirates

Industrial market countries

Australia, Austria, Belgium, Canada, Denmark, Finland, France, Germany, West, Ireland, Italy, Japan, Netherlands, New Zealand, Norway, Spain, Sweden, Switzerland, United Kingdom, United States

Nonmember countries

Albania, Angola, Bulgaria, Cuba, Czechoslovakia, Germany, East, Korea, North, Mongolia, USSR

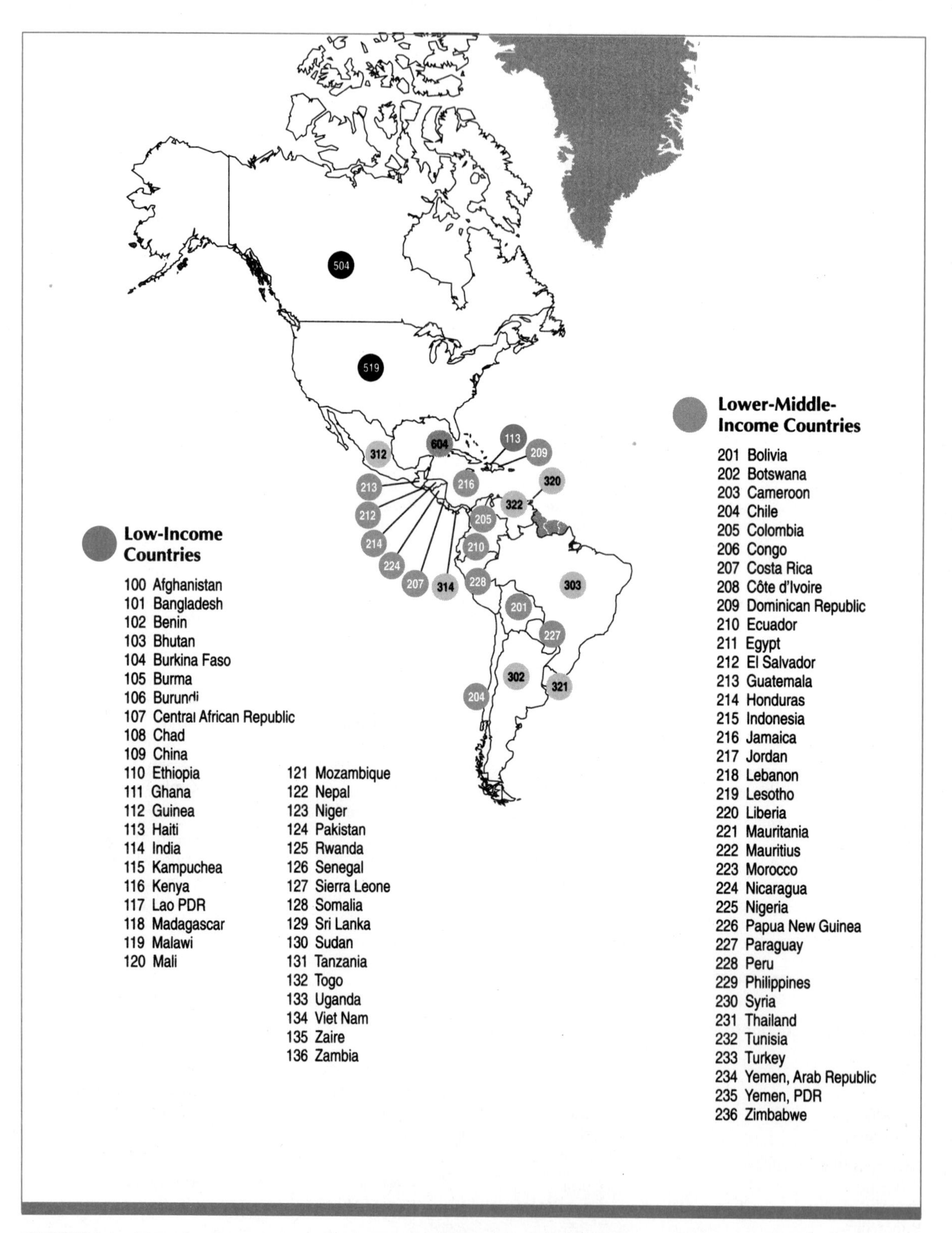

FIGURE 1-1 Map showing name and economic category for 128 countries with populations of 1 million or more as identified in the *World Development Report* 1987 (World Bank, 1987)

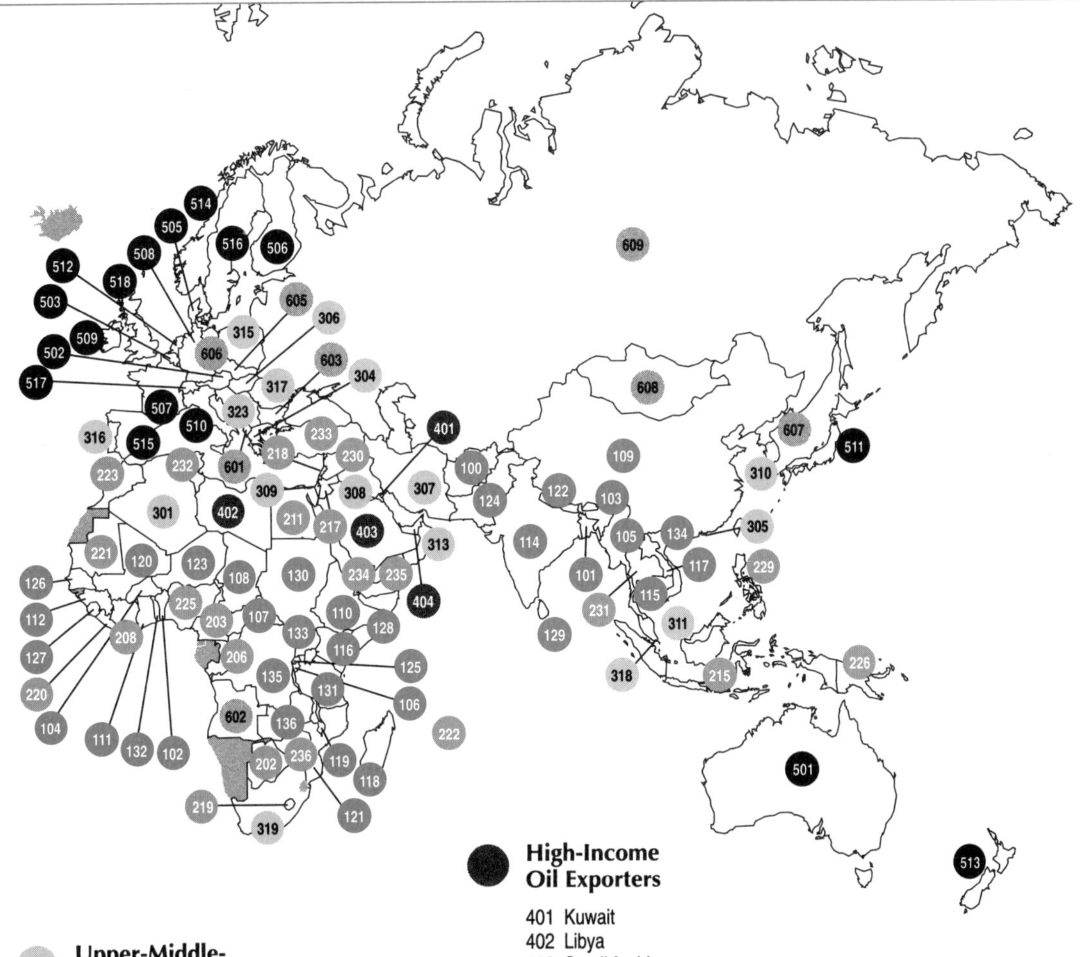

High-Income Oil Exporters

401 Kuwait
402 Libya
403 Saudi Arabia
404 United Arab Emirates

Upper-Middle-Income Countries

301 Algeria
302 Argentina
303 Brazil
304 Greece
305 Hong Kong
306 Hungary
307 Iran
308 Iraq
309 Israel
310 Korea, South
311 Malaysia
312 Mexico
313 Oman
314 Panama
315 Poland
316 Portugal
317 Romania
318 Singapore
319 South Africa
320 Trinidad & Tobago
321 Uruguay
322 Venezuela
323 Yugoslavia

Industrial Market Countries

501 Australia
502 Austria
503 Belgium
504 Canada
505 Denmark
506 Finland
507 France
508 Germany, West
509 Ireland
510 Italy
511 Japan
512 Netherlands
513 New Zealand
514 Norway
515 Spain
516 Sweden
517 Switzerland
518 United Kingdom
519 United States

Nonmember Countries

601 Albania
602 Angola
603 Bulgaria
604 Cuba
605 Czechoslovakia
606 Germany, East
607 Korea, North
608 Mongolia
609 USSR

has two possible values, yes and no, which can be represented by numbers such as 1 and 0. Since the numbers merely represent categories, these are qualitative variables.

Sometimes a qualitative variable has ordered categories. Four economic categories are ordered by level of economic development: low income, lower-middle income, upper-middle income, industrial market. A country might be classified by birth rate as low, moderate, or high. Such categories are ordered, but without exact numerical meaning. Such a variable is often called an *ordinal qualitative variable.*

An **ordinal qualitative variable** has categories with a natural ordering, but not exact numerical values.

Unit of measurement is the basic unit used for measuring and recording the value of a variable. Life expectancy for the United States is 76 years; year is the unit of measurement for life expectancy. Per capita gross national product is reported in U. S. dollars; U. S. dollar is the unit of measurement for gross national product. Units of measurement should always be included in reports of numerical information.

The **unit of measurement** of a variable is the basic unit (such as inches, years, dollars) used to measure and record the values of the variable.

With this introduction to the World Bank data set and some terminology, we can talk about how to get started in data analysis.

1-3 Questions to Ask Before Starting Data Analysis

We should ask some questions before starting any data analysis. The answers will help us decide whether an analysis is worthwhile. They will also help to guide us as we begin. Here we address some of these questions in terms of the World Bank data set.

What are the goals of our analysis? What are we trying to learn? There are times when we know very little about the variables and relationships between the variables. Then we might state a vague goal such as: Learn about the variables and about relationships between these variables. It is important to be more specific when possible. We can be more specific in describing the goals of our analysis of the World Bank indicators.

First, we want to study each variable separately, looking at characteristics of the variable for the entire group of countries. Consider the variable life expectancy: How do life expectancies vary across countries? Are the life expectancies clustered around some center value? Are there two or more clusters of countries with similar life expectancies? Are life expectancies scattered across a wide range?

Second, we want to study each variable separately within economic categories. We also want to compare results across economic categories. For life expectancy: How do life expectancies vary across the low-income countries? How do life expectancies for the low-income countries differ from life expectancies among the industrial market countries?

Finally, we want to look at relationships between variables, for all countries and within economic categories. Is there any relationship between female primary school enrollment and birth rates? We want to answer such questions overall and within economic categories.

Can we use the data set to meet the goals of our analysis? This is a question too often overlooked by investigators and data analysts. Many times, the variables recorded cannot legitimately meet the goals of the investigation. Or the data are so poorly collected or full of errors that no analysis is worthwhile. We should always carefully consider what questions our collection of data can really answer, setting limits on interpretations and reported results.

There are many limitations of our data, some clearly stated in the *World Development Report* 1987 (World Bank, 1987). Many indicators are based on information supplied by individual countries. Methods and quality of collecting and reporting vary greatly across countries. Some deliberate misreporting may occur. These problems make comparisons across countries hard to interpret.

Even under the best conditions, some indicators may be difficult to determine with accuracy. It cannot be easy, for example, to estimate the percentage of married women of childbearing age who use contraception.

Our indicators are countrywide averages. Conditions can vary widely within a country. Life expectancy, for example, is reported as 56 years for India and 76 years for the United States. This suggests that lifespans are shorter in India than in the United States. However, there are well-to-do people in India with advantages similar to those available to well-off Americans; life expectancies for these people are probably much longer than 56 years. Likewise, there are Americans who do not have access to these advantages; their life expectancies may be much less than 76 years. Thus, even when we trust the quality of our information, we have to be careful when making interpretations.

Recognizing the limitations of our data, we can set limits on our interpretations of analyses. We will use the indicators to compare countries in a general way. But we will not attach great importance to exact differences. We will also recognize that conditions within countries may vary greatly.

We must also be cautious because there are *missing values* for many variables in the World Bank data set. The worst case is for percentage of married women of childbearing age using contraception in 1984. Thirty-three, or about one-fourth, of the countries have no value recorded for this variable. Three countries—Afghanistan, Kampuchea (Cambodia), and Lebanon—have missing values for most of the variables. We must recognize that any results we obtain apply only to countries with available information. When a variable has many missing values, comparisons across economic categories will be even more suspect and interpretations more tentative than they would be otherwise.

A case has a **missing value** for a variable if no information on the variable is available for that case.

Each of the World Bank indicators corresponds to a single year or short time period. Therefore, we can make no analyses of possible time trends in the variables. All of our results will apply only to the year or time period for which the variables are recorded.

How do we get a data set ready for analysis? Data must be collected and compiled before we can analyze it. Each variable should be measured and recorded correctly. We must avoid mistakes in transcribing information from data collection forms to a computer file. All this care involves proofreading and error checking. Although tedious, proofreading and error checking are important in statistics and data analysis. Analysis may be worthless if there are mistakes in the data.

Since our data come from tables compiled by the World Bank, error checking in collection and compilation is out of our control; all we can do is be aware of possible inadequacies. We should be careful with transcribing, as from tables to computer files, from computer printouts to reports. We will also be cautious when interpreting the results of our analysis.

Some Suggested Reading and a Caution

In the next four chapters, we will use some simple data analysis techniques to study the World Bank data. The chapters describe and report the results of some analyses; other analyses are left to the exercises. Hopefully, these chapters not only introduce some useful concepts in data analysis, but also describe some interesting aspects of the World Bank data we are exploring. For more on using data analysis to learn about a data set, see *Exploratory Data Analysis* (Tukey, 1977); *Data Analysis and Regression* (Mosteller and Tukey, 1977); *Applications, Basics and Computing of Exploratory Data Analysis* (Velleman and Hoaglin, 1981); and *Understanding Robust and Exploratory Data Analysis* (Hoaglin, Mosteller, and Tukey, editors, 1983).

The World Bank data set provides an extended example for use in Chapters 2–5. We look at some World Bank indicators again in Section 15-1 when we consider the correlation coefficient as a descriptive statistic measuring linear association between two quantitative variables. You may notice that we do not mention the World Bank data set anywhere else in Part III, when we discuss statistical inference. The reason is this: Formal statistical inference involves using a carefully selected sample to learn about a larger population. The World Bank data set contains information *on an entire population* of countries: 128 nations with 1 million or more inhabitants. Data analysis is very appropriate for this collection of World Bank indicators, but statistical inference is not.

1–4 Using a Computer Statistical Software Package: Minitab

You might feel overwhelmed by the idea of studying the World Bank data set. The collection includes many indicators measured on some or all of 128 countries. Counting, tabulating, calculating, and graphing by hand would be a time-consuming task for such a large data set. Fortunately, many computer programs in the form of statistical software packages are available for use on the computer to help us with such an analysis.

As an example of the use of a statistical package, we include in this text a discussion of the use of a particular package, called Minitab. This discussion is intended as an introduction to using a computer for statistical analysis. Minitab is useful for such an introduction because it is relatively simple to use. Also, Minitab is a widely used package that has been the subject of much error checking. Since errors can arise in statistical computing, we can have more confidence in a program whose developers check for errors and make changes in the program when necessary. A major drawback of using Minitab in this presentation is that it is not widely used for statistical analysis outside of schools, partly because its graphical and analysis capabilities are not as extensive as those of some other very good statistical packages. However, once you have learned to use one statistical package, you will probably find that it is not too hard to master others.

Most chapters in this book have a Minitab appendix that illustrates the use of Minitab commands for statistical procedures described in the chapter. A glossary at the end of the book provides a short definition of each command we cover in these appendices. If you do not want to worry about statistical packages at first reading, you can skip all the Minitab appendices; the chapter discussions do not depend on any of this material.

The Minitab appendix at the end of Chapter 1 describes some of the Minitab commands for reading, writing, and editing data files.

Summary of Chapter 1

Data analysis involves the use of tables, graphs, and descriptive statistics to study variables and relationships between variables in a data set. It is extremely important to formulate as precisely as possible what it is we are trying to learn from a given data set, stating goals and recognizing limitations.

Statistical inference involves more formal reasoning than data analysis. We use a carefully selected sample to make probability statements about a larger population. The validity of these statements depends on certain assumptions being met. Careful experimental design allows us to collect samples of observations that allow valid inferences about the population sampled.

Appendix to Chapter 1: The World Bank Indicators

The data come from the World Development Indicators section of the *World Development Report* 1987 (World Bank, 1987). Unless stated otherwise, variables are for 1985.

Birth rate: Birth rate is the number of live births per 1,000 population. Units are live births per 1,000 population.

Calorie supply: Daily calorie supply per capita is the calorie equivalent of a country's food supply in 1985 divided by its population size. Units are calories per person.

Child death rate: Child death rate is the estimated number of deaths of 1- to 4-year-old children per 1,000 children in this age group. Units are deaths of 1- to 4-year-olds per 1,000 1- to 4-year-old children.

Contraception use: Percentage of married women of childbearing age using contraception in 1984 covers women who are practicing contraception and women whose husbands are practicing contraception. Childbearing age is generally defined as 15–44 years, although some countries use intervals 18–44, 15–49, or 19–49. Unit of measurement is percent.

Death rate: Death rate is the number of deaths per 1,000 population. Units are deaths per 1,000 population.

Fertility rate: Total fertility rate is the number of children born to a woman if she lives to the end of her childbearing years and if the number of children she bears at each age corresponds to the current age-specific fertility rate for her country. Fertility rate estimates the average number of children born per woman in a country. Units are children per woman.

Gross domestic product: Gross domestic product, reported in millions of U. S. dollars, is an estimate of the total market value of all goods and services produced by a country.

Gross national product: Per capita gross national product, reported in U. S. dollars, is an estimate of the average market value of all goods and services produced per person, with an adjustment for income from abroad. The method of calculating per capita gross national product is described in the *World Development Report* 1987, with emphasis on difficulties in comparing values of this indicator across countries.

Higher education enrollment: Number enrolled in higher education as percentage of age group is the number of people enrolled in higher education divided by the number of 20- to 24-year-olds in 1984, times 100. Unit of measurement is percent.

Infant mortality rate: Infant mortality rate is the estimated number of infants who die during their first year of life, per 1,000 live births. The units are deaths of infants in their first year per 1,000 live births.

Life expectancy: Life expectancy at birth is the estimated number of years a baby would be expected to live if mortality patterns prevailing in 1985 continued throughout his or her life. Unit of measurement is year. Male life expectancy is the estimate for males only; female life expectancy is the estimate for females only.

Number of cities of over 500,000 *people in* 1980*:* The unit of measurement is city.

Population growth: Percentage average annual population growth for 1980–1985 is calculated from mid-year population totals. Unit of measurement is percent.

Population size: Population size is the estimated total population of a country in the middle of 1985. Refugees who are not permanently settled are credited to their country of origin in the population counts. Population totals are given in millions of people.

Primary school enrollment: Number enrolled in primary school as percentage of age group is the number of people enrolled in primary school divided by the number of 6- to 11-year-olds in 1984, times 100. Countries differ in ages and duration of primary schooling. It is possible for this variable to have a value greater than 100% if some children in primary school are above or below the standard primary school age. Unit of measurement is percent. Male primary school enrollment is estimated for males only; female primary school enrolіment is estimated for females only.

Minitab® Appendix for Chapter 1

Minitab is a statistical package, a computer program that performs a number of statistical procedures. Many schools have a version of Minitab on their mainframe computers. Minitab is also available for use on personal computers (the IBM PC, XT, AT series, or compatibles; and the IBM Personal System/2, with MS-DOS or IBM PC-DOS). At this writing, the Student Edition of Minitab for the personal computer costs about $50 and is available from Addison-Wesley Publishing Company, Route 128, Reading, Massachusetts 01867.

For a complete listing of Minitab commands and discussion of Minitab capabilities, consult the manual for the student edition of Minitab (Schaefer and Anderson, 1989) or the Minitab handbook (Ryan, Joiner, and Ryan, 1985). The commands are the same for all the versions of Minitab, both mainframe and personal computer.

Our discussion here assumes that you have accessed the Minitab program. Whether you are using Minitab on a mainframe or a personal computer, you must learn from your instructor, computing center, or manual how to load Minitab. Once Minitab is loaded, the computer will give you a special prompt, such as:

```
MTB>
```

When this special prompt appears, the Minitab program is waiting for your instructions. We will describe some of the Minitab commands for reading, writing, and editing data files.

The examples in this appendix use the data in Exercise 5-35 at the end of Chapter 5. As explained in that exercise, the United Nations ranks countries according to mortality rate for children under 5 years of age (Grant, 1987). The 20 countries with the highest under-5 mortality rate and the 20 countries with the lowest under-5 mortality rate are listed, along with values for four variables: under-5 mortality rate or child mortality rate (MORT), life expectancy (LIFE), percentage of infants born weighing less than 2,500 grams (WEIGHT), and daily per capita calorie supply as percentage of requirements (CALORIE). A missing value is denoted in Exercise 5-35 by −9.

Creating a Data File or Worksheet

Suppose we want to create a computer data file containing these four variables, along with an identification (ID) code for each of the 40 countries. A data file is an array of numbers that looks similar to the list in Exercise 5-35.

Some versions of Minitab and other statistical packages allow alphabetic values for variables. We will use numerical values, appropriate for any statistical package, throughout this discussion. So, instead of country names in this example, we will use ID codes 1−40.

Each case (in this example, each country) takes up a row in the array. We will have five columns, for ID, MORT, LIFE, WEIGHT, and CALORIE values. The missing value code in Minitab is an asterisk, so we will type * instead of −9 for missing values. The first four lines in the data file, corresponding to Afghanistan, Mali, Sierra Leone, and Malawi, will look like this:

```
1  329  38  20    *
2  302  43  13   68
3  302  35  14   91
4  275  46  10   95
```

The last four lines in the data file, for Japan, Switzerland, Finland, and Sweden, will look like this:

```
37    9  77   5  113
38    9  76   5  129
39    8  74   4  114
40    8  76   4  116
```

There are two ways to create this data file in Minitab (in what is called a worksheet). We can enter the data by rows, using the READ command, or we can enter the data by columns, using the SET command.

Entering Data with the READ Command

To enter the values row-wise into five columns, we use the READ command. When the Minitab prompt (we will call it MTB>) appears, we type **READ C1, C2, C3, C4, C5** or, more briefly, **READ C1-C5**. (Note that columns need not be entered in sequential order. We could, for instance, type **READ C10 C5 C1-C3**.) We can use either capital letters or lowercase letters in all commands. In the illustrations, boldface denotes the commands and numbers that we type. We type the READ command after the prompt like this:

```
MTB> read c1-c5
```

C1 stands for column 1 in the data file (or worksheet), C2 for column 2, and so on. Minitab will respond with a data prompt (such as DATA>) and accept five numbers per row. Minitab will continue to accept rows of numbers until we type the word **END**. For our example, we type 40 rows of numbers, and then type **END**. After each typed line, we push the ENTER (or RETURN) key, telling Minitab to process that line.

```
MTB>    read c1-c5
DATA>   1   329  38  20   *
DATA>   2   302  43  13   68
  .               .
  .               .
  .               .
DATA>   39    8  74   4  114
DATA>   40    8  76   4  116
DATA>   end
```

The dots indicate that we have typed all 40 lines of values, even though they are not all shown here. We leave at least one space between separate data values.

Using "&" to Indicate More Than One Line per Row of Values

If we want to use more than one line to enter a row of values, we can use the ampersand (&) to indicate this to Minitab. For instance, using two lines to enter the first row of our data set would look like this:

```
MTB>    read c1-c5
DATA>   1  329  38  &
CONT>   20  *
DATA>
```

The ampersand is useful when we have many variables (columns) per case.

Entering Data with the SET Command

To enter the values column-wise into 40 rows, we use the SET command, one column (variable) at a time. To enter the 40 ID numbers into column 1 of the worksheet, we type **SET C1** after the prompt:

```
MTB>   set c1
```

Minitab will respond with a data prompt, accepting numbers until we type **END**:

```
MTB>    set c1
DATA>   1  2  3  4  5  6  7  8  9  10  11  12  13  14  15  16  17 18  19  20  21  22
DATA>   23  24  25  26  27  28  29  30  31  32  33  34  35  36  37  38  39  40
DATA>   end
```

To enter the values for child mortality, we type **SET C2**. After the DATA> prompt, we type the 40 values for MORT and then type **END**. We follow a similar procedure for the other columns (or variables).

Displaying Data with the PRINT Command

After entering our file, we may want to display it, for proofreading or other checks. To do this, we use the PRINT command:

```
MTB> print c1-c5
```

Minitab will respond by displaying the 40 rows of the worksheet:

```
ROW  c1   c2   c3  c4   c5

  1   1  329   38  20    *
  2   2  302   43  13   68
      .    .    .   .    .
      .    .    .   .    .
      .    .    .   .    .
 39  39    8   74   4  114
 40  40    8   76   4  116
```

The dots again indicate that Minitab will type all 40 rows of values. Note that with any command filling up the display, Minitab will ask you whether you want to continue (Y/N). Type Y to continue displaying output; type N to stop the output.

Correcting Mistakes with the LET Command

Suppose we notice a mistake in row 39 (for Finland). The line is incorrectly entered as

```
39  8  75  4  114
```

with 75 instead of 74 as life expectancy in column 3. The incorrect value is in column 3, row 39. To change it, we use the LET command:

```
MTB> let c3(39)=74
```

Now row 39 will look like this:

```
39  8  74  4  114
```

Deleting Rows with the DELETE Command

Suppose we inadvertently enter row 2 twice, so the worksheet looks like this:

```
1  329  38  20   *
2  302  43  13  68
2  302  43  13  68
3  302  35  14  91
:   :    :   :   :
```

The extra entry for the second country (Mali) is in row 3. To delete row 3, we use the DELETE command:

```
MTB>  delete row 3
```

The worksheet now looks like this:

```
1  329  38  20   *
2  302  43  13  68
3  302  35  14  91
:   :    :   :   :
```

Erasing a Column with the ERASE Command

To delete a column, we use the ERASE command. Suppose we do not want the ID codes in column 1 of the worksheet. Then we type **ERASE C1**:

```
MTB>  erase c1
```

The new file contains only the four columns for MORT, LIFE, WEIGHT, and CALORIE.

Naming Columns with the NAME Command

Consider now our completed worksheet, with 5 variables (columns) for each of 40 cases (rows). It can be very helpful for labeling output to name the columns. To do this, we use the NAME command. For instance:

```
MTB>    name c1 'id'  c2 'mort'  &
CONT>   c3 'life'
MTB>    name c4 'weight'  c5 'calorie'
```

Note that we can use the continue line symbol, &, as in data input. Names are included in single quotes and have eight or fewer characters. A name cannot begin or end with a blank and may not contain the single quote (same as the apostrophe) character.

Summarizing Worksheet Contents with the INFO Command

The INFORMATION command summarizes the contents of the Minitab worksheet we are currently working with. The worksheet consists of the data file and constants, plus names of variables that we are currently using. For the worksheet we have just created, if we type `INFO`:

```
MTB>  info
```

we get the following information:

```
COLUMN  NAME      COUNT  MISSING

c1      id         40
c2      mort       40
c3      life       40
c4      weight     40       4
c5      calorie    40       3

CONSTANTS USED: NONE
```

CONSTANTS USED refers to stored constants, and we have none. (Note that to Minitab, INFO is the same as INFORMATION. We need type only the first four letters of *any* command name.)

Saving a Minitab Worksheet

When we leave Minitab, on purpose or because the computer goes down, we lose any worksheets we have not saved. To save a Minitab worksheet (including names for columns and any stored constants, which we will discuss later), we use the SAVE command. On a personal computer, we use:

```
MTB>  save 'b:example'
```

to save the worksheet in a file called EXAMPLE.MTW on drive B. Minitab automatically adds the extension .MTW to the file name we specify in the SAVE command in the personal computer version; the particular extension used on mainframe computers may depend on the installation. To save the file named EXAMPLE.MTW on hard disk drive C, we use

```
MTB>  save 'c:example'
```

On a mainframe computer, we use

```
MTB>  save 'example'
```

to save the worksheet on a file named EXAMPLE.MTW. We should choose names for files that remind us what is in the files, and we must be careful to choose new file names. If we use the SAVE command with an existing file name, the new file replaces the old one.

Retrieving a Saved Minitab Worksheet

The RETRIEVE command inputs data we have saved using the SAVE command. If we have just entered Minitab on the personal computer and we want to input the worksheet on file EXAMPLE.MTW from drive B, we use

```
MTB> retrieve 'b:example'
```

We use C instead of B in the command if the file is saved on drive C. If we have just entered Minitab on a mainframe computer, we use

```
MTB> retrieve 'example'
```

Suppose we have been working in Minitab for a while with other data, and now we want to use the worksheet in file EXAMPLE.MTW. Because there are five columns of data in file EXAMPLE.MTW, we must ERASE columns 1–5 in our worksheet:

```
MTB> erase c1-c5
```

Now we can retrieve file EXAMPLE.MTW as explained above.

Saving Data in an ASCII File

The SAVE command creates a file only the Minitab program can read. We can save data in an ASCII file that other programs can read, using the WRITE command. If we are using a personal computer and we want to save our data on a file named EXERCISE.DAT on drive B, we use

```
MTB> write 'b:exercise' c1-c5
```

Then our five columns of data will be saved on drive B in a file called EXERCISE.DAT. Minitab automatically adds .DAT at the end of the file name when we use the WRITE command. To do the same thing using Minitab on a mainframe computer, we type

```
MTB> write 'exercise' c1-c5
```

Inputting Data from an ASCII File

To input data from an ASCII file (perhaps we saved it using the WRITE command or created it using the text editor), we use the READ command. The command for the personal computer looks like this:

```
MTB> read 'b:exercise' c1-c5
```

if columns C1–C5 are cleared on our worksheet and we want to input data from file EXERCISE.DAT on drive B. On the mainframe, the command looks like this:

```
MTB> read 'exercise' c1-c5
```

Using OUTFILE to Save a Minitab Work Session

The command OUTFILE saves all our commands and Minitab's responses on a disk file that we can access later for editing or printing. To save this output in a file called SAMPLE.LIS, we use

```
MTB> outfile 'sample'
```

Minitab automatically appends .LIS to the file name specified in the OUTFILE command. The NOOUTFILE command stops the recording of commands and Minitab responses:

```
MTB> nooutfile
```

We can alternate the OUTFILE and NOOUTFILE commands to be selective in what we store.

Sending a Minitab Work Session to a Printer

The PAPER command sends the Minitab session directly to a printer. The command is

```
MTB> paper
```

To stop routing output to the printer, we use the NOPAPER command:

```
MTB> nopaper
```

Again, we can alternate PAPER and NOPAPER commands to select output for printing.

Only one of the OUTFILE and PAPER commands can be in effect at one time. If OUTFILE is in effect and we type PAPER, we stop the OUTFILE command; typing OUTFILE when PAPER is in effect stops the PAPER command.

Creating a Missing Value Code with the CODE Command

Suppose we are using Minitab on a personal computer and we want to read the data for Exercise 5-35 directly from a data disk. The file name is E5X35 (for

Exercise 5-35) and the variables ID, MORT, LIFE, WEIGHT, and CALORIE are in columns 1–5. On this ASCII file, the seven missing values are coded as −9, because not all users may be using Minitab. Every statistical package recognizes numbers, but missing value codes vary. The following sequence of commands reads the file E5X35 from the data disk in drive B, changes the seven occurrences of −9 to * (the Minitab missing value code), names the variables, and then saves the worksheet on drive C in a file called M5X35 (for Minitab worksheet of Exercise 5-35).

```
MTB>  read 'b:e5x35' c1-c5
MTB>  code(-9)'*' c4 c4
MTB>  code(-9)'*' c5 c5
MTB>  name c1 'id' c2 'mort' c3 'life' c4 'weight' c5 'calorie'
MTB>  save 'c:m5x35'
```

The first CODE command changes all values of −9 in column 4 to an asterisk and then puts the coded values back into column 4. The second CODE command does the same for column 5. Columns 4 and 5 were the only columns with missing values in this file.

A missing value code is useful in a statistical package. Minitab excludes missing values from plots and certain statistical procedures. Whenever Minitab encounters an asterisk, it will treat the value as missing.

Recoding Data Values with the CODE Command

Suppose now we have a Minitab worksheet containing the data from Exercise 5-35, with columns named as above. We might find it useful to create a new column that distinguishes the 20 countries with the highest values of child mortality from the 20 countries with the lowest values for this variable. We can do this using the CODE command. Let's create a new variable called MORTCODE in column 6 of our worksheet. We will let MORTCODE equal 1 for the countries with low values of MORT, 2 for the countries with high values of MORT. Low values of MORT are from 0 to 20, high values from 200 to 400. One way to create MORTCODE in column 6 is this:

```
MTB>   code(0:20) 1, (200:400) 2 &
CONT>  in 'MORT' store results in c6
MTB>   name c6 'mortcode'
```

We could have typed C2 instead of 'MORT' to convey the same information to Minitab. Note that we used the continuation symbol, &, to show Minitab that our CODE command took up more than one line. Column 6 is now named MORTCODE. This column contains a 2 for the 20 countries with IDs 1–20 and a 1 for countries with IDs 21–40.

A coded variable can have more than two values. We might create a variable CALCODE in column 7 that equals 1 for countries with CALORIE values in the range from 51 to 75, 2 for countries with CALORIE values from 76 to

100, 3 for CALORIE values from 101 to 125, and 4 for CALORIE values from 126 to 150:

```
MTB>   code(51:75) 1, (76:100) 2, (101:125) 3, (126:150) 4 &
CONT>  in 'calorie' store results in c7
MTB>   name c7 'calcode'
```

We could have typed C5 instead of 'CALORIE' in this command. Column 7 is now named CALCODE. It has values 1, 2, 3, and 4 plus * for the three missing values.

If we type the PRINT command

```
MTB>  print c1-c7
```

Minitab will print something like this:

```
ROW  id  mort  life  weight  calorie  mortcode  calcode

  1   1   329    38      20        *         2        *
  2   2   302    43      13       68         2        1
  3   3   302    35      14       91         2        2
  .   .     .     .       .        .         .        .
  .   .     .     .       .        .         .        .
 38  38     9    76       5      129         1        4
 39  39     8    74       4      114         1        3
 40  40     8    76       4      116         1        3
```

The dots here indicate that all 40 rows would be printed.

Creating New Columns with the COPY Command

For some analyses, we will want to consider the 20 countries with MORTCODE = 1 separately from the 20 countries with MORTCODE = 2. We might want the variables for the two sets of countries in separate groups of columns. Let's put the values of ID, MORT, LIFE, WEIGHT, CALORIE, and CALCODE for the countries with MORTCODE = 1 in columns 11–16; for the countries with MORTCODE = 2, in columns 21–26. Here is one way to accomplish this, using the COPY command:

```
MTB>   copy c1-c5, c7 to c11-c16;
SUBC>  use only 'mortcode'=1.
MTB>   copy c1-c5, c7 to c21-c26;
SUBC>  use only 'mortcode'=2.
```

The COPY command copies the first set of columns into the second set of columns, subject to the conditions after the semicolon. The semicolon indicates that a subcommand follows. A command that includes a subcommand ends with a period. Columns C11–C16 all have 20 rows, one for each of the 20 countries with low values of MORT (MORTCODE = 1). Columns C21–C26 also have 20 rows, one for each of the 20 countries with high values of MORT (MORTCODE = 2). We could have obtained the same results with these commands:

```
MTB>   copy c1-c5, c7 to c11-c16;
SUBC>  use only 'mort'=0:20.
MTB>   copy c1-c5, c7 to c21-c26;
SUBC>  use only 'mort'=200:400.
```

Then columns C11–C16 include only countries with values of MORT in the range of 0 to 20 (or MORTCODE = 1). Columns C21–C26 include only countries with values of MORT from 200 to 400 (or MORTCODE = 2).

Our major limitation in producing new columns is that the student edition of Minitab allows a worksheet containing at most 2,000 numbers.

We might want to name our new columns. For instance, we could use these commands:

```
MTB>   name c11 'id1'  c12 'mort1'  c13 'life1' &
CONT>  c14 'weight1' c15 'calorie1'  c16 'calcode1'
MTB>   name c21 'id2'  c22 'mort2'  c23 'life2' c24 'weight2' &
CONT>  c25 'calorie2'  c26 'calcode2'
```

To display a list of the data on our worksheet, we can use the PRINT command.

Rearranging a Data File Using the SORT Command

The countries on our file are listed in decreasing order of child mortality (MORT). We can use the SORT command to rearrange a data file in increasing order for a single variable. If we want our data file arranged in increasing order of child mortality, we can use the command

```
MTB>  sort c2 c1, c3-c7 c2 c1, c3-c7
```

Processing this command, Minitab rearranges the rows for columns 1–7 in increasing order of values in C2, which contains the values for child mortality, MORT. The rearranged rows will be put back in columns C1–C7. (If we use columns with data already in them, Minitab writes over the original data. These data are lost unless they have been saved somewhere.) Now countries with IDs 39 and 40 (Finland and Sweden) will be at the top of our file and Afghanistan (ID = 1) will be at the bottom.

If we want to leave the original columns alone and put the sorted columns in C31–C37, we could use

```
MTB>  sort c2 c1, c3-c7 c32, c31, c33-c37
```

Columns 1–7 are unchanged, as are all our other original columns. Columns 31–37 contain the variables for countries sorted on values of MORT in C2.

These two examples illustrate how the SORT command works. Minitab divides the number of columns listed after SORT in half. It sorts the first half of the columns, based on the values in the first column listed. The sorted contents of these columns are then stored in the second set of columns, in the order listed.

We have used the SORT command to sort several variables (including ID) based on MORT. The values of the variables for a single country are still in one row within the specified columns. If we wanted only to sort the values of MORT without worrying about the other variables, we could use

```
MTB>  sort c2 c9
```

Column 2 remains unchanged. Column 9 contains the 40 values of MORT listed in increasing order. The sorted values of MORT in C9 do not correspond to the country IDs in C1.

Ranking the Values in a Single Column with the RANK Command

We can also assign ranks based on a variable without rearranging the worksheet, using the RANK command. If we use

```
MTB>  rank 'calorie' c10
```

Minitab will put the rank for each CALORIE value in C10. The smallest value receives rank 1, the next smallest receives rank 2, the largest receives rank 40. Tied values of CALORIE receive the average of the ranks they share.

CHAPTER 2

Studying One Variable at a Time: Lists, Tables, and Plots

IN THIS CHAPTER

Data list
Ordered (ranked) data
Dot plot
Stem-and-leaf plot
Frequency table
Frequency plot, histogram
Quantiles
Box plot, box graph
Symmetrical, positively skewed, and negatively skewed distributions
Unimodal, bimodal, and multimodal distributions

What can we learn about per capita calorie supplies in the World Bank countries? How do life expectancies vary across these countries? Can we identify different groups of countries based on birth rates? What is a good way to summarize primary school enrollments?

All these questions are about individual variables. We consider ways to study individual variables in Chapters 2 and 3. Chapter 3 discusses numbers, called descriptive statistics, used to summarize location and spread in a set of values. Chapter 2 looks at lists, tables, and plots. We use these tools to learn about the distribution of a variable—that is, how the values of the variable are distributed along the number line.

2-1 Lists and Dot Plots

A *list* can be useful if the number of cases is not too large: we simply list all of the values of a variable. We will see many data lists in the examples and exercises throughout this book.

A **data list** is a listing of the values of a variable in a data set.

Consider the life expectancies listed in Table 2-1 for each of the World Bank countries. This alphabetical list is useful for looking up the life expectancy for a country. Three countries (Afghanistan, Kampuchea, and Lebanon) have no 1985 life expectancy information available; these three countries must be excluded from analyses involving the life expectancies.

When looking at Table 2-1, you may find yourself wondering which countries have the longest and shortest life expectancies. Another (perhaps more interesting) way to look at the life expectancies is to order their values from smallest to largest, as in Table 2-2. We say the observations are *sorted* or *ordered* by life expectancy. The values of the ordered variables in a sorted list may go from smallest to largest or from largest to smallest. Either way, extreme values are readily apparent, and we can scan intermediate values more easily than in an unordered list. We see from Table 2-2 that the shortest life expectancy is 40 years, for Guinea and Sierra Leone. Australia and France share the longest life expectancy, 78 years. Afghanistan, Kampuchea, and Lebanon are shown at the bottom of the table as having no 1985 life expectancy information available.

Values are **ordered, sorted,** or **ranked** if they are listed in order of magnitude.

Suppose we just want to get a feel for life expectancies; for the moment we do not care which life expectancies go with which countries. Then the ordered list of life expectancies in Table 2-2 is much more useful than the scrambled list of numbers in Table 2-1. Creating such an ordered list of values is often the first step in data analysis.

Can we do better than simply listing the values in order of magnitude? We may be able to interpret a plot or graph of some kind more easily than a

TABLE 2-1 Life expectancy at birth in 1985 for 128 countries. Countries are listed alphabetically.

Country	Life expectancy at birth (years)	Country	Life expectancy at birth (years)	Country	Life expectancy at birth (years)
Afghanistan	Missing	Guatemala	60	Oman	54
Albania	70	Guinea	40	Pakistan	51
Algeria	61	Haiti	54	Panama	72
Angola	44	Honduras	62	Papua New Guinea	52
Argentina	70	Hong Kong	76	Paraguay	66
Australia	78	Hungary	71	Peru	59
Austria	74	India	56	Philippines	63
Bangladesh	51	Indonesia	55	Poland	72
Belgium	75	Iran	60	Portugal	74
Benin	49	Iraq	61	Romania	72
Bhutan	44	Ireland	74	Rwanda	48
Bolivia	53	Israel	75	Saudi Arabia	62
Botswana	57	Italy	77	Senegal	47
Brazil	65	Jamaica	73	Sierra Leone	40
Bulgaria	71	Japan	77	Singapore	73
Burkina Faso	45	Jordan	65	Somalia	46
Burma	59	Kampuchea	Missing	South Africa	55
Burundi	48	Kenya	54	Spain	77
Cameroon	55	Korea, North	68	Sri Lanka	70
Canada	76	Korea, South	69	Sudan	48
Central African Republic	49	Kuwait	72	Sweden	77
		Lao PDR	45	Switzerland	77
Chad	45	Lebanon	Missing	Syria	64
Chile	70	Lesotho	54	Tanzania	52
China	69	Liberia	50	Thailand	64
Colombia	65	Libya	60	Togo	51
Congo	58	Madagascar	52	Trinidad and Tobago	69
Costa Rica	74	Malawi	45	Tunisia	63
Côte d'Ivoire	53	Malaysia	68	Turkey	64
Cuba	77	Mali	46	Uganda	49
Czechoslovakia	70	Mauritania	47	USSR	70
Denmark	75	Mauritius	66	United Arab Emirates	70
Dominican Republic	64	Mexico	67	United Kingdom	75
Ecuador	66	Mongolia	63	United States	76
Egypt	61	Morocco	59	Uruguay	72
El Salvador	64	Mozambique	47	Venezuela	70
Ethiopia	45	Nepal	47	Viet Nam	65
Finland	76	Netherlands	77	Yemen Arab Republic	45
France	78	New Zealand	74	Yemen, PDR	46
Germany, East	59	Nicaragua	59	Yugoslavia	72
Germany, West	75	Niger	44	Zaire	51
Ghana	53	Nigeria	50	Zambia	52
Greece	68	Norway	77	Zimbabwe	57

TABLE 2-2 Life expectancy at birth in 1985 for 128 countries. Countries are ordered from smallest to largest life expectancy.

Country	expectancy at birth (years)	Country	expectancy at birth (years)	Country	expectancy at birth (years)
Guinea	40	Cameroon	55	Albania	70
Sierra Leone	40	Indonesia	55	Argentina	70
Angola	44	South Africa	55	Chile	70
Bhutan	44	India	56	Czechoslovakia	70
Niger	44	Botswana	57	Sri Lanka	70
Burkina Faso	45	Zimbabwe	57	USSR	70
Chad	45	Congo	58	United Arab Emirates	70
Ethiopia	45	Burma	59	Venezuela	70
Lao PDR	45	Germany, East	59	Bulgaria	71
Malawi	45	Morocco	59	Hungary	71
Yemen Arab Republic	45	Nicaragua	59	Kuwait	72
Mali	46	Peru	59	Panama	72
Somalia	46	Guatemala	60	Poland	72
Yemen, PDR	46	Iran	60	Romania	72
Mauritania	47	Libya	60	Uruguay	72
Mozambique	47	Algeria	61	Yugoslavia	72
Nepal	47	Egypt	61	Jamaica	73
Senegal	47	Iraq	61	Singapore	73
Burundi	48	Honduras	62	Austria	74
Rwanda	48	Saudi Arabia	62	Costa Rica	74
Sudan	48	Mongolia	63	Ireland	74
Benin	49	Philippines	63	New Zealand	74
Central African Republic	49	Tunisia	63	Portugal	74
Uganda	49	Dominican Republic	64	Belgium	75
Liberia	50	El Salvador	64	Denmark	75
Nigeria	50	Syria	64	Germany, West	75
Bangladesh	51	Thailand	64	Israel	75
Pakistan	51	Turkey	64	United Kingdom	75
Togo	51	Brazil	65	Canada	76
Zaire	51	Colombia	65	Finland	76
Madagascar	52	Jordan	65	Hong Kong	76
Papua New Guinea	52	Viet Nam	65	United States	76
Tanzania	52	Ecuador	66	Cuba	77
Zambia	52	Mauritius	66	Italy	77
Bolivia	53	Paraguay	66	Japan	77
Côte d'Ivoire	53	Mexico	67	Netherlands	77
Ghana	53	Greece	68	Norway	77
Haiti	54	Korea, North	68	Spain	77
Kenya	54	Malaysia	68	Sweden	77
Lesotho	54	China	69	Switzerland	77
Oman	54	Korea, South	69	Australia	78
		Trinidad and Tobago	69	France	78

Missing: Afghanistan, Kampuchea, Lebanon

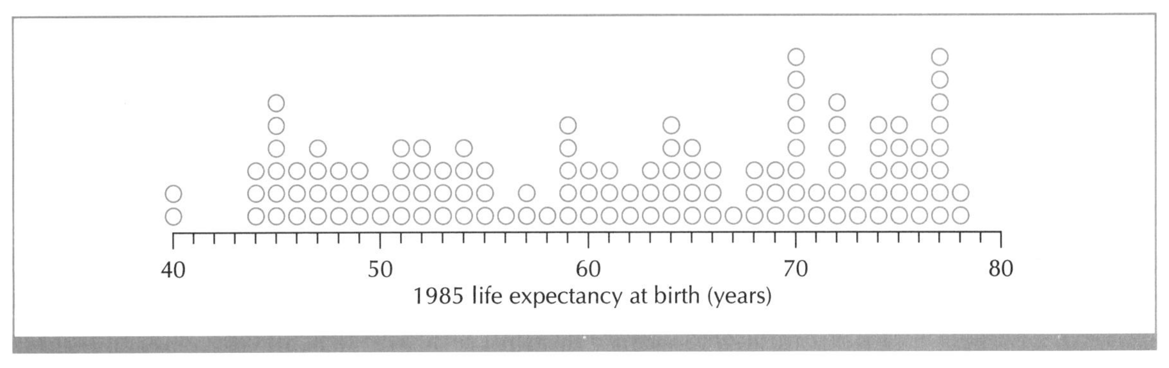

FIGURE 2-1 Dot plot of 1985 life expectancy at birth for 125 countries. Three countries are excluded because of missing values.

list of numbers. Often we can *see* more in a plot than in a list. In fact, much of modern data analysis (or exploratory data analysis) involves use of graphical techniques to examine a data set. One simple graphical technique is the dot plot.

A *dot plot* uses dots or circles to represent values, arranged along a line or axis showing the scale and units of measurement. The dot plot shown in Figure 2-1 is a picture representation of the ordered life expectancies in Table 2-2. Only the 125 countries with nonmissing values are included in the plot.

A **dot plot** is a graphical display, with dots representing values positioned along an axis or number line.

We stack dots corresponding to cases with the same value. Eight stacked dots, for instance, indicate that eight countries have a life expectancy of 70 years. Figure 2-1 shows us that 11 countries have life expectancies of 45 years or less. There is a fairly even distribution of countries with life expectancies in the range from 46 to 69 years. A large group of countries have life expectancies from 70 to 78 years.

We will consider life expectancies again later. In Section 2-2, we use stem-and-leaf plots to look at fertility rate and calorie supply.

2-2

Stem-and-Leaf Plots

A *stem-and-leaf plot* is especially useful for getting a quick picture of a set of values when graphing by hand rather than by computer. The plot gets its name from the way it is constructed: A fixed number of the leftmost digits of a value form the stem; one or more digits to the right of the stem form the leaf.

A **stem-and-leaf plot** uses the values taken on by a variable in a graphical display showing the frequency of values in specified intervals.

TABLE 2-3 Total fertility rate in 1985 for 128 countries.

Fertility rate	Country	Fertility rate	Country
1.3	Germany (West)	4.7	Ecuador, Egypt, Haiti
1.4	Denmark	4.9	Mongolia, Morocco, South Africa
1.5	Italy, Netherlands, Switzerland	5.2	El Salvador, Kuwait
1.6	Belgium	5.4	Papua New Guinea
1.7	Canada, Finland, Hungary, Norway, Singapore, Sweden	5.6	Central African Republic, Iran, Nicaragua
1.8	Germany (East), Hong Kong, Japan, United Kingdom, United States	5.7	Bangladesh, Chad, Guatemala
2.0	Australia, Bulgaria, Cuba, France, Greece, Portugal, Spain	5.8	Lesotho
2.1	Austria, Czechoslovakia, New Zealand, Romania, Yugoslavia	5.9	Bolivia, United Arab Emirates
2.3	China, Poland, USSR	6.0	Guinea, Honduras, Yemen PDR
2.4	Korea (South)	6.1	Pakistan, Zaire
2.5	Chile, Mauritius	6.2	Bhutan, Ethiopia, Jordan, Zimbabwe
2.6	Ireland, Uruguay	6.3	Algeria, Congo, Mauritania, Mozambique, Nepal
2.8	Jamaica, Trinidad and Tobago	6.4	Angola, Ghana, Lao PDR
2.9	Israel	6.5	Benin, Burkina Faso, Burundi, Côte d'Ivoire, Madagascar, Mali, Sierra Leone, Togo
3.2	Panama, Sri Lanka, Thailand	6.6	Sudan
3.3	Argentina, Colombia, Costa Rica	6.7	Botswana, Iraq, Oman, Senegal, Syria
3.4	Albania	6.8	Cameroon, Somalia, Yemen Arab Republic, Zambia
3.6	Brazil	6.9	Liberia, Nigeria, Uganda
3.7	Malaysia	7.0	Niger, Tanzania
3.8	Korea (North)	7.1	Saudi Arabia
3.9	Burma, Turkey, Venezuela	7.2	Libya
4.0	Dominican Republic	7.6	Malawi
4.1	Indonesia	7.8	Kenya
4.3	Mexico, Peru, Philippines	8.0	Rwanda
4.4	Paraguay		
4.5	India		
4.6	Tunisia, Viet Nam		

Missing: Afghanistan, Kampuchea, Lebanon

Let's construct a stem-and-leaf plot for the variable fertility rate. Fertility rate is an estimate of the average number of children born per woman in a country. As we can see from the listing in Table 2-3, three countries (Afghanistan, Kampuchea, and Lebanon) have no information available on fertility rate. We must exclude them from analyses involving this variable.

Since each fertility rate has just two digits, the choice of stem and leaf is easy: The number to the left of the decimal point is the stem; the number to the right of the decimal point forms the leaf. Fertility rates in Table 2-3 range from 1.3 to 8.0. Thus, the stems go from 1 to 8, as shown in Figure 2-2a.

The first fertility rate in Table 2-3 is 1.3, for West Germany. This number has a stem of 1 and a leaf of 3 and is plotted in Figure 2-2b. The next fertility rate is 1.4, for Denmark. This number has a stem of 1 and a leaf of 4, added to the plot in Figure 2-2c. By the time we add the value 2.0 for Spain, the plot looks like Figure 2-2d.

A completed stem-and-leaf plot of fertility rates is displayed in Figure 2-3. The legend of the plot explains how to interpret the stems and leaves. We see a peak in the distribution at the row for stem 6; there is a concentration of countries with fertility rates near 6. The plot shows another concentration of countries around fertility rate 2.

FIGURE 2-2 Steps in constructing a stem-and-leaf plot of 1985 total fertility rates. The stem is an integer and the leaf is a decimal value.

Stem	Leaf
1	
2	
3	
4	
5	
6	
7	
8	

a. The stems go from 1 to 8.

Stem	Leaf
1	3
2	
3	
4	
5	
6	
7	
8	

b. The value 1.3 for West Germany is plotted.

Stem	Leaf
1	3 4
2	
3	
4	
5	
6	
7	
8	

c. The value 1.4 for Denmark is added to the plot.

Stem	Leaf
1	3 4 5 5 5 6 7 7 7 7 7 7 8 8 8 8 8
2	0 0 0 0 0 0 0
3	
4	
5	
6	
7	
8	

d. After the value 2.0 for Spain is added, the plot looks like this.

FIGURE 2-3 Stem-and-leaf plot of 1985 total fertility rates for 125 countries. The stem is an integer and the leaf is a decimal value. Three countries are excluded because of missing values.

Stem	Leaf
1	3 4 5 5 5 6 7 7 7 7 7 7 8 8 8 8 8
2	0 0 0 0 0 0 0 1 1 1 1 1 3 3 3 4 5 5 6 6 8 8 9
3	2 2 2 3 3 3 4 6 7 8 9 9 9
4	0 1 3 3 3 4 5 6 6 7 7 7 9 9 9
5	2 2 4 6 6 6 7 7 7 8 9 9
6	0 0 0 1 1 2 2 2 2 3 3 3 3 3 4 4 4 5 5 5 5 5 5 5 5 6 7 7 7 7 7 8 8 8 8 9 9 9
7	0 0 1 2 6 8
8	0

The purpose of a stem-and-leaf plot is to provide insight into the distribution of a set of values. We might decide that the rows are too long in Figure 2-3, that the plot is too condensed. Variations of the stem-and-leaf plot use two or more rows of leaves for each stem, in effect stretching the plot. One such variation of the stem-and-leaf plot of fertility rates is shown in Figure 2-4, with two rows for each stem: one for leaves 0–4 and the other for leaves 5–9. We can see that there is a large group of countries with fertility rates between 1.5 and 2.5. Another large group has values between 5.5 and 7.0. Three countries have fertility rates greater than 7.5 children per woman.

Another variation on the plot in Figure 2-3 might use five rows for each stem: for leaves 0 and 1, leaves 2 and 3, leaves 4 and 5, leaves 6 and 7, leaves 8 and 9. You might try constructing this variation yourself for the fertility rate values, and compare it with the plots in Figures 2-3 and 2-4.

A stem-and-leaf plot is more condensed than a dot plot. Each stack of dots in a dot plot corresponds to a single value. Each row or stem in a stem-and-leaf plot corresponds instead to a range of values. The stem-and-leaf plot summarizes the information in a dot plot, while still showing us every data value. We determine the stems, leaves, and number of rows per stem to give a plot that is more informative than a simple list, but not as stretched out as a dot plot.

How we define stems and leaves depends in large part on the values of a variable. Figure 2-5a shows a stem-and-leaf plot of calorie supplies. Estimated daily calorie supply ranges from 1,504 to 3,831 calories per person. The stems in the plot are in hundreds of calories, ranging from 15 hundred to 38 hundred calories. Leaves are in calories. The smallest value, 1,504, has a stem of 15 and a leaf of 04. The largest value, 3,831, has a stem of 38 and a leaf of 31.

Figure 2-5b is a variation of the stem-and-leaf plot in Figure 2-5a, with single digits as leaves. To obtain a leaf for Figure 2-5b, we *cut* the rightmost (units) digit from the corresponding calorie supply value in Figure 2-5a, so leaves in Figure 2-5b are in tens of calories. The two stem-and-leaf plots convey almost the same information. However, in his book *Exploratory Data Analysis* (1977), John Tukey suggests that single-digit leaves are generally preferable, making plots easier to look at and interpret. Do you find the plot in Figure 2-5b easier to examine and interpret than the one in Figure 2-5a?

FIGURE 2-4 Stem-and-leaf plot of 1985 total fertility rates for 125 countries. The stem is an integer and the leaf is a decimal value. Each stem takes up two rows, one for leaves 0–4 and the other for leaves 5–9. Three countries are excluded because of missing values.

Stem	Leaf
1	3 4
1	5 5 5 6 7 7 7 7 7 7 8 8 8 8 8
2	0 0 0 0 0 0 0 1 1 1 1 1 3 3 3 4
2	5 5 6 6 8 8 9
3	2 2 2 3 3 3 4
3	6 7 8 9 9 9
4	0 1 3 3 3 4
4	5 6 6 7 7 7 9 9 9
5	2 2 4
5	6 6 6 7 7 7 8 9 9
6	0 0 0 1 1 2 2 2 2 3 3 3 3 3 4 4 4
6	5 5 5 5 5 5 5 5 6 7 7 7 7 7 8 8 8 8 9 9 9
7	0 0 1 2
7	6 8
8	0

FIGURE 2-5 Stem-and-leaf plots of 1985 calorie supply per capita for 124 countries. Four countries are excluded because of missing values.

Stem	Leaf
15	04
16	78 81
17	28 37 47 88
18	17 55 99
19	19 24 69
20	34 38 50 54 54 72 78 83 89
21	16 37 46 48 51 54 59 71 73 81 89
22	11 19 28 36 40 50 50 94
23	11 35 37 41 42 58 85
24	19 25 48 61 62 69
25	05 33 47 49 71 74 83 85
26	02 02 33 77 78 84 95 98
27	26 40 71 96
28	03 07 36 41 56
29	26 47 79
30	06 26 60 97
31	22 22 28 31 38 51 61 67 68 77
32	21 39 63 80
33	43 58 59 85 86 89
34	32 32 40 65 74 82
35	14 38 47
36	02 12 25 63 63 79
37	21 91
38	31

a. The stem is in hundreds of calories; the leaf is in calories.

Stem	Leaf
15	0
16	78
17	2348
18	159
19	126
20	335557788
21	13445557788
22	11234559
23	1334458
24	124666
25	03447788
26	00377899
27	2479
28	00345
29	247
30	0269
31	2223356667
32	2368
33	455888
34	334678
35	134
36	012667
37	29
38	3

b. The stem is in hundreds of calories; the leaf is in tens of calories.

The stem-and-leaf plots in Figure 2-5 show a peak and large concentration of values near the stem for 21 hundred calories. There is another, somewhat smaller, peak and concentration of values near 31 hundred calories. The World Bank countries seem to fall into two major groups: countries with adequate per capita calorie supplies and countries with inadequate per capita calorie supplies.

With stems representing hundreds of calories in Figure 2-5, we get a plot with 24 rows or stems. Different choices for stems and leaves are possible. Figure M2-6 in the Minitab Appendix for Chapter 2 shows a stem-and-leaf plot for these calorie supplies with five rows for each stem value. The single-digit stems are in thousands of calories and the single-digit leaves are in hundreds of calories (obtained after cutting off the two rightmost digits of each calorie supply value).

The way a stem-and-leaf plot is constructed makes it easy to read the actual data values. The plot could be constructed with stems across the bottom and leaves as columns, but the numbers are not as easy to read that way.

So far we have considered lists and displays of all the values of a variable. If the number of cases is large, such lists and displays can get unwieldy. An important goal of data analysis is data reduction: We want fewer numbers to look at. A good summary provides insights into the data, highlighting features and characteristics that we might otherwise have missed. We will discuss some simple ways to summarize information about a variable in Sections 2-3 and 2-4.

2-3 Frequency Tables, Frequency Plots, and Histograms

We might like to summarize the values of a variable, without necessarily listing each value. A *frequency table* provides such a summary. If a variable takes on relatively few values, a frequency table shows the number of occurrences of each value. Otherwise, a frequency table displays the number of occurrences of values within specified intervals. We will use the first type of frequency table for number of cities of over 500,000 persons, and the second type of frequency table for primary school enrollments.

When a variable takes on relatively few values, a **frequency table** lists the number of occurrences of each of these values.

When a variable takes on relatively many values, a **frequency table** lists the number of occurrences of values over specified intervals.

The number of cities of over 500,000 persons (large cities) in 1980 is shown in Table 2-4. One way to summarize this information is to count the countries with no large cities, with one large city, and so on. Then we can display these counts in a frequency table, such as Table 2-5. Table 2-5 is based on the 123 countries with nonmissing information, the last column showing the percentage of these 123 countries with each number of large cities. Twenty-seven (22.0%) countries had no large cities; 51 (41.5%) countries had one such

TABLE 2-4 Number of cities of over 500,000 persons in 1980 for 128 countries.

Number of large cities	Countries
0	Albania, Bhutan, Burkina Faso, Burundi, Central African Republic, Chad, Congo, El Salvador, Honduras, Kuwait, Lao PDR, Lesotho, Liberia, Malawi, Mali, Mauritania, Mongolia, Nepal, Niger, Papua New Guinea, Rwanda, Sierra Leone, Somalia, Togo, Trinidad and Tobago, Yemen Arab Republic, Yemen PDR
1	Afghanistan, Algeria, Angola, Austria, Benin, Bolivia, Bulgaria, Cameroon, Canada, Chile, Costa Rica, Côte d'Ivoire, Cuba, Czechoslovakia, Denmark, Dominican Republic, Ethiopia, Finland, Guatemala, Guinea, Haiti, Hong Kong, Hungary, Ireland, Israel, Jamaica, Jordan, Kenya, Lebanon, Libya, Madagascar, Malaysia, Mozambique, New Zealand, Nicaragua, Panama, Paraguay, Portugal, Romania, Senegal, Singapore, Sri Lanka, Sudan, Switzerland, Tanzania, Thailand, Tunisia, Uganda, Uruguay, Zambia, Zimbabwe
2	Belgium, Burma, Ecuador, Egypt, Ghana, Greece, Korea (North), Peru, Philippines, Saudi Arabia, Syria, Zaire
3	Bangladesh, Germany (East), Iraq, Netherlands, Sweden, Yugoslavia
4	Colombia, Morocco, Turkey, Venezuela, Viet Nam
5	Argentina, Australia
6	France, Iran, Spain
7	Korea (South), Mexico, Pakistan, South Africa
8	Poland
9	Indonesia, Italy, Japan, Nigeria, Norway
11	Germany (West)
14	Brazil
17	United Kingdom
36	India
50	USSR
65	United States
78	China
Missing:	Botswana, Kampuchea, Mauritius, Oman, United Arab Emirates

city. Three (2.4%) countries had 50 or more cities of over 500,000 persons in 1980. Note that because of rounding, the listed percentages do not sum to exactly 100%.

A *frequency plot* is a graphical presentation of the number of occurrences of each value of a variable. The frequency plot in Figure 2-6 displays the same information as Table 2-5. The horizontal axis corresponds to values of the variable, number of large cities (the left-hand column of Table 2-5). The

TABLE 2-5 Number of cities of over 500,000 persons (large cities) in 1980, summarized for 123 countries. Five countries have missing values.

Number of large cities	Number of countries	Percentage of countries
0	27	22.0
1	51	41.5
2	12	9.8
3	6	4.9
4	5	4.1
5	2	1.6
6	3	2.4
7	4	3.3
8	1	.8
9	5	4.1
11	1	.8
14	1	.8
17	1	.8
36	1	.8
50	1	.8
65	1	.8
78	1	.8
Total	123	100.0

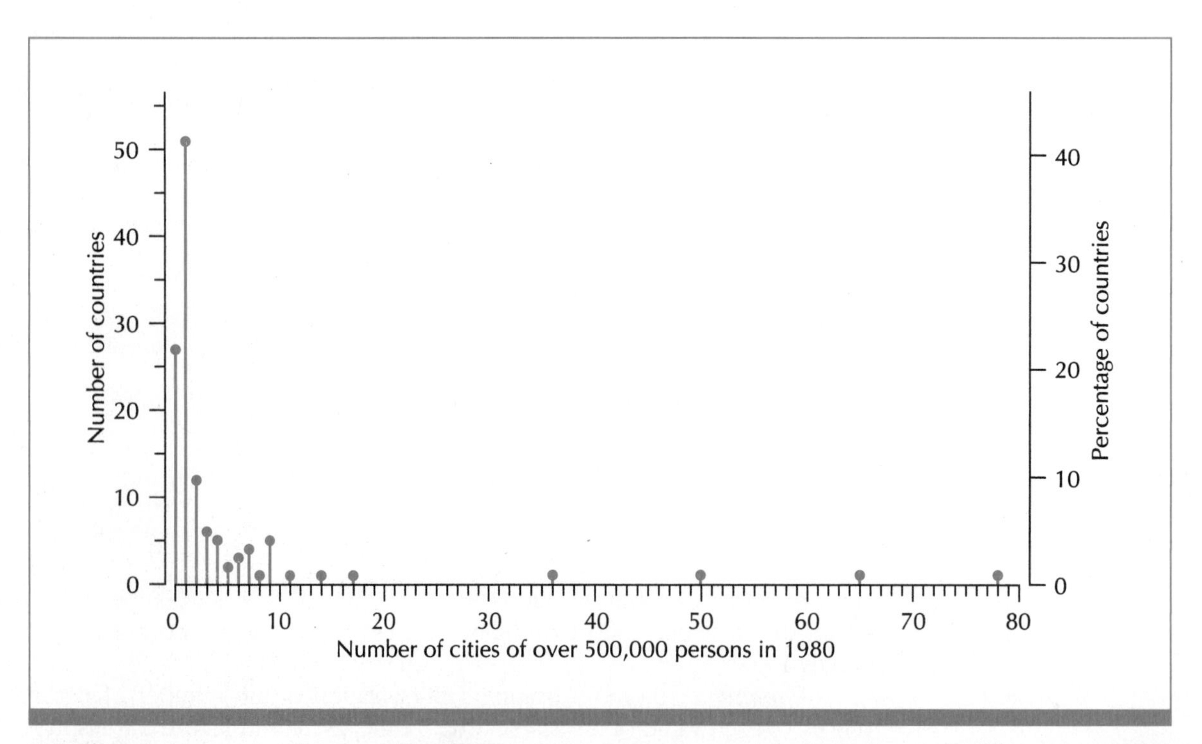

FIGURE 2-6 Frequency plot of number of cities of over 500,000 persons in 1980, summarized for 123 countries. Five countries are excluded because of missing values.

vertical axis on the left refers to the frequency or number of countries with any given value (the center column of Table 2-5). The vertical axis on the right refers to the percentage of countries with any given value (the right-hand column of Table 2-5). The heights of the vertical lines show the relative frequencies of the data values. Other names for a frequency plot are *bar chart* and *bar graph*. Frequency plots are related to histograms, to be discussed shortly.

When a variable takes on relatively few values, a **frequency plot** is a graphical way of displaying the number of occurrences of each of these values.

Values for primary school enrollment are listed in Table 2-6. Primary school enrollment is the number of children enrolled in primary school as the percentage of 6- to 11-year-olds in the country. Some countries have children outside of this age interval in primary school, so percentages may be greater than 100. The list in Table 2-6 is ordered from smallest to largest primary school enrollment. Certainly this ordered list is easier to study than an unordered list (such as enrollments for countries listed in alphabetical order). But with so many different values, we might prefer a table that has fewer numbers to look at, a summary of the enrollment figures in Table 2-6.

How can we summarize the numbers in Table 2-6? In a summary, we may not feel the need to discriminate between individual enrollment figures that are very close, such as 76 and 77. Perhaps we can look at intervals of numbers without losing too much information. Such a grouping of values is shown in Table 2-7, a frequency table that groups values of the variable into intervals. Primary school enrollments are grouped into intervals of length 10: 20–29, 30–39, and so on. The table shows that four countries have enrollments in the range from 20 to 29, for instance. Primary school enrollments range from values in the 20's to well above 100%. A majority of countries have primary school

TABLE 2-6 Number enrolled in primary school in 1984 as percentage of 6–11-year age group, for 119 countries. Nine countries have missing information.

25	62	80	97	99	102	106	113
25	62	83	97	99	102	106	113
28	62	83	97	99	103	107	114
29	64	84	97	99	103	107	115
32	66	87	97	99	103	107	116
32	67	87	97	99	103	107	116
37	67	90	97	100	104	107	116
38	68	90	98	100	105	107	118
42	70	91	98	101	105	107	118
45	76	92	98	101	105	108	119
49	76	94	98	101	105	108	120
49	76	95	98	101	106	109	121
55	77	96	98	101	106	109	131
57	77	97	98	101	106	111	134
61	77	97	98	102	106	112	

TABLE 2-7 Number enrolled in primary school in 1984 as percentage of 6–11-year age group for 119 countries. Nine countries have missing information.

Primary school enrollment	Number of countries	Percentage of countries
20–29	4	3.4
30–39	4	3.4
40–49	4	3.4
50–59	2	1.7
60–69	9	7.6
70–79	7	5.9
80–89	6	5.0
90–99	30	25.2
100–109	37	31.1
110–119	12	10.1
120–129	2	1.7
130–139	2	1.7
Total	119	100.0

enrollments from 90% to 110%. The percentages listed in the right-hand column of Table 2-7 do not sum to 100 exactly, because of rounding.

It is clear in Table 2-7 where each value should go. A value of 29% goes in the first interval, a value of 30% goes in the second interval, and so on. However, such intervals are commonly labeled somewhat differently, as shown in Table 2-8. The intervals designated in Table 2-8 have overlapping endpoints, which can be convenient for presenting frequency information. But how does the reader interpret the endpoints of the intervals? With the labeling in Table 2-8, for instance, the reader does not know whether a value of 30% should go in the interval 20–30 or in the interval 30–40. For compiling the frequency table, it does not matter whether an endpoint is counted in the corresponding upper or lower interval, just so the counting is consistent (to ensure that the intervals all have the same length and are comparable). If 30% goes in the interval 30–40, for example, then 40% goes in the interval 40–50. The convention should be clearly noted somewhere in the legend. The legend for Table 2-8 explains that intervals include the lower endpoint, not the upper endpoint. Therefore, the intervals in Table 2-8 are the same as those in Table 2-7. Note that in Table 2-8, the missing observations are included in the tabulation. Also, the column for percentage of countries is not included in Table 2-8.

We can display the information in a frequency table such as Table 2-7 graphically via a histogram. A *histogram* displays frequency information for a variable, summarized over intervals. Figure 2-7 is a histogram displaying the same information as Table 2-7. The horizontal axis shows the scale and units for the variable. The vertical axes show frequencies on the left and percentages on the right. It is helpful to show both frequencies and percentages, but many people draw histograms showing just one or the other. Most histograms are

TABLE 2-8 Number enrolled in primary school in 1984 as percentage of 6–11-year-olds for 128 countries. Intervals include the lower endpoint, not the upper endpoint.

Primary school enrollment	Number of countries
20–30	4
30–40	4
40–50	4
50–60	2
60–70	9
70–80	7
80–90	6
90–100	30
100–110	37
110–120	12
120–130	2
130–140	2
Missing	9
Total	128

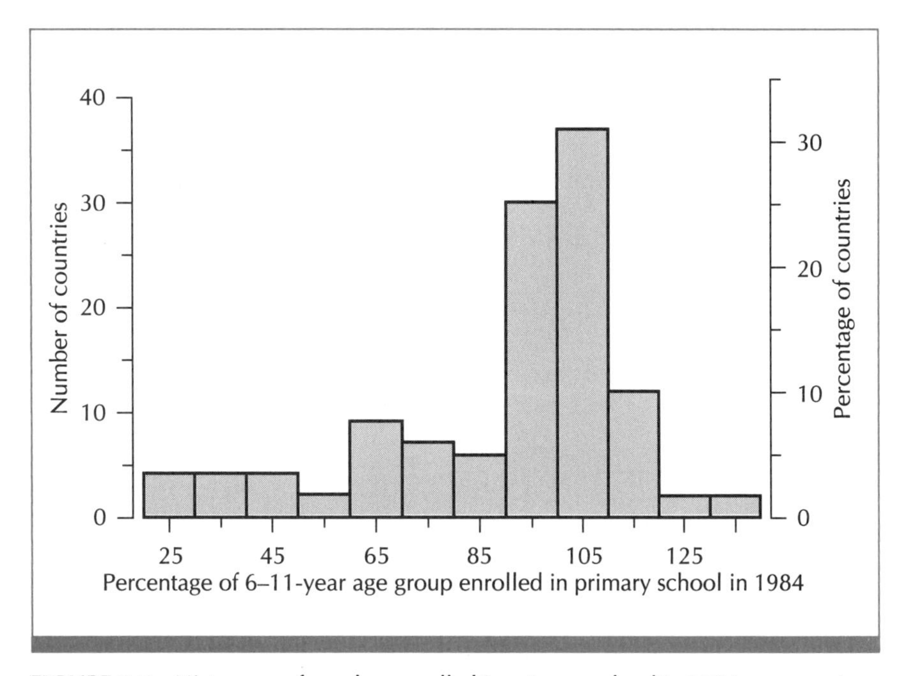

FIGURE 2-7 Histogram of number enrolled in primary school in 1984 as percentage of 6–11-year age group, for 119 countries. Intervals include the lower endpoint, not the upper endpoint. Nine countries are excluded because of missing values.

set up like the one in Figure 2-7. However, they can be oriented differently. Bars might go from left to right, for instance, as a stem-and-leaf plot is usually drawn.

> When a variable takes on relatively many values, a **histogram** is a graphical way of displaying the number of occurrences of values over specified intervals.

Figure 2-8 is a histogram with intervals of length 5: 25–30, 30–35, and so on. The endpoints of the intervals are labeled in this histogram; in Figure 2-7, we labeled the midpoints of the intervals. Either way is acceptable.

The histograms in Figures 2-7 and 2-8 show us that a majority of the countries had primary school enrollments between 90% and 110%. A few countries had enrollments greater than 110%. A large portion of the countries had relatively low primary school enrollments.

Since these two histograms provide essentially the same visual information, how can we decide between them? In general, how do we choose the

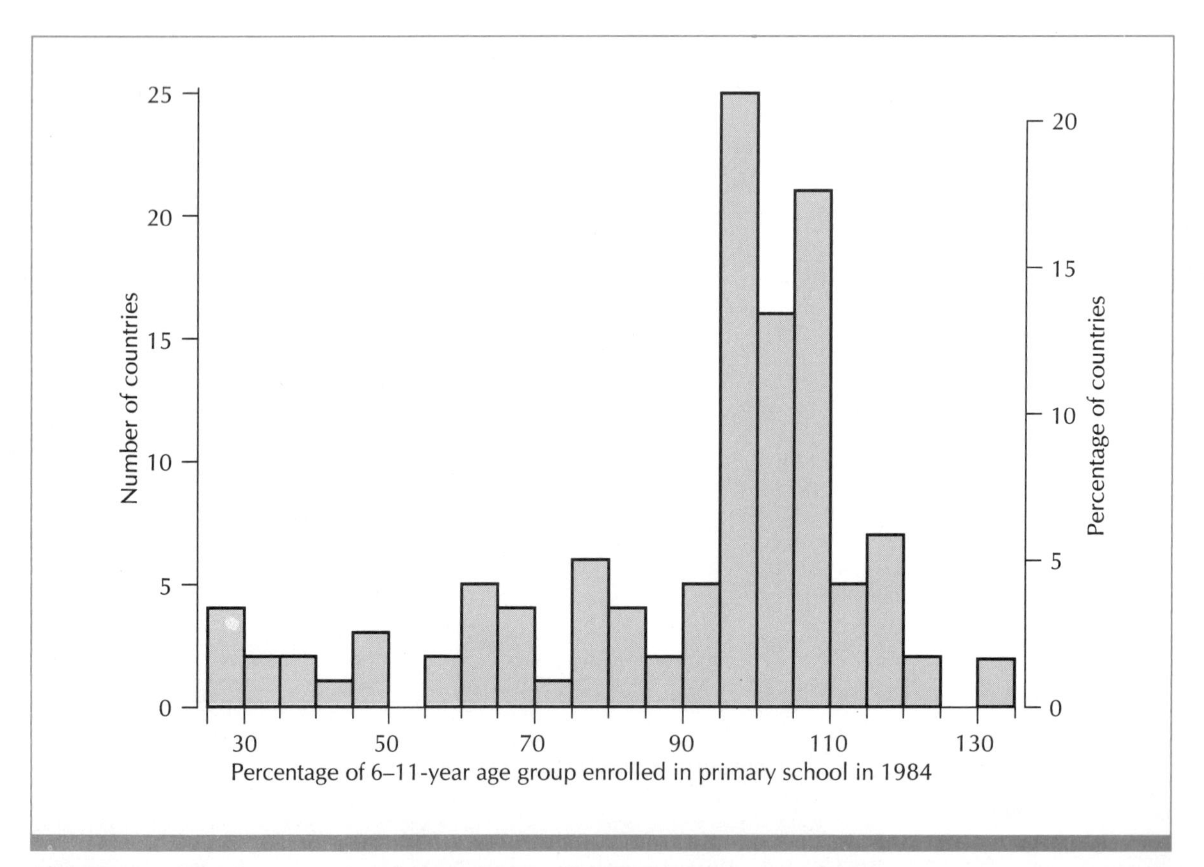

FIGURE 2-8 Histogram of number enrolled in primary school in 1984 as percentage of 6–11-year age group, for 119 countries. Intervals include the lower endpoint, not the upper endpoint. Nine countries are excluded because of missing values.

interval widths for a frequency table or histogram? If the interval widths are too great, there are few intervals and the histogram is of limited use in summarizing the variable. The extreme case is when the intervals are so wide that all values are contained in a single interval. The other extreme is when intervals are so narrow that at most a single value is contained in any interval; the resulting histogram conveys the same information as a dot plot. There is no set way to choose interval widths. We must choose a satisfactory middle ground between too many intervals and too few.

2-4 Describing the Shape of a Distribution

The *distribution* of a variable describes how the values of the variable are positioned along the number line. A dot plot shows the exact distribution of a set of values. A stem-and-leaf plot and a histogram each provide an abbreviation or summary of a distribution.

> The **distribution** of a variable is a description of how the values of the variable are positioned along an axis or number line.

As we will see later when we discuss statistical inference, the shape of the distribution of a set of observations can help us decide on a method of formal analysis. There are many shapes that a distribution can have. Three major types are symmetric distributions, distributions that are skewed to the left (negatively skewed), and distributions that are skewed to the right (positively skewed).

A distribution is *symmetric* if the values to the right of some center point form a mirror image of the values to the left of that point. Seldom will we find a real set of data values that is exactly symmetrical. But we will often encounter sets of values that are roughly or approximately symmetrical.

> A distribution is **symmetric** if the values to the right of some center point form a mirror image of the values to the left of that point.

Figure 2-9 shows a dot plot of gross national product for 19 industrial market countries. These 19 values are roughly symmetrical about a center point somewhere between 10,000 and 11,000 dollars.

Primary school enrollments are displayed in Figure 2-10 for 22 upper-

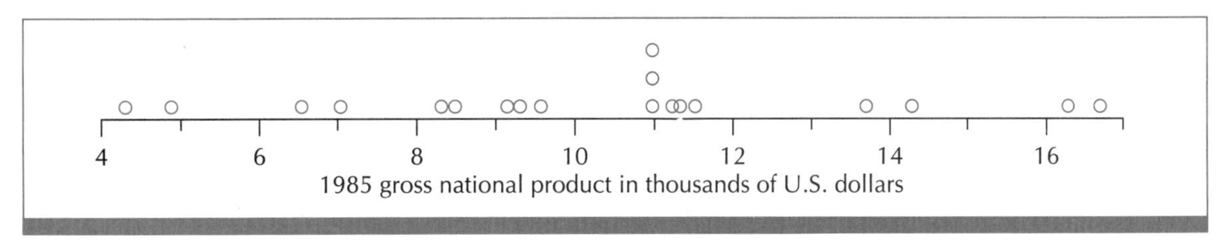

FIGURE 2-9 Dot plot of per capita gross national product in 1985 for 19 industrial market countries.

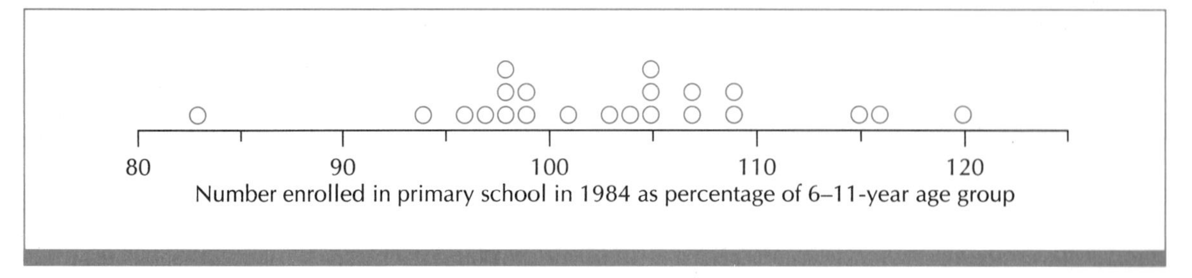

FIGURE 2-10 Dot plot of number enrolled in primary school in 1984 as percentage of 6–11-year age group for 22 upper-middle-income countries. One country is excluded with a missing value.

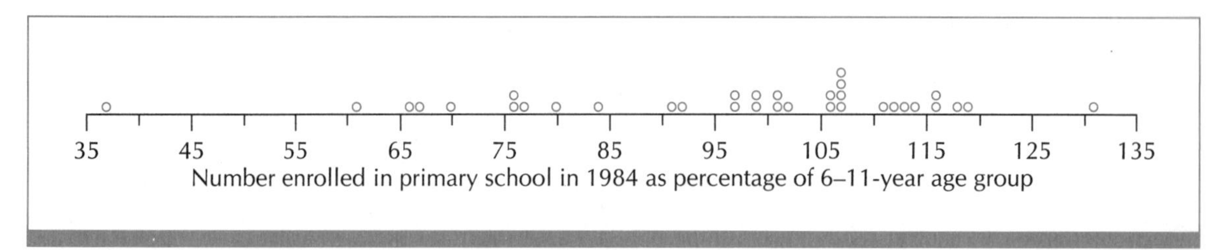

FIGURE 2-11 Dot plot of number enrolled in primary school in 1984 as percentage of 6–11-year age group for 34 lower-middle-income countries. Two countries are excluded because of missing values.

middle-income countries. These 22 values are approximately symmetrical about a center point between 100% and 105%.

Consider the histogram of primary school enrollments for 119 countries in Figure 2-8. There is a concentration of relatively high values in the 95–110 range. The distribution tails off to the left across a wide range of relatively low values. We say such a distribution is *skewed to the left* or *negatively skewed.*

> A distribution is **negatively skewed** or **skewed to the left** if there is a concentration of relatively large values, with some scatter over a range of smaller values.

Another negatively skewed distribution is illustrated in Figure 2-11. This dot plot shows primary school enrollments for 34 lower-middle-income countries. Again, there is a concentration of values on the right, and a tailing off of values to the left.

Figure 2-12 shows higher education enrollments for 119 countries. There is a concentration of relatively low values in the range from 0 to 5 or 10. The distribution tails off to the right across a wide range of relatively high values. We say this distribution is *skewed to the right* or *positively skewed.*

> A distribution is **positively skewed** or **skewed to the right** if there is a concentration of relatively small values, with some scatter over a range of larger values.

Child death rates are plotted in Figure 2-13 for 124 countries. A value of 0 indicates a country with a death rate among 1–4-year-olds of less than .5 per

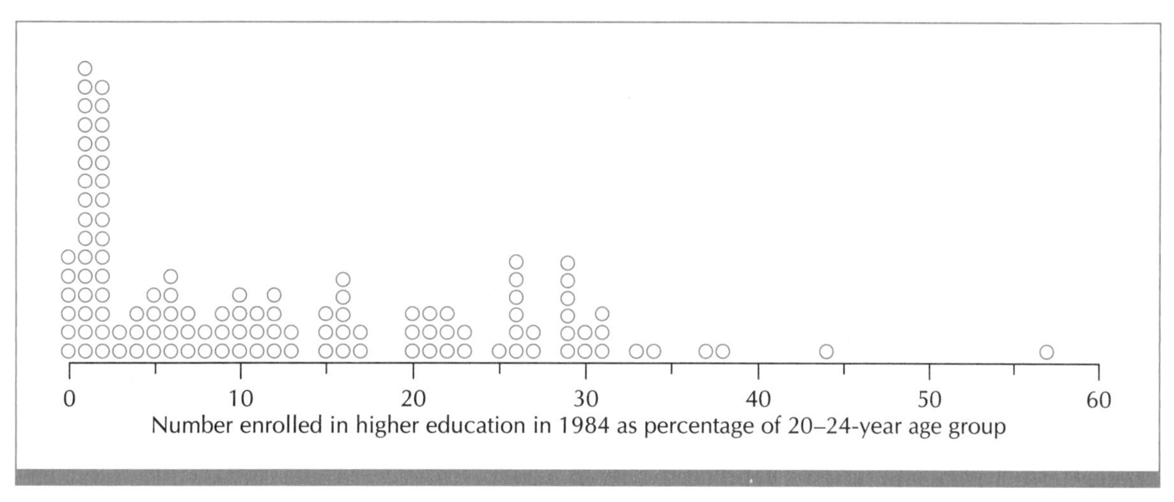

FIGURE 2-12 Dot plot of number enrolled in higher education in 1984 as percentage of 20–24-year age group for 119 countries. Nine countries are excluded because of missing values.

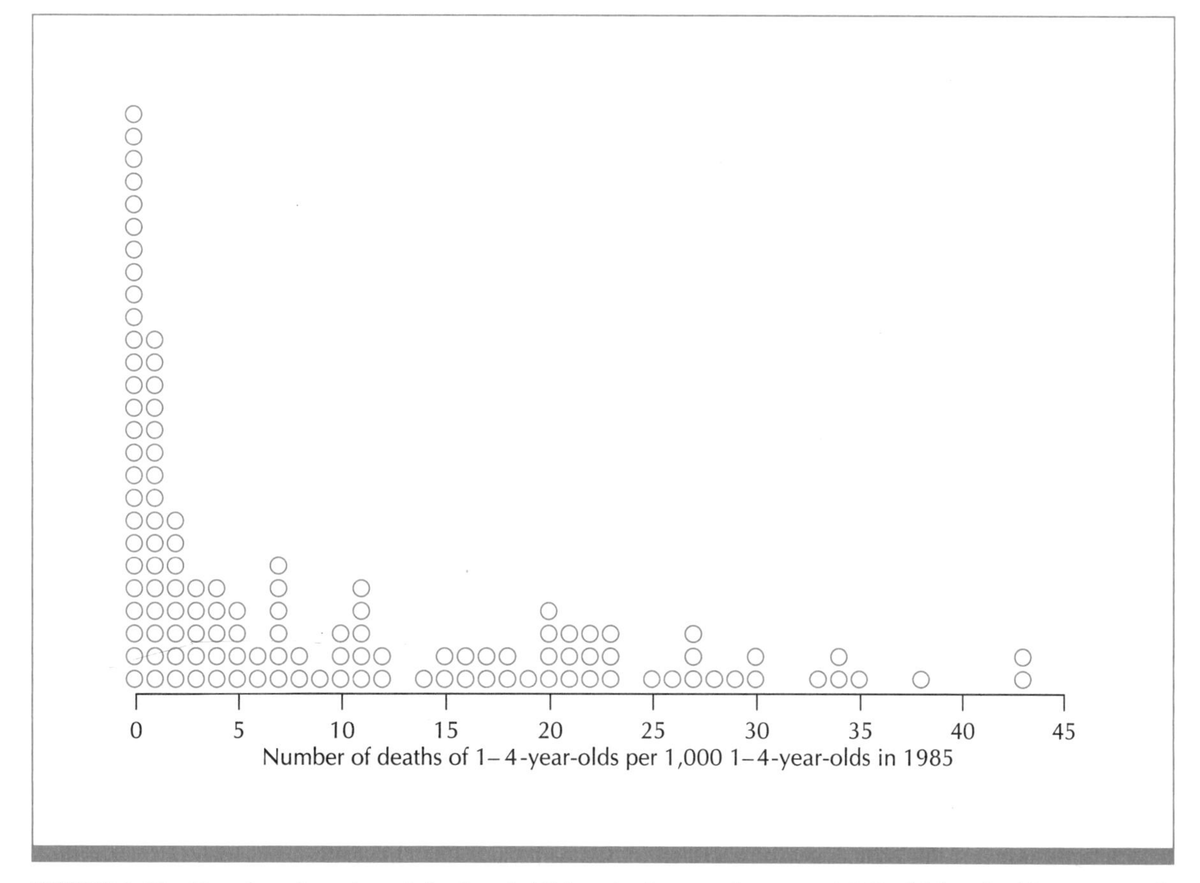

FIGURE 2-13 Dot plot of number of deaths of children 1–4 years of age per 1,000 children in this age group in 1985 for 124 countries. Four countries are excluded because of missing values.

1,000. This distribution is skewed to the right or positively skewed. A majority of the values are less than 10. The others range from 10 to 43 deaths per 1,000 1–4-year-olds.

The term *mode* or *peak* refers to a major concentration of values in a distribution. Describing the number and locations of such modes or peaks is an important part of studying a distribution, because we may be able to identify subgroups based on the modes that we see.

A **mode** or **peak** in a distribution is a major concentration of values.

Consider the plot of death rates in Figure 2-14. This distribution has a major concentration of values (a mode or peak) in the interval from 7 to 10 deaths per 1,000 population. Because there is only one major peak, this is a *unimodal distribution.* The primary school enrollments in Figure 2-8 form a unimodal distribution, as do the higher education enrollments in Figure 2-12 and the child death rates in Figure 2-13.

A **unimodal distribution** has one major peak or concentration of values.

A distribution may have more than one major peak or concentration of values. Consider the dot plot of fertility rates in Figure 2-15. (We saw stem-and-leaf plots of these values in Figures 2-3 and 2-4.) There is a major peak or mode in the interval from 1.7 to 2.1 children per woman. Another major peak is in the interval from 6.2 to 6.8 children per woman. Because there are two major peaks, this is a *bimodal distribution.* The two peaks suggest that we can consider two distinct groups of countries, differentiated by high and low values of fertility rate. The stem-and-leaf plot of per capita calorie supplies in Figure 2-5 also looks bimodal. As we noted when discussing that plot, there seem to

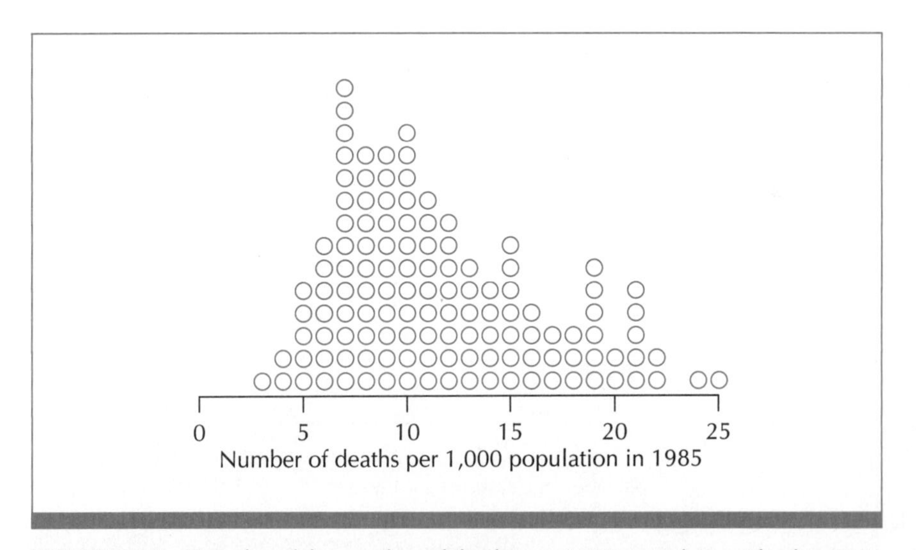

FIGURE 2-14 Dot plot of the number of deaths per 1,000 population (death rate) in 1985 for 125 countries. Three countries are excluded because of missing values. This is a unimodal distribution.

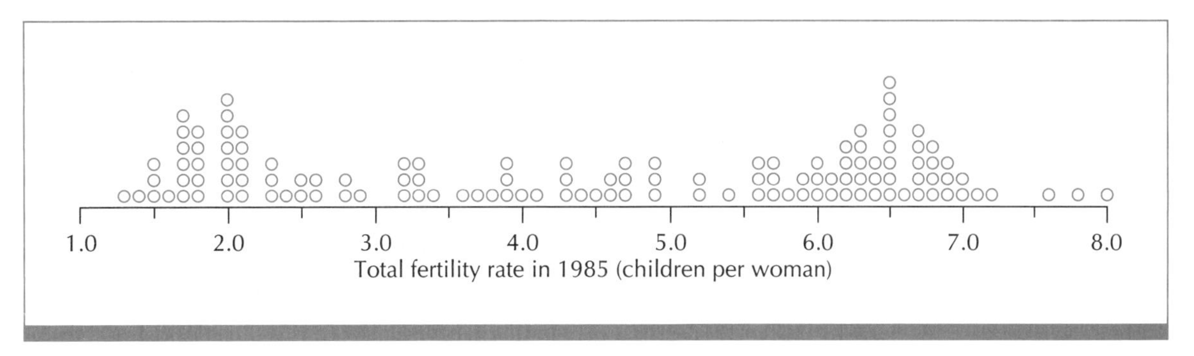

FIGURE 2-15 Dot plot of total fertility rate in 1985 for 125 countries. Three countries are excluded because of missing values. This is a bimodal distribution.

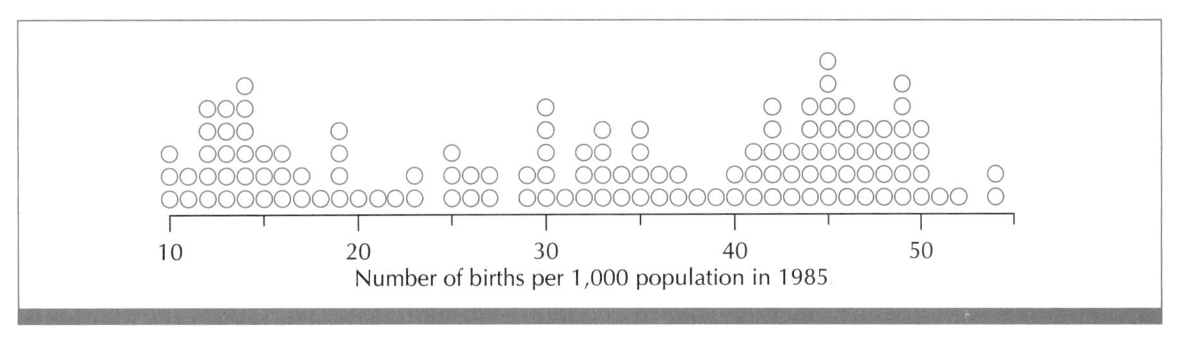

FIGURE 2-16 Dot plot of number of births per 1,000 population (birth rate) in 1985 for 125 countries. Three countries are excluded because of missing values. This is a multimodal distribution.

be two groups of countries, corresponding to adequate and inadequate per capita calorie supplies.

> A distribution is **bimodal** if it has two major peaks or concentrations of values.

We say a distribution with several peaks or modes is *multimodal*. The birth rates illustrated in Figure 2-16 form such a distribution. Values are concentrated in three intervals: from 12 to 16, from 30 to 35, and from 42 to 50 births per 1,000 population. We might classify the countries into three distinct groups based on birth rates. The life expectancies in Figure 2-1 also form a multimodal distribution; countries could be separated into several groups based on life expectancy.

> A distribution is **multimodal** if it has several major peaks or concentrations of values.

In Section 2-5 we use box plots and box graphs to summarize the distribution of a variable and to compare distributions. To discuss box plots and box graphs, we first have to introduce quantiles.

2-5

Quantiles, Box Plots, and Box Graphs

Numbers that divide an ordered list of values into equal or approximately equal sized groups are called *quantiles*. Because they summarize the information in a set of values, quantiles are descriptive statistics. Descriptive statistics are the subject of the next chapter, so you might wonder why we are discussing them here. We do so because quantiles are used in the construction of box plots and box graphs, graphical tools useful for summarizing the distribution of a variable and for comparing two or more distributions. Since graphical tools are the main subject of Chapter 2, box plots and box graphs (and therefore quantiles) are introduced here rather than in Chapter 3.

Quantiles divide an ordered list of values into equal or approximately equal sized groups. If two values satisfy the definition of a quantile, we let the quantile equal the average of the two values.

Commonly used quantiles include the *median, quartiles, deciles,* and *percentiles*. The median divides an ordered list of values in half. Quartiles, deciles, and percentiles divide an ordered list of values into 4, 10, and 100 groups, respectively (if the data set is large enough).

The **median** divides an ordered list of values in half: At least half the values are less than or equal to the median and at least half are greater than or equal to the median.

Quartiles divide an ordered list of values into 4 groups of equal or approximately equal size. At least one-fourth of the values are less than or equal to the **first quartile** and at least three-fourths are greater than or equal to the first quartile. The **second quartile** is the same as the median. At least three-fourths of the values are less than or equal to the **third quartile** and at least one-fourth are greater than or equal to the third quartile.

Deciles divide an ordered list of values into 10 groups of equal or approximately equal size, provided there are enough values. At least $10x$% of the values are less than or equal to the xth decile and at least $(100 - 10x)$% of the values are greater than or equal to the xth decile.

Percentiles divide an ordered list of values into 100 groups of equal or approximately equal size, provided there are enough values. At least x% of the values are less than or equal to the xth percentile and at least $(100 - x)$% of the values are greater than or equal to the xth percentile.

The median, which is the same as the 50th percentile, the fifth decile, and the second quartile, divides a set of ordered values into two groups. The median of an odd number of values is the middle value. For the three numbers 1, 2, and 5, the median is the middle value, 2. The median of an even number of values is typically defined to be the average of the two middle values. For instance, the median of the four numbers 1, 2, 5, and 8 is 3.5, the average of the middle values 2 and 5.

Box plots and *box graphs* are graphical displays based on quantiles (Tukey, 1977). The *box plot* or *box-and-whisker plot* is defined by five numbers: the minimum, the first quartile, the median, the third quartile, and the maximum value for a variable. The first and third quartiles define the extremes of the box. The median is indicated within the box. The minimum and maximum values determine the whiskers. Figure 2-17 shows the structure of a box plot.

> A **box plot** is a graphical summary of the distribution of a variable. The box plot is constructed from five descriptive statistics: the minimum (or smallest value), the first quartile, the median, the third quartile, and the maximum (or largest value).

The box plot provides a brief summary of location, variation, and extreme values of a distribution. Box plots are especially useful for comparing two or more distributions. For example, Figure 2-18 shows box plots of life expectancy for 35 low-income countries and 19 industrial market countries. Before discussing the information contained in this figure, let's find the numbers we need to construct these two box plots.

Consider first the 35 nonmissing values of life expectancy for low-income countries: 40, 40, 44, 44, 45, 45, 45, 45, 45, 46, 46, 47, 47, 47, 48, 48, 48, 49, 49, 49, 51, 51, 51, 51, 52, 52, 52, 53, 54, 54, 56, 59, 65, 69, 70. To find the median, we can start counting at the smallest value and continue until we come to the middle value. (The middle in this case is the 18th value. To find how far to count, divide the sample size by 2 and then round up to the nearest integer. In this case, the sample size is 35, and 35/2 = 17.5, which we round up to 18.) Alternatively, we could start counting at the largest value, until we come to the middle. Either way, we find that the median of these 35 life expectancies is 49.

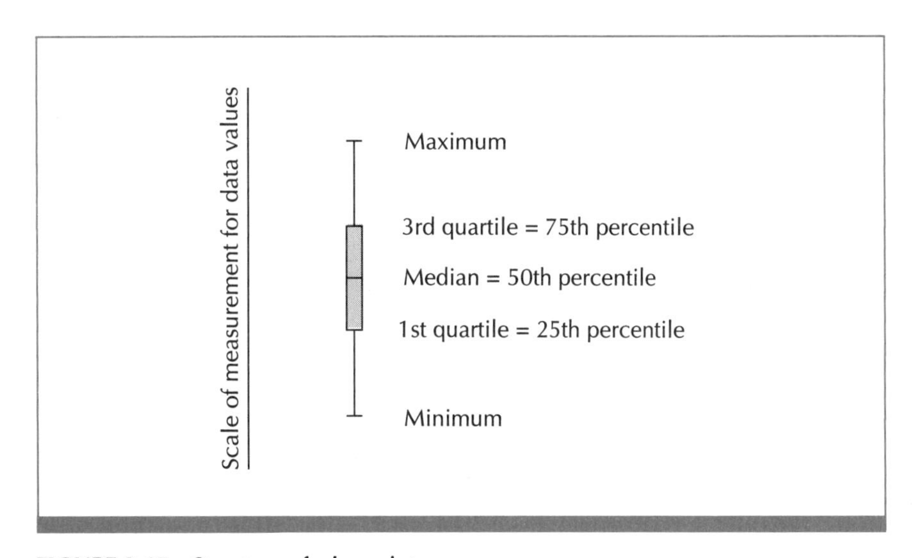

FIGURE 2-17 Structure of a box plot

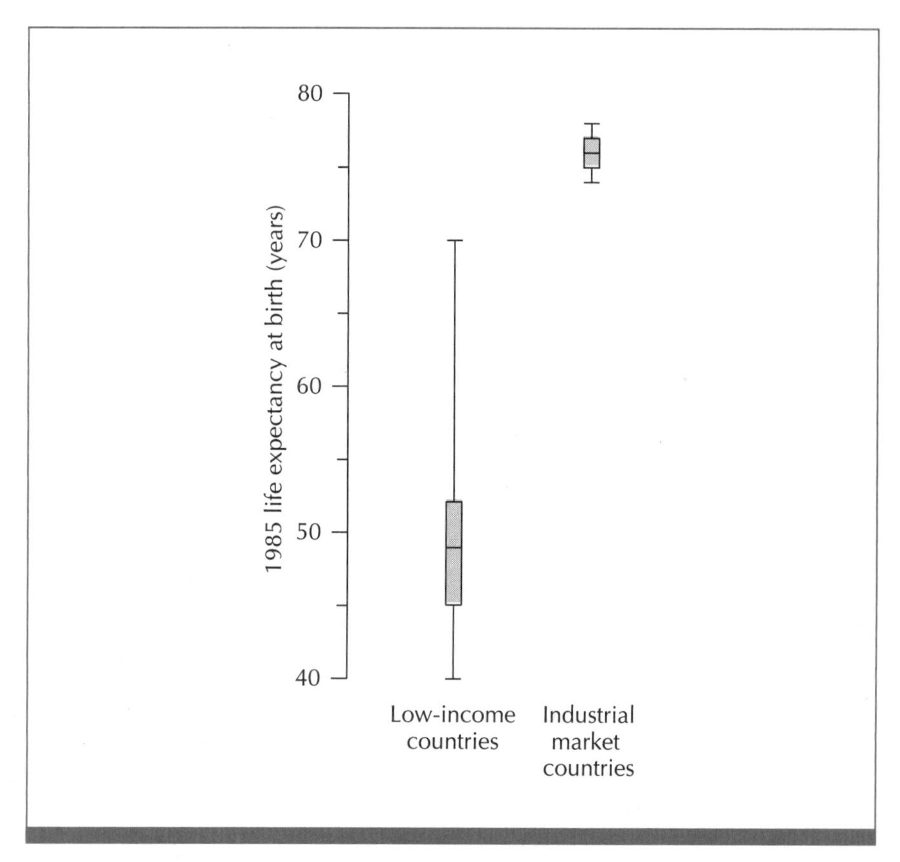

FIGURE 2-18 Box plots of 1985 life expectancy at birth, for 35 low-income countries and 19 industrial market countries. Two low-income countries are excluded because of missing values.

The first quartile, or 25th percentile, is the number with at least one-quarter of the values below and at least three-quarters above. Since there are 35 life expectancies for low-income countries, we start counting at the smallest and continue until we get to the 9th smallest value. (To find how far to count, divide the sample size by 4 and then round up to the nearest integer. In this case, 35/4 = 8.75, which we round to 9.) Doing this, we find the first quartile to be 45. The third quartile, or 75th percentile, is the number with at least three-quarters of the values below and at least one-quarter above. Because there are 35 values, we can start counting at the largest and continue until we get to the 9th largest value. If we do this, we find the third quartile to be 52.

The extreme life expectancies for the low-income countries are easy to see in the ordered list above. The five summary numbers for constructing the box plot of life expectancy for the low-income countries are the minimum, 40; the first quartile, 45; the median, 49; the third quartile, 52; and the maximum, 70. Since these summary numbers are based on quartiles, we know that approximately one-fourth of the life expectancies for these low-income countries are in each of these intervals: 40–45, 45–49, 49–52, and 52–70.

The life expectancies for the 19 industrial market countries are: 74, 74, 74, 75, 75, 75, 75, 76, 76, 76, 77, 77, 77, 77, 77, 77, 77, 78, 78. You can show for yourself that the five summary numbers for constructing the box plot of life expectancy for the industrial countries are the minimum, 74; the first quartile, 75; the median, 76; the third quartile, 77; and the maximum, 78.

The simple plots in Figure 2-18 illustrate that life expectancies are very different for the low-income and industrial market countries. Not only are the centers of the two sets of values widely separated, but also there is no overlap between the two ranges of values. The longest life expectancy among the low-income countries (maximum = 70 years) is 4 years less than the shortest life expectancy among the industrial market countries (minimum = 74 years). Life expectancies for the industrial market countries appear to be fairly symmetrical about the median. The distribution of life expectancies for the low-income countries appears less symmetrical. Dot plots (or stem-and-leaf plots or histograms) would provide more information on shape and symmetry of these two distributions.

Box Graphs

A *box graph* is similar to a box plot, with more information (Tukey, 1977; Cleveland, 1985). As illustrated in Figure 2-19, the box is the same as in a box

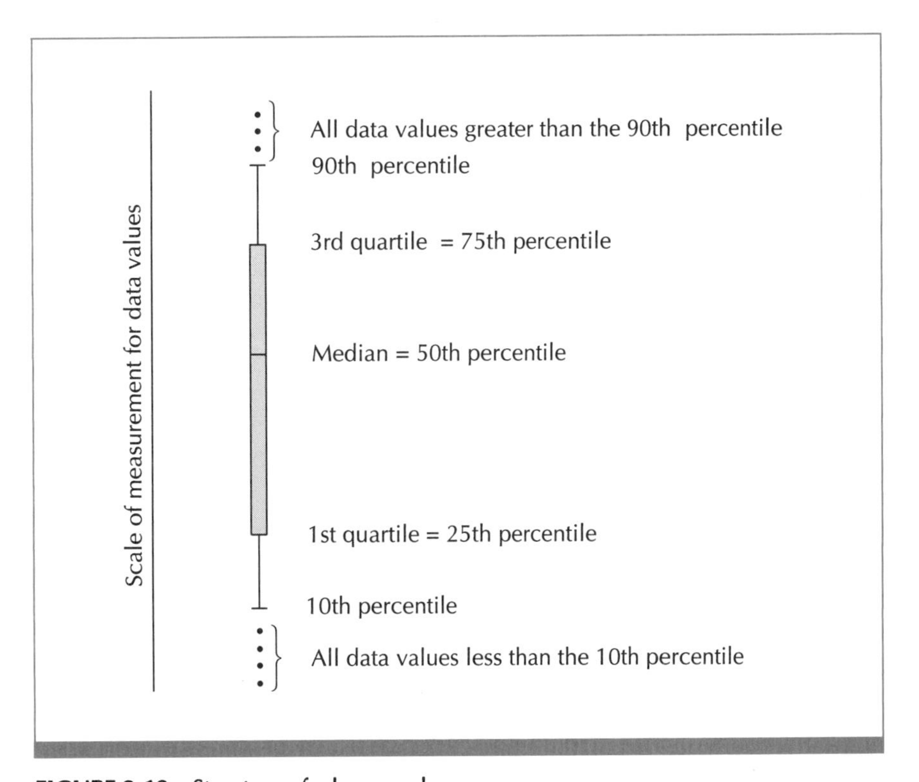

FIGURE 2-19 Structure of a box graph

plot. The 10th and 90th percentiles are the endpoints of the whiskers. In addition, values greater than the 90th percentile and less than the 10th percentile are plotted.

A **box graph** is a graphical summary of the distribution of a variable. The box graph is constructed using the 10th percentile, the first quartile, the median, the third quartile, and the 90th percentile, plus all values less than the 10th percentile and all values greater than the 90th percentile.

Box graphs of population growth are displayed in Figure 2-20 for 35 low-income countries and 19 industrial market countries. The 35 nonmissing values of population growth for the low-income countries are: 1.2, 1.4, 1.8, 2.0, 2.0, 2.2, 2.2, 2.2, 2.3, 2.3, 2.4, 2.4, 2.5, 2.5, 2.6, 2.6, 2.6, 2.6, 2.7, 2.7, 2.9, 2.9, 3.0, 3.0, 3.0, 3.1, 3.1, 3.1, 3.2, 3.2, 3.3, 3.3, 3.5, 3.5, 4.1. For the box-and-whisker portion of the graph, we find the 10th percentile, 2.0; the first quartile, 2.3; the

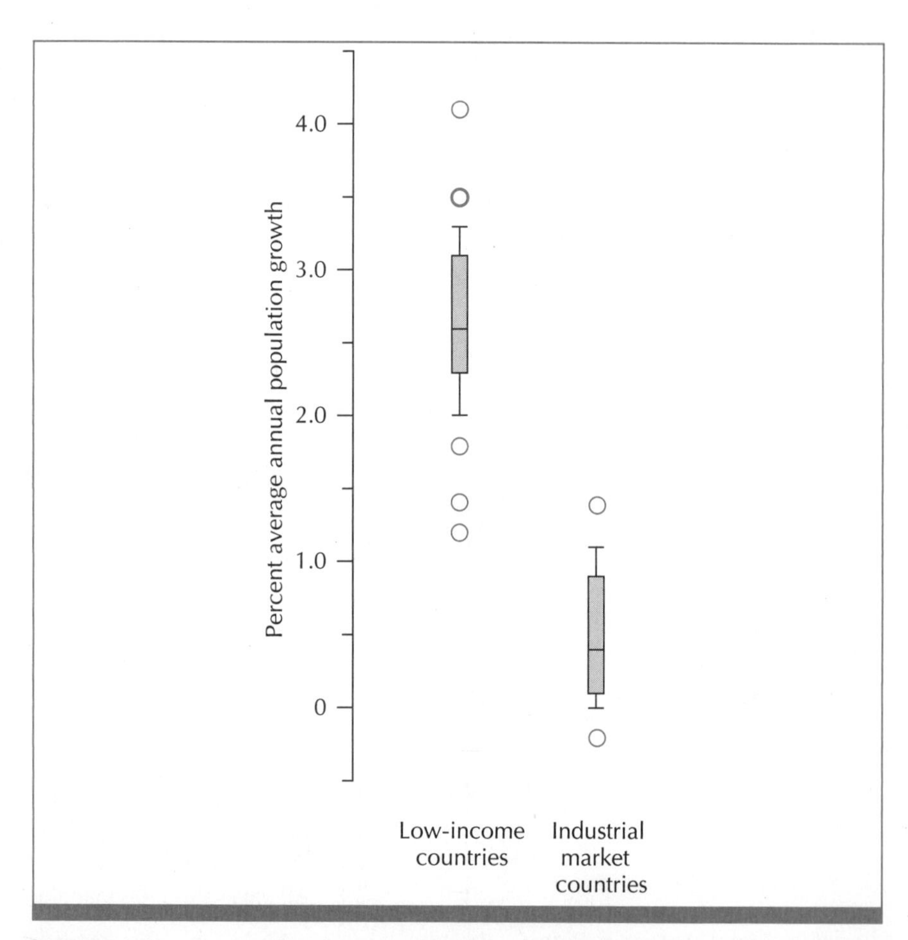

FIGURE 2-20 Box graphs of percent average annual growth of population over the period 1980–1985, for 35 low-income countries and 19 industrial market countries. Two low-income countries are excluded because of missing values.

median, 2.6; the third quartile, 3.1; and the 90th percentile, 3.3. In addition, we plot the values less than the 10th percentile value of 2.0: 1.2, 1.4, and 1.8. We also plot the values greater than the 90th percentile value of 3.3: 3.5, 3.5, and 4.1.

The 19 values of population growth for the industrial market countries are: −.2, 0, .1, .1, .1, .1, .2, .3, .3, .4, .5, .6, .7, .7, .9, .9, 1.0, 1.1, 1.4. To draw the box-and-whisker portion of the graph, we need the 10th percentile, 0; the first quartile, .1; the median, .4; the third quartile, .9; and the 90th percentile, 1.1. We also plot the value −.2 that is less than the 10th percentile and the value 1.4 that is greater than the 90th percentile.

Comparing the box graphs in Figure 2-20 shows that the two groups of countries have very different levels of population growth. There is only slight overlap between the two sets of values. The range from the minimum to the maximum is somewhat larger for the low-income countries than for the industrial market nations. However, the spread is very similar in the box-and-whisker portions of the two graphs. These two distributions appear to have different locations but similar variation.

In Chapter 3 we will discuss descriptive statistics that summarize location or central tendency and descriptive statistics that summarize spread or variation in a set of values.

Summary of Chapter 2

One way to study the values of a variable is to list all of them. A list in which the values are ordered from smallest to largest (or vice versa) is especially useful. A dot plot represents values as points along a number line, providing a clear picture of the distribution of values of the variable. A stem-and-leaf plot is more compact than a dot plot; it uses the actual values taken on by a variable in a graphical display that summarizes the shape of the distribution of values.

It is often useful to summarize the values of a variable rather than to list or plot all values. Frequency tables, frequency plots, and histograms are useful for this purpose.

Plots such as dot plots, stem-and-leaf plots, and histograms help us see the symmetry or skewness of a distribution of values. We also look for modes or peaks, which are concentrations of values in a distribution.

Box plots and box graphs are simple graphical displays that provide information about the location, spread, and extremes of a set of values. These plots are very simple and are especially useful for comparing distributions.

Minitab Appendix for Chapter 2

The examples in this appendix are based on the Minitab worksheet we described in the Minitab Appendix for Chapter 1, using the data in Exercise 5-35.

Creating a Dot Plot

To produce a dot plot, we use the command DOTPLOT. For a dot plot of life expectancies for all 40 countries, we use

```
MTB> dotplot 'life'
```

or

```
MTB> dotplot c3
```

to produce the plot in Figure M2-1.

To display a dot plot of life expectancies for the 20 countries with low child mortality, we can use the data in column 13:

```
MTB> dotplot 'life1'
```

or

```
MTB> dotplot c13
```

to produce the plot in Figure M2-2.

With the command

```
MTB> dotplot 'life2'
```

we get the plot in Figure M2-3. We could type C23 instead of 'LIFE2' to get the same results.

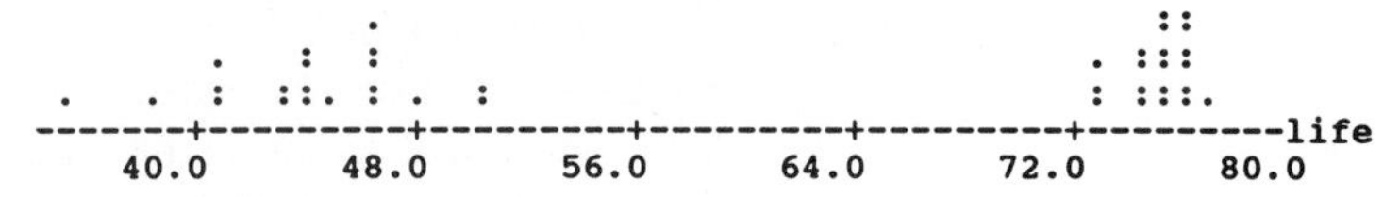

FIGURE M2-1 Dot plot of life expectancies for the 40 countries in Exercise 5-35

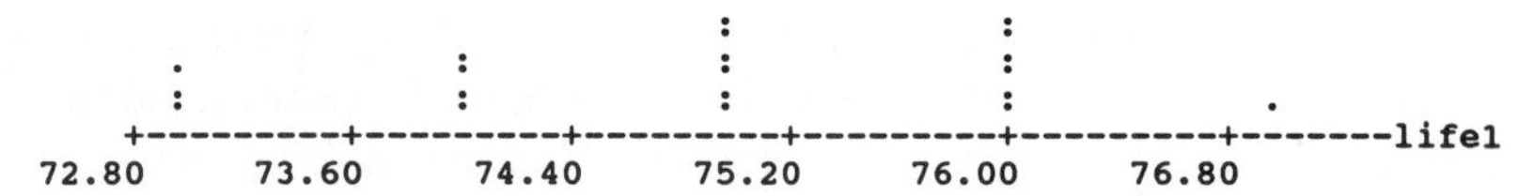

FIGURE M2-2 Dot plot of life expectancies for the 20 countries with low child mortality in Exercise 5-35

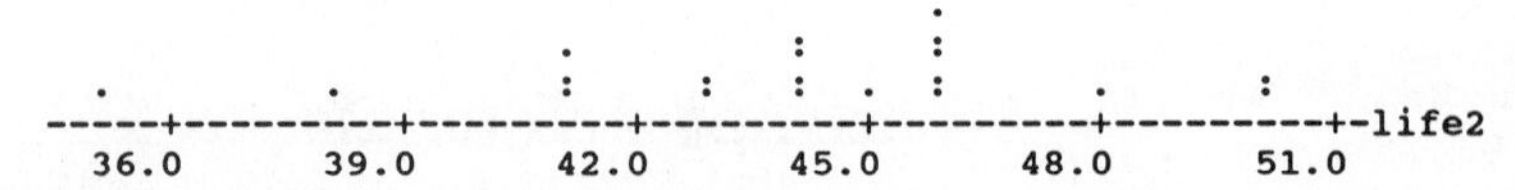

FIGURE M2-3 Dot plot of life expectancies for the 20 countries with high child mortality in Exercise 5-35

Creating Several Dot Plots with the Same Scale

To produce the dot plots in Figures M2-2 and M2-3 using the same scale, we can use the SAME subcommand with the DOTPLOT command:

```
MTB>   dotplot 'life1' 'life2';
SUBC>  same.
```

The semicolon indicates that a subcommand is to follow; we type a period at the end of the subcommand. We could use C13 instead of 'LIFE1' and C23 instead of 'LIFE2'. The resulting plots are displayed in Figure M2-4.

We can obtain the same plots as in Figure M2-4 using the LIFE data in C3 and the BY subcommand with MORTCODE, which is in C6:

```
MTB>   dotplot 'life';
SUBC>  by 'mortcode'.
```

The BY subcommand performs the main command within each level of the specified variable. The variable following a BY subcommand must be what Minitab calls a classification variable, having only integer values between −9999 and +9999.

Creating a Stem-and-Leaf Plot

We use the STEM-AND-LEAF command to produce stem-and-leaf plots. The command

```
MTB>  stem 'calorie'
```

produces the plot in Figure M2-5.

The display shows that 37 countries are included in the plot, with three missing values. The middle column gives the stems, the values to the right being the leaves. In the leftmost column, the number in parentheses is the number of values in the row (stem) containing the median. For these 37 values, the median or 50th percentile is 114. There are two values of CALORIE in the row containing the median, 114. For rows above the median row, the number on the left shows the number of values in that row and above. There are 17 values of CALORIE in rows for stems 6–10, for instance. For rows below the

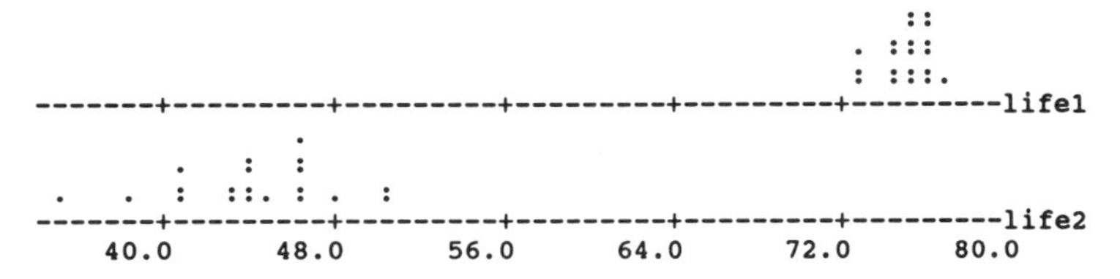

FIGURE M2-4 Dot plots of life expectancies for the 20 countries with low child mortality (LIFE1) and for the 20 countries with high child mortality (LIFE2)

```
Stem-and-leaf of calorie     N  = 37
Leaf Unit = 1.0              N* =  3

   2     6 88
   3     7 1
   3     7
   4     8 4
   7     8 579
  11     9 1123
  15     9 5778
  17    10 22
  17    10
 (2)    11 34
  18    11 5556
  14    12 2
  13    12 899
  10    13 0012
   6    13 79
   4    14 0023
```

FIGURE M2-5 Stem-and-leaf plot of per capita calorie supply as percentage of requirements for the 40 countries in Exercise 5-35

```
Stem-and-leaf of calspply  N  = 124
Leaf Unit = 100

   1     1 5
   7     1 667777
  13     1 888999
  33     2 00000000011111111111
  48     2 222222223333333
  62     2 44444455555555
  62     2 666666667777
  50     2 88888999
  42     3 00001111111111
  28     3 2222333333
  18     3 444444555
   9     3 66666677
   1     3 8
```

FIGURE M2-6 Stem-and-leaf plot of 1985 calorie supply per capita for 124 countries. The stem is in thousands of calories and the leaf is in hundreds of calories. These are the same data plotted in Figure 2-5.

median row, the number on the left shows the number of values in that row and below. For example, there are 10 values in the rows for stems 13 and 14. To obtain the original data values from the stem-and-leaf display, multiply each value in the plot by the leaf unit (1 in this example).

We can use the BY subcommand with STEM-AND-LEAF. The command

```
MTB>   stem 'calorie';
SUBC>  by 'mortcode'.
```

results in two stem-and-leaf plots, one for countries with MORTCODE = 1 and one for countries with MORTCODE = 2.

In Minitab, stems and leaves must be single digits, so we cannot directly produce a stem-and-leaf plot such as the one in Figure 2-5. Suppose the nonmissing World Bank values of per capita calorie supply, for 124 nations, are in a column named CALSPPLY on our worksheet. If we ask for a stem-and-leaf plot of CALSPPLY, we get the output in Figure M2-6. Note that the single-digit stems represent thousands of calories and the single-digit leaves represent hundreds of calories. For these 124 values of CALSPPLY, Minitab prints a stem-and-leaf plot with five rows per stem: for leaves 0 and 1, 2 and 3, 4 and 5, 6 and 7, 8 and 9.

Creating a Histogram

Let's return now to our worksheet for Exercise 5-35. For a histogram of the CALORIE variable, we use

```
MTB> histogram 'calorie'
```

to obtain the histogram in Figure M2-7. Note that in Minitab figures and tables, N denotes the number of nonmissing values for a variable, while N* denotes the number of missing values.

We can control the construction of the histogram with the INCREMENT and START subcommands. INCREMENT specifies the interval width and START specifies the midpoint of the first interval. The command

```
MTB>   histogram 'calorie';
SUBC>  increment=10;
SUBC>  start=55.
```

will produce the output in Figure M2-8. Note that the order in which we give the subcommands is not important.

```
Histogram of calorie   N = 37   N* = 3

Midpoint   Count
      70       3  ***
      80       1  *
      90       7  *******
     100       6  ******
     110       2  **
     120       5  *****
     130       7  *******
     140       6  ******
```

FIGURE M2-7 Histogram of CALORIE for 40 countries in Exercise 5-35

```
Histogram of calorie   N = 37   N* = 3

Midpoint   Count
    55.0       0
    65.0       2  **
    75.0       1  *
    85.0       4  ****
    95.0       8  ********
   105.0       2  **
   115.0       6  ******
   125.0       4  ****
   135.0       6  ******
   145.0       4  ****
```

FIGURE M2-8 Histogram of CALORIE for the 40 countries in Exercise 5-35, using HISTOGRAM with INCREMENT and START subcommands

We can use the BY and SAME subcommands with the HISTOGRAM command, as we can with DOTPLOT.

The HISTOGRAM output includes a frequency table. For discrete (classification or qualitative) data, an appropriate choice of interval width will produce a frequency table and frequency plot showing individual values.

Producing a Frequency Table with the TALLY Command

We can use the TALLY command to obtain a frequency table for a classification variable (a variable that takes only integer values between −9999 and +9999 or missing values). The command

```
MTB>  tally 'calcode'
```

produces a frequency table for the CALCODE variable:

```
calcode    COUNT
   1          3
   2         12
   3          9
   4         13
   N=        37
   *=         3
```

If we use the subcommand ALL, the output includes counts, cumulative counts, percentages, and cumulative percentages. Thus, the command

```
MTB>   tally 'calcode';
SUBC>  all.
```

yields the output

```
calcode    COUNT    CUMCNT    PERCENT    CUMPCT
   1          3         3       8.11       8.11
   2         12        15      32.43      40.54
   3          9        24      24.32      64.86
   4         13        37      35.14     100.00
   N=        37
   *=         3
```

Creating a Box Plot

The BOXPLOT command produces box plots. (We cannot get box graphs with Minitab.) If we use

```
MTB>  boxplot 'weight'
```

we get the box plot in Figure M2-9.

```
             -----------------------------
          ---I        +                  I-------------------------
             -----------------------------
      +---------+---------+---------+---------+---------+------weight
     3.5       7.0      10.5      14.0      17.5      21.0
```

FIGURE M2-9 Box plot of the variable WEIGHT for 40 countries in Exercise 5-35

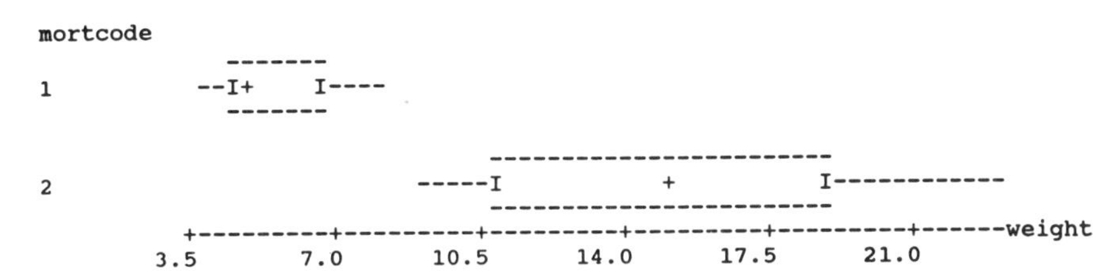

FIGURE M2-10 Box plots of WEIGHT for the 20 countries with MORTCODE = 1 and for the 20 countries with MORTCODE = 2

We can use the BY subcommand, as follows:

```
MTB>   boxplot 'weight';
SUBC>  by 'mortcode'.
```

Two box plots will be displayed, one for countries with MORTCODE = 1 and one for countries with MORTCODE = 2, as in Figure M2-10.

Exercises for Chapter 2

In the exercises, answer the following questions: What would you need to know about the sample to be willing to use it to make inferences about a larger population? What is that larger population (if any)? What limitations do you see in the sample?

For each figure and table, include a legend that completely describes its contents. Note the number of cases included and the number excluded if there are missing values.

EXERCISE 2-1

The numbers of doctoral degrees awarded in the United States in 1984–1985 are listed here by race/ethnic group (American Council on Education, 1987):

Group	Number of doctoral degrees	Percent of total
American Indian	119	.4
Asian/Pacific Islander	1,106	3.4
Black	1,154	3.6
Hispanic	677	2.1
Nonresident Alien	5,317	16.5
White	23,934	74.1
Total	32,307	100.0

Construct a frequency plot of this information. Do you think you can evaluate the information more easily using the table or the plot, or do you think there is no big difference?

EXERCISE 2-2

In a study of domestic water usage conducted from July 1981 to June 1982 in Perth, Western Australia, average toilet flush volume (in liters) was determined for 147 households (James and Knuiman, 1987):

4.0	4.0	4.5	5.4	6.0	6.0	6.1	6.1	6.4	6.5	6.5
6.6	6.9	7.0	7.0	7.0	7.0	7.0	7.0	7.0	7.2	7.3
7.4	7.5	7.5	7.6	7.7	7.8	8.0	8.0	8.0	8.0	8.0
8.0	8.1	8.2	8.2	8.2	8.2	8.4	8.5	8.5	8.6	8.7
8.7	8.8	8.9	8.9	9.0	9.0	9.0	9.0	9.0	9.2	9.2
9.2	9.3	9.4	9.5	9.5	9.6	9.7	9.7	9.7	9.8	9.8
9.9	10.0	10.0	10.0	10.0	10.0	10.0	10.0	10.0	10.0	10.0
10.0	10.0	10.0	10.0	10.0	10.1	10.1	10.1	10.4	10.5	10.5
10.5	10.6	10.6	10.6	10.6	10.6	10.7	10.8	10.9	11.0	11.0
11.0	11.0	11.0	11.0	11.0	11.0	11.0	11.0	11.0	11.1	11.2
11.2	11.4	11.4	11.4	11.4	11.4	11.5	11.5	11.5	11.5	11.6
11.8	11.8	11.8	11.9	12.0	12.0	12.0	12.0	12.0	12.2	12.3
12.4	12.5	12.5	13.0	13.0	13.0	13.1	13.4	13.5	14.1	14.6
15.0	15.0	16.0	20.0							

a. Construct a stem-and-leaf plot of these 147 toilet flush volumes.

b. How many peaks does this distribution have? (Is the distribution unimodal, bimodal, or multimodal?)

c. Would you describe this distribution as symmetrical, negatively skewed, positively skewed, or none of these?

d. What can you say about this sample of toilet flush volumes based on the stem-and-leaf plot?

e. Why do you think there are so many 0 leaves in this sample? (*Hint:* Volumes were recorded by human chart readers.)

EXERCISE 2-3

In a study of domestic water usage conducted in Perth, Western Australia, from July 1981 to June 1982, average shower flow rate (in liters/minute) was determined for 129 houses (James and Knuiman, 1987):

2.2	2.3	3.2	3.3	3.4	3.4	3.5	3.6	3.7	3.7	3.8
3.9	4.0	4.1	4.3	4.5	4.6	4.8	4.8	4.9	5.0	5.0
5.0	5.0	5.1	5.1	5.1	5.4	5.4	5.5	5.5	5.6	5.6
5.6	5.7	5.8	5.9	6.0	6.0	6.0	6.0	6.1	6.2	6.2
6.2	6.2	6.3	6.3	6.4	6.4	6.4	6.5	6.6	6.6	6.6
6.7	6.7	6.8	6.9	6.9	6.9	6.9	7.0	7.0	7.0	7.1
7.2	7.2	7.3	7.3	7.4	7.5	7.5	7.5	7.5	7.5	7.6
7.6	7.8	8.0	8.2	8.2	8.3	8.3	8.4	8.4	8.8	9.0
9.1	9.2	9.2	9.3	9.3	9.3	9.3	9.5	9.6	9.6	9.6

9.7	9.8	9.8	10.2	10.3	10.4	10.4	10.4	10.5	10.5	10.6
10.8	10.8	11.2	11.3	11.3	11.5	11.9	11.9	11.9	12.3	12.7
13.8	14.3	14.6	15.0	15.0	15.3	15.5	18.9			

a. Construct a stem-and-leaf plot of these 129 shower flow rates.

b. How many peaks does this distribution have? (Is the distribution unimodal, bimodal, or multimodal?)

c. Would you describe this distribution as symmetrical, negatively skewed, positively skewed, or none of these?

d. What can you say about this sample of shower flow rates from what you see in the stem-and-leaf plot?

e. Can you use this sample of shower flow rates to infer characteristics of a larger population of households? What would you need to know about the sample to feel comfortable doing this?

EXERCISE 2-4

The coldest temperature on record (°F) is shown here for each state (*USA Today,* October 22, 1987, page 10A; from National Climatic Data Center):

Alabama	−27	Louisiana	−16	Ohio	−39
Alaska	−80	Maine	−48	Oklahoma	−27
Arizona	−40	Maryland	−40	Oregon	−54
Arkansas	−29	Massachusetts	−35	Pennsylvania	−42
California	−45	Michigan	−51	Rhode Island	−23
Colorado	−61	Minnesota	−59	South Carolina	−19
Connecticut	−32	Mississippi	−19	South Dakota	−58
Delaware	−17	Missouri	−40	Tennessee	−32
Florida	−2	Montana	−70	Texas	−23
Georgia	−17	Nebraska	−47	Utah	−69
Hawaii	12	Nevada	−50	Vermont	−50
Idaho	−60	New Hampshire	−46	Virginia	−30
Illinois	−35	New Jersey	−34	Washington	−48
Indiana	−35	New Mexico	−50	West Virginia	−37
Iowa	−47	New York	−52	Wisconsin	−54
Kansas	−40	North Carolina	−34	Wyoming	−63
Kentucky	−34	North Dakota	−60		

a. Construct a frequency table and draw a histogram using each of three widths (say, widths 5, 10, and 20). Which interval width do you prefer for the histogram? Why?

b. Is the distribution symmetrical, negatively skewed, positively skewed, or none of these?

c. Is the distribution unimodal, bimodal, or multimodal?

EXERCISE 2-5

Stress loads at which 41 graphite beams fractured (bend stress, MPa $\times$ 10^{-2}) are listed below (Cheng, 1987; from Cooper, 1984):

2.7555	2.9890	3.0065	3.0649	3.1233	3.1525	3.1525
3.1817	3.2225	3.2284	3.2692	3.2984	3.3276	3.3276
3.3743	3.3743	3.3860	3.3860	3.3860	3.4152	3.4152
3.4152	3.4444	3.4619	3.4736	3.4736	3.5028	3.5028
3.5320	3.5436	3.5611	3.5611	3.5728	3.5903	3.6195
3.6779	3.7071	3.7363	3.7363	3.7363	4.0282	

a. Construct a frequency table and histogram using each of three widths (say, widths of .1, .2, and .5). Which interval width do you prefer for the histogram? Why?

b. Is the distribution unimodal, bimodal, or multimodal?

c. Construct a box plot of these observations.

d. Would you describe this distribution as symmetrical, negatively skewed, positively skewed, or none of these?

e. Compare the information about these stress loads that you obtain from the histograms and the box plot.

EXERCISE 2-6

A stem-and-leaf plot can provide a quick way to look at a distribution when you are doing some data analysis by hand, rather than on the computer. Consider the following unordered list of strength measurements (in pounds per square inch) on 100 samples of yarn (Duncan, 1974, page 67; from U.S. Department of Agriculture, 1945):

66	117	132	111	107	85	89	79	91	97	138	103	111
86	78	96	93	101	102	110	95	96	88	122	115	92
137	91	84	96	97	100	105	104	137	80	104	104	106
84	92	86	104	132	94	99	102	101	104	107	99	85
95	89	102	100	98	97	104	114	111	98	99	102	91
95	111	104	97	98	102	109	88	91	103	94	105	103
96	100	101	98	97	97	101	102	98	94	100	98	99
92	102	87	99	62	92	100	96	98				

a. Start a stem-and-leaf plot with stems in tens and leaves in units. Use two rows per stem. Your stems will go from 6 to 13 (two rows each).

b. Now write in the leaves as you come to them in the list of yarn strengths. The first leaf is 6, in the second row for stem 6, since the first yarn strength is 66. Reading across, the next yarn strength is 117. The leaf will be 7, in the second row for stem 11. And so on. You will end up with a stem-and-leaf plot in which the leaves are not listed in order of magnitude. From this plot you can easily rearrange the leaves in order to get a stem-and-leaf plot with ordered leaves.

c. Is this distribution of yarn strengths unimodal, bimodal, or multimodal?

d. Would you describe the distribution as symmetrical, negatively skewed, positively skewed, or none of these?

EXERCISE 2-7

Survival times (in days from diagnosis) are listed below for 43 patients with chronic granulocytic leukemia (Hollander and Wolfe, 1973, pages 251–252; Siddiqui and Gehan, 1966):

7	47	58	74	177	232	273	285	317
429	440	445	455	468	495	497	532	571
579	581	650	702	715	779	881	900	930
968	1,077	1,109	1,314	1,334	1,367	1,534	1,712	1,784
1,877	1,886	2,045	2,056	2,260	2,429	2,509		

a. What was the median survival time for this group of patients?

b. Construct a box graph for these survival times.

c. Describe the distribution in terms of shape, location, variation, and symmetry or skewness.

EXERCISE 2-8

Numbers of pregnancies per 1,000 girls 15–19 years of age in 1980 are listed below by state (*USA Today,* December 11, 1986, page 7A; from Alan Guttmacher Institute):

Alabama	117.3	Louisiana	118.1	Ohio	101.3
Alaska	124.2	Maine	86.9	Oklahoma	119.5
Arizona	123.2	Maryland	122.5	Oregon	118.7
Arkansas	117.2	Massachusetts	85.7	Pennsylvania	90.3
California	140.2	Michigan	102.4	Rhode Island	83.1
Colorado	113.7	Minnesota	77.0	South Carolina	113.7
Connecticut	80.7	Mississippi	125.0	South Dakota	86.4
Delaware	105.6	Missouri	106.4	Tennessee	113.0
Florida	131.2	Montana	93.3	Texas	137.0
Georgia	130.9	Nebraska	80.7	Utah	94.6
Hawaii	105.6	Nevada	144.0	Vermont	94.8
Idaho	96.4	New Hampshire	80.7	Virginia	107.4
Illinois	100.6	New Jersey	95.8	Washington	122.3
Indiana	101.9	New Mexico	125.6	West Virginia	103.6
Iowa	79.0	New York	100.7	Wisconsin	84.8
Kansas	101.0	North Carolina	110.3	Wyoming	126.6
Kentucky	110.7	North Dakota	74.8		

a. Construct a frequency table and draw a histogram summarizing these pregnancy rates.

b. Is the distribution symmetrical, negatively skewed, positively skewed, or none of these?

c. Is the distribution unimodal, bimodal, or multimodal?

EXERCISE 2-9

Consider the following list of shear strengths of welds of stainless steel, in pounds per weld (Duncan, 1974, page 67):

2,385	2,280	2,330	2,360	2,350	2,350	2,370	2,310	2,280
2,310	2,310	2,330	2,280	2,290	2,190	2,280	2,270	2,260
2,250	2,260	2,270	2,270	2,305	2,310	2,340	2,330	2,340
2,350	2,360	2,360	2,340	2,280	2,290	2,350	2,330	2,280
2,285	2,250	2,340	2,330	2,350	2,275	2,190	2,240	2,230
2,210	2,220	2,190	2,230	2,160	2,270	2,400	2,350	2,360
2,360	2,300	2,350	2,340	2,290	2,250	2,270	2,340	2,310
2,360	2,300	2,430	2,340	2,440	2,370	2,340	2,360	2,340
2,330	2,380	2,350	2,360	2,390	2,360	2,400	2,320	2,360
2,350	2,340	2,320	2,350	2,330	2,320	2,300	2,280	2,230
2,290	2,270	2,290	2,270	2,270				

a. Construct a stem-and-leaf plot of these weld strengths. Use the first two digits as the stem and the last two digits as the leaf. Your stems will go from 21 to 24. Also, use two rows per stem. Construct the plot from the weld strengths as they are listed. You can always rearrange the leaves later to get a stem-and-leaf plot with ordered leaves.

b. Use the information in the stem-and-leaf plot to construct a histogram. Compare the information provided by the two graphs.

c. Is the distribution of weld strengths unimodal, bimodal, or multimodal?

d. Would you describe this distribution as symmetrical, negatively skewed, positively skewed, or none of these?

EXERCISE 2-10

Our ability to read a sign at night depends on the light intensity near the sign. The following are measurements of light intensity (in candela per square meter) for 30 highway signs in a city (Milton and Arnold, 1986, page 249; based on "Use of Retroreflectors in the Improvement of Nighttime Highway Visibility," by H. Waltman in *Color*):

10.9	1.7	9.5	2.9	9.1	3.2	9.1	7.4	13.3	13.1
6.6	13.7	1.5	7.4	9.9	13.6	17.3	3.6	4.9	13.1
7.8	10.3	10.3	9.6	5.7	6.3	2.6	15.1	2.9	16.2

a. Construct a stem-and-leaf plot of these intensity measurements.

b. Describe the distribution in terms of location, variability, peaks, and symmetry or skewness.

EXERCISE 2-11

Consider the following measurements of rainfall (in acre-feet) from 26 seeded clouds (Devore, 1987, page 30; from "A Bayesian Analysis of a Multiplicative Treatment Effect in Weather Modification," *Technometrics,* 1975, pages 161–166):

4.1	7.7	17.5
31.4	32.7	40.6
92.4	115.3	118.3
119.0	129.6	198.6

200.7	242.5	255.0
274.7	274.7	302.8
334.1	430.0	489.1
703.4	978.0	1,656.0
1,697.8	2,745.6	

a. Construct a box graph of these rainfall measurements.

b. Discuss the information provided in this box graph.

EXERCISE 2-12

The following are measurements of sulfur dioxide concentrations (in micrograms per cubic meter) from a damaged Bavarian forest (Milton and Arnold, 1986, page 226; based on Roberts, 1983):

33.4	38.6	41.7	43.9	44.4	45.3	46.1	47.6	50.0	52.4	52.7	53.9
54.3	55.1	56.4	56.5	60.7	61.8	62.2	63.4	65.5	66.6	70.0	71.5

a. Construct a dot plot and a box plot of these measurements. Compare the information provided by the two plots.

b. Is the distribution unimodal, bimodal, or multimodal?

c. Would you describe this distribution as symmetric, negatively skewed, positively skewed, or none of these?

d. The average concentration of sulfur dioxide in undamaged parts of the country was 20 micrograms per cubic meter. Does the evidence here suggest that the damage in the forest might be related to acid rain? Or would you hesitate to suggest such a connection? Explain your answer.

EXERCISE 2-13

Shear strengths of welds in pounds per weld are shown below for 95 samples of the same material (Duncan, 1974, page 69):

146	148	146	140	150	146	146	142	152	148	158	152
146	156	152	150	150	146	148	152	156	162	160	146
156	150	146	154	150	148	148	148	150	152	150	152
152	152	148	152	146	152	150	144	152	154	138	154
160	106	152	150	150	150	152	158	152	156	156	152
154	154	144	148	154	152	152	154	154	158	158	156
160	162	154	148	152	146	158	162	160	160	158	154
160	162	164	150	150	152	144	150	152	154	152	

a. Construct a dot plot and a box graph of these observations. Compare the information provided in these two plots.

b. Is this distribution unimodal, bimodal, or multimodal?

c. Would you describe the distribution as symmetrical, negatively skewed, positively skewed, or none of these?

EXERCISE 2-14

The following list shows percentages of tanks with leaks, from daily inspections of fuel tanks (Duncan, 1974, page 68; from Kauffman, 1945, page 17):

32.6	35.5	44.0	43.3	40.8	40.0	49.4	45.5	46.8	45.9	45.9
49.5	46.4	50.5	50.5	50.0	53.8	52.8	53.3	53.6	57.6	55.7
56.4	58.7	59.3	59.5	59.6	55.9	57.8	55.1	62.7	61.4	60.5
67.3	72.0	72.4	73.3	70.1	72.7	73.8	75.0	75.5	82.3	80.5
80.0	80.5	86.8	54.6							

a. Construct a dot plot and box graph for these observations. Compare the information provided by the two plots.

b. Describe the distribution with respect to location, variation, peaks, and symmetry or skewness.

EXERCISE 2-15

Errors (in meters) made using a lightweight handheld laser range finder to locate an object from 500 meters are shown below (Milton and Arnold, 1986, page 252; from *Civil Engineering,* February 1983, page 52). A "+" indicates an overestimate and a "−" indicates an underestimate.

−.10	−.05	+.01	+.03	+.06	−.07
−.03	+.01	+.03	+.09	−.06	−.02
+.02	+.05	+.10			

a. Construct a dot plot and box plot of these errors. Compare the information provided by these two plots.

b. Describe the distribution of errors in terms of location, variation, and symmetry or skewness.

EXERCISE 2-16

In a study of X-ray microanalysis as a method of chemical analysis, a chemical was analyzed that was in theory 26.6% potassium by weight. Listed here are the percentages of potassium found by X-ray microanalysis of 27 samples (Milton and Arnold, 1986, pages 249–250; based on Kiss, 1983).

21.9	23.1	24.0	24.6	24.9	25.4	26.7	22.0	23.4	24.1	24.7
25.1	25.5	27.2	22.1	23.7	24.2	24.8	25.2	26.5	27.8	22.1
23.8	24.5	24.8	25.3	26.5						

a. Construct a stem-and-leaf plot and box plot of these observations. Compare the information provided in these two plots.

b. Describe the distribution with respect to location, spread, peaks, and symmetry or skewness.

c. How well did X-ray microanalysis seem to do in this experiment?

EXERCISE 2-17

Lifespan (in kilometers driven) was determined for each of 70 aluminum air batteries being developed for electric cars (Milton and Arnold, 1986, page 547; from "Aluminum-Air Battery Development: Toward an Electric Car," *Energy and Technology Review,* June 1983, pages 20–33):

1,625	1,726	2,498	1,942	2,216	1,631	2,101	1,820	2,239
2,037	2,618	2,173	1,902	2,415	1,698	2,587	1,810	2,612
1,733	2,245	1,947	1,622	2,016	1,867	2,831	2,357	1,747

2,417	2,021	2,639	1,650	2,702	1,929	2,381	1,719	2,291
2,093	1,603	3,150	2,109	1,913	1,727	1,672	2,071	2,815
1,871	2,750	2,280	1,763	2,470	2,353	1,893	2,897	2,539
2,150	1,702	1,802	2,925	2,918	1,635	3,070	1,988	2,306
1,616	2,178	1,750	3,200	1,690	2,592	2,072		

a. Construct a frequency table and histogram for these battery lifespans.

b. Describe the distribution in terms of location, spread, peaks, and symmetry or skewness.

EXERCISE 2-18

Numbers of defects found in 29 100-yard pieces of wool cloth are listed below (Duncan, 1974, page 42; data provided by the Bendix Radio Division of Bendix Aviation Corporation):

2	0	0	1	1	2	2	1	0	1	1	0	2	1	
1	5	0	3	1	1	1	2	2	1	0	0	0	1	4

Construct a frequency table and a frequency plot summarizing this information. Discuss the results.

EXERCISE 2-19

Base pay of the state's governor in 1986 is shown here (in dollars) for each state (*USA Today,* December 11, 1986, page 9C; data supplied by Council of State Governments).

Alabama	63,839	Louisiana	73,400	Ohio	65,000
Alaska	85,728	Maine	35,000	Oklahoma	70,128
Arizona	62,500	Maryland	75,000	Oregon	72,000
Arkansas	35,000	Massachusetts	75,000	Pennsylvania	75,000
California	49,100	Michigan	85,800	Rhode Island	49,500
Colorado	60,000	Minnesota	84,560	South Carolina	60,000
Connecticut	65,000	Mississippi	63,000	South Dakota	55,120
Delaware	70,000	Missouri	81,000	Tennessee	68,200
Florida	78,757	Montana	50,542	Texas	94,350
Georgia	79,356	Nebraska	40,000	Utah	60,009
Hawaii	59,400	Nevada	65,000	Vermont	60,000
Idaho	50,000	New Hampshire	62,880	Virginia	75,000
Illinois	58,000	New Jersey	85,000	Washington	63,000
Indiana	65,988	New Mexico	60,000	West Virginia	72,000
Iowa	64,000	New York	100,000	Wisconsin	75,337
Kansas	65,000	North Carolina	98,196	Wyoming	70,000
Kentucky	61,200	North Dakota	60,862		

a. Construct a frequency table and draw a histogram summarizing these governors' salaries.

b. Is the distribution symmetrical, negatively skewed, positively skewed, or none of these?

c. Is the distribution unimodal, bimodal, or multimodal?

EXERCISE 2-20

Forty-six U.S. universities accepted less than half of applicants in 1985 and had an average freshman SAT score of at least 1200. The average yearly cost (in dollars; tuition, room and board, books, supplies, any out-of-state surcharge) for each of these universities is shown below (*USA Today,* December 15, 1986, page 2D).

U.S. Naval Academy	0	Tufts	17,060
U.S. Military Academy	0	Virginia	9,320
Stanford	17,458	Penn	17,210
Harvard/Radcliffe	17,395	Lafayette	14,600
Princeton	17,555	Haverford	15,930
Yale	17,400	Wesleyan (Connecticut)	16,565
Brown	17,264	William & Mary	10,084
Cooper Union	1,300	Colgate	15,680
Amherst	15,920	Trinity (Connecticut)	15,370
Dartmouth	17,285	Bates	15,070
USAF Academy	0	Colby	16,000
Bowdoin	15,620	Bucknell	14,965
Georgetown	15,830	Chicago	17,310
Duke	14,340	Cal Tech	16,385
Williams	15,498	Northwestern	16,175
Columbia	17,175	Hamilton (New York)	15,100
Middlebury	14,440	Claremont/McKenna	15,300
Cornell	16,490	Notre Dame	12,240
Rice	9,770	Carleton	13,575
Swarthmore	16,200	Wellesley	15,980
MIT	17,700	Harvey Mudd	16,030
Davidson	12,470	UNC, Chapel Hill	7,470
Washington & Lee	11,780	Vassar	15,498

a. Construct a frequency plot and histogram of these college costs.

b. Describe the distribution in terms of peaks and skewness.

EXERCISE 2-21

In the 1980 Wisconsin Restaurant Survey, information on number of full-time employees was obtained on 265 restaurants (Ryan, Joiner, and Ryan, 1985, pages 77, 321–328). Listed below are number of full-time employees (number of restaurants): 0 (58), 1 (25), 2 (29), 3 (24), 4 (17), 5 (11), 6 (16), 7 (10), 8 (8), 9 (1), 10 (16), 11 (1), 12 (3), 13 (3), 14 (1), 15 (4), 16 (1), 18 (3), 20 (10), 25 (5), 26 (1), 28 (1), 30 (6), 32 (1), 35 (2), 36 (1), 40 (3), 42 (1), 51 (1), 80 (1), 250 (1).

a. Construct a frequency table and frequency plot for number of full-time employees. For the frequency plot, you may wish to exclude the largest value or show a break in the scale on the horizontal axis.

b. Describe the distribution of number of full-time employees for these 265 restaurants.

c. These 265 restaurants were part of 1,000 restaurants surveyed. With only 26.5% response, would you be willing to use this sample to draw inferences about all Wisconsin restaurants?

EXERCISE 2-22

Information was obtained on the length of stay (in days) for all patients voluntarily committed to the acute psychiatric unit of a Wisconsin health care center during the first half of 1981 (Ryan, Joiner, and Ryan, 1985, page 82). Listed below are the lengths of stay (number of patients): 0 (3), 1 (11), 2 (5), 3 (2), 4 (3), 5 (5), 6 (2), 7 (2), 8 (3), 9 (2), 10 (1), 11 (3), 12 (1), 13 (2), 14 (1), 15 (1), 18 (1), 19 (2), 25 (4), 35 (2), 45 (1), 75 (1).

a. Construct a frequency table and frequency plot for these lengths of stay.

b. Describe the distribution of length of stay for this group of patients.

EXERCISE 2-23

Numbers of cancers diagnosed in Western Australia in 1982 are shown below for each of the ten leading sites, separately for males and females (Hatton and Clarke-Hundley, 1984, page 21).

Site	Number of males	Site	Number of females
Lung	360	Breast	384
Prostate	234	Melanoma	158
Colon	147	Colon	143
Melanoma	142	Lung	109
Bladder	129	Cervix	98
Stomach	102	Rectum	81
Rectum	100	Uterus	73
All leukemias	46	Ovary	66
Kidney	44	Bladder	49
Pancreas	40	All leukemias	48

Calculate the percentage of the 1,344 males in each site category. Calculate the percentage of 1,209 females in each site category. Construct two frequency plots using this information, one plot for males and one for females. Note that each of these frequency plots summarizes frequency information about a qualitative or categorical variable: site of cancer. You may want to construct each frequency plot with the cancer sites listed in the order shown above, along the vertical axis (for easier reading). Frequencies and percentages can then be indicated on horizontal axes, say frequencies on the bottom horizontal axis and percentages on the top. A frequency plot arranged in this way for a qualitative variable is a form of dot chart, discussed in Chapter 4. If you want to compare the plots for males and females, how should you construct the scales for the two frequency plots?

EXERCISE 2-24 In a study of preovulatory estrogen levels and basal body temperature, investigators estimated the mid-cycle fertile period for each of 24 women (Carter and Blight, 1981).

Length of fertile period (days)	Number of women
4	1
5	4
6	3
7	3
8	6
9	5
10	1
11	1
Total	24

a. Construct a frequency plot based on this frequency table.

b. Calculate any descriptive statistics you think are appropriate.

c. Discuss your findings.

EXERCISE 2-25 The percentage of dwelling units with lead paint is shown here for 23 Massachusetts communities (*The Boston Sunday Globe,* January 25, 1987, page 29).

Community	Percentage of dwelling units with lead paint	Community	Percentage of dwelling units with lead paint
Arlington	63.6	Malden	79.4
Boston	80.7	Medford	83.6
Brockton	63.6	New Bedford	83.5
Brookline	72.6	Newton	72.8
Chelsea	93.6	Pittsfield	71.1
Chicopee	52.8	Quincy	76.3
Fall River	86.4	Somerville	48.6
Framingham	33.8	Springfield	69.4
Haverhill	81.9	Waltham	60.6
Lawrence	82.9	Weymouth	49.9
Lowell	76.1	Worcester	77.3
Lynn	82.1		

Display and/or summarize these observations in any ways that you find useful.

CHAPTER 3

Studying One Variable at a Time: Descriptive Statistics

IN THIS CHAPTER

Measures of central tendency or location
Mean, trimmed mean, weighted mean
Median
Measures of spread or variation
Range, interquartile range
Standard deviation, for a population and for a sample

How can we describe central tendency of life expectancies in a group of countries? How much variation is there among those life expectancies? We might address these questions in terms of *descriptive statistics,* numbers used to describe or summarize a set of values.

A **descriptive statistic** is a number used to describe or summarize a set of values.

One class of descriptive statistics that we have already encountered consists of the quantiles, numbers that divide a set of ordered values into equal or approximately equal sized groups. We use quantiles to construct box plots and box graphs.

Now we are going to discuss some descriptive statistics that measure location or central tendency and some that measure spread or variation, for a single variable. The median is a descriptive statistic that we have seen already in our discussion of quantiles. The median, which divides an ordered list of values in half, is a measure of location or central tendency. The range from the minimum to the maximum is a summary measure of variation or spread in a set of values. We begin our discussion with measures of central tendency and then consider measures of variation.

3-1 Measures of Central Tendency

A number describing the location of a set of values is a *measure of central tendency* or *measure of location.* The mean and the median are the most common measures of central tendency. We have already defined the *median* as a number that divides an ordered list of values in half. Now we will look at the mean, then consider variations of the mean and compare them with the median as measures of location of a set of values.

A **measure of central tendency** or **location** is a descriptive statistic that summarizes central tendency or location of a distribution of values.

The *mean* or *average* of a set of values is the arithmetic average of the values. When calculated from a sample, it is called the *sample mean* or *sample average.* To calculate the mean, add up the values and then divide by the number of values:

$$\textbf{Mean or average} = \frac{\text{Sum of the values}}{\text{Number of values}}$$

In statistics we commonly let n denote the number of data values. We let x_1 denote the first value, x_2 the second value, and so on, to x_n, the nth value. The capital Greek letter sigma, Σ, denotes summation in mathematics. Putting these symbols together, we can write the calculation formula for a mean as

$$\text{Mean of } x_1 \text{ through } x_n = \frac{1}{n}\sum_{i=1}^{n} x_i$$

These symbols tell us to add up the values and then divide by the number of values. When calculating an average of sample values, we often denote the resulting sample mean by $\bar{x}$.

The letter representing the values in the formula for the mean is arbitrary. We could let y_1, for example, denote the first value, y_2 denote the second, and so on. (If calculated for a sample, we would then let $\bar{y}$ denote the sample mean.) We can write the calculation formula for the mean of y_1 through y_n as

$$\text{Mean of } y_1 \text{ through } y_n = \frac{1}{n}\sum_{i=1}^{n} y_i$$

Let's calculate the mean life expectancy for several groups of countries.

Life expectancies are shown in Table 3-1 for four high-income oil exporters. We calculate the mean life expectancy for these four countries by adding the four life expectancies and then dividing by 4:

$$\frac{72 + 60 + 62 + 70}{4} = 66 \text{ years}$$

The average is exactly 66. The mean life expectancy for these four high-income oil exporting nations is 66 years. Note that because of the symmetry of these four values of life expectancy, the median (the average of the middle values 62 and 70) also equals 66 years.

Recall that we listed the life expectancies for the industrial market countries in Section 2-5. The average of these 19 life expectancy values is 76.05 years, to two decimal places. If we round to the nearest year, we say the mean life expectancy for the industrial market nations is about 76 years. Because the distribution of these 19 life expectancies is fairly symmetrical, the mean is very close to the median, which we found to be 76 years in Section 2-5.

Thirty-five of the 37 low-income countries have nonmissing values for life expectancy, listed in Section 2-5. The average of these 35 life expectancies is 50.2, or about 50 years. Recall that the median life expectancy for these low-income countries is 49 years: About half the values are greater than or equal to 49 and about half are less than or equal to 49. The mean is larger than the median because the distribution of these 35 life expectancies is somewhat

TABLE 3-1 Life expectancy at birth in 1985 for four high-income oil exporting nations.

Country	1985 life expectancy at birth (years)
Kuwait	72
Libya	60
Saudi Arabia	62
United Arab Emirates	70

positively skewed; three low-income countries have relatively long life expectancies compared with the others (see Figure 3-3). These three values contribute to making the mean larger than the median.

We can interpret the mean or arithmetic average as a balance point or center of gravity. Figure 3-1 is a dot plot of life expectancies for the four high-income oil exporters. Each dot represents a single country and is positioned along the axis at the point corresponding to the country's life expectancy. Think of the axis as a plank of wood for a seesaw and the dots as children of identical size and weight sitting perfectly still on the seesaw. Where should the fulcrum be placed so that the seesaw will remain balanced in the horizontal position? It should be placed at the center of gravity—the mean or arithmetic average of the positions along the seesaw. For the four values plotted in Figure 3-1, the mean of 66 years is the arithmetic average of the positions along the axis. It is the point where we should place the fulcrum to balance the seesaw.

The four points in Figure 3-1 are symmetrically placed about the mean. The mean of 66 years is a good measure of the center of these values. However, 66 is not really close to any of the four values. As we can see from this example, a measure of the center of a set of values need not be particularly close to any of the values.

A dot plot of life expectancies for the 19 industrial market countries is shown in Figure 3-2. The arithmetic mean or balance point is 76.05 years.

Life expectancies for 35 low-income countries are plotted in Figure 3-3. The balance point is at the arithmetic average of 50.2 years.

When we average the life expectancies for a group of countries to calculate the mean, we weight each country equally. This is illustrated in Figures 3-1, 3-2, and 3-3, where dots represent countries and each dot is the same size. Does it make sense to treat countries equally in describing the center of the life expectancies, when some countries have more people than others?

If we consider a country as a single entity with its associated life expectancy, then the mean makes sense. However, if we want to describe life expectancies for the people in a group of countries, then we do not want a simple average across countries. We might instead weight each country's life expectancy by its population size in calculating an average. Then we have a **weighted mean.**

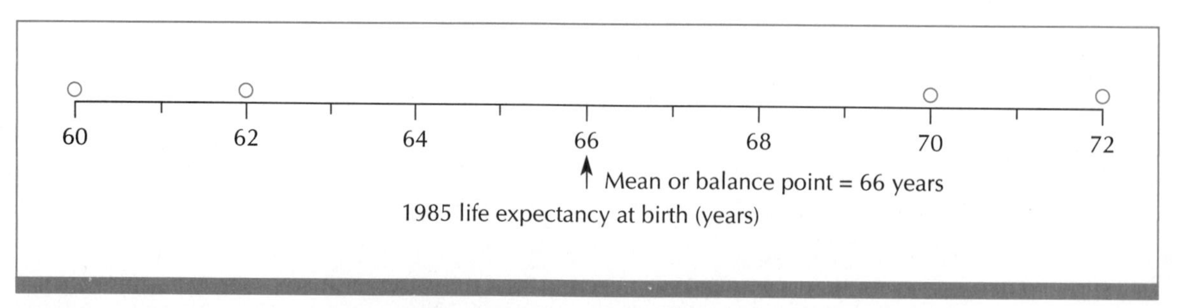

FIGURE 3-1 Dot plot of 1985 life expectancy at birth for four high-income oil exporting nations.

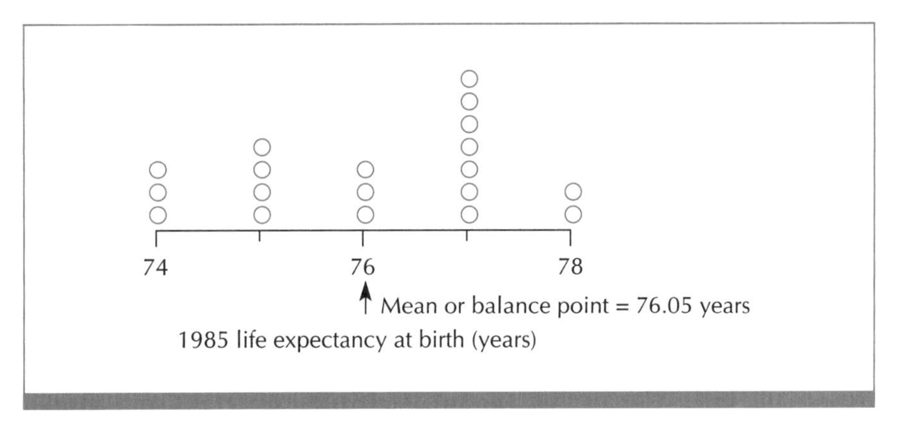

FIGURE 3-2 Dot plot of 1985 life expectancy at birth for 19 industrial market countries.

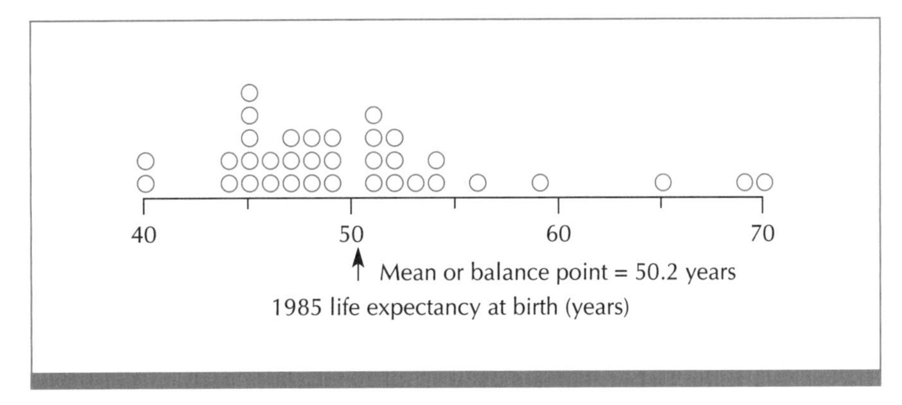

FIGURE 3-3 Dot plot of 1985 life expectancy at birth for 35 low-income countries. Two countries are excluded because of missing values.

To calculate a **weighted mean** or **weighted average,** multiply each value by an appropriate weight, add these products, and then divide by the sum of the weights:

$$\text{Weighted mean} = \frac{\text{Sum of the weighted data values}}{\text{Sum of the weights}}$$

Suppose we let x_1 denote the first data value, x_2 the second value, and so on, through x_n. Let w_1, w_2, through w_n denote a set of weights. Then we can write the corresponding weighted mean as

$$\text{Weighted mean} = \frac{\sum_{i=1}^{n} w_i x_i}{\sum_{j=1}^{n} w_j}$$

which is the same as the definition above. (Note that this formula for the weighted mean uses different subscript letters in the numerator and denomi-

TABLE 3-2 Population in mid-1985 and life expectancy at birth in 1985 for four high-income oil exporters.

Country	Population (millions) in mid-1985	1985 life expectancy at birth (years)	Product of population and life expectancy
Kuwait	1.7	72	122.4
Libya	3.8	60	228.0
Saudi Arabia	11.5	62	713.0
United Arab Emirates	1.4	70	98.0
Sum or Total	18.4	264	1,161.4

Mean life expectancy: $\frac{264}{4} = 66$ years

Weighted mean life expectancy, life expectancy weighted by population size: $\frac{1{,}161.4}{18.4} = 63.12$, or 63 years

nator. This is to remind us that these sums are calculated separately, and then we find the quotient.)

Table 3-2 shows population size and life expectancy for the four high-income oil exporters. Let's calculate a weighted average life expectancy, with weights equal to population sizes. We multiply the life expectancy for each country by the population size for that country. The resulting products are shown in the last column of Table 3-2. We add these products, getting a sum of 1,161.4. We also add the weights. The sum of the weights is 18.4, the total population (in millions) for the four high-income oil exporters. We then divide 1,161.4 by 18.4 to get a weighted average life expectancy of about 63 years.

The weighted average life expectancy is 3 years less than the unweighted average. Examining Table 3-2, we see why. The two countries with the largest populations, Saudi Arabia and Libya, have the shorter life expectancies. Kuwait and United Arab Emirates have longer life expectancies, but relatively small population sizes. The two countries with the larger populations (and shorter life expectancies) contribute more to the weighted mean than to the unweighted mean. The weighted mean life expectancy of 63 years is a measure of the middle of life expectancies for the combined populations of these four high-income oil exporting nations. Exercises 3-2 and 3-3 ask you to find weighted mean life expectancy for the industrial market and low-income countries, respectively.

Trimmed Means as Measures of Central Tendency

The problem with the mean as a measure of central tendency is that it can be greatly influenced by a few extreme values. One very large value can make the mean much larger than it would be if that value were excluded (see Exercise 3-1), and similarly for an extremely small value. In such cases, the mean or average may not be a good measure of center.

TABLE 3-3 Population in mid-1985 for 35 low-income countries (values missing for Afghanistan and Kampuchea).

Country	Population (millions) in mid-1985	Country	Population (millions) in mid-1985
Bhutan	1.2	Madagascar	10.2
Central African Republic	2.6	Ghana	12.7
Togo	3.0	Mozambique	13.8
Lao PDR	3.6	Uganda	14.7
Sierra Leone	3.7	Sri Lanka	15.8
Benin	4.0	Nepal	16.5
Burundi	4.7	Kenya	20.4
Chad	5.0	Sudan	21.9
Somalia	5.4	Tanzania	22.2
Haiti	5.9	Zaire	30.6
Rwanda	6.0	Burma	36.9
Guinea	6.2	Ethiopia	42.3
Niger	6.4	Viet Nam	61.7
Senegal	6.6	Pakistan	96.2
Zambia	6.7	Bangladesh	100.6
Malawi	7.0	India	765.1
Mali	7.5	China	1,040.3
Burkina Faso	7.9		

Measures of central tendency based on 35 nonmissing values (in millions of people):
Mean: 69.0
5% *trimmed mean*: 19.6
15% *trimmed mean*: 13.5
Median: 7.9

Consider, for example, the population sizes shown in Table 3-3 for 35 low-income countries. The mean is 69.0 million people. But only 4 of the 35 countries have populations greater than 69.0 million; most have populations much smaller than 69.0 million. The mean is so large mainly because two countries have very large populations: China (1,040.3 million) and India (765.1 million). The mean does not provide a very satisfactory measure of the center of these 35 population sizes.

One way around this problem is to exclude very large and very small values before calculating a mean. The resulting measure of central tendency is called a trimmed mean. To calculate a 5% *trimmed mean,* we exclude the largest 5% and the smallest 5% of the values and calculate the mean of the remaining values. For a 15% *trimmed mean,* we calculate the mean after excluding the largest 15% and the smallest 15% of the values. In general, for an x% *trimmed mean* we exclude the smallest x% and the largest x% of the values and calculate the mean of the remaining $(100 - 2x)$% of the values. Typical values for the percent x excluded from each end are integers from 1 to 15. It is easiest to find a trimmed mean from an ordered list of values (such as we might have in a stem-and-leaf plot).

A **trimmed mean** of a set of values is a mean with a specified percentage of the largest and smallest values excluded from the calculation. An ***x*% trimmed mean** is a mean calculated after the largest *x*% and the smallest *x*% of the values have been excluded.

Let's find the 5% trimmed mean and the 15% trimmed mean for the 35 population sizes in Table 3-3. Five percent of 35 is 1.75, or about 2. For the 5% trimmed mean we exclude the two largest population sizes (for China and India) and the two smallest population sizes (for Bhutan and Central African Republic). The mean of the remaining 31 values is 19.6 million, the 5% trimmed mean. Eleven of the 35 countries have populations larger than 19.6 million, and 24 have populations smaller than 19.6 million.

Fifteen percent of 35 is 5.25, or about 5. For the 15% trimmed mean we exclude the five largest population sizes (for China, India, Bangladesh, Pakistan, and Viet Nam) and the five smallest population sizes (for Bhutan, Central African Republic, Togo, Lao PDR, and Sierra Leone). The mean of the remaining 25 population sizes is 13.5 million, the 15% trimmed mean. Fifteen of the 35 countries have populations greater than 13.5 million, and 20 have populations smaller than 13.5 million.

Only 4 of the 35 countries in Table 3-3 have population sizes greater than the mean value of 69.0 million people. Eleven countries have populations greater than the 5% trimmed mean of 19.6 million. Fifteen countries have populations greater than the 15% trimmed mean of 13.5 million. The 5% trimmed mean seems closer to the center than the mean. The 15% trimmed mean seems closer to the center than the 5% trimmed mean. Following this line of thought, we might say that the center of the population sizes is a number that divides the set of values in half—that is, the *median*.

The median population size for these low-income countries is 7.9 million, the middle of the 35 ordered values in Table 3-3. Eighteen of the 35 population sizes are greater than or equal to 7.9 million and 18 are less than or equal to 7.9 million. As a measure of central tendency, the median has two satisfying characteristics. It really is the center of a set of values in the sense that it divides the data set in half. Also, the median is not influenced by extremely small or large values. We can think of the median as the most extreme instance of a trimmed mean. If the number of values is odd, we trim all but the middle value, which is the median. If the number of values is even, we trim all but the middle two values and average them to obtain the median.

Measures of Central Tendency: An Example

Let's now consider measures of central tendency for the percentage of married women of childbearing age using contraception in 1984. We will look at three groups of countries: low-income countries, industrial market countries, and lower-middle-income countries.

A dot plot of contraception use is shown in Figure 3-4 for the 28 low-income countries with nonmissing values. This distribution is positively skewed: Most of the values are concentrated between 0 and 8%, while two countries

(Sri Lanka and China) have much higher levels of contraception use (57% and 69%, respectively). Four measures of central tendency are shown in Figure 3-4. The mean is greater than the median, inflated by the few large values. This is true in general for a positively skewed distribution. Eliminating extreme values from the calculation of the mean dampens this effect; the trimmed means are closer to the median. Note that trimming does not affect the median at all.

Figure 3-5 illustrates levels of contraception use for 13 industrial market countries with nonmissing values. This distribution is negatively skewed: Two countries have relatively low values compared with the others. The mean is less than the median for these 13 values; this is in general true for a negatively

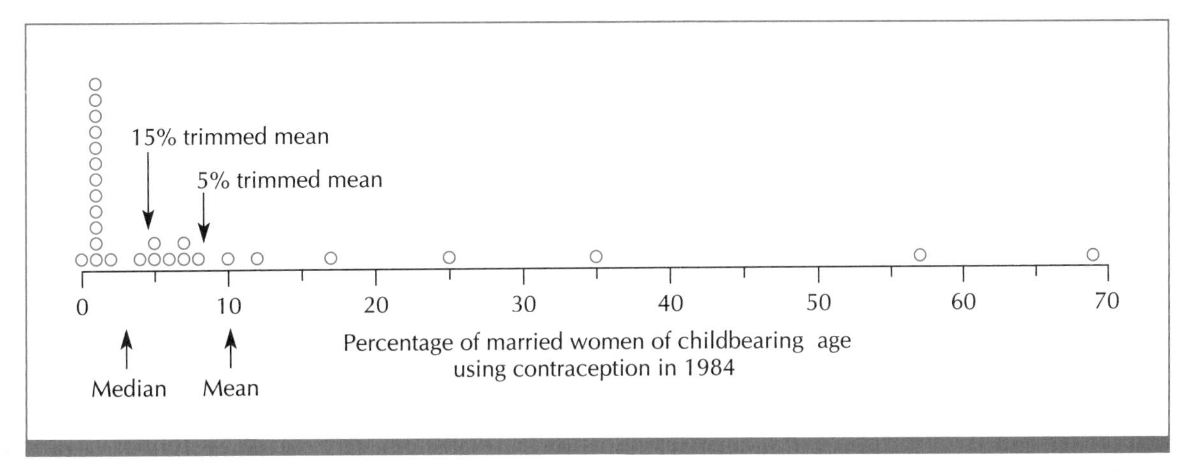

FIGURE 3-4 Dot plot of percentage of married women of childbearing age using contraception in 1984, for 28 low-income countries. Nine countries are excluded because of missing values.

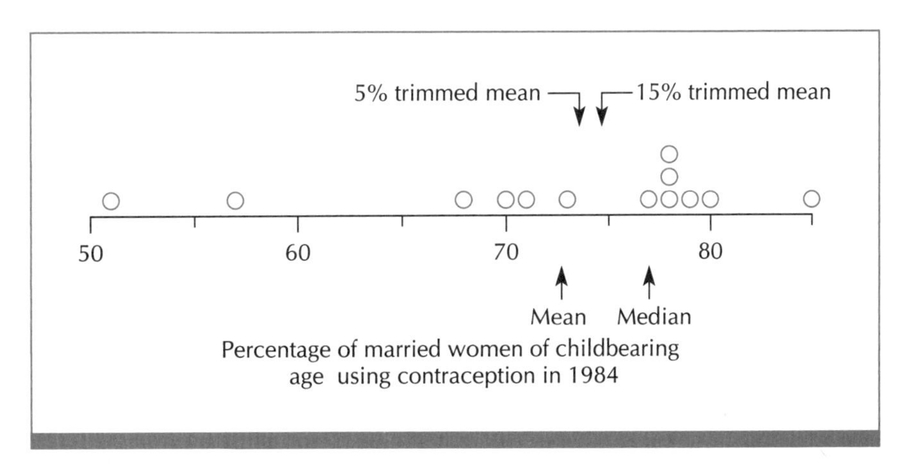

FIGURE 3-5 Dot plot of percentage of married women of childbearing age using contraception in 1984, for 13 industrial market countries. Six countries are excluded because of missing values.

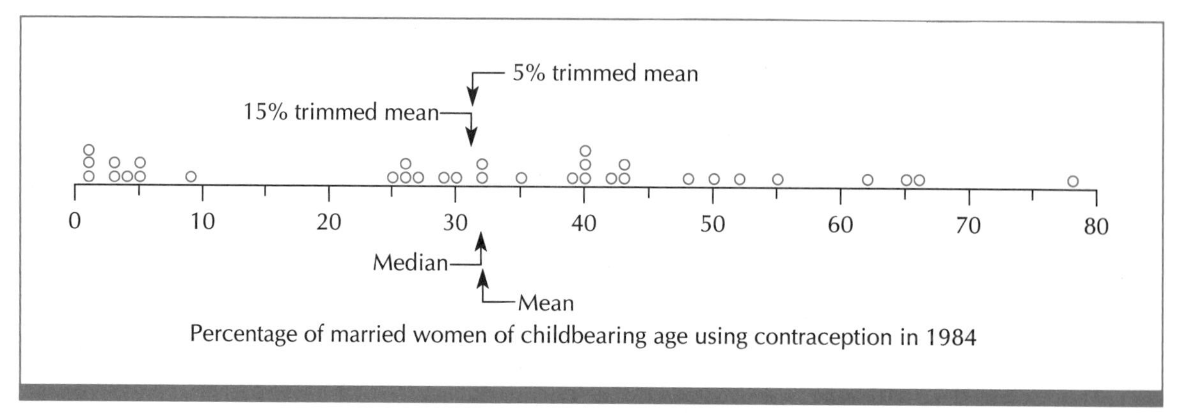

FIGURE 3-6 Dot plot of percentage of married women of childbearing age using contraception in 1984, for 33 lower-middle-income countries. Three countries are excluded because of missing values.

skewed distribution. The trimmed means lie between the median and the mean. Again, the median is the same for the trimmed and untrimmed sets of values.

Estimated levels of contraception use are plotted in Figure 3-6 for 33 lower-middle-income countries with nonmissing values. This dot plot is more symmetrical than the ones in Figures 3-4 and 3-5. When a distribution is nearly symmetric, measures of central tendency are similar. For these 33 levels of contraception use, the measures of central tendency illustrated in Figure 3-6 all equal either 31% or 32%.

The median contraception use is 3% for the low-income countries, 32% for the lower-middle-income countries, and 77% for the industrial market countries with available information. The lower-middle-income countries fall between the low-income countries and the industrial market countries with respect to contraception use, but there is a fair amount of overlap in the three distributions. Nine of the lower-middle-income countries, for example, have levels of contraception use less than 10%, typical of the low-income group. One lower-middle-income country (Mauritius) has a recorded contraception use of 78%, a typical value for the industrial market group.

We turn now to measures of variation.

3-2 Measures of Variation

As we know, measures of central tendency help describe the location of a set of values along a number line. But location alone provides a very incomplete description of a distribution. Compare Figures 3-2 and 3-3, for instance. We know that the center of the life expectancies for the industrial market countries (Figure 3-2) is different from the center for the low-income countries (Figure 3-3). But the spread in the values is also very different for the two groups of countries. The distribution of life expectancies for the industrial market countries is very compact; the life expectancies are not far from each other. On the other hand, the distribution for the low-income countries is spread out; there is a great deal of variation among these values. This comparison suggests that we will have a more complete summary of a distribution if we describe variation as well as location.

One way to study the variation in a set of values is to look at the extremes, the *maximum* (or largest value) and the *minimum* (or smallest value). These two extremes define the *range*. Sometimes we think of the range as the difference between the maximum and minimum values. For the life expectancies of low-income countries, the range is from a minimum of 40 years to a maximum of 70 years; we say these life expectancies span a range of 30 years. On the other hand, the life expectancies for the industrial market countries span a range of 4 years, from a minimum of 74 to a maximum of 78 years. Certainly the range helps summarize the difference in variation between these two sets of values.

The **range** is defined by the minimum (smallest) and the maximum (largest) values. The difference between these two extremes is a measure of spread or variation in a set of values.

The range depends only on the extreme values of a distribution; for this reason it may not provide a good summary of the variation among the bulk of the values. Instead of comparing the minimum and maximum values, we might compare the first and third quartiles. The first and third quartiles define the *interquartile range*. We know that about one-fourth of the values are below the first quartile and about one-fourth are above the third quartile. Therefore, the interquartile range from the first to the third quartile contains the middle 50% of the ordered data values. The interquartile range describes variation in the middle half of the distribution, whereas the range describes variation only between extremes of the distribution.

The **interquartile range** is defined by the first and third quartiles. The difference between the third and first quartiles describes spread or variation in the middle half of a distribution of values.

We found the quartiles for life expectancies of low-income countries and for industrial market countries in Section 2-5. The middle 50% of the nonmissing life expectancies for low-income countries lie in the interquartile range from 45 to 52 years. For comparison, the middle 50% of the life expectancies for industrial market countries are in the interquartile range from 75 to 77 years. We might say the interquartile range is 7 years for the low-income countries, compared with 2 years for the industrial market countries. Note that the interquartile range is the length of the box portion of a box plot or box graph (see Figure 2-18, for instance).

The Standard Deviation as a Measure of Variation

The most commonly used measure of spread or variation is the *standard deviation,* often abbreviated SD. (We use the sample standard deviation in a number of formal statistical procedures for making inferences about population means and variances, as we will see in Part III.) In contrast to the range and interquartile range, we use all the data values to calculate the standard deviation, which measures variation of those values about the mean.

When the standard deviation measures variation in a population, we call it the *population standard deviation* and denote it by the Greek symbol σ; when it measures variation in a sample, we call it the *sample standard deviation* and denote it by the letter s. The computing formulas for the population standard deviation and the sample standard deviation differ slightly, as shown below. The sample standard deviation is used extensively in formal statistical inference. For purposes of data analysis we will follow common practice and use the calculation formula for the sample standard deviation; this is the formula used in computer statistical packages for calculating the standard deviation as a descriptive statistic.

To calculate a **population standard deviation,** we take the difference between each value and the mean, and square it. We add up the squared differences, and then divide by the number of data values. Then we take the square root:

$$\sigma = \text{Population standard deviation} = \sqrt{\frac{\text{Sum of (data value} - \text{mean)}^2}{\text{Number of values}}}$$

To calculate the **sample standard deviation,** we follow the same steps, except that we divide by 1 less than the number of data values:

$$s = \text{Sample standard deviation} = \sqrt{\frac{\text{Sum of (data value} - \text{mean)}^2}{\text{Number of values} - 1}}$$

We can write the calculation formula for the sample standard deviation, s, this way:

$$s = \sqrt{\frac{\sum_{i=1}^{n} (x_i - \bar{x})^2}{n - 1}} = \sqrt{\frac{\sum_{i=1}^{n} x_i^2 - n(\bar{x})^2}{n - 1}}$$

where n denotes the number of values; x_1, x_2, through x_n represent the data values; and $\bar{x}$ denotes the sample mean. The second formula given above for the sample standard deviation is algebraically equivalent to the first, and is easier to use for hand calculations.

The standard deviation is always greater than or equal to 0. It tends to be large when values are very spread out about the mean, and to be small when values are close to the mean. If all the values are the same, then the standard deviation is 0. However, if the values are not all equal, the standard deviation will always be greater than 0.

Let's calculate the standard deviation of the life expectancies for the four high-income oil exporters, plotted in Figure 3-1. Using the formula for the sample standard deviation, Table 3-4 shows the steps in the calculation. The standard deviation is 5.9, or about 6 years.

TABLE 3-4 Calculating the standard deviation of 1985 life expectancy at birth for four high-income oil exporters, using the formula for the sample standard deviation for purposes of data analysis.

	Value Life expectancy (years)	Value − Mean Life expectancy − 66	(Value − Mean)^2 (Life expectancy − 66)^2
	60	−6	36
	62	−4	16
	70	4	16
	72	6	36
Sum	264	0	104

Mean: $\frac{264}{4} = 66$ years

Standard deviation: $\sqrt{\frac{104}{4 - 1}} = 5.9$ years

The second column of Table 3-4 contains some interesting information—distances of the individual values from the mean. These distances are examples of residuals. In general, a *residual* is the difference between an observation and a summary value. (Later, when we discuss modeling experimental results as part of statistical inference, we will see that this summary value is an estimate of the mean, or a predicted value, of the observation based on a probability model we assume for the experiment.)

> A **residual** is the difference between an observation and a summary value for the observation.

The summary value we use in this simple example is the mean of the four observations. A residual is then the difference between an observation and this mean. The residuals in column 2 of Table 3-4 show us that the life expectancies range from 6 years below the mean to 6 years above the mean. If we plot these residuals, we get a dot plot that has the same appearance as the dot plot of life expectancies in Figure 3-1, but the values are shifted, with a mean of 0. These shifted values (or residuals) give us a feel for the variation about the mean. Statisticians often look at differences such as these; we find residuals very useful for many analyses in formal statistical inference.

The standard deviation of the 35 life expectancies for low-income countries (Figure 3-3) is 6.9 years; recall that the range is 30 years and the interquartile range is 7 years. The standard deviation of life expectancies for the 19 industrial market countries (Figure 3-2) is 1.3 years; the range is 4 years and the interquartile range is 2 years. For simple data analysis, we may prefer to use the interquartile range and the range as measures of variation; the variation they represent is easier to visualize than for the standard deviation.

Recall that we used box graphs to compare population growth for low-income and industrial market countries in Figure 2-20. We noticed that the variation in the two sets of values appeared to be about the same. This observation is supported when we compare measures of variation. For the low-income countries, the range in population growth values is 2.9, the interquartile range is .8, and the standard deviation is .6. The range in population growth values for the industrial market countries is 1.6, the interquartile range is .8, and the standard deviation is .4.

In Chapters 2 and 3, we have considered ways to look at a single variable. In Chapters 4 and 5, we look at ways to study relationships between two or more variables.

Summary of Chapter 3

Measures of central tendency describe the location of a set of values. The mean or average can be greatly influenced by extreme values. Trimmed means are less affected by extremes, whereas the median is not affected at all. Sometimes we find it useful to calculate a weighted mean, in which a value's contribution to the calculation is determined by a weight assigned to that value. (When we weight the values equally, we get the mean.)

We consider three measures of spread or variation. The range is defined by the extremes, while the interquartile range is defined by the middle 50% of the ordered values. The standard deviation is a measure of variation about the mean calculated from all the data values. The calculation formulas differ slightly for the population standard deviation and the sample standard deviation. When calculating the standard deviation as a descriptive statistic, we follow common practice and use the calculation formula for the sample standard deviation. We use the sample standard deviation extensively in classical statistical inference. For simple data analysis, the interquartile range and range are easier to interpret than the standard deviation.

Minitab Appendix for Chapter 3

Calculating Descriptive Statistics

The DESCRIBE command will provide all of the descriptive statistics in Chapter 3 except the weighted mean. Referring to the data for Exercise 5-35, the command

```
MTB> describe 'calorie'
```

produces the output in Figure M3-1.

The N column gives the number of nonmissing observations; N*, the

```
                N       N*     MEAN   MEDIAN   TRMEAN    STDEV   SEMEAN
calorie        37        3   110.00   114.00   110.58    22.38     3.68

              MIN      MAX       Q1       Q3
calorie     68.00   143.00    91.50   130.00
```

FIGURE M3-1 Descriptive statistics for the variable CALORIE in Exercise 5-35.

```
            mortcode        N       N*      MEAN    MEDIAN    TRMEAN
mort               1       20        0    10.950    11.000    11.000
                   2       20        0    248.10    239.50    245.72
life               1       20        0    74.900    75.000    74.889
                   2       20        0    44.050    44.000    44.222
weight             1       19        1     5.579     5.000     5.529
                   2       17        3     15.24     15.00     15.13
calorie            1       20        0    128.00    129.50    128.00
                   2       17        3     88.82     91.00     89.33

            mortcode    STDEV   SEMEAN       MIN       MAX        Q1        Q3
mort               1    1.638    0.366     8.000    13.000    10.000    12.000
                   2    32.43     7.25    210.00    329.00    225.00    258.50
life               1    1.165    0.261    73.000    77.000    74.000    76.000
                   2    3.663    0.819    35.000    50.000    41.500    46.000
weight             1    1.346    0.309     4.000     8.000     4.000     7.000
                   2     4.44     1.08      9.00     23.00     10.50     19.50
calorie            1    10.40     2.32    113.00    143.00    115.25    138.50
                   2    10.80     2.62     68.00    102.00     84.50     97.00
```

FIGURE M3-2 Descriptive statistics for four variables in Exercise 5-35, separately for the countries with low child mortality (MORTCODE = 1) and the countries with high child mortality (MORTCODE = 2).

number of missing observations. MEAN (arithmetic average) and MEDIAN are as we described them in Chapter 3. MIN is the smallest and MAX the largest value. TRMEAN is the 5% trimmed mean. Q1 and Q3 are the first and third quartiles, respectively. STDEV is the standard deviation. SEMEAN is the standard error of the mean, equal to the standard deviation divided by the square root of the number of nonmissing observations N. We will see the standard error of the mean in Part III.

The BY subcommand with the DESCRIBE command instructs Minitab to describe variables within levels of a classification variable (with integer values between −9999 and +9999 or missing). Recall that in the Minitab Appendix for Chapter 1, we created a variable MORTCODE equal to 1 for the 20 countries with low child mortality, and equal to 2 for the 20 countries with high child mortality. If we use the command

```
MTB>   describe c2-c5;
SUBC>  by 'mortcode'.
```

we obtain descriptive statistics for the variables in columns 2–5, separately for the two sets of countries, as shown in Figure M3-2. The command

```
MTB>   describe 'mort', 'life', 'weight', 'calorie';
SUBC>  by 'mortcode'.
```

will produce the same display.

Calculating a Weighted Mean

We must do more work to calculate a weighted mean. Suppose we have the population and life expectancy data for the four high-income oil exporters (Table 3-2) in our Minitab worksheet. Population (POP) is in column 1 and life expectancy (LIFE) is in column 2:

```
ROW     pop    life
 1      1.7     72
 2      3.8     60
 3     11.5     62
 4      1.4     70
```

We want to calculate weighted mean life expectancy, with life expectancies weighted by population size. The following LET command instructs Minitab to multiply the life expectancy by the population size for each country and put the results in column 3:

```
MTB> let c3=c1*c2
```

or, equivalently:

```
MTB> let c3='pop'*'life'
```

Now we want to sum the values in column 3 and store the sum as the stored constant K1. We also want to sum the weights (population sizes) in column 1 and store this sum as the stored constant K2. The weighted mean equals K1 divided by K2. The following sequence of commands calculates the weighted mean from the column of products (C3) and the column of weights (POP in C1):

```
MTB> sum c3 k1
MTB> sum 'pop' k2
MTB> let k3=k1/k2
MTB> print k3
K3    63.1196
```

The printed value of K3, 63.1196, is the weighted mean we found in Table 3-2.

Storing Constants in Our Worksheet

We can store constants in our worksheet with designations such as K1, K2, and K3, as we used above. Such stored constants can be useful in calculations, as for a weighted mean.

Summarizing Data by Columns

SUM is an example of a command that summarizes the data in columns. SUM adds the nonmissing values in a column. Other column-wise commands are COUNT (number of values in a column, missing and nonmissing); N (number of nonmissing values in a column); NMISS (number of missing values in a column); MEAN (average of nonmissing values in a column); STDEV (standard deviation of nonmissing values in a column); MEDIAN (median of nonmissing values in a column); MINIMUM (smallest nonmissing value in a column); MAXIMUM (largest nonmissing value in a column); and SSQ (sum of the squares of the nonmissing values in a column).

We can use a column command just to print a value. In the previous example for the four high-income oil exporters, we could use the MEAN command this way:

```
MTB> mean 'life'
   MEAN = 66.000
```

Minitab prints MEAN = 66.000, the mean of the four life expectancies. If we want to store that mean as a stored constant, say K4, we use

```
MTB> mean 'life' k4
   MEAN = 66.000
```

Minitab prints MEAN = 66.000, and saves this value as the stored constant K4.

Summarizing Data by Rows

Sometimes we find it useful to summarize data row-wise. For example, the command

```
MTB> rsum c11-c14 c20
```

calculates a sum of four numbers (the numbers in columns 11–14) for each row in the worksheet. These sums are stored in the corresponding rows of column 20. (If there are any missing values in columns 11–14, Minitab calculates the sum of the nonmissing values.)

All row-wise operations perform a function row-wise over all but the last column listed, where the results are stored. Row-wise operations in addition to RSUM are RCOUNT (total number of values), RN (number of nonmissing values), RNMISS (number of missing values), RMEAN (average of values), RSTDEV (standard deviation of values), RMEDIAN (median value), RMINIMUM (smallest value), RMAXIMUM (largest value), and RSSQ (sum of the squared values).

Row operations can involve constants and stored constants. For example, the command

```
MTB> rsum 10 k1 c2 c5
```

will add 10 and the value in the stored constant K1 to each value in column 2 and place the results in corresponding positions of column 5. If there are missing values in any positions of column 2, the sum of 10 and the value in K1 will be placed in the corresponding position in column 5.

Calculating Functions of Columns of Data Using the LET Command

The LET command is useful for calculating functions of columns of data. For instance, the command

```
MTB> let c5=absolute(c4)
```

puts the absolute value of elements of column 4 into column 5. Besides ABSOLUTE, other functions are SQRT (take the square root), LOGTEN (take the logarithm base 10), LOGE (take the natural logarithm), EXPO (calculate the mathematical constant e to the power given), ANTILOG (calculate 10 to the power given), ROUND (round to the nearest integer), SIGNS (assign $+1$ for a positive value, -1 for a negative value, 0 for a zero value).

Performing Arithmetic on Columns of Data Using the LET Command

The LET command is also useful for arithmetic operations on columns of data. We use the symbols + for addition, − for subtraction, * for multiplication, / for division, and ** for exponentiation. Minitab carries out arithmetic with the usual algebraic order of precedence. Expressions within parentheses are evaluated first. Within parentheses, functions are evaluated first, then exponentiation, then multiplication and division, and then addition and subtraction. If two or more operations have the same precedence, Minitab carries them out from left to right. The expression

```
MTB> let c10=(c1-k6)**2
```

tells Minitab to subtract the stored constant K6 from each value in column 1, square the results, and put them in column 10. The expression

```
MTB> let c5=c2+c3/4
```

tells Minitab to divide each element in column 3 by the number 4, add the results to the corresponding elements of column 2, and store in column 5.

To illustrate use of the LET command, suppose for each of the four high-income oil exporters, we want to calculate population as percentage of total population for the four countries. The following sequence of commands accomplishes this:

```
MTB> sum 'pop' k2
MTB> let c3=('pop'/k2)*100
MTB> name c3 'prcntpop'
```

If we now use the print command

```
MTB> print 'pop' 'prcntpop'
```

we get this output:

```
ROW    pop   prcntpop
 1     1.7     9.2391
 2     3.8    20.6522
 3    11.5    62.5000
 4     1.4     7.6087
```

Exercises for Chapter 3

In the exercises, answer the following questions: What would you need to know about the sample to be willing to use it to make inferences about a larger population? What is that larger population (if any)? What limitations do you see in the sample?

For each figure and table, include a legend that completely describes its contents. Note the number of cases included and the number excluded if there are missing values.

EXERCISE 3-1

Consider this simple (hypothetical) example of how different a mean and median can be. Suppose that while looking for a job, you discover that the average salary of employees at Company A last year was $135,000. Intrigued, you do some investigation. You discover that Company A has a president, plus seven other employees. You are able to determine last year's salaries for these other seven: $8,000; $8,000; $9,000; $9,000; $11,000; $15,000; $20,000. You are surprised at how low these salaries are, until you find last year's salary for the president: $1,000,000. Do you want to work for Company A? Compare the mean with the median and a trimmed mean as measures of central tendency for these eight salaries. Which measure, if any, conveys a more correct impression of a typical salary?

EXERCISE 3-2

Population in mid-1985 (millions), life expectancy at birth in 1985 (years) and their product are shown below for 19 industrial market countries.

Country	Population	Life expectancy	Product
Australia	15.8	78	1,232.4
Austria	7.6	74	562.4
Belgium	9.9	75	742.5
Canada	25.4	76	1,930.4
Denmark	5.1	75	382.5
Finland	4.9	76	372.4
France	55.2	78	4,305.6
Germany, West	61.0	75	4,575.0
Ireland	3.6	74	266.4
Italy	57.1	77	4,396.7
Japan	120.8	77	9,301.6
Netherlands	14.5	77	1,116.5
New Zealand	3.3	74	244.2
Norway	4.2	77	323.4
Spain	38.6	77	2,972.2
Sweden	8.4	77	646.8
Switzerland	6.5	77	500.5
United Kingdom	56.5	75	4,237.5
United States	239.3	76	18,186.8
Sum	737.7	1,445	56,295.8

a. Find the mean life expectancy for these 19 countries.

b. Find the 10% trimmed mean life expectancy for these countries.

c. Find the weighted mean life expectancy, with life expectancies weighted by population size.

d. Compare the three measures of central tendency from parts (a), (b), and (c).

e. How do these three measures compare with the life expectancy of the United States, the industrial market country with the largest population?

EXERCISE 3-3

Population in mid-1985 (millions), life expectancy at birth in 1985 (years), and their product are shown below for 35 low-income nations.

Country	Population	Life expectancy	Product
Bangladesh	100.6	51	5,130.6
Benin	4.0	49	196.0
Bhutan	1.2	44	52.8
Burkina Faso	7.9	45	355.5
Burma	36.9	59	2,177.1
Burundi	4.7	48	225.6
Central African Republic	2.6	49	127.4
Chad	5.0	45	225.0
China	1,040.3	69	71,780.7
Ethiopia	42.3	45	1,903.5
Ghana	12.7	53	673.1
Guinea	6.2	40	248.0
Haiti	5.9	54	318.6
India	765.1	56	42,845.6
Kenya	20.4	54	1,101.6
Lao PDR	3.6	45	162.0
Madagascar	10.2	52	530.4
Malawi	7.0	45	315.0
Mali	7.5	46	345.0
Mozambique	13.8	47	648.6
Nepal	16.5	47	775.5
Niger	6.4	44	281.6
Pakistan	96.2	51	4,906.2
Rwanda	6.0	48	288.0
Senegal	6.6	47	310.2
Sierra Leone	3.7	40	148.0
Somalia	5.4	46	248.4
Sri Lanka	15.8	70	1,106.0
Sudan	21.9	48	1,051.2
Tanzania	22.2	52	1,154.4
Togo	3.0	51	153.0
Uganda	14.7	49	720.3
Viet Nam	61.7	65	4,010.5
Zaire	30.6	51	1,560.6
Zambia	6.7	52	348.4
Sum	2,415.3	1,757	146,424.4

a. Find the mean life expectancy for these 35 low-income countries.

b. Find the 15% trimmed mean life expectancy for these countries.

c. Find the weighted mean life expectancy, with life expectancy weighted by population size.

d. Compare the three measures of central tendency in parts (a), (b), and (c).

e. Now consider just China and India. Find the mean life expectancy for these two countries. Find the weighted mean life expectancy, with life expectancy weighted by population size. Compare these two measures of central tendency.

f. Consider the 33 low-income countries that remain after you exclude China and India. Find the mean life expectancy for these 33 countries. Find the weighted mean life expectancy, with life expectancy weighted by population size. Compare these two measures of central tendency.

g. Why are the mean and weighted mean life expectancy you calculated for the 33 countries in part (f) closer in value than the mean and weighted mean life expectancy for all 35 countries from parts (a) and (c)?

EXERCISE 3-4

The percentage of married women of childbearing age using contraception in 1984 is shown below for each of 28 low-income countries with information available.

Country	Contraception use	Country	Contraception use
Somalia	0	Sierra Leone	4
Burkina Faso	1	Burma	5
Burundi	1	Sudan	5
Chad	1	Benin	6
Guinea	1	Haiti	7
Malawi	1	Nepal	7
Mali	1	Pakistan	8
Niger	1	Ghana	10
Rwanda	1	Senegal	12
Tanzania	1	Kenya	17
Uganda	1	Bangladesh	25
Zaire	1	India	35
Zambia	1	Sri Lanka	57
Ethiopia	2	China	69

a. Find the median, mean, 5% trimmed mean, and 15% trimmed mean of contraception use for these 28 low-income countries.

b. Find the weighted mean contraception use, with contraception use weighted by population size. Population sizes for these countries are given in Exercise 3-3.

c. Compare the measures of central tendency you found in parts (a) and (b).

EXERCISE 3-5

As part of a study of vegetation damage, researchers measured the oxidant content of dew water in 12 samples collected at Port Burwell, Ontario, from August 25 to August 30, 1960. The 12 measurements in parts per million (ppm) ozone are (Hollander and Wolfe, 1973, page 57; from Cole and Katz, 1966):

.08 .11 .15 .17 .17 .20 .21 .22 .28 .31 .32 .35

a. Construct a dot plot of these observations.

b. Find the mean, 15% trimmed mean, and median. Locate these three measures of central tendency on the plot. How do they compare?

c. Find the range, interquartile range, and standard deviation. What does each of these descriptive statistics measure?

d. Describe the distribution of these ozone measurements.

EXERCISE 3-6

Operating hours until first failure of air-conditioning equipment are shown below for 13 Boeing 720 airplanes (Hollander and Proschan, 1984, page 176; from Proschan, 1963).

23 50 50 55 74 90 97 102 130 194 359 413 487

a. Construct a dot plot of these observations.

b. Is the distribution unimodal, bimodal, or multimodal?

c. Would you describe this distribution as symmetrical, negatively skewed, positively skewed, or none of these?

d. Calculate the mean, 15% trimmed mean, and median of these values. Locate these three measures of central tendency on your plot and compare them.

e. Find and discuss these three measures of variation: range, interquartile range, standard deviation.

EXERCISE 3-7

The following are errors (in inches) made by a robot in applying an adhesive (Milton and Arnold, 1986, page 250; based on Hegland, 1983):

.001 .002 .003 .004 .006 .001 .002 .003 .004 .006 .001
.003 .003 .006 .001 .003 .003 .005 .007 .002 .003 .004
.005 .008 .004

These values range from .001 to .008 inch. We often find it helpful in calculations to use coded or transformed values. If we change our units to thousandths of an inch, then we multiply each number above by 1,000. The transformed values then range from 1 to 8. You may wish to use these new units in your calculations below.

a. Construct a dot plot. Describe the distribution in terms of location, spread, peaks, and symmetry or skewness.

b. Calculate the mean, 15% trimmed mean, and median of these values. Locate these three measures of central tendency on your plot and compare them.

c. Calculate and discuss these three measures of variation: range, interquartile range, standard deviation.

EXERCISE 3-8

A new chip can be reprogrammed without removing it from the microcomputer. Times (in seconds) to reprogram a byte of memory on this chip are shown below (Milton and Arnold, 1986, page 258; from *Design News,* April 1983, page 26).

11.6	12.3	12.5	12.9	13.0	13.1	13.2	13.3
13.3	13.4	13.8	14.2	14.7	15.1	15.3	

a. Construct a dot plot of these programming times.

b. Calculate the mean, 15% trimmed mean, and median of these values. Locate these three measures of central tendency on your plot and compare them.

c. Calculate the range, interquartile range, and standard deviation. What do these three descriptive statistics measure?

d. Describe this distribution in terms of location, spread, peaks, and symmetry or skewness.

e. A company has claimed that a byte of memory on this chip can be reprogrammed in less than 14 seconds. Does this claim seem reasonable?

EXERCISE 3-9

Consider the following measurements of leaf protein (mg/g fresh weight) from six plants of a variety of soybean (Devore, 1982, page 23; from *Science,* volume 199, page 974):

4.9	5.1	6.1	11.7	14.0	16.1

a. Construct a dot plot of these six measurements.

b. Describe this distribution in terms of location, spread, peaks, and symmetry or skewness.

c. Calculate the mean, 15% trimmed mean, and median of these values. Locate these three measures of central tendency on the plot and compare them.

d. Find the range, interquartile range, and standard deviation. Discuss these three measures of variation.

EXERCISE 3-10

Researchers measured pulmonary compliance ($cm^3/cm\ H_2O$) for each of 16 construction workers who had been exposed over a long period to asbestos. Pulmonary compliance is a measure of how well the lungs expand and contract. These measurements were taken 8 months after asbestos exposure (Devore, 1987, page 275; from "Acute Effects of Chrysotile Asbestos Exposure on Lung Function," *Environ. Research,* 1978, pages 360–372).

167.9	180.8	184.8	189.8	194.8	200.2	201.9	206.9
207.2	208.4	226.3	227.7	228.5	232.4	239.8	258.6

a. Construct a dot plot of these measurements.

b. Calculate the mean, 5% trimmed mean, and median of these values. Locate these three descriptive statistics on your plot and compare them.

c. Find the range and interquartile range.

d. Describe this distribution in terms of location, variation, peaks, and symmetry or skewness.

EXERCISE 3-11

Researchers measured levels of the amino acid alanine (in mg/100 ml) for six normal baby boys on an isoleucine-free diet (Devore, 1987, page 275; from "The Essential Amino Acid Requirements of Infants," *Amer. J. Nutrition,* 1964, pages 322–330):

1.44 2.70 2.80 2.84 2.94 3.54

a. Construct a dot plot of these measurements.

b. Is this distribution symmetrical, negatively skewed, or positively skewed?

c. Calculate the mean, 15% trimmed mean, and median of these values. Locate these three measures of location on your plot and compare them.

d. Find the range, interquartile range, and standard deviation. Discuss these three measures of variation.

EXERCISE 3-12

Investigators measured radiation levels (in milliroentgens per hour) in the television display areas of 10 department stores (Devore, 1987, page 301; from "Many Set Color TV Lounges Show Highest Radiation," *J. Environmental Health,* 1969, pages 359–360):

.15 .16 .36 .40 .48 .50 .50 .60 .80 .89

You may wish to change units by multiplying each value by 100 before doing any calculations.

a. Construct a dot plot of these radiation levels.

b. Would you describe this distribution as symmetrical, negatively skewed, positively skewed, or none of these?

c. Calculate the mean, 10% trimmed mean, and median of these values. Locate these three descriptive statistics on your plot and compare them.

d. Find the range, interquartile range, and standard deviation. Discuss these three measures of variation.

e. The limit recommended for such radiation exposure is .50 milliroentgen per hour. Based on these sample measurements, discuss the safety of department store television display areas at the time this study was done.

EXERCISE 3-13

A softball player wanted to compare his hitting distance with two types of bats (wooden and aluminum) and two brands of balls. He hit four balls pitched from a pitching machine for each combination of bat and ball type. The distances he hit the softballs are shown below (Shaughnessy, 1988).

Bat, ball type	Distances hit (feet)			
Wooden bat, Dudley Thunder	230	242	242	250
Wooden bat, Worth Red Dot	258	264	265	275
Aluminum bat, Dudley Thunder	265	270	277	282
Aluminum bat, Worth Red Dot	290	302	310	318

a. Draw a dot plot for each of the four sets of distances. Use the same scale for each plot.

b. Find the mean, standard deviation, and range for each of the four sets of distances.

c. Using the wooden bat, on average what was the player's hitting advantage using the Worth Red Dot ball rather than the Dudley Thunder?

d. Using the aluminum bat, on average what was the player's hitting advantage using the Worth Red Dot rather than the Dudley Thunder?

e. Looking at your answers to parts (c) and (d), is the hitting advantage of the Worth Red Dot softball the same for the wooden and aluminum bats?

f. Using the Worth Red Dot softball, on average what was the player's hitting advantage using the aluminum rather than the wooden bat?

g. Using the Dudley Thunder softball, on average what was the player's hitting advantage using the aluminum rather than the wooden bat?

h. Looking at your answers to parts (f) and (g), is the hitting advantage of the aluminum bat the same for the two brands of softball?

i. Compare the range and standard deviation across the four sets of distances. What can you say about the variation in distances hit under the four sets of conditions?

CHAPTER 4

Studying Two Variables at a Time

IN THIS CHAPTER

Two-way frequency table or contingency table

Plots of a quantitative variable over levels of a qualitative variable

Dot chart

Scatterplot

Logarithmic transformation

Looking at individual variables can be interesting, but analysis should not stop there. In fact, we went beyond individual variables in Chapters 2 and 3 when we compared plots and descriptive statistics for life expectancy and population growth across economic categories. Such analyses address our second goal: to study each indicator separately within economic categories and compare results across economic categories. We continue now by discussing a number of ways to study relationships between two variables. Let's begin with two-way frequency tables, for studying relationships between two qualitative variables.

4-1 Two-Way Frequency Tables for Studying the Relationship Between Two Qualitative Variables

In making projections of future population sizes, the World Bank classifies countries as having primary school enrollments for females greater than or less than 70% (World Bank, 1987, page 281). Analysts make separate projections for the two groups of countries. To see how many countries are in each group, we might look at a frequency table such as Table 4-1. This table shows the number of countries with high female primary school enrollments, low enrollments, and missing enrollment information.

From female primary school enrollment, we have created an ordinal qualitative variable with two possible values: low (less than or equal to 70%) and high (greater than 70%). About 22% of the 128 countries have low female primary school enrollments as defined by this new variable; about 11% have missing values. Do these percentages vary across economic categories? To find out, we can count the number of countries with low female primary school enrollments, the number with high enrollments, and the number with missing information, within each economic category. We might then arrange these counts in a table that has columns defined by level of female primary school enrollment and rows defined by economic categories. (Alternatively, columns could be defined by economic category and rows by enrollment level.)

In general, when we count the number of cases within each combination of values for two qualitative variables and arrange these counts in a two-dimen-

TABLE 4-1 Countries classified by number of females enrolled in primary school in 1984 as percentage of 6–11-year age group.

1984 primary school enrollment for females	Number of countries	Percentage of countries
Less than or equal to 70%	28	22
Greater than 70%	86	67
Missing	14	11
Total	128	100

sional table with rows defined by one variable and columns by another, we create a *two-way frequency table.* We use such a table to look for association between the two variables.

A **two-way frequency table**, or two-dimensional contingency table, is a tabular display of the number of cases within each combination of categories of two qualitative variables.

If the distribution of frequencies across categories of one variable depends on the category of the other variable, we say the two variables are *associated.* If the distribution of frequencies across categories of one variable is about the same for each category of the other variable, we say the two variables are *independent,* or not associated.

We say two qualitative variables are **associated** if the distribution across categories of one variable depends on the category of the other variable.

We say two qualitative variables are **independent** or **not associated** if the distribution across categories of one variable is about the same for all categories of the other variable.

Table 4-2 is a two-way frequency table for economic category and female primary school enrollment. Examining the table, we see that the distribution of counts across levels of female primary school enrollment depends on economic category. Equivalently, the distribution of counts across economic categories is different for the different levels of female primary school enrollment. Based on this table, we say there is an association between the two variables: Lower female primary school enrollments and more missing values are associated with lower economic categories; higher female primary school enrollments and fewer missing values are associated with greater economic development as measured by economic category.

Do lower female primary school enrollments tend to be associated with lower levels of contraception use among the World Bank countries? We address this question by examining the two-way frequency table in Table 4-3. In

TABLE 4-2 Countries classified by economic category and number of females enrolled in primary school in 1984 as percentage of 6–11-year age group.

	1984 primary school enrollment for females			
Economic category	≤70%	>70%	Missing	Total
Low income	18	13	6	37
Lower-middle income	9	24	3	36
Upper-middle income	0	22	1	23
High-income oil exporting	1	2	1	4
Industrial market	0	18	1	19
Nonmember	0	7	2	9
Total	28	86	14	128

TABLE 4-3 Countries classified by number of females in primary school in 1984 as percentage of age group and percentage of married women of child-bearing age using contraception in 1984. (Thirty-nine countries are excluded because of missing values on one or both of these variables.) The number in parentheses to the right of a frequency is the percentage of the row total. The number in parentheses below a frequency is the percentage of the column total.

Percent contraception use	Female primary school enrollment		
	≤70%	>70%	Total
≤35%	23 (53)	20 (47)	43 (100)
	(96)	(31)	(48)
>35%	1 (2)	45 (98)	46 (100)
	(4)	(69)	(52)
Total	24 (27)	65 (73)	89 (100)
	(100)	(100)	(100)

this table, female school enrollment has two categories: low (≤70%) and high (>70%). Percent contraception use also has two categories: low (≤35%) and high (>35%). The cutoff value of 35% defining the categories for contraception use divides the countries with nonmissing information into approximately equal sized groups. Note that Table 4-3 is based on the 89 countries with nonmissing information on both variables.

For interpreting a two-way frequency table, it can be helpful to show each frequency as a percentage of its row total and as a percentage of its column total. In Table 4-3, these percentages are displayed along with the frequencies. The number in parentheses to the right of a frequency is the percentage of the row total. The number in parentheses below a frequency is the percentage of the column total. These row and column percentages help us see the strength of the association between female primary school enrollment and contraception use in Table 4-3. As the first column shows, all but one of the countries with lower enrollments also have lower contraception use. The second column shows that 69% of countries with higher enrollments also have greater contraception use. Looking at the table row-wise, we see that more than half the countries with lower contraception use have lower enrollments. All but one country with greater contraception use also have higher female primary school enrollments.

We have to be careful interpreting Table 4-3. Thirty-nine (30%) of the 128 countries have missing information on one or both variables. Also, we have taken two quantitative variables and created two new variables, each with just a low and a high category. *How we select the dividing point for low and high can affect what we see in the frequency table.* We might choose several ways to divide the variables and see if the same relationship holds in each analysis. Or, we might decide not to categorize the indicators at all, and instead examine the relationship between them using scatterplots, as discussed in Section 4-3.

We will look at female primary school enrollment and contraception use again. But first, we consider ways to study the relationship between a quantitative variable and a qualitative variable.

4-2

Tables and Graphs for Studying the Relationship Between a Quantitative Variable and a Qualitative Variable

We saw two modes or peaks when we looked at the plot of fertility rates in Figure 2-15. (Recall that fertility rate is an estimate of the average number of children per woman in a country.) One peak was around 2 children per woman and another around 6.5 children per woman. Are lower fertility rates associated with higher levels of economic development? Let's see how the quantitative variable fertility rate is related to the qualitative variable economic category.

Table 4-4 shows some descriptive statistics for fertility rates, overall and within economic categories. We see that both the mean and median fertility rate decrease as economic category increases from low-income to lower-middle-income to upper-middle-income to industrial market. The high-income oil exporters are on a par with the low-income countries. The nonmembers are close to the upper-middle-income group.

Figure 4-1 shows a dot plot of fertility rates for each economic category. The axes are lined up, with the same scale for each plot, making visual comparisons easy. These plots clearly show the relationship between fertility rate and economic category. The industrial market countries have the lowest fertility rates and the least spread in the values. The other economic categories all show much higher fertility rates in general, and greater variation among the values.

We might choose to show a box plot of the quantitative variable for each level of the qualitative variable. Such a display allows easy comparisons of

TABLE 4-4 Descriptive statistics for 1985 total fertility rates, overall and within economic categories. The summary statistics are the number in the group, *n*; the mean; median; standard deviation, SD; minimum; maximum. Two low-income countries (Afghanistan, Kampuchea) and one lower-middle-income country (Lebanon) are excluded because of missing values.

Economic category	*n*	Mean	Median	SD	Minimum	Maximum
Low income	35	6.1	6.4	1.2	2.3	8.0
Lower-middle income	35	5.1	5.4	1.4	2.5	6.9
Upper-middle income	23	3.4	2.9	1.6	1.7	6.7
High-income oil exporting	4	6.4	6.5	1.0	5.2	7.2
Industrial market	19	1.8	1.7	.3	1.3	2.6
Nonmembers	9	3.2	2.3	1.6	1.8	6.4
Overall	125	4.5	4.7	2.0	1.3	8.0

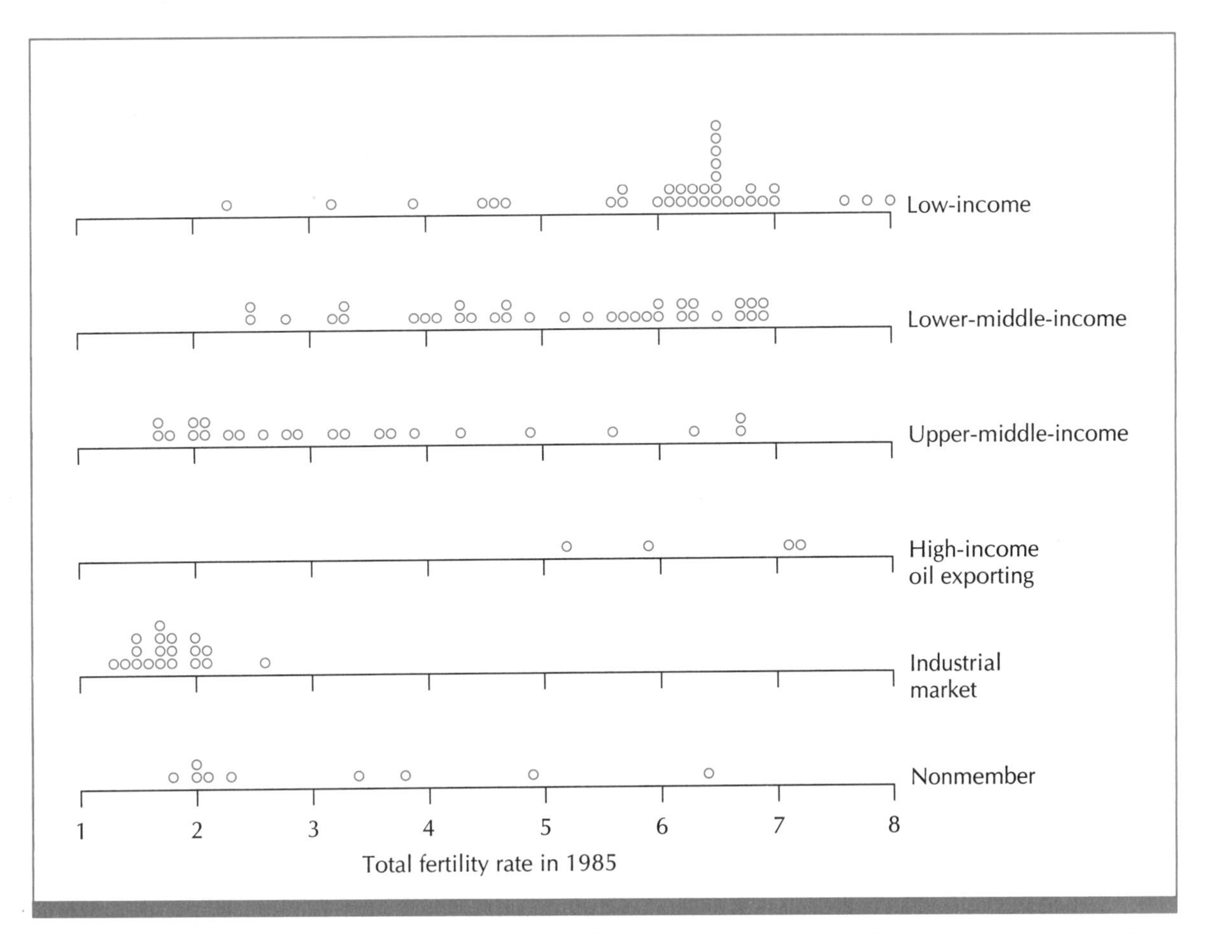

FIGURE 4-1 Dot plot of 1985 total fertility rate for each economic category. Two low-income countries and one lower-middle-income country are excluded because of missing values.

center, spread, and extreme values of the quantitative variable within each level of the qualitative variable. Figure 4-2 displays a box plot of fertility rate for each economic category.

A less informative (but often used) type of plot is shown in Figure 4-3. Here the mean fertility rate, the mean plus one standard deviation, and the mean minus one standard deviation are plotted for each economic category. This type of plot can be misleading. It suggests symmetry when the distribution of values may not be symmetrical at all (as we see in Figures 4-1 and 4-2). We get no information about skewness or concentrations of values. Displays such as those in Figures 4-1 and 4-2 are much more informative.

Even less informative and potentially more misleading than Figure 4-3 is a graph that shows only the means of the quantitative variable across levels of the qualitative variable (Figure 4-4). Some authors connect the means to the

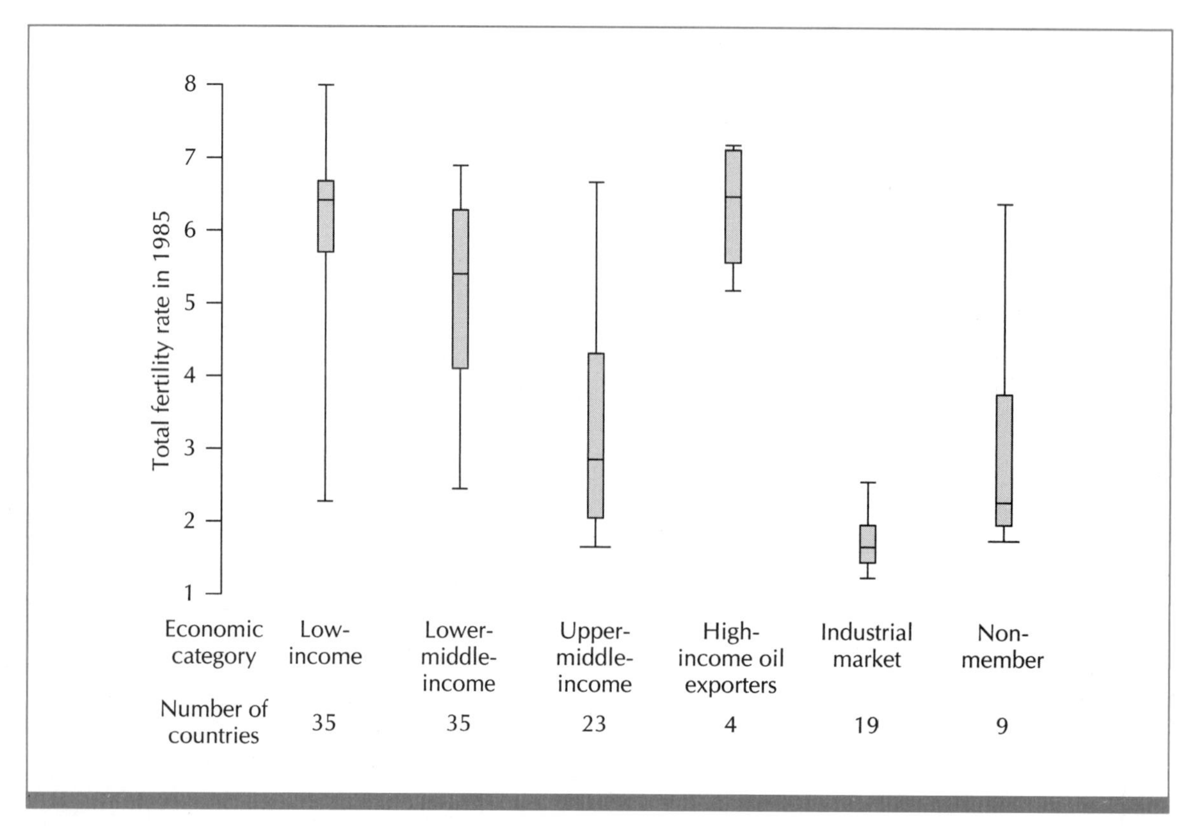

FIGURE 4-2 Box plot of 1985 total fertility rate for each economic category. Two low-income countries and one lower-middle-income country are excluded because of missing values.

horizontal axis with a line or bar, making the graph look like a histogram or frequency plot. The visual impression of differences between means is completely dependent on the scale chosen for the quantitative variable. Since no measures of variation are provided, we cannot make meaningful comparisons across categories. This is not a useful graph.

Dot Charts

Using box plots, histograms, and dot plots, we can compare the distribution of a quantitative variable across levels of a qualitative variable. Sometimes we want to compare totals (or percentages) of a quantitative variable across levels of a qualitative variable. A good graphical tool for such a comparison is the dot chart.

A *dot chart* is a graphical display of totals (or percentages) for a quantitative variable at each level of a qualitative variable (Cleveland, 1985). Examples of dot charts are given in Figures 4-5 and 4-6. Levels of the qualitative

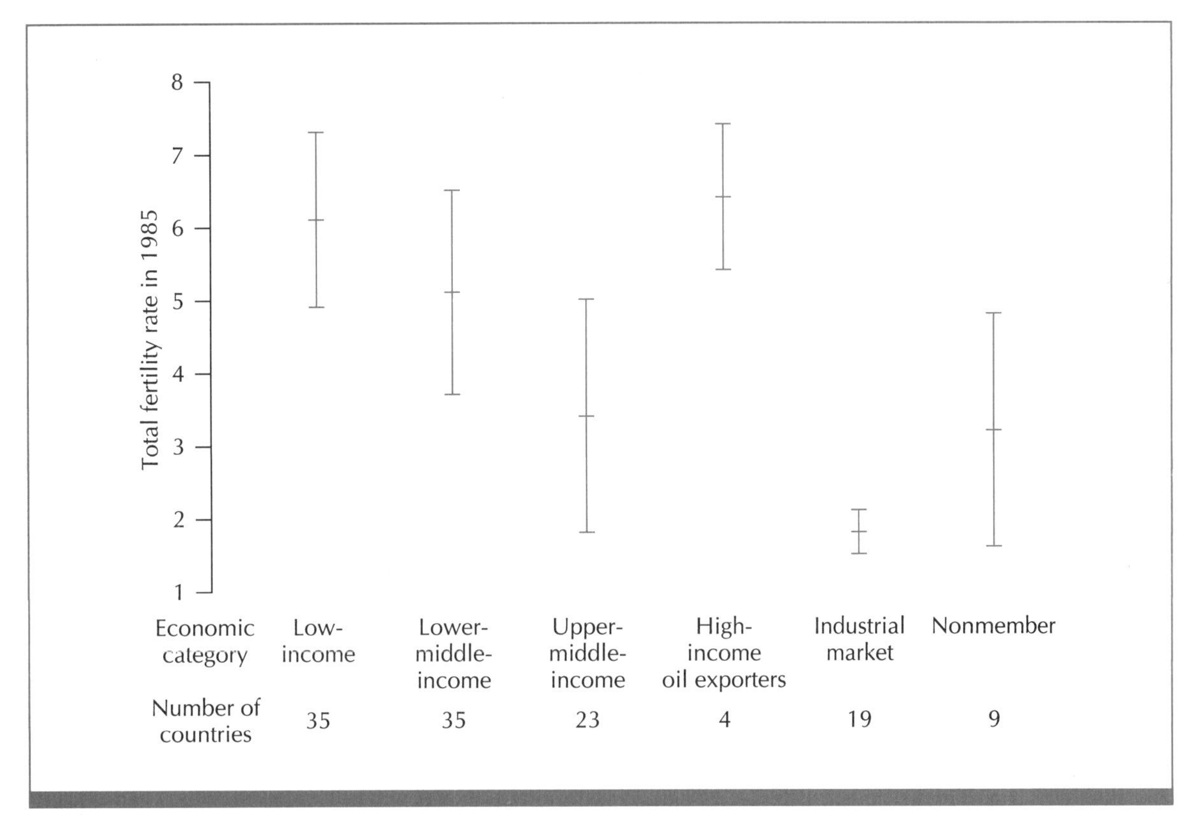

FIGURE 4-3 Mean total fertility rate in 1985, mean plus one standard deviation, and mean minus one standard deviation for each economic category. (Two low-income countries and one lower-middle-income country are excluded because of missing values.) *This type of plot is of limited usefulness.*

variable are listed at the left of the chart. For each category, a horizontal line illustrates the value (and relative value or percentage of total) of the quantitative variable. The axis at the bottom of the chart shows the scale for the quantitative variable; the axis at the top shows percentage of total.

> A **dot chart** is a graphical tool for comparing totals (or percentages) for a quantitative variable across levels of a qualitative variable. We make visual comparisons along a straight axis.

Figure 4-5 displays population size for each of five economic categories (nonmember nations are excluded). The lower axis shows population in millions of people and the upper axis is percentage of total population among the countries included in the chart. A dot indicates the population size for each economic category. A line helps us connect the economic category label with the corresponding dot.

The dot chart in Figure 4-6 displays total gross domestic product (an estimate of the value of goods and services produced by a country, in millions

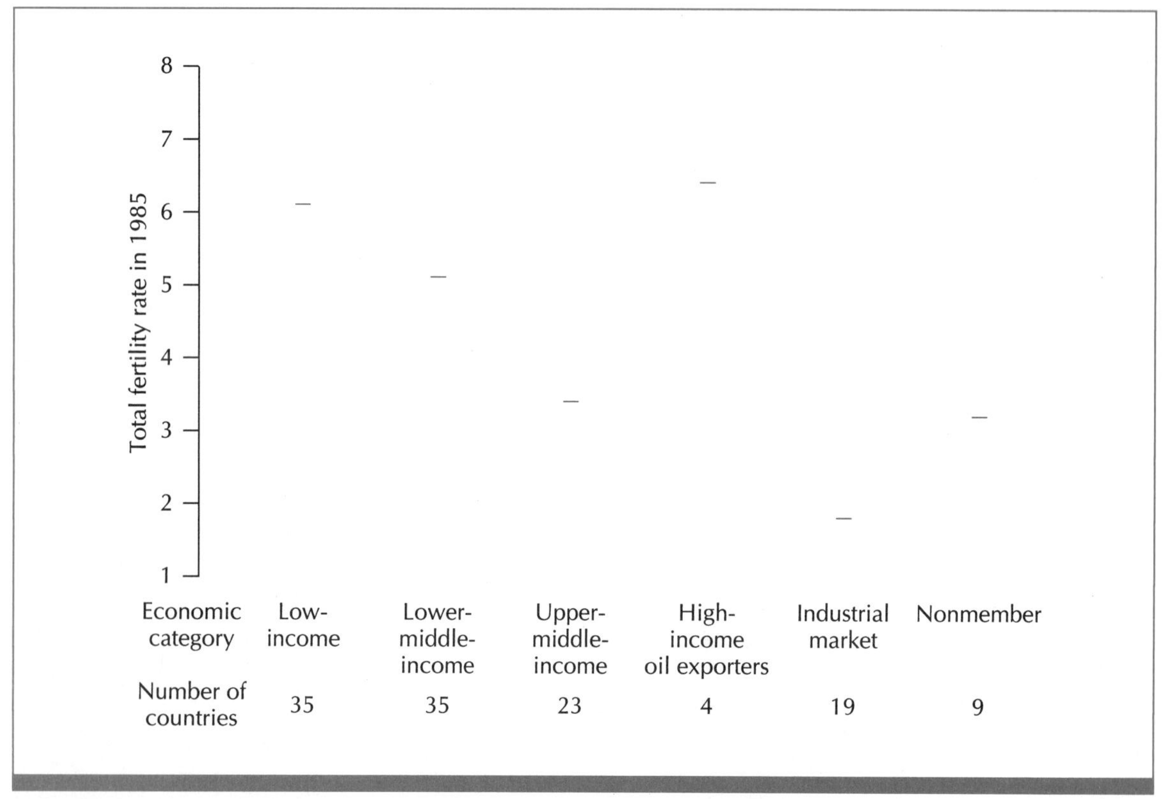

FIGURE 4-4 Mean total fertility rate in 1985 for each economic category. (Two low-income countries and one lower-middle-income country are excluded because of missing values.) *Please do not use this type of graph!*

of U.S. dollars) for each of five economic categories, again excluding nonmembers. The scale for gross domestic product is shown at the bottom. The upper axis gives percentage of total gross domestic product among the countries included in the plot.

Figure 4-5 shows that the low-income countries comprise over 50% of the total population of countries included in the chart; the industrial market countries represent less than 20% of the population total. Figure 4-6 shows that the low-income countries account for less than 10% of the total gross domestic product of countries included in the chart, while the industrial market countries account for over 70% of this total. We must be careful in comparing the two graphs because they are based on different numbers of countries. (The two charts could have been based only on countries with nonmissing information on both variables.) In addition, gross domestic product is difficult to compare across countries, especially across economic categories. However, these two dot charts give an overwhelming impression that the low-income countries include most of the population and the industrial market countries most of the material wealth.

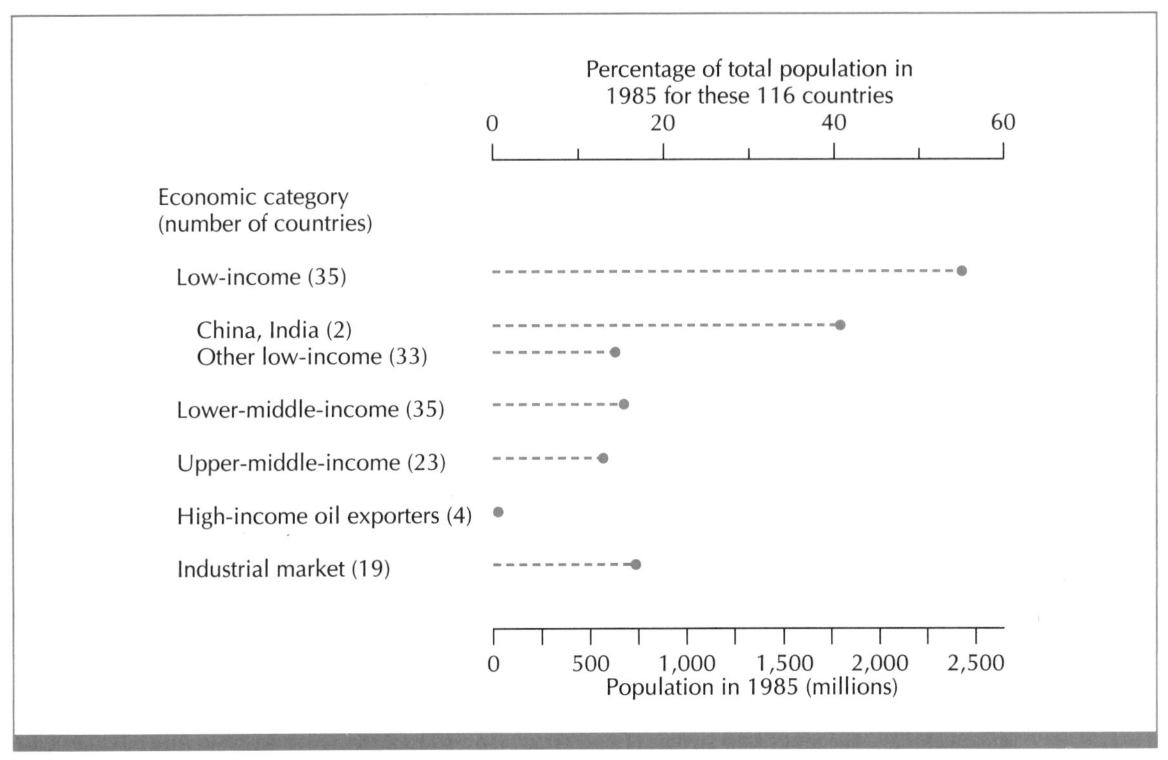

FIGURE 4-5 Dot chart showing population in 1985 and percentage of total, by economic category.

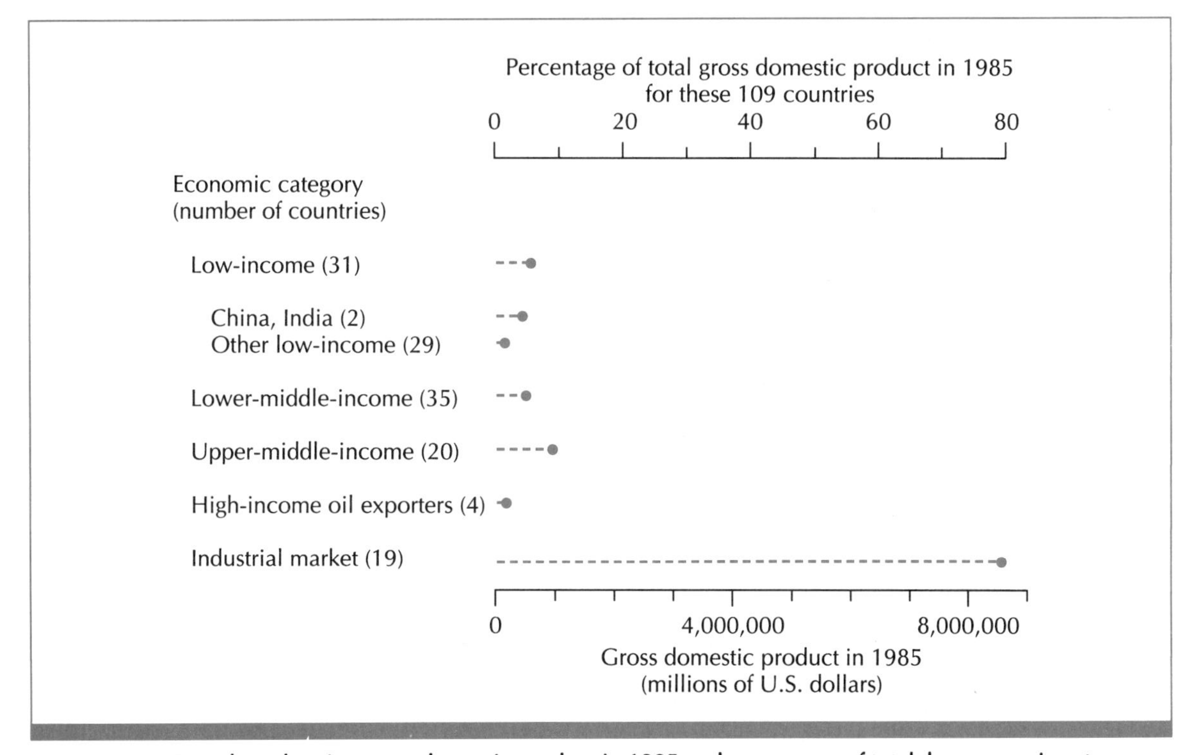

FIGURE 4-6 Dot chart showing gross domestic product in 1985 and percentage of total, by economic category.

When used to display frequencies and percentages across categories of a single qualitative variable, a dot chart is the same as a frequency plot (see Exercise 2-23). Dot charts present the same kind of information commonly displayed in pie charts for many business and social science applications. As a graphical tool, the dot chart is easier to interpret because we make comparisons of totals or percentages along a straight axis. A pie chart requires more difficult visual comparisons of angles. For a discussion of the advantages of the dot chart over the pie chart for visual interpretations of data, see Cleveland (1985).

Section 4-3 looks at ways to examine the relationship between two quantitative variables.

4-3 Scatterplots for Studying the Relationship Between Two Quantitative Variables

Life expectancies are shorter among low-income countries than among industrial market nations (Figure 2-18). How do life expectancies in general vary with gross national product among the World Bank countries? We can examine the relationship between these two quantitative variables using a scatterplot.

A *scatterplot* or scattergram is a two-dimensional graphical display of two quantitative variables. The scale for one variable is on the vertical axis, the scale for the other variable on the horizontal axis. Each case is represented by a point on the plot positioned according to the values of the two variables.

> A **scatterplot** or scattergram is a two-dimensional graphical display of two quantitative variables, the scale for one variable on the vertical axis and the scale for the other on the horizontal axis. Each case is represented by a point on the plot positioned according to the values of the two variables.

Gross national product per capita and life expectancy are listed in Table 4-5 for the four high-income oil exporters. We see that the two countries with greater per capita gross national products also have longer life expectancies. Figure 4-7 shows a scatterplot for these data. Each of the four points on the graph represents a high-income oil exporting nation. Since we are interested in seeing how life expectancies vary with per capita gross national product, we follow tradition and plot per capita gross national product along the horizontal axis and life expectancy along the vertical axis. (The variable on the horizontal axis is often called the independent variable; the variable on the vertical axis, the dependent variable. *Such designations are merely conventions; they do not imply any cause-and-effect relationship between the two variables.*)

The scatterplot in Figure 4-7 uses a *range frame* (Tufte, 1983, page 130): Each axis extends from the minimum to the maximum value of the corresponding variable. The vertical axis extends from 60 to 72 years, the horizontal axis from 7,170 to 19,270 U.S. dollars. Note that the axes do not intersect. Such

TABLE 4-5 Per capita gross national product in 1985 and life expectancy at birth in 1985 for four high-income oil exporting nations.

Nation	Per capita gross national product in 1985 (U.S. dollars)	Life expectancy at birth in 1985 (years)
Libya	7,170	60
Saudi Arabia	8,850	62
Kuwait	14,480	72
United Arab Emirates	19,270	70

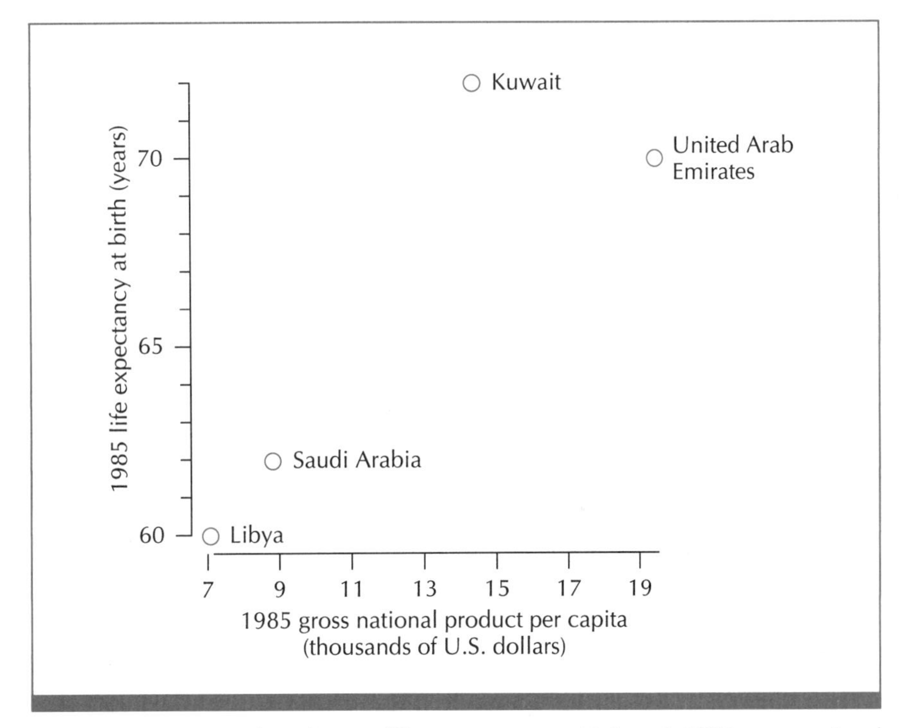

FIGURE 4-7 Scatterplot of 1985 life expectancy at birth and 1985 gross national product per capita for four high-income oil exporting nations. The range of values for each variable is indicated by the line on the corresponding axis.

a range frame allows efficient use of the plotting portion of the graph, while clearly showing the range of each variable. With a range frame, we avoid the often misleading perception of an origin where axes meet (see Exercise 4-1).

> In a **range frame** for a scatterplot, each axis extends from the minimum to the maximum value for the corresponding variable.

Sometimes on a scatterplot we provide some identifying information near each plotted point. In Figure 4-7, country name is indicated near each point. Such labels can be useful in plots without too many points (when labeling might lead to confusing clutter).

Our overall impression from Figure 4-7 is that high-income oil exporters with greater per capita gross national products tend to have longer life expectancies. This is a very small group of countries. We would not necessarily expect to see the same relationship within another group of countries or in our data set as a whole. Figure 4-8 shows a scatterplot of life expectancy and gross

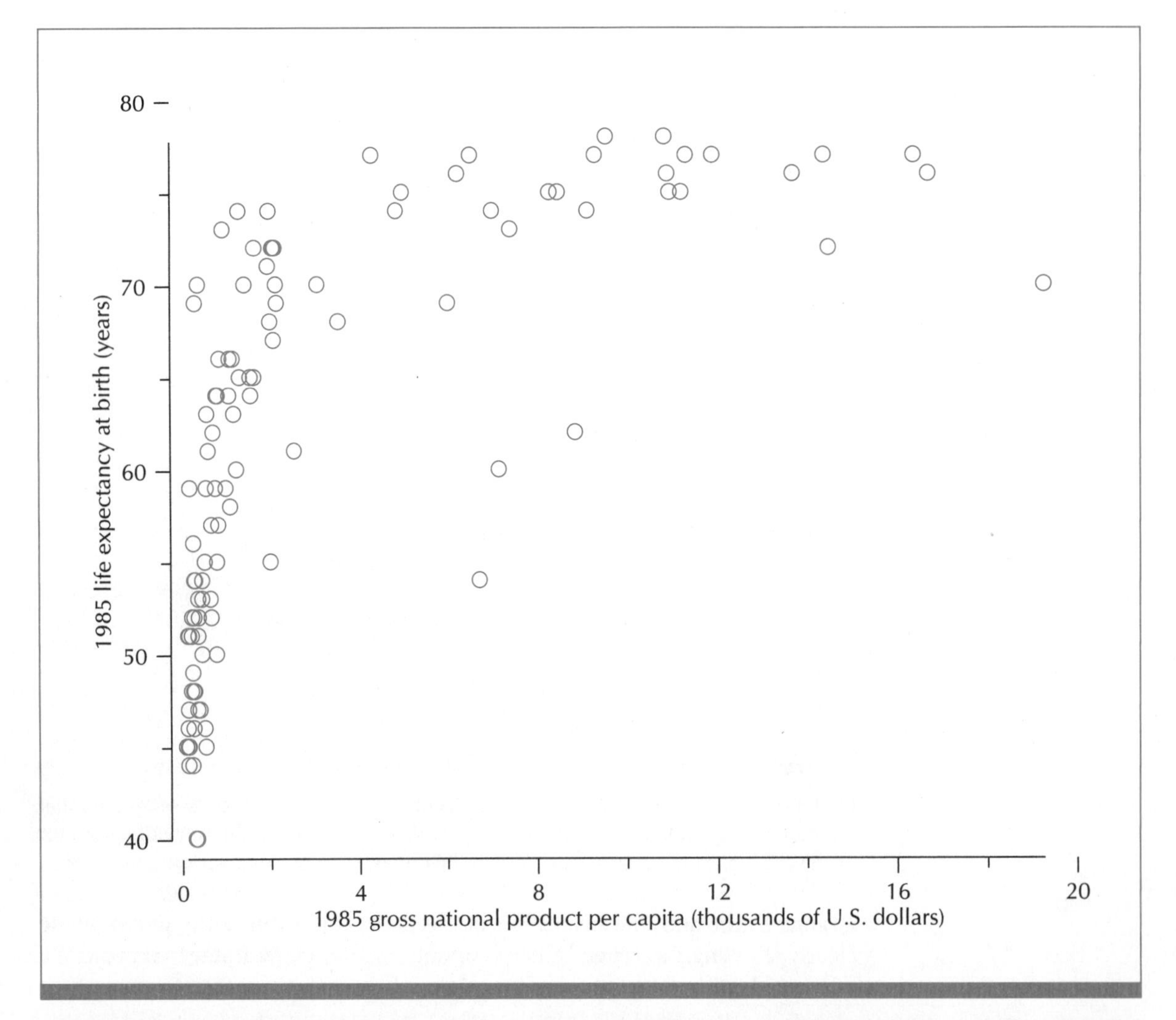

FIGURE 4-8 Scatterplot of 1985 life expectancy at birth and 1985 gross national product per capita for 109 countries. Nineteen countries are excluded because of missing values. The range of values for each variable is indicated by the line on the corresponding axis.

national product, based on the 109 countries with information on both variables. Each country in this plot is represented by an open circle. Darker circles indicate more than one case with the same (or very similar) values. For suggestions on dealing with overlap in scatterplots, see Cleveland (1985).

Figure 4-8 shows that countries with greater per capita gross national products tend to have longer life expectancies. For countries with per capita gross national products less than $2,000, small differences in income are associated with much larger differences in life expectancy than is true for other countries. Also, more than half the countries have per capita incomes less than $2,000. (The median per capita gross national product is $1,010.) This creates congestion in the left portion of the plot. It would be nice if we could stretch out the gross national product scale for lower incomes and compress the scale for higher incomes. One such change of scale uses the logarithm of per capita gross national product. We call such a change of scale a *transformation*. When we transform the scale of a variable, we hope to make interpretations easier in our analysis.

A **transformation** of a variable is a mathematical manipulation of each value of the variable. When we make a transformation, we transform the original scale of measurement for the variable to a new scale. For example, the logarithmic transformation involves taking the logarithm of each value of a variable, the square root transformation involves taking the square root of each value of a variable, and a power transformation involves taking a power (such as the square or cube) of each value of a variable.

We will use logarithm base-10 to transform the values of per capita gross national product to a new scale. Recall that the logarithm base-10 of a constant is the number you raise 10 to in order to get the constant back. For example, the logarithm base-10 of 1,000, denoted $\log_{10}(1{,}000)$, is 3 since $10^3 = 1{,}000$. Similarly, $\log_{10}(10{,}000) = 4$ since $10^4 = 10{,}000$ and $\log_{10}(5{,}000) = 3.69897$ since $10^{3.69897} = 5{,}000$.

To make the **logarithmic transformation** of a variable, take the logarithm of each value of the variable.

The **logarithm base-10,** denoted $\log_{10}$, of a constant is the number you raise 10 to in order to get the constant back. For example, $\log_{10}(10{,}000) = 4$ since $10^4 = 10{,}000$.

A scatterplot of life expectancy and the logarithm of per capita gross national product is shown in Figure 4-9. The scale for the logarithm base-10 of per capita gross national product is shown on the top of the plot. The same scale is shown at the base of the plot, but with actual per capita gross national products indicated on the logarithmic scale.

What does Figure 4-9 show? We see that on the logarithmic scale, the distance between gross national products of $100 and $1,000 is the same as the distance between $1,000 and $10,000. We have indeed stretched out the lower incomes and compressed the higher incomes. This relieves the congestion we saw in Figure 4-8 and gives us a clearer picture of the relationship between

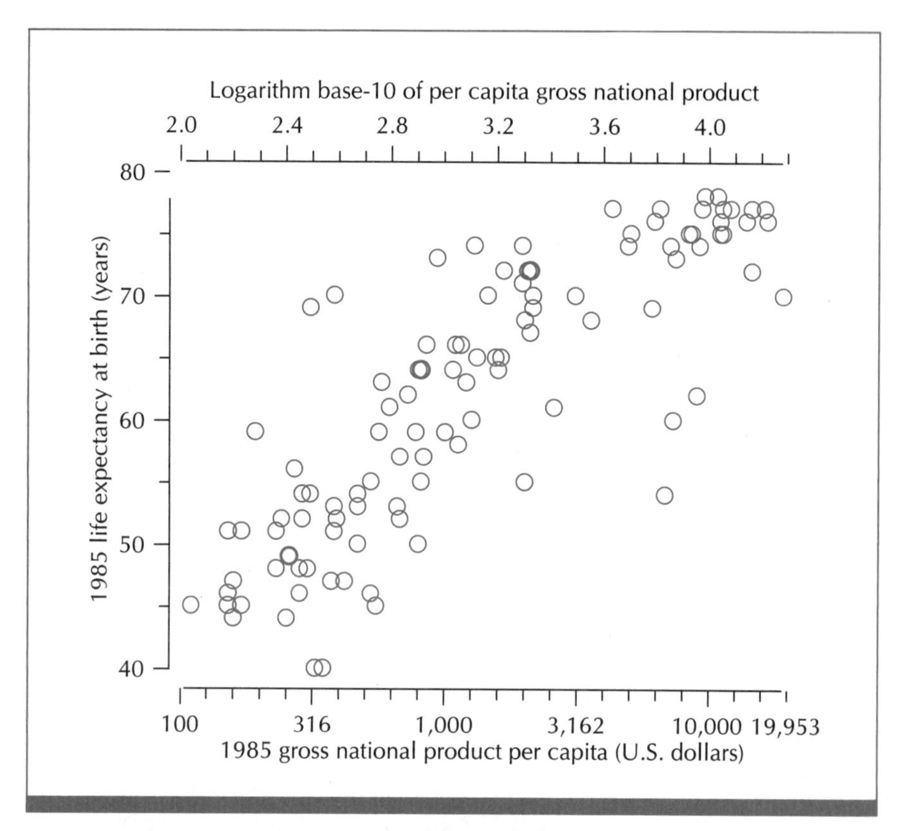

FIGURE 4-9 Scatterplot of 1985 life expectancy at birth and the logarithm of 1985 gross national product per capita for 109 countries. Nineteen countries are excluded because of missing values. The range of values for each variable is indicated by the line on the corresponding axis.

gross national product and life expectancy. We see that average life expectancy increases in roughly a linear fashion as the logarithm of per capita gross national product increases.

In Figure 4-10, we have added a dot plot for life expectancy on the left vertical axis. We have also added a dot plot for gross national product (on the logarithmic scale) on the bottom axis. This is a variation on the dot-dash-plot suggested by Tufte (1983, page 133). We get a visual impression of the distribution of each variable separately as well as the relationship between the two variables.

With a transformation, we get a clearer picture of the relationship between gross national product and life expectancy. Sometimes it is helpful to transform both variables, making two changes of scale in a scatterplot (see Exercise 4-26). Trial and error determines whether a transformation of one or both variables helps to illustrate the relationship between two variables.

As another example, Figure 4-11 shows a scatterplot of primary school enrollment and gross national product. We again see congestion of points at low income levels. Transforming gross national product by taking the logarithm base-10, we obtain the scatterplot in Figure 4-12. We see an increasing relationship between gross national product and primary school enrollments for countries with per capita gross national products less than \$1,000. This increasing relationship does not exist for countries with per capita gross national products over \$1,000. For these countries, primary school enrollments level off, fluctuating around 100%.

The examples so far have shown positive or increasing relationships

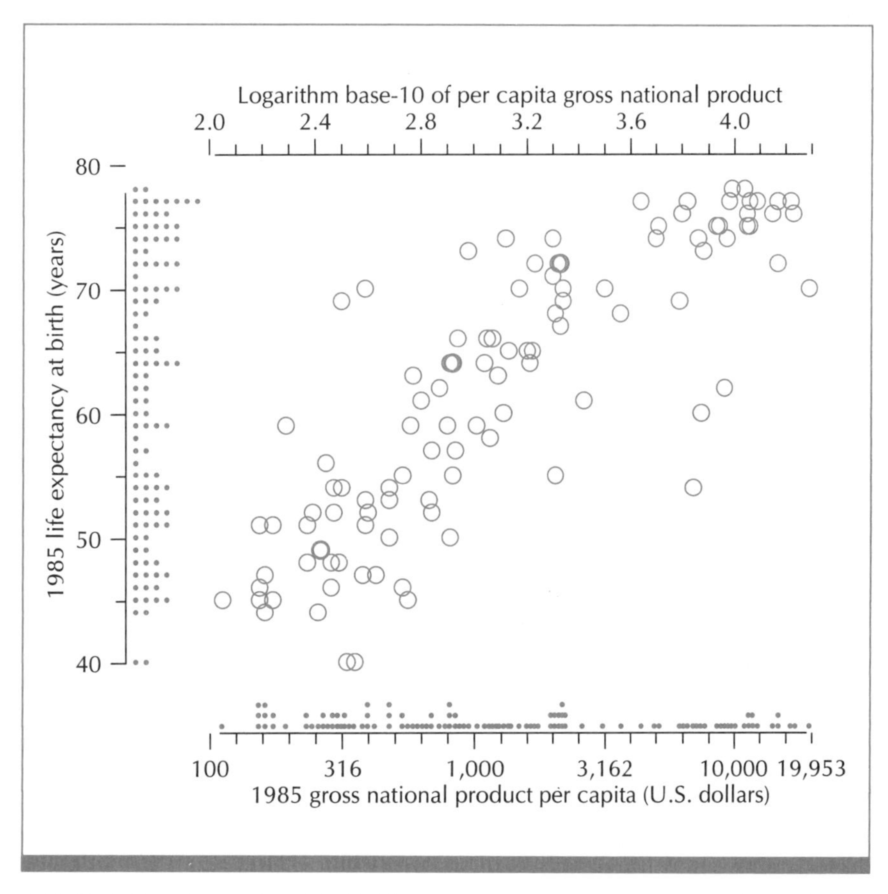

FIGURE 4-10 Scatterplot of 1985 life expectancy at birth and the logarithm of 1985 gross national product per capita for 109 countries. Nineteen countries are excluded because of missing values. The range of values for each variable is indicated by the line on the corresponding axis. A dot plot of the 109 life expectancies is on the left vertical axis. A dot plot of the 109 values of per capita gross national product is on the bottom axis.

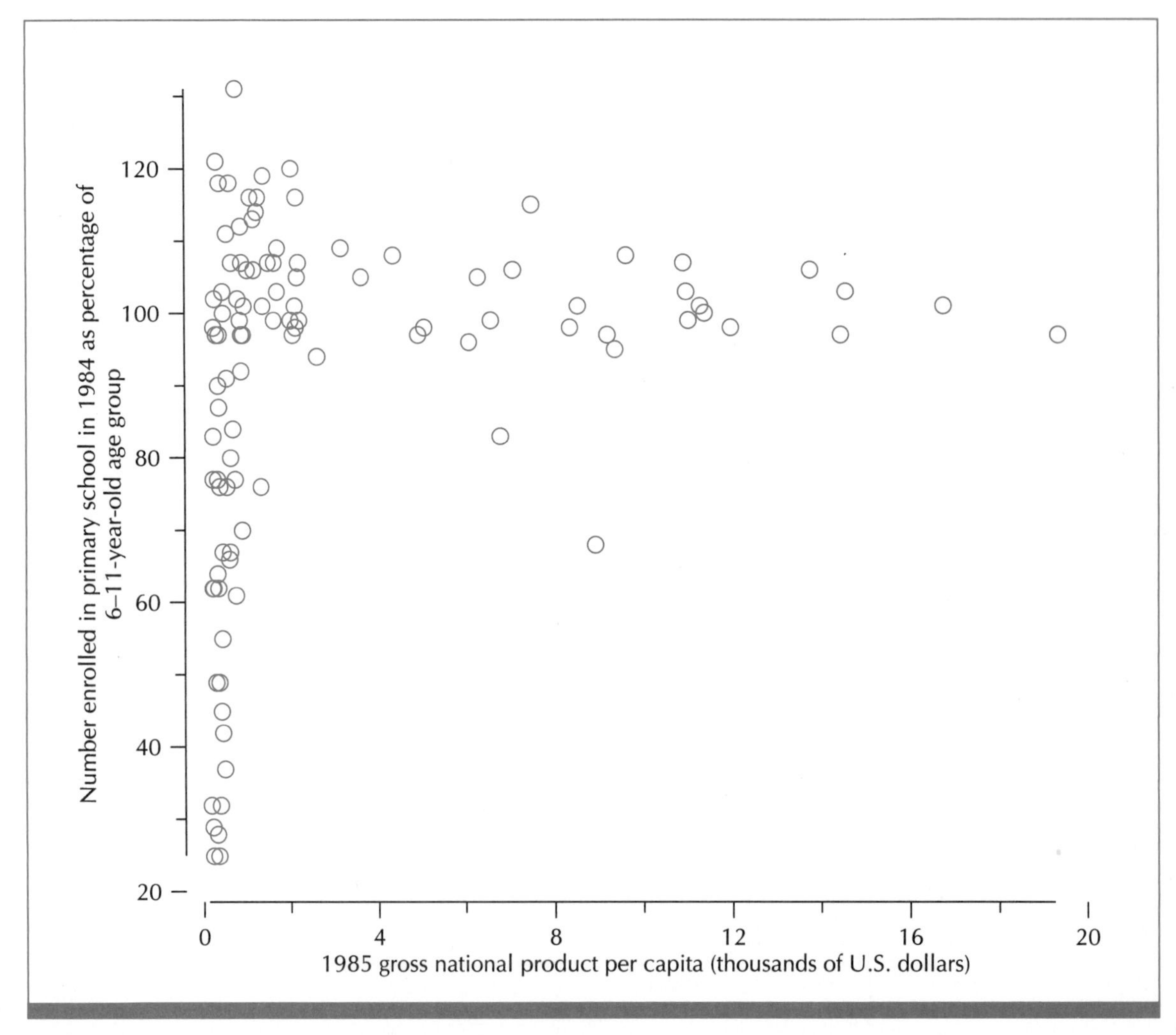

FIGURE 4-11 Scatterplot of number enrolled in primary school in 1984 as percentage of 6–11-year age group and 1985 gross national product per capita for 104 countries. Twenty-four countries are excluded because of missing values. The range of values for each variable is indicated by the line on the corresponding axis.

between two variables. Two quantitative variables can also have a negative or decreasing relationship. Figure 4-13 shows that birth rates decrease with increasing gross national product for the four high-income oil exporters.

Sometimes we see little or no relationship between two variables. Figure 4-14 reveals no meaningful relationship between calorie supply and gross national product for the four high-income oil exporters.

In this chapter we have considered a number of ways to study two variables at a time. In Chapter 5 we examine relationships among three or more variables.

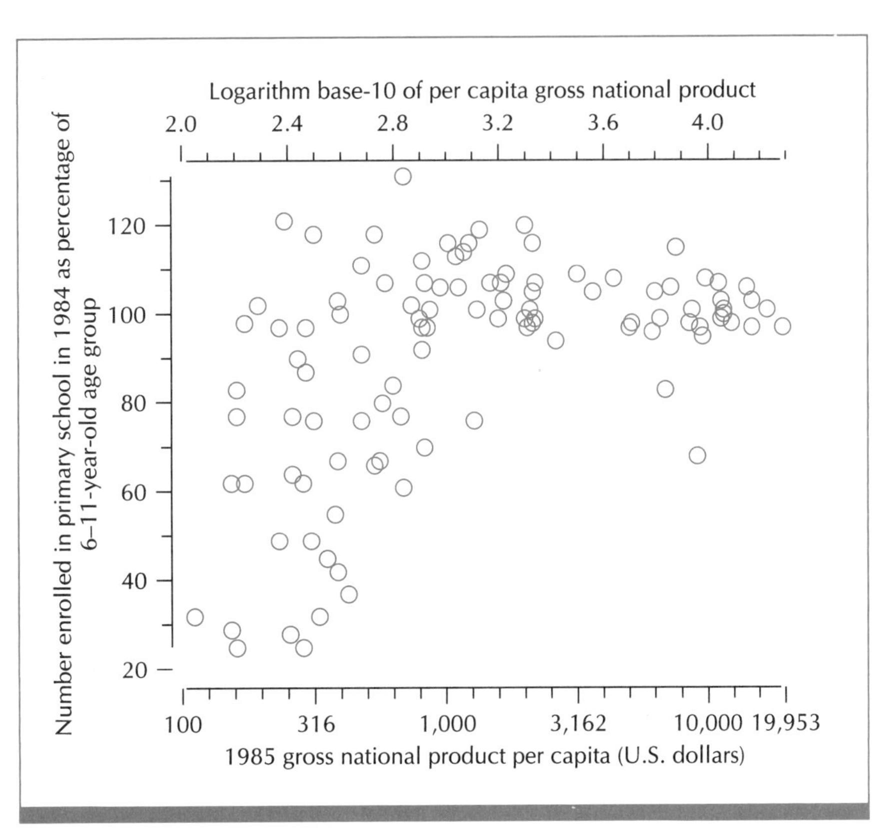

FIGURE 4-12 Scatterplot of number enrolled in primary school in 1984 as percentage of 6–11-year age group and the logarithm of 1985 gross national product per capita for 104 countries. Twenty-four countries are excluded because of missing values. The range of values for each variable is indicated by the line on the corresponding axis.

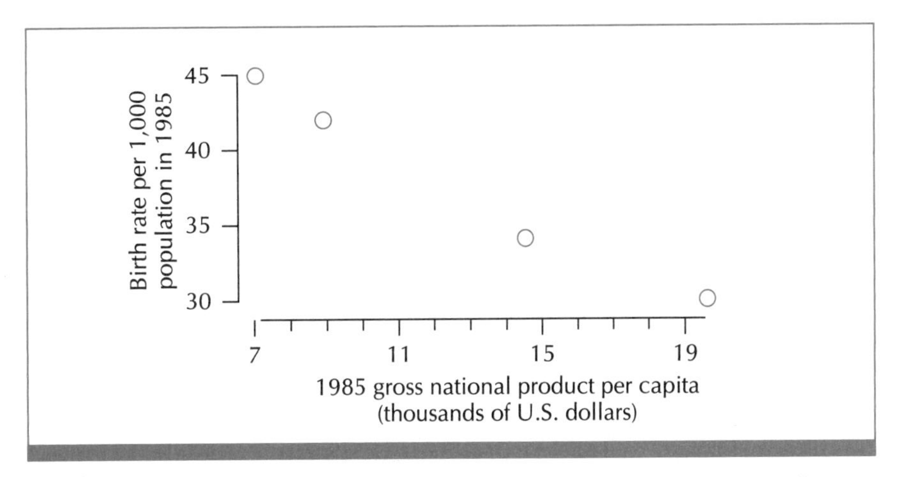

FIGURE 4-13 Scatterplot of birth rate in 1985 and 1985 gross national product per capita for four high-income oil exporting nations. The range of values for each variable is indicated by the line on the corresponding axis.

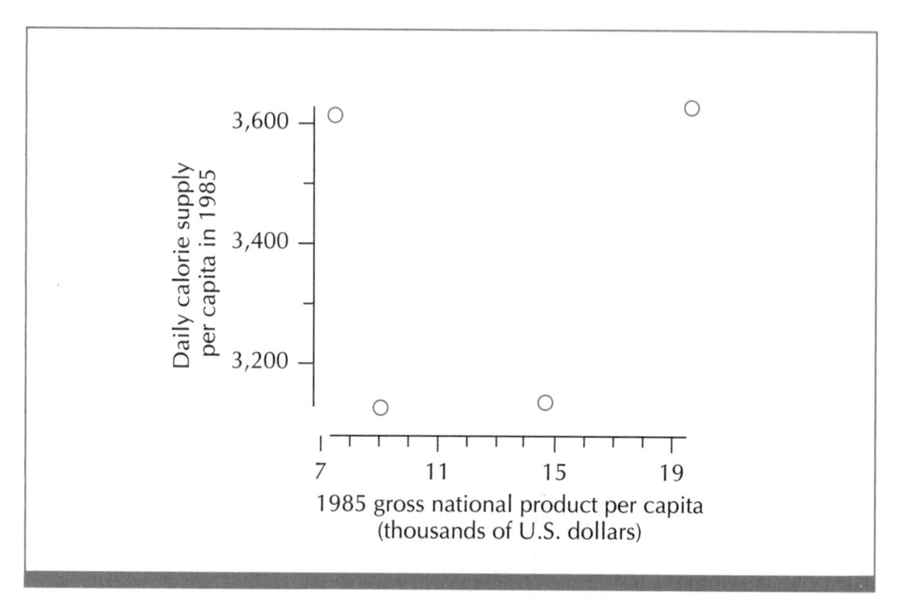

FIGURE 4-14 Scatterplot of daily calorie supply per capita and gross national product per capita in 1985 for four high-income oil exporting nations. The range of values for each variable is indicated by the line on the corresponding axis.

Summary of Chapter 4

Two-way frequency tables (or contingency tables) are tools for examining the relationship between two qualitative variables.

There are a number of ways we can study a quantitative variable at different levels of a qualitative variable. We can tabulate descriptive statistics for the quantitative variable across categories of the qualitative variable. Box plots, histograms, and dot plots allow us to compare the distribution of a quantitative variable across levels of a qualitative variable.

A dot chart displays totals (or percentages) of a quantitative variable for each level of a qualitative variable.

The scatterplot is a valuable graphical tool for looking at the relationship between two quantitative variables. With a range frame, we make efficient use of all the space in a scatterplot. Taking a transformation of a variable can lead to a scatterplot that is easier to look at, or shows a simpler relationship between two variables. Sometimes it is useful to transform both variables before plotting.

Minitab Appendix for Chapter 4

Creating a Two-Way Frequency Table

Recall that in the Minitab Appendix for Chapter 1, we worked with the data for Exercise 5-35. We created a coded variable for child mortality in column 6

(MORTCODE = 1 for countries with low child mortality, MORTCODE = 2 for countries with high child mortality). We created a coded variable for calorie supply in column 7 (CALCODE = 1, 2, 3, or 4 for values of CALORIE in the ranges 51–75, 76–100, 101–125, or 126–150, respectively). To create a two-dimensional frequency table with categories of MORTCODE for rows and categories of CALCODE for columns, we use the TABLE command:

```
MTB>  table 'mortcode' 'calcode'
```

to produce the display in Figure M4-1. We could have specified C6 instead of 'MORTCODE' and C7 instead of 'CALCODE'.

To get row percentages, we use the subcommand ROWPERCENTS. The subcommand COLPERCENTS provides column percentages for each cell in the table. The command

```
MTB>   table 'mortcode' 'calcode';
SUBC>  rowpercents;
SUBC>  colpercents.
```

results in the display in Figure M4-2. The row percentage is the cell frequency as a percentage of the row total. The column percentage is the cell frequency as a percentage of the column total.

```
ROWS: mortcode       COLUMNS: calcode

             1         2         3         4       ALL

 1           0         0         7        13        20
 2           3        12         2         0        17
ALL          3        12         9        13        37

  CELL CONTENTS --
                  COUNT
```

FIGURE M4-1 Two-dimensional frequency table of MORTCODE and CALCODE in Exercise 5-35.

```
ROWS: mortcode       COLUMNS: calcode

             1         2         3         4       ALL

 1          --        --     35.00     65.00    100.00
            --        --     77.78    100.00     54.05

 2       17.65     70.59     11.76        --    100.00
        100.00    100.00     22.22        --     45.95

ALL       8.11     32.43     24.32     35.14    100.00
        100.00    100.00    100.00    100.00    100.00

  CELL CONTENTS --
                  % OF ROW
                  % OF COL
```

FIGURE M4-2 Row and column percentages for the two-dimensional frequency table of MORTCODE and CALCODE.

Calculating Descriptive Statistics Within Categories

We know from the Minitab Appendix for Chapter 3 that we can use the DESCRIBE command with the BY subcommand to get descriptive statistics of one or more quantitative variables within levels of a categorical (classification) variable. The command

```
MTB>   describe 'mort' 'life' 'weight';
SUBC>  by 'calcode'.
```

gives the display in Figure M4-3. We get descriptive statistics for child mortality, life expectancy, and percent low birth weights for each level of CALCODE (codes 1, 2, 3, 4, and missing).

Producing Dot Plots Within Categories

We can use the DOTPLOT command with the BY subcommand to display dot plots of a quantitative variable within levels of a qualitative variable. The command

```
MTB>   dotplot 'weight';
SUBC>  by 'calcode'.
```

produces the output in Figure M4-4.

	calcode	N	N*	MEAN	MEDIAN	TRMEAN
mort	1	3	0	262.0	252.0	262.0
	2	12	0	246.08	243.50	244.10
	3	9	0	57.2	11.0	57.2
	4	13	0	11.538	12.000	11.636
	*	3	0	259.0	232.0	259.0
life	1	3	0	45.00	46.00	45.00
	2	12	0	43.67	44.00	43.90
	3	9	0	69.00	75.00	69.00
	4	13	0	74.692	75.000	74.727
	*	3	0	42.67	44.00	42.67
weight	1	3	0	13.33	13.00	13.33
	2	11	1	15.82	17.00	15.78
	3	8	1	6.125	5.500	6.125
	4	12	1	5.583	5.500	5.600
	*	2	1	17.50	17.50	17.50

	calcode	STDEV	SEMEAN	MIN	MAX	Q1	Q3
mort	1	36.1	20.8	232.0	302.0	232.0	302.0
	2	26.17	7.55	210.00	302.00	225.25	258.50
	3	94.1	31.4	8.0	231.0	8.5	113.5
	4	1.391	0.386	9.000	13.000	10.000	13.000
	*	61.1	35.3	216.0	329.0	216.0	329.0
life	1	1.73	1.00	43.00	46.00	43.00	46.00
	2	3.92	1.13	35.00	50.00	41.00	46.00
	3	12.62	4.21	44.00	77.00	61.50	76.00
	4	1.032	0.286	73.000	76.000	74.000	75.500
	*	4.16	2.40	38.00	46.00	38.00	46.00
weight	1	2.52	1.45	11.00	16.00	11.00	16.00
	2	4.87	1.47	9.00	23.00	10.00	20.00
	3	2.295	0.811	4.000	10.000	4.000	8.000
	4	1.084	0.313	4.000	7.000	5.000	6.750
	*	3.54	2.50	15.00	20.00	*	*

FIGURE M4-3 Descriptive statistics for MORT, LIFE, and WEIGHT within levels of CALCODE.

```
calcode
1                                   .     .        .
        +---------+---------+---------+---------+---------+-------weight
calcode
2
                         .  :         .  .        .  .  .  .  .     .
        +---------+---------+---------+---------+---------+-------weight
calcode  .
3        :  .  .     :     .
        +---------+---------+---------+---------+---------+-------weight
calcode     :  .  .
 4       :  :  :  :
        +---------+---------+---------+---------+---------+-------weight
       3.5       7.0      10.5      14.0      17.5      21.0
```

FIGURE M4-4 Dot plots of the WEIGHT variable for the four nonmissing values of CALCODE.

```
mortcode

              -------
1           --I+     I----
              -------

                                    -----------------------
2                              -----I           +          I------------
                                    -----------------------
         +---------+---------+---------+---------+---------+------weight
       3.5       7.0      10.5      14.0      17.5      21.0
```

FIGURE M4-5 Box plots of the WEIGHT variable for the two levels of MORTCODE.

Producing Box Plots Within Categories

We can use the BOXPLOT command with the BY subcommand to display a box plot of a quantitative variable for each level of a classification variable. The command

```
MTB>   boxplot 'weight';
SUBC>  by 'mortcode'.
```

produces two box plots of the WEIGHT variable—one for countries with low child mortality and one for countries with high child mortality, as shown in Figure M4-5.

Creating Scatterplots

The PLOT command provides scatterplots. If we use the command

```
MTB>  plot 'weight' 'calorie'
```

we get the scatterplot in Figure M4-6, with WEIGHT on the vertical axis and CALORIE on the horizontal axis.

To get scatterplots with range frames, we use the YSTART and XSTART subcommands:

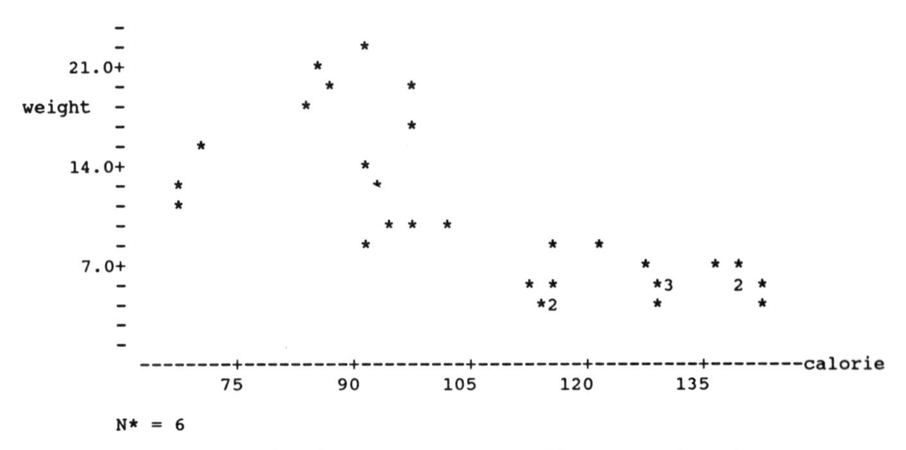

FIGURE M4-6 Scatterplot of WEIGHT versus CALORIE in Exercise 5-35.

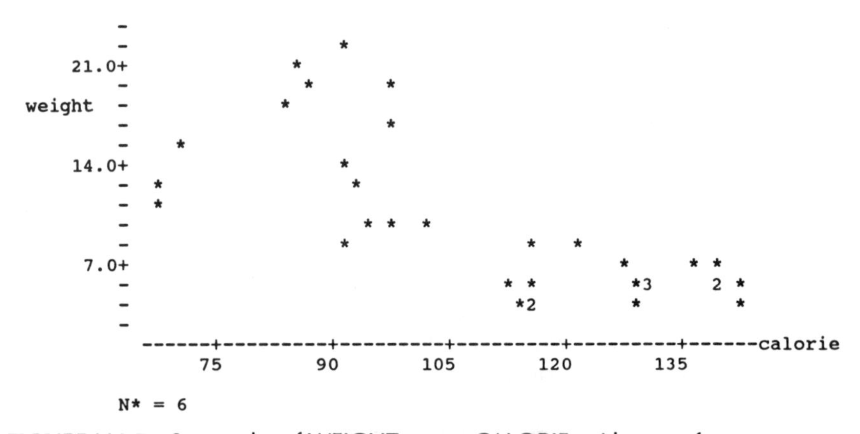

FIGURE M4-7 Scatterplot of WEIGHT versus CALORIE, with range frame.

```
MTB>   plot 'weight' 'calorie';
SUBC>  ystart 4 23;
SUBC>  xstart 68 143.
```

The YSTART subcommand tells Minitab we want the vertical axis to go from 4 (the minimum value of WEIGHT) to 23 (the maximum value of WEIGHT). The XSTART subcommand tells Minitab we want the horizontal axis to go from 68 (the minimum value of CALORIE) to 143 (the maximum value of CALORIE). We get the plot in Figure M4-7. The scatterplots in Figures M4-6 and M4-7 are very similar in appearance since Minitab automatically creates scatterplots with axes that are nearly range frames.

Minitab uses the asterisk as the plotting symbol in scatterplots. If two or more cases have the same plotted point, the number of coinciding points is plotted instead of an asterisk.

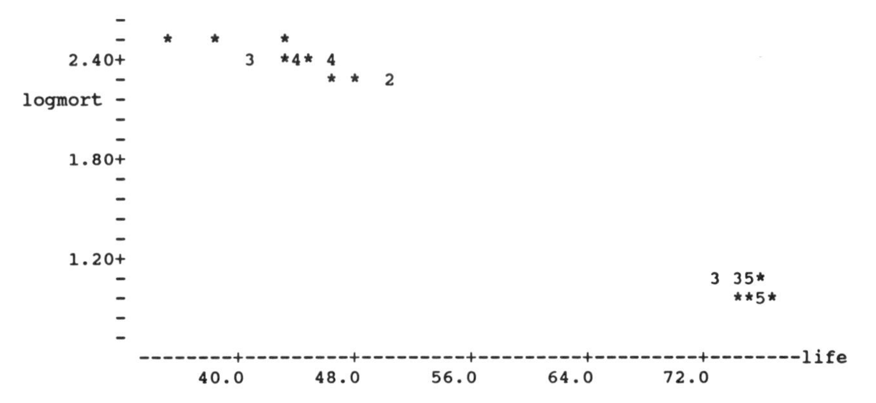

FIGURE M4-8 Scatterplot of LOGMORT versus LIFE.

Transforming Data with the LET Command

We can use the LET command to transform data before plotting. The sequence of commands

```
MTB> let c29=logten(c2)
MTB> name c29 'logmort'
MTB> plot 'logmort' 'life'
```

will produce a scatterplot with the logarithm base-10 of child mortality on the vertical axis and life expectancy on the horizontal axis, as shown in Figure M4-8.

Minitab does not have a command for producing a dot chart.

Exercises for Chapter 4

In the exercises, answer the following questions: What would you need to know about the sample to be willing to use it to make inferences about a larger population? What is that larger population (if any)? What limitations do you see in the sample?

For each figure and table, include a legend that completely describes its contents. Note the number of cases included and the number excluded if there are missing values.

Use range frames when drawing scatterplots. When we say plot variable 1 versus variable 2, we mean plot variable 1 on the vertical axis and variable 2 on the horizontal axis.

EXERCISE 4-1

Consider the values of 1985 gross national product per capita and 1985 life expectancy at birth for four high-income oil exporters, shown in Table 4-5.

A scatterplot of life expectancy versus gross national product is displayed in Figure 4-7 for these four countries. In that scatterplot, we used a range frame. Each axis extends from the minimum to the maximum value for the corresponding variable. The axes do not meet. Compare this scatterplot with two other ways to draw a plot:

a. Plot life expectancy versus per capita gross national product for the four high-income oil exporters. In this plot, let each axis extend from 0 to the maximum value for the variable. Let the axes meet at 0.

b. Plot life expectancy versus per capita gross national product for the four high-income oil exporters. Let the horizontal axis go from \$7,000 to \$20,000. Let the vertical axis go from 55 to 75 years. Let the axes meet (at 55 for life expectancy and \$7,000 for per capita gross national product).

c. Discuss the advantages of the scatterplot with range frame in Figure 4-7 over those you drew in parts (a) and (b).

EXERCISE 4-2

The Boston Red Sox spend an estimated \$36,754 per minor-league player per year ("The Cost to Develop a Major Leaguer," *The Boston Globe,* June 7, 1989, page 47):

Category of expense	Cost (\$)
Player contracts (including meals, equipment, expenses)	14,215
Signing bonuses	6,085
Scouts (salaries)	6,000
Scouts (expenses)	3,723
Managers, coaches, instructors (salaries)	2,885
Minor-league spring training	2,192
Instructors (travel)	554
Player insurance	508
Statistics	269
National association fees	208
Medical	115

Construct a dot chart to show the breakdown of yearly expenses per minor-league player. Show percentage of total as well as dollar costs.

EXERCISE 4-3

Researchers investigated the cancer risk associated with saccharin use in this study. For 2 years, the experimenters fed male rats diets with percent sodium saccharin ranging from 0 to 7.5. A response was development of a bladder tumor in a rat living past the time the first tumor appeared. Dose (percent sodium saccharin in the diet) and response information are shown below (Van Ryzin and Rai, 1987; from Carlborg, 1985).

Dose	Number of responses/ Number of animals tested	Proportion of responses
0	0/324	.000
1	5/658	.008
3	8/472	.017
4	12/189	.063
5	15/120	.125
6.25	20/120	.167
7.5	37/118	.314

Plot the proportion of responses versus dose. How would you describe the dose–response curve based on these plotted points?

EXERCISE 4-4

The accompanying frequency table summarizes the results of a study of 1,224 diabetic people, with each person classified by duration of diabetes and presence of the eye disease retinopathy (Knuiman and Speed, 1988).

Duration of patient's diabetes (years)	Patient has retinopathy			Proportion yes
	Yes	No	Total	
0–2	46	290	336	.14
3–5	52	211	263	.20
6–8	44	134	178	.25
9–11	54	91	145	.37
12–14	38	53	91	.42
15–17	39	42	81	.48
18–20	23	23	46	.50
21+	52	32	84	.62

a. Plot proportion yes versus duration of diabetes. (You will have to pick a single value to represent a duration interval. You might reasonably use 1 for the interval 0–2, for instance.)

b. Describe the relationship between duration of diabetes and the proportion of diabetic patients with retinopathy.

EXERCISE 4-5

Times to failure (time units not given in the source) are listed below for tires manufactured with three different methods of production (Hsieh, 1986; from Bain, 1978).

Method 1: 10.03 10.47 10.58 11.48 11.60 12.41 13.03
13.51 14.48 16.96 17.08 17.27 17.90 18.21
19.30 20.10 21.51 21.78 21.79 25.34

Method 2:	10.10	11.01	11.20	12.95	13.19	14.81	16.03
	17.01	18.96	24.10	24.15	24.52	26.05	26.44
	28.59	30.24	31.03	33.51	33.61	40.68	
Method 3:	19.07	19.51	19.62	20.47	20.78	21.37	22.08
	22.61	23.47	26.02	26.23	26.47	27.07	27.43
	28.28	29.10	29.66	30.67	30.81	34.36	

a. Construct a dot plot of failure times for each production method. Use the same scale for each plot.

b. Compare the three distributions of failure times. In particular, compare location and variation for the three sets of failure times.

EXERCISE 4-6

Field-goal ratio and point-after-touchdown ratio for the 1983 regular season are listed here for 29 National Football League kickers (Berry and Berry, 1985).

Kicker (Team)	Point-after-touchdown Conversions/Attempts (%)	Field goals Goals/Attempts (%)
Allegre (Baltimore)	22/24 (92)	30/35 (86)
Anderson (Pittsburgh)	38/39 (97)	27/31 (87)
Karlis (Denver)	33/34 (97)	21/25 (84)
Lowery (Kansas City)	44/45 (98)	24/30 (80)
Wersching (San Francisco)	51/51 (100)	25/30 (83)
Haji-Sheikh (New York Giants)	22/23 (96)	35/42 (83)
M. Bahr (Cleveland)	38/40 (95)	21/24 (88)
Septien (Dallas)	57/59 (97)	22/27 (81)
Kempf (Houston)	33/34 (97)	17/21 (81)
Murray (Detroit)	38/38 (100)	25/32 (78)
Stenerud (Green Bay)	52/52 (100)	21/26 (81)
Luckhurst (Atlanta)	43/45 (96)	17/22 (77)
C. Bahr (Los Angeles Raiders)	51/53 (96)	21/27 (78)
M. Andersen (New Orleans)	37/38 (97)	18/23 (78)
Johnson (Seattle)	49/50 (98)	18/25 (72)
Ricardo (Minnesota)	33/34 (97)	25/33 (76)
Moseley (Washington)	62/63 (98)	33/47 (70)
von Schamann (Miami)	45/48 (94)	18/27 (67)
Leahy (New York Jets)	36/37 (97)	16/24 (67)
O'Donoghue (St. Louis)	45/47 (96)	15/28 (54)
Benirschke (San Diego)	43/45 (96)	15/24 (63)
Breech (Cincinnati)	39/41 (95)	16/23 (70)
Franklin (Philadelphia)	24/27 (89)	15/26 (58)
Danelo (Buffalo)	33/34 (97)	10/20 (50)
Thomas (Chicago)	35/38 (92)	14/25 (56)
Smith (New England)	12/15 (80)	3/6 (50)
Capece (Tampa Bay)	23/26 (88)	10/20 (50)
Steinfort (New England and Buffalo)	17/18 (94)	7/21 (33)
Nelson (Los Angeles Rams)	33/37 (89)	5/11 (45)

a. Construct two dot plots: one for percent successful point-after-touchdown attempts and one for percent successful field-goal attempts. Describe and compare the two distributions in terms of location, variation, peaks, and symmetry or skewness.

b. Plot percent successful field-goal attempts versus percent successful point-after-touchdown attempts. Describe the relationship between the two variables shown in the scatterplot.

EXERCISE 4-7

For the 1973 entering class of 15 American law schools, the class average scores on an examination known as the LSAT and the class average of undergraduate grades are shown below (Efron and Tibshirani, 1986).

LSAT	Grades	LSAT	Grades	LSAT	Grades
576	3.39	578	3.03	555	3.00
605	3.13	545	2.76	635	3.30
666	3.44	661	3.43	653	3.12
572	2.88	558	2.81	580	3.07
651	3.36	575	2.74	594	2.96

a. Construct a dot plot of class average LSAT scores and a dot plot of class average undergraduate grades. Describe the distribution of each set of values.

b. Construct a scatterplot of class average LSAT versus class average undergraduate grades. Discuss the information provided in the plot.

EXERCISE 4-8

Wire is wound around plastic spools used in electric motors. When current passes through the wire, the temperature of the spool rises. The accompanying table shows two measurements of temperature rise (°C) made on each of 12 such plastic spools (Nelson, 1986, page 21).

Spool:	1	2	3	4	5	6	7	8	9	10	11	12
First measurement:	45.0	45.1	45.4	45.9	45.9	46.0	46.2	46.5	46.5	46.8	47.0	50.6
Second measurement:	44.9	44.7	45.8	45.3	45.8	45.2	45.2	45.5	46.0	46.1	45.5	50.0

a. Construct a dot plot of the first set of temperature rise measurements. Using the same scale, construct a dot plot of the second set of temperature rise measurements. Describe the two distributions and compare them.

b. Construct a scatterplot of the second measurement versus the first measurement. Sketch the relationship you would expect if the two measurements were perfectly consistent. Do the two measurements appear to be consistent? Discuss the information about the two sets of measurements revealed in the scatterplot.

EXERCISE 4-9

A researcher rated the readability of unpublished reports and published articles written by engineers. A report with a low score is more easily understood. Results are shown below (Milton and Arnold, 1986, page 320; from "Engineers' English" by W. H. Emerson, *CME,* June 1983, pages 54–56).

Unpublished reports: 2.39 2.56 2.36 2.62 2.51 2.29 2.58 2.41 2.86 2.49 2.33 1.94 2.14

Published articles: 1.79 1.87 1.62 1.96 1.75 1.74 2.06 1.69 1.67 1.94 1.33 1.70 1.65

a. Construct a stem-and-leaf plot of scores for unpublished reports and another for published articles. Use the first digits as the stem and the last digit as the leaf.

b. Describe the distribution of each set of values. How do the two distributions compare?

EXERCISE 4-10

Researchers measured airborne bacteria (number of colonies per cubic foot) in eight carpeted hospital rooms and eight uncarpeted hospital rooms (Devore, 1982, page 315; from "Microbial Air Sampling in a Carpeted Hospital," *J. Environmental Health,* 1968, page 405).

Carpeted:	7.1	8.2	10.1	10.8	11.8	13.0	14.0	14.6
Uncarpeted:	3.8	7.2	8.3	10.1	11.1	12.0	12.1	13.7

a. Construct dot plots of the two sets of values, using the same scales for each.

b. Describe the distribution of bacterial counts for the carpeted rooms and for the uncarpeted rooms. Compare the two distributions.

EXERCISE 4-11

Researchers measured carbon monoxide concentration (parts per million) and benzo(a)pyrene concentration (μg per 1,000 cubic meters) in 16 different air samples from Herald Square in New York City (Devore, 1982, page 457; from "Carcinogenic Air Pollutants in Relation to Automobile Traffic in New York City," *Environmental Science and Technology,* 1971, pages 145–150).

Carbon monoxide	Benzo(a)-pyrene	Carbon monoxide	Benzo(a)-pyrene
2.8	.5	5.5	1.3
15.5	.1	12.0	5.7
19.0	.8	5.6	1.5
6.8	.9	19.5	6.0
5.5	1.0	11.0	7.3
5.6	1.1	12.8	8.1
9.6	3.9	5.5	2.2
13.3	4.0	10.5	9.5

a. Construct a dot plot for each of these two variables. Describe each distribution in terms of location, variation, concentrations of values, and symmetry or skewness.

b. Construct a scatterplot of carbon monoxide concentration versus benzo(a)pyrene concentration. Discuss the relationship between the two variables shown in the plot.

EXERCISE 4-12

Sodium content and potassium content (no units given) in perspiration of 10 healthy women are shown here (Oja and Nyblom, 1989; from Johnson and Wichern, 1982, page 182).

Woman:	1	2	3	4	5	6	7	8	9	10
Sodium:	48.5	65.1	47.2	53.2	55.5	36.1	24.8	33.1	47.4	54.1
Potassium:	9.3	8.0	10.9	12.2	9.7	7.9	14.0	7.6	8.5	11.3

a. Construct a dot plot for each of these two sets of values. Describe each distribution.

b. Draw a scatterplot of sodium content versus potassium content. What is the relationship between the two variables revealed in this plot?

EXERCISE 4-13

Maximal oxygen uptake ($ml \cdot kg^{-1} \cdot min^{-1}$) is a measure of lung function and capacity for work. Values of maximal oxygen uptake are listed below for male world-class athletes in their 20's. Several athletes were tested in each of five different sports (Wilmore, 1984).

Wrestling:	58.3	50.4	60.9	64.0	54.3	
Weightlifting:	40.1	42.6	49.5	50.7	46.3	41.5
Shot/discus:	49.5	42.8	42.6	47.5		
Ice hockey:	61.5	54.6	53.6			
Cross-country skiing:	63.9	73.9	78.3	73.0		

a. Using the same scales, construct a dot plot of maximal oxygen uptake for athletes in each of the five sports. Line up the plots under one another for easy comparisons. Discuss and compare these distributions.

b. Find the mean and the median of each of the five sets of values. Find the range, interquartile range, and standard deviation for each set of values. Construct a table showing measures of central tendency and variation, as well as sample size, for each sport. Discuss these measures within each sport and compare sports.

EXERCISE 4-14

Maximal oxygen uptake is a measure of lung function and capacity for work. Values of maximal oxygen uptake (in $ml \cdot kg^{-1} \cdot min^{-1}$) are listed below for female world-class athletes in their teens and early 20's. Several athletes were tested in each of four sports (Wilmore, 1984).

Basketball:	42.3	42.9	49.6
Swimming:	46.2	43.4	40.5

Distance running:	63.2	50.8	57.5	
Volleyball:	43.5	56.0	41.7	50.6

a. Using the same scales, construct a dot plot of maximal oxygen uptake for athletes in each of the four sports. Line up the plots under one another for easy comparisons. Discuss and compare these distributions.

b. Find the mean, median, range, interquartile range, and standard deviation for each of the four sets of values. Construct a table showing measures of location and variation, as well as sample size, for each sport. Discuss these descriptive statistics within each sport and compare sports.

EXERCISE 4-15

Is cigarette smoking associated with delayed conception? To answer this question, investigators studied 586 women who were pregnant with planned pregnancies and had gotten pregnant within 24 cycles of trying. Since oral contraceptives are associated with delayed conception, women whose most recent method of birth control had been the pill were not included. A woman was classified as a smoker if she smoked on average at least one cigarette a day during at least the first cycle she was trying to get pregnant. The accompanying frequency table classifies women by their smoking status and number of cycles to pregnancy (Weinberg and Gladen, 1986; Baird and Wilcox, 1985):

	Nonsmokers		Smokers	
Cycle	Number	(Percent)	Number	(Percent)
1	198	(40.7)	29	(29)
2	107	(22.0)	16	(16)
3	55	(11.3)	17	(17)
4	38	(7.8)	4	(4)
5	18	(3.7)	3	(3)
6	22	(4.5)	9	(9)
7	7	(1.4)	4	(4)
8	9	(1.9)	5	(5)
9	5	(1.0)	1	(1)
10	3	(.6)	1	(1)
11	6	(1.2)	1	(1)
12	6	(1.2)	3	(3)
More than 12	12	(2.5)	7	(7)
Total	486	(100)	100	(100)

a. Construct a frequency plot showing the number of nonsmokers in each cycle category. Using the same scale for frequencies, construct a frequency plot showing the number of smokers in each cycle category.

b. Construct a frequency plot showing the percentage of nonsmokers in each cycle category. Using the same scale for percentages, construct a frequency plot showing the percentage of smokers in each cycle category.

c. What information is provided in the two plots you constructed in part (a)? Compare this with the information provided in the two plots from part (b). What do these plots suggest about the relationship between smoking and delayed conception?

d. Suppose you are concerned that some women reporting pregnancy in the first cycle might have become pregnant by accident. Exclude the first cycle category from your plots. Do your interpretations change?

EXERCISE 4-16

Researchers are investigating an antibody known as 64K autoantibody as a possible early warning of Type I diabetes (*Science News,* volume 133, June 18, 1988, page 389). In one study, researchers found 64K autoantibodies in 18 of 20 patients newly diagnosed with Type I diabetes. They found the antibodies in none of 18 controls (people without diabetes). Display these results in a two-way frequency table. What do these results suggest?

EXERCISE 4-17

Will substances that cause cancer in people cause cancer in mice? Will substances that cause cancer in mice cause cancer in people? These questions are important because scientists routinely use animal studies to evaluate possible carcinogenicity of chemicals in humans (*Statistical Science,* volume 3, 1988, pages 3–56). An evaluation of 266 chemicals tested in rats and in mice by the National Cancer Institute and the National Toxicology Program yielded the following results (*Statistical Science,* volume 3, 1988, page 34; from "Species Correlation in Long-Term Carcinogenicity Studies," by J. K. Haseman and J. E. Huff, *Cancer Lett.,* volume 37, 1987, pages 125–132):

	Rats	
Mice	Carcino-genic	Not car-cinogenic
Carcinogenic	67	36
Not carcinogenic	32	131

a. For what percentage of the 266 chemicals do the rats and mice agree with respect to carcinogenicity (or lack of carcinogenicity)?

b. Of the chemicals carcinogenic for at least one species, what percentage was carcinogenic in both?

c. What do these results suggest about agreement between humans and mice (or rats) with respect to potential cancer-causing agents?

EXERCISE 4-18

In the United States in 1984–1985, bachelor's degrees were received by 12,402 Hispanic men, 13,472 Hispanic women, 23,018 black men, 34,455 black women, 405,085 white men, 421,021 white women, 13,554 Asian/Pacific Islander men, 11,841 Asian/Pacific Islander women, 1,998 American Indian men, 2,248 American Indian women, 20,091 nonresident alien men, and 9,126 nonresident alien women (American Council on Education, 1987, page 19).

a. Arrange this information in a two-way frequency table showing number of degrees by race and sex.

b. Construct two frequency plots (or dot charts) of bachelor's degrees by race—one for men and one for women. Use scales that allow the best comparisons of the two plots.

c. Display the data in any other way you find informative.

d. Discuss your findings.

EXERCISE 4-19

Researchers in Martinique found that 10 of 17 patients with tropical spastic paraparesis (a common paralytic disease in the tropics) had antibodies to HTLV-I (human T-cell lymphotropic virus I, a virus associated with some lymphomas and leukemias). Twelve of 303 controls (people without tropical spastic paraparesis) had antibodies to HTLV-I. (Numbers with antibodies were calculated from percentages reported in *Science,* volume 236, May 29, 1987, page 1059.)

a. Arrange these results in a two-way frequency table showing patient group and presence/absence of HTLV-I antibodies.

b. Discuss the possible implications of these findings.

EXERCISE 4-20

The snowberry fly has markings on its wings that make it resemble one of its predators, the zebra spider. Do these markings discourage zebra spiders from attacking snowberry flies? Investigators carried out an experiment to determine the reaction of zebra spiders to four types of potential prey: another spider, a housefly, a snowberry fly, and a snowberry fly with blackened wings (special markings obscured). The table shows the number of trials in which a predator spider pounced on the potential prey. (The numbers in the last column were calculated from percentages given in Mather and Roitberg, 1987.)

Prey	Number of trials	Number of trials in which predator spider pounced on prey
Spider	40	2
Housefly	40	24
Snowberry fly	76	15
Blackened wing snowberry fly	33	13

a. Arrange these results in a two-way frequency table.

b. Do these findings suggest that the markings of the snowberry fly are protective against attacks by zebra spiders?

EXERCISE 4-21

A study of Rhode Island 12th graders identified 59 females and 74 males who were academically prepared for science studies, with course work including calculus and physics. When interviewed, 11 of the 59 females and 47 of the 74 males expressed an interest in a career in engineering, science, or technology (*Science,* volume 236, May 8, 1987, page 660).

a. Display these results in a two-way frequency table.

b. Discuss the relationship between sex and science career interest in this group of students.

EXERCISE 4-22

A large group of domestic cats (in the United States and Canada) considered to be at high risk of infection with feline immunodeficiency virus were tested for presence of the virus. Of 663 female cats tested, 51 were positive for the virus. Of 855 male cats tested, 168 were positive (Yamamoto et al., 1989).

a. Arrange these results in a two-way frequency table.

b. Does there appear to be an association between sex and presence of the virus in this group of high-risk cats?

EXERCISE 4-23

Is obesity related to abnormal liver function tests? Researchers studied 39 people who were at least 11% above their ideal body weight and had abnormal results of liver function tests (*Science News,* volume 135, May 27, 1989, page 332). None of the volunteers had problems such as alcohol abuse that might contribute to liver problems. The researchers gave the volunteers a diet and exercise program to follow. At the end of the study period (about a year and a half), four volunteers had gained weight; all four still had abnormal liver function tests. Eighteen volunteers had lost less than 10% of their body weight; 11 of the 18 still had abnormal liver function tests. Seventeen volunteers had lost more than 10% of their body weight; 4 of the 17 still had abnormal liver function tests.

a. Arrange these results in a two-way frequency table to show the relationship between extent of weight loss and presence of abnormal liver function tests.

b. What do these results suggest about the relationship between obesity and evidence of abnormal liver function?

EXERCISE 4-24

Injuries treated at the University of Rochester Section of Sports Medicine from May 1975 to July 1983 are classified below by sex of the injured person and site of the injury (DeHaven and Lintner, 1986):

Site	Males	Females
Knee	1,157	401
Ankle	326	89
Shoulder/upper arm	224	15
Hand/finger	151	20
Hip/thigh	126	11
Elbow/forearm	107	28
Foot/toe	72	20

a. Tabulate and/or plot these results in any reasonable way.

b. Discuss the relationship between sex and site of injury in this group of patients.

EXERCISE 4-25

In a study of smokeless tobacco use among high school students in two Arkansas communities, researchers classified 901 students by use and by grade (Marty, McDermott, and Williams, 1986):

	Grade 10	Grade 11	Grade 12
Uses smokeless tobacco	45	68	58
Does not use smokeless tobacco	281	262	187

Of the 171 users of smokeless tobacco, 162 provided information on duration of use:

Duration of use (years)	<1	1–2	2–3	3–4	4–5	>5
Number of users	26	35	26	22	28	25

One hundred seventy provided information on frequency of use:

Frequency of use (days/week)	≤1	2–3	4–5	6–7
Number of users	31	20	20	99

One hundred seventy provided information on extent of daily use:

Number of dips or chews/day	1	2–3	4–5	6–7	8–9	≥10
Number of users	30	51	36	24	8	21

a. Tabulate and/or plot these results in any reasonable way.

b. Is there an association between grade in school and use of smokeless tobacco among this group of students?

c. Construct a frequency table and frequency plot of each of these variables: duration of use, frequency of use, and number of dips or chews per day. Discuss your findings.

EXERCISE 4-26

Use the information in the accompanying table to look for a possible association between the age of the mother and the likelihood of a baby with Down's syndrome. (Data contributed by P. A. P. Moran to the collection of problems in Andrews and Herzberg, 1985, pages 221–222. The data, for births in Australia from 1942 to 1952, originally appeared in "A Survey of Mongoloid Births in Victoria, Australia," by R. D. Collman and A. Stoller, *American Journal of Public Health,* volume 57, 1962, pages 813–829.)

Age of mother (years)	"Center" of age interval (years)	Number of births	Number of mothers of babies with Down's syndrome	Down's ratio
20 or younger	17.5	35,555	15	.00042
Over 20, under 25	22.5	207,931	128	.00062
At least 25, under 30	27.5	253,450	208	.00082
At least 30, under 35	32.5	170,970	194	.00113
At least 35, under 40	37.5	86,046	297	.00345
At least 40, under 45	42.5	24,498	240	.00980
45 or older	47.5	1,707	37	.02168

Let age of mother refer to the "center" of the age interval in column 2 of the table. Let Down's ratio refer to the number of mothers of babies with Down's syndrome divided by the total number of births, shown in the last column. You may wish to multiply each Down's ratio by 10,000 (or 100,000) before answering the following questions.

a. Construct a scatterplot of Down's ratio versus age of mother.

b. Construct a scatterplot of the logarithm of Down's ratio versus age of mother.

c. Construct a scatterplot of Down's ratio versus the logarithm of age of mother.

d. Construct a scatterplot of the logarithm of Down's ratio versus the logarithm of age of mother.

e. Discuss the apparent relationship between maternal age and Down's ratio, as shown in parts (a)–(d).

f. Does taking the logarithm of one or both variables help to illustrate the relationship between the two variables?

g. Discuss limitations of this data set and the scatterplots you constructed.

EXERCISE 4-27

In this experiment investigators studied the response of male beetles to an airborne sex pheromone (Nordheim, Tsiatis, and Shapas, 1983). The experimenters considered four dose rates (units not given), with 30 beetles exposed at each dose rate. The number of beetles responding within 60 seconds is shown below for each group:

Dose rate	Number of beetles responding within 60 seconds
10^{-6}	2
10^{-5}	10
10^{-4}	17
10^{-3}	25

Display these experimental results in a two-way frequency table. Plot the data in an informative way. Discuss your findings.

EXERCISE 4-28

In 1982 in Western Australia, 1,317 males and 854 females died of ischaemic heart disease, 1,119 males and 828 females died of cancer, 371 males and 460 females died of cerebral vascular disease, and 346 males and 147 females died from accidents (Hatton and Clarke-Hundley, 1984, page 44). Display these results in tabular and/or graphical form. Discuss your findings.

CHAPTER 5

Studying More Than Two Variables at a Time

IN THIS CHAPTER

- Multidimensional frequency table (or contingency table)
- Studying two quantitative variables within levels of a qualitative variable
- Scatterplot matrix
- Framed rectangles on a map

Is the relationship between female primary school enrollment and contraception use the same for all economic categories? How does the difference between female and male life expectancy vary with overall life expectancy within economic categories? What is a good way to display scatterplots involving birth rate, calorie supply, contraception use, and female primary school enrollment? How should we display quantitative information on a map? We will approach each of these questions with a tool for studying more than two variables at a time. Then we will discuss some general principles for constructing effective graphical displays. Let's begin by examining the relationship among three qualitative variables.

5-1 Multidimensional Frequency Tables for Several Qualitative Variables

We used a two-way frequency table (Table 4-3) to look at the relationship between female primary school enrollment and contraception use among the 89 countries with information on both variables. We found an association between female primary school enrollment and contraception use among these countries. Most of the countries with low enrollments also have low contraception use; more than half of the countries with higher enrollments also have greater contraception use.

Is the relationship between these two variables similar for each economic category? To answer this question, we can look at a three-way frequency table (or three-dimensional contingency table). A three-way frequency table displays counts of cases within each combination of three qualitative variables.

> A **multiway frequency table** or multidimensional contingency table displays the number of cases within each combination of categories of several qualitative variables.

Table 5-1 shows the number of countries within each combination of levels of economic category, contraception use, and female primary school enrollment, as a two-way display of female primary school enrollment and contraception use for each economic category. The data could be arranged differently, but separate displays by economic category seem reasonable here.

We can see a strong association among the three variables in Table 5-1. Lower enrollments and lower contraception use are associated with lower economic categories. Thirty-nine countries have missing information on either female primary school enrollment or contraception use, or both. It might be informative to display missing categories for these two variables in a breakdown by economic category. We will not do that here. However, we will be cautious in our interpretations because of the large number of missing values.

TABLE 5-1 Countries classified by economic category, percentage of married women of childbearing age using contraception in 1984, and number of females in primary school in 1984 as percentage of 6–11-year age group. Thirteen low-income countries, four lower-middle-income countries, six upper-middle-income countries, all four high-income oil exporters, seven industrial market countries, and five nonmembers are excluded because of missing information on female primary school enrollment or contraception use or both.

Economic category	Percent contraception use in 1984	1984 primary school enrollment for females		
		≤ 70%	> 70%	Total
Low-income	≤ 35%	16	6	22
	> 35%	0	2	2
	Total	16	8	24
Lower-middle-income	≤ 35%	7	11	18
	> 35%	1	13	14
	Total	8	24	32
Upper-middle-income	≤ 35%	0	2	2
	> 35%	0	15	15
	Total	0	17	17
Industrial market	≤ 35%	0	0	0
	> 35%	0	12	12
	Total	0	12	12
Nonmembers	≤ 35%	0	1	1
	> 35%	0	3	3
	Total	0	4	4

5-2

Scatterplots for Studying Two Quantitative Variables Within Levels of a Qualitative Variable

Instead of categorizing values of contraception use and female primary school enrollment as either low or high, we might prefer to study these two indicators as quantitative variables. Let's see how they are related within economic categories.

Figure 5-1 is a scatterplot of contraception use and female primary school enrollment, with economic categories designated by different plotting symbols. A difficulty with this approach is that overlapping plotting symbols result when two or more countries have similar values of the two variables. Contrasting colors to designate levels of the qualitative variable can be very effective (Cleveland, 1985, pages 205–207).

We see an overall increasing relationship between female primary school enrollment and contraception use. The lower left-hand portion of the plot corresponds to lower contraception use and lower female primary school enrollments. These countries are predominantly low-income and lower-middle-income. The upper right-hand portion of the plot corresponds to greater con-

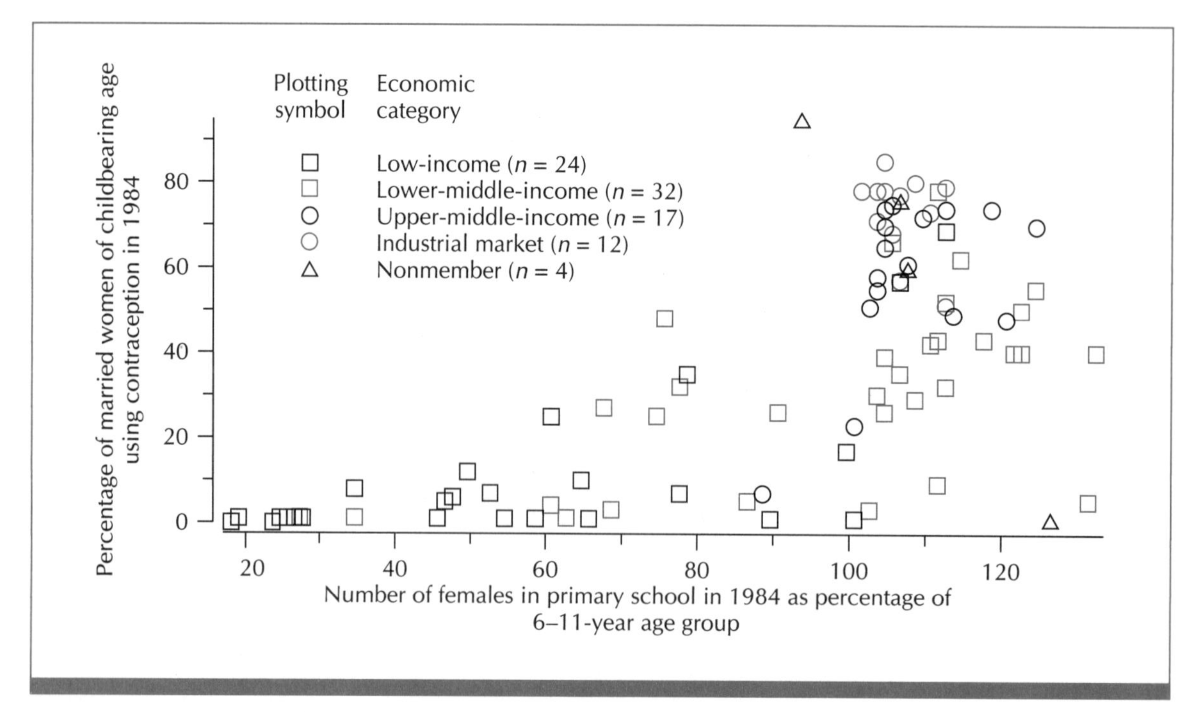

FIGURE 5-1 Scatterplot of percentage of married women of childbearing age using contraception in 1984 and number of females in primary school in 1984 as percentage of 6–11-year age group, with economic categories distinguished by plotting symbols. Thirty-nine countries are excluded because of missing values on at least one of the two indicators: 13 low-income, 4 lower-middle-income, 6 upper-middle-income, all 4 high-income oil exporters, 7 industrial market, and 5 nonmember countries.

traception use and higher female primary school enrollments. These countries are primarily industrial market and upper-middle-income. Figure 5-1 gives us a strong visual impression of an increasing relationship between female primary school enrollment and contraception use. Both these indicators are positively associated with economic category.

A scatterplot such as Figure 5-1 can be confusing if there are many overlapping plotting symbols. An alternative is to construct separate plots of the two quantitative variables for each level of the qualitative variable. This is done in Figure 5-2 for economic category and two variables related to life expectancy.

Figure 5-2 shows scatterplots of the difference between female and male life expectancy (vertical axis) and overall life expectancy (horizontal axis) for each of the six economic categories. The differences between female and male life expectancies over the six graphs range from −2 (life expectancy 2 years longer for males than females) to 9 (life expectancy 9 years longer for females than males). Overall life expectancies range from 40 to 78 years. All scatterplots have the same scales, making visual comparisons easier. Both plotted variables increase with economic category. Also, the relationship between the two plotted variables depends on economic category.

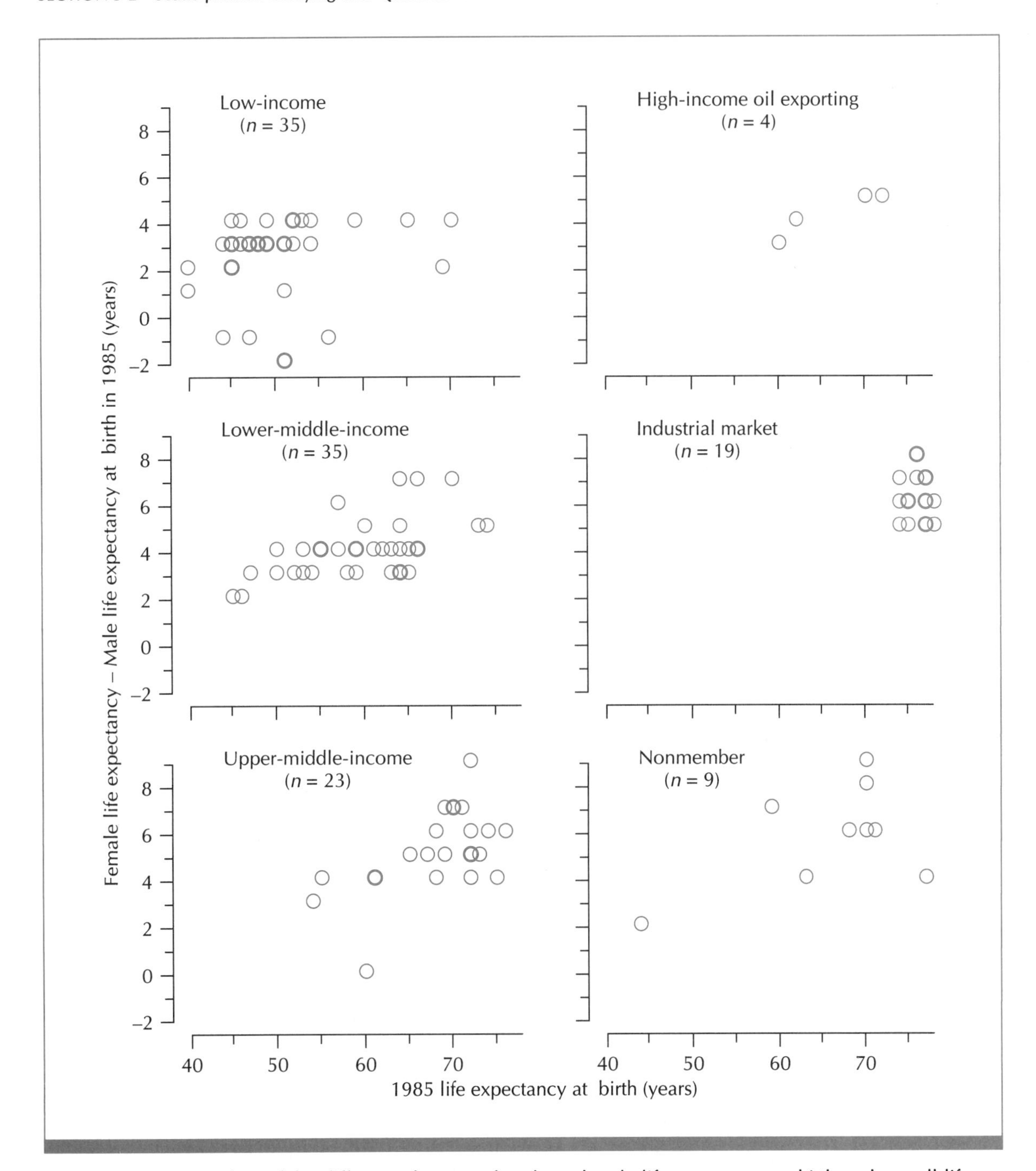

FIGURE 5-2 Scatterplots of the difference between female and male life expectancy at birth and overall life expectancy at birth in 1985 for each economic category. Two low-income countries and one lower-middle-income country are excluded because of missing information on life expectancy.

For low-income countries, life expectancies tend to be short and the differences between female and male life expectancies are relatively small. Among the lower- and upper-middle-income nations, there is more variation in life expectancies. There is also a striking positive relationship between overall life expectancy and the difference between female and male life expectancies. Countries with greater economic development have longer life expectancies and greater differences between female and male life expectancies. However, an increasing relationship between the two plotted variables is not apparent among nations with life expectancies greater than 70 years. (We might hope that the difference between female and male life expectancies will decline with continued economic development!)

In Section 5-3, we group several scatterplots into a scatterplot matrix.

5-3 The Scatterplot Matrix for Several Quantitative Variables

We saw a positive association between female primary school enrollment and contraception use in Figure 5-1. Suppose we would like to consider, in addition, birth rate and calorie supply. We can construct a scatterplot for each pair of variables. If we arrange these plots as in Figure 5-3, we have what is known as a *scatterplot matrix* (Cleveland, 1985; Chambers et al., 1983).

> A **scatterplot matrix** displays scatterplots for pairs of quantitative variables. In the upper right-hand portion are scatterplots for each pair of variables. For each scatterplot in the upper right, there is a corresponding scatterplot in the lower left-hand portion, in which the same two variables are plotted, on opposite axes. The variable names are shown in the body of the scatterplot matrix.

There are six possible pairs of the four indicators. The upper right-hand portion of Figure 5-3 shows six scatterplots resulting from these six pairings. For each plot in the upper right of Figure 5-3, there is a corresponding graph in the lower left-hand portion, with the same two variables plotted on opposite axes. We show variable names in the body of the scatterplot matrix. Scales for the axes lie outside the matrix. Each axis approximately spans the range of values for the corresponding variable.

The 89 World Bank countries with nonmissing values for each of the four indicators are represented in Figure 5-3. Among these 89 nations, female primary school enrollment ranges from a little more than 10% to almost 130%. Contraception use ranges from 0 to over 80%. Birth rate extends from just over 5 to just under 55 births per 1,000 population. Calorie supply per capita ranges from about 1,500 to 3,800 calories per day.

Perhaps the most striking relationship is the strong negative association between contraception use and birth rate. Both contraception use and birth rate are strongly associated with daily calorie supply per capita. Countries with lower calorie supplies tend to have lower levels of contraception use and higher birth rates.

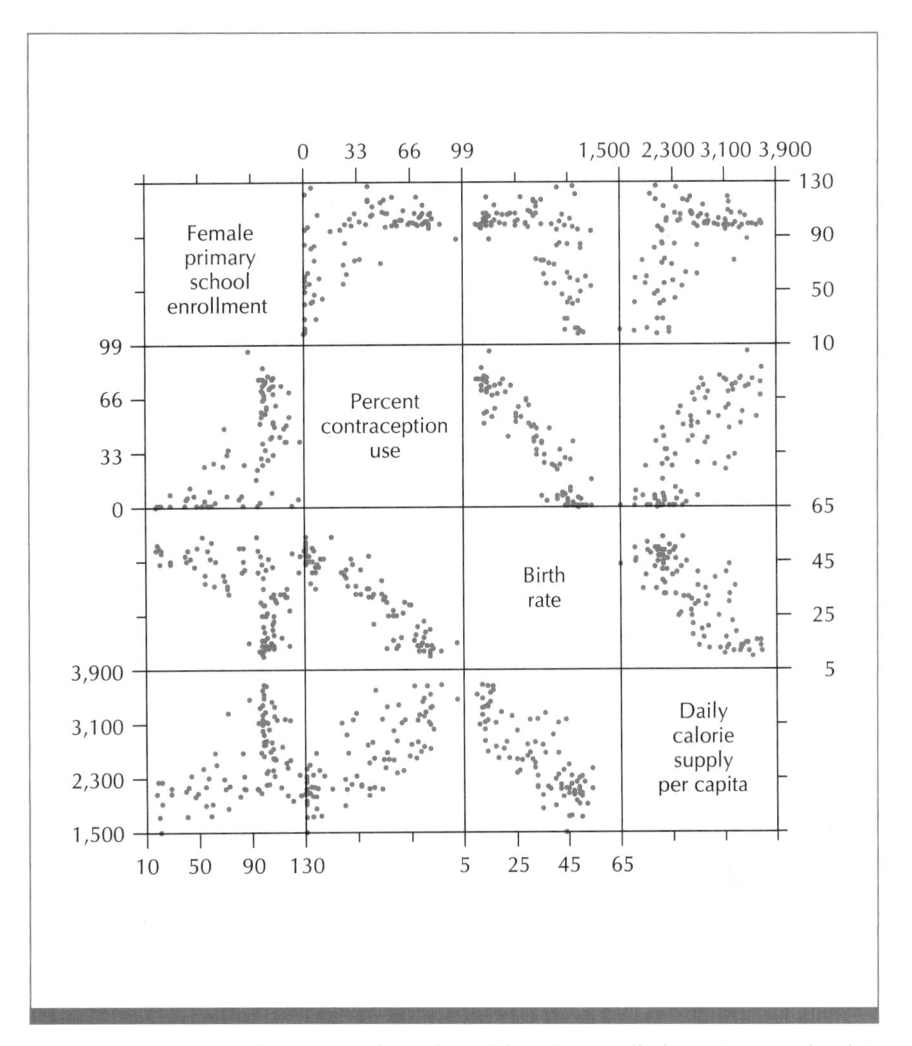

FIGURE 5-3 Scatterplot matrix of number of females enrolled in primary school in 1984 as percentage of 6–11-year age group, percentage of married women of child-bearing age using contraception in 1984, birth rate per 1,000 population in 1985, and daily calorie supply per capita in 1985. Plots are based on the 89 World Bank countries with nonmissing information on all four indicators. Thirty-nine countries are excluded because of missing values on at least one of the four indicators: 13 low-income, 4 lower-middle-income, 6 upper-middle-income, all 4 high-income oil exporters, 7 industrial market, and 5 nonmember countries.

Female primary school enrollment has an increasing relationship with contraception use and calorie supply, and a decreasing relationship with birth rate. Countries with lower female primary school enrollments tend to have lower contraception use, higher birth rates, and lower calorie supplies per capita. Among countries with female primary school enrollments above 90%, there is a great deal of variation in values of the other three variables.

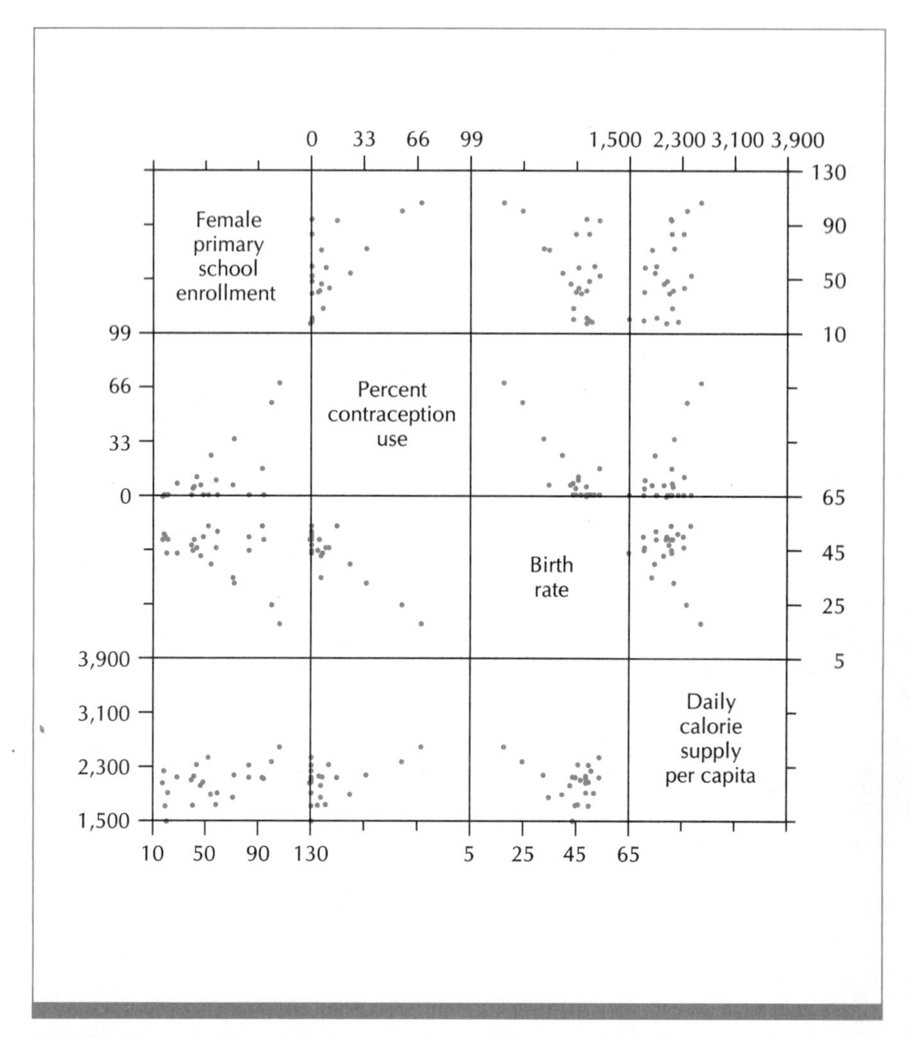

FIGURE 5-4a Scatterplot matrix based on 24 low-income countries with nonmissing information on all four indicators. Thirteen countries are excluded because of missing values.

Now let's look at a separate scatterplot matrix for each of four economic categories: low-income, lower-middle income, upper-middle income, and industrial market. We use the same scales as in Figure 5-3, for easier comparisons.

In Figure 5-4a, two countries have relatively high levels of contraception use and low birth rates. These two countries are China and Sri Lanka, as shown in Table 5-2. Excluding these two nations, we see among the remaining low-income countries low levels of contraception use and calorie supply, high birth rates, and generally low female primary school enrollments.

Figure 5-4d reveals very little variation in values of the four indicators

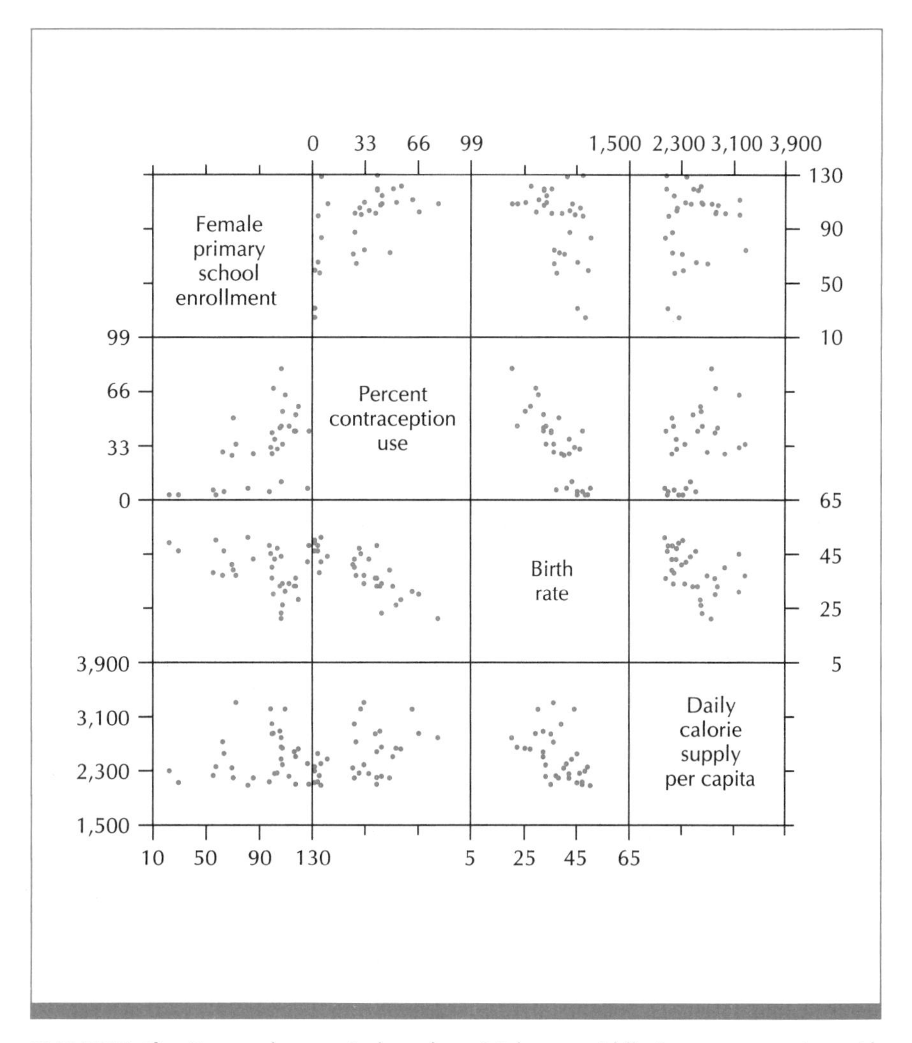

FIGURE 5-4b Scatterplot matrix based on 32 lower-middle-income countries with nonmissing information on all four indicators. Four countries are excluded because of missing values.

TABLE 5-2 Number of females enrolled in primary school in 1984 as percentage of 6–11-year age group, percentage of married women of child-bearing age using contraception in 1984, birth rate per 1,000 population in 1985, and daily calorie supply per capita in 1985 for two low-income countries

Country	Female primary school enrollment	Percent contraception use	Birth rate	Calorie supply
China	107	69	18	2,602
Sri Lanka	101	57	25	2,385

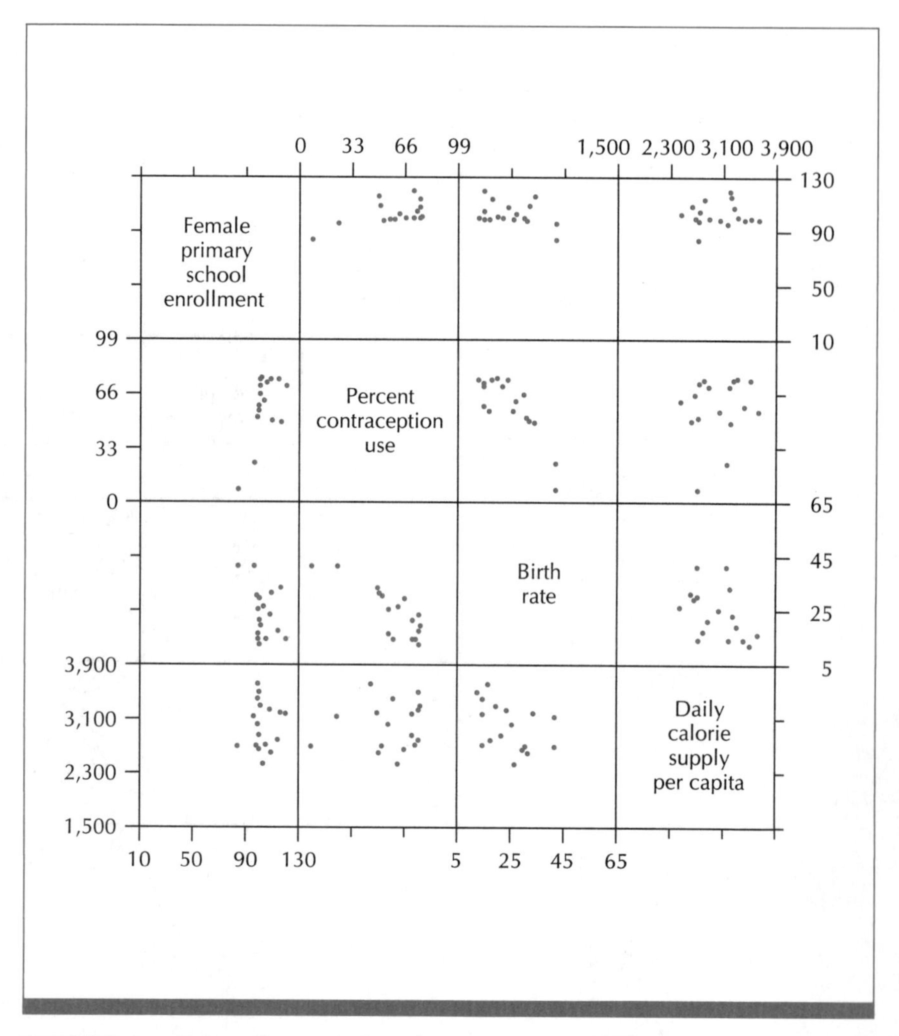

FIGURE 5-4c Scatterplot matrix based on 17 upper-middle-income countries with nonmissing information on all four indicators. Six countries are excluded because of missing values.

among the industrial market countries. These 12 industrial market nations have high female primary school enrollments and daily calorie supplies per capita, relatively high levels of contraception use, and low birth rates.

A comparison of the scatterplot matrices in Figures 5-4a–d shows that the middle-income countries lie between the low-income and industrial market nations as far as these four indicators are concerned. The lower-middle-income countries are closer to the low-income group. The upper-middle-income countries are closer to the industrial market group.

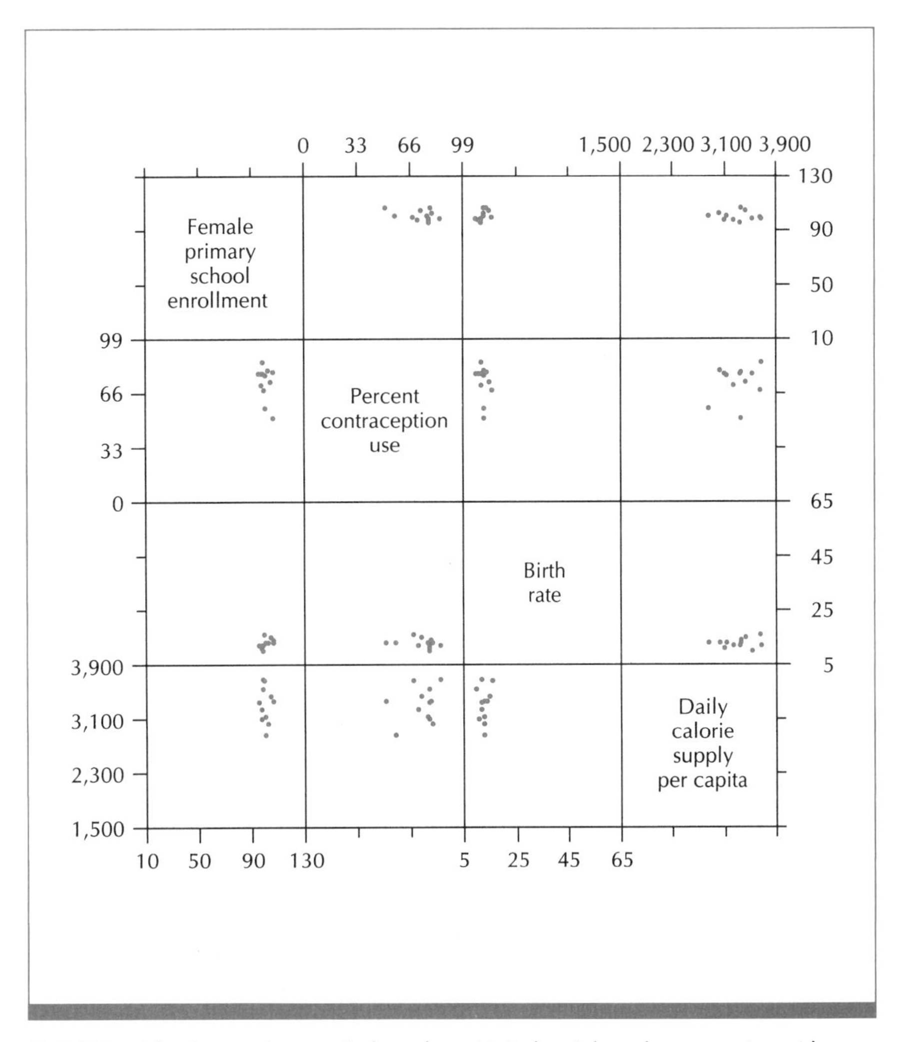

FIGURE 5-4d Scatterplot matrix based on 12 industrial market countries with non-missing information on all four indicators. Seven countries are excluded because of missing values.

we see a strong negative association between contraception use and birth rate. Female primary school enrollment and calorie supply are positively associated with contraception use and negatively associated with birth rate.

In Figure 5-4c, two countries have relatively low levels of contraception use and high birth rates compared with the other upper-middle-income countries represented in the plot. These two countries are Algeria and Iran. Even with these two nations excluded, we still see a decreasing relationship between contraception use and birth rate. There is little association for any other pair of indicators.

All four indicators in Figures 5-3 and 5-4 are related to economic category. Birth rates decrease with economic category. Female primary school enrollment, contraception use, and calorie supply all increase with economic category.

Figure 5-3 is based on the 89 World Bank countries with nonmissing information on all four indicators. Eighty-five countries are included in Figure 5-4. The four extra countries in Figure 5-3 are in the nonmember economic category, which is not included in Figure 5-4. We must be careful when comparing plots, or other tools of data analysis, when varying numbers of cases are included. For this reason we included in each scatterplot matrix only countries with nonmissing information on each of the indicators. Note that in all our figures and tables, we document numbers of countries included and excluded.

5-4

Displaying a Quantitative Variable by Geographic Location: Framed Rectangles on a Map

We often find it useful to convey information related to geographic location by means of a map. In Chapter 1, we displayed a map of the world in Figure 1-1 showing the location and economic category of the World Bank countries. We can think of geographic location as represented by two quantitative variables (corresponding to longitude and latitude, for example.) Figure 1-1 was thus a graphical representation of four variables at a time: the qualitative variable country name, the qualitative variable economic category, and the two quantitative variables corresponding to geographic location.

In this section we discuss how to display a quantitative variable by geographic location. We use framed rectangles on a map to see how calorie supply is related to geographic location among countries on the African continent.

A quantitative variable may be displayed by geographic location using **framed rectangles on a map**.

A **framed rectangle** is a shaded rectangle within a rectangular frame. The lower and upper ends of the frame define lower and upper bounds for the quantitative variable. Tick marks placed outside the frame mark the halfway point between these extremes. The value of the variable is indicated by the height of the shaded rectangle.

We use framed rectangles on a map in Figure 5-5 to display calorie supply for 39 African nations. We can compare calorie supplies across countries by the heights of the shaded rectangles; the frame makes visual comparisons easy. Framed rectangles allow easier interpretations than the shading commonly used on maps, since gradations in shading are hard to differentiate visually. Also, with shading, larger geographic regions tend to make a stronger visual impact than smaller ones. Both of these problems are avoided when we use framed rectangles on a map. Refer to Cleveland (1985) for further discussion of maps for display of quantitative information by geographic location.

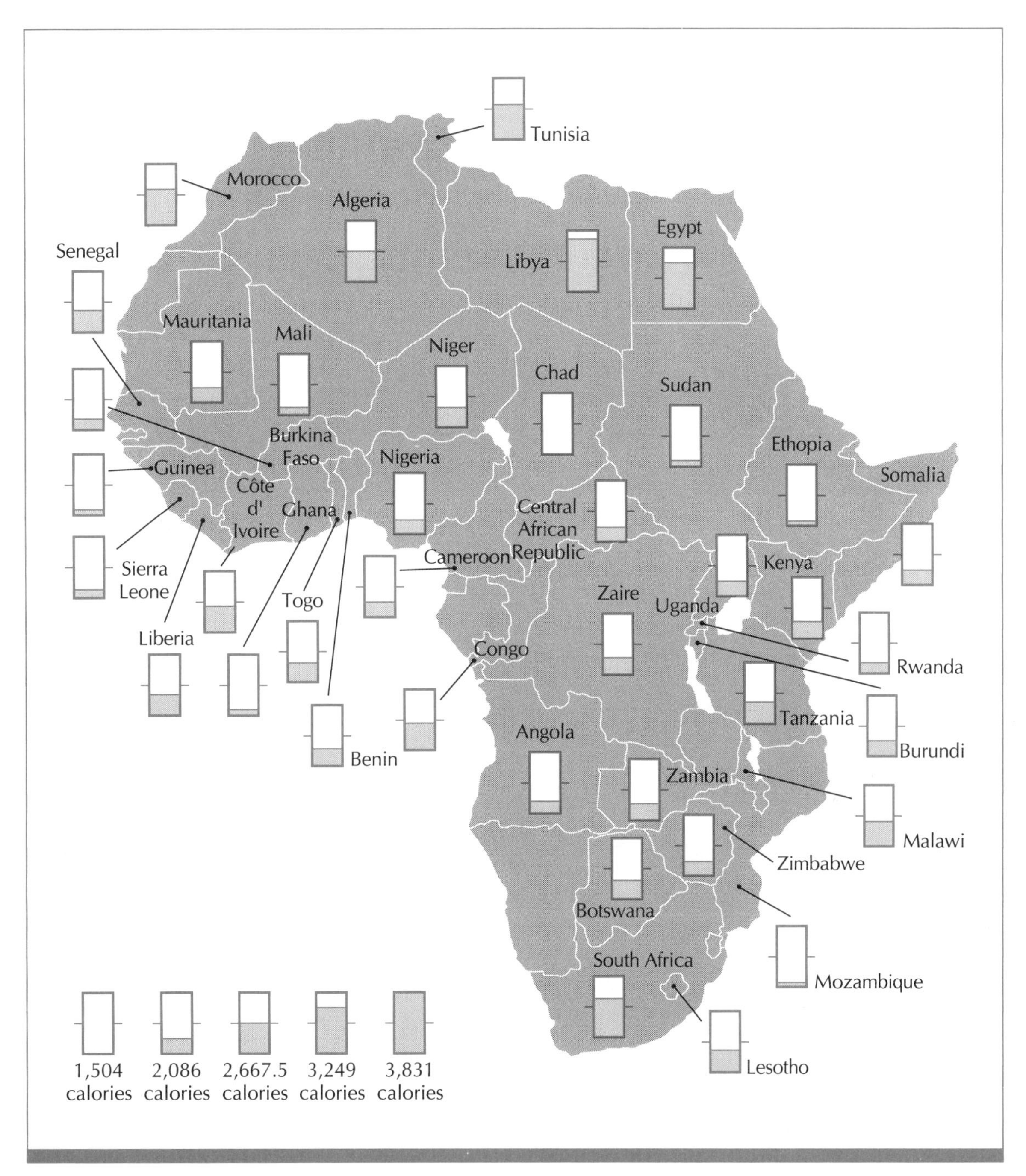

FIGURE 5-5 Framed rectangles on a map of Africa display daily calorie supply per capita in 1985 for 39 African nations. The top of each frame corresponds to the maximum (3,831 calories) and the bottom of each frame to the minimum (1,504 calories) daily calorie supply per capita in 1985 among the 124 World Bank countries with non-missing values.

The bottom of each frame on the map in Figure 5-5 corresponds to the minimum daily calorie supply per capita in 1985 among all 124 World Bank countries with information on this indicator (1,504 calories). The top of each frame corresponds to the maximum daily calorie supply per capita in 1985 among these 124 countries (3,831 calories). The tick marks on the outsides of the frames are halfway between these two values (2,667.5 calories). For reference, the median daily calorie supply per capita in 1985 among the 124 World Bank countries is 2,594 calories, slightly below the tick marks. The mean is 2,675 calories, just above the tick marks. We could have chosen different extremes for the frames. The ones we have used allow comparisons of the African countries with the minimum and maximum over 124 nations.

Looking at the map, we see immediately that Chad has a calorie supply equal to the minimum among World Bank countries. Two neighboring Saharan countries, Sudan and Ethiopia, have calorie supplies very close to this minimum. Most of the 39 countries have low values for calorie supply. Only six countries have calorie supplies above the median (2,594 calories) of 124 World Bank countries. These are the five northernmost countries (Morocco, Algeria, Tunisia, Libya, and Egypt) and the southernmost country (South Africa).

The map in Figure 5-5 gives us a strong overall impression of the relationship between geographic location and calorie supply among these 39 African nations. With framed rectangles, a lot of information is provided in a concise and visually effective fashion. (You might construct a map of Africa, using different shadings to represent relative calorie supplies. Compare the visual effectiveness of your shaded map with the map in Figure 5-5.)

5-5 Effective Graphs

An effective graph summarizes quantitative information in a way that aids visual interpretations. The map in Figure 5-5, for example, helps us see geographical trends in calorie supply for 39 African nations. No extraneous elements that might interfere with visual interpretations are included in the map.

In his book *The Visual Display of Quantitative Information,* Edward Tufte makes a number of suggestions for construction of effective graphical displays. One of these is to maximize the space used for presentation of data and minimize space that is either empty or filled with nonessential elements (such as unnecessary lines, dots, or words). We have tried to follow this advice in graphical displays contained in this book. In scatterplots, for example, the axis for each variable in most cases spans the range of values for the variable. Plotted points then fill the graph as much as possible, minimizing uninformative blank space.

Some types of graphical displays are preferable to others because they are easier to interpret. A dot chart allows more accurate visual comparisons of percentages or proportions than does a pie chart, for example. For excellent discussions of principles of graphing data, see Edward Tufte's *The Visual Display of Quantitative Information* (Tufte, 1983), Tufte's *Envisioning Information* (Tufte, 1990), and William Cleveland's *The Elements of Graphing Data* (Cleveland, 1985).

We now leave data analysis for a while. In Chapters 6, 7, and 8, we discuss probability. Then in Part III, we will use data analysis and probability as we consider ideas in experimental design and statistical inference.

Summary of Chapter 5

Multiway frequency tables (multidimensional contingency tables) allow us to look at relationships among several qualitative variables. Scatterplots of two quantitative variables at each level of a qualitative variable can be useful for studying the relationship between the two quantitative variables within levels of the qualitative variable. The scatterplot matrix provides a way to display several scatterplots within a single figure. Framed rectangles on a map effectively illustrate how a quantitative variable depends on geographic location; framed rectangles are easier than shadings to interpret visually. Graphical displays should be designed to allow easy visual interpretation, with maximum use of data and minimal use of blank space and extraneous material.

Minitab Appendix for Chapter 5

Creating Multiway Frequency Tables

To produce multidimensional frequency tables, we use the TABLE command. The TABLE command can be followed by up to 10 classification variables. (A classification variable takes integer values between −9999 and +9999 or missing values.) Minitab uses the first variable for rows of the table and the second variable for columns. Minitab prints a separate table for each combination of values for any other variables listed. Consider the example based on Exercise 5-35. In the Minitab Appendix for Chapter 1, we created two classification variables, CALCODE and MORTCODE. We will create another classification variable based on WEIGHT, using the CODE command:

```
MTB>  code (0:5)1 (6:30)2 'weight' c30
MTB>  name c30 'wtcode'
```

The variable WTCODE in column 30 equals 1 for countries with values of WEIGHT less than or equal to 5, equals 2 for countries with values of WEIGHT greater than or equal to 6. The TABLE command

```
CONTROL: mortcode =  1
ROWS: wtcode       COLUMNS: calcode

             1         2         3         4        ALL

 1           0         0         4         6         10
 2           0         0         3         6          9
ALL          0         0         7        12         19

CONTROL: mortcode =  2
ROWS: wtcode       COLUMNS: calcode

             1         2         3         4        ALL

 1           0         0         0         0          0
 2           3        11         1         0         15
ALL          3        11         1         0         15
  CELL CONTENTS --
                    COUNT
```

FIGURE M5-1 Three-dimensional frequency table of WTCODE, CALCODE, and MORTCODE produced by the TABLE command

```
MTB> table 'wtcode' 'calcode' 'mortcode'
```

produces two frequency tables of WTCODE by CALCODE, one for each of the two levels of MORTCODE, as shown in Figure M5-1.

Superimposing Scatterplots on the Same Graph

Recall that in the Minitab Appendix for Chapter 1, we produced columns named MORT1 and LIFE1 that contained the child mortality and life expectancy data, respectively, for the 20 countries with low child mortality. The columns named MORT2 and LIFE2 contained the child mortality and life expectancy data, respectively, for the 20 countries with high child mortality. Suppose we want to plot child mortality versus life expectancy for each set of countries. We can produce separate scatterplots using the PLOT command twice, once for MORT1 versus LIFE1 and once for MORT2 versus LIFE2. We can get the same scales for both plots by using the YSTART and XSTART subcommands. To superimpose these two scatterplots on the same graph, we can use the MPLOT command:

```
MTB> mplot 'mort1' 'life1' 'mort2' 'life2'
```

Minitab will produce the two scatterplots on the same graph, using different plotting symbols for the two sets of points, as shown in Figure M5-2. Numerals (2, 3, 4, etc.) indicate the number of cases with the same plotted points.

We can achieve the same result using the LPLOT command, instructing Minitab to plot MORT versus LIFE by MORTCODE:

```
MTB> lplot 'mort' 'life' by 'mortcode'
```

The output is shown in Figure M5-3.

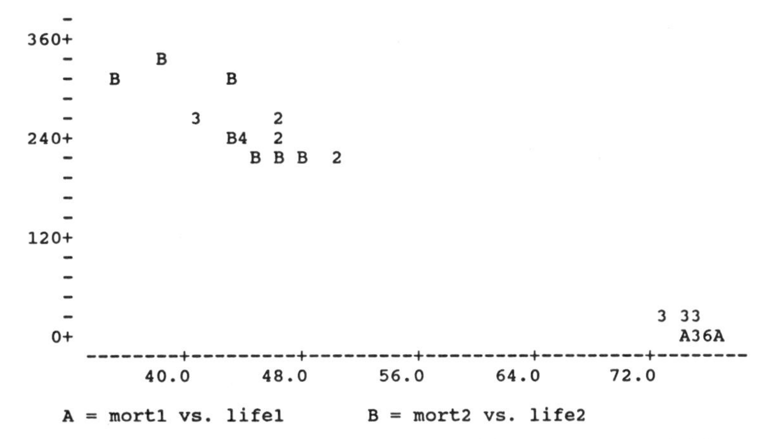

FIGURE M5-2 Scatterplot of child mortality versus life expectancy, with different plotting symbols for low- and high-child-mortality countries, using the MPLOT command

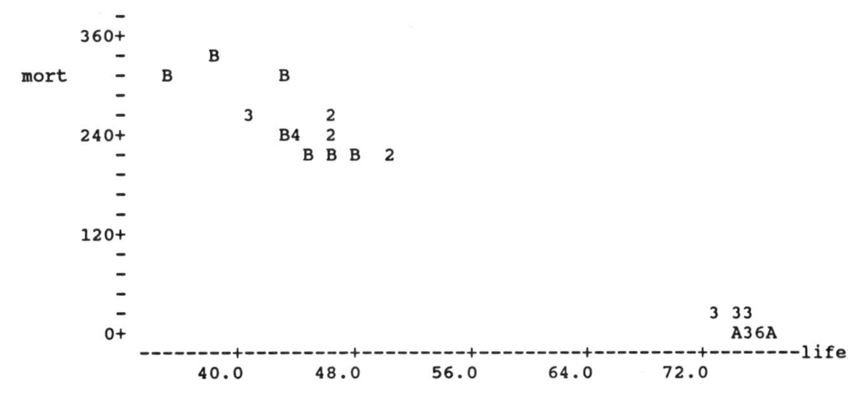

FIGURE M5-3 Scatterplot of child mortality versus life expectancy, with different plotting symbols for low- and high-child-mortality countries, using the LPLOT command

Creating a Scatterplot Matrix

Minitab will not produce a scatterplot matrix directly (some statistical packages will). To construct a scatterplot matrix, we can use the PLOT command to produce scatterplots, controlling the scales using the YSTART and XSTART subcommands. We can then assemble these plots into a scatterplot matrix by hand or by using a word-processing program.

We cannot use Minitab to produce a map with framed rectangles.

Exercises for Chapter 5

In the exercises, answer the following questions: What would you need to know about the sample to be willing to use it to make inferences about a larger population? What is that larger population (if any)? What limitations do you see in the sample?

For each figure and table, include a legend that completely describes its contents. Note the number of cases included and the number excluded if there are missing values.

EXERCISE 5-1

How does survival of mice exposed to DDT and urethane compare with survival of unexposed mice? The accompanying table summarizes results of a study designed to address this question (Breslow, 1988; *J. Nat. Cancer Inst.*, volume 52, 1974, pages 233–239). The table shows the percentage of mice alive at five times, by sex and exposure groups.

	Control	2 ppm DDT	10 ppm DDT	50 ppm DDT	250 ppm DDT	Urethane
Males						
Number of animals	348	362	367	396	372	315
Percentage alive at						
0 weeks	100	100	100	100	100	100
70 weeks	80.7	83.7	81.7	84.6	72.3	72.1
90 weeks	63.5	57.2	57.2	52.8	28.5	39.7
110 weeks	32.5	22.4	25.1	21.5	1.3	8.9
130 weeks	13.8	3.0	8.2	3.5	0.0	0.1
Females						
Number of animals	363	354	370	349	334	248
Percentage alive at						
0 weeks	100	100	100	100	100	100
70 weeks	79.1	77.4	83.8	80.2	66.5	72.2
90 weeks	60.0	55.1	64.9	56.4	41.3	40.7
110 weeks	35.0	23.7	42.4	26.9	9.9	12.1
130 weeks	14.0	5.6	15.4	6.3	1.5	1.2

a. For the male mice, plot percentage alive versus time for each treatment group. You may want to use different plotting symbols for each treatment group on the same graph. Or you may wish to plot a separate graph for each treatment group.

b. Repeat part (a) for female mice.

c. Separately for males and females, compare survival across treatment groups.

d. Separately for each treatment, compare survival of males and females.

e. Discuss all of your results.

EXERCISE 5-2

When scientists say that two treatments have a synergistic effect, they mean that the effect of the two treatments taken together is greater than the sum of their separate effects. In this experiment, researchers explored possible synergistic antitumor effects of two agents that stimulate the activity of an animal's immune system. The experimenters injected mice with lymphoma cells and then treated them with interferon or monoclonal antibody or a combination of the two agents. The number of mice surviving 90 days after injection with the tumor cells is shown below, along with the number of animals treated with each treatment combination (Piegorsch, Weinberg, and Margolin, 1988; Basham et al., 1986). The table shows number of animals alive at 90 days/number of animals treated.

Interferon (units/mouse/day)	Monoclonal antibody (units/mouse/day)				
	0	.1	1	10	100
0	0/10	0/10	3/10	2/10	9/40
10^4	1/40	1/10	6/10	7/10	8/10

Do the experimental results suggest a synergistic antitumor effect of these two forms of immunotherapy? Graph the data in any way that helps to answer this question.

EXERCISE 5-3

Consider the following information on body weight and kidney weight for 25 normal mice and 9 diabetic mice, from a study by Dr. E. Jones of the Children's Cancer Research Foundation in Boston, Massachusetts (Hill and Padmanabhan, 1984; from Bishop, 1973).

Body weight (g)	Kidney weight (mg)	Body weight (g)	Kidney weight (mg)
	Normal Mice		
34	810	37	780
43	480	38	660
35	680	32	750
33	920	36	780
34	650	32	670
26	650	32	670
30	650	38	700
31	560	42	720
31	620	36	800
27	740	44	830
28	600	33	640
27	640	38	800
30	690		
	Diabetic Mice		
42	1,030	46	1,100
44	1,240	34	1,040
38	1,150	44	1,080
52	1,280	38	870
48	1,240		

a. Construct two dot plots of body weight—one for the normal mice and one for the diabetic mice. Use the same scale for each plot. Describe and compare the two distributions.

b. Repeat part (a) for kidney weight.

c. Plot kidney weight versus body weight for the normal mice. Using the same scales, plot kidney weight versus body weight for the diabetic mice.

d. Is the relationship between body weight and kidney weight the same for the normal and diabetic mice? Discuss your findings.

EXERCISE 5-4

The accompanying table shows forearm tremor frequency (in Hz) for each of five weights (in pounds) applied at the wrist, for six volunteers (Hollander and Wolfe, 1973, page 175; based on Fox and Randall, 1970). Each value is the average of five measurements.

Volunteer	0 lb	1.25 lb	2.5 lb	5 lb	7.5 lb
1	3.01	2.85	2.62	2.63	2.58
2	3.47	3.43	3.15	2.83	2.70
3	3.35	3.14	3.02	2.71	2.78
4	3.10	2.86	2.58	2.49	2.36
5	3.41	3.32	3.08	2.96	2.67
6	3.07	3.06	2.85	2.50	2.43

Construct one or more scatterplots showing forearm tremor frequency versus weight applied to determine whether the relationship between the two variables is the same for each volunteer. Discuss your findings.

EXERCISE 5-5

Percent minority enrollment in grades kindergarten through 12 in 1980, percent minority teachers in 1982, and percent minority new hires in 1982 are shown below for 17 states with high minority enrollments (American Council on Education, 1987, page 27).

State	Percent minority enrollment in public schools in 1980	Percent minority teachers in 1982	Percent minority new hires in 1982
Alabama	33.6	27.0	13.0
Arizona	33.7	11.6	10.9
Arkansas	23.5	19.4	12.8
California	42.9	18.9	25.2
Delaware	28.8	16.8	9.0
Florida	32.2	21.3	11.8
Georgia	34.3	26.4	14.7
Illinois	28.6	6.5	4.8
Louisiana	43.4	30.7	9.1

State	Percent minority enrollment in public schools in 1980	Percent minority teachers in 1982	Percent minority new hires in 1982
Maryland	33.5	26.5	10.4
Mississippi	51.6	38.3	19.8
New Mexico	57.0	28.3	25.4
New York	32.0	9.9	17.8
North Carolina	31.9	21.6	16.6
South Carolina	43.5	25.5	16.7
Texas	45.9	24.0	18.1
Virginia	27.5	20.3	14.7

a. Construct a scatterplot matrix with these three variables. Discuss your findings.

b. Use framed rectangles on a map of the United States to investigate the relationship between these variables and geography. Discuss your findings.

EXERCISE 5-6

Engine displacement, city and expressway gasoline usage, and weight are shown here for ten cars weighing 1,000 kg or less (Ramsay, 1988; from 1986 *Consumer Reports*).

Car	Engine displacement (liters)	City gas (liters/ 100 km)	Expressway gas (liters/ 100 km)	Weight (100 kg)
Chevrolet Chevette	1.6	13.3	6.8	10.0
Chevrolet Spectrum	1.5	10.1	5.3	8.7
Dodge Colt	1.5	11.0	5.6	9.9
Dodge Omni	1.6	11.5	5.6	9.5
Honda Civic	1.5	11.5	6.5	9.2
Nissan Sentra	1.6	10.5	5.6	9.5
Renault Alliance	1.4	12.0	5.6	9.1
Toyota Tercel	1.5	11.0	5.5	9.7
Honda Civic CRX	1.5	9.4	5.6	9.0
Nissan Pulsar NX	1.6	9.7	5.1	9.2

Construct a scatterplot matrix using these four variables. What observations can you make about relationships between pairs of variables for these small cars?

EXERCISE 5-7

Estimated radiocarbon dates ($\pm$ 1 standard deviation) are shown by depth of samples taken from sites of two different archeological digs in Tasmania (Cosgrove, 1989). The units for estimated radiocarbon dates are years before A.D. 1950. In this reference, the term *standard deviation* indicates an estimate of errors in counting of the modern radiocarbon standard, background and sample.

Shannon River Valley		Bluff Cave, Florentine River Valley	
Depth (cm)	Estimated radiocarbon date ± 1 SD	Depth (cm)	Estimated radiocarbon date ± 1 SD
5	2,450 ± 70	5	11,630 ± 200
15	10,440 ± 160	10	13,100 ± 110
22	18,480 ± 200	15	13,830 ± 220
23	17,660 ± 250	20	16,120 ± 180
25	19,080 ± 280	30	21,410 ± 240
45	30,840 ± 480	35	24,190 ± 410
50	16,200 ± 590	42	27,770 ± 420
		50	28,000 ± 720
		53	23,640 ± 310
		55	30,750 ± 1,340
		60	30,420 ± 690

a. Construct a scatterplot of estimated radiocarbon date versus depth of the sample for each of the two sites. Use the same scales for each plot.

b. Is the relationship between depth of the sample and radiocarbon age the same for the two sites? Discuss the information provided in your two plots.

EXERCISE 5-8

Investigators studied eight hot springs in the Cascade Range in north central Oregon (Ingebritsen, Sherrod, and Mariner, 1989). They measured discharge temperature (°C) and concentration (mg/liter) of calcium (Ca), sodium (Na), and chlorine (Cl) in hot springs water at each site.

Hot spring	Discharge temperature	Ca	Na	Cl
Austin	86	35	305	390
Bagby	58	3.3	53	14
Breitenbush	84	95	745	1,200
Bigelow	59	195	675	1,250
Belknap	73	210	660	1,200
Foley	79	510	555	1,350
Kahneeta	83	13	400	240
Unnamed (on Rider Creek)	46	215	405	790

a. Construct a dot plot of each of the four variables. Describe each distribution in terms of location, variation, concentrations of values, and symmetry or skewness.

b. Construct a scatterplot matrix using these four variables. Discuss the relationships between pairs of variables you see in the plots.

EXERCISE 5-9

As part of a study of movements and survival of black ducks, U.S. Fish and Wildlife Service workers captured and examined 50 female black ducks in

November and December 1983 (Pollock, Winterstein, and Conroy, 1989). Thirty-one were hatch-year ducks, born during the previous breeding season. The other 19 were after-hatch-year ducks, at least 1 year old. The workers recorded body weight and wing length for each duck. They also calculated a condition index (body weight divided by wing length) for each duck.

Weight (gm)	Wing length (mm)	Condition index (gm/mm)	Weight (gm)	Wing length (mm)	Condition index (gm/mm)
		Hatch-year ducks			
1,140	266	4.29	1,070	267	4.01
1,160	264	4.39	1,270	276	4.60
1,120	262	4.27	1,080	260	4.15
1,070	268	3.99	1,150	271	4.24
940	252	3.73	1,030	265	3.89
1,240	271	4.58	1,160	275	4.22
1,120	265	4.23	1,180	263	4.49
1,010	272	3.71	1,050	271	3.87
1,040	270	3.85	1,050	275	3.82
1,200	276	4.35	1,160	266	4.36
1,280	270	4.74	1,150	263	4.37
1,250	272	4.59	1,220	268	4.55
1,090	275	3.96	1,140	262	4.35
1,040	255	4.08	1,140	270	4.22
1,130	268	4.22	1,120	274	4.09
1,180	259	4.56			
		After-hatch-year ducks			
1,160	277	4.19	1,250	276	4.53
1,260	280	4.50	1,050	275	3.82
1,080	267	4.04	1,320	285	4.63
1,140	277	4.11	1,260	269	4.68
1,200	283	4.24	1,110	270	4.11
1,100	264	4.17	1,280	281	4.55
1,420	270	5.26	1,270	270	4.70
1,120	272	4.12	1,370	275	4.98
1,110	271	4.10	1,220	265	4.60
1,340	275	4.87			

a. For the variable body weight, construct two box plots, one for the hatch-year ducks and another for the after-hatch-year ducks. Display these two box plots on the same graph to allow comparison between the two groups of ducks. In the same way, construct box plot displays to compare wing length and condition index between the two groups of ducks.

b. For each variable, discuss the information provided in the plots for each age group. Use the plots to compare the two age groups.

c. For the 31 hatch-year ducks, construct a scatterplot matrix using the three variables body weight, wing length, and condition index. Discuss the relationships between pairs of variables revealed in the plots.

d. Repeat part (c) for the 19 after-hatch-year ducks.

e. Compare the scatterplot matrices for the two age groups. Are relationships between pairs of variables different for the two groups?

EXERCISE 5-10

Researchers measured crying activity in 38 4–7-day-old babies, 20 females and 18 males. They tested these children 3 years later to measure speech and intellectual development. Sex, infant cry count (units not given), and 3-year Stanford–Binet IQ score are shown below for each child (Hollander and Proschan, 1984, pages 150–151; from Karelitz et al., 1964).

Girls		Boys	
Cry count	IQ score	Cry count	IQ score
10	87	20	90
12	94	17	94
16	100	12	97
19	103	12	103
14	106	9	103
10	109	23	103
15	112	13	104
15	114	16	106
9	119	27	108
12	119	18	109
19	120	18	109
16	124	18	112
20	132	23	113
15	133	21	114
22	135	16	118
31	135	12	120
16	136	17	141
22	157	30	155
33	159		
13	162		

a. Construct a stem-and-leaf plot of the girls' cry counts. Use tens as stems and units as leaves, two rows per stem. Construct a similar plot for the boys. Describe these two distributions of cry counts. Compare the distributions for the girls and boys.

b. Construct a stem-and-leaf plot of the girls' IQ scores. Use the units as leaves, other digits as stems. Construct a similar plot for the boys. Describe the two distributions of IQ scores. Compare the distributions for the girls and boys.

c. For the girls, construct a scatterplot of IQ score versus cry count. Discuss the relationship between the two variables shown in the plot.

d. Construct a scatterplot of IQ score versus cry count for the boys. Use the same scales as for the girls. Describe the information provided in the plot.

e. Compare the two scatterplots. Is the relationship between infant cry count and age-3 IQ score the same for the girls and the boys?

EXERCISE 5-11

Researchers recorded obesity and blood pressure for each person in a random sample of 58 Mexican–American women and a random sample of 44 Mexican–American men aged 35–60 years in a small California town. (A sample is a random sample if each member of the population had an equal and independent chance of being included in the sample.) Obesity is recorded as actual weight divided by ideal weight (based on New York Metropolitan life tables). Blood pressure (BP) is systolic blood pressure in millimeters of mercury (mm Hg). The results are shown here (from a study by J. W. Farquhar and associates discussed in Hollander and Proschan, 1984, pages 147 and 150).

Obesity	BP	Obesity	BP	Obesity	BP
		Women			
1.50	140	1.59	150	1.43	130
1.63	132	2.39	150	1.50	112
.92	138	1.17	116	1.33	124
1.09	112	1.24	116	1.44	110
1.23	160	1.50	140	1.34	124
2.04	138	1.13	118	1.11	104
1.38	114	1.35	138	1.42	170
1.55	144	1.33	108	1.22	108
1.07	98	.97	112	1.26	100
1.65	120	1.01	118	1.54	130
1.43	128	1.74	128	1.51	118
1.36	110	1.37	148	1.67	162
1.03	128	1.32	108	1.56	116
1.33	104	1.56	122	1.25	98
1.24	110	1.27	118	1.57	116
1.30	118	1.32	138	1.41	142
1.21	124	1.20	120	1.15	118
1.43	122	1.24	112	1.28	126
1.75	138	2.20	136	1.64	136
1.73	208				
		Men			
1.31	130	1.31	148	1.19	146
1.11	122	1.34	140	1.17	146
1.56	132	1.18	110	1.04	124
1.03	150	.88	120	1.29	114
1.26	136	1.16	118	1.32	190
1.37	118	1.25	130	1.48	112
1.58	126	.93	162	1.29	124
1.06	126	1.19	134	.96	110
1.13	118	1.19	110	.81	94
1.11	118	1.29	140	1.29	128
1.28	126	1.20	140	1.02	124
1.09	104	1.08	134	1.04	130
1.14	124	1.13	110	1.16	134
1.57	144	1.07	116	1.04	118
1.37	118	1.26	132		

a. Construct box plots of obesity separately for men and women. Display the two box plots in the same graph to allow comparisons. Describe the distribution of obesity values for men and women. Compare the distributions for men and women.

b. Repeat part (a) for blood pressure.

c. Construct a scatterplot of blood pressure versus obesity for women. Construct a similar scatterplot for men, using the same scales. Describe the relationship between obesity and blood pressure for men and for women. Is the relationship the same for the two sexes?

EXERCISE 5-12 The table shows concentrations (nanograms per milliliter) of the brain metabolite homovanillic acid (HVA) in cerebrospinal fluid, full-scale intelligence quotient (IQ), memory quotient (MQ), and IQ − MQ for nine patients with a disorder known as Korsakoff's psychosis (Dietz, 1989; from McEntee and Mair, 1978).

HVA	IQ	MQ	IQ − MQ
21	89	60	29
23	90	59	31
25	122	102	20
25	87	64	23
26	89	61	28
31	106	79	27
40	104	80	24
48	106	80	26
75	127	88	39

The difference between IQ and MQ is a measure of memory impairment; greater impairment is associated with larger differences.

a. Construct a scatterplot matrix using these four variables.

b. Discuss the relationships between pairs of variables shown in the scatterplot matrix.

EXERCISE 5-13 In this study of energy requirements of grazing Merino wether sheep in Australia, researchers determined outdoor maintenance requirements (in Mcal/sheep/day) by radioassay of urinary CO_2. They carried out four separate experiments, in one location. Animal weights (in kg) and energy requirements are shown for each of these experiments (Wallach and Goffinet, 1987; from Young and Corbett, 1972).

Weight	Requirements	Weight	Requirements	Weight	Requirements	Weight	Requirements
			Experiment 1				
22.1	1.31	30.0	1.23	33.8	1.46	49.2	2.53
25.1	1.46	30.2	1.01	34.3	1.14	51.8	1.87
25.1	1.00	30.2	1.12	34.9	1.00	51.8	1.92
25.7	1.20	33.2	1.25	42.6	1.81	52.5	1.65
25.9	1.36	33.2	1.32	43.7	1.73	52.6	1.70
26.2	1.27	33.2	1.47	44.9	1.93	53.3	2.66
27.0	1.21	33.9	1.03	49.0	1.78		

Weight	Requirements	Weight	Requirements	Weight	Requirements	Weight	Requirements
Experiment 2				**Experiment 3**			
23.9	1.37	32.1	1.80	46.7	2.21	28.6	2.13
25.1	1.29	32.6	1.75	37.1	2.11	29.2	1.80
26.7	1.26	33.1	1.82	31.8	1.39	26.2	1.05
27.6	1.39	34.1	1.36	36.1	1.79		
28.4	1.27	34.2	1.59	**Experiment 4**			
28.9	1.74	44.4	2.33	45.9	2.36	34.4	1.63
29.3	1.54	44.6	2.25	36.8	2.31	26.4	1.27
29.7	1.44	52.1	2.67	34.4	1.85	27.5	.94
31.0	1.47	52.4	2.28				
31.0	1.50	52.7	3.15				
31.8	1.60	53.1	2.73				
32.0	1.67	52.6	3.73				

a. Plot energy requirement versus animal weight for each of the four experiments. Use the same scales for each plot.

b. What is the relationship between animal weight and outdoor maintenance energy requirements?

c. Do the four experiments demonstrate the same relationship between animal weight and energy requirements? Discuss your findings.

EXERCISE 5-14

Researchers have studied patterns of recovery of stroke patients over time. Such patterns provide useful baselines for evaluating individual patients. In this study, researchers obtained recovery information on 368 patients who survived at least 8 weeks from initial examination. The number (percentage) of these 368 patients past each of three recovery milestones at the initial examination, as well as 1, 2, 4, 6, and 8 weeks later are shown in the table (Partridge, Johnston, and Edwards, 1987).

Recovery milestone	Initial examination	1 week later	2 weeks later	4 weeks later	6 weeks later	8 weeks later
Maintain sitting balance for two minutes	217 (59.0)	280 (76.1)	315 (85.6)	334 (90.8)	337 (91.6)	338 (91.8)
Stand up to free-standing position	105 (28.5)	158 (42.9)	193 (52.4)	230 (62.5)	243 (66.0)	260 (70.7)
Independent walking inside	53 (14.4)	101 (27.4)	138 (37.5)	166 (45.1)	181 (49.2)	196 (53.3)

a. On the same scatterplot, plot the percentage of patients past the milestone versus time, for each of the three milestones. You may want to use different plotting symbols (or colors) for the three milestones.

b. Discuss the recovery patterns over time for these three milestones.

EXERCISE 5-15

Values of maximal oxygen uptake (in ml · kg^{-1} · min^{-1}) are listed here for world-class athletes in their teens and 20's. (Maximal oxygen uptake is a measure of lung function and capacity for work.) Several male and female athletes were tested in each of four sports (Wilmore, 1984).

Basketball	*Females:*	42.3	42.9	49.6		
	Males:	41.9	45.9	50.0		
Speed skating	*Females:*	52.0	46.1			
	Males:	56.1	72.9	64.6		
Cross-country skiing	*Females:*	61.5	68.2	56.9		
	Males:	63.9	73.9	78.3	73.0	
Distance running	*Females:*	63.2	50.8	57.5		
	Males:	65.5	72.2	77.4	78.1	73.2

Display these values in any way(s) that will allow comparisons of maximal oxygen uptake across sports for each sex, and between males and females for each sport.

EXERCISE 5-16

Age, height, weight, and maximal oxygen capacity are listed below for 13 male world-class distance runners (Wilmore, 1984). Maximal oxygen capacity is a measure of the lungs' capacity for work.

Age (years)	Height (cm)	Weight (kg)	Maximal oxygen capacity (ml · kg^{-1} · min^{-1})
10	144.3	31.9	56.6
26	176.1	64.5	72.2
26	178.9	63.9	77.4
26	177.0	66.2	78.1
27	178.7	64.9	73.2
32	177.3	64.3	70.3
35	174.0	63.1	66.6
36	177.3	69.6	65.1
40–49	180.7	71.6	57.5
55	174.5	63.4	54.4
50–59	174.7	67.2	54.4
60–69	175.7	67.1	51.4
70–75	175.6	66.8	40.0

a. Construct a scatterplot matrix using these four variables. (When age is listed as an interval, use a reasonable value such as the midpoint of the interval in the plots that include age.)

b. Discuss the relationships between pairs of variables that you see in these plots.

EXERCISE 5-17

The number of female mosquitos captured coming to bite in a yard with an electrocuting device and the number killed in the device are listed here for each of five different 2-hour sessions at each of two sites (Nasci, Harris, and Porter, 1983).

Trial	1	2	3	4	5
Site 1					
Electrocuting device	31	44	129	15	11
Human bait	94	146	194	54	39
Site 2					
Electrocuting device	49	151	30	12	17
Human bait	90	172	219	60	21

a. Using the same scales, construct four dot plots: for each site, a plot of mosquito numbers killed in the electrocuting device and a plot of numbers captured by the person.

b. How do the numbers captured vary by site? How do the numbers captured vary between the electrocuting device and human bait?

c. For each site, draw a scatterplot of number of mosquitos captured by the person versus number killed in the electrocuting device. What is the relationship between the two variables for each site? Compare sites.

EXERCISE 5-18

The human immunodeficiency virus (HIV) is the virus associated with the acquired immune deficiency syndrome (AIDS). In this study, investigators tested residents of six African countries for presence of HIV (Kanki et al., 1987). They classified residents into three groups. The risk group included prostitutes and people visiting outpatient clinics for sexually transmitted diseases. The disease group consisted of people with tuberculosis and patients hospitalized in infectious disease or internal medicine wards. The control group included healthy adults from the same geographic regions as the risk and disease groups. The numbers testing positive for HIV antibodies in serum samples are shown below by country and group.

Burkina Faso: 1 positive of 22 tested in the disease group, 45 positive of 340 tested in the risk group, 2 positive of 416 tested in the control group

Ivory Coast: 4 positive of 40 tested in the disease group, 46 positive of 232 tested in the risk group, 38 positive of 1,067 tested in the control group

Guinea: 1 positive of 131 tested in the disease group, 0 positive of 13 tested in the risk group, 2 positive of 314 tested in the control group

Guinea Bassau: 0 positive of 273 tested in the disease group, 0 positive of 39 tested in the risk group, 0 positive of 151 tested in the control group

Senegal: 2 positive of 178 tested in the disease group, 3 positive of 422 tested in the risk group, 0 positive of 426 tested in the control group

Mauritania: 2 positive of 35 tested in the disease group, 0 positive of 9 tested in the risk group, 0 positive of 140 tested in the control group

a. Display these results in one or more frequency tables.

b. Discuss any differences between groups with respect to detection of antibodies to HIV.

c. Discuss any differences among countries.

d. Use framed rectangles on a map of Africa to look for geographic trends. Discuss your findings.

EXERCISE 5-19

In 1975–1976: 499,602 men and 418,786 women received bachelor's degrees in the United States, 165,474 men and 143,789 women received master's degrees, 26,010 men and 7,777 women received doctorates, and 52,365 men and 9,720 women received first professional degrees.

In 1984–1985: 482,528 men and 496,949 women received bachelor's degrees, 143,390 men and 142,861 women received master's degrees, 21,700 men and 11,243 women received doctorates, and 50,455 men and 24,608 women received first professional degrees (American Council on Education, 1987, page 187).

a. Arrange this information in one or more frequency tables.

b. Discuss the relationship between sex and degree, separately for the two academic years. Is the relationship the same for the two academic years?

c. Discuss the relationship between year and degree, separately for men and women. Is the relationship the same for men and women?

EXERCISE 5-20

Of 22,632,000 White 18–24-year-olds in the United States in 1985, 18,916,000 had completed high school and 6,500,000 had enrolled in college. Of 3,716,000 Black 18–24-year-olds, 2,810,000 had completed high school and 734,000 had enrolled in college. Of 2,221,000 Hispanic 18–24-year-olds, 1,396,000 had completed high school and 375,000 had enrolled in college (American Council on Education, 1987, page 17).

a. Display these numbers in one or more frequency tables. Plot the data in any helpful way.

b. Discuss the relationship between race and high school completion.

c. Discuss the relationship between race and college entrance.

EXERCISE 5-21

Participants in the 1974 General Social Surveys were asked three questions, each beginning with:

> "Please tell me whether or not you think it should be possible for a pregnant woman to obtain a legal abortion . . ."

The three questions ended with:

A: ". . . if she is married and does not want any more children."
B: ". . . if the family has a very low income and cannot afford any more children."
C: ". . . if she is not married and does not want to marry the man."

Of 1,060 respondents, 413 said yes to all three questions and 430 said no to all three. Of the remaining 217 respondents, 29 said yes to A and B but no to C; 16 said yes to A and C but no to B; 18 said yes to A but no to B and C; 60 said yes to B and C but no to A; 57 said yes to B but no to A and C; 37 said yes to C but no to A and B (Tanner and Wong, 1987; from Haberman, 1979).

a. Arrange these results in one or more frequency tables.

b. Interpret the attitudes toward abortion reflected by respondents in this survey.

EXERCISE 5-22

Are adults with lifelong exposure to malaria less susceptible to the infection than are children? In a study of malaria in a region of Kenya, researchers treated 83 adults and 62 children (aged 6 months to 5 years) for malaria, achieving what is called a radical cure. By 56 days after radical cure, 57 of the 62 children and 13 of the 83 adults had developed malaria infections. By 84 days after radical cure, all 62 of the children and 48 of the 83 adults had developed malaria infections [*Science,* volume 237, August 7, 1987, pages 639–642).

a. Arrange these observations in a frequency table. Construct any plots that seem helpful.

b. Do adults seem to be less susceptible to malaria infection than children?

EXERCISE 5-23

A study of alcoholism in Sweden included men of known paternity, born to single women and adopted by nonrelatives at a young age. Researchers defined two types of alcoholism. Type I alcoholism is associated with onset after age 25, ability to abstain, infrequent fights when drinking, guilt about alcohol, and psychological dependence. Type II alcoholism is associated with onset before age 25, inability to abstain, frequent fights when drinking, little guilt about drinking, and little dependence. The researchers classified the 862 male adoptees by whether they had severe Type I alcoholism, whether they had a Type I genetic background, and whether they were raised in an environment that would contribute to excessive drinking:

Type I genetic back-ground	Contributing environment	Number with severe Type I abuse/Number in category
No	No	16/376
No	Yes	3/72
Yes	No	22/328
Yes	Yes	10/86

The researchers classified the 862 male adoptees in a similar fashion with respect to Type II alcoholism:

Type II genetic back-ground	Contributing environment	Number with severe Type II abuse/Number in category
No	No	11/567
No	Yes	8/196
Yes	No	12/71
Yes	Yes	5/28

Numbers of adoptees with alcohol abuse were calculated from percentages given in Cloninger (1987).

a. Tabulate and/or plot these results in any way that seems reasonable.

b. Discuss the relative contributions of genetic and environmental background to Type I and Type II alcoholism suggested by this study.

EXERCISE 5-24

Investigators classified sports injuries among children in a French health care district in 1981 and 1982 by age and sex of the child and type of injury. (These frequencies were calculated from percentages reported in Tursz and Crost, 1986.)

Type of injury	Boys 6–11 years	Boys ≥ 12 years	Girls 6–11 years	Girls ≥ 12 years
Contusions	77	119	49	87
Cuts, lacerations	61	32	21	7
Sprains, strains	22	45	12	43
Fractures	43	52	32	36

a. Tabulate and/or plot the results in any way that seems helpful.

b. Discuss the relationships among age, sex, and type of injury in this group of children.

EXERCISE 5-25

The accompanying table shows by species the number of animals testing positive for rabies and the number tested in Maryland, in 1982, 1983, and 1984. (Numbers positive for rabies were calculated from percentages reported in Beck, Felser, and Glickman, 1987.)

	Number positive for rabies/Number tested		
	1982	1983	1984
Raccoon	119/1,484	736/3,134	964/1,691
Skunk	13/80	28/120	32/69
Fox	0/79	5/116	19/91
Bat	17/753	51/1,169	46/1,098
Groundhog	0/72	5/215	13/445
Deer	0/11	1/24	1/36
Rabbit	0/64	0/102	2/202
Mouse/rat	0/100	0/86	1/144

	Number positive for rabies/Number tested		
	1982	1983	1984
Opossum	0/99	0/256	2/510
Chipmunk/squirrel	0/197	0/260	2/597
Ferret/mink	0/24	0/22	0/28
Beaver/muskrat	0/6	2/26	0/12
Horse	0/8	0/19	1/27
Cattle	1/27	3/72	2/103
Cat	0/609	7/1,069	15/1,503
Dog	0/603	0/750	1/801
Goat/sheep/pig	0/12	0/23	0/34

a. Tabulate and/or plot these results in any way that seems helpful.

b. Discuss these results from the point of view of a Maryland public health worker.

EXERCISE 5-26

Schistosomiasis is a parasitic infection common in the tropics. The parasites are carried by snails and passed to humans through contact with water (as in lakes, rivers, and irrigation canals). In a study of schistosomiasis in an Egyptian village, residents were examined and tested for presence of two forms of the parasite, Schistosoma mansoni and Schistosoma haematobium (Ismail et al., 1988). Researchers classified a total of 1,031 villagers by presence of one or both parasites:

Infection with Schistosoma haematobium	Infection with Schistosoma mansoni	
	Yes	No
Yes	119	46
No	213	653

The villagers were also classified by presence of infection and occupation:

Occupation	Number examined	Number infected with Schisto-soma haema-tobium	Number infected with Schisto-soma mansoni
Small child (not yet in school)	169	3	10
Student	382	99	162
Housewife/girl	241	15	34
Farmer	222	47	124
Nonfarm worker	17	0	2

In addition, the villagers were classified by age, sex, and presence of infection:

Age (years)	Number examined	Number infected with Schistosoma haematobium	Number infected with Schistosoma mansoni	Age (years)	Number examined	Number infected with Schistosoma haematobium	Number infected with Schistosoma mansoni
	Females				**Males**		
< 5	63	2	3	< 5	75	0	4
5–15	183	19	48	5–15	235	89	134
16–25	107	13	18	16–25	82	30	57
26–45	101	2	6	26–45	84	5	37
> 45	47	1	1	> 45	54	4	24

a. Tabulate and/or plot these results in any way that seems helpful.

b. Does there appear to be an association between infection with one schistosomiasis parasite and infection with the other?

c. For each of the two parasites, do the infection rates differ for different occupations? Is the association between infection and occupation different for the two parasites?

d. For each of the two parasites, are the infection rates different for males and females? Is the association between infection and sex different for the two parasites?

e. For each of the two parasites, do infection rates depend on age? Is the association between infection and age different for the two parasites?

f. For each of the two parasites, is the age effect on infection different for males and females? This is the same as asking if the sex effect on infection is different for different age groups.

EXERCISE 5-27

Self-reported cigarette smoking among Rhode Island physicians is shown here by specialty and year. The table shows number of smokers/number of physicians (percent smokers). (Numbers of smokers were calculated from percentages reported in Buechner et al., 1986.)

	1963	1968	1973	1978	1983
Internal medicine	31/113 (27.4)	24/158 (15.2)	21/229 (9.2)	31/352 (8.8)	30/496 (6.0)
General and family practice	59/171 (34.5)	59/274 (21.5)	37/215 (17.2)	21/167 (12.6)	19/227 (8.4)
Surgery	32/100 (32.0)	33/130 (25.4)	38/150 (25.3)	23/164 (14.0)	13/154 (8.4)
Pediatrics	14/60 (23.3)	13/78 (16.7)	19/92 (20.7)	12/119 (10.1)	10/140 (7.1)
Obstetrics and gynecology	26/57 (45.6)	21/65 (32.3)	23/82 (28.0)	18/94 (19.1)	16/111 (14.4)
Orthopedic surgery	11/27 (40.7)	12/50 (24.0)	11/58 (19.0)	6/64 (9.4)	9/72 (12.5)

a. Plot percent cigarette smokers versus year for each of the six specialties.

b. What is the trend in percent cigarette smokers over time?

c. Does the time trend vary with specialty?

EXERCISE 5-28

Indicators related to quality of life and children's health are listed for 29 countries (Grant, 1987). Malnutrition is the percentage of children under 5 years of age suffering from mild to moderate malnutrition (60% to 80% of desirable weight), 1980–1984. Water is the percentage of the population with access to drinking water in 1983. Polio is the percentage of 1-year-old children fully immunized against polio, 1984–1985. Low weight is the percentage of infants born weighing less than 2,500 grams (5.5 pounds) in 1982–1983. Breastfeeding is the percentage of mothers wholly or partially breastfeeding their babies for at least 6 months, 1980–1984.

Country	Malnu-trition	Water	Polio	Low weight	Breast-feeding
Sierra Leone	24	23	12	14	94
Malawi	30	51	55	10	95
Niger	17	34	4	20	30
Rwanda	29	60	50	17	98
Yemen	54	31	8	9	76
Yemen, Dem.	32	50	14	12	73
Burundi	30	26	29	14	95
Bangladesh	63	42	2	50	97
Sudan	53	48	9	15	86
Bolivia	49	43	46	10	91
Haiti	65	33	24	17	85
Uganda	16	16	8	10	70
Pakistan	62	39	32	27	96
Ghana	23	43	18	17	70
Egypt	46	75	67	7	91
Peru	42	52	48	9	72
Indonesia	27	33	25	14	97
Congo	30	29	59	15	97
Kenya	30	28	57	13	84
Honduras	29	69	75	9	28
Brazil	55	76	86	9	19
Burma	50	25	2	20	90
El Salvador	52	51	55	13	77
Philippines	40	54	53	14	58
Colombia	43	81	61	10	58
Thailand	29	65	65	12	47
Panama	48	62	71	8	48
Chile	10	85	91	9	28
Costa Rica	46	93	74	9	20

a. Construct a stem-and-leaf plot for each of the five variables. Describe the distribution of each variable.

b. Construct a scatterplot matrix with these five variables.

c. Discuss the relationships between pairs of variables that you see in these plots.

EXERCISE 5-29

Baseball statistics summarizing the 1987 season are listed for 26 major-league teams (*USA Today,* October 6, 1987, page 4c; and October 7, 1987, page 5c). St. Louis, San Francisco, Detroit, and Minnesota were division winners in 1987. St. Louis and Minnesota were playoff winners. Minnesota won the World Series. League is a coded variable: 1 = American League, 2 = National League. Div is a coded variable: 1 = Eastern Division, 2 = Western Division. DivWin shows division winners: 1 = yes, 0 = no. POWin shows the playoff winners: 1 = yes, 0 = no. WSWin shows the World Series winner: 1 = yes, 0 = no. Win% is the percentage of games won by the team, to three decimal places. Runs is the number of runs scored by the team. RunAllow is the number of runs allowed by the team. HR is the number of homeruns hit by the team. Walks is the number of walks or bases on balls received by the team. BatAve is the proportion of at-bats that resulted in hits for the team. ERA is the team's earned run average, the average number of runs allowed by the team's pitching staff per nine innings of play. HitA is the number of hits allowed by the team. HRAllow is the number of homeruns allowed by the team. WalkA is the number of walks or bases on balls allowed by the team. SO is the number of batters struck out by the team's pitching staff.

Team	League	Div	DivWin	POWin	WSWin	Win%	Runs	RunAllow
St. Louis	2	1	1	1	0	.586	798	693
NY Mets	2	1	0	0	0	.568	823	698
Montreal	2	1	0	0	0	.562	741	720
San Francisco	2	2	1	0	0	.556	783	669
Cincinnati	2	2	0	0	0	.519	783	752
Philadelphia	2	1	0	0	0	.494	702	749
Pittsburgh	2	1	0	0	0	.494	723	744
Chicago Cubs	2	1	0	0	0	.472	720	801
Houston	2	2	0	0	0	.469	648	678
Los Angeles	2	2	0	0	0	.451	635	675
Atlanta	2	2	0	0	0	.429	747	829
San Diego	2	2	0	0	0	.401	668	763
Detroit	1	1	1	0	0	.605	896	735
Toronto	1	1	0	0	0	.593	845	655
Milwaukee	1	1	0	0	0	.562	862	817
NY Yankees	1	1	0	0	0	.549	788	758
Minnesota	1	2	1	1	1	.525	786	806
Kansas City	1	2	0	0	0	.512	715	691
Oakland	1	2	0	0	0	.500	806	789
Boston	1	1	0	0	0	.481	842	825
Seattle	1	2	0	0	0	.481	760	801
Chicago W. Sox	1	2	0	0	0	.475	748	746
Texas	1	2	0	0	0	.463	823	849
California	1	2	0	0	0	.463	770	803
Baltimore	1	1	0	0	0	.414	729	880
Cleveland	1	1	0	0	0	.377	742	957

a. Select a few of these variables that interest you. Plot these variables overall, by league, and by division within league.

b. Use a scatterplot matrix to examine relationships between pairs of variables.

c. Discuss your findings.

EXERCISE 5-30

Information on racial composition, fires and thefts, age of housing, homeowners insurance availability, and median family income is presented for each of 47 zip code areas of Chicago. The data were collected as part of a study to investigate possible discrimination in homeowners insurance underwriting practices in Chicago. From a report published by the U.S. Commission on Civil Rights (U.S. Commission on Civil Rights, 1979), this data set was contributed by S. E. Fienberg to a collection of problems edited by Andrews and Herzberg (1985, pages 407–411). Zip is the zip code area. Minor is the percentage of the population of minority racial background (provided by the U.S. Bureau of the Census). Fires gives the number of fires per 1,000 housing units during 1975 (provided by the Chicago Fire Department). Thefts gives the number of thefts per 1,000 population in 1975 (provided by the Chicago Police Department). Old is the percentage of housing units built before 1940 (provided by the U.S. Bureau of the Census). Vol is the number of new (voluntary) homeowners policies and renewals less nonrenewals and cancellations, per 100 housing

Team	HR	Walks	BatAve	ERA	HitA	HRAllow	WalkA	SO
St. Louis	94	644	.263	3.91	1,484	129	533	873
NY Mets	192	592	.268	3.84	1,407	135	510	1,032
Montreal	120	501	.265	3.92	1,428	145	446	1,012
San Francisco	205	511	.260	3.68	1,407	146	547	1,038
Cincinnati	192	514	.266	4.24	1,486	170	485	919
Philadelphia	169	587	.254	4.18	1,453	167	587	877
Pittsburgh	131	535	.264	4.20	1,377	164	562	914
Chicago Cubs	209	504	.264	4.55	1,524	159	628	1,024
Houston	122	526	.253	3.84	1,363	141	525	1,137
Los Angeles	125	445	.252	3.72	1,415	130	565	1,097
Atlanta	152	641	.258	4.63	1,529	163	587	837
San Diego	113	577	.260	4.27	1,402	175	602	897
Detroit	225	653	.272	4.02	1,430	180	563	976
Toronto	215	555	.269	3.74	1,323	158	567	1,064
Milwaukee	163	598	.276	4.62	1,548	169	529	1,039
NY Yankees	196	604	.262	4.36	1,475	179	542	900
Minnesota	196	523	.261	4.63	1,465	210	564	990
Kansas City	168	523	.262	3.86	1,424	128	548	923
Oakland	199	593	.260	4.32	1,442	176	531	1,042
Boston	174	606	.278	4.77	1,584	190	517	1,034
Seattle	161	500	.272	4.49	1,503	199	497	919
Chicago W. Sox	173	487	.258	4.30	1,436	189	537	792
Texas	194	567	.266	4.63	1,388	199	760	1,103
California	172	590	.252	4.38	1,481	212	504	941
Baltimore	211	524	.258	5.01	1,555	226	547	870
Cleveland	187	489	.263	5.28	1,566	219	606	849

Zip	Minor	Fires	Thefts	Old	Vol	Invol	Income
60626	10.0	6.2	29	60.4	5.3	.0	11,744
60640	22.2	9.5	44	76.5	3.1	.1	9,323
60613	19.6	10.5	36	73.5	4.8	1.2	9,948
60657	17.3	7.7	37	66.9	5.7	.5	10,656
60614	24.5	8.6	53	81.4	5.9	.7	9,730
60610	54.0	34.1	68	52.6	4.0	.3	8,231
60611	4.9	11.0	75	42.6	7.9	.0	21,480
60625	7.1	6.9	18	78.5	6.9	.0	11,104
60618	5.3	7.3	31	90.1	7.6	.4	10,694
60647	21.5	15.1	25	89.8	3.1	1.1	9,631
60622	43.1	29.1	34	82.7	1.3	1.9	7,995
60631	1.1	2.2	14	40.2	14.3	.0	13,722
60646	1.0	5.7	11	27.9	12.1	.0	16,250
60656	1.7	2.0	11	7.7	10.9	.0	13,686
60630	1.6	2.5	22	63.8	10.7	.0	12,405
60634	1.5	3.0	17	51.2	13.8	.0	12,198
60641	1.8	5.4	27	85.1	8.9	.0	11,600
60635	1.0	2.2	9	44.4	11.5	.0	12,765
60639	2.5	7.2	29	84.2	8.5	.2	11,084
60651	13.4	15.1	30	89.8	5.2	.8	10,510
60644	59.8	16.5	40	72.7	2.7	.8	9,784
60624	94.4	18.4	32	72.9	1.2	1.8	7,342
60612	86.2	36.2	41	63.1	.8	1.8	6,565
60607	50.2	39.7	147	83.0	5.2	.9	7,459
60623	74.2	18.5	22	78.3	1.8	1.9	8,014
60608	55.5	23.3	29	79.0	2.1	1.5	8,177
60616	62.3	12.2	46	48.0	3.4	.6	8,212
60632	4.4	5.6	23	71.5	8.0	.3	11,230
60609	46.2	21.8	4	73.1	2.6	1.3	8,330
60653	99.7	21.6	31	65.0	.5	.9	5,583
60615	73.5	9.0	39	75.4	2.7	.4	8,564
60638	10.7	3.6	15	20.8	9.1	.0	12,102
60629	1.5	5.0	32	61.8	11.6	.0	11,876
60636	48.8	28.6	27	78.1	4.0	1.4	9,742
60621	98.9	17.4	32	68.6	1.7	2.2	7,520
60637	90.6	11.3	34	73.4	1.9	.8	7,388
60652	1.4	3.4	17	2.0	12.9	.0	13,842
60620	71.2	11.9	46	57.0	4.8	.9	11,040
60619	94.1	10.5	42	55.9	6.6	.9	10,332
60649	66.1	10.7	43	67.5	3.1	.4	10,908
60617	36.4	10.8	34	58.0	7.8	.9	11,156
60655	1.0	4.8	19	15.2	13.0	.0	13,323
60643	42.5	10.4	25	40.8	10.2	.5	12,960
60628	35.1	15.6	28	57.8	7.5	1.0	11,260
60627	47.4	7.0	3	11.4	7.7	.2	10,080
60633	34.0	7.1	23	49.2	11.6	.3	11,428
60645	3.1	4.9	27	46.6	10.9	.0	13,731

Adapted from Figure 67.1 of *Data: A Collection of Problems from Many Fields for the Student and Research Worker*, by D. F. Andrews and A. M. Herzberg.

units, from December 1977 through February 1978 (provided by insurance companies to the Illinois Department of Insurance). Invol is the number of new fair-plan (involuntary) homeowners insurance policies and renewals per 100 housing units, from December 1977 through May 1978 (provided by the Illinois Department of Insurance); most fair-plan policyholders were rejected for voluntary policies by homeowners insurance companies. Income is the median family income for the area as estimated by the U.S. Bureau of the Census.

a. Plot each of the variables (other than zip code). Describe the distribution of each variable.

b. Select a few of these variables that interest you. Construct a scatterplot matrix using these variables. Discuss the relationship between pairs of variables revealed in your scatterplot matrix.

c. Using the accompanying map of Chicago zip code areas, use framed rectangles to display by zip code area any variable(s) you wish. You may wish to use separate graphs for different variables. Are there geographic trends?

d. Display these variables in any other way you find useful. Discuss your results.

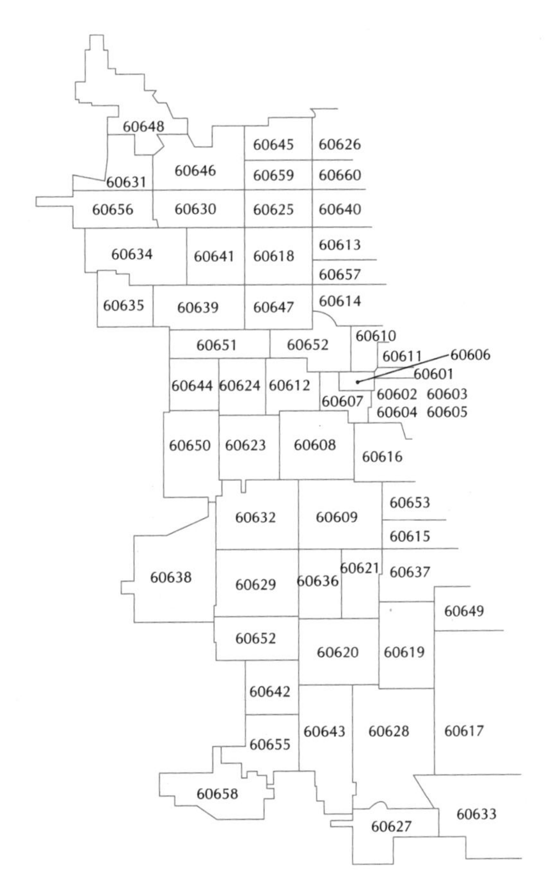

EXERCISE 5-31 Use framed rectangles on a map of the United States to display the pregnancy rates for teenage girls given in Exercise 2-8. Discuss any geographic trends you see.

EXERCISE 5-32 Use framed rectangles on a map of the United States to display the record cold temperatures listed in Exercise 2-4. Discuss any geographic trends you see.

EXERCISE 5-33 Use framed rectangles on a map of the United States to display the governors' salaries listed in Exercise 2-19. Discuss any geographic trends you see.

EXERCISE 5-34 District scores for third graders in reading, mathematics, and science are shown here for 27 Massachusetts urbanized centers (*The Boston Sunday Globe,* November 30, 1986, page 96).

Urbanized center	Reading	Mathematics	Science
Attleboro	1290	1250	1300
Boston	1190	1210	1180
Brockton	1200	1190	1180
Cambridge	1250	1250	1230
Chelsea	1110	1130	1110
Chicopee	1280	1230	1250
Everett	1190	1220	1190
Fall River	1170	1190	1190
Fitchburg	1260	1240	1270
Gloucester	1240	1250	1260
Greenfield	1300	1340	1360
Haverhill	1260	1240	1250
Holyoke	1150	1170	1150
Lawrence	1100	1110	1110
Lowell	1220	1250	1230
Lynn	1200	1210	1190
Malden	1260	1230	1250
New Bedford	1210	1220	1190
Pittsfield	1270	1270	1280
Quincy	1290	1270	1290
Revere	1260	1220	1220
Salem	1240	1300	1260
Somerville	1190	1190	1190
Springfield	1210	1220	1210
Waltham	1310	1310	1290
Watertown	1340	1360	1330
Worcester	1250	1260	1250

Graph these observations in any ways that you find helpful. Discuss your findings.

EXERCISE 5-35 The United Nations ranks countries according to mortality rate for children under 5 years of age (Grant, 1987). The 20 countries with the highest under-5 mortality rates and the 20 with the lowest under-5 mortality rates are listed, along with values for four variables. Mort is under-5 mortality rate in 1985:

annual number of deaths of children under 5 years of age per 1,000 live births. Life is life expectancy: the number of years newborn children could be expected to live based on prevailing mortality risks in 1985. Weight is percentage of infants born weighing less than 2,500 grams (5.5 pounds) in 1982–1983. Calorie is the daily calorie supply per capita as percentage of requirements in 1983. A −9 denotes a missing value.

Country	Mort	Life	Weight	Calorie
Afghanistan	329	38	20	−9
Mali	302	43	13	68
Sierra Leone	302	35	14	91
Malawi	275	46	10	95
Guinea	259	41	18	84
Ethiopia	257	41	13	93
Somalia	257	41	−9	89
Mozambique	252	46	16	71
Burkina Faso	245	46	21	85
Angola	242	43	19	87
Niger	237	44	20	97
Central African Republic	232	44	23	91
Chad	232	46	11	68
Guinea-Bissau	232	44	15	−9
Senegal	231	44	10	102
Mauritania	223	45	10	97
Kampuchea	216	46	−9	−9
Liberia	215	50	−9	102
Rwanda	214	48	17	98
Yemen	210	50	9	92
Belgium	13	74	5	140
German Dem. Rep.	13	73	6	142
Italy	13	75	7	140
USA	13	75	7	137
Germany, Federal Republic of	12	74	5	130
Ireland	12	73	4	143
Singapore	12	73	8	115
Spain	12	75	−9	132
United Kingdom	12	74	7	128
Australia	11	75	6	115
France	11	75	5	139
Hong Kong	11	76	8	122
Canada	10	76	6	130
Denmark	10	75	6	131
Netherlands	10	76	4	129
Norway	10	76	4	115
Japan	9	77	5	113
Switzerland	9	76	5	129
Finland	8	74	4	114
Sweden	8	76	4	116

Display these observations in any ways that you find helpful. Discuss your findings.

CHAPTER 6

Some Ideas in Probability Needed for Statistical Inference

IN THIS CHAPTER

Experiment, outcome, sample space, event
Probability function
Conditional probability, Bayes' rule
Independence, independent events, dependent events
Random variable, probability distribution of a random variable
Mean or expected value, variance, standard deviation of a random variable

So far we have used tools of data analysis to learn about a collection of information. In particular, we applied these tools in an exploratory study of the World Bank indicators. For such an analysis to make any sense, we have to make some assumptions about the data set. We must assume that the variables (such as the World Bank indicators) are measured and recorded correctly, for instance. When we know or suspect that these assumptions are not met (as in possible deliberate misreporting of World Bank indicators by member countries), we must be careful in our interpretations of results.

In formal statistical inference, we go beyond the goals of data analysis. We want to use samples to learn about larger populations. The tools of data analysis can help us do this. In addition, however, we want to use our observations to make probability statements about the larger populations. To make such probability statements, we must assume more about the observations than is necessary for data analysis.

Suppose, for example, that we are medical researchers. We want to compare a new treatment with a standard treatment in curing patients with a disease. We treat ten patients with the new treatment and ten patients with the standard treatment. Eight patients receiving the new treatment are cured, while three patients receiving the standard treatment are cured. How can we use the results of this experiment to make a statement comparing the cure rates for the two treatments? (By cure rate, we mean the proportion of patients cured.)

Since the numerical results of this experiment are so simple, data analysis amounts to reporting that eight of ten patients were cured under the new treatment and three of ten patients were cured under the standard treatment. (We could present this information in a frequency table.) In formal statistical inference, we might ask this question: If there were really no difference in cure rates between the new and standard treatments, would we be surprised by our observed experimental results? Or: How *likely* is it that we would *by chance* have observed a difference at least as extreme as eight out of ten cured versus three out of ten cured, if the two treatments really had the same cure rate? The answer to this question helps us make a statement comparing the two treatments. (Formal statistical analysis of this type of experiment is the subject of Sections 16-4 and 16-5.)

When we use the words *likely* and *by chance,* as in the question above, we are dealing with probabilities. Even the term *cure rate* refers to the likelihood or probability of cure in a group of patients.

We must make some assumptions about our experiment for any statements involving probabilities to make sense. We must assume, for instance, that the patients receiving the new treatment are similar to the patients receiving the standard treatment. Then, if there were really no difference in cure rates between the two treatments, we would expect a similar number of patients cured in each group. If the experimental results show a striking difference between proportions cured in the two groups, we might doubt that the two treatments are really similar.

In general, statistical inference involves making probability statements about populations based on what we observe in our samples. All formal statistical analysis depends on making assumptions that justify these inferences. When we design an experiment to meet the assumptions needed later for formal analysis of the results, we are in the realm of experimental design, an important part of statistics that we will address in Part III. The ideas in probability that we need for formal statistical inference are the subject of discussion here in Part II.

6-1

Probability as Chances

Probability is a part of everyday life. Buying a state lottery ticket, we like to know our chance of winning. We may guess the odds that the hometown team will win the next game. If a screening test comes up positive for the acquired immune deficiency syndrome (AIDS), we would like to know the probability that we really do have AIDS.

On radio and television weather reports, we hear statements such as "partly cloudy today with a 50% chance of showers" or "the probability of rain today is 90%." What does it mean to say the chance of rain is 90%? We address this question in Example 6-1.

EXAMPLE 6-1

When weather forecasters say there is a 90% probability of rain, they mean that the chances are 9 in 10 that at least .01 inch (or .2 millimeter) of rain will fall during the forecast period (usually 12 hours) at any given point in the forecast area (National Weather Service *Operations Manual,* 1984; kindly provided by Joe Bocchieri and Keith Seitter). If you live in an area receiving a 90% probability of rain forecast, then over the next 12 hours there are 9 chances in 10 that your location will receive at least .01 inch of rainfall.

A 50% chance of snow means that over the forecast period, any given point in the forecast area has 1 chance in 2 of receiving at least .01 inch water-equivalent of snow.

Ranges for probability of precipitation forecasts are from near 100% down to 20%. With probabilities below 20%, precipitation is usually not mentioned in the main part of a weather forecast.

The *probability* of an event is the chance or likelihood of the event occurring. A weather forecaster estimates the chance of precipitation. A lottery player wants to know the likelihood that her selection will win. A cancer researcher may estimate the probability that a male cigarette smoker will develop lung cancer. Figuring the chance of a serious nuclear reactor accident might be the task of a government regulatory commission. An oddsmaker assesses the likelihood that the Lakers will win four of a possible seven games in the National Basketball Association championship finals.

The **probability** of an event is the chance or likelihood of the event occurring.

Some definitions that help us discuss probability more formally are given in Section 6-2. We will illustrate the terms defined with some simple examples in probability.

6-2

Experiment, Outcome, Sample Space, Events

By an *experiment* we will mean a process leading to a well-defined observation, called the experimental *outcome*. The *sample space* is the set of all possible outcomes of the experiment.

An **experiment** is a process leading to a well-defined observation or outcome.

The **sample space** is the set of all possible outcomes of the experiment.

EXAMPLE 6-2

In one version of a state lottery called Megabucks, a player purchases a ticket after selecting six different integers from 1 to 36. Part of the ticket price goes into the state treasury and the other part into the pot. On drawing day, state officials select six integers from 1 to 36. Players holding tickets with the winning six numbers share the pot. This experiment involves selection of six different integers from 1 to 36. An observation or outcome is a particular combination of six numbers. The sample space is the set of all possible combinations of six integers from 1 to 36.

EXAMPLE 6-3

A researcher carries out this experiment as part of a study of cigarette smoking and lung cancer. She selects a male cigarette smoker at random from among all male cigarette smokers. She then keeps in touch with him until he either develops lung cancer or dies with no evidence of lung cancer. The sample space for this experiment contains two possible outcomes: either the man develops lung cancer or the man dies with no evidence of lung cancer. (The idea of random selection is one we will encounter when we discuss experimental design and statistical inference. *Random selection* of an individual from a population means that each member of the population has an equal chance of being selected.)

EXAMPLE 6-4

The New York Knicks and the Phoenix Suns are meeting in the National Basketball Association championship series. The team that wins four of a possible seven games wins the championship. The experiment involves the two teams playing until one team has won four games. An observation or outcome is a particular sequence of wins by the two teams, with one team winning four games. The sample space consists of all possible outcomes of the series.

We will consider two types of sample spaces: finite sample spaces and continuous sample spaces. A *finite sample space* contains a finite number of outcomes. The sample spaces in Examples 6-2, 6-3, and 6-4 are all finite sample spaces.

A **finite sample space** is a sample space that contains a finite number of outcomes.

A *continuous sample space* equals an interval of values. The next three examples describe some continuous sample spaces.

A **continuous sample space** is a sample space that equals an interval of values.

EXAMPLE 6-5

For this experiment, we time the length of a Friday morning statistics class. The class is scheduled to meet for 50 minutes on Fridays beginning at 10:30 A.M. The class must be out of the room by 11:25 A.M. at the latest. Therefore, the outcome or length of the class meeting can be from 0 to 55 minutes. The sample space is the interval [0, 55] consisting of all real numbers from 0 to 55, a continuous sample space. (In interval notation, brackets indicate that the endpoints are included in the interval.)

EXAMPLE 6-6

In a life-testing experiment, an engineer tests a computer chip until failure. The outcome is the time in hours from start of testing until failure. In theory, any number greater than or equal to 0 is possible. Therefore, the sample space consists of all real numbers greater than or equal to 0, denoted by the interval $[0, \infty)$. (In interval notation, a parenthesis indicates that the endpoint is not included in the interval.) This is a continuous sample space.

EXAMPLE 6-1
(continued)

You measure the rainfall at your home over a 12-hour period. The outcome is the amount of rainfall, in inches. The sample space consists of numbers greater than or equal to 0, a continuous sample space.

Some sample spaces are neither finite nor continuous according to our definitions. Suppose, for instance, that our experiment consists of counting the number of defects on a copper sheet for use in personal computer hardware. The sample space contains all possible outcomes of this experiment: 0, 1, 2, 3, and so on. Since this sample space contains all the nonnegative integers, it is a discrete, countably infinite sample space. Although very useful for many applications, we will not consider such sample spaces in this book.

An *event* is a subset of the sample space. One event in the Megabucks lottery, Example 6-2, consists of the outcomes with all six numbers greater than 3. Another event consists of all outcomes that include the number 30. In the basketball championship, Example 6-4, one event consists of outcomes with the series ending in five games. Another event consists of all outcomes in which the Knicks win the series.

An **event** is a subset of the sample space.

When the sample space is continuous, we generally consider events that are continuous, or intervals, as well. For the statistics class in Example 6-5, one event includes all outcomes when the class runs long. This event includes class times longer than 50 minutes, represented by the interval $(50, 55]$. The interval $[49, 51]$ denotes the event that the class lasts from 49 to 51 minutes. (Recall, in interval notation, a parenthesis indicates that the endpoint is not included in the interval; a bracket, that the endpoint is included in the interval.)

One event in the life-testing experiment in Example 6-6 is that the computer chip lasts more than 2,000 hours, denoted by the interval $(2{,}000, \infty)$. The event that the computer chip fails within the first 1,000 hours is represented by the interval $[0, 1{,}000]$.

If you measure less than half an inch of rainfall in Example 6-1, the event is denoted by the interval $[0, .5)$.

In Section 6-3, we define a probability function for a finite sample space. We consider probability functions for continuous sample spaces in Chapter 8 and in Part III. Probability functions formalize our intuitive notions about probability, and help us build the probability models we need for statistical inference.

6-3

Probability Functions

A *probability function* is a rule that describes how we assign probabilities or chances to events.

> A **probability function** assigns a unique number or probability to each outcome in a finite sample space S. The probability of each outcome is greater than or equal to 0. The sum of the probabilities of all the outcomes in S equals 1. The probability of an event E, denoted by $P(E)$, is the sum of the probabilities of all the outcomes in E.

From the definition of a probability function, we see that the probability of the entire sample space is 1:

$$\text{If } S \text{ denotes the sample space, then } P(S) = 1.$$

The probability that an event E does not occur equals 1 minus the probability that the event E does occur:

$$\text{For any event } E,\ P(\text{not } E) = 1 - P(E).$$

We sometimes find it useful to consider the probability that at least one of several events occur. The probability that at least one of several events occur equals 1 minus the probability that none of the events occur.

For any collection of events:
$P(\text{at least one of the events occur}) = 1 - P(\text{none of the events occur})$

Also, the probability that at least one of several events occur is less than or equal to the sum of the individual probabilities of the events:

For any collection of events:
P(at least one of the events occur)
$\leq$ sum of the individual probabilities of the events

Suppose that several events have no outcomes in common; the experiment can result in at most one of these events. Then the probability that at least one of the events occur equals the sum of the individual probabilities of the events:

If several events have no outcomes in common, then:
P(at least one of the events occur)
= sum of the individual probabilities of the events

Note that if several events have no outcomes in common, then *at least one occurs* is the same as *exactly one occurs.*

A sample space S and a probability function P for S provide a *probability model* for an experiment. Examples 6-7 and 6-8 illustrate two different probability models for the same experiment.

A sample space S and a probability function P defined for S provide a **probability model** for an experiment.

EXAMPLE 6-7

The experiment consists of tossing a coin three times and noting for each toss whether the coin comes up heads or tails. For a single toss, let H indicate that the coin lands heads up and T that the coin lands tails up. Assume that when the coin is tossed it will come up either heads or tails and not, for example, land on its side and roll away where we cannot see it.

We can denote an outcome of the three coin tosses by an ordered triple. The first, second, and third elements of the ordered triple show the results of the first, second, and third tosses, respectively. For instance, the ordered triple (H, T, H) denotes one possible outcome of the experiment: heads on the first toss, tails on the second toss, heads on the third toss.

The sample space S contains eight possible outcomes, listed in the first column of Table 6-1. Suppose we think of these outcomes as resulting from independent tosses of a fair coin. (By a *fair coin,* we mean heads and tails are equally likely. By *independent tosses,* we mean the outcome of one toss does not in any way affect the outcome of another toss.) Then under the resulting probability model, each outcome in S is equally likely. For this model, the probability function P assigns the number $\frac{1}{8}$ to each of the eight outcomes in S.

Let E denote the event that all three tosses result in heads. Then E contains exactly one outcome and we write $E = \{(H, H, H)\}$, using braces to enclose the outcomes in the event. According to the probability model in Table 6-1, $P(E) = \frac{1}{8}$.

If event E does not occur, then there is at least one tail in the three tosses. From the relation $P(\text{not } E) = 1 - P(E)$, we see that

$$P(\text{at least one tail}) = 1 - P(\text{no tails}) = 1 - P(E) = 1 - \frac{1}{8} = \frac{7}{8}$$

Suppose A_1 denotes the event that heads come up on the first two tosses, A_2 the event that heads come up on the second and third tosses, and A_3 the event that heads come up on the first and third tosses. Table 6-2 shows the outcomes in these events. Under the probability model in Table 6-1, each of events A_1, A_2, and A_3 has probability $\frac{1}{4}$. The probability that at least one of these three events occur is less than or equal to the sum of the individual probabilities of the three events:

$$P(\text{at least one of events } A_1, A_2, \text{ and } A_3 \text{ occur}) \le P(A_1) + P(A_2) + P(A_3)$$

The event that at least one of A_1, A_2, and A_3 occur is the same as the event that at least two tosses result in heads. Therefore,

$$P(\text{at least two heads}) \le P(A_1) + P(A_2) + P(A_3) = \frac{1}{4} + \frac{1}{4} + \frac{1}{4} = \frac{3}{4}$$

In this simple example, it is easy to find the exact probability of at least two heads to be $\frac{1}{2}$.

TABLE 6-1 With this probability model, each possible outcome of three coin tosses is equally likely.

Outcome	Probability of the outcome
(H, H, H)	$\frac{1}{8}$
(H, H, T)	$\frac{1}{8}$
(H, T, H)	$\frac{1}{8}$
(T, H, H)	$\frac{1}{8}$
(H, T, T)	$\frac{1}{8}$
(T, H, T)	$\frac{1}{8}$
(T, T, H)	$\frac{1}{8}$
(T, T, T)	$\frac{1}{8}$

TABLE 6-2 Some possible events resulting from three coin tosses and their probabilities based on the probability model in Table 6-1

Description	Event	Probability of the event
Heads on the first two tosses	A_1 = {(H, H, H), (H, H, T)}	$P(A_1) = \frac{1}{4}$
Heads on the second and third tosses	A_2 = {(H, H, H), (T, H, H)}	$P(A_2) = \frac{1}{4}$
Heads on the first and third tosses	A_3 = {(H, H, H), (H, T, H)}	$P(A_3) = \frac{1}{4}$
Exactly one tail	B_1 = {(H, H, T), (H, T, H), (T, H, H)}	$P(B_1) = \frac{3}{8}$
Exactly two tails	B_2 = {(T, T, H), (T, H, T), (H, T, T)}	$P(B_2) = \frac{3}{8}$
Three tails	B_3 = {(T, T, T)}	$P(B_3) = \frac{1}{8}$

Let B_1 denote the event that exactly one tail comes up, B_2 the event that exactly two tails come up, and B_3 the event that exactly three tails come up. The outcomes in B_1, B_2, and B_3 are listed in Table 6-2. Under the probability model in Table 6-1, each of B_1 and B_2 has probability $\frac{3}{8}$ and B_3 has probability $\frac{1}{8}$.

The three events B_1, B_2, and B_3 have no outcomes in common. Therefore, the probability that at least one of these three events occur is equal to the sum of the individual probabilities of the three events:

$$P(\text{at least one of } B_1, B_2, \text{and } B_3 \text{ occur}) = P(B_1) + P(B_2) + P(B_3)$$

The event that at least one of B_1, B_2, and B_3 occur is the same as the event that there is at least one tail. Therefore,

$$P(\text{at least one tail}) = P(B_1) + P(B_2) + P(B_3) = \frac{3}{8} + \frac{3}{8} + \frac{1}{8} = \frac{7}{8}$$

This is the same answer we obtained for the probability of at least one tail in our earlier calculation.

Let's consider now an alternative probability model for the three-coin-toss experiment.

EXAMPLE 6-8

We toss a coin three times. This time, however, the tosses are not all independent of one another. If the first two tosses result in one head and one tail, then the three-toss outcome has probability $\frac{1}{8}$. However, if heads come up on the first two tosses, then heads are sure to come up on the third toss. Similarly, if tails come up on the first two tosses, then tails will come up on the third. The probability model for this experiment is shown in Table 6-3.

Two of the eight outcomes in Table 6-3 have probability 0. We can define the sample space by including only outcomes with probabilities greater than 0, as in Table 6-4.

TABLE 6-3 The probability model for the three-coin-toss experiment described in Example 6-8

Outcome	Probability of the outcome
(H, H, H)	$\frac{1}{4}$
(H, H, T)	0
(H, T, H)	$\frac{1}{8}$
(H, T, T)	$\frac{1}{8}$
(T, H, H)	$\frac{1}{8}$
(T, H, T)	$\frac{1}{8}$
(T, T, H)	0
(T, T, T)	$\frac{1}{4}$

TABLE 6-4 An alternative description of the probability model for the three-coin-toss experiment in Example 6-8

Outcome	Probability of the outcome
(H, H, H)	$\frac{1}{4}$
(H, T, H)	$\frac{1}{8}$
(H, T, T)	$\frac{1}{8}$
(T, H, H)	$\frac{1}{8}$
(T, H, T)	$\frac{1}{8}$
(T, T, T)	$\frac{1}{4}$

As Example 6-8 illustrates, we can often write the sample space for an experiment in more than one way. Also, the outcomes in a finite sample space need not be equally likely.

Odds or odds ratios are used in a variety of fields, including sports, gambling, and public health. In Section 6-4, we define what we mean by the odds of an event occurring. This section is optional; it is not required to understand any other concepts we will discuss.

6-4 The Odds of an Event (Optional)

Odds ratios depend on the probability model for an experiment. The *odds of an event* is the probability that the event occurs divided by the probability that the event does not occur:

$$\text{For any event } E, \textbf{ odds of event } \boldsymbol{E} = \frac{P(E)}{P(\text{not } E)} = \frac{P(E)}{1 - P(E)}$$

We illustrate the use of the odds ratio with two examples.

EXAMPLE 6-7 *(continued)*

We toss a fair coin three times, with the probability model shown in Table 6-1. Let A denote the event that at least one head comes up in the three tosses. Then A contains seven outcomes and $P(A) = \frac{7}{8}$. The odds of at least one head appearing is the odds of event A:

$$\text{Odds of at least one head} = \frac{P(A)}{1 - P(A)} = \frac{7/8}{1/8} = 7$$

We can denote this odds ratio by 7:1. The interpretation is that over many repetitions of the three-coin-toss experiment, we expect at least one head about seven times for every one time that no heads come up.

If B is the event that a head appears on the first toss, then B contains four outcomes and $P(B) = \frac{1}{2}$. The odds of a head on the first toss is the odds of event B:

$$\text{Odds of a head on the first toss} = \frac{P(B)}{1 - P(B)} = \frac{1/2}{1/2} = 1$$

This 1:1 ratio indicates that in three independent tosses of a fair coin, a head on the first toss is as likely as a tail on the first toss.

EXAMPLE 6-4 *(continued)*

The Houston Rockets and the Chicago Bulls are meeting in the National Basketball Association championship finals. An oddsmaker lays the odds of the Bulls winning the series as 3:2 or 3 to 2 in favor of the Bulls. According to this oddsmaker, what is the probability that the Bulls will win the championship? We can write the 3:2 odds ratio as

$$\frac{P(\text{Bulls win})}{P(\text{Rockets win})} = \frac{3}{2}$$

The oddsmaker is predicting that the Bulls have 3 chances of winning, compared with 2 chances for the Rockets. Since the Bulls are given 3 chances out of a total of 5 (three chances for the Bulls plus two chances for the Rockets), he believes that

$$P(\text{Bulls win}) = \frac{3}{5} = .6 \quad \text{and} \quad P(\text{Rockets win}) = \frac{2}{5} = .4$$

The oddsmaker believes there is a 60% chance that the Bulls will win the championship.

6-5 Conditional Probability, Independent and Dependent Events

Sometimes we want to revise a probability based on new information. A conditional probability is an updated probability given this new information. Related to conditional probability is the idea of independence (independence of events and independence of observations). We use independence to develop probability models for many experimental situations.

Consider updating probabilities with new information. Suppose you are a project leader at work and your boss has promised to send you a new assistant. In your experience, about 75% of the assistants your boss sends are women. So, if asked, you are likely to say that there is a 75% chance or probability $\frac{3}{4}$ that your new assistant is a woman. At lunch the day before your assistant is to arrive, a friend tells you that your new assistant is bald. With this additional information, you may revise your guess. Since many more bald people are male than female, you might decide the probability of a female assistant is a good deal less than $\frac{3}{4}$. This revised probability is the conditional probability that your assistant is a woman, given that the assistant is bald. We say the event that the assistant is a woman and the event that the assistant is bald are dependent events, because knowing that one of the events occurred alters the estimated probability of the other event.

The *conditional probability* of an event A given an event B is the revised probability of event A occurring, based on the information that event B has occurred. We define such a conditional or revised probability this way:

> The **conditional probability of event *A* given event *B***, denoted $P(A|B)$, is the probability that events A and B occur together, divided by the probability of event B:
>
> $$P(A|B) = \frac{P(A \text{ and } B)}{P(B)}$$
>
> provided that $P(B)$ is greater than 0.

The conditional probability of event A given event B is defined only when event B is possible, so $P(B)$ must be greater than 0. Sometimes we find it useful to write the definition of conditional probability in the following way:

$$\text{For any events } A \text{ and } B \text{ with } P(B) \text{ greater than } 0,$$
$$P(A \text{ and } B) = P(B) \times P(A|B).$$

The probability that both A and B occur equals the probability of B times the conditional probability of A given B. This relation gives us a way to find the probability of both A and B occurring when we know the probability of B and the conditional probability of A given B. We will find this useful in some examples that follow.

Two events are *dependent* if knowing that one of the events occurred changes the calculated probability that the other event occurred. Formally, we say:

> Two events C and D are **dependent events** if $P(C|D)$ does not equal $P(C)$. This is the same as saying that $P(D|C)$ does not equal $P(D)$, and the same as saying that $P(C \text{ and } D)$ does not equal $P(C) \times P(D)$.

Let's illustrate the ideas of conditional probability and dependent events with an example.

EXAMPLE 6-9

The following frequency table classifies each of 2,475 people serving time in Georgia prisons for murder, according to race of the convict and race of the convict's victim:

Race of the convict	Race of the victim		
	Black	White	Total
Black	1,438	228	1,666
White	64	745	809
Total	1,502	973	2,475

These data appear, along with additional information, in an article discussing possible racial biases in the Georgia death penalty system ("Supreme Court Ruling on Death Penalty," *Chance,* 1988).

We might use this information to construct a probability model for classifying a convicted murderer in Georgia by race and race of the victim:

Outcome	Probability of the outcome
(prisoner black, victim black)	$\frac{1{,}438}{2{,}475} = .58$
(prisoner black, victim white)	$\frac{228}{2{,}475} = .09$
(prisoner white, victim black)	$\frac{64}{2{,}475} = .03$
(prisoner white, victim white)	$\frac{745}{2{,}475} = .30$

Let B denote the event that a convicted murderer is white. Listing the outcomes in B within braces, we can write: B = {(prisoner white, victim black), (prisoner white, victim white)}. According to this probability model, the probability that a convicted murderer is white equals

$$P(B) = \frac{64}{2{,}475} + \frac{745}{2{,}475} = \frac{809}{2{,}475} = .33$$

Let C be the event that the murder victim was white: C = {(prisoner black, victim white), (prisoner white, victim white)}. According to this probability model,

$$P(C) = \frac{228}{2{,}475} + \frac{745}{2{,}475} = \frac{973}{2{,}475} = .39$$

If events B and C both occur, then the convicted murderer is white and the victim was white. Therefore,

$$P(B \text{ and } C) = P(\text{prisoner white, victim white}) = \frac{745}{2{,}475} = .30$$

The conditional probability that the convicted murderer is white given that the victim was white is

$$P(\text{prisoner is white}|\text{victim was white}) = \frac{P(\text{prisoner white and victim white})}{P(\text{victim white})}$$

and we can write this as

$$P(B|C) = \frac{P(B \text{ and } C)}{P(C)} = \frac{745/2{,}475}{973/2{,}475} = \frac{745}{973} = .77$$

Without additional information, we would say there is about a 33% chance that a convicted murderer in Georgia is white. Knowing that the victim was white, we revise this estimate and say there is about a 77% chance that the convicted murderer is white.

The event B that the convicted murderer is white and the event C that the victim was white are dependent events, since $P(B|C)$ does not equal $P(B)$. The race of the convicted murderer and the race of the victim are *dependent* or *associated characteristics.* (We will see this idea again in Sections 16-3 and 16-5 when we discuss formal statistical procedures to test for independence of two qualitative variables.)

Two events are *independent* if knowing that one of the events occurred does not change the calculated probability that the other event occurred. The probability that the two independent events both occur is the product of the individual probabilities of the two events.

> Two events A and B are **independent events** if $P(A|B) = P(A)$. This is the same as saying that $P(B|A) = P(B)$, and the same as saying that $P(A \text{ and } B)$ equals $P(A) \times P(B)$.

We can use the idea of independence to construct probability models for experiments, as illustrated in Example 6-10.

EXAMPLE 6-10

A person is Rh positive (Rh+) if a substance called the Rh+ antigen is on the surface of the red blood cells. If this Rh+ antigen is not there, a person is Rh negative (Rh−).

Each year at City Hospital, about 51% of the babies born are boys. About 85% of the babies are Rh+, regardless of sex, the rest being Rh−. Sex and presence of the Rh+ antigen are *independent* or *unassociated characteristics* since males are as likely as females to be Rh positive. (In Sections 16-3 and 16-5 we will address the problem of deciding whether two qualitative variables are independent or unassociated.)

Suppose we classify a baby born at City Hospital by sex and by presence of the Rh+ antigen. Then there are four possible outcomes in the sample space. We can use the given information to construct a probability function for this sample space.

The outcome (girl, Rh+) represents the result that the baby is a girl and is Rh+. Since sex and presence of the Rh+ antigen are independent characteristics, we calculate the probability of this outcome as

$$P(\text{girl, Rh+}) = P(\text{girl}) \times P(\text{Rh+}) = .49 \times .85 = .4165$$

We calculate the probabilities of the other outcomes in a similar way (Exercise 6-5).

We can extend the definition of independence to any finite number of events. Consider events A_1 through A_k. It is not enough to say that any pair of these events are independent (see Exercise 6-28). Instead, we say the following:

> Events A_1 through A_k are **independent events** if for any subcollection of two or more of these events, the probability that each of the events in the subcollection occurs is equal to the product of the probabilities of these events taken separately.

This definition of independence tells us that if A_1 through A_k are independent events, then the probability that A_1 through A_k all occur is the product of the probabilities $P(A_1)$ through $P(A_k)$.

EXAMPLE 6-7
(continued)

We toss a fair coin three times and the tosses are independent of one another. That is, the result of one toss does not affect the result of another toss. What is the probability of three heads in a row?

Let E_1 be the event that a head appears on the first toss, E_2 the event that a head appears on the second toss, and E_3 the event that a head appears on the third toss. Because the coin is fair, the probability of each of these events is $\frac{1}{2}$. From the description of the experiment, we know that E_1, E_2, and E_3 are independent events.

The probability of three heads in a row is the probability that each of the events E_1, E_2, and E_3 occurs. Therefore, the probability of three heads in three tosses equals

$$\begin{aligned} P(\text{three heads in a row}) &= P(E_1, E_2, \text{ and } E_3 \text{ each occur}) \\ &= P(E_1) \times P(E_2) \times P(E_3) = \frac{1}{2} \times \frac{1}{2} \times \frac{1}{2} = \frac{1}{8} \end{aligned}$$

Similar calculations give the entire probability model for this experiment, shown in Table 6-1.

In Example 6-7, we use the independence of the coin tosses to build a probability model for the experiment. In general, we use the concept of independence to develop probability models for experiments in formal statistical inference. An important assumption for building these probability models is that the observations are independent, as we will see in Part III.

We can use a formula called Bayes' rule for calculating a conditional probability when we know certain other probabilities and conditional probabilities. We discuss Bayes' rule in Section 6-6. That material is optional; the information on Bayes' rule is not necessary for understanding any topics that follow.

6-6 Bayes' Rule (Optional)

Bayes' rule is a formula for calculating a conditional probability when certain other probabilities and conditional probabilities are known.

For any events A and E with positive probability, the following relationship (called **Bayes' rule**) is true:

$$P(E|A) = \frac{P(E) \times P(A|E)}{P(E) \times P(A|E) + P(\text{not } E) \times P(A|\text{not } E)}$$

Among other applications, we can use Bayes' rule to study the effectiveness of medical screening tests, as illustrated in the next example.

EXAMPLE 6-11

Some people have proposed large-scale screening programs for the acquired immune deficiency syndrome (AIDS). Is such a program a good idea? In particular, what is the likelihood that a person testing positive for AIDS really has AIDS? If a person tests negative for AIDS, what is the chance he or she really does not have the disease?

To answer these questions, we need an estimate of the proportion of people affected with the AIDS virus in the population to be tested. We also need some information on how good the screening test is at identifying presence and absence of the virus.

Suppose that 1.5 million of roughly 250 million Americans are infected with the AIDS virus. Then our estimate of the proportion infected is 1.5 million divided by 250 million, or .006 ("Random Testing for AIDS?" *Chance,* 1988).

Assume the probability that a person with AIDS is correctly diagnosed (called the *sensitivity* of the screening test) is .98. Assume also the probability that a person without AIDS is correctly diagnosed (called the *specificity* of the screening test) is .93. (These values are based on results reported in Weiss et al., 1985.)

The **sensitivity** of a diagnostic test is the probability that a person with the condition under study will test positive.

The **specificity** of a diagnostic test is the probability that a person without the condition under study will test negative.

Suppose that a single person is randomly selected from the United States population and tested for AIDS. The sample space for this experiment has four outcomes, corresponding to the four possible combinations of disease status (infected with the AIDS virus or not infected) and test result (positive or negative).

Let A denote the event that the person has the AIDS virus. Let T denote the event that the person tests positive for AIDS. Then the probabilities we have estimated above can be written as

$$P(A) = .006$$
$$P(T|A) = .98$$
$$P(\text{not } T|\text{not } A) = .93$$

We would like to find $P(A|T)$, the probability that a person testing positive really has AIDS. We can find this conditional probability using Bayes' rule:

$$P(A|T) = \frac{P(A) \times P(T|A)}{P(A) \times P(T|A) + P(\text{not } A) \times P(T|\text{not } A)}$$
$$= \frac{.006 \times .98}{(.006 \times .98) + (.994 \times .07)} = .0779$$

Here, $P(\text{not } A) = .994 = 1 - .006$ is the probability the person does not have AIDS and $P(T|\text{not } A) = .07 = 1 - .93 = 1 - P(\text{not } T|\text{not } A)$ is the probability that a person without AIDS tests positive.

What have we shown here? $P(A|T)$ is the probability that a person testing positive for AIDS is really infected with the virus. This conditional probability is .0779, meaning that of every 10,000 people testing positive for AIDS, we would expect only 779 to actually have the disease! The other 9,221 people are *false positives:* they test positive for AIDS but really do not have the disease.

We can use Bayes' rule again to find the probability that a person testing negative really does not have AIDS, $P(\text{not } A|\text{not } T)$:

$$P(\text{not } A|\text{not } T) = \frac{P(\text{not } A) \times P(\text{not } T|\text{not } A)}{P(\text{not } A) \times P(\text{not } T|\text{not } A) + P(A) \times P(\text{not } T|A)}$$
$$= \frac{.994 \times .93}{(.994 \times .93) + (.006 \times .02)} = .99987$$

This conditional probability tells us that of every 100,000 people testing negative for AIDS, we would expect 99,987 to really be free of the virus. The other 13 people are *false negatives:* they test negative for AIDS but really have the disease. They are infected people missed by the screening procedure.

We can use a tree diagram such as the one in Figure 6-1 to illustrate how Bayes' rule works. Imagine that 100,000 people selected at random from the general United States population are tested for AIDS. We assume the proportion of people in the general United States population with AIDS is .006. Therefore, as the tree diagram in Figure 6-1 illustrates, we expect about 600 of the 100,000 people tested to have AIDS and about 99,400 to be free of the disease.

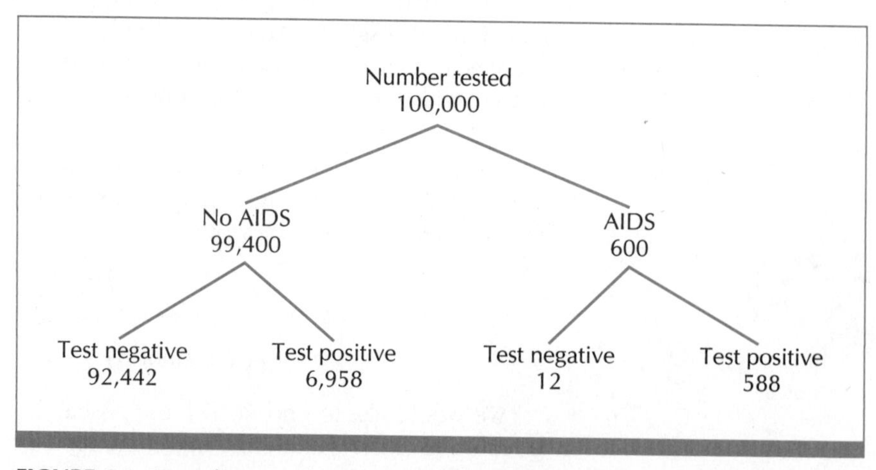

FIGURE 6-1 Tree diagram illustrating the use of Bayes' rule in Example 6-11

We assume a 98% chance that a person with AIDS tests positive for the disease. Therefore, of the 600 people we expect with AIDS, about $.98 \times 600 = 588$ will test positive for AIDS; about $.02 \times 600 = 12$ will test negative.

Similarly, we assume a 93% chance that a person without AIDS will test negative for the disease. Therefore, of the 99,400 people we expect to be free of the AIDS virus, about $.93 \times 99{,}400 = 92{,}442$ will test negative for AIDS; about $.07 \times 99{,}400 = 6{,}958$ will test positive. These expected numbers are shown along the bottom of the diagram in Figure 6-1.

To find the probability that a person testing positive for AIDS really has the disease, we look at the expected numbers in our diagram. We expect 6,958 people without AIDS and 588 people with AIDS to test positive for the disease. Therefore, we calculate

$$P(A|T) = \frac{588}{6{,}958 + 588} = .0779$$

which is the same answer we found using Bayes' rule directly.

In a similar fashion we can find the probability that a person testing negative really is free of the AIDS virus. We expect 92,442 people without AIDS and 12 people with AIDS to test negative for the disease. Therefore,

$$P(\text{not } A|\text{not } T) = \frac{92{,}442}{92{,}442 + 12} = .99987$$

which is again the same answer we found using Bayes' rule directly.

What do these calculations tell us? If the screening procedure comes up negative, there is a very large likelihood the tested individual is in fact free of AIDS. However, if the test results are positive, there is less than an 8% chance that the person really has AIDS! Used as a general screening device, this test would make a lot of people without AIDS very nervous, at least until they could be rechecked. In practice, a screening procedure might not do as well as we have described (Barnes, 1987). What do these calculations suggest to you about the practicality of a large-scale screening procedure such as this?

The probability of a false positive and the probability of a false negative depend on how well the test identifies presence and absence of the condition under study. These probabilities also depend on the proportion of people with the condition in the population. (See Exercises 6-9 through 6-12.) Note that for this type of application, it is *conditional probabilities* (of disease given a positive test result, for instance) that are of main public interest.

6-7 Random Variables

A random variable represents a numerical observation resulting from an experiment. We will be using random variables in our discussion of statistical inference.

A *random variable* is a rule that assigns a number to each outcome in the sample space. A *finite random variable* is a random variable that takes on a finite number of values. A *continuous random variable* is a random variable that takes values in an interval of numbers.

> A **random variable** is a rule that assigns a number to each outcome in the sample space.
>
> A **finite random variable** is a random variable that takes on a finite number of values.
>
> A **continuous random variable** is a random variable that takes on values in an interval of numbers.

It is possible to have a random variable that is neither finite nor continuous. For instance, if a random variable X denotes the number of defects on a sheet of copper used in personal computer hardware, then X has a countably infinite set of possible values: 0, 1, 2, and so on. Although very useful, we will not consider such random variables in this book.

Examples 6-12, 6-13, and 6-14 illustrate three finite random variables.

EXAMPLE 6-12

You apply for a job and you are either offered the job or not. We can write a sample space as $S = \{\text{success, failure}\}$, where success indicates you are offered the job and failure indicates that you are not. Let the random variable X represent the rule that assigns the number 1 to success and the number 0 to failure. We can write this rule as

$$X(\text{success}) = 1$$
$$X(\text{failure}) = 0$$

X is a finite random variable that takes on two values, 0 and 1. The random variable X codes the outcomes success and failure as numbers, 1 and 0, respectively. This kind of coding is common. For instance, a worker may code the answer to a survey question as a number before entering it on a computer file.

EXAMPLE 6-13

A pollster interviews a couple with two children. He asks the couple the gender of each child. A possible sample space is $S = \{(\text{G, G}), (\text{G, B}), (\text{B, G}), (\text{B, B})\}$. The outcome (G, B) indicates, for example, that the firstborn child is a girl and the secondborn child is a boy.

Perhaps the pollster is really interested only in the number of boys, not birth order. Then he might record their answer using the following rule, denoted by W:

$$W(\text{G, G}) = 0$$
$$W(\text{G, B}) = 1$$
$$W(\text{B, G}) = 1$$
$$W(\text{B, B}) = 2$$

W is a finite random variable that takes on three values: 0, 1, and 2. The pollster uses the rule W to count the number of male children the couple has.

EXAMPLE 6-14 You are interested in your letter grade for an introductory statistics course. We might write a sample space as $S = \{A, B, C, D, F\}$. To see how your grade contributes to your grade point average, you must convert your grade to a number. The random variable Z is a rule assigning a number to each letter grade:

$$Z(A) = 4$$
$$Z(B) = 3$$
$$Z(C) = 2$$
$$Z(D) = 1$$
$$Z(F) = 0$$

Z is a finite random variable with five possible values: 0, 1, 2, 3, and 4.

Example 6-15 illustrates two continuous random variables and one finite random variable.

EXAMPLE 6-15 We select one freshman at random from among all freshmen at State University. Each freshman represents a possible outcome of the experiment; the sample space consists of all freshmen at State University.

We record height, weight, and sex for the freshman selected. Let the random variable X be the height in inches of the student. We can think of X as a rule that assigns to each student in the sample space the number corresponding to the student's height in inches. Let the random variable Y be the weight in kilograms of the student selected. Then Y is a rule that assigns to each student in the sample space the number that is the student's weight in kilograms. X and Y are continuous random variables, since height and weight can take values over intervals of numbers.

Let the random variable Z code the sex of the student selected. Z equals 1 if the student is a woman and 2 if the student is a man. Since it has two possible values, Z is a finite random variable.

As this example illustrates, we can define more than one random variable on the same sample space. This makes sense because in real data collection situations, many numerical variables may be of interest for each individual or outcome observed.

We are often more interested in the numbers assigned by a random variable than in the actual outcomes in the sample space. Then we want to assign probabilities to events defined by the random variable.

For instance, the pollster in Example 6-13 may be interested in the probability that a couple with two children has one boy and one girl. We write this probability as $P(W = 1)$, where W is the random variable that counts the number of male children.

In Example 6-15 we may be interested in the probability that the height of the freshman selected is from 66 to 72 inches. We write this probability as $P(66 \leq X \leq 72)$, where X is the random variable corresponding to the height in inches of the student selected. Because a student is selected at random from among the freshmen at State University, the probability that X takes a value from 66 to 72 is the same as the proportion of freshmen at State University who are from 66 to 72 inches tall.

We denote the probability that the student selected is female by $P(Z = 1)$, where Z is 1 for a female and 2 for a male. Because the freshman was selected at random, $P(Z = 1)$ is the same as the proportion of women in the freshman class at State University.

We can write the probability that the student selected is female and from 66 to 72 inches in height as $P(Z = 1, 66 \leq X \leq 72)$. Again, because of the random selection, this probability corresponds to the proportion of freshmen at State University who are females from 66 to 72 inches in height.

The probability that the student selected weighs more than 100 kilograms is $P(Y > 100)$, where Y represents the student's weight in kilograms. We denote the probability that the freshman selected is male and weighs more than 100 kilograms by $P(Z = 2, Y > 100)$. We can write $P(Z = 2, 66 \leq X \leq 72, Y > 100)$ to denote the probability that the freshman selected is male, from 66 to 72 inches tall, and weighs more than 100 kilograms.

The Probability Distribution of a Random Variable

A probability function for the sample space determines the probability of an event defined by a random variable. The *probability distribution* of a random variable X refers to the collection of probabilities assigned to events defined by X.

> The **probability distribution** of a random variable is the collection of probabilities assigned to events defined by the random variable.

We can use the probability distribution of a random variable to find the mean or expected value of the random variable, as well as its variance and standard deviation. Many problems in statistical inference involve asking questions about the mean and standard deviation of a random variable.

We define expected value, variance, and standard deviation for finite random variables in Section 6-8. We can give similar definitions for continuous random variables. In Chapter 8 and Part III, we discuss some continuous random variables.

6-8 Mean, Variance, and Standard Deviation of a Finite Random Variable

The mean of a random variable is a measure of location, a measure of the center of the probability distribution of the random variable. The *mean* of a finite random variable X is a weighted average of the values the random variable takes on, each value weighted by the probability that X equals that value.

The **mean** of a finite random variable X equals $\Sigma\, xP(X = x)$, where the sum is over all numbers x with $P(X = x)$ greater than 0.

Some symbols for the mean of a random variable X are μ and μ_X.

The mean of a random variable X is also called the *expected value of X*, sometimes denoted by EX or $E(X)$. The term *expected value* refers to the average value of X we *expect* to see over many observations. The notation EX or $E(X)$ reminds us that the mean is an expected average in this sense.

The variance of a random variable is a measure of the variation of the random variable about its mean. The *variance* of a random variable X equals $E(X - \mu)^2$, the expected squared deviation of X about its mean.

The **variance** of a finite random variable X equals $\Sigma\,(x - \mu)^2P(X = x)$, where μ denotes the mean of X and the sum is over all numbers x with $P(X = x)$ greater than 0.

If a random variable X takes on only one value, then the variance of X equals 0. Otherwise, the variance of X will always be greater than 0 (see Exercise 6-25). Some symbols for the variance of X are σ^2, σ_X^2, and $\text{Var}(X)$.

The standard deviation is also a measure of variation in a random variable. The *standard deviation* of a random variable X is the positive square root of the variance of X. We denote the standard deviation of X by σ, σ_X, or $\sqrt{\text{Var}(X)}$.

The **standard deviation** of a random variable is the positive square root of the variance of the random variable.

The units of the mean and standard deviation of a random variable X are the same as the units of X. The variance of X is in squared units.

In a sampling situation, we have a sample of observed values of a random variable. We use the sample mean and sample standard deviation to estimate the mean and standard deviation, respectively, of the random variable. When we discuss statistical inference, we see how to use the (observed) sample mean and sample standard deviation to make inferences about the (unknown) mean and standard deviation of a random variable.

Example 6-16 illustrates how to find the mean, variance, and standard deviation of a finite random variable. We will consider a real taste-test experiment similar to the one described in Example 6-16 when we introduce the main ideas of statistical inference in Chapter 9.

EXAMPLE 6-16

Three statistics students volunteer for a taste test comparing Coke and Pepsi. Each student tastes samples in two identical-looking cups and decides which beverage he or she prefers. How many students do we expect to pick Pepsi? We can answer this question only after we specify a probability model for the problem. We will describe one possible probability model.

Suppose the students make selections independently of one another. Suppose also that the probability of picking Pepsi is $\frac{3}{5}$ and the probability of picking Coke is $\frac{2}{5}$ for all three students. We will refer to the tasters as the first, second, and third students. One possible outcome of the experiment is that the first and second students pick Pepsi and the third student picks Coke. We denote this outcome by (Pepsi, Pepsi, Coke). There are eight possible outcomes in the sample space, listed in Table 6-5.

Let the random variable Y equal the number of Pepsi selections for any given outcome of the experiment. The values of Y for the eight outcomes in the sample space are shown in column 2 of Table 6-5.

We assume the students make independent selections. Thus, the probability for each outcome is calculated as shown in column 3 of the table. For example,

$$P(\text{Pepsi, Pepsi, Coke}) = \frac{3}{5} \times \frac{3}{5} \times \frac{2}{5} = \frac{18}{125} = .144$$

Under our probability model, the probability that the first two students pick Pepsi and the third student picks Coke is $\frac{18}{125}$ or .144.

We see that the random variable Y takes on four values: 0, 1, 2, and 3. The probability distribution for Y is given by:

$$P(Y = 3) = \frac{27}{125}$$

$$P(Y = 2) = \frac{18}{125} + \frac{18}{125} + \frac{18}{125} = \frac{54}{125}$$

$$P(Y = 1) = \frac{12}{125} + \frac{12}{125} + \frac{12}{125} = \frac{36}{125}$$

$$P(Y = 0) = \frac{8}{125}$$

The mean or expected value of Y is

$$\begin{aligned} E(Y) &= 3 \times P(Y = 3) + 2 \times P(Y = 2) \\ &\quad + 1 \times P(Y = 1) + 0 \times P(Y = 0) \\ &= 3 \times \frac{27}{125} + 2 \times \frac{54}{125} + 1 \times \frac{36}{125} + 0 \times \frac{8}{125} \\ &= \frac{225}{125} = 1.8 \text{ Pepsi selections} \end{aligned}$$

The expected value of Y is 1.8, the expected number of Pepsi selections among the three students. This expected value does not equal any of the values Y can take on. How should we interpret this number? We can think of the expected value as the average number of Pepsi selections among three students, if we were able to do the experiment many times under the same conditions. Suppose many groups of three students participate and the probability model above applies to each group. If we divide the total number of Pepsi selections by the number of student groups, the result will be close to 1.8. In

TABLE 6-5 The value of the random variable Y and the probability of each outcome in the sample space for Example 6-16

Outcome	Value of Y (number of Pepsi selections)	Probability of the outcome
(Pepsi, Pepsi, Pepsi)	3	$\frac{3}{5} \times \frac{3}{5} \times \frac{3}{5} = \frac{27}{125} = .216$
(Pepsi, Pepsi, Coke)	2	$\frac{3}{5} \times \frac{3}{5} \times \frac{2}{5} = \frac{18}{125} = .144$
(Pepsi, Coke, Pepsi)	2	$\frac{3}{5} \times \frac{2}{5} \times \frac{3}{5} = \frac{18}{125} = .144$
(Coke, Pepsi, Pepsi)	2	$\frac{2}{5} \times \frac{3}{5} \times \frac{3}{5} = \frac{18}{125} = .144$
(Coke, Coke, Pepsi)	1	$\frac{2}{5} \times \frac{2}{5} \times \frac{3}{5} = \frac{12}{125} = .096$
(Coke, Pepsi, Coke)	1	$\frac{2}{5} \times \frac{3}{5} \times \frac{2}{5} = \frac{12}{125} = .096$
(Pepsi, Coke, Coke)	1	$\frac{3}{5} \times \frac{2}{5} \times \frac{2}{5} = \frac{12}{125} = .096$
(Coke, Coke, Coke)	0	$\frac{2}{5} \times \frac{2}{5} \times \frac{2}{5} = \frac{8}{125} = .064$

some groups, all three students may pick Pepsi, in other groups one or two students may pick Pepsi, and in other groups no students may pick Pepsi. But the average number of Pepsi selections per group will be about 1.8.

The variance of the random variable Y in this example is

$$\begin{aligned}\text{Var}(Y) &= (3 - 1.8)^2P(Y = 3) + (2 - 1.8)^2P(Y = 2) \\ &\quad + (1 - 1.8)^2P(Y = 1) + (0 - 1.8)^2P(Y = 0) \\ &= 1.44 \times \frac{27}{125} + .04 \times \frac{54}{125} + .64 \times \frac{36}{125} + 3.24 \times \frac{8}{125} \\ &= .72 \text{ (Pepsi selections)}^2\end{aligned}$$

The standard deviation of Y is $\sqrt{\text{Var}(Y)} = \sqrt{.72} = .85$ Pepsi selection.

Note that the sample space for the taste test in Example 6-16 (the eight outcomes are listed in Table 6-5) looks like the sample space for a three-coin-toss experiment such as we considered in Example 6-7. Each student in the taste test has two possible choices (Pepsi or Coke), while each coin toss has two possibilities (head or tail). When we have independent repetitions of a two-outcome experiment and the probabilities of the two outcomes are the same for each repetition, we say we have a *binomial experiment*. We will discuss probability models for binomial experiments in Chapter 7. Many real experiments involve repetitions having two possible outcomes (for example, cured versus not cured in a medical experiment, alive versus dead in an animal toxicity study, product A versus product B in a market research survey, acceptable versus defective in an industrial setting). We will discuss formal statistical analysis of such binomial experiments in Sections 10-2, 10-5, 16-1, and 16-2.

Example 6-17 is a simple application of probabilities and expected values.

EXAMPLE 6-17

How do we make decisions in situations involving risk, when we are not certain of the consequences of possible choices or courses of action? Knowledge, prejudice, and assessment of costs and benefits may come into play. But our decisions in risky situations may also be influenced by the way we perceive these situations (Allman, 1985), as the following experiment illustrates.

Experimenters presented two groups of physicians with a hypothetical problem regarding an impending outbreak of a rare disease (Kahneman and Tversky, 1982). If untreated, the disease will kill 600 people.

The experimenters offered physicians in the first group two possible programs for fighting the disease. With program A, 200 of the disease victims will be saved. With program B, there is a $\frac{1}{3}$ chance that all 600 disease victims will be saved and a $\frac{2}{3}$ chance that none of them will be saved. The expected number of disease victims saved under program B is 200 (Exercise 6-26). Therefore, as far as *expected* number of victims saved is concerned, programs A and B are equivalent. A majority of the physicians in this group chose program A; they preferred to save 200 disease victims for sure rather than risk saving none of the victims.

The experimenters also presented physicians in the second group with two programs for dealing with the disease. With program C, 400 disease victims will die. With program D, there is a $\frac{2}{3}$ chance that all 600 disease victims will die and a $\frac{1}{3}$ chance that none of them will die. The expected number of deaths among disease victims is 400 for program D (Exercise 6-27), the same as for program C. However, a majority of physicians in this group chose program D. These physicians preferred to take a chance and try to prevent all the disease victims from dying rather than accept the certainty of 400 dying.

The results of the programs offered the two groups of physicians are the same. Programs A and C both guarantee 200 victims saved, with 400 deaths. Programs B and D offer a $\frac{1}{3}$ chance that all disease victims are saved (none die) and a $\frac{2}{3}$ chance that none are saved (all the disease victims die). The choices offered the two groups were similar in content but not in presentation. The physicians in the first group chose between two gains. These physicians tended to prefer a certain gain (in saved lives) to an uncertain situation with possible large gain and possible no gain. The physicians in the second group chose between two losses. These physicians tended to reject the certain loss (deaths) and choose the uncertain course with possible no loss as well as possible large loss.

How we evaluate and react to a situation involving uncertainty and risk seems to depend not only on our background, experience, knowledge, and so on, but also on whether we consider the problem in terms of gains or losses, benefits or costs. (When making decisions in the face of uncertainty, it matters whether we think of the proverbial glass as half full or half empty!)

Summary of Chapter 6

A probability function defined on a sample space formalizes the way we assign probabilities or chances to events.

The conditional probability of event A given event B is the probability that events A and B both occur divided by the probability of event B. Bayes' rule provides a way of calculating a conditional probability when certain other probabilities and conditional probabilities are known. A useful application of Bayes' rule is in evaluation of medical screening procedures.

Two events are independent if knowing that one of the events occurred does not change the calculated probability that the other event occurred. The concept of independence is useful in building many probability models. Two events are dependent if knowing that one of the events occurred does change the calculated probability that the other event occurred.

A random variable is a rule that assigns a number to each outcome in a sample space, representing numerical observations made during an experiment. The probability distribution of a random variable specifies the probability of each event defined by the random variable. The method of analysis we select in statistical inference depends on the probability distribution of the variable observed.

The mean or expected value of a random variable is a measure of the center of the values taken on by the random variable. The variance and standard deviation measure the spread or variation in the values taken on by the random variable. Much of statistical inference involves asking questions about the mean and variance of a random variable. In a classical analysis, we base our inferences about the unknown mean and variance of a random variable on the sample mean and sample variance calculated from the observations.

Exercises for Chapter 6

EXERCISE 6-1

In 1984, researchers asked a group of 1,060 Vermont 4th and 5th graders about their experiences with cigarette smoking (Haugh et al., 1986). Of the 515 girls, 379 said they had never smoked, 114 said they had tried a few cigarettes, 7 reported they were regular smokers, and 15 said they used to smoke. Among the 545 boys, 322 said they had never smoked, 187 said they had tried a few cigarettes, 20 described themselves as regular smokers, and 16 as former smokers.

a. Arrange these findings in a two-way frequency table. Use this information to construct a probability model for classifying a Vermont 4th or 5th grader by sex and smoking history.

b. Write down an appropriate sample space and probability function for this problem.

c. Find the probability of the following events: the child is a girl; the child has never tried cigarettes; the child has tried a few cigarettes; the child is a regular smoker; the child is a boy who smokes regularly; the child is a girl who has never smoked.

d. Find the following conditional probabilities: the probability the child is a girl, given the child is a regular smoker; the probability the child is a regular smoker, given the child is a girl; the probability the child is a boy, given the child has tried a few cigarettes; the probability the child has tried a few cigarettes, given the child is a boy.

e. Are sex and smoking status independent or dependent under this probability model?

EXERCISE 6-2

In a 1984 survey for the Veterans Administration, researchers interviewed 2,033 women veterans with recent wartime service (World War II, Korean War, Vietnam War). The researchers classified these women by exposure to combat (exposed or not) and by job (nurse or not). Of the 396 nurses, 94 were exposed to combat and 302 were not. Of the 1,637 women veterans who were not nurses, 67 were exposed to combat and 1,570 were not (Dienstfrey, 1986).

a. Arrange these results in a two-way frequency table. Use this information to construct a probability model for classifying women veterans by exposure to combat and job.

b. Write down an appropriate sample space and probability function for this problem.

c. Are exposure to combat and job status independent or dependent under this probability model?

EXERCISE 6-3

An obstetrical nurse has observed over long experience that 12.75% of the babies she has helped deliver were boys with no hair, 25.5% were boys with very little hair, 12.75% were boys with lots of hair, 12.25% were girls with no hair, 24.5% were girls with very little hair, and 12.25% were girls with lots of hair.

a. If a newborn baby is classified as above by sex and by quantity of hair, construct an appropriate sample space. Use the nurse's observations to define a probability function for this sample space.

b. Under this probability model, are sex and quantity of hair of a newborn baby independent?

EXERCISE 6-4

At a toy factory, employees work around the clock producing miniature cars. Standards for the toys have been set, so that a quality control inspector can classify any given miniature car as ready for shipment or not acceptable for shipment. One week 9,000 miniature cars are produced. These cars can be classified by acceptability for shipment and by the shift that produced them, as follows:

Shift	Not acceptable	Ready for shipment	Total
Day shift, 7 A.M.–3 P.M.	100	2,900	3,000
Evening shift, 3 P.M.–11 P.M.	300	2,700	3,000
Night shift, 11 P.M.–7 A.M.	600	2,400	3,000
Total	1,000	8,000	9,000

The quality control inspector selects at random one miniature car for inspection from the 9,000 cars and classifies it according to the shift that produced it and its acceptability for shipment.

a. Write down a sample space and probability function for this experiment.

b. Is the chance that a toy car is not acceptable independent of the shift producing it?

EXERCISE 6-5

Refer to the experiment in Example 6-10 to answer parts (a)–(e). Note that part (c) uses optional material on odds ratios.

a. Write down a sample space and probability function for this experiment.

b. Find the following probabilities: the probability the baby is a boy; the probability the baby is Rh+; the probability the baby selected is a boy who is Rh+.

c. Find the following odds: the odds the baby is a girl; the odds the baby is Rh+; the odds the baby is a girl who is Rh+.

d. Find the conditional probability that the baby is a girl, given the baby is Rh+. Find the probability the baby is a girl. Why are these two probabilities equal?

e. Find the conditional probability that the baby is Rh+, given the baby is a girl. Why does this conditional probability equal the probability that the baby selected is Rh+?

EXERCISE 6-6

Last year at City Hospital, 510 of the babies born were boys and 490 were girls. Of the 510 boys, 31 were born with a condition affecting color vision called red–green colorblindness; 2 of the girls were born with this condition. Use this information to construct a probability model for classifying a newborn baby by sex and presence or absence of red–green colorblindness. [Parts (c) and (e) use optional material on odds ratios.]

a. Write down a sample space for this experiment. Find a probability function for this sample space based on the given information.

b. Find the following probabilities: the probability that the baby selected is a girl; the probability that the baby selected has red–green colorblindness; the probability that the baby selected is a girl and has red–green colorblindness.

c. Find the following odds: the odds the baby has red–green colorblindness; the odds the baby is a boy; the odds the baby is a boy with red–green colorblindness.

d. Find the following conditional probabilities: the probability the baby is a boy, given the baby has red–green colorblindness; the probability the baby has red–green colorblindness, given the baby is a boy.

e. Find the following odds: for a girl baby, the odds she has red–green colorblindness; for a baby boy, the odds he has red–green colorblindness; for a baby with red–green colorblindness, the odds the baby is a boy; for a baby without red–green colorblindness, the odds the baby is a boy.

f. Are the event that the baby is a boy and the event that the baby has red–green colorblindness independent or dependent events?

g. Red–green colorblindness is known in genetics as a sex-linked characteristic. Discuss what this means in light of your answers to parts (a)–(f) of this problem.

EXERCISE 6-7

A unit of a large machine consists of two components A and B. The probability that component A fails during operation is .01 and the probability that component B fails during operation is .05. The components A and B operate independently of one another in the sense that the functioning or failure of one of the components does not affect the likelihood of failure for the other component. Suppose that during a test run, components A and B are each tested and classified as success if functional, failure if nonfunctional.

a. Write down a sample space for this experiment. Find the probability function for this sample space. That is, based on the assumptions above, write down the probability of each outcome in the sample space.

b. Suppose the components A and B are wired in parallel, as shown here:

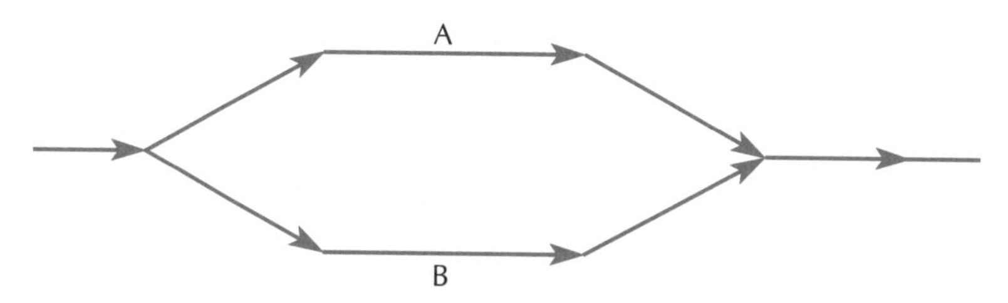

The unit of the machine functions if either component A or component B is functional. What is the probability that the unit is functional during operation? What is the probability that the unit fails during operation?

c. Suppose that instead components A and B are wired in series, as shown here:

The unit of the machine functions if and only if both components A and B are functional. What is the probability that the unit is functional during operation? What is the probability the unit fails during operation?

EXERCISE 6-8

Three Boston sports teams played on May 4, 1988. From the $2\frac{1}{2}$-point line favoring the Celtics over the Knicks, a basketball fan estimated that the Celtics had probability .55 of winning that day. (This exercise uses optional material on odds ratios.)

a. *The Boston Globe* gave the odds of the Bruins beating the Devils in the Stanley Cup hockey playoffs as $8\frac{1}{2}$ to 10 in favor of the Bruins. Based on these odds, what is the estimated probability of the Bruins winning the game?

b. *The Boston Globe* favored the Red Sox with 10:12 odds over the White Sox. Based on this odds ratio, what is the estimated probability of the Red Sox winning the baseball game?

c. Let an outcome record a win or loss that day for each of the three Boston teams. Write down an appropriate sample space.

d. What does it mean to assume that the results for the three Boston teams that day were independent of one another? Assuming independence and using the estimated probabilities from above, write down a probability function for this sample space.

e. Find the probability of each of the following events: all three Boston teams win; at least one Boston team wins; no Boston team wins.

Note: That day, all three Boston teams lost. (Odds and point line from *The Boston Globe,* May 4, 1988, pages 52, 57.)

EXERCISE 6-9

Suppose a screening procedure for AIDS is not as good as the one described in Example 6-11. This screening test is positive for 90% of people with the disease and negative for 85% of people without the disease. Suppose that in the population tested, the proportion infected with the AIDS virus is .006. (This exercise uses optional material on Bayes' rule.)

a. Find the probability that a person testing positive really has AIDS. What is the probability of a false positive?

b. Find the probability that a person testing negative really does not have AIDS. What is the probability of a false negative?

c. Compare your results with those we found in Example 6-11.

EXERCISE 6-10

Suppose a screening procedure for AIDS is better than the one described in Example 6-11. This screening test is positive for 99% of people with the disease and negative for 99% of people without the disease. Suppose that in the population tested, the proportion infected with the AIDS virus is .006. (This exercise uses optional material on Bayes' rule.)

a. Find the probability that a person testing positive really has AIDS. What is the probability of a false positive?

b. Find the probability that a person testing negative really does not have AIDS. What is the probability of a false negative?

c. Compare your results with those we found in Example 6-11 and Exercise 6-9.

EXERCISE 6-11

Suppose the screening procedure described in Example 6-11 is applied to a population at high risk for AIDS—say, 10% of people in this population are infected with the AIDS virus. Suppose, as in Example 6-11, that the screening test is positive for 98% of people with the disease and negative for 93% of people without the disease. (This exercise uses optional material on Bayes' rule.)

a. Find the probability that a person testing positive really has AIDS. What is the probability of a false positive?

b. Find the probability that a person testing negative really does not have AIDS. What is the probability of a false negative?

c. Compare your results with those we found in Example 6-11.

EXERCISE 6-12

A screening procedure for AIDS is applied to a population at high risk for AIDS; 10% of this population have AIDS. This screening test is positive for 90% of people with the disease and negative for 85% of people without the disease. (This exercise uses optional material on Bayes' rule.)

a. Find the probability that a person testing positive really has AIDS. What is the probability of a false positive?

b. Find the probability that a person testing negative really does not have AIDS. What is the probability of a false negative?

c. Compare your results with those in Exercise 6-9 and Exercise 6-11.

EXERCISE 6-13

Sometimes people are automatically tested more than once in a screening program. Suppose in Example 6-11, each person is tested twice and a person is said to be positive if and only if both test results come up positive. (The same result would be achieved by retesting only people who are positive on the first test.) How does this change the probability of a false positive and the probability of a false negative? (This exercise uses optional material on Bayes' rule.)

a. Find the probability that a person with a positive result really does have AIDS. What is the probability of a false positive?

b. Find the probability that a person with a negative result really does not have AIDS. What is the probability of a false negative?

c. Compare your results with those in Example 6-11.

EXERCISE 6-14

In a country with a stable population, about .8% of all adult men develop lung cancer and 30% of all adult men are smokers. Studies of lung cancer patients have shown that 56.3% of all adult male lung cancer patients are cigarette smokers. It is also known that of the adult men who will never develop lung cancer, 29.8% are smokers. Answer the following questions for this country. (This exercise uses optional material on odds ratios and Bayes' rule.)

a. What is the probability that an adult male cigarette smoker will develop lung cancer? What are the odds of an adult male cigarette smoker developing lung cancer?

b. What is the probability of an adult male nonsmoker developing lung cancer? What are the odds of an adult male nonsmoker developing lung cancer?

c. Compare your answers to parts (a) and (b).

d. Why is it important for these calculations to assume that the population of the country is stable? (That is, very few people are moving in and out of the country.)

e. What other simplifying assumptions are necessary for these calculations to be valid?

EXERCISE 6-15

A new *in utero* diagnostic procedure has been developed to test for a particular type of birth defect. A preliminary investigation showed the test results to be positive for 95% of the pregnant women who subsequently gave birth to babies having the birth defect under study. The results were negative for 97% of the pregnant women who later gave birth to babies not having the birth defect.

Suppose that the incidence of the birth defect in the United States is 100 per 100,000 births. That is, for every 100,000 babies born in the United States, 100 are found to have the birth defect. (This exercise uses optional material on Bayes' rule.)

a. If a woman has a positive test result, what is the probability her baby will have the birth defect? What is the probability of a false positive?

b. If a woman has a negative test result, what is the probability her baby will not have the birth defect? What is the probability of a false negative?

c. Discuss the advantages and disadvantages of using this diagnostic procedure.

EXERCISE 6-16

How useful is the lie detector or polygraph test in assessing a person's guilt or innocence? Based on studies involving 120 guilty and 120 innocent people (Gastwirth, 1987, page 217), we will estimate the probability that a guilty person is detected as guilty to be .88. We will estimate the probability that an innocent person is classified as innocent to be .86. (This exercise uses optional material on Bayes' rule.)

a. Suppose the polygraph is used among people indicted of a felony. We will assume that past experience has shown that about 90% of people indicted of a felony are in fact guilty of the felony.

(i) Find the probability that a person with a positive (guilty) lie detector result is really guilty. What is the probability of a false positive?

(ii) Find the probability that a person with a negative (not guilty) lie detector result is really innocent. What is the probability of a false negative?

b. Suppose the polygraph is used to screen for illicit drug use among employees of a large corporation. We will assume that experience with similar companies has shown that about 3% of employees are involved in illicit drug use.

(i) Find the probability that an employee with a positive (drug use) lie detector result really is an illicit drug user. What is the probability of a false positive?

(ii) Find the probability that a person with a negative (no drug problem) lie detector result is really not an illicit drug user. What is the probability of a false negative?

c. Taking all the probabilities in parts (a) and (b) at face value, discuss the importance of the proportion guilty in the test population to interpretation of lie detector results.

d. We applied the lie detector test to two different populations: people indicted of a felony and people working for a large corporation. Our assessment of the lie detector was based on baseline studies of 120 guilty people and 120 innocent people. How might the makeup of this baseline test group affect our interpretation of the results in our two applications? Discuss.

EXERCISE 6-17

A cancer researcher wants to test a new combination of chemotherapy and radiation on skin tumors in laboratory mice. The researcher administers the treatment to each of four laboratory mice having the type of skin tumor under study. After a week of treatment, the researcher records failure or success for each mouse, depending on whether or not skin tumor cells are observed on the animal.

a. Write down a sample space for this problem, where an outcome shows success or failure for each of the four animals in the experiment.

b. Suppose that the treatment combination has no effect on the skin tumors. However, there is a .1 probability that the tumor will disappear spontaneously over a week of observation. If the mice are independent of one another with regard to disappearance of the skin tumor, find the probability function for the sample space in this problem.

c. With the probability model above, what is the probability that three or four of the mice will be free of skin tumors at the end of the week of treatment? What is the probability that one or none of the mice will be free of skin tumors at the end of the week of treatment?

d. Define a random variable X to be the number of mice that are free of skin tumors at the end of the week of treatment. Find the values X takes on and the probability that X takes on each of these values. Find the expected value, variance, and standard deviation of X.

EXERCISE 6-18

The experiment is the same as in Exercise 6-17. Suppose now that the treatment does have an effect on the skin tumor. Assume there is a probability of .8 that the skin tumor will disappear by the end of the week of treatment.

a. If mice are independent of one another with respect to disappearance of skin tumors, find the probability function for the sample space in this problem.

b. With this probability model, what is the probability that three or four of the mice will be free of skin tumors at the end of the week of treatment? What is the probability that one or none of the mice will be free of skin tumors at the end of the week of treatment?

c. Define a random variable X to be the number of mice that are free of skin tumors at the end of the week of treatment. Find the values X takes on and the probability that X takes on each of these values. Find the expected value, variance, and standard deviation of X.

d. How do the answers to this exercise compare with the answers to Exercise 6-17?

EXERCISE 6-19

Ralph's history teacher gives a surprise quiz one day in class. The quiz consists of four multiple-choice questions. Each question has six answers to choose from, one being the correct answer and the other five incorrect. Suppose that Ralph has not studied history all semester and that he guesses on this quiz in such a way that he is equally likely to choose any one of the six answers to any question and his answer to one question does not in any way influence his answer to another question. (Maybe he rolls a fair die to decide on his answer.) Suppose an experiment consists of Ralph taking the quiz. An outcome shows whether Ralph's answer was correct or incorrect for each of the four questions.

a. Write down a sample space and find the probability of each outcome in the sample space.

b. Define a random variable W that counts the number of correct answers Ralph guesses on the quiz. List the possible values for W and the probability that W takes on each of these values.

c. Find the expected value, variance, and standard deviation of W.

d. What is the probability that Ralph guesses at least two answers correctly? That is, find $P(W \geq 2)$.

EXERCISE 6-20

Vin throws a dart at a target, with each throw scored in the following way. A shot hitting the bull's eye in the center of the target is worth 6 points. Hits in concentric circles about the center are worth from 5 points down to 1 point, 5 points within the circle closest to the center and 1 point for the circle farthest from the center. A shot that misses the target is worth 0 points. Vin is a good dart player and experience has shown that under ideal conditions his likelihood of achieving any given score is the following:

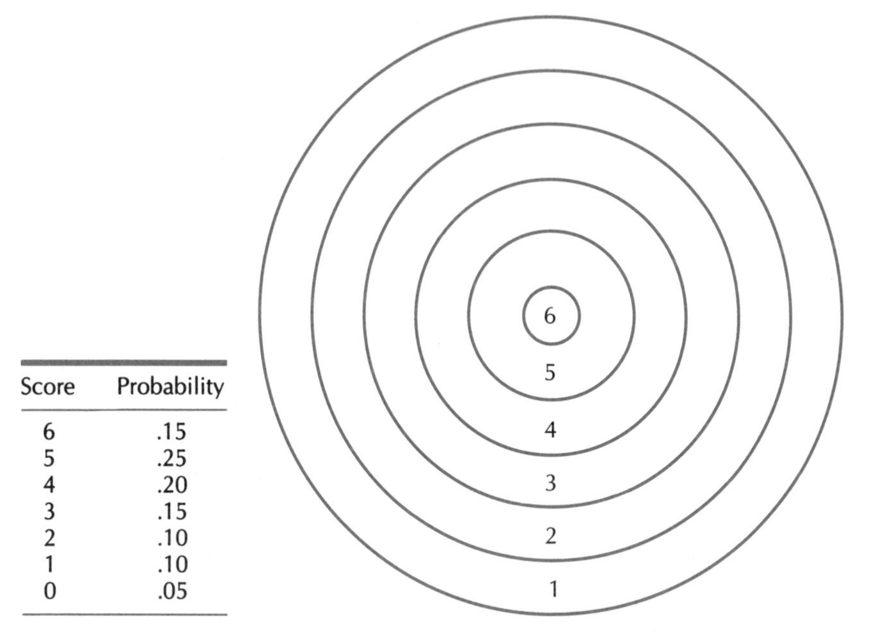

Score	Probability
6	.15
5	.25
4	.20
3	.15
2	.10
1	.10
0	.05

Let the random variable Y represent the score Vin gets with one throw under ideal conditions.

a. Find $P(Y \geq 4)$, the probability that Vin gets 4 or more points.

b. Find $P(Y < 3)$, the probability that Vin gets less than 3 points.

c. Find $P(3 \leq Y \leq 5)$, the probability that Vin gets 3, 4, or 5 points.

d. Find the expected value, variance, and standard deviation of Y. How would you interpret each of these values?

EXERCISE 6-21 An experiment consists of rolling a six-sided die once and noting how many dots are face up at the end of the roll. This is a fair die, so that each of the six sides is equally likely to be face up at the end of the roll. [Part (b) uses optional material on odds ratios.]

a. Write down the sample space for this experiment and the probability function for this sample space.

b. What are the odds that 1, 2, or 3 dots will be face up at the end of the roll?

Suppose a gambler wins \$.50, \$1.50, or \$2.50 if the die comes up 4, 5, or 6, respectively. He loses \$.50, \$1.50, or \$2.50 if the die comes up 3, 2, or 1, respectively. Let the random variable W denote the gambler's winnings at the end of one die roll, where a loss is the same as a negative gain.

c. Find the values that W takes on and the probability that W takes on each of these values.

d. Find $P(W > 0)$, the probability that the gambler wins some money.

e. Calculate the gambler's expected gain, which is the same as the expected value of *W*.

f. Find the variance and standard deviation of *W*.

EXERCISE 6-22

An experiment consists of rolling a six-sided die once and noting how many dots are face up at the end of the roll. For this die, the chance of a particular side being face up at the end of a roll is proportional to the number of dots on that side. [Part (b) uses optional material on odds ratios.]

a. Write down the sample space S for this experiment. Find the probability function for S. (*Hint:* Let $P(i) = c \times i$, where c is a constant and i denotes the number of dots face up at the end of the roll. Show that the probabilities of the outcomes in S sum to 1 if $c = \frac{1}{21}$.)

b. What are the odds that 1, 2, or 3 dots will be face up at the end of the roll?

Suppose a gambler wins $.50, $1.50, or $2.50 if the die comes up 4, 5, or 6, respectively. He loses $.50, $1.50, or $2.50 if the die comes up 3, 2, or 1, respectively. Let the random variable *W* denote the gambler's winnings at the end of one die roll, where a loss is the same as a negative gain.

c. Find the values that *W* takes on and the probability that *W* takes on each of these values.

d. Find $P(W > 0)$, the probability that the gambler wins some money.

e. Calculate the gambler's expected gain, which is the same as the expected value of *W*.

f. Find the variance and standard deviation of *W*.

g. Compare your answers to this exercise with your answers to Exercise 6-21.

EXERCISE 6-23

A six-sided die is rolled once and each outcome is equally likely. A gambler wins $1 for each dot that appears face up at the end of the roll. Let the random variable *Y* denote the gambler's winnings at the end of one die roll.

a. Find the values *Y* takes on and the probability that *Y* takes on each of these values.

b. Find $P(Y > 0)$, the probability that the gambler wins some money.

c. Find the expected value of *Y*, the gambler's expected winnings.

d. Find the variance and standard deviation of *Y*.

e. Compare these answers with the answers to Exercise 6-21.

EXERCISE 6-24

A six-sided die is rolled once and the probabilities of the outcomes are those you found in Exercise 6-22. A gambler wins $1 for each dot that appears face up at the end of the roll. Let the random variable *Y* denote the gambler's winnings at the end of one die roll.

a. Find the values *Y* takes on and the probability that *Y* takes on each of these values.

b. Find $P(Y > 0)$, the probability that the gambler wins some money.

c. Find the expected value of Y, the gambler's expected winnings.

d. Find the variance and standard deviation of Y.

e. Compare these answers with the answers to Exercises 6-22 and 6-23.

EXERCISE 6-25 Suppose X is a finite random variable. Show that if $P(X = c) = 1$ for some constant c, then the variance of X is 0. Show that if X takes on at least two different values with positive probability, then the variance of X is greater than 0.

EXERCISE 6-26 If no treatment program is implemented, an outbreak of a rare disease is expected to kill 600 people. With one possible treatment plan, call it program B, there is a $\frac{1}{3}$ probability that all 600 of the disease victims will be saved and a $\frac{2}{3}$ probability that none of them will be saved. Find the expected number of disease victims saved under this treatment plan. (*Hint:* Let the random variable X denote the number of disease victims saved under the program B treatment plan. Find the expected value of X.)

EXERCISE 6-27 If no treatment program is implemented, an outbreak of a rare disease is expected to kill 600 people. With one possible treatment plan, call it program D, there is a $\frac{2}{3}$ probability that all 600 disease victims will die and a $\frac{1}{3}$ probability that none of them will die. Find the expected number of deaths under this treatment plan. (*Hint:* Let the random variable Y denote the number of disease victims who die under the program D treatment plan. Find the expected value of Y.)

EXERCISE 6-28 This problem illustrates why a special definition of independence for several events is necessary. Suppose a fair die is rolled twice, and the two rolls are independent. (A fair die is one with each of its six sides equally likely to be face up at the end of a roll.) Let an outcome be represented by the number of dots face up on each of the two rolls. For instance, (3, 5) indicates that 3 dots came up on the first roll and 5 dots came up on the second roll.

a. Write down a sample space and probability function for this experiment.

Let A be the event that the first roll results in 1, 2, or 3 dots face up. Let B be the event that the second roll results in 4, 5, or 6 dots face up. Let C be the event that the total number of dots face up on the two rolls is 3 or 11. Let D be the event that the second roll results in 2, 5, or 6 dots face up.

b. Find the probability of each of these events: A, B, C, A and B, A and C, B and C, A and B and C.

c. Use your answer to part (b) to show that A, B, and C are pairwise independent but the probability of their intersection does not equal the product of their separate probabilities.

d. Find the probability of each of these events: *A, C, D, A* and *C, A* and *D, C* and *D, A* and *C* and *D*.

e. Use your answer to part (d) to show that the probability of *A* and *C* and *D* equals the product of their separate probabilities. Show that *A, C,* and *D* are *not* pairwise independent.

(*Note:* To say several events are pairwise independent means that for any two of the events, the probability of their intersection equals the product of their separate probabilities.)

EXERCISE 6-29

Joe DiMaggio had a lifetime batting average of .325. Suppose .325 was the probability of his getting a hit in any single time at bat. Suppose also that his times at bat were independent of each other as far as his getting a hit was concerned. With these assumptions, calculate the probability that Joe DiMaggio would have at least one hit in four at-bats during a single game. How reasonable are the assumptions you made in order to calculate this probability? (This exercise is based on an article by Tom Short and Larry Wasserman, "Should We Be Surprised by the Streak of Streaks?" *Chance,* Volume 2, Spring 1989, page 13.)

EXERCISE 6-30

In a simple genetic model for a characteristic, an individual has two genes for the characteristic. A gene can be one of two types in this model: dominant or recessive. If the individual has at least one dominant-type gene, then the dominant form of the characteristic is expressed. If the individual has two recessive-type genes, then the recessive form of the characteristic is expressed. Phenotype refers to the form of the characteristic expressed. If we denote the dominant form by **B** and the recessive form by **b**, then the gene combinations (or genotypes) **BB**, **Bb**, and **bB** result in phenotype **B**. The genotype **bb** results in phenotype **b**.

a. In a hybrid cross in genetics, both parents have one dominant gene and one recessive gene for the characteristic. An offspring receives one gene for the characteristic from each parent. In the simplest Mendelian genetics model of independent selection, we assume that each of a parent's two genes is equally likely to be passed on to the offspring. We also assume that the parents are independent of one another with regard to the gene passed on to the offspring. With these assumptions, develop a probability model for the genotype of the offspring. Show that under this model we expect a 3:1 ratio of dominant to recessive phenotypes among the offspring of the hybrid cross.

b. In a dihybrid cross, both parents have one dominant and one recessive gene for each of two characteristics. Say, each parent has a dominant gene **B** and a recessive gene **b** for the first characteristic, and a dominant gene **C** and a recessive gene **c** for the second characteristic. In the simplest Mendelian genetics model of independent selection, we assume that for each characteristic, each of a parent's two genes is equally likely to be passed on to the offspring. We also assume independence between characteristics and between parents with regard to the genes passed on to the offspring. With these assumptions, develop a probability model for the genotype of the offspring for these two characteristics. Show that under this model, we expect a 9:3:3:1 ratio of the phenotypes **BC**:**Bc**:**bC**:**bc** (dominant for both characteristics: dominant for the first and recessive for the second: recessive for the first and dominant for the second: recessive for both).

CHAPTER 7

Finite Probability Models Based on Counting Techniques

IN THIS CHAPTER

Permutations, combinations
Binomial distribution
Hypergeometric distribution

We mentioned in Chapter 6 that binomial experiments are commonly used by researchers in many fields. Recall that a binomial experiment consists of independent repetitions of a two-outcome experiment, with the probabilities of the two outcomes constant for each repetition. In Example 6-16, we considered a taste test of Coke versus Pepsi as a binomial experiment. Formal statistical analysis of a binomial experiment is discussed later in several sections on statistical inference (in particular, Sections 10-2, 10-5, 16-1, and 16-2). In this chapter, we develop the probability model for a binomial experiment.

In another common type of investigation, results can be summarized in a 2×2 frequency table. Our introductory example in Chapter 6 was such a study; patients were classified according to the treatment they received (new or standard) and their response to treatment (cured or not cured). One possible probability model for such an experiment leads to a probability distribution called a hypergeometric distribution. Formal statistical analysis of a 2×2 frequency table based on a hypergeometric distribution is discussed in Sections 11-5 and 16-5. We will develop the probability model that leads to a hypergeometric distribution in this chapter.

Both the binomial and the hypergeometric probability distributions involve techniques in counting. In Section 7-1, we discuss two counting problems that lead to some important distributions in statistics. One counting problem asks us to find the number of ways to arrange a finite number of objects in order (permutations). The other asks us to find the number of ways to select some of the items in a group when order of selection does not matter (combinations). After discussing permutations and combinations, we will use them to develop the binomial distribution and the hypergeometric distribution.

7-1 Permutations and Combinations

A *permutation* is an ordered arrangement of a finite number of items. How many possible ways are there to arrange n items in order? Consider first the special case in the following example.

EXAMPLE 7-1

Three United States gold medal winners in the 1988 Seoul Summer Olympics agree to be photographed for a magazine story: Theresa Edwards (basketball), Janet Evans (swimming), and Florence Griffith Joyner (track). If the three athletes stand in a row for the photograph, how many ways are there for them to line up?

The photographer has three athletes to choose from for the left position: Edwards, Evans, and Griffith Joyner. Once that athlete has been selected, the photographer has two athletes to choose from for the center spot. When this second athlete has been chosen, the third must automatically stand on the right. Therefore, there are $3 \times 2 \times 1 = 6$ possible permutations or ordered arrangements of the three women:

Edwards, Evans, Griffith Joyner
Edwards, Griffith Joyner, Evans
Evans, Edwards, Griffith Joyner
Evans, Griffith Joyner, Edwards
Griffith Joyner, Edwards, Evans
Griffith Joyner, Evans, Edwards

A shorthand notation for the product $3 \times 2 \times 1$ is $3!$. We read the symbol $3!$ as "3 factorial."

In general, for any positive integer n, the symbol $n!$ for n factorial denotes the product of all the integers from 1 to n. By convention, we say 0 factorial equals 1: $0! = 1$.

If we want to arrange n items in a row, we have n items to choose from for the first position. With that position filled, there are $n - 1$ remaining items to choose from for the second position, and so on. Therefore, the number of possible permutations or ordered arrangements of n items is the product of the integers from 1 to n, or $n!$.

A **permutation** is an ordered arrangement of a finite number of objects. The number of possible permutations of n objects equals the product of the integers from 1 to n. This product is called n factorial and denoted $n!$.

A group of items selected from a larger collection without regard to the order of selection is a *combination*. Suppose a collection contains n items and k of these items are chosen. How many possible combinations of size k are there? This is the same as asking: How many ways are there to divide the n items into two groups, the first group with k items and the second group with $n - k$ items? Consider first the following special case.

EXAMPLE 7-2

Six United States gold medal winners in the 1988 Seoul Summer Olympics agree to be photographed for a magazine story: the three female athletes noted in Example 7-1, plus Carl Lewis (track), Greg Louganis (diving), and Kenny Monday (wrestling). The editor wants to divide the six athletes into two groups, one group of three to be photographed for the magazine cover, the other three to be photographed for the accompanying story. How many possible ways are there to divide the athletes into two groups of size three, a cover group and a story group?

We are not concerned here about the order of the athletes in a particular group. We care only about which athletes will be photographed for the magazine cover and which for the story inside. There are 20 ways to divide the athletes into two groups of size three, a cover group and a story group. The 20 possibilities are listed in Table 7-1.

Can we find the number 20 without writing down all 20 possible arrangements? Yes, we can. Imagine lining up the six athletes in a row and having the three athletes on the left appear in the cover photograph, the three athletes on the right in the story photograph.

TABLE 7-1 Arrangements of six athletes into two groups of size three, one group for a cover photograph and one group for a story photograph. The order within the two groups does not matter to us.

Arrangement	Cover group	Story group
1	Edwards, Evans, Griffith Joyner	Lewis, Louganis, Monday
2	Edwards, Evans, Lewis	Griffith Joyner, Louganis, Monday
3	Edwards, Evans, Louganis	Griffith Joyner, Lewis, Monday
4	Edwards, Evans, Monday	Griffith Joyner, Lewis, Louganis
5	Edwards, Griffith Joyner, Lewis	Evans, Louganis, Monday
6	Edwards, Griffith Joyner, Louganis	Evans, Lewis, Monday
7	Edwards, Griffith Joyner, Monday	Evans, Lewis, Louganis
8	Edwards, Lewis, Louganis	Evans, Griffith Joyner, Monday
9	Edwards, Lewis, Monday	Evans, Griffith Joyner, Louganis
10	Edwards, Louganis, Monday	Evans, Griffith Joyner, Lewis
11	Evans, Griffith Joyner, Lewis	Edwards, Louganis, Monday
12	Evans, Griffith Joyner, Louganis	Edwards, Lewis, Monday
13	Evans, Griffith Joyner, Monday	Edwards, Lewis, Louganis
14	Evans, Lewis, Louganis	Edwards, Griffith Joyner, Monday
15	Evans, Lewis, Monday	Edwards, Griffith Joyner, Louganis
16	Evans, Louganis, Monday	Edwards, Griffith Joyner, Lewis
17	Griffith Joyner, Lewis, Louganis	Edwards, Evans, Monday
18	Griffith Joyner, Lewis, Monday	Edwards, Evans, Louganis
19	Griffith Joyner, Louganis, Monday	Edwards, Evans, Lewis
20	Lewis, Louganis, Monday	Edwards, Evans, Griffith Joyner

There are $6! = 6 \times 5 \times 4 \times 3 \times 2 \times 1 = 720$ ways to line up the six athletes. But some of these arrangements will be the same as far as the cover/story division is concerned. For example, the arrangement (Edwards, Evans, Griffith Joyner, Lewis, Louganis, Monday) puts the three women on the cover and the men in the story photograph. But the arrangements (Evans, Griffith

Joyner, Edwards, Louganis, Lewis, Monday) and (Griffith Joyner, Evans, Edwards, Monday, Lewis, Louganis) do, too.

Once three athletes have been selected for the three leftmost positions, any of the $3! = 6$ permutations of these three athletes gives the same cover group. Likewise, the three athletes in the rightmost positions can be arranged in any of $3! = 6$ ways and still make up the same story group. Therefore, the $6! = 720$ permutations of the six athletes must be divided by the $3!$ ways to arrange the three leftmost athletes and still get the same cover group, and also divided by the $3!$ ways to arrange the three rightmost athletes and still get the same story group. So we find that there are

$$\frac{6!}{3!3!} = 20$$

ways to select three athletes for the cover, a photograph of the remaining three athletes going with the story. A common notation for the number of ways to divide six elements into two distinct groups of size three is $\binom{6}{3}$:

$$\binom{6}{3} = \frac{6!}{3!3!} = 20$$

In general, suppose a collection of n items is divided into two distinct groups, the first group containing k items and the second group containing $n - k$ items, where k is an integer from 0 to n. The number of ways to do this is

$$\binom{n}{k} = \frac{n!}{k!(n-k)!}$$

The symbols $\binom{n}{k}$ and $\binom{n}{n-k}$ denote the same number.

A **combination** is a group of objects selected from a larger collection without regard to order of selection. There are

$$\binom{n}{k} = \frac{n!}{k!(n-k)!}$$

combinations or ways to select k objects from among n objects without regard to order of selection. This is also the number of ways to divide n objects into two distinct groups, the first of size k and the second of size $n - k$.

There is exactly one way to arrange n items into two distinct groups, the first group containing n items and the second group containing no items. So, 1 must equal $\binom{n}{n}$ and $\binom{n}{0}$:

$$1 = \binom{n}{n} = \binom{n}{0} = \frac{n!}{n!0!}$$

For this reason we say $0!$ equals 1.

In Example 7-3, we use counting techniques to figure out how many ways a seven-game series can end.

EXAMPLE 7-3

The Atlanta Hawks and the Los Angeles Lakers are meeting in the National Basketball Association championship final series. How many possible outcomes are there in this best-of-seven game series?

Let H stand for a game won by the Hawks and L for a game won by the Lakers. Then any outcome of the series can be represented by a sequence of H's and L's. For instance, the sequence (H, L, H, H, H) denotes the outcome in which the Hawks win the championship in five games, the Lakers winning only the second game.

There are two ways the series can end in four games. These outcomes can be denoted by (H, H, H, H) and (L, L, L, L) since the series ends in four games only if one team wins the first four games.

How many ways can the series end in six games? Suppose the Hawks win the championship in six games. This means the Hawks win three of the first five games and then win the sixth, thus ending the series. (If the Hawks win four of the first five games, the series ends before the sixth game.) How many ways are there for the Hawks to win three of the first five games? The five games are divided into two groups, a group of three games won by the Hawks and a group of two games won by the Lakers. There are

$$\binom{5}{3} = \frac{5!}{3!2!} = 10$$

ways to divide the first five games into three games won by the Hawks and two games won by the Lakers. So, there are 10 ways the Hawks could win the series in six games. Similarly, there are 10 ways the Lakers could win the series in six games. Altogether there are 20 ways the best-of-seven series could end in six games, as listed in Table 7-2.

Exercise 7-6 asks you to find the number of ways the series can end in

TABLE 7-2 The number of ways a seven-game series can end in six games

Hawks win the championship	Lakers win the championship
(H, H, H, L, L, H)	(L, L, L, H, H, L)
(H, H, L, H, L, H)	(L, L, H, L, H, L)
(H, L, H, H, L, H)	(L, H, L, L, H, L)
(L, H, H, H, L, H)	(H, L, L, L, H, L)
(H, H, L, L, H, H)	(L, L, H, H, L, L)
(H, L, H, L, H, H)	(L, H, L, H, L, L)
(L, H, H, L, H, H)	(H, L, L, H, L, L)
(H, L, L, H, H, H)	(L, H, H, L, L, L)
(L, H, L, H, H, H)	(H, L, H, L, L, L)
(L, L, H, H, H, H)	(H, H, L, L, L, L)

five games and the number of ways the series can end in seven games. Putting all these totals together will give you the number of possible outcomes of such a best-of-seven series.

In Section 7-2 we use combinations to derive the binomial probability distributions. A binomial probability distribution provides the probability model for independent repetitions of a two-outcome experiment, when the probabilities of the two outcomes are the same for each repetition.

7-2

The Binomial Distributions

A coin toss is an example of an experiment with exactly two possible outcomes (heads or tails). Other examples include a taste test in which a taster must choose one of two possible products (product A or product B), an animal's survival in a toxicity study (alive or dead), a test of a product (functional or not), and a student's performance on an examination (pass or fail). We sometimes call an experiment with exactly two possible outcomes a *Bernoulli experiment* or *Bernoulli trial.*

A **Bernoulli experiment** or **Bernoulli trial** is an experiment that has exactly two possible outcomes. For convenience, we refer to these two outcomes as success and failure.

Suppose the random variable X counts the number of successes in several independent repetitions of a Bernoulli experiment. If the probability of the success outcome is the same for each repetition, we say X has a binomial probability distribution, and we call X a binomial random variable.

In the next example, we find the probability distribution of a particular binomial random variable. Then we will discuss binomial distributions in general and look at another example.

EXAMPLE 7-4

We roll a fair six-sided die four times. (A die is *fair* if each side is equally likely to be face up at the end of a roll.) We win \$1 for each result that is divisible by 3. We win nothing otherwise. What is the probability we win at least \$2? What is our expected gain (the amount we expect to be ahead at the end of the game)? We can phrase these questions in terms of a random variable with a binomial probability distribution.

Let the random variable X equal the number of results that are divisible by 3 in the four rolls. If success refers to 3 or 6 dots coming up and failure to 1, 2, 4, or 5 dots coming up in a single roll, then X counts the number of successes in four rolls of a fair die. The probability of success equals $\frac{1}{3}$ for each roll. If the rolls are independent, then X has a binomial distribution. Let's find that binomial distribution.

Our experiment consists of four rolls of a fair die. If we assume the four rolls are independent of one another, then we can find the probability of each outcome of the experiment. The 16 possible outcomes are listed in Table 7-3, along with the value of the random variable X and the probability of each outcome.

The random variable X has five possible values: 0, 1, 2, 3, and 4. From the probabilities in Table 7-3, we can find the probability distribution for X. For example, the only way X can equal 4 is when each roll results in success (3 or 6 dots face up). Therefore, $P(X = 4) = P(\text{S, S, S, S}) = \frac{1}{81}$.

X equals 3 when the four rolls result in three successes and one failure. There are four ways the experiment could end in three successes and one failure, and each of these four outcomes has probability $\frac{2}{81}$. Therefore, $P(X = 3) = 4 \times \frac{2}{81} = \frac{8}{81}$. In a similar fashion we can find the entire probability distribution for X, shown in Table 7-4.

The probability we win at least \$2 is the probability that the random variable X is greater than or equal to 2:

$$P(X \geq 2) = P(X = 2) + P(X = 3) + P(X = 4) = \frac{24}{81} + \frac{8}{81} + \frac{1}{81} = \frac{33}{81}$$

Our expected gain is the same as the expected value of X:

$$\begin{aligned} E(X) &= \sum_{k=0}^{4} kP(X = k) \\ &= 0 \times \frac{16}{81} + 1 \times \frac{32}{81} + 2 \times \frac{24}{81} + 3 \times \frac{8}{81} + 4 \times \frac{1}{81} \\ &= \frac{108}{81} = 1\tfrac{1}{3} \text{ dollars, or about } \$1.33 \end{aligned}$$

A *binomial experiment* consists of n independent repetitions of an experiment with two possible outcomes, called success and failure for convenience. The probability p of success is the same for each repetition and $0 < p < 1$. If the random variable X counts the number of successes in these n repetitions, then we say X has a binomial distribution, denoted Binomial(n, p). A binomial random variable X takes on integer values from 0 to n. Let's find $P(X = k)$ where k is an integer from 0 to n.

A **binomial experiment** consists of a finite number of independent repetitions of an experiment with two possible outcomes, called success and failure, with the probability of success the same for each repetition.

If X equals k, then k of the n repetitions of the experiment resulted in the success outcome and the other $n - k$ repetitions resulted in the failure outcome. Because the n repetitions are independent of one another, the probability of a particular sequence of k successes and $n - k$ failures is $p^k(1 - p)^{n-k}$. For example, the outcome consisting of k successes followed by $n - k$ failures has this probability. The outcome consisting of $n - k$ failures followed by k successes also has this probability. Likewise, any outcome that includes k successes and $n - k$ failures has probability $p^k(1 - p)^{n-k}$.

TABLE 7-3 The outcomes of the experiment in Example 7-4, the value of the random variable X, and the probability of each outcome

Outcome	Value of X (number of successes)	Probability
(S, S, S, S)	4	$\frac{1}{3} \times \frac{1}{3} \times \frac{1}{3} \times \frac{1}{3} = \frac{1}{81}$
(S, S, S, F)	3	$\frac{1}{3} \times \frac{1}{3} \times \frac{1}{3} \times \frac{2}{3} = \frac{2}{81}$
(S, S, F, S)	3	$\frac{1}{3} \times \frac{1}{3} \times \frac{2}{3} \times \frac{1}{3} = \frac{2}{81}$
(S, F, S, S)	3	$\frac{1}{3} \times \frac{2}{3} \times \frac{1}{3} \times \frac{1}{3} = \frac{2}{81}$
(F, S, S, S)	3	$\frac{2}{3} \times \frac{1}{3} \times \frac{1}{3} \times \frac{1}{3} = \frac{2}{81}$
(S, S, F, F)	2	$\frac{1}{3} \times \frac{1}{3} \times \frac{2}{3} \times \frac{2}{3} = \frac{4}{81}$
(S, F, S, F)	2	$\frac{1}{3} \times \frac{2}{3} \times \frac{1}{3} \times \frac{2}{3} = \frac{4}{81}$
(F, S, S, F)	2	$\frac{2}{3} \times \frac{1}{3} \times \frac{1}{3} \times \frac{2}{3} = \frac{4}{81}$
(S, F, F, S)	2	$\frac{1}{3} \times \frac{2}{3} \times \frac{2}{3} \times \frac{1}{3} = \frac{4}{81}$
(F, S, F, S)	2	$\frac{2}{3} \times \frac{1}{3} \times \frac{2}{3} \times \frac{1}{3} = \frac{4}{81}$
(F, F, S, S)	2	$\frac{2}{3} \times \frac{2}{3} \times \frac{1}{3} \times \frac{1}{3} = \frac{4}{81}$
(F, F, F, S)	1	$\frac{2}{3} \times \frac{2}{3} \times \frac{2}{3} \times \frac{1}{3} = \frac{8}{81}$
(F, F, S, F)	1	$\frac{2}{3} \times \frac{2}{3} \times \frac{1}{3} \times \frac{2}{3} = \frac{8}{81}$
(F, S, F, F)	1	$\frac{2}{3} \times \frac{1}{3} \times \frac{2}{3} \times \frac{2}{3} = \frac{8}{81}$
(S, F, F, F)	1	$\frac{1}{3} \times \frac{2}{3} \times \frac{2}{3} \times \frac{2}{3} = \frac{8}{81}$
(F, F, F, F)	0	$\frac{2}{3} \times \frac{2}{3} \times \frac{2}{3} \times \frac{2}{3} = \frac{16}{81}$

TABLE 7-4 The binomial probability distribution for the random variable X in Example 7-4

k	$P(X = k)$
0	$1 \times \frac{16}{81} = \frac{16}{81}$
1	$4 \times \frac{8}{81} = \frac{32}{81}$
2	$6 \times \frac{4}{81} = \frac{24}{81}$
3	$4 \times \frac{2}{81} = \frac{8}{81}$
4	$1 \times \frac{1}{81} = \frac{1}{81}$

The number of ways the experiment can end in k successes and $n - k$ failures is

$$\binom{n}{k} = \frac{n!}{k!(n - k)!}$$

This is the number of ways to divide n repetitions into two distinct groups—a group of k successes and a group of $n - k$ failures.

The event that X equals k contains the $\binom{n}{k}$ experimental outcomes corresponding to k successes and $n - k$ failures. Each of these $\binom{n}{k}$ outcomes has probability $p^k(1 - p)^{n-k}$. The probability that X equals k is the sum of these $\binom{n}{k}$ probabilities, so

$$P(X = k) = \binom{n}{k} p^k(1 - p)^{n-k}$$

for any integer k from 0 to n.

Suppose a random variable X counts the number of successes in n independent repetitions of a Bernoulli experiment, with probability p of success on each repetition. Then X has a **binomial probability distribution**, denoted **Binomial(n, p).** Probabilities for X have the form

$$P(X = k) = \binom{n}{k} p^k(1 - p)^{n-k}$$

for integer values of k from 0 to n.

We can check that this formula gives us the probabilities we found for the random variable X in Example 7-4. For instance,

$$P(X = 2) = \binom{4}{2}\left(\frac{1}{3}\right)^2\left(\frac{2}{3}\right)^2 = \frac{4!}{2!2!} \times \left(\frac{1}{3}\right)^2 \times \left(\frac{2}{3}\right)^2 = 6 \times \frac{1}{9} \times \frac{4}{9} = \frac{24}{81}$$

which is the same as the value we listed in Table 7-4.

We can find the mean and variance of a binomial random variable as follows:

If X has a Binomial(n, p) probability distribution, then the expected value of X is $n \times p$, or simply np:

$$E(X) = \sum_{k=0}^{n} kP(X = k) = \sum_{k=0}^{n} k \binom{n}{k} p^k(1 - p)^{n-k} = np$$

The variance of X is $n \times p \times (1 - p)$ or $np(1 - p)$:

$$\begin{aligned}\text{Var}(X) &= \sum_{k=0}^{n} (k - E(X))^2\, P(X = k) \\ &= \sum_{k=0}^{n} (k - np)^2 \binom{n}{k} p^k(1 - p)^{n-k} = np(1 - p)\end{aligned}$$

The formula for the expected value of a binomial random variable makes intuitive sense. If we conduct n independent Bernoulli trials with probability p of success on each trial, then we expect, on average, the proportion p of these n trials (that is, np trials) to be successes.

In Example 7-5, we find a probability distribution for a random variable that we will see again when we discuss formal statistical analysis of a taste-test experiment in Chapter 9.

EXAMPLE 7-5

Twelve members of a statistics class participate in a taste test comparing Coke and Pepsi. Each student receives two identical-looking cups, one filled with Coke and the other with Pepsi. A student tastes samples from each cup and decides which beverage he or she prefers.

Suppose that Coke and Pepsi are equally likely to be preferred. Then how many students would we expect to express a preference for Coke? Would we be surprised if nine or more students made the same selection?

To answer these questions, we must build a probability model for our experiment. Suppose the students make independent selections, so one student's choice does not in any way affect another student's choice. Suppose also that each student has the same probability p of choosing Coke. (This second assumption may greatly simplify reality, because of physiological differences between people.) If Coke and Pepsi are equally likely to be preferred, then $p = \frac{1}{2}$.

Let the random variable Y denote the number of students who choose Coke. Then under the model assumptions above, Y has a Binomial(12, $\frac{1}{2}$) distribution. We know the probability distribution of Y without having to write down all possible outcomes in the sample space. This is fortunate; it would be extremely tedious to write down all possible sets of preferences for the 12 students, since there are $2^{12} = 4{,}096$ of them! Table 7-5 shows the probability distribution for the random variable Y.

If Coke and Pepsi are equally likely to be preferred, then we would expect half of the 12 students to express a preference for Coke and half to express a preference for Pepsi. The number of students we expect to prefer Coke is equal to the expected value of Y: $E(Y) = 12 \times \frac{1}{2} = 6$ students. The variance of Y is $\text{Var}(Y) = 12 \times \frac{1}{2} \times \frac{1}{2} = 3$ students^2 and the standard deviation of Y is $\sqrt{\text{Var}(Y)} = 1.7$ students.

Would we be surprised to see nine or more students express the same preference? The probability that nine or more students choose Coke is the probability that Y is greater than or equal to 9:

$$P(Y \geq 9) = \sum_{k=9}^{12} \binom{12}{k} \left(\frac{1}{2}\right)^k \left(\frac{1}{2}\right)^{n-k} = .073$$

TABLE 7-5 Binomial probability distribution for the random variable Y in Example 7-5. The probabilities listed have been rounded to 4 decimal places; this is why they do not sum exactly to 1.

k	$P(Y = k)$
0	$\binom{12}{0}(\frac{1}{2})^0(\frac{1}{2})^{12} = 1 \times (\frac{1}{2})^{12} = .0002$
1	$\binom{12}{1}(\frac{1}{2})^1(\frac{1}{2})^{11} = 12 \times (\frac{1}{2})^{12} = .0029$
2	$\binom{12}{2}(\frac{1}{2})^2(\frac{1}{2})^{10} = 66 \times (\frac{1}{2})^{12} = .0161$
3	$\binom{12}{3}(\frac{1}{2})^3(\frac{1}{2})^9 = 220 \times (\frac{1}{2})^{12} = .0537$
4	$\binom{12}{4}(\frac{1}{2})^4(\frac{1}{2})^8 = 495 \times (\frac{1}{2})^{12} = .1208$
5	$\binom{12}{5}(\frac{1}{2})^5(\frac{1}{2})^7 = 792 \times (\frac{1}{2})^{12} = .1934$
6	$\binom{12}{6}(\frac{1}{2})^6(\frac{1}{2})^6 = 924 \times (\frac{1}{2})^{12} = .2256$
7	$\binom{12}{7}(\frac{1}{2})^7(\frac{1}{2})^5 = 792 \times (\frac{1}{2})^{12} = .1934$
8	$\binom{12}{8}(\frac{1}{2})^8(\frac{1}{2})^4 = 495 \times (\frac{1}{2})^{12} = .1208$
9	$\binom{12}{9}(\frac{1}{2})^9(\frac{1}{2})^3 = 220 \times (\frac{1}{2})^{12} = .0537$
10	$\binom{12}{10}(\frac{1}{2})^{10}(\frac{1}{2})^2 = 66 \times (\frac{1}{2})^{12} = .0161$
11	$\binom{12}{11}(\frac{1}{2})^{11}(\frac{1}{2})^1 = 12 \times (\frac{1}{2})^{12} = .0029$
12	$\binom{12}{12}(\frac{1}{2})^{12}(\frac{1}{2})^0 = 1 \times (\frac{1}{2})^{12} = .0002$

Under the equal-preference model, there is about a 7% chance of seeing 9 or more of the 12 students select Coke. Similarly, under the equal-preference model, there is about a 7% chance of seeing 9 or more of the 12 students select Pepsi. So, under the equal-preference model, there is about a 14% chance that 9 or more students express the same preference. Whether we find such an event surprising depends on whether we think an event with a 14% chance of occurring to be likely or unlikely. (It would no doubt depend as well on whether we had a financial tie with one or the other beverage company!) As unbiased observers, we may not be too surprised to see 9 or more of the 12 students make the same selection, even if we believe in the equal-preference model.

Probabilities for Binomial(n, p) distributions are listed in Table A at the end of the book, for selected values of n and p.

In Section 7-3, we discuss another group of probability distributions based on combinations, the hypergeometric distributions. A hypergeometric distribution is sometimes used to model experimental results that can be summarized in a 2×2 frequency table.

7-3

The Hypergeometric Distributions

Suppose the random variable X counts the number of Type 1 objects in a sample selected at random from a finite collection of objects, each classified as either Type 1 or Type 2. Then we say X has a hypergeometric probability distribution, and we call X a hypergeometric random variable.

In the next example, we find the probability distribution of a particular hypergeometric random variable. Then we will discuss the hypergeometric distribution in general and apply it to another example.

EXAMPLE 7-6

A paper bag contains six spark plugs—four good ones and two bad ones. You are changing the spark plugs in your car, unaware that there are any bad ones in your bag. Without looking, you reach into the bag and select four spark plugs. What is the probability that at least one of the spark plugs you select is bad? What is the expected number of bad spark plugs in your sample? We can phrase these questions in terms of a random variable that has a hypergeometric probability distribution.

Let the random variable X equal the number of bad spark plugs among the four you select. If the four spark plugs in the sample were randomly selected from among the six in the bag, then X has a hypergeometric distribution. Let's find this hypergeometric distribution.

When we say that four spark plugs are randomly selected from the six in the bag, we mean that each sample of size four is equally likely to be selected. We know that the number of ways to select four items from among six items is

$$\binom{6}{4} = \frac{6!}{4!2!} = 15$$

Fifteen is the number of ways to divide the six spark plugs into two distinct groups: a group of four that is taken from the bag and a group of two that stays in the bag.

There are 15 ways to select four spark plugs from the six in the bag, and each of these outcomes is equally likely. Let's designate the four good spark plugs in the bag with the letters a, b, c, and d, and the two bad spark plugs with the letters u and v. Then we can list the 15 possible outcomes of the experiment as in Table 7-6.

The random variable X has three possible values: 0, 1, and 2. From the probabilities in Table 7-6, we can find the probability distribution for X. For instance, X equals 0 only when all four spark plugs selected are good ones, so $P(X = 0) = P(\text{a, b, c, d selected}) = \frac{1}{15}$. X equals 1 when one bad spark plug and three good ones are selected. We see in Table 7-6 that there are eight ways to select one bad spark plug and three good ones. Each of these eight outcomes has probability $\frac{1}{15}$, so $P(X = 1) = 8 \times \frac{1}{15} = \frac{8}{15}$. Similarly, we find that the probability that X equals 2 is $\frac{6}{15}$. The probability distribution for X is summarized in Table 7-7.

TABLE 7-6 The outcomes of the experiment in Example 7-6, the value of the random variable X, and the probability of each outcome. The letters a, b, c, and d represent good spark plugs; u and v represent bad spark plugs. The order within the two groups (selected and not selected) does not matter.

Outcome		Value of X, number of bad spark plugs selected	Probability
Selected	Not selected		
a, b, c, d	u, v	0	$\frac{1}{15}$
a, b, c, u	d, v	1	$\frac{1}{15}$
a, b, c, v	d, u	1	$\frac{1}{15}$
a, b, d, u	c, v	1	$\frac{1}{15}$
a, b, d, v	c, u	1	$\frac{1}{15}$
a, c, d, u	b, v	1	$\frac{1}{15}$
a, c, d, v	b, u	1	$\frac{1}{15}$
b, c, d, u	a, v	1	$\frac{1}{15}$
b, c, d, v	a, u	1	$\frac{1}{15}$
a, b, u, v	c, d	2	$\frac{1}{15}$
a, c, u, v	b, d	2	$\frac{1}{15}$
a, d, u, v	b, c	2	$\frac{1}{15}$
b, c, u, v	a, d	2	$\frac{1}{15}$
b, d, u, v	a, c	2	$\frac{1}{15}$
c, d, u, v	a, b	2	$\frac{1}{15}$

TABLE 7-7 The hypergeometric probability distribution for the random variable X in Example 7-6

k	$P(X = k)$
0	$\frac{1}{15}$
1	$\frac{8}{15}$
2	$\frac{6}{15}$

The probability that you select at least one bad spark plug is the same as the probability that the random variable X equals 1 or 2: $P(X \geq 1) = \frac{8}{15} + \frac{6}{15} = \frac{14}{15}$. We obtain the same answer by noting that the probability of at least one bad spark plug equals 1 minus the probability of no bad spark plugs selected:

$$P(X \geq 1) = 1 - P(X = 0) = 1 - \frac{1}{15} = \frac{14}{15}$$

The expected number of bad spark plugs you select is the same as the expected value of X:

$$\begin{aligned} E(X) &= 0 \times P(X = 0) + 1 \times P(X = 1) + 2 \times P(X = 2) \\ &= 0 \times \frac{1}{15} + 1 \times \frac{8}{15} + 2 \times \frac{6}{15} = \frac{20}{15} \end{aligned}$$

or $1\frac{1}{3}$ spark plugs.

Consider now a general hypergeometric probability distribution. Suppose that in a group of N objects, m_1 are Type 1 and $m_2 = N - m_1$ are Type 2. We select a sample of size n at random from among the N objects. If the random variable X counts the number of Type 1 objects in the sample, then we say X has a hypergeometric distribution.

To specify the distribution of X, we will consider three conditions: the sample size n is less than or equal to both m_1 and m_2; the sample size n is greater than the number m_1 of Type 1 objects; the sample size n is greater than the number m_2 of Type 2 objects. (It is possible for n to be greater than both m_1 and m_2.)

Let's consider first the case that the sample size n is less than or equal to the number m_1 of Type 1 objects and the number m_2 of Type 2 objects. (This would be the case if, for example, you were selecting $n = 4$ spark plugs from a bag containing $m_1 = 5$ bad ones and $m_2 = 15$ good ones.) In this case, X can take any integer value k from 0 to n. Because of the random selection, $P(X = k)$ equals the number of ways to select k of m_1 Type 1 objects and $n - k$ of m_2 Type 2 objects, divided by the number of ways to select n of N objects.

There are $\binom{m_1}{k}$ ways to select k of m_1 Type 1 objects. There are $\binom{m_2}{n-k}$ ways to select $n - k$ of m_2 Type 2 objects. The number of ways to select k of m_1

Type 1 objects *and* $n - k$ of m_2 Type 2 objects is $\binom{m_1}{k} \times \binom{m_2}{n-k}$. The number of ways to select n objects from a total of $N = m_1 + m_2$ is $\binom{N}{n}$. Therefore, $P(X = k)$ is

$$P(X = k) = \frac{\binom{m_1}{k}\binom{m_2}{n-k}}{\binom{N}{n}}$$

for integer values of k from 0 to n.

Let's consider next the second condition: the sample size n is greater than the number m_1 of Type 1 objects. (This is the case in Example 7-6, where you select $n = 4$ spark plugs from a bag containing $m_1 = 2$ bad ones and $m_2 = 4$ good ones.) Since X counts the number of Type 1 objects in the sample, X cannot be greater than m_1. The largest possible value for X is the minimum of n and m_1, denoted minimum(n, m_1). In Example 7-6, the largest possible number of bad spark plugs in the sample is 2, the minimum of $n = 4$ and $m_1 = 2$.

Now suppose the sample size n is larger than the number m_2 of Type 2 objects. Then the number of Type 1 objects in the sample is at least $n - m_2$. (This would be the case if you were selecting $n = 4$ spark plugs from a bag containing $m_1 = 5$ bad ones and $m_2 = 3$ good ones. There would have to be at least one bad spark plug in the sample, with $n - m_2 = 4 - 3 = 1$.) The smallest possible value for X in this case is the maximum of 0 and $n - m_2$, denoted by maximum$(0, n - m_2)$.

Suppose a random variable X counts the number of Type 1 objects in a sample of size n selected at random from a collection of N objects, m_1 of Type 1 and $m_2 = N - m_1$ of Type 2. Then X has a **hypergeometric probability distribution,** with probabilities of the form

$$P(X = k) = \frac{\binom{m_1}{k}\binom{m_2}{n-k}}{\binom{N}{n}}$$

where k is an integer from maximum$(0, n - m_2)$ to minimum(n, m_1).

We can check that the formula for $P(X = k)$ works for the hypergeometric random variable in Example 7-6. You are selecting $n = 4$ spark plugs from a bag containing $m_1 = 2$ bad ones and $m_2 = 4$ good ones. The random variable X counts the number of bad spark plugs in the sample. The smallest value X can take on is maximum$(0, n - m_2)$ = maximum$(0, 4 - 4) = 0$. The largest value X can take on is minimum(n, m_1) = minimum$(4, 2) = 2$. So, X can take the values 0, 1, and 2. Using the formula for hypergeometric probabilities, we find

$$P(X = 0) = \frac{\binom{2}{0}\binom{4}{4}}{\binom{6}{4}} = \frac{1 \times 1}{15} = \frac{1}{15}$$

$$P(X = 1) = \frac{\binom{2}{1}\binom{4}{3}}{\binom{6}{4}} = \frac{2 \times 4}{15} = \frac{8}{15}$$

$$P(X = 2) = \frac{\binom{2}{2}\binom{4}{2}}{\binom{6}{4}} = \frac{1 \times 6}{15} = \frac{6}{15}$$

This is the same probability distribution we found for X in Table 7-7.

We can find the mean and variance of a hypergeometric random variable as follows:

The expected value of a hypergeometric random variable X equals nm_1/N:

$$E(X) = \sum kP(X = k) = \sum k \frac{\binom{m_1}{k}\binom{m_2}{n-k}}{\binom{N}{n}} = \frac{nm_1}{N}$$

where the sum is over all integers k from maximum$(0, n - m_2)$ to minimum(n, m_1). That is, the expected number of Type 1 objects in the sample equals the sample size n times the proportion of Type 1 objects in the total collection.

The variance of a hypergeometric random variable X is

$$\begin{aligned} \text{Var}(X) &= \sum (k - E(X))^2 P(X = k) \\ &= \sum \left(k - \frac{nm_1}{N}\right)^2 \frac{\binom{m_1}{k}\binom{m_2}{n-k}}{\binom{N}{n}} \\ &= \frac{nm_1}{N} \times \left(1 - \frac{m_1}{N}\right) \times \left(\frac{N-n}{N-1}\right) \end{aligned}$$

where the sum is again over integers k from maximum$(0, n - m_2)$ to minimum(n, m_1).

In Example 7-7, we use a hypergeometric distribution in a quality control application involving acceptance sampling.

EXAMPLE 7-7

Suppose you are in charge of quality control at Rocky's Rocking Horse Company. Each day, 90 rocking horses are produced, and you select a small number of these for careful inspection. You classify each rocking horse in your sample as either acceptable or unacceptable for shipment to toy outlets. If you find too many unacceptable rocking horses in your sample, you then inspect the entire day's production. Otherwise, you pronounce the day's production as acceptable for shipment, with no further inspection.

You know that the probability a sample passes inspection depends on the proportion of unacceptable rocking horses produced that day, as well as on the number of rocking horses you inspect. How can you use this knowledge to find an appropriate sampling and decision scheme?

Let's find out how your inspection routine will perform if you randomly select 5 rocking horses from the 90 produced; you accept the day's production lot only if all 5 are acceptable. Let the random variable W count the number of unacceptable rocking horses in your sample of 5. Then W has a hypergeometric distribution that depends on the number of unacceptable horses produced that day.

Suppose that 10% of the production is unacceptable. Then 9 of the 90 rocking horses are unacceptable and 81 are acceptable. The probability the sample passes inspection is the probability W equals 0:

$$P(W = 0) = \frac{\binom{9}{0}\binom{81}{5}}{\binom{90}{5}} = .58$$

The probability is .58 that the sample passes inspection, when 10% of the lot is unacceptable.

If 20% of the lot is unacceptable, then 18 rocking horses are unacceptable and 72 are acceptable. The probability you accept the lot is the probability W equals 0:

$$P(W = 0) = \frac{\binom{18}{0}\binom{72}{5}}{\binom{90}{5}} = .32$$

The probability is .32 that the sample passes inspection, when 20% of the lot is unacceptable.

We can continue to calculate the probability a sample passes inspection, assuming different proportions of unacceptable rocking horses in the lot. Table 7-8 lists some of these probabilities.

For a given sampling and decision rule, a plot of the probability a sample passes inspection for different proportions unacceptable in the lot is called an *operating characteristic curve*. In Example 7-7, the sampling and decision rule

TABLE 7-8 Probability of accepting the day's production lot in Example 7-7, for different proportions of unacceptable rocking horses in the lot. The decision rule is to accept the lot if no unacceptable rocking horses are found in a sample of 5 selected at random from the 90 rocking horses produced in a day.

Proportion defective in the lot	Probability of accepting the lot
$\frac{0}{90} = 0$	$\frac{\binom{0}{0}\binom{90}{5}}{\binom{90}{5}} = 1$
$\frac{9}{90} = .1$	$\frac{\binom{9}{0}\binom{81}{5}}{\binom{90}{5}} = .58$
$\frac{18}{90} = .2$	$\frac{\binom{18}{0}\binom{72}{5}}{\binom{90}{5}} = .32$
$\frac{27}{90} = .3$	$\frac{\binom{27}{0}\binom{63}{5}}{\binom{90}{5}} = .16$
$\frac{36}{90} = .4$	$\frac{\binom{36}{0}\binom{54}{5}}{\binom{90}{5}} = .07$
$\frac{45}{90} = .5$	$\frac{\binom{45}{0}\binom{45}{5}}{\binom{90}{5}} = .03$
$\frac{54}{90} = .6$	$\frac{\binom{54}{0}\binom{36}{5}}{\binom{90}{5}} = .009$
$\frac{63}{90} = .7$	$\frac{\binom{63}{0}\binom{27}{5}}{\binom{90}{5}} = .002$
$\frac{72}{90} = .8$	$\frac{\binom{72}{0}\binom{18}{5}}{\binom{90}{5}} = .0002$
$\frac{81}{90} = .9$	$\frac{\binom{81}{0}\binom{9}{5}}{\binom{90}{5}} = .000003$
$\frac{90}{90} = 1$	0

is: Select 5 rocking horses at random from among the 90 produced; accept the lot if there are no unacceptable rocking horses in the sample. Figure 7-1 shows a sketch of the operating characteristic curve for this sampling and decision rule. Some of the calculations for this curve are shown in Table 7-8. Note that the operating characteristic curve for this example is not continuous because there are just 91 possible values for the proportion of unacceptable items in the lot ($0 = \frac{0}{90}, \frac{1}{90}, \frac{2}{90}, \ldots, \frac{89}{90}, \frac{90}{90} = 1$).

For a sampling and decision rule in an acceptance sampling inspection problem, an **operating characteristic curve** is a plot of the probability of accepting the lot for different proportions of defectives in the lot.

The probability of accepting the lot depends on both the size of the sample inspected and the decision rule. Each sampling and decision rule has its own operating characteristic curve (see Exercise 7-9).

In acceptance sampling for quality control, we compare the operating characteristic curves for several sampling and decision rules. We then select an inspection scheme that seems to best balance the opposing goals of minimizing sampling costs and accepting only lots with small proportions defective.

Note that such acceptance sampling procedures serve the *defensive* purpose of preventing unacceptable lots from being shipped to consumers (Schilling, 1982). Taking the *offensive* in quality control, we should use good experimental design during product development to design quality into the product (see Box, Hunter, and Hunter, 1978; and Part III of this book) and

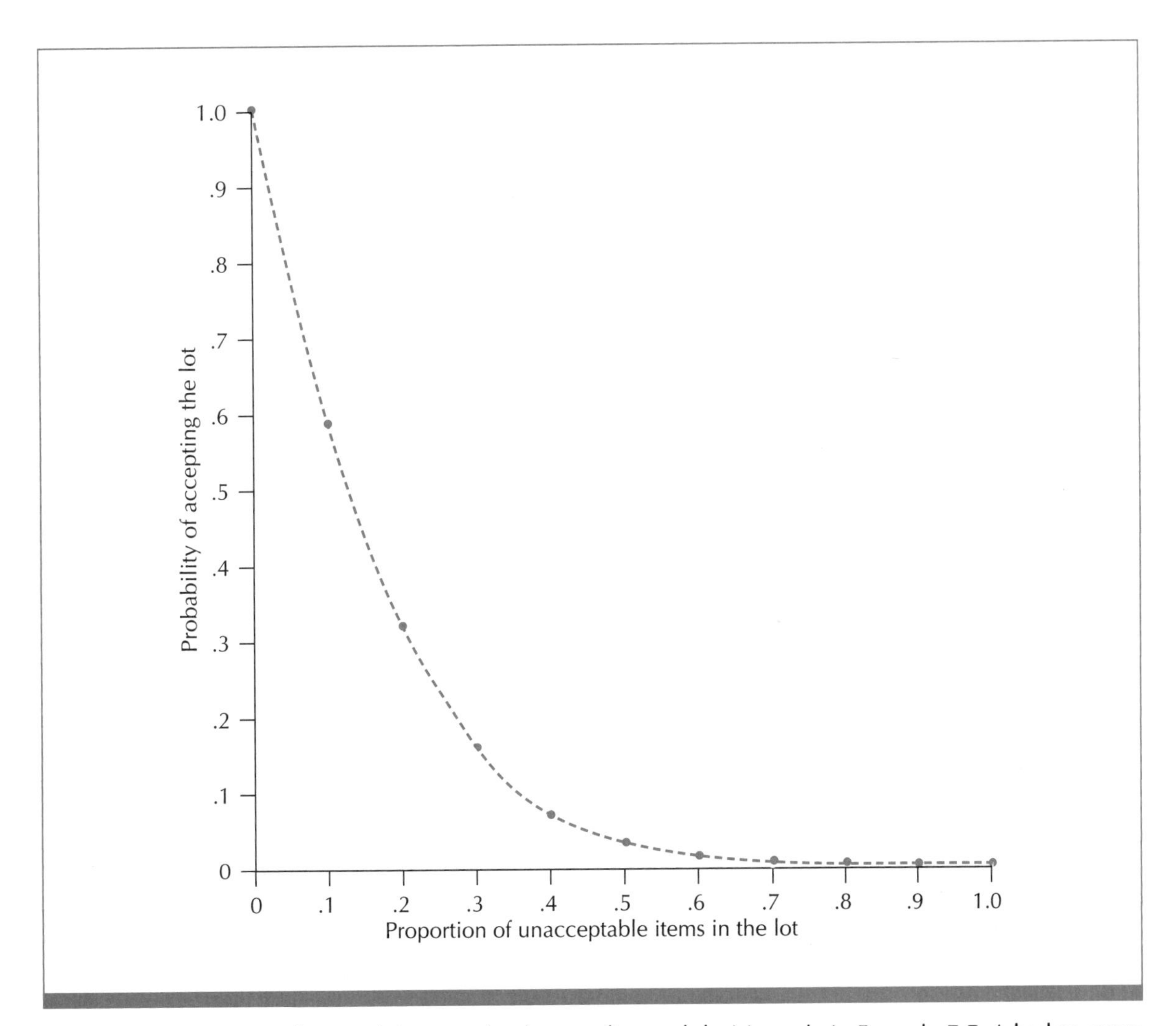

FIGURE 7-1 Operating characteristic curve for the sampling and decision rule in Example 7-7. A broken curve connects the points calculated in Table 7-8; the possible values for the proportion of unacceptable items in the lot are $0 = \frac{0}{90}, \frac{1}{90}, \frac{2}{90}, \ldots, \frac{89}{90}, \frac{90}{90} = 1$.

charting tools of statistical process control (as discussed in Ryan, 1989) to monitor quality of the product.

In Chapters 6 and 7, we have considered finite sample spaces and finite random variables. We can extend these ideas to experiments that have a countably infinite set of possible outcomes. For instance, we might count the telephone calls coming through a switchboard in a fixed interval of time, or we might count the dents in a piece of sheet metal. Then we can think of our count (or random variable) as having any nonnegative integer as a possible value. In repeated independent Bernoulli trials, if we count the number of failures before the first success, we again have a random variable that can take on any nonnegative integer value. Although interesting and useful, we will not consider probability distributions for countably infinite random variables. (See, for example, Larsen and Marx, 1986, Chapter 4; or Rice, 1988, Chapter 2.)

We will use some continuous probability distributions when we discuss statistical inference in Part III. Recall that a continuous random variable takes values in an interval of numbers. We call the probability distribution for a continuous random variable a **continuous probability distribution.** In Chapter 8, we will discuss a special group of continuous probability distributions, the Gaussian (or normal) distributions. Random variables with Gaussian probability distributions form the basis for much of classical statistical inference.

Summary of Chapter 7

Permutations and combinations form the basis for two important types of probability distributions: the binomial distributions and the hypergeometric distributions.

Suppose we have n independent repetitions of a two-outcome (success, failure) experiment, where p is the probability of success on each repetition. A random variable that counts the number of successes in these n repetitions has a binomial distribution, denoted Binomial(n, p).

Suppose a sample is selected at random from a finite collection of objects, each object classified as either Type 1 or Type 2. A random variable that counts the number of Type 1 objects in the sample has a hypergeometric probability distribution.

Minitab Appendix for Chapter 7

Finding Binomial Probabilities

Minitab provides probabilities for binomial distributions using the PDF (probability function or probability density function) command with the BINOMIAL subcommand. To display probabilities for the binomial distribution with sample size $n = 12$ and probability of success $p = .5$, we use the command

```
MTB>   pdf;
SUBC>  binomial 12 0.5.
```

Minitab will print the probabilities, as shown in Figure M7-1.

The command

```
MTB>   pdf;
SUBC>  binomial 4 0.6.
```

results in a listing of probabilities for the Binomial(4, .6) distribution, as shown in Figure M7-2.

We can also use the PDF command to find the probability that a binomial random variable takes on a single value. For instance,

```
MTB>   pdf 11;
SUBC>  binomial 12 0.5.
        k          P(X = k)
      11.00         0.0029
```

causes Minitab to print $P(X = 11)$, where X is a random variable having the Binomial(12, .5) distribution. We see that $P(X = 11)$ equals .0029.

```
BINOMIAL WITH N =   12   P = 0.500000
   K            P( X = K)
   0              0.0002
   1              0.0029
   2              0.0161
   3              0.0537
   4              0.1208
   5              0.1934
   6              0.2256
   7              0.1934
   8              0.1208
   9              0.0537
  10              0.0161
  11              0.0029
  12              0.0002
```

FIGURE M7-1 Probabilities for the Binomial(12, .5) distribution

```
BINOMIAL WITH N =    4   P = 0.600000
   K            P( X = K)
   0              0.0256
   1              0.1536
   2              0.3456
   3              0.3456
   4              0.1296
```

FIGURE M7-2 Probabilities for the Binomial(4, .6) distribution

```
BINOMIAL WITH N =    4  P = 0.600000
   K  P( X LESS OR = K)
   0            0.0256
   1            0.1792
   2            0.5248
   3            0.8704
   4            1.0000
```

FIGURE M7-3 Cumulative probabilities for the Binomial(4, .6) distribution

Finding Cumulative Binomial Probabilities

Minitab will also provide cumulative probabilities with the CDF (cumulative distribution function) command. The command

```
MTB>   cdf;
SUBC>  binomial 4 0.6.
```

results in a display of cumulative probabilities of the form $P(X \leq c)$, where X has the Binomial(4, .6) distribution. These cumulative probabilities are displayed in Figure M7-3.

We can also have Minitab print a single cumulative probability, $P(X \leq c)$, for a specified value of c. The command

```
MTB>   cdf 3;
SUBC>  binomial 12 0.5.
     k        P(X LESS OR = k)
   3.00             0.0730
```

will cause Minitab to print the probability $P(X \leq 3)$, where X has the Binomial(12, .5) distribution. We see that $P(X \leq 3)$ equals .0730.

Finding Values Corresponding to Cumulative Binomial Probabilities

The INVCDF command with the BINOMIAL subcommand provides the number c such that $P(X \leq c)$ equals a specified probability, where X has the indicated binomial distribution. If no value of c works exactly, Minitab will print two values of c, corresponding to cumulative probabilities surrounding the specified value. For instance,

```
MTB>   invcdf 0.025;
SUBC>  binomial 12 0.5.
     k  P(X LESS OR = k)        k  P(X LESS OR = k)
     2        0.0193            3        0.0730
```

Minitab does not print probabilities for hypergeometric distributions.

Exercises for Chapter 7

EXERCISE 7-1

Four floats are lining up for the homecoming parade. How many ways are there for the four floats to line up?

EXERCISE 7-2 The starting five players of the high school basketball team are lining up for a yearbook picture. How many different ways are there to line up the players in a row? If the five players line up in a completely random fashion, what is the probability that the tallest player will be in the center position for the photograph?

EXERCISE 7-3 The final examination in your statistics class consists of eight questions. You may pick any five of the questions to answer. How many ways can you select five questions from the eight questions on the test?

EXERCISE 7-4 You and eight other students are taking a beginning statistics class. The instructor has decided that she will randomly select six of you to pass the course, and the other three to fail. How many ways are there for the instructor to choose the six students who will pass the course? What is the probability that you will pass the course?

EXERCISE 7-5 How many ways are there to select 6 elements from a set containing 36 elements, without regard to order of selection? In the Megabucks game described in Example 6-2, assume the winning numbers are selected at random. If a worthy citizen purchases one ticket with one six-number combination, what is the probability the citizen will win?

EXERCISE 7-6 Referring to Example 7-3, find the number of ways a best-of-seven game series can end in five games. Find the number of ways such a series can end in seven games. Find the total number of ways a best-of-seven game series can end.

EXERCISE 7-7 Your softball team has made it to the statewide finals. League organizers are trying to decide whether to have a five-game series or a seven-game series for the playoffs.

a. Suppose whether your team wins or loses any one game is independent of whether you win or lose any other game. What does this mean?

b. Decide whether your team has a better chance of winning a five-game series or a seven-game series if the probability you beat your opponent in any single game is: .4, .5, or .6.

EXERCISE 7-8 Suppose in Example 7-2 the editor uses a random process to choose which three of the six Olympic gold medal winners will appear on the magazine cover. Then the 20 possible outcomes in the sample space, listed in Table 7-1, are equally likely. Under this probability model, find the following probabilities:

a. The probability that the athletes on the cover will all be the same sex.

b. The probability that Florence Griffith Joyner will be on the cover. To find this probability without counting the outcomes in Table 7-1, create a hypergeometric random variable in the following way. Call Florence Griffith Joyner a Type 1 athlete and call the other five athletes Type 2. Let the random variable W count the number of Type 1 athletes on the cover. Find $P(W = 1)$.

c. The probability that Florence Griffith Joyner and Carl Lewis will be on the cover. To find this probability without counting the outcomes in Table 7-1, create a hypergeometric random variable in the following way. Call Florence Griffith Joyner and Carl Lewis Type 1 athletes and call the other four athletes Type 2. If X counts the number of Type 1 athletes in the sample, find $P(X = 2)$.

d. The probability that at least one female athlete will be in the cover photograph.

EXERCISE 7-9

You are in charge of quality control at a company that makes expensive sports cars. Factory workers produce ten cars a day. You subject all ten cars to some quality inspection. In addition, you select a sample of the ten cars for exhaustive testing. Each car tested is classified as either acceptable or not acceptable for shipment.

One day, all ten cars pass the preliminary inspection. Construct a table similar to Table 7-8 and plot the operating characteristic curve if the sampling and decision rule is:

a. Accept the day's production with no exhaustive testing.

b. Accept the ten cars produced that day if one car selected at random is found to be acceptable.

c. Accept the ten cars produced that day if two cars selected at random are both found to be acceptable.

d. Accept the ten cars produced that day if five cars selected at random are all found to be acceptable.

e. Accept the ten cars produced that day if all ten cars are found to be acceptable after exhaustive testing.

f. Compare the operating characteristic curves you found in parts (a)–(e). Discuss how each sampling and decision rule balances the opposing goals of minimizing inspection costs and accepting the day's production only when the proportion of unacceptable cars in the lot is small.

Construct a table similar to Table 7-8 and plot the operating characteristic curve if the sampling and decision rule is:

g. Accept the ten cars produced that day if, of two cars selected at random, one or none is found to be unacceptable.

h. Accept the ten cars produced that day if, among five cars selected at random, one or none is found to be unacceptable.

i. Accept the day's production if one or none of the ten cars produced is found to be unacceptable.

j. Compare the sampling and decision rules in parts (g)–(i) with each other and with the ones in parts (c)–(e).

EXERCISE 7-10

A couple plans to have two children. Suppose that they are able to carry out their plan.

a. Assume that the sex of one child is independent of the sex of the other. What does this mean?

b. Assume also that the probability p of a girl is the same for both births. Let an outcome note the sex of the firstborn child and the sex of the second-born child. Write down the sample space and probability function for this experiment.

c. If the random variable Y counts the number of female children the couple has, write down the probability distribution of Y.

d. Find the following probabilities: the probability the couple has one girl and one boy; the probability the couple has two girls; the probability the couple has two boys; the probability the couple has two children of the same sex.

e. Find the number of girls the couple can expect to have when the probability p that a baby is a girl is: $\frac{12}{25}$, $\frac{1}{2}$, or $\frac{13}{25}$.

EXERCISE 7-11

Your small cousin went trick-or-treating and has a bag of miniature candy bars. He says you can pick three without looking. You love Yummy candy bars the best. Let X denote the number of Yummy candy bars among the three you select. Find the probability distribution of X if:

a. There are five Yummy bars among ten candy bars in the bag.

b. There are two Yummy bars among ten candy bars in the bag.

c. There are eight Yummy bars among ten candy bars in the bag.

EXERCISE 7-12

An American swimmer is competing in three events in the summer Olympics. Experts give her probability p of winning, for each of the events.

a. Assume that her winning or losing one event is independent of her result in any other event. What does this mean?

b. Let an outcome note whether or not she wins the gold medal, for each of the three events. Write down the sample space and probability function for this experiment.

c. If the random variable X counts the number of gold medals the swimmer wins among the three events, write down the probability distribution of X.

d. Find the following, when the probability p she wins any single event is .4, .5, or .6: the probability the swimmer wins no gold medals; the probability she wins at least one gold medal; the probability she wins two or three gold medals; the probability she wins all three gold medals; the number of gold medals she can expect to win.

EXERCISE 7-13

At a factory there are three shifts. The numbers of men and women working each shift are summarized below:

Shift	Men	Women
Day	10	10
Evening	10	5
Night	9	3

Management announces that it will randomly select four workers from each shift for intensive performance review. When two men and two women are selected from each shift, the women claim sex discrimination. Answer the following for this problem:

a. If there really is random selection, find the probability that two or more of the four workers selected from the day shift will be women.

b. If there really is random selection, find the probability that two or more of the four workers selected from the evening shift will be women.

c. If there really is random selection, find the probability that two or more of the four workers selected from the night shift will be women.

d. If the selection is independent for different shifts, use your answers to parts (a)–(c) to find the probability under random selection that two or more women from the day shift *and* two or more women from the evening shift *and* two or more women from the night shift will be selected.

e. Use your answer to part (d) to discuss the women's claim that the selection process was *not* random, but biased toward selection of women.

f. How would your answers to parts (a)–(e) change if three women from each shift had been selected?

g. How would your answers to parts (a)–(e) change if one woman from each shift had been selected?

EXERCISE 7-14

You are a consulting statistician at a medical research center. Physicians at the center have developed a potential new treatment for a disease and wish to try it out in a preliminary study on humans. They will use the results of this study to decide whether to continue research on this treatment. The physicians come to you for advice on designing their study. To develop a probability model for their experiment, you make two simplifying assumptions:

Response to treatment is independent for different patients.
Each patient has the same probability p of being cured.

a. Discuss the reasonableness of these simplifying assumptions for this experimental situation.

If the physicians treat a total of n patients, you suggest the following decision rule. Select a number c from 0 to n. Let X denote the number of patients cured. Then decide to:

Continue research if $X \geq c$; discontinue research if $X < c$.

b. What probability distribution does X have?

c. Find $P(X \geq c)$ for the following situations:

(i) $n = 10, c = 6, p = \frac{3}{4}$
(ii) $n = 10, c = 8, p = \frac{3}{4}$
(iii) $n = 10, c = 6, p = \frac{9}{10}$
(iv) $n = 10, c = 8, p = \frac{9}{10}$
(v) $n = 20, c = 12, p = \frac{3}{4}$
(vi) $n = 20, c = 16, p = \frac{3}{4}$
(vii) $n = 20, c = 12, p = \frac{9}{10}$
(viii) $n = 20, c = 16, p = \frac{9}{10}$

d. The probability that the physicians will continue research on the treatment equals $P(X \geq c)$. Write a report to the physicians explaining how this probability depends on: the sample size n, the criterion number c selected, and the probability p of cure for any given patient. (The physicians must decide on the smallest value of p that makes the treatment worth further investigation.)

EXERCISE 7-15

Complete parts (a), (b), and (c) for each combination of sample size n and probability of success p listed in (i)–(iv):

(i) $n = 2$ and $p = \frac{1}{10}, \frac{1}{4}, \frac{1}{2}, \frac{3}{4}, \frac{9}{10}$
(ii) $n = 3$ and $p = \frac{1}{10}, \frac{1}{4}, \frac{1}{2}, \frac{3}{4}, \frac{9}{10}$
(iii) $n = 4$ and $p = \frac{1}{10}, \frac{1}{4}, \frac{1}{2}, \frac{3}{4}, \frac{9}{10}$
(iv) $n = 5$ and $p = \frac{1}{10}, \frac{1}{4}, \frac{1}{2}, \frac{3}{4}, \frac{9}{10}$

a. Write down the Binomial(n, p) probability distribution.

b. Let X represent a random variable having a Binomial(n, p) distribution. Graph the probability function in a plot similar to a frequency plot (Section 2-3), with possible values k on the horizontal axis and probabilities $P(X = k)$ on the vertical axis.

c. Find the expected value and variance of each distribution from the definitions given in Section 6-8 and from the formulas given in Section 7-2.

CHAPTER 8

The Gaussian (Normal) Distributions

IN THIS CHAPTER

Gaussian (normal) distributions
Standard Gaussian distribution
Central Limit Theorem

We need to learn about the Gaussian distributions for a couple of reasons. One very good reason is that much of classical statistical analysis (including many techniques developed early in this century) depends on the assumption that we have independent observations from a Gaussian distribution. For instance, in Section 10-3 we will discuss the t test, a tool for making inferences about a population mean. For the probability statements associated with a t test to be valid, we must assume that we have independent observations from a Gaussian distribution. The two-sample t test, used for comparing two population means, is the subject of Section 11-3. For valid probability statements using a two-sample t test, we must assume that we have two independent random samples from Gaussian distributions. In Chapters 12 and 13 we discuss experiments that might be analyzed using analysis of variance techniques, and in Chapter 15 we discuss fitting linear regression models; all of these classical statistical techniques depend on the assumption that observations come from a Gaussian distribution. In Sections 8-1 and 8-2, we will discuss what it means for observations to come from a Gaussian distribution.

Another reason to learn about Gaussian distributions is that a number of large-sample procedures allow us to make inferences based on a particular Gaussian distribution, the *standard Gaussian distribution,* even if the observations themselves do not come from a Gaussian distribution. For instance, in Section 10-1 we discuss large-sample inferences about a population mean. With a large enough sample size (plus a random sample), these inferences can be based on the standard Gaussian distribution even if the observations do not come from a Gaussian distribution. Large-sample inference about a proportion is the subject of Section 10-2. Again, if the (random) sample size is large enough, we can base our inferences on the standard Gaussian distribution. In Section 11-1 we consider large-sample inferences about two population means and in Section 11-2, large-sample inference about two proportions. With large enough sample sizes (plus independent random samples), inferences can be based on the standard Gaussian distribution. Justification for these large-sample inferences based on the standard Gaussian distribution comes from the Central Limit Theorem, a result we discuss in Section 8-3.

If a random variable has a Gaussian probability distribution, then that variable represents a continuous-type observation or measurement such as height, weight, thickness, strength, temperature, or pressure. How can we find probabilities associated with such a random variable? We need this knowledge because, as mentioned above, many probability statements in statistical inference are based on a Gaussian distribution.

Also mentioned above, much of classical statistical inference depends on the assumption that a sample of observations comes from a Gaussian distribution. If we collect a sample of independent observations from such a distribution, what will a plot of the sample values look like? To address this question, we might use a computer program to simulate a sample of independent observations from a Gaussian distribution. (The RANDOM command in Minitab will accomplish this, as described in the appendix to this chapter.)

As an illustration, consider a particular Gaussian distribution, with mean

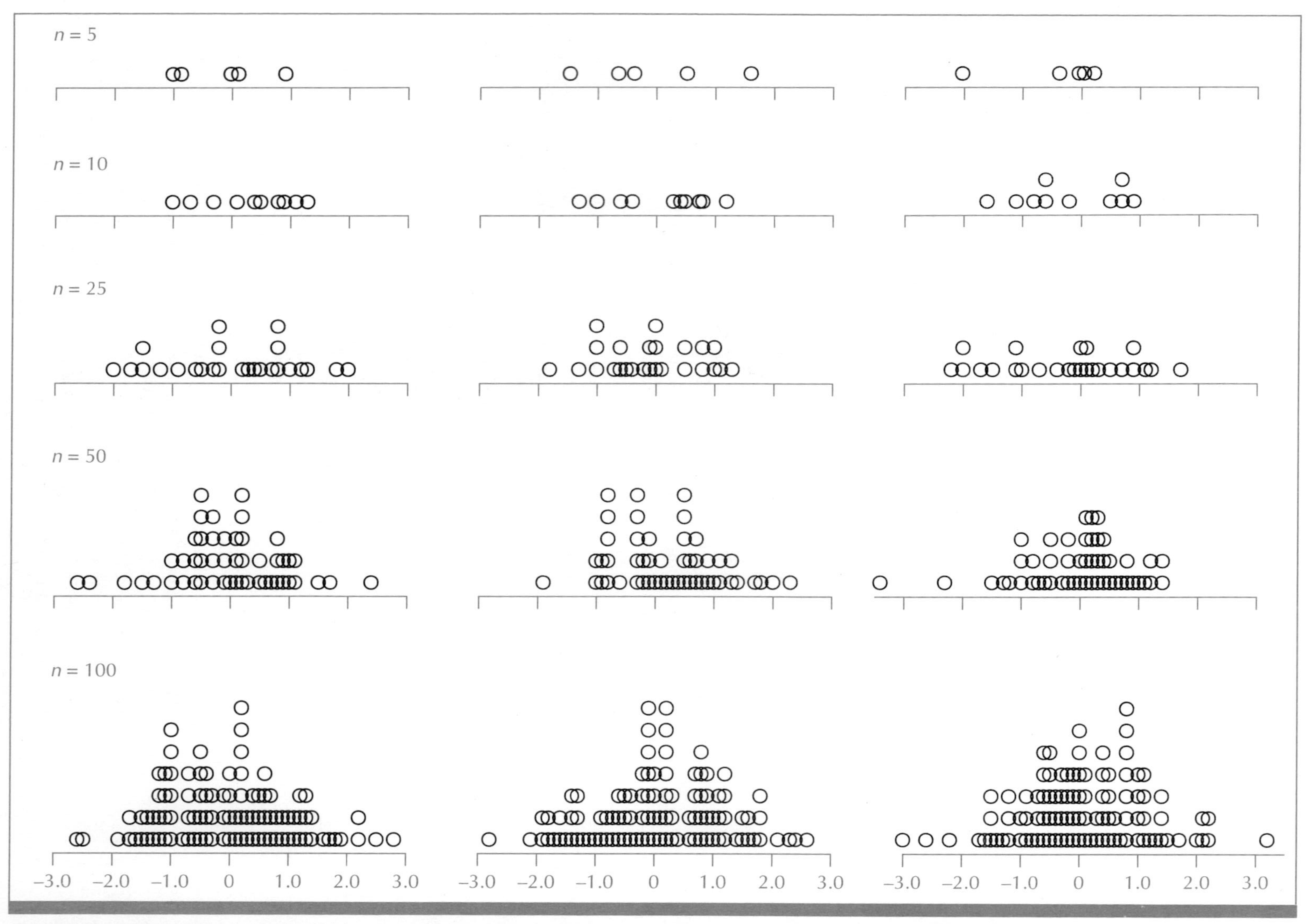

FIGURE 8-1 Dot plots of simulated samples of independent observations from the standard Gaussian distribution (obtained using the RANDOM command in Minitab). Three samples were simulated for each of these sample sizes: 5, 10, 25, 50, and 100.

0 and standard deviation 1, called the *standard Gaussian distribution*. Figure 8-1 shows plots of 15 different simulated samples of independent observations from the standard Gaussian distribution. We can think of each sample as a set of independent observations of a random variable having a standard Gaussian distribution. The first row in the figure shows dot plots of values from three samples, each of size 5. Rows 2–5 show dot plots for three samples of, respectively, size 10, 25, 50, and 100.

Looking at the plots in Figure 8-1, you might guess the standard Gaussian distribution is a unimodal distribution that is symmetric about 0. If you guessed this, you would be right. Notice the differences between plots of different samples from the same distribution. The standard Gaussian distribution has mean 0, but the *sample* means are not all 0. Just as there is variation in the sample values, there is variation among the sample means. Similarly, the standard deviation of the standard Gaussian distribution is 1, but the sample standard deviations do not all equal 1.

We expect variation among different sets of observations of a random variable. We cannot predict exactly what values of a random variable we will observe. However, we can use the distribution of the random variable to make *probability statements* about one or more observed values of the variable. In Section 8-1, we discuss how to make probability statements regarding a single observation of a random variable that has a Gaussian distribution. These ideas can then be extended to making probability statements about several independent observations from a Gaussian distribution.

In Section 8-2, we check to see if the distribution of a real data set can be approximated by a Gaussian distribution. Such a check in an experimental situation can help us decide if it is reasonable to use a statistical procedure that depends on assuming Gaussian observations. Then in Section 8-3, we discuss the Central Limit Theorem. This theorem and related results are useful in statistical inference based on large samples.

Before discussing the Gaussian distributions, a comment on their name is in order. The common term for these distributions is the *normal distributions*. The traditional designation "normal" is unfortunate because it helps to perpetuate the extremely common and *incorrect* notion that measurements "normally" follow a normal distribution. Many sets of measurements do not follow a Gaussian distribution, as we will see. To avoid incorrect interpretations, we will refer to these distributions as the Gaussian distributions, after the great mathematician Karl Friedrich Gauss, who described them.

8-1 The Gaussian Distributions

If a random variable Y has a Gaussian distribution, then Y can in theory take any real number value. How can we find the probability that a Gaussian random variable Y is in an interval of numbers? For instance, what is the probability that Y has a value from 10 to 25, $P(10 \leq Y \leq 25)$? Or, what is the probability that Y is greater than 50, $P(Y > 50)$? Before addressing such questions

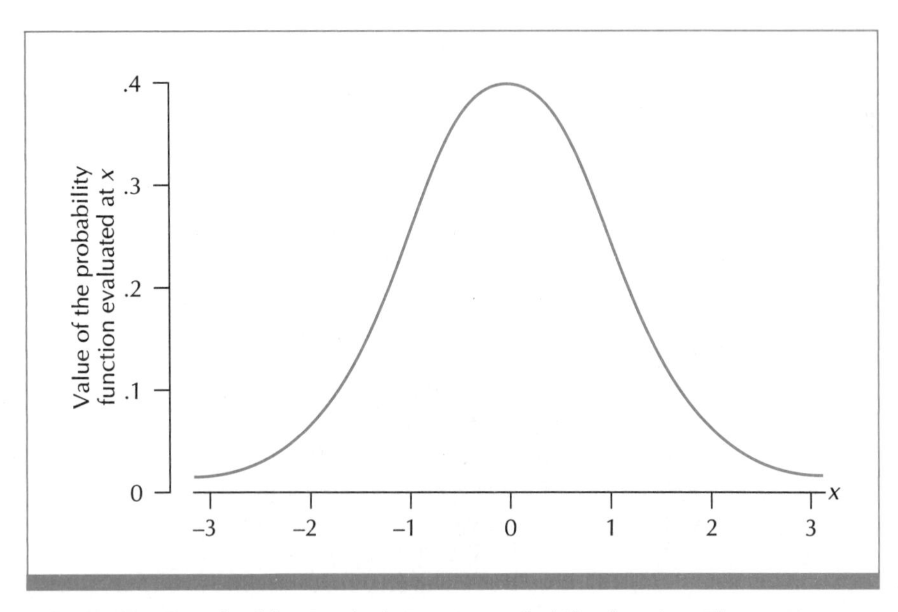

FIGURE 8-2 Graph of the standard Gaussian probability function. The area between the curve and the horizontal axis equals 1.

for a general Gaussian distribution, we will consider a special case, the standard Gaussian distribution.

The standard Gaussian distribution is defined in terms of the curve graphed in Figure 8-2, the standard Gaussian probability function. The area under this curve (between the curve and the horizontal axis) equals 1.

Let Z denote a random variable having the standard Gaussian distribution. The probability that Z takes a value from c to d, $P(c \le Z \le d)$, is the area under the standard Gaussian probability function between the numbers c and d. Such an area is shaded in Figure 8-3a.

We might want to find the probability that Z is greater than or equal to c, $P(Z \ge c)$. Then we find the area under the standard Gaussian probability function to the right of c. The shaded portion of Figure 8-3b shows such an area.

For the probability $P(Z \le c)$ that Z is less than or equal to c, we find the area under the standard Gaussian probability function to the left of c. Such an area is shaded in Figure 8-3c.

Now suppose we want to find the probability that Z equals a number c, $P(Z = c)$. We find the area under the standard Gaussian probability function at the number c, as illustrated in Figure 8-3d. But this is the area of a line segment, a rectangle with width 0. Therefore, we say $P(Z = c)$ equals 0. [This is true for all continuous random variables. If X is a continuous random variable, then $P(X = c) = 0$ for any number c.] From this, we see that $P(Z \ge c) = P(Z > c)$ and $P(c \le Z \le d) = P(c < Z < d)$, and so on.

The **standard Gaussian distribution** is a continuous probability distribution. The probability a standard Gaussian random variable is in an inter-

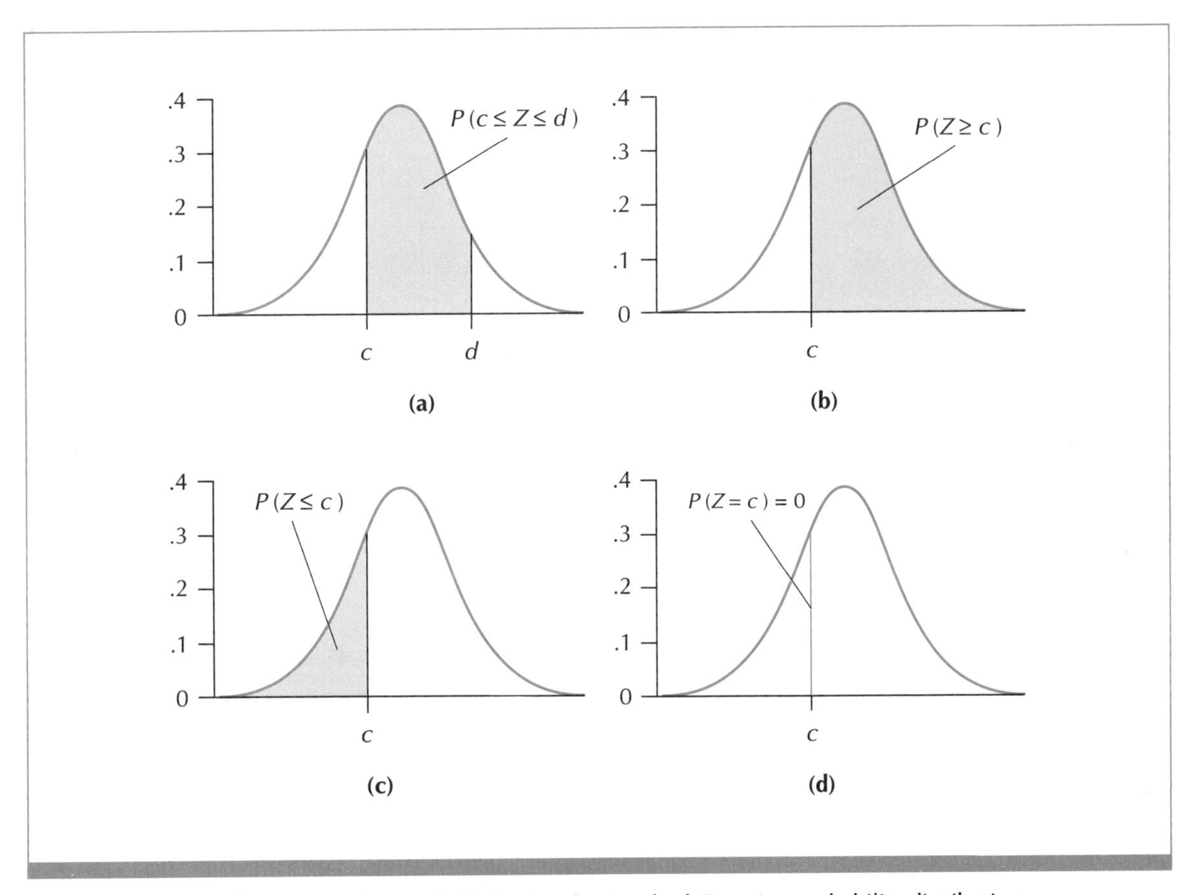

FIGURE 8-3 Let Z denote a random variable having the standard Gaussian probability distribution.
(a) The area under the graph between c and d equals $P(c \leq Z \leq d)$.
(b) The area under the curve to the right of c equals $P(Z \geq c)$.
(c) The area under the curve to the left of c equals $P(Z \leq c)$.
(d) Because Z is a continuous random variable, $P(Z = c)$ equals 0 for any number c.

val of numbers equals the area over that interval under the graph of the standard Gaussian probability function, illustrated in Figure 8-2.

Table B at the back of the book lists cumulative probabilities of the form $P(Z \leq c)$ for selected nonnegative values of c. For instance, $P(Z \leq 1) = .8413$ and $P(Z \leq 2.51) = .9940$. We can find a tail probability (so called because it represents the relatively small probability for a *tail* of the distribution) such as $P(Z \geq 2.51)$ by subtraction: $P(Z \geq 2.51) = 1 - P(Z \leq 2.51) = .0060$.

A **cumulative probability** has the form $P(X \leq c)$ where X is a random variable and c is a constant.

A **tail probability** for a random variable X is a probability that is small (less than .5) and has the form $P(X \geq c)$ or $P(X \leq c)$ for some number c.

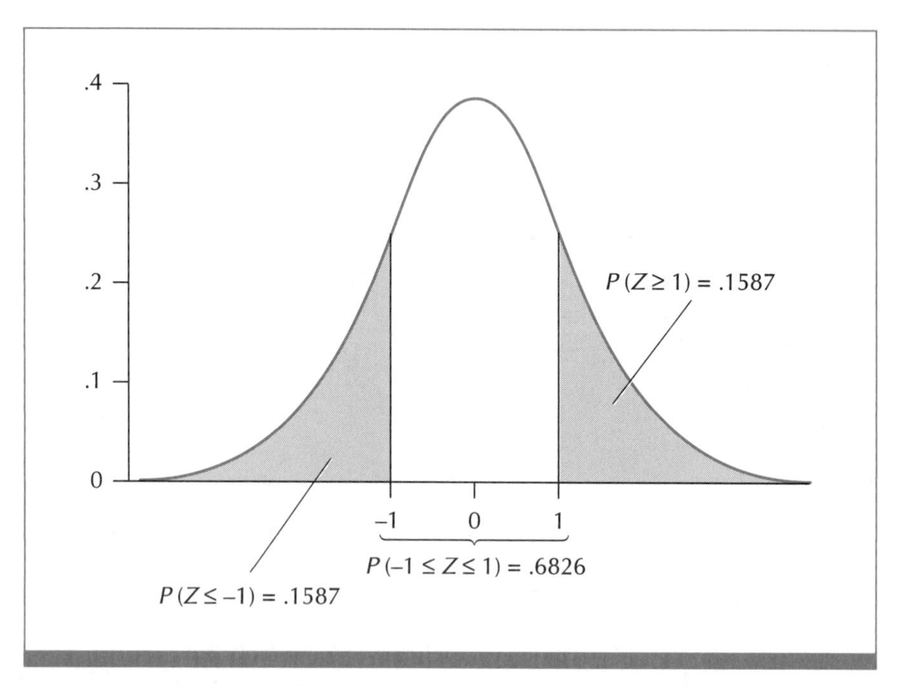

FIGURE 8-4 The standard Gaussian probability function is symmetric about 0. If Z has a standard Gaussian distribution, then $P(Z \le -1) = P(Z \ge 1) = .1587$.

We see from Figure 8-2 that the standard Gaussian probability function is symmetric about 0. So, for any positive number c, $P(Z \ge c)$ equals $P(Z \le -c)$. For example, Figure 8-4 illustrates the fact that $P(Z \le -1) = P(Z \ge 1) = 1 - P(Z \le 1) = .1587$. With this information, we can calculate $P(-1 \le Z \le 1)$:

$$\begin{aligned} P(-1 \le Z \le 1) &= 1 - P(Z \le -1) - P(Z \ge 1) \\ &= 1 - .1587 - .1587 = .6826 \end{aligned}$$

A picture helps us visualize the probability we want to calculate. For instance, suppose we want to find $P(-2 \le Z \le -1)$. The corresponding area is shaded in Figure 8-5. Because the standard Gaussian curve is symmetric about 0, $P(-2 \le Z \le -1) = P(1 \le Z \le 2)$. Also, $P(1 \le Z \le 2) = P(Z \le 2) - P(Z < 1)$. We can read these last two probabilities from Table B. Therefore,

$$\begin{aligned} P(-2 \le Z \le -1) = P(1 \le Z \le 2) &= P(Z \le 2) - P(Z < 1) \\ &= .9772 - .8413 = .1359 \end{aligned}$$

Suppose we want to find $P(-1 \le Z \le 2)$. From Figure 8-6, we see that the probability we want is $P(Z \le 2) - P(Z < -1)$. Using Table B, we find that $P(Z \le 2) = .9772$ and $P(Z < -1) = P(Z > 1) = 1 - P(Z \le 1) = .1587$. Therefore,

$$P(-1 \le Z \le 2) = .9772 - .1587 = .8185$$

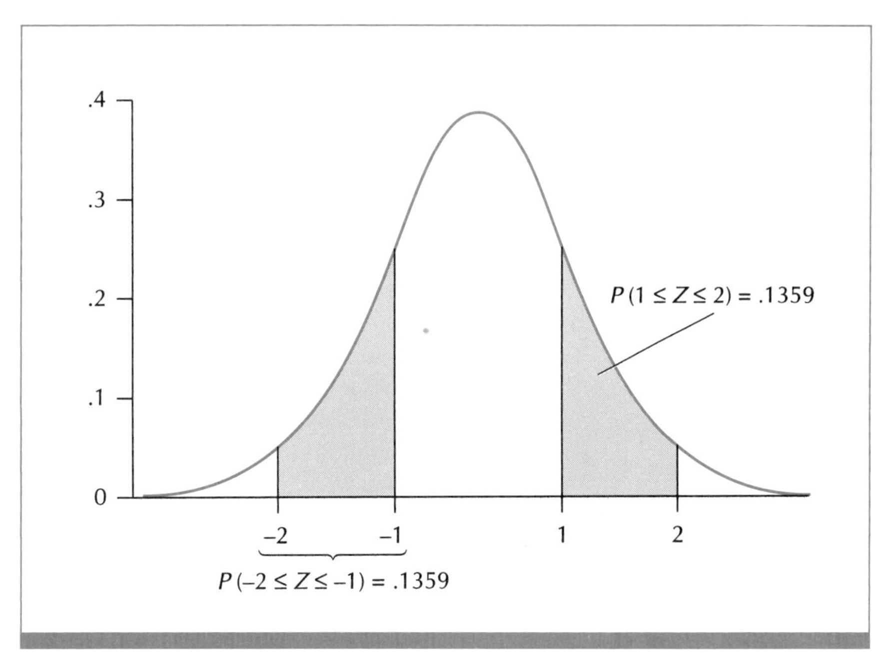

FIGURE 8-5 If Z has the standard Gaussian distribution, then $P(-2 \leq Z \leq -1) = P(1 \leq Z \leq 2) = P(Z \leq 2) - P(Z < 1) = .9772 - .8413 = .1359$.

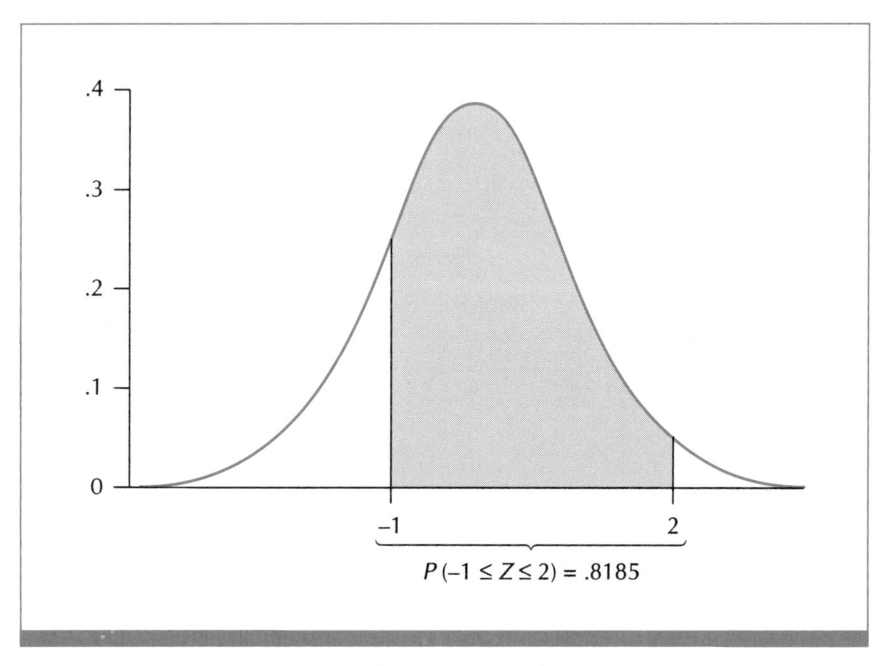

FIGURE 8-6 If Z has the standard Gaussian distribution, then $P(-1 \leq Z \leq 2) = P(Z \leq 2) - P(Z < -1) = .9772 - .1587 = .8185$.

Again, let Z denote a random variable having the standard Gaussian distribution. Because this distribution is symmetric about 0, the mean or expected value of Z is 0. The variance of Z equals 1. We summarize this information as follows:

If Z is a standard Gaussian random variable,
then $E(Z) = 0$ and $\text{Var}(Z) = 1$.

Thus, as noted in the introduction to this chapter, the standard Gaussian distribution is a Gaussian distribution with mean 0 and variance 1. Now let σ denote a positive number and let μ denote any number. If $X = \sigma Z$, then X has a Gaussian distribution with mean 0 and variance σ^2, standard deviation σ. If $Y = X + \mu = \sigma Z + \mu$, then Y has a Gaussian distribution with mean μ and variance σ^2. That is, $E(Y) = \mu$ and $\text{Var}(Y) = \sigma^2$.

Suppose the random variable Y has a Gaussian distribution with mean μ and variance σ^2. We want to find the probability that Y takes a value between two numbers c and d, $P(c \leq Y \leq d)$. To find this probability, we standardize the random variable Y.

Suppose a random variable X has mean μ and variance σ^2 (or standard deviation σ). We **standardize *X*** by subtracting the mean μ and then dividing by the standard deviation σ. The **standardized random variable**

$$\frac{X - \mu}{\sigma}$$

has mean 0 and standard deviation 1.

When we standardize a random variable, we are converting the variable into standard units. A value of the standardized variable shows the distance, in standard deviations, that the corresponding value of the original variable lies above or below its mean.

If the random variable Y has a Gaussian distribution with mean μ and variance σ^2, then the standardized variable $Z = (Y - \mu)/\sigma$ has the standard Gaussian distribution, with mean 0 and variance 1.

A **Gaussian distribution** is a continuous probability distribution. If a random variable Y has a Gaussian distribution with mean μ and variance σ^2 (or standard deviation σ), then

$$\frac{Y - \mu}{\sigma}$$

has the standard Gaussian distribution.

If we write Y as $Y = \sigma Z + \mu$, then we see that

$$P(c \leq Y \leq d) = P(c \leq \sigma Z + \mu \leq d) = P\left(\frac{c - \mu}{\sigma} \leq Z \leq \frac{d - \mu}{\sigma}\right)$$

Therefore, we can use a table of probabilities for the standard Gaussian distribution, such as Table B, to find $P(c \leq Y \leq d)$. That is, we can use a table of

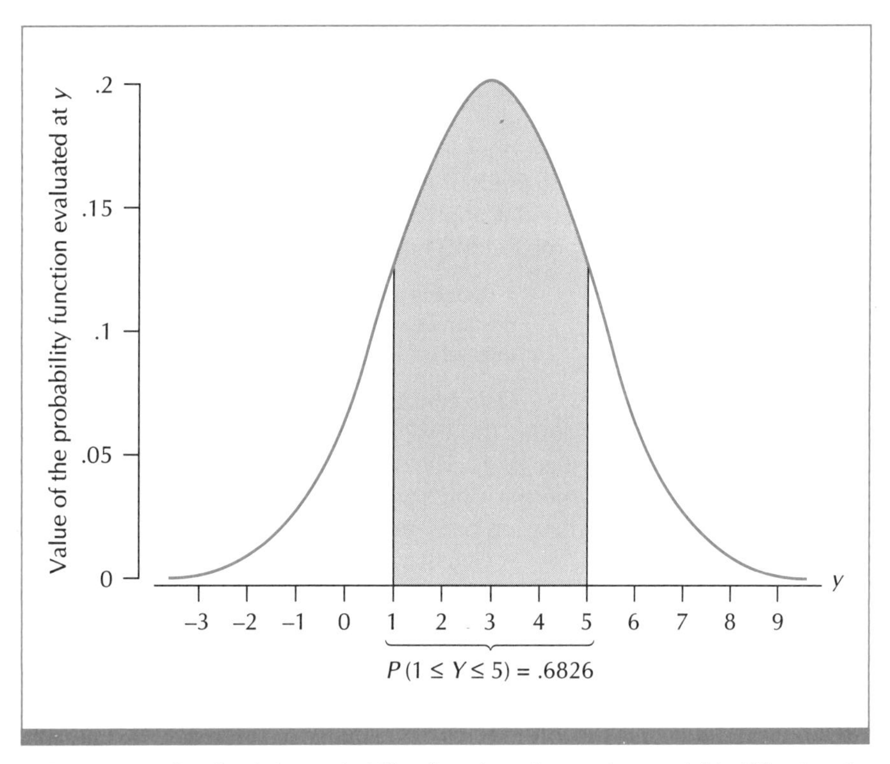

FIGURE 8-7 Graph of the probability function of a random variable Y having the Gaussian distribution with mean 3 and variance 4. The shaded area corresponds to $P(1 \leq Y \leq 5)$.

probabilities for the standard Gaussian distribution to make probability statements about any Gaussian random variable.

Suppose, for example, that Y has a Gaussian distribution with mean 3 and variance 4. Let's find the probability that Y takes a value from 1 to 5, $P(1 \leq Y \leq 5)$. The corresponding area is shaded in Figure 8-7. Let Z denote a standard Gaussian random variable. Then

$$P(1 \leq Y \leq 5) = P\left(\frac{1-3}{\sqrt{4}} \leq \frac{Y-3}{\sqrt{4}} \leq \frac{5-3}{\sqrt{4}}\right)$$
$$= P(-1 \leq Z \leq 1) = .6826$$

We can use the ideas in this section to decide whether the distribution of a set of values can be approximated by a Gaussian distribution. In Section 8-2, we evaluate whether the distribution of a set of baseball statistics can be approximated by a Gaussian distribution. In an experimental situation, we use such an evaluation to decide if a statistical procedure based on the assumption of Gaussian observations is justified.

8-2

Approximating a Distribution of Values by a Gaussian Distribution

What does it mean to say that a set of values approximately follows a Gaussian distribution? It means that the proportion of values between any two numbers c and d approximately equals the area between c and d under that Gaussian probability function.

A **distribution of data values is approximately Gaussian** if the proportion of values in any interval approximately equals the area over that interval under the appropriate Gaussian curve.

Let's consider an example. Table 8-1 shows the number of hits allowed during the 1987 season for each of the 26 major-league baseball teams. We see that $\frac{15}{26} = .577$ is the proportion of teams with 1,400 to 1,500 hits allowed. Is there a Gaussian distribution with an area about .577 under the probability function between 1,400 and 1,500? Let's see if we can find one.

A Gaussian distribution is symmetric about its mean. So, for a Gaussian approximation to work, the data values should be symmetrically distributed. Figure 8-8 shows a dot plot and box plot of the 26 values of hits allowed. The plots look reasonably symmetric, so we will continue.

We must specify the mean and standard deviation of the Gaussian distribution we want to use. It is reasonable to use the mean and standard deviation of the data values. The mean number of hits allowed per team in 1987 was 1,457.5 and the standard deviation was 66.7. Can we approximate the distribu-

TABLE 8-1 Hits allowed during the 1987 season by the 26 major-league baseball teams (*The Sporting News*, October 12, 1987, pages 37, 42; *USA Today*, October 6, 1987, page 4c; *USA Today*, October 7, 1987, page 5c; kindly provided by Lee Panas)

Team	Hits allowed	Team	Hits allowed
Blue Jays	1,323	Phillies	1,453
Astros	1,363	Twins	1,465
Pirates	1,377	Yankees	1,475
Rangers	1,388	Angels	1,481
Padres	1,402	Cardinals	1,484
Giants	1,407	Reds	1,486
Mets	1,407	Mariners	1,503
Dodgers	1,415	Cubs	1,524
Royals	1,424	Braves	1,529
Expos	1,428	Brewers	1,548
Tigers	1,430	Orioles	1,555
White Sox	1,436	Indians	1,566
Athletics	1,442	Red Sox	1,584

Mean = 1,457.5 Standard deviation = 66.7

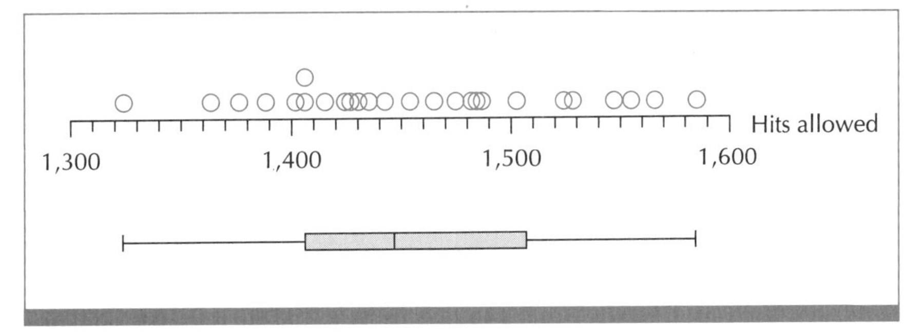

FIGURE 8-8 Dot plot and box plot of the number of hits allowed in 1987 by the 26 major-league baseball teams

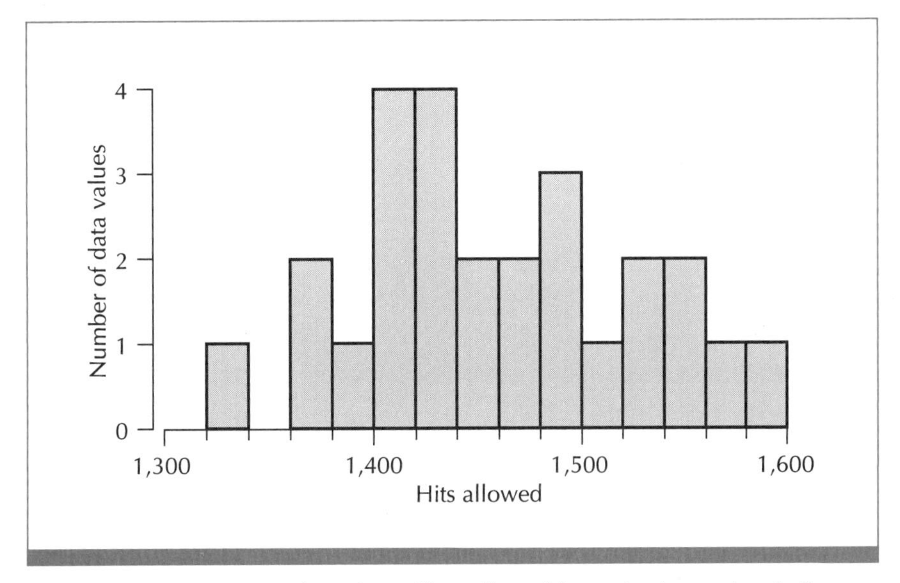

FIGURE 8-9 Histogram of number of hits allowed by major-league baseball teams in 1987

tion of 1987 hits allowed by the Gaussian distribution with mean 1,457.5 and standard deviation 66.7?

A histogram of hits allowed is shown in Figure 8-9. The proportion of values between two interval endpoints is the proportion of the total area of the histogram between those two endpoints. For instance, the proportion of teams with 1,420 to 1,480 hits allowed is $\frac{8}{26}$ = .308. The area between 1,420 and 1,480 is $\frac{8}{26}$ or roughly 31% of the total area of the histogram.

The probability function for the Gaussian distribution with mean 1,457.5 and standard deviation 66.7 is graphed in Figure 8-10. Does the area under this curve between two points approximate the proportion of the area of the histo-

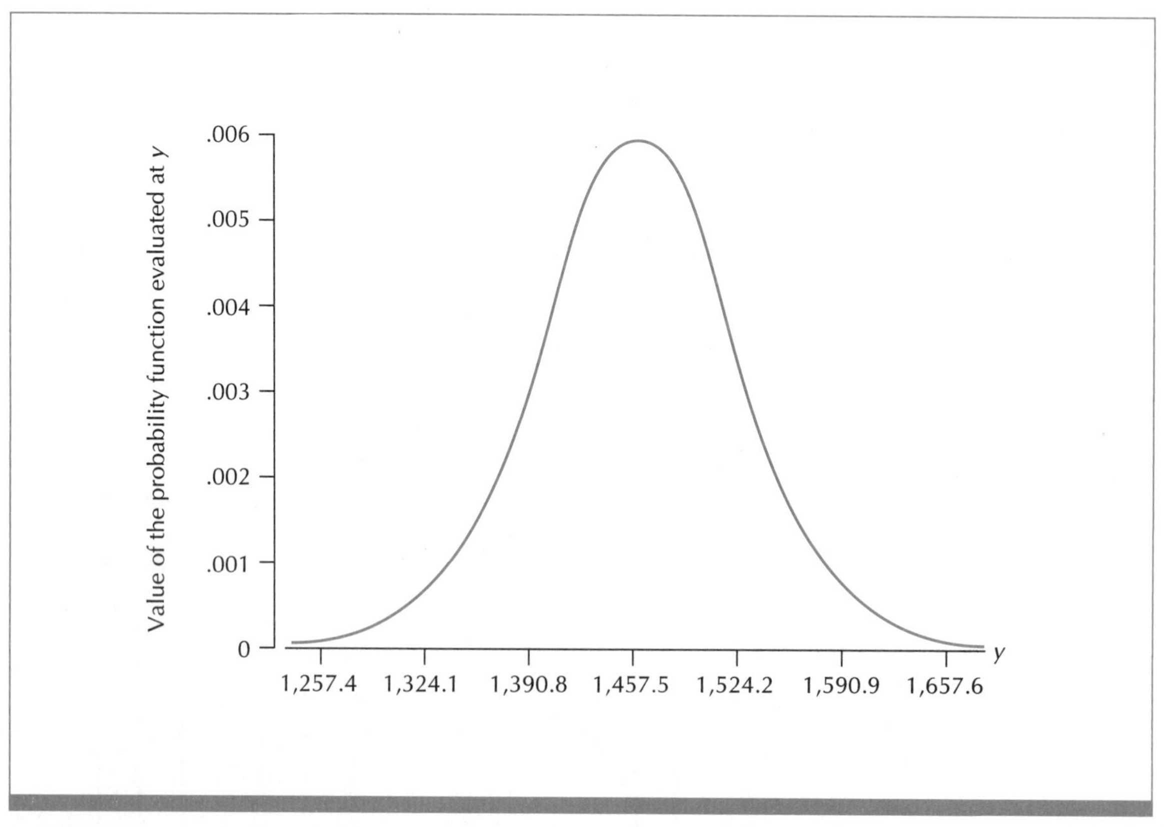

FIGURE 8-10 Graph of the probability function for the Gaussian distribution with mean 1,457.5 and standard deviation 66.7

gram between the two points? Let's find, for instance, the area under the curve between 1,420 and 1,480.

Let Y denote a random variable having the Gaussian distribution with mean 1,457.5 and standard deviation 66.7. Let Z denote a standard Gaussian random variable. The area under the curve in Figure 8-10 between 1,420 and 1,480 equals $P(1{,}420 \le Y \le 1{,}480)$. Then

$$
\begin{aligned}
P(1{,}420 \le Y \le 1{,}480) & \\
&= P\left(\frac{1{,}420 - 1{,}457.5}{66.7} \le \frac{Y - 1{,}457.5}{66.7} \le \frac{1{,}480 - 1{,}457.5}{66.7}\right) \\
&= P(-.56 \le Z \le .34) \\
&= P(Z \le .34) - P(Z \le -.56) \\
&= .6331 - .2877 = .3454
\end{aligned}
$$

The area under the curve in Figure 8-10 between 1,420 and 1,480 is .3454. This is not too far from the observed proportion .308 of teams with 1,420 to 1,480

TABLE 8-2 Comparison of observed frequencies with those expected based on the Gaussian distribution with mean 1,457.5 and standard deviation 66.7, for number of hits allowed in 1987

Interval of values for a Gaussian distribution with mean 1,457.5 and standard deviation 66.7	Interval of values for a standard Gaussian distribution	Area under the curve over this interval	Expected frequency in this interval	Observed frequency in this interval
1,457.5 to 1,490.9	0 to .5	.1915	.1915 × 26 = 4.98	5
1,424.1 to 1,457.5	−.5 to 0	.1915	.1915 × 26 = 4.98	5
1,390.8 to 1,524.2	−1 to 1	.6826	.6826 × 26 = 17.75	17
Greater than 1,524.2	Greater than 1	.1587	.1587 × 26 = 4.13	5
Less than 1,390.8	Less than −1	.1587	.1587 × 26 = 4.13	4
1,324.1 to 1,590.9	−2 to 2	.9544	.9544 × 26 = 24.81	25
Greater than 1,590.9	Greater than 2	.0228	.0228 × 26 = .59	0
Less than 1,324.1	Less than −2	.0228	.0228 × 26 = .59	1
1,257.3 to 1,657.7	−3 to 3	.9974	.9974 × 26 = 25.93	26
Greater than 1,657.7	Greater than 3	.0013	.0013 × 26 = .03	0
Less than 1,257.3	Less than −3	.0013	.0013 × 26 = .03	0
1,400 to 1,500	−.86 to .64	.5440	.5440 × 26 = 14.14	15

hits allowed. Using this Gaussian distribution, we would expect .3454 × 26 = 8.98 of the 26 teams to have 1,420 to 1,480 hits allowed. This agrees pretty well with the observed frequency of 8.

Table 8-2 summarizes similar calculations. An entry in column 1 shows an interval of values for hits allowed. Column 2 shows the same interval standardized by subtracting the sample mean of 1,457.5 and then dividing by the sample standard deviation of 66.7. Column 3 shows the area under the standard Gaussian curve over the interval shown in column 2; column 3 is the proportion expected in the interval in column 1 using this Gaussian distribution. This proportion times 26 gives the expected number of data values in the interval, listed in column 4. For comparison, column 5 gives the observed number of values of hits allowed in the interval.

Consider, for example, the interval that extends from the sample mean to half a standard deviation above the sample mean: from 1,457.5 to 1,457.5 + .5 × 66.7, or from 1,457.5 to 1,490.9. This interval is the first entry in Table 8-2. If we standardize by subtracting 1,457.5 and then dividing by 66.7, this interval becomes 0 to .5. Using Table B at the back of the book, we see that the probability that a standard Gaussian random variable Z is between 0 and .5 is

$$P(0 \leq Z \leq .5) = P(Z \leq .5) - P(Z \leq 0) = .6915 - .5000 = .1915$$

If the Gaussian approximation is reasonable, we expect about 19% of the data values in the interval from 1,457.5 to 1,490.9. Nineteen percent of 26 is about 5, and 5 is in fact the observed number of values of hits allowed in this interval. For each of the intervals in Table 8-2, we see that there is close agreement between the observed frequencies and those expected based on the Gaussian approximation.

TABLE 8-3 Comparison of observed proportions with those expected based on the Gaussian distribution with mean 1,457.5 and standard deviation 66.7, for number of hits allowed by major-league baseball teams in 1987

Interval of values	Proportion expected by Gaussian approximation	Proportion observed
1,457.5 to 1,490.9	.1915	$\frac{5}{26}$ = .19
1,424.1 to 1,457.5	.1915	$\frac{5}{26}$ = .19
1,390.8 to 1,524.2	.6826	$\frac{17}{26}$ = .65
Greater than 1,524.2	.1587	$\frac{5}{26}$ = .19
Less than 1,390.8	.1587	$\frac{4}{26}$ = .15
1,324.1 to 1,590.9	.9544	$\frac{25}{26}$ = .96
Greater than 1,590.9	.0228	$\frac{0}{26}$ = 0
Less than 1,324.1	.0228	$\frac{1}{26}$ = .04
1,257.3 to 1,657.7	.9974	$\frac{26}{26}$ = 1
Greater than 1,657.7	.0013	$\frac{0}{26}$ = 0
Less than 1,257.3	.0013	$\frac{0}{26}$ = 0
1,400 to 1,500	.5440	$\frac{15}{26}$ = .58

Column 1 of Table 8-3 lists the intervals shown in the first column of Table 8-2. For each interval, column 2 shows the proportion of data values expected in the interval based on the Gaussian approximation (column 3 of Table 8-2). Column 3 shows the proportion of data values observed in the interval (column 5 of Table 8-2, divided by 26). We have checked only a few intervals, but the similarity between these observed and expected proportions suggests that the Gaussian distribution with mean 1,457.5 and standard deviation 66.7 gives a reasonable approximation to the distribution of hits allowed by major-league baseball teams in 1987.

Standardized Data Values

We indicated previously that a standardized random variable shows standard deviations above or below the mean for a random variable. Similarly, a *standardized data value* shows how many standard deviations above or below the mean a data value is. We standardize a data value by subtracting the mean of the values and then dividing by the standard deviation:

> We obtain a **standardized data value** by subtracting the mean of the set of data values and dividing by the standard deviation.

For instance, in 1987 the Rangers allowed 1,388 hits. When we subtract the sample mean and divide by the sample standard deviation, we get a standardized value of

$$\frac{1{,}388 - 1{,}457.5}{66.7} = -1.04$$

The Rangers were about one standard deviation below the mean number of hits allowed among the major-league teams. The Orioles allowed 1,555 hits in 1987. Subtracting the sample mean and dividing by the sample standard deviation, we get this standardized value:

$$\frac{1{,}555 - 1{,}457.5}{66.7} = 1.46$$

The Orioles were about one and a half standard deviations above the mean number of hits allowed among the major-league teams in 1987.

Referring to Table 8-2, we find that 17 of the 26 values of hits allowed were within one standard deviation of the mean, 5 were more than one standard deviation above the mean, and 4 were more than one standard deviation below the mean. We can interpret all of the intervals in column 1 of Table 8-2 in terms of standard deviations above or below the sample mean. This is the meaning of the standardized intervals in column 2 of the table. (For more discussion of standard units and Gaussian approximations, see Freedman, Pisani, and Purves, 1978, Chapter 5.)

More on Gaussian Approximations

The assumption that we can approximate the distribution of a set of data values by a Gaussian distribution is the basis for many techniques in classical statistical inference. However, many distributions *cannot* be approximated by a Gaussian distribution. Consider, for instance, the data displayed in Figure 8-11. (We saw this plot in Chapter 2.) The figure shows a dot plot of number enrolled in higher education in 1984 as percentage of 20–24-year-old age group, for 119

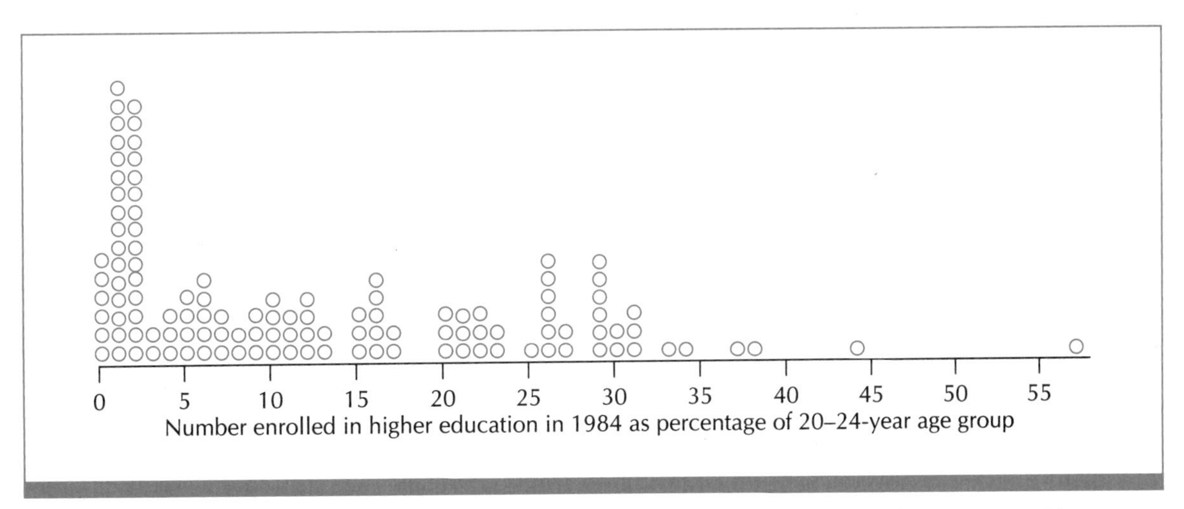

FIGURE 8-11 Dot plot of number enrolled in higher education in 1984 as percentage of 20–24-year-old age group for 119 countries (*World Development Report 1987*, pages 262, 263)

countries. The plot is skewed to the right, not at all symmetrical. The distribution of the 119 data values cannot be well approximated by a Gaussian distribution (see Exercise 8-16).

If the proportion of data values in any interval approximately equals the area over that interval under a Gaussian curve, then we say we can approximate the distribution of data values using that Gaussian distribution. Similarly, if the probability that a random variable is in any interval approximately equals the area over that interval under a Gaussian curve, then we say the random variable has approximately that Gaussian distribution.

> The **distribution of a random variable is approximately Gaussian** if the probability that the random variable is in any interval approximately equals the area over that interval under a Gaussian curve.

In Section 8-3 we discuss a famous result known as the Central Limit Theorem. This result states that if we have a large number of independent observations from the same distribution, then the average of these observations has approximately a Gaussian distribution. This is very useful in large-sample statistical analysis, when we can base our inferences on a Gaussian distribution.

8-3 The Central Limit Theorem

We say random variables X_1, X_2 through X_n are *independent* if they represent independent numerical observations made during an experiment. Observations are *independent* if the result of any one observation does not in any way affect the results for the other observations.

Let X_1, X_2 through X_n denote independent random variables with the same probability distribution. Then we say X_1 through X_n represent a *random sample in the probability sense.*

> A **random sample in the probability sense** is a collection of independent random variables with the same probability distribution.

For our purposes, a random sample in the probability sense represents numerical observations on the same variable made during independent and identical repetitions of an experiment. For example, suppose we have a group of mice with the same genetic background, each with a tumor of the same type and size. We house, feed, and care for the animals in the same way and treat them all with the same antitumor regimen. At the end of the treatment period, we plan to measure the tumor size on each mouse. Under these experimental conditions, we can think of the collection of tumor measurements as a random sample in the probability sense.

Suppose the random variables X_1 through X_n represent a random sample in the probability sense. Let μ denote the mean and σ^2 the variance of each X_i. Let Y denote the sum and $\bar{X} = Y/n$ the average of X_1 through X_n. The *Central Limit Theorem* states that if the sample size n is large enough, then Y and $\bar{X}$

are each approximately Gaussian distributed. In particular, the Central Limit Theorem tells us that the standardized version of $\bar{X}$ (or Y):

$$\frac{\bar{X} - \mu}{\sigma/\sqrt{n}} = \frac{Y - n\mu}{\sqrt{n}\sigma}$$

has approximately the standard Gaussian distribution if the sample size n is large enough.

A version of the Central Limit Theorem: Suppose X_1 through X_n are independent random variables with the same probability distribution. Let μ denote the mean and σ the standard deviation of each X_i. Let $\bar{X}$ denote the average of X_1 through X_n. If the sample size n is large enough, the distribution of

$$\frac{\bar{X} - \mu}{\sigma/\sqrt{n}}$$

is approximately the standard Gaussian distribution. The larger the sample size, the better the approximation.

How large does the sample size n have to be for the Gaussian approximation to apply to the distribution of $\bar{X}$ or Y? The answer depends on how different the probability distribution of the X_i's is from a Gaussian distribution. Let's illustrate this idea.

Suppose we toss a fair coin many times. The tosses are independent of one another in that the result of one toss does not affect in any way the result of another. We record a 1 for each head and a 0 for each tail. Let X_1 denote the result of the first toss, X_2 the result of the second toss, and so on. Then the X_i's are independent random variables, each with the same probability distribution defined by the two probabilities:

$$P(X_i = 0) = \frac{1}{2} \quad \text{and} \quad P(X_i = 1) = \frac{1}{2}$$

Let $Y_1 = X_1$, $Y_2 = X_1 + X_2$, and so on. Since Y_n is the sum of X_1 through X_n, counting the number of heads in n independent tosses of a fair coin, Y_n has the Binomial(n, $\frac{1}{2}$) distribution. The left-hand side of Figure 8-12 shows plots of the Binomial(n, $\frac{1}{2}$) distribution for several values of n. When the probability of success is $\frac{1}{2}$, a binomial distribution is symmetrical for any sample size n. For sample size 10, the shape of the binomial probability distribution closely resembles that of a Gaussian distribution. For sample size 25, the similarity in shape between the binomial and Gaussian distributions is striking (see Exercise 8-17).

Suppose now the coin we toss is not fair. Instead, the probability of a head is $\frac{1}{10}$. Then Y_n counts the number of successes in n independent trials with the probability of success equal to $\frac{1}{10}$ on each trial, and so has a Binomial(n, $\frac{1}{10}$) distribution. The right-hand side of Figure 8-12 shows plots of the Binomial(n, $\frac{1}{10}$) distribution for several values of n. For small sample sizes, the distribution is very asymmetrical. The binomial distribution is more symmetrical for sample size 25 than for smaller sample sizes. The binomial distri-

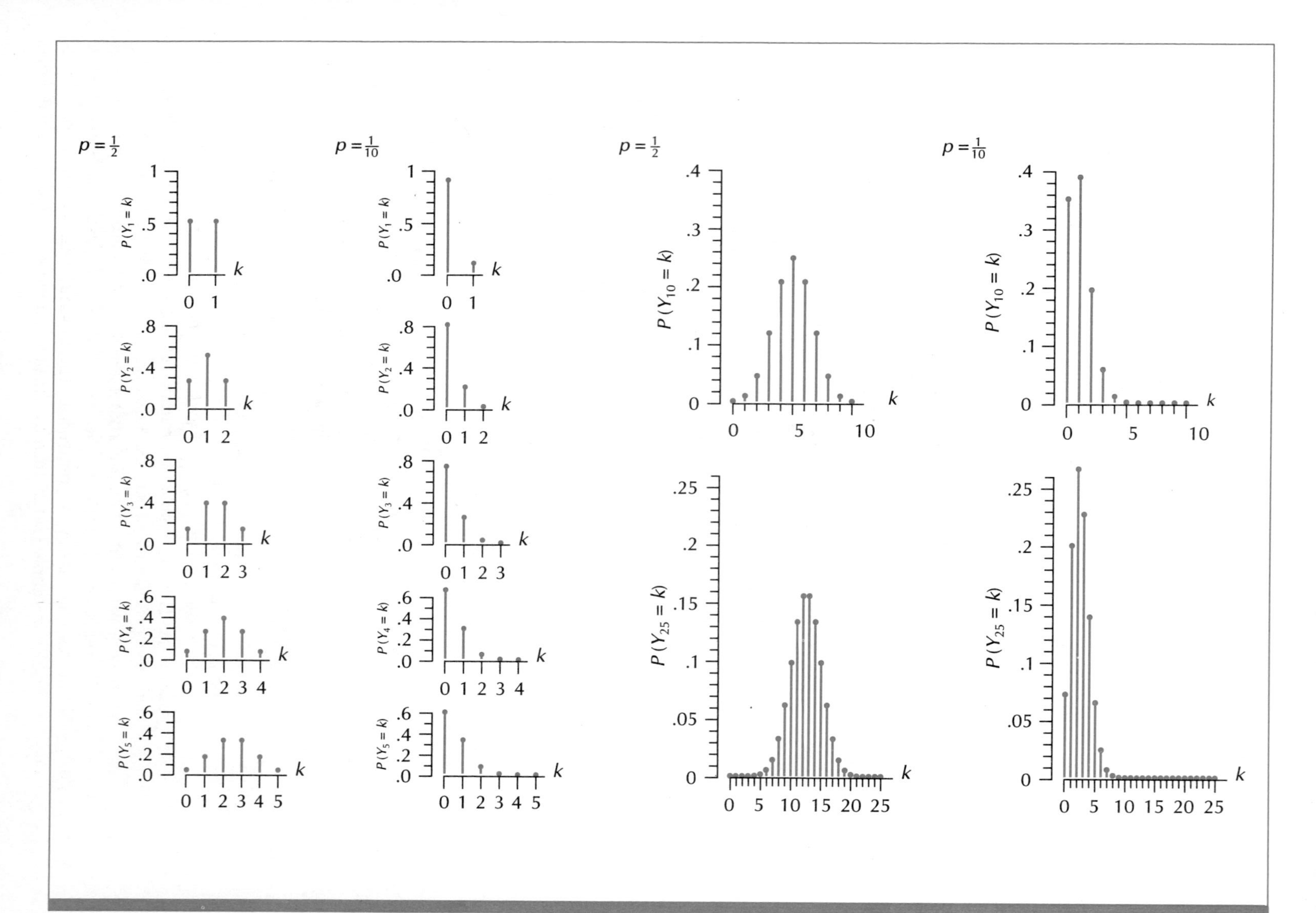

FIGURE 8-12 Plot of the Binomial(n, p) distribution for $p = \frac{1}{2}$ and $p = \frac{1}{10}$ and several values of n

bution will more and more resemble a Gaussian distribution as the sample size increases. For a comparison of the Binomial(25, $\frac{1}{10}$) distribution with a Gaussian distribution, see Exercise 8-18.

As Figure 8-12 illustrates, the sample size necessary for $\bar{X}$ (or Y) to have approximately a Gaussian distribution depends on the shape of the distribution of the random variables X_1 through X_n. If the observations have a symmetrical distribution, a sample of size 30 may be enough. For skewed or multimodal distributions, the sample size must be larger; how much larger depends on how far from symmetrical and unimodal is the distribution of the individual observations.

Another Large-Sample Result

We will find a result related to the Central Limit Theorem useful in large-sample statistical inference. As before, suppose X_1 through X_n are independent random variables with the same probability distribution. Let μ denote the mean and σ the standard deviation of each X_i. Let $\bar{X}$ and s denote the sample mean and sample standard deviation, respectively, of X_1 through X_n. The Central Limit Theorem says that if the sample size n is large enough, then

$$\frac{\bar{X} - \mu}{\sigma/\sqrt{n}}$$

has approximately the standard Gaussian distribution. Suppose we use the sample standard deviation s to estimate σ. A related result states that if n is large enough, then

$$\frac{\bar{X} - \mu}{s/\sqrt{n}}$$

has approximately the standard Gaussian distribution. Sometimes we call $s/\sqrt{n}$ the *standard error of* $\bar{X}$ and denote it by SE, so SE $= s/\sqrt{n}$.

> Suppose X_1 through X_n represent a random sample in the probability sense. Let $\bar{X}$ denote the average and s the sample standard deviation of X_1 through X_n. The **standard error of the mean**, SE $= s/\sqrt{n}$, is the estimated standard deviation of the random variable $\bar{X}$.

The large-sample result stated above says that if n is large enough, $(\bar{X} - \mu)/\text{SE}$ has approximately the standard Gaussian distribution.

> **A large-sample result related to the Central Limit Theorem:** Suppose X_1 through X_n are independent random variables having the same probability distribution, with mean μ. Let $\bar{X}$ denote the sample average and s the sample standard deviation of X_1 through X_n. Let SE $= s/\sqrt{n}$ denote the standard error of the mean. If the sample size n is large enough, then
>
> $$\frac{\bar{X} - \mu}{s/\sqrt{n}} = \frac{\bar{X} - \mu}{\text{SE}}$$
>
> has approximately the standard Gaussian distribution. The larger the sample size, the better the approximation.

What is the usefulness of this large-sample approximation? We are often interested in asking questions about the mean μ of a population. Suppose we want to use a sample of independent observations from the population, say X_1 through X_n, to learn about the population mean. The sample mean $\bar{X}$ is a good guess or estimate of the unknown population mean μ. In large-sample statistical analysis, we use $\bar{X}$ as the basis of our inferences about μ. That is, we use the distribution of $\bar{X}$ to make probability statements about μ. The various versions of the Central Limit Theorem say that if the sample size is large enough, the probability distribution of $\bar{X}$ is approximately Gaussian, *no matter what the probability distribution of the individual random variables* X_1 through X_n. This amazing result tells us that if we have a random sample in the probability sense and if the sample size is large enough, we can base inferences about μ on a Gaussian distribution; we do not have to worry about the distribution of the individual observations.

We will see an illustration of the use of this large-sample result in Example 9-3. Section 10-1 gives a more general discussion of large-sample inference about a population mean, based on a Gaussian distribution. In Section 10-2 we look at large-sample inferences about a proportion. These large-sample procedures are based on the Central Limit Theorem result described above. This Central Limit Theorem result can be extended to differences between two sample means. The large-sample procedures discussed in Sections 11-1 and 11-2 are based on this extension to two large samples. In Section 11-1 we consider large-sample inferences about two means and in Section 11-2, large-sample inferences about two proportions.

Summary of Chapter 8

We introduce the standard Gaussian probability distribution and show how to find probabilities for a general Gaussian distribution using probabilities for the standard Gaussian distribution. We discuss the idea of approximating a distribution with a Gaussian distribution.

A version of the Central Limit Theorem states that if X_1 through X_n are independent and identically distributed random variables (with finite mean and variance), then for large enough sample size n, the sample mean $\bar{X}$ has approximately a Gaussian distribution. This result is extremely useful for making inferences about a population mean when the sample size is large.

Minitab Appendix for Chapter 8

Finding Cumulative Probabilities for the Standard Gaussian Distribution

Let Z denote a random variable having the standard Gaussian distribution (also called the standard normal distribution). To find a cumulative probability of the form $P(Z \leq c)$ we use the CDF command with the NORMAL subcommand. To find $P(Z \leq 1.5)$, we use

```
MTB>   cdf 1.5;
SUBC>  normal.
 1.5000  0.9332
```

Minitab prints the value 1.5000, plus the cumulative probability 0.9332.

Finding Cumulative Probabilities for a Gaussian Distribution That Is Not Standardized

If we want a cumulative probability for a Gaussian distribution with mean μ and standard deviation σ, we can specify μ and σ as part of the NORMAL subcommand. To find $P(Y \leq 3.4)$, where Y has the Gaussian distribution with mean 1 and standard deviation 2, we use the command

```
MTB>   cdf 3.4;
SUBC>  normal 1 2.
 3.4000  0.8849
```

Minitab prints the value 3.4000, plus the cumulative probability 0.8849.

We can instruct Minitab to print $P(Y \geq 3.4)$ with the following commands

```
MTB>   cdf 3.4 k1;
SUBC>  normal 1 2.
MTB>   let k2=1-k1
MTB>   print k2
K2     0.115070
```

In the CDF command, we specified that we wanted the cumulative probability stored in constant K1. We can then use this cumulative probability later by referring to the stored constant K1.

Finding Values of a Gaussian Probability Function

If we want to print the value of a Gaussian probability function at a particular point, we use the PDF command with the NORMAL subcommand. The command

```
MTB>   pdf 0;
SUBC>  normal.
 0.0000  0.3989
```

causes Minitab to print the value 0.3989 of the standard Gaussian probability function evaluated at 0 (note that this is *not* a probability). If we want to save this value in the stored constant K2, we use the command

```
MTB>    pdf 0 k2;
SUBC>   normal.
```

We can then use the value stored in K2 for later commands.

To print the value of the probability function for the Gaussian distribution with mean 1 and standard deviation 2, evaluated at 3.4, we use the command

```
MTB>    pdf 3.4;
SUBC>   normal 1 2.
 3.4000  0.0971
```

Minitab prints the value 0.0971 of the appropriate probability function evaluated at 3.4 (again, this is *not* a probability).

Finding Values Corresponding to Gaussian Cumulative Probabilities

The INVCDF command with the NORMAL subcommand prints the value c for which $P(X \leq c)$ equals a specified probability, where X has the Gaussian distribution indicated in the subcommand. The command

```
MTB>    invcdf 0.95 k3;
SUBC>   normal.
MTB>    print k3
K3      1.64485
```

prints the value of c for which $P(Z \leq c) = .95$, where Z has the standard Gaussian distribution. This value of c equals 1.64485 and is stored as constant K3. The command

```
MTB>    invcdf 0.025;
SUBC>   normal 1 2.
 0.0250   -2.9199
```

prints the value of c for which $P(Y \leq c) = .025$ where Y has the Gaussian distribution with mean 1 and standard deviation 2. This value of c equals -2.9199.

Simulating Random Samples Using the RANDOM Command

The RANDOM command generates simulated random samples of numbers from a specified distribution. This command can be useful for seeing what distributions look like. Recall that in Section 8-2 we looked at hits allowed in 1987 by the 26 major-league baseball teams. The average of the 26 values of hits allowed was 1,457.5, and the standard deviation was 66.7. Suppose we want Minitab to simulate 26 random values from the Gaussian distribution with

```
Histogram of C1   N = 26

Midpoint   Count
  1330.0       0
  1350.0       1  *
  1370.0       3  ***
  1390.0       0
  1410.0       2  **
  1430.0       3  ***
  1450.0       3  ***
  1470.0       2  **
  1490.0       4  ****
  1510.0       3  ***
  1530.0       1  *
  1550.0       1  *
  1570.0       1  *
  1590.0       1  *
  1610.0       1  *
```

FIGURE M8-1 Histogram of 26 random values simulated from the Gaussian distribution with mean 1,457.5 and standard deviation 66.7

mean 1,457.5 and standard deviation 66.7, and put these values in column 1. We use the command

```
MTB>   random 26 c1;
SUBC>  normal 1457.5 66.7.
```

We can then use the histogram command on column 1 to look at the distribution of these 26 values:

```
MTB>   histogram c1;
SUBC>  increment=20;
SUBC>  start=1330.
```

The INCREMENT subcommand specifies intervals of width 20. The START subcommand specifies the midpoint of the first interval as 1330. These subcommands provide a histogram set up like the one we used in Figure 8-9 to look at the distribution of hits allowed. The Minitab display is shown in Figure M8-1.

We may want to look at two more sets of simulated random values to get a feel for how the distributions of sample values can vary. The commands

```
MTB>   random 26 c2 c3;
SUBC>  normal 1457.5 66.7.
MTB>   histogram c2 c3;
SUBC>  increment=20;
SUBC>  start=1330.
```

simulate two more sets of 26 random values from the Gaussian distribution with mean 1,457.5 and standard deviation 66.7, and plot them in the histograms shown in Figure M8-2.

```
Histogram of C2   N = 26          Histogram of C3   N = 26
                                  2 Obs. below the first class
Midpoint   Count
  1330.0     0                    Midpoint   Count
  1350.0     1  *                   1330.0     0
  1370.0     3  ***                 1350.0     1  *
  1390.0     1  *                   1370.0     1  *
  1410.0     2  **                  1390.0     1  *
  1430.0     1  *                   1410.0     3  ***
  1450.0     6  ******              1430.0     4  ****
  1470.0     6  ******              1450.0     2  **
  1490.0     0                      1470.0     4  ****
  1510.0     3  ***                 1490.0     2  **
  1530.0     0                      1510.0     3  ***
  1550.0     1  *                   1530.0     1  *
  1570.0     2  **                  1550.0     2  **
```

FIGURE M8-2 Histograms of two more sets of 26 random values simulated from the Gaussian distribution with mean 1,457.5 and standard deviation 66.7

The RANDOM command with the BINOMIAL subcommand simulates random values from the specified binomial distribution. For instance,

```
MTB>   random 30 c1;
SUBC>  binomial 5 0.25.
```

simulates 30 random values from the Binomial(5, .25) distribution and puts them in column 1.

We can investigate the Central Limit Theorem result using the RANDOM command. As an example, we might use the BERNOULLI subcommand to simulate random values from a Bernoulli distribution (the same as a binomial distribution with sample size 1). The command

```
MTB>   random 20 c1-c40;
SUBC>  bernoulli 0.2.
```

simulates 40 sets of 20 random values from the Bernoulli distribution with probability of success .2, the Binomial(1, .2) distribution, one set for each of columns 1–40. Now we will calculate a row-wise sum of the first five columns of observations and put it in column 41. C41 will contain 20 values, each a sum of five Bernoulli(.2) observations:

```
MTB>  rsum c1-c5 c41
```

Similarly, we will calculate row-wise sums of C1–C10, C1–C20, and C1–C40:

```
MTB>  rsum c1-c10 c42
MTB>  rsum c1-c20 c43
MTB>  rsum c1-c40 c44
```

Column 42 now contains 20 values, each a sum of ten simulated values from the Bernoulli(.2) distribution. Similarly, column 43 contains 20 values, each a sum of 20 values from the Bernoulli(.2) distribution. Column 44 contains 20 values, each a sum of 40 values from the Bernoulli(.2) distribution.

```
Histogram of C1   N = 20

Midpoint   Count
       0      14  **************
       1       6  ******

Histogram of C41   N = 20

Midpoint   Count
       0       6  ******
       1      10  **********
       2       1  *
       3       2  **
       4       1  *

Histogram of C42   N = 20

Midpoint   Count
       0       1  *
       1       7  *******
       2       7  *******
       3       0
       4       4  ****
       5       0
       6       0
       7       1  *

Histogram of C43   N = 20

Midpoint   Count
       2       2  **
       3       5  *****
       4       5  *****
       5       3  ***
       6       2  **
       7       1  *
       8       1  *
       9       1  *

Histogram of C44   N = 20

Midpoint   Count
       3       1  *
       4       0
       5       2  **
       6       1  *
       7       2  **
       8       2  **
       9       3  ***
      10       5  *****
      11       2  **
      12       1  *
      13       1  *
```

FIGURE M8-3 Histogram of 20 observations on the sum of 1, 5, 10, 20, and 40, respectively, random values simulated from the Bernoulli(.2) distribution

We can display histograms of C1 and C41–C44 to get a feel for the relationship between sample size and the Central Limit Theorem Gaussian approximation to the distribution of a sum (or average):

```
MTB> histogram c1, c41-c44
```

The histograms are shown in Figure M8-3. You may try similar manipulations to investigate the Central Limit Theorem result.

Exercises for Chapter 8

EXERCISE 8-1

Let Z be a standard Gaussian random variable. Draw a sketch of the appropriate area under the standard Gaussian curve and use Table B to find the following probabilities:

a. $P(-2.5 \leq Z \leq -1.5)$

b. $P(Z \leq -2.4)$

c. $P(-1.5 \leq Z \leq .5)$

d. $P(.5 \leq Z \leq 1.5)$

e. $P(Z \geq 2.55)$

EXERCISE 8-2

Suppose the random variable Y has a Gaussian distribution with mean 20 and variance 9. Draw a sketch of the appropriate area under the Gaussian curve and find the following probabilities:

a. $P(Y \leq 15)$

b. $P(10 \leq Y \leq 30)$

c. $P(10 \leq Y < 15)$

d. $P(Y \geq 18)$

e. $P(25 \leq Y \leq 30)$

f. $P(14 \leq Y \leq 26)$

g. $P(Y > 23)$

h. $P(Y < 17)$

i. $P(Y \geq 28)$

EXERCISE 8-3

Mercury concentration (in parts per million) is listed for each of 115 swordfish inspected for mercury (Lee and Krutchkoff, 1980).

.05	.07	.07	.13	.13	.19	.24	.25
.28	.32	.39	.45	.46	.53	.54	.56
.60	.60	.61	.62	.65	.71	.72	.75
.76	.79	.81	.81	.82	.82	.82	.83
.83	.83	.84	.85	.89	.90	.91	.92
.92	.93	.95	.95	.97	.97	.98	1.00
1.00	1.01	1.02	1.04	1.05	1.05	1.08	1.10
1.12	1.12	1.14	1.14	1.15	1.16	1.20	1.20
1.20	1.20	1.20	1.21	1.22	1.25	1.25	1.26
1.27	1.27	1.29	1.29	1.29	1.29	1.30	1.31
1.32	1.32	1.37	1.37	1.39	1.39	1.40	1.40
1.41	1.42	1.43	1.44	1.45	1.54	1.54	1.58
1.58	1.60	1.60	1.62	1.62	1.66	1.66	1.68
1.69	1.72	1.74	1.85	1.89	1.96	2.06	2.10
2.23	2.25	2.72					

a. Plot these observations.

b. Can the distribution of these observations be approximated by a Gaussian distribution?

c. At the time these measurements were made, the Food and Drug Administration stated that seafood with more than 1 part per million mercury contamination should not be eaten. Discuss this safety limit in relation to this data set.

EXERCISE 8-4

Serum total cholesterol concentration (mg/dl) is listed below for each of 23 24-month-old baboons, raised on a high cholesterol, saturated fat diet (McMahan, 1981).

141	135	127	200	184	122	219	114	136	253
243	188	239	135	165	140	186	134	110	103
144	252	169							

Plot these observations. Can the distribution of these observations be approximated by a Gaussian distribution?

EXERCISE 8-5

In this study of the association between hyperglycemia and relative hyperinsulinemia, investigators administered standard glucose tolerance tests to 13 control patients and 20 obese patients on the Pediatric Clinical Research Ward, University of Colorado Medical Center. As part of the study, workers determined plasma inorganic phosphate levels from blood samples zero hours after a standard-dose oral glucose challenge. The recorded values of plasma inorganic phosphate (mg/dl) are shown below (Zerbe, 1979).

Control patients:	4.3	3.7	4.0	3.6	4.1	3.8	3.8	4.4
	5.0	3.7	3.7	4.4	4.7			
Obese patients:	4.3	5.0	4.6	4.3	3.1	4.8	3.7	5.4
	3.0	4.9	4.8	4.4	4.9	5.1	4.8	4.2
	6.6	3.6	4.5	4.6				

a. Plot these observations.

b. Do the plasma inorganic phosphate measurements for the control patients appear to be approximately Gaussian distributed?

c. Do the plasma inorganic phosphate measurements for the obese patients appear to be approximately Gaussian distributed?

d. Compare center and variation for these two distributions.

EXERCISE 8-6

In this experiment, investigators studied a method of determining aflatoxin levels in contaminated peanuts. (This problem is important in sampling inspection. Inspectors want to protect consumers while not rejecting too many good peanuts.) They ground the peanuts into meal and then blended a sample of the meal in a chemical solution. They divided this blend equally among 16 centrifuge bottles. The determination of aflatoxin concentration for each bottle (units not given) is shown below (Quesenberry, Whitaker, and Dickens, 1976; from Waltking, Bleffert, and Kiernan, 1968).

121.23	71.69	117.91	91.09	104.86	151.00	125.40
83.94	137.53	83.49	116.78	90.72	100.54	74.59
137.19	146.25					

Plot these observations. Do these observations appear to come from a Gaussian distribution?

EXERCISE 8-7

Mental retardation is associated with some metabolic diseases. In this experiment, investigators studied metabolism of tyrosine among 36 mentally handicapped patients (Geertsema and Reinecke, 1984). The measured response was the total amount of tyrosine catabolites excreted in the urine (in μmoles per 100 ml urine).

a. Ten separate measurements on one patient resulted in the following observations:

.325 .317 .375 .325 .508 .117 .150 .317 .275 .383

Plot these observations. Do they appear to come from a Gaussian distribution?

b. The average of ten observations on each of the 36 patients is shown below. These averages are listed in increasing order.

.309	.328	.355	.368	.379	.381	.383	.391
.393	.411	.440	.444	.447	.464	.505	.521
.554	.593	.613	.620	.628	.650	.674	.697
.699	.715	.725	.754	.818	.835	.868	.995
1.099	1.115	1.185	1.693				

Plot these averages. Can the distribution of these averages be approximated by a Gaussian distribution?

c. The investigators selected 1.0 μmole per 100 ml urine as a cutoff for classifying a patient's tyrosine metabolism as clinically negative (less than 1.0, no metabolic problem) or clinically positive (greater than 1.0, a metabolic disorder). Considering the average determinations in part (b), what can you say about these patients in light of this cutoff?

EXERCISE 8-8

Life expectancy at birth in 1985 is plotted for 125 countries in Figure 2-1. Can the distribution of these 125 life expectancies be approximated by a Gaussian distribution?

EXERCISE 8-9

Total fertility rate in 1985 is plotted for 125 countries in Figures 2-4 and 2-15. Can the distribution of these 125 fertility rates be approximated by a Gaussian distribution?

EXERCISE 8-10

A frequency plot of number of cities of over 500,000 persons in 1980, summarizing information for 123 countries, is shown in Figure 2-6. Can the distribution of these 123 data values be approximated by a Gaussian distribution?

EXERCISE 8-11

Number enrolled in primary school in 1984 as percentage of 6–11-year age group is summarized for 119 countries with a histogram in Figure 2-7, with a dot plot in Figure 2-12. Can the distribution of these 119 percentages be approximated by a Gaussian distribution?

EXERCISE 8-12

Number of deaths of children 1–4 years of age per 1,000 children in this age group in 1985 is plotted for 124 countries in Figure 2-13. Can the distribution of these 124 data values be approximated by a Gaussian distribution?

EXERCISE 8-13

In men's golf, all 19 top money winners for 1986 made over $300,000 that year. The average drive for each of these 19 winners is listed below (*USA Today*, January 15, 1987, page 6C).

Player	Average drive (yards)	1986 earnings (dollars)
Greg Norman	277.5	653,295
Bob Tway	268.2	652,780
Payne Stewart	266.4	535,389
Andy Bean	273.9	491,937
Dan Pohl	273.1	463,629
Hal Sutton	262.2	429,433
Tom Kite	253.9	394,164
Ben Crenshaw	261.4	388,168
Ray Floyd	258.6	380,508
B. Langer	259.0	379,799
John Mchaffey	262.9	378,172
Calvin Peete	248.6	374,953
Fuzzy Zoeller	266.4	358,115
Joey Sindelar	277.7	341,230
Jim Thorpe	266.2	326,086
Ken Green	267.0	317,834
Larry Mize	254.9	314,050
Doug Tewell	258.1	310,285
Corey Pavin	260.3	304,557

a. Plot average drive for these 19 golfers.

b. Find the mean and standard deviation of the 19 average drives.

c. Let W be a Gaussian random variable with mean and standard deviation that you calculated in part (b). Find the following probabilities:

(i) $P(255 \leq W \leq 260)$
(ii) $P(260 \leq W \leq 265)$
(iii) $P(265 \leq W \leq 270)$
(iv) $P(W < 255)$
(v) $P(W \geq 270)$

d. Find the proportion of the 19 average drives listed above that are

(i) from 255 to 260
(ii) from 260 to 265
(iii) from 265 to 270
(iv) less than 255
(v) greater than or equal to 270

Compare these proportions with the probabilities you found in part (c).

e. Can the distribution of these 19 average drives be approximated by a Gaussian distribution?

f. Can the distribution of the 19 earnings be approximated by a Gaussian distribution?

EXERCISE 8-14 The percentage of mothers wholly or partially breastfeeding their babies for at least 6 months in 1980–1984 is listed below for 29 countries (Grant, James P., 1987, pages 130–131).

Country	Percentage of mothers breastfeeding their babies	Country	Percentage of mothers breastfeeding their babies
Sierra Leone	94	Peru	72
Malawi	95	Indonesia	97
Niger	30	Congo	97
Rwanda	98	Kenya	84
Yemen	76	Honduras	28
Yemen, Dem.	73	Brazil	19
Burundi	95	Burma	90
Bangladesh	97	El Salvador	77
Sudan	86	Philippines	58
Bolivia	91	Colombia	58
Haiti	85	Thailand	47
Uganda	70	Panama	48
Pakistan	96	Chile	28
Ghana	70	Costa Rica	20
Egypt	91		

a. Plot these data.

b. Find the mean and the standard deviation for these 29 data values.

c. Let Y be a Gaussian random variable with mean and standard deviation that you calculated in part (b). Find the following probabilities:

(i) $P(Y < 30)$
(ii) $P(Y > 90)$
(iii) $P(50 \leq Y \leq 80)$

d. Find the proportion of data values that are:

(i) less than 30
(ii) greater than 90
(iii) from 50 to 80

Compare these proportions with the probabilities you found in part (c).

e. Can the distribution of these 29 percentages be approximated by a Gaussian distribution?

EXERCISE 8-15 Base pay of the governor in 1986 is listed in Exercise 2-19 for each of the 50 states (*USA Today,* December 11, 1986, page 9C).

a. Plot these salaries.

b. Find the mean and the standard deviation for these 50 salaries.

c. Let W be a Gaussian random variable with mean and standard deviation that you calculated in part (b). Find the following probabilities:

(i) $P(58{,}000 \le W \le 68{,}000)$
(ii) $P(70{,}000 \le W \le 76{,}000)$
(iii) $P(W > 80{,}000)$
(iv) $P(W < 50{,}000)$

d. Find the proportion of the 50 salaries that are:

(i) from \$58,000 to \$68,000
(ii) from \$70,000 to \$76,000
(iii) greater than \$80,000
(iv) less than \$50,000

Compare these proportions with the probabilities you found in part (c).

e. Can the distribution of these 50 salaries be approximated by a Gaussian distribution?

EXERCISE 8-16 Number enrolled in higher education in 1984 as percentage of 20–24-year-old age group is plotted in Figure 8-11 for 119 countries. Can the distribution of these 119 data values be approximated by a Gaussian distribution?

EXERCISE 8-17 Suppose the random variable Y has a Binomial$(25, \frac{1}{2})$ distribution. This distribution is illustrated in Figure 8-12.

a. Find the mean and variance of Y. (See Section 7-2.)

b. Find the following probabilities:

(i) $P(10 \le Y \le 15)$
(ii) $P(Y \ge 15)$
(iii) $P(Y \le 10)$
(iv) $P(8 \le Y \le 17)$
(v) $P(Y \le 7)$
(vi) $P(Y \ge 18)$

c. Let W be a Gaussian random variable with the same mean and variance as Y, which you found in part (a). Find the following probabilities:

(i) $P(10 \le W \le 15)$
(ii) $P(W \ge 15)$
(iii) $P(W \le 10)$
(iv) $P(8 \le W \le 17)$
(v) $P(W \le 7)$
(vi) $P(W \ge 18)$

d. Compare the probabilities you found in part (b) with those you found in part (c). Does the binomial distribution of Y seem to be well approximated by the Gaussian distribution of W?

EXERCISE 8-18 Suppose the random variable X has a Binomial(25, $\frac{1}{10}$) distribution. This distribution is illustrated in Figure 8-12.

a. Find the mean and variance of X. (See Section 7-2.)

b. Find the following probabilities:

(i) $P(1 \leq X \leq 4)$
(ii) $P(X \geq 4)$
(iii) $P(X \leq 1)$
(iv) $P(0 \leq X \leq 5)$
(v) $P(X \geq 6)$

c. Let U be a Gaussian random variable with the same mean and variance as X, which you found in part (a). Find the following probabilities:

(i) $P(1 \leq U \leq 4)$
(ii) $P(U \geq 4)$
(iii) $P(U \leq 1)$
(iv) $P(0 \leq U \leq 5)$
(v) $P(U \geq 6)$

d. Compare the probabilities you found in part (b) with those you found in part (c). Does the binomial distribution of X seem to be well approximated by the Gaussian distribution of U?

EXERCISE 8-19 Fifty college students participated in a coin-tossing experiment (Alex Olsen, personal communication, 1986). For the first stage of the experiment, each student tossed a coin once and recorded whether a head or tail landed face up. For the second stage, each student tossed a coin twice and recorded the number of heads and tails in the two tosses. Similarly, each student recorded the number of heads in five other stages of the experiment, involving 3, 4, 5, 6, and 10 tosses of a coin. The results are shown below:

One toss:	Number of heads	0	1									
	Number of students	22	28									
Two tosses:	Number of heads	0	1	2								
	Number of students	11	23	16								
Three tosses:	Number of heads	0	1	2	3							
	Number of students	7	15	22	6							
Four tosses:	Number of heads	0	1	2	3	4						
	Number of students	4	18	16	10	2						
Five tosses:	Number of heads	0	1	2	3	4	5					
	Number of students	1	7	21	16	4	1					
Six tosses:	Number of heads	0	1	2	3	4	5	6				
	Number of students	1	2	16	17	9	5	0				
Ten tosses:	Number of heads	0	1	2	3	4	5	6	7	8	9	10
	Number of students	0	0	1	8	9	15	10	2	3	2	0

a. Make a frequency plot of the results of each of the seven stages of the experiment.

b. What assumptions must you make to discuss this experiment in terms of the Central Limit Theorem? Discuss the reasonableness of these assumptions for this experiment.

c. Suppose the assumptions you discussed in part (b) hold for this experiment. Discuss how the results of the experiment relate to the Central Limit Theorem.

CHAPTER 9

Basic Ideas in Statistics

IN THIS CHAPTER

Population, parameter
Sample, random sample
Statistical inference
Experimental design
Hypothesis testing, null and alternative hypotheses
Test statistic
Observed significance level, *p*-value
Significance level, power
Point estimate, interval estimate, confidence interval
Parametric, nonparametric

The realm of statistics includes both data analysis and statistical inference. In data analysis, we use graphical tools and descriptive statistics to explore a data set. We want to learn as much as we can about variables and relationships between variables, within a particular sample. With statistical inference, we move beyond the goals of data analysis. We want to use a sample to learn about (make inferences about) the population from which the sample came.

We use the techniques of statistical inference to make probability statements based on our observations. The validity of these statements depends on certain assumptions about the observations (regarding independence or probability distributions, for instance). To assure the reasonableness of these assumptions, we must take care in how we obtain the observations. Collecting samples that allow valid inferences is the realm of experimental design. We should not talk about statistical inference without discussing design of experiments.

Consider, for example, a taste test of Coke and Pepsi. What are the goals of a taste test? How can we conduct the test in order to meet these goals? Should we give some tasters Coke and some other tasters Pepsi and try to determine which group is more satisfied? Should we hand each taster a can of Coke and a can of Pepsi and ask him to decide which he likes best? Or should we do something else? Our willingness to assess the results of the taste test will depend on whether we think the experimental design allowed a fair comparison of the two soft drinks. We will see how one group of students approached this problem, in Example 9-1.

Now consider this question: Will you pedal farther in 15 minutes on an exercise bicycle while listening to music than if you do not listen to music? One student designed an experiment to answer this question for herself. As we will see in Example 9-2, she tried to control as many extraneous variables (such as clothing type and time of day) as possible to allow a valid assessment of her experimental results.

As another example, the engineer in charge of an electroplating process is concerned that too much gold is being plated onto components. Thousands of components are plated each day and she cannot examine all of them. So she decides to use a sample of the components plated one day to learn about the population of all components plated that day. She would like to use the sample to estimate the average thickness of gold plated on all components that day. She would also like to use the sample to decide if the average gold thickness on components plated that day was above the target value. How should she select components for her sample to give her the most confidence in her inferences? We will consider this problem in Example 9-3.

These three examples, introduced in Section 9-2, form the basis of our introductory discussion of statistical inference. Before considering them, however, we need to define some basic terms that will make the discussion clearer.

9-1

Some Definitions Related to Statistical Inference

In *statistical inference,* we use a sample to learn about a population:

> **Statistical inference** refers to the process of drawing conclusions about a population based on a sample from that population.

The *population* is the group or collection we want to learn about. For our purposes, we assume the population is very large compared with the size of the sample.

> The **population** is the set or collection we are interested in learning about. We learn about the population by studying a sample of the population.

Sometimes the population is hypothetical. For instance, to get a feel for whether a coin is fair, we might flip it 200 times and count the number of heads. We make inferences about the fairness of the coin (or the probability of heads) based on the sample of 200 tosses. The population is the hypothetical collection of results we would get if we could toss the coin indefinitely.

A similar situation occurs when an investigator carries out a controlled experiment. Say, he divides some animals into two groups. He subjects all the animals in one group to one experimental condition or treatment, and subjects all the animals in the other group to another condition. He then compares the responses across the two experimental groups. The sample consists of all the animals involved in the experiment. The population is the hypothetical collection of similar animals that might be subjected to the same experimental conditions, if time and money permitted unlimited investigation.

Other times the population has substance. In a quality control setting, a sample of items is used to draw inferences about the acceptability of a large collection of items produced. Or, a sample of 1-year-old babies may be examined so that inferences can be made about physical characteristics (such as height and weight) of a large population of 1-year-old babies.

When we can study the entire population, we are in the realm of data analysis (as when we studied the World Bank data set). Statistical inference is appropriate when we cannot observe the entire population, but can observe only a subset or *sample* of the population. We use this sample to learn about the population.

> A **sample** is a subset of the population. We use the observations in the sample to make inferences about the population.

In formal statistical analysis, the subject of the remaining chapters, we want our sample to be a *random sample in the probability sense:*

> By a **random sample in the probability sense,** we mean a sample with observations that are independent and have the same probability distribution.

We want the probability distribution of the observations in the sample to be the same as that of the population. That is, we want a sample that is *representative* of the population.

> We say a sample is **representative** of the population if it is similar to the population with respect to characteristics we want to study.

The best way to ensure getting a random sample in the probability sense is to collect a *random sample in the experimental sense:*

> When sampling from a population, we say we have a **random sample in the experimental sense** if each member of the population had an equal and independent chance of being included in the sample.

Sometimes, as in a quality control setting, a true random sample from the population may not be hard to get. Other times, as in examining 1-year-old babies, a true random sample from the population may be difficult, or costly, or even impossible to obtain. In such cases, we must be especially careful when interpreting results.

In formal statistical analysis, we select a sample from a population. We make assumptions about the distribution of the observations, then use the sample to make probability statements about the population. Sometimes our statistical inference is *parametric:*

> In **parametric statistical inference,** we assume a specific type of probability distribution (for example, Gaussian) for the observations.

Other times our inference is *nonparametric:*

> In **nonparametric statistical inference,** our assumptions do not specify a particular type of probability distribution for the observations.

We will use both parametric and nonparametric statistical analyses in the chapters that follow.

The descriptions "parametric" and "nonparametric" come from the idea of a *parameter:*

> A number that characterizes the distribution of a population is a **parameter.**

The mean and median of a population are parameters that describe the center of the population values. The standard deviation of a population is a parameter that describes the spread in the population values. One goal of statistical inference is to use a sample to estimate population parameters.

One way to estimate a population parameter is with a *point estimate:*

> A **point estimate** is a single number calculated from the sample to estimate a population parameter.

We often use the sample mean as a point estimate of the population mean. We may use the sample standard deviation as a point estimate of the population standard deviation.

Because of sampling variation, we would not expect to get the same point estimate of a parameter from different samples. Point estimates have variation associated with them. So in addition to a point estimate, we might like an *interval estimate:*

An **interval estimate** for a population parameter is a range of reasonable values for the parameter.

Our interval estimates will take the form of *confidence intervals,* with probability interpretations. We will calculate a confidence interval for a population mean (average gold thickness) in Example 9-3. Confidence intervals have different forms depending on the parameter we estimate and the assumptions we make about the sample observations. We will see a number of different forms of confidence intervals in the chapters that follow.

A **confidence interval** is an interval estimate of a parameter, with a probability interpretation.

In *hypothesis testing,* we formalize the way we ask questions about one or more populations. We ask whether the sample is consistent with a specific hypothesis about the population(s).

Hypothesis testing is a formal strategy for comparing two statements about the state of nature in an experimental situation.

We want to compare two statements about the state of nature. The first statement is generally a "status quo" or "no difference" statement. We call it the *null hypothesis,* or H_0.

The **null hypothesis** is a statement about the state of nature in an experimental situation, generally a "status quo" or "no difference" statement. We often denote the null hypothesis by H_0.

The other statement is an alternative to the null hypothesis. We call it the *alternative hypothesis,* or H_a.

The **alternative hypothesis** is a statement about the state of nature, providing an alternative to that specified in the null hypothesis. We often denote the alternative hypothesis by H_a.

As we will see when we discuss the general strategy of hypothesis testing in Section 9-3, we build a probability model for our experiment under the null hypothesis and use this model to make probability statements based on our sample.

The general strategy of hypothesis testing is the same for all hypothesis testing situations. Before describing this general strategy, let's consider three examples.

9-2 Three Examples

We will refer to the following examples later in the chapter as we discuss experimental design and formal statistical analysis.

EXAMPLE 9-1

On the first day of a statistics course, I walked into the classroom with a package of paper cups and several cans of the diet, caffeine-free versions of Coke and Pepsi. I asked the class to design and carry out a taste test comparing these two soft drinks. In particular, I asked the class to:

- State a specific goal. Write this goal in terms of hypotheses to be compared.
- Design an experiment to meet this goal.
- State assumptions and describe how the experiment will be analyzed.
- Carry out the experiment.
- Analyze and interpret the results.

The class enthusiastically accepted my challenge. The following is a summary of their discussion and results.

State a specific goal. The class discussed two goals. The first was to determine whether tasters could tell a difference between Coke and Pepsi. The second was to see, more specifically, if they had a preference for one of these soft drinks. The students decided the second goal was more interesting and relevant for a taste test.

Goal: To see if tasters have a preference for Coke or Pepsi.

They then wrote this goal in terms of two statements to be compared. One statement, called the null hypothesis, says there is no preference. The other statement, called the alternative hypothesis, says there is a preference. (As noted earlier, we sometimes denote the null hypothesis by H_0 and the alternative hypothesis by H_a.)

Null hypothesis: Tasters have no preference for Coke or Pepsi.
Alternative hypothesis: Tasters have a preference for either Coke or Pepsi.

Design an experiment to meet this goal. One student, Karen, could not taste either soft drink for dietary reasons, so she carried out the mechanics of the taste test. The class decided that each of the other 12 students should taste both Coke and Pepsi. The students did not want any taster to know the brand of beverage tasted at any stage. (Why?) They also did not want Karen to know the identity of beverages tasted. (Why?) So the students chose a **double-blind experiment.**

A **double-blind experiment** is one in which neither the participants nor the experimenter know the identity of the treatments as they are administered and evaluated.

The students wrote a letter C on the bottoms of 12 paper cups and a letter P on the bottoms of 12 more. They filled all the cups about three-quarters full—those labeled C, with Coke, and those labeled P, with Pepsi. Then they paired the cups, one C cup and one P cup per pair. They switched the cups within a pair around, so no one knew which was the C cup and which was the P cup.

Everyone except Karen left the room. Karen placed one pair of cups on each desk, then the others returned. A taster sampled from each cup as often as desired, leaving enough beverage in a cup so that the letter C or P on the bottom was not visible. Each of the 12 tasters chose a preferred beverage. If a taster really felt no preference, he or she had to choose one of the cups anyway. After everyone had made a selection, each taster gave Karen the cup of the preferred drink; she checked the letter on the bottom of the cup and wrote it down.

State assumptions and describe how the experiment will be analyzed. All the students had already had a beginning statistics course, so they knew they needed to make some simplifying assumptions in order to analyze the results of the experiment. The assumptions they made were:

Experimental assumptions

The tasters make choices independently of one another.
Each taster has the same probability p of choosing Coke.

The students believed the first assumption was reasonable. However, they thought the second assumption was most likely a great simplification of reality; we will discuss this later in the chapter.

Let Y denote the number of tasters who choose Coke. Under the two assumptions listed above, Y has a Binomial(12, p) distribution. We can rewrite the null and alternative hypotheses as

$$H_0:\ p = \frac{1}{2} \quad \text{and} \quad H_a:\ p \neq \frac{1}{2}$$

The null hypothesis says that a taster is just as likely to select Pepsi as Coke. This seems reasonable when a taster really has no preference, but must select one of the cups anyway. Under the null hypothesis, the random variable Y has a Binomial(12, $\frac{1}{2}$) distribution. We found this probability distribution in Example 7-5. The probabilities, listed in Table 7-5, are shown again in Table 9-1.

Extreme values of Y, near 0 or 12, cause us to doubt the null hypothesis, suggesting that there is a preference for Pepsi or Coke. Moderate values of Y, not too far from 6, are consistent with the null hypothesis of no preference. With this in mind, the students chose the following decision rule:

Decision rule

If $3 \leq Y \leq 9$, say the experimental results are consistent with the no-preference null hypothesis.

If $Y \leq 2$ or $Y \geq 10$, say the experimental results are inconsistent with the no-preference null hypothesis, suggesting there is a preference.

Using Table 9-1, we can calculate the probability of saying the results are inconsistent with the null hypothesis, when the null hypothesis is really true:

$$P(Y \leq 2 \text{ or } Y \geq 10 \text{ when } H_0 \text{ is true}) = .038$$

We call .038 the *significance level* of the test. This is the probability the students will say the experimental results are inconsistent with what they would expect

TABLE 9-1 Probabilities for a random variable Y having a Binomial(12, $\frac{1}{2}$) distribution. The probabilities have been rounded to 4 decimal places; this is why they do not sum exactly to 1.

k	$P(Y = k)$
0	.0002
1	.0029
2	.0161
3	.0537
4	.1208
5	.1934
6	.2256
7	.1934
8	.1208
9	.0537
10	.0161
11	.0029
12	.0002

under the null hypothesis, when the null hypothesis is really true. With this decision rule, there is about a 4% chance the students will say there is a preference for Coke or Pepsi when there really is none.

Carry out the experiment. The class carried out the experiment as described above. Nine of the students selected Coke as their preferred beverage and three selected Pepsi.

Analyze and interpret the results. Since nine students selected Coke, $Y = 9$. Using the decision rule given above, the class decided the experimental results were consistent with the no-preference null hypothesis.

The students then calculated the probability under the null hypothesis of seeing results as extreme as or more extreme than the observed result. This probability is called the *observed significance level* or *p-value*. We can find the p-value using the probabilities in Table 9-1:

$$p\text{-value} = P(Y \leq 3 \text{ or } Y \geq 9 \text{ when } H_0 \text{ is true}) = .146$$

Note that $Y \leq 3$ is just as extreme in the direction of a Pepsi preference as $Y \geq 9$ is in the direction of a Coke preference. If there were really no preference, there would be slightly better than a 14% chance of seeing a result at least as extreme as the one observed.

The class decided that seeing 9 of 12 students prefer Coke was not extreme enough to discredit the no-preference null hypothesis; they were unwilling to say there was a preference. In addition, they noted that the 12 tasters (graduate students at a northeastern university) were unlikely to be representative of the population of American soft drink consumers. Therefore, they could not interpret the experimental results in terms of any population beyond the confines of the class.

EXAMPLE 9-2

Will you pedal farther in 15 minutes on an exercise bicycle while listening to a Walkman or not? As part of a class project, a student tried to answer this question for herself (Walsh, 1988). The student, Michele, used an exercise bicycle and a Walkman equipped with a tape of five "pop rock" songs. She planned to use her results to compare two statements:

Null hypothesis: The median distance pedaled is the same with and without the Walkman.

Alternative hypothesis: The median distance pedaled is different with than without the Walkman.

Michele carried out her experiment on four consecutive weekdays, between 8:00 and 8:30 A.M. Each morning she pedaled for 15 minutes, timed by a clock with an alarm. The odometer on the exercise bicycle was covered for each trial, so she could not see the distance pedaled. Two of the four runs were with the Walkman and two without. All other conditions (clothing, warm-up, room temperature) were constant. By a random process such as drawing numbers from a hat, Michele determined that she would make the runs on the first and fourth days wearing the Walkman. She made the runs on the second and third days without the Walkman.

This experiment was not double-blind as the taste test was. Michele was both the experimenter and the participant in her investigation. She obviously knew which treatment (Walkman or no Walkman) she used in each trial. Since she knew why she was conducting the experiment, her preconceptions could influence her results. This is a major drawback of a study that is not blinded. Can you think of some ways Michele could have improved her experimental design?

The results of Michele's experiment are shown in Table 9-2. Let's use her results to compare her null and alternative hypotheses.

We should always start by plotting data in any way that seems reasonable. Figure 9-1 shows a plot of Michele's experimental results. We see that she pedaled farther in both runs with the Walkman than she did in either run without the Walkman.

These results suggest that Michele tends to pedal farther wearing the Walkman. Let's do a formal statistical analysis of the results. As in many experimental situations, there are several approaches we could take. For this illustration, we will keep things simple, using a method of analysis known as the median test (see Section 11-5).

The median of Michele's four observations is 3.2. We classify each observation by treatment and by position relative to the overall median. This information is displayed in Table 9-3.

We know that half the experimental runs were made with the Walkman and half without. We also know that half the observations are above the overall median and half below. Would we be surprised, if the null hypothesis were true, to see both the Walkman results above the overall median and both the no-Walkman results below? To answer this question, we need to develop a probability model for the experiment under the null hypothesis.

TABLE 9-2 Results of an experiment to determine the effect of wearing a Walkman on distance pedaled on an exercise bicycle

Day	Experimental condition	Miles pedaled
Tuesday	Walkman	3.5
Wednesday	No Walkman	2.9
Thursday	No Walkman	2.7
Friday	Walkman	3.8

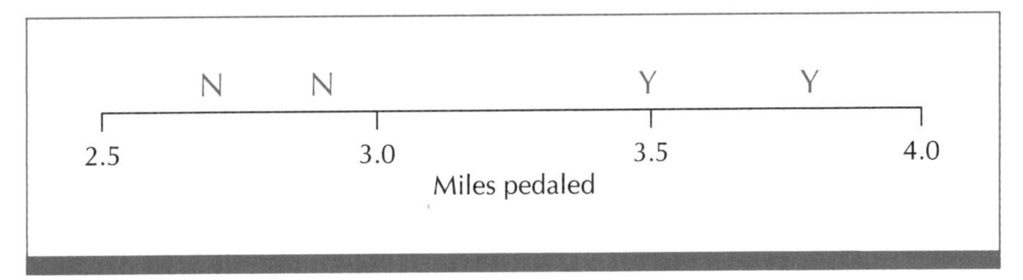

FIGURE 9-1 Plot of the results of the exercise bicycle experiment in Example 9-2: Y = Walkman, N = No Walkman

TABLE 9-3 The observations in Example 9-2 classified by treatment and position relative to the overall median

	Treatment		
	Walkman	No Walkman	Total
Above median	2	0	2
Below median	0	2	2
Total	2	2	4

Let's assume that the results of the four trials are independent of one another:

Experimental assumptions

The results of the four trials are independent.

Under the null hypothesis, we expect a random distribution of the two below-median results across the four trials, regardless of treatment. Let X equal the number of Walkman results below the overall median. Then under the null hypothesis, X has a hypergeometric distribution:

$$P(X = k \text{ when } H_0 \text{ is true}) = \frac{\binom{2}{k}\binom{2}{2-k}}{\binom{4}{2}} \qquad \text{for } k = 0, 1, 2$$

The hypergeometric probability $P(X = k \text{ when } H_0 \text{ is true})$ equals the number of ways we could select k Walkman trials and $2 - k$ no-Walkman trials to be below the overall median, divided by the number of ways we could select two of the four trials to be below the overall median. (We discussed hypergeometric distributions in Section 7-3.)

The observed value of X is 0. This value is as extreme as possible in supporting the statement that the median distance pedaled is greater with the Walkman. A value of 2 is just as extreme in supporting the statement that the median distance pedaled is greater without the Walkman. The alternative hypothesis covers both these possibilities. Therefore, the observed significance level or p-value is

$$P(X = 0 \text{ or } X = 2 \text{ when } H_0 \text{ is true}) = \frac{\binom{2}{0}\binom{2}{2}}{\binom{4}{2}} + \frac{\binom{2}{2}\binom{2}{0}}{\binom{4}{2}} = \frac{1}{6} + \frac{1}{6} = \frac{1}{3}$$

If the null hypothesis were true, there would be 1 chance in 3 of seeing results at least as extreme as those actually observed. This is a pretty good chance; based on the p-value, we would say the experimental results are consistent with the null hypothesis.

This conclusion seems to contradict the observations. Both distances pedaled with the Walkman are greater than both distances pedaled without the Walkman. An explanation of this seeming contradiction is that we have a very small sample size. Stronger evidence in favor of either hypothesis requires larger sample sizes. Also, in our analysis, we ignored the actual distances pedaled, noting only whether each observation was above or below the overall median. In Chapter 11 we discuss ways to use the relative magnitudes (ranks) as well as the actual values of the observations to analyze the results of this type of experiment.

EXAMPLE 9-3

In a high-technology factory, workers make hardware for personal computers. At one stage in the production process, gold is electroplated onto small components somewhat resembling sewing needles. The target thickness for the gold is 1.000 unit. The minimum acceptable thickness is .980 unit. While thicknesses greater than 1.000 unit are acceptable for use, managers want to avoid the use of too much gold.

Emily is in charge of the electroplating process. She knows that the process is controlled to keep thicknesses above the minimum acceptable value of .980 unit, but she is concerned that perhaps too much gold is being used. Emily decides to design an experiment to answer the question: Is too much gold being plated onto these components?

Out of a day's production lot consisting of many thousands of these components, Emily selects 100 at random for inspection. (She selects a *random sample,* meaning that each component in the lot has an equal and independent chance of being included in the sample.) She plans to use her results to compare two statements:

Null hypothesis: The average thickness of gold plated onto the components is 1.000 unit.
Alternative hypothesis: The average thickness of gold plated onto the components is greater than 1.000 unit.

The alternative hypothesis is one-sided, allowing for an average thickness greater than 1.000 unit, but not for an average less than 1.000 unit. This one-sided alternative corresponds to Emily's concern that too much gold is being used.

Let μ denote the average thickness of gold on components in the day's production lot. We can write the null and alternative hypotheses as

$$H_0: \ \mu = 1.000 \qquad \text{and} \qquad H_a: \ \mu > 1.000$$

Let $\bar{X}$ and s denote the sample mean and standard deviation, respectively, of the 100 thicknesses observed in the sample. The estimated standard deviation of $\bar{X}$ is called the *standard error* (SE) of $\bar{X}$. The standard error of $\bar{X}$ for this example is SE $= s/\sqrt{100}$.

The **standard error** of a variable quantity is an estimate of the standard deviation of the quantity.

For example, if $\bar{X}$ and s represent the sample average and standard deviation of a random sample of size n, then the standard error of $\bar{X}$, called the **standard error of the mean,** is SE $= s/\sqrt{n}$.

To carry out a statistical analysis, Emily makes three assumptions about her experiment:

Experimental assumptions

The observations are independent.
The observations come from the same distribution.
The sample size of 100 is large enough that $(\bar{X} - \mu)$/SE has approximately the standard Gaussian distribution (see the Central Limit Theorem results in Section 8-3).

Because she selects 100 components at random from a large production lot, Emily believes these assumptions are reasonable for her experiment.

Under the null hypothesis, $\mu = 1.000$ and $(\bar{X} - 1.000)$/SE has approximately the standard Gaussian distribution. Emily decides that values of this quantity near 0 are consistent with the null hypothesis. Values much greater than 0 are inconsistent with her null hypothesis, suggesting that too much gold is being electroplated. With this in mind, she chooses the following decision rule:

Decision rule

If $(\bar{X} - 1.000)$/SE < 1.65, say the results are consistent with the null hypothesis.
If $(\bar{X} - 1.000)$/SE ≥ 1.65, say the results are inconsistent with the null hypothesis.

Using a table of probabilities for the standard Gaussian distribution (such as Table B), we see that

$$P\left(\frac{\bar{X} - 1.000}{\text{SE}} \geq 1.65 \text{ when } H_0 \text{ is true}\right) \doteq .0495$$

or about .05. (The symbol $\doteq$ means two quantities are close in value or approximately equal.) There is about a 5% chance that Emily will say her experimental results are inconsistent with the null hypothesis, when the null hypothesis is really true. We call this probability the *significance level* of her test of hypothesis.

Besides asking whether too much gold is being used, Emily wants to estimate the mean thickness μ of gold on components in the production lot. She will use the sample mean $\bar{X}$ as a point estimate of μ. But, because it is based on a sample, she does not expect $\bar{X}$ to equal μ exactly. She would like to use her sample to find a range of reasonable values for the average thickness of gold plated onto the components that day. Like the test of hypotheses, her range of reasonable values for μ is based on the standard Gaussian distribution.

Referring to a table of probabilities for the standard Gaussian distribution (such as Table B), Emily makes the following approximate probability statement:

$$P\left(-1.96 \leq \frac{\bar{X} - \mu}{\text{SE}} \leq 1.96\right) \doteq .95$$

Using some algebra, she rewrites this as

$$P(\bar{X} - 1.96\text{SE} \leq \mu \leq \bar{X} + 1.96\text{SE}) \doteq .95$$

From this probability statement, Emily decides to use the interval from $\bar{X}$ − 1.96SE to $\bar{X}$ + 1.96SE as a range of reasonable values (or interval estimate) of μ.

We say that the interval ($\bar{X}$ − 1.96SE, $\bar{X}$ + 1.96SE) is an approximate 95% *confidence interval* for μ. The correct probability interpretation of this confidence interval is: Suppose Emily selects 100 components at random from the production lot. Based on the 100 measured thicknesses, she constructs a confidence interval as above. She puts these 100 components back into the lot and then selects another 100 components at random, again constructing a confidence interval based on the 100 measured thicknesses. If she repeats this experiment many times, she obtains many confidence intervals for μ, one from each experiment. She would expect about 95% of these confidence intervals to contain μ, about 5% not to contain μ. The probability interpretation of a confidence interval is based on this idea of repeated sampling. Since Emily actually takes only one sample, she calculates just one confidence interval for μ; this interval either contains μ or it does not.

If Emily had started with a different probability statement, she would have obtained a different confidence interval. For instance, ($\bar{X}$ − 2.58SE, $\bar{X}$ + 2.58SE) is an approximate 99% confidence interval for μ.

FIGURE 9-2 Stem-and-leaf plot of the measured thickness of gold for 100 components in Example 9-3. The stem is the thickness to one-hundredth of a unit; the leaf gives the nearest thousandth of a unit.

Stem	Leaf
.97	8 9
.98	4 5 6 6 7 9 9
.99	4 4 5 6 6 7 7 7 7 8 8 9 9 9
1.00	0 0 0 1 1 1 1 2 2 3 3 3 4 5 5 6 6 8 9
1.01	0 0 0 0 0 1 1 1 2 2 2 3 3 3 4 5
1.02	1 1 3 4 5 5 6 6 8 9
1.03	2 3 4 7 8
1.04	5 6 9
1.05	
1.06	0 3
1.07	4 6 8
1.08	1 2 5 6
1.09	6 7
1.10	4 6 7
1.11	
1.12	5
1.13	1 4
1.14	
1.15	2 5 9
1.16	5
1.17	
1.18	3
1.19	
1.20	1
1.21	
1.22	7

Having decided on her analysis, Emily goes ahead and randomly selects 100 components from the day's production lot. The 100 measured thicknesses are displayed in the stem-and-leaf plot in Figure 9-2. The distribution is positively skewed. We see that two of the 100 observations are below the acceptable limit of .980 unit, most thicknesses are clustered around the target value of 1.000 unit, and there is a long tail toward larger values. This plot supports Emily's concern that too much gold is being used in the plating process. (Note, however, that 2% of units with gold thickness below the .980 unit specification may be unacceptably large. This is a consideration in quality control that we are not addressing here.)

Emily calculates the sample mean, standard deviation, and standard error based on her 100 observations:

$$\bar{X} = 1.0345 \text{ units} \qquad s = .0537 \text{ unit} \qquad \text{SE} = .00537 \text{ unit}$$

To compare her hypotheses, Emily then calculates

$$\frac{\bar{X} - 1.000}{\text{SE}} = \frac{1.0345 - 1.000}{.00537} = 6.42$$

Since 6.42 is larger than 1.65, she decides the results are inconsistent with the null hypothesis; the average thickness μ of gold on the components seems to be greater than 1.000 unit.

Emily decides to calculate the probability of seeing a value of $(\bar{X} - 1.000)/\text{SE}$ greater than or equal to 6.42 under the null hypothesis:

$$P\left(\frac{\bar{X} - 1.000}{\text{SE}} \geq 6.42 \text{ when } H_0 \text{ is true}\right)$$

This probability is the *p-value* or *observed significance level* of her experimental results. This p-value approximately equals $P(Z \geq 6.42)$, where Z has the standard Gaussian distribution. From Table B, we see that $P(Z \geq 6.42)$ is less than .0002. If the average gold thickness were really 1.000 unit, the probability is less than .0002 that a random sample of size 100 would result in a value of $(\bar{X} - 1.000)/\text{SE}$ greater than or equal to 6.42. Such a small p-value strongly discredits the null hypothesis; we say the experimental results are extremely inconsistent with the null hypothesis.

Emily estimates the average thickness μ of gold plated onto components that day with the sample mean, $\bar{X} = 1.0345$ units. She calculates a 95% confidence interval for μ:

$$(\bar{X} - 1.96\text{SE}, \bar{X} + 1.96\text{SE})$$
$$= (1.0345 - 1.96 \times .00537, 1.0345 + 1.96 \times .00537) = (1.0240, 1.0450)$$

Emily uses the interval 1.024 to 1.045 units as her range of reasonable values for μ. This interval is illustrated in Figure 9-3, along with a box plot of the observations. Because of the large sample size, the interval estimate of the mean is narrow compared with the range of the observations.

Emily estimates that the average gold thickness on components in her lot is .024 to .045 unit greater than the target value. Her next step is to decide, with other members of management, whether this excess justifies the expense of modifications to the gold plating process. If so, she will try to modify the process so that less gold is used while still achieving acceptable minimum

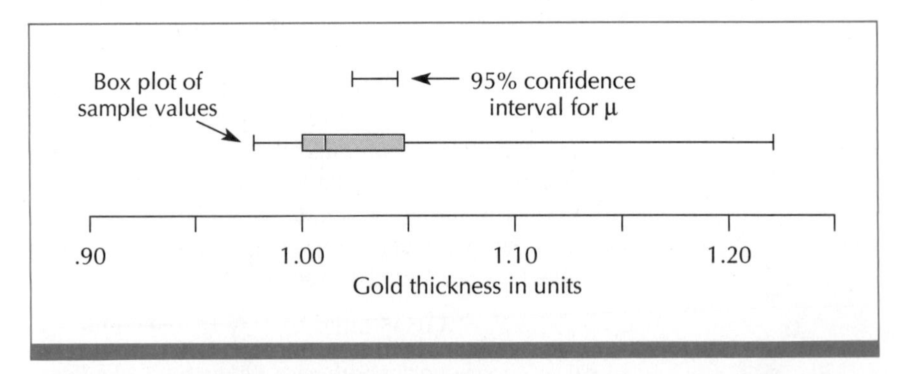

FIGURE 9-3 An approximate 95% confidence interval for the mean thickness μ of gold plated onto components in Example 9-3. Also shown is a box plot summarizing the 100 sample values.

thickness. Do you think, looking at Figure 9-2, that the plating process has more problems with *variability* than with *average* thickness of gold plated? (We will discuss problems in variability in Chapter 14.)

We now leave these three examples briefly to outline the general strategy of hypothesis testing, in Section 9-3. We then return to the examples in Section 9-4, discussing them in terms of ideas in hypothesis testing and estimation.

9-3

The General Strategy of Hypothesis Testing

We want to use the sample to compare two statements about the state of nature. The first statement is generally a "status quo" or "no difference" statement, called the *null hypothesis.* The other statement is an alternative to the null hypothesis, called the *alternative hypothesis.* We will discuss two approaches to hypothesis testing: the *significance level approach* and the *p-value approach.* Let's outline the significance level approach first.

The significance level approach to hypothesis testing

1. State a null hypothesis and an alternative hypothesis to be compared.
2. Formulate a *test statistic* that measures how far the observations differ from what we would expect under the null hypothesis.
3. Make assumptions about the sample, specifying the probability distribution the test statistic would have if the null hypothesis were true.
4. Select a number α, called the *significance level* of the test. The significance level is a probability. Typical values of α are .05 and .01.
5. Define a set of possible values of the test statistic called the *acceptance region.* Values in the acceptance region are consistent with the null hypothesis, and

   $$P(\text{test statistic is in the acceptance region when } H_0 \text{ is true}) = 1 - \alpha$$

   The remaining possible values of the test statistic make up the *rejection region.* Values in the rejection region are inconsistent with the null hypothesis, and

   $$P(\text{test statistic is in the rejection region when } H_0 \text{ is true}) = \alpha$$

6. Based on the regions defined in step 5, formulate a *decision rule:*

   If the test statistic is in the acceptance region, say the results are consistent with the null hypothesis.

   If the test statistic is in the rejection region, say the results are inconsistent with the null hypothesis.

7. Collect a random sample in the probability sense that will allow us to compare the hypotheses in step 1. Calculate the test statistic based on the sample. Use the decision rule in step 6 to decide whether the observations are consistent with the null hypothesis. Draw conclusions based on the experimental results.

Using the formal strategy of hypothesis testing, we use the test statistic to decide whether the experimental results seem to be consistent with the null hypothesis.

> A **test statistic** is a measure of how much the sample observations differ from what we would expect if the null hypothesis were true.

When our test statistic falls in the acceptance region, we say the experimental results are consistent with the null hypothesis; using the terminology of classical statistics, we fail to reject the null hypothesis. When the test statistic falls in the rejection region, we say the experimental results are inconsistent with the null hypothesis; in the words of classical statistics, we reject the null hypothesis. The classical terminology implies a finality to a hypothesis testing situation that seldom exists. A single experiment rarely decides an issue conclusively. Rather, it adds to our body of knowledge in some way. We make decisions and choose courses of action based on experimental results, doing the best we can with the information available, aware that later results may lead us to different conclusions. We should keep this in mind whenever we interpret the results of an experiment.

We can see from steps 4, 5, and 6 above that the significance level is the probability of saying the results are inconsistent with the null hypothesis, when the null hypothesis is really true.

> Using the significance level approach to hypothesis testing, the **significance level** (or level) is the probability of saying the observations are inconsistent with the null hypothesis, when the null hypothesis is really true.

The significance level is the probability of making what we call a *Type I error*.

> In hypothesis testing, we make a **Type I error** if we say the experimental results are inconsistent with the null hypothesis, when the null hypothesis is really true.

We make a *Type II error* if we say the results are consistent with the null hypothesis when the alternative is really true.

> In hypothesis testing, we make a **Type II error** if we say the experimental results are consistent with the null hypothesis, when the alternative is really true.

Whether we make an error depends on the decision we make and on which hypothesis is true, as summarized in Table 9-4.

Related to significance level is the *power of the test:*

> Using the significance level approach to hypothesis testing, the **power of the test** is the probability of saying the results are inconsistent with the null hypothesis, when an alternative to the null hypothesis is really true.

For a specific alternative, the power of the test is 1 minus the probability of making a Type II error. We will discuss how to calculate the power of a test in Section 9-4.

TABLE 9-4 Whether we make an error in hypothesis testing depends on the decision we make and on the state of nature.

	State of nature	
Decision	Null hypothesis is true	Alternative hypothesis is true
Results consistent with null hypothesis	No error	Type II error
Results inconsistent with null hypothesis	Type I error	No error

Another approach to hypothesis testing is the p-value approach, outlined below.

The p-value approach to hypothesis testing

1. State a null hypothesis and an alternative hypothesis to be compared.
2. Formulate a test statistic that measures how far the observations differ from what we would expect under the null hypothesis.
3. Make assumptions about the sample, specifying the probability distribution the test statistic would have if the null hypothesis were true.
4. Collect a random sample in the probability sense that will allow us to compare the hypotheses stated in step 1. Calculate the test statistic based on the sample.
5. Calculate the *observed significance level* or *p-value.* The p-value is the probability under the null hypothesis of seeing a test statistic as extreme as or more extreme (in the direction of the alternative) than the one actually observed.
6. If the p-value is large, say the results are consistent with the null hypothesis. If the p-value is small, say the results are inconsistent with the null hypothesis.

A large p-value means a test statistic at least as extreme as that observed would not be surprising if the null hypothesis were true; so, we say the results are consistent with the null hypothesis. A small p-value means we would not be likely to see a test statistic as extreme as or more extreme than the one observed, if the null hypothesis were true. Therefore, we say the results are inconsistent with the null hypothesis.

The ***p*-value,** or observed significance level, is the probability of seeing a test statistic as extreme as or more extreme (in the direction of the alternative) than the one observed, if the null hypothesis were really true.

There are many questions we can ask at this point. How do we decide what the hypotheses should be? What assumptions do we make about the sample? What if the assumptions we make do not apply to the actual sampling process? How do we select a test statistic? Should we use the significance level or p-value approach to hypothesis testing? When is a p-value "small" and when is it "large"? As we will see in Section 9-4, the answer to all of these questions is: It depends on the situation.

9-4 Some Comments on Hypothesis Testing

Let's address the questions we just posed in terms of the examples in Section 9-2. Then we will make some additional observations on hypothesis testing.

How do we decide what the hypotheses should be? The null hypothesis is generally a statement of no difference or status quo. In Example 9-1, the null hypothesis states that the students have no preference for Coke or Pepsi. The null hypothesis in Example 9-2 is that the median distance pedaled is the same with and without the Walkman. In Example 9-3, the null hypothesis says that the average thickness of gold plated onto the components is equal to the target value of 1.000 unit.

The alternative hypothesis is an alternative to the null hypothesis. In Example 9-1, the alternative states that the students have a preference for either Coke or Pepsi. This alternative allows for deviations from the no-preference null hypothesis in the direction of either a preference for Coke or a preference for Pepsi. Such a two-sided alternative is appropriate when we do not know what to expect from the experiment. We also use a two-sided alternative when we want to allow for extremes on either side of the null hypothesis, even when we do have some prior expectation of how the experiment will turn out.

> A **two-sided alternative hypothesis** allows for extremes in two directions, on either side of the state of nature specified in the null hypothesis.

The alternative hypothesis in Example 9-2 is two-sided. It states that the median distance pedaled is different with than without the Walkman. This alternative allows the possibility that the median distance pedaled is greater with the Walkman, as well as the possibility that the median distance pedaled is greater without the Walkman.

A one-sided alternative makes sense in Example 9-3. The manager is concerned that too much gold is being used. Her alternative hypothesis says that the average thickness of gold plated onto the components is greater than the target value of 1.000 unit.

> A **one-sided alternative hypothesis** allows for extremes in just one direction, on just one side of the state of nature specified in the null hypothesis.

Null and alternative hypotheses must be developed specially for each experimental situation. They should be written down *before* the data are examined. The reason is that after looking at the sample, we may be able to state hypotheses that fit what we see, defeating the point of hypothesis testing. In fact, null and alternative hypotheses should be developed during the experimental design stage, before the sample is selected. With clearly stated hypotheses, we can then determine whether our planned experiment will be adequate to test the hypotheses that interest us.

What assumptions do we make about the sample? The assumptions we make depend on what we know about the experimental design, on how the observations were obtained. They also depend on what we know or are willing to assume about the distribution of values in the population. The major assumption that is common to all the hypothesis testing situations we will discuss is that our observations were made independently and came from a common population. Our aim is to use this sample of independent observations to infer characteristics of the larger population.

The students in Example 9-1 assumed that the tasters made independent beverage selections. They also assumed that each taster had the same probability p of choosing Coke. Such assumptions are simplifying, perhaps unrealistic. There may be biological differences that make some tasters more likely to prefer Coke, others more likely to prefer Pepsi. We might think of p as the proportion of tasters who will choose Coke. (In a large-scale test, this is the most reasonable interpretation of p. Then the null hypothesis states that half the population prefers Coke; the alternative states that either more than half the population prefers Coke or more than half prefers Pepsi.) We then simplify the situation by supposing that each taster has probability p of choosing Coke. Under the no-preference null hypothesis, this probability p is equal to $\frac{1}{2}$.

For the bicycling experiment in Example 9-2, we assumed the four observations were independent. Under the null hypothesis, we assumed a random distribution of results across the four trials, regardless of treatment (Walkman or no Walkman).

In the sampling inspection problem of Example 9-3, the experimenter assumed the observations were independent, with the same distribution. These assumptions seemed reasonable because the components in the sample were selected at random from a large production lot. She made an additional assumption—that the sample size was large enough to use a Gaussian approximation to the distribution of the sample mean.

In the chapters that follow, we will see how different sets of assumptions lead to different probability models. For a given experiment, the assumptions we make about the sampling process determine the hypothesis testing procedure we should use.

What if the assumptions we make do not apply to the actual sampling process? In most hypothesis testing situations, we must make some simplifying assumptions. We then build a probability model we think will be good enough to make reasonable inferences about the state of nature. If our assumptions provide too poor a model of reality, the inferences we make may be incorrect; how incorrect depends on how badly our model approximates reality.

For instance, in Example 9-1, there were 12 tasters. We assumed the tasters made independent selections, each taster having probability p of choosing Coke. There were men and women in the class. Students represented several racial backgrounds and national origins. If some factors (such as sex, race, or cultural background) make some students tend to have similar responses, the model assumptions break down. Then it is not clear how the results of the test of hypotheses apply to the actual experiment.

How do we select a test statistic? The test statistic measures how far the observed sample differs from what we would expect if the null hypothesis were true. It must have a known probability distribution under the null hypothesis. This probability distribution allows us to make probability statements based on the experimental results.

The test statistic Y in Example 9-1 equals the number of tasters choosing Coke. With the assumptions we make, Y has a Binomial(12, $\frac{1}{2}$) distribution under the null hypothesis.

In Example 9-2, the test statistic X equals the number of Walkman results below the overall median. With independent observations, X has a hypergeometric distribution under the null hypothesis.

The test statistic is $(\bar{X} - 1.000)/\text{SE}$ in Example 9-3. With the assumptions we make, this test statistic has approximately the standard Gaussian distribution under the null hypothesis.

In the chapters that follow, we will discuss tests of hypotheses for a number of different experimental situations. We will see how the choice of test statistic depends on the experimental design, on the type of observations (such as continuous or categorical), and on the assumptions we make about the sampling process.

How do we define an acceptance region and a rejection region when using the significance level approach? The acceptance region contains possible values of the test statistic near what is expected under the null hypothesis. The rejection region contains values far from what is expected under the null hypothesis.

In Example 9-1, the test statistic Y is the number of tasters choosing Coke. We expect values of Y near 6 under the null hypothesis. Values close to 0 or 12 are far from what we expect under the null hypothesis. Three reasonable choices of acceptance and rejection regions are shown in Table 9-5, along with the significance level associated with each choice.

TABLE 9-5 Three possible choices of acceptance and rejection regions in Example 9-1

Choice	Acceptance region	Rejection region	Significance level, α
1	{2, 3, 4, 5, 6, 7, 8, 9, 10}	{0, 1, 11, 12}	.006
2	{3, 4, 5, 6, 7, 8, 9}	{0, 1, 2, 10, 11, 12}	.038
3	{4, 5, 6, 7, 8}	{0, 1, 2, 3, 9, 10, 11, 12}	.146

The students in Example 9-1 selected choice 2, with significance level .038. With this choice, the probability of saying the results are inconsistent with the null hypothesis, when the null hypothesis is really true, is .038.

The test statistic in Example 9-3 is $(\bar{X} - 1.000)/\text{SE}$, where $\bar{X}$ is the sample mean gold thickness. We expect values of the test statistic near 0 under the null hypothesis, values greater than 0 under the alternative. The acceptance region is $(-\infty, c)$ and the rejection region is $[c, \infty)$. Here, c is a constant that gives the desired significance level.

The manager wanted a significance level α close to .05. She knew that if Z is a standard Gaussian random variable, then $P(Z \geq 1.65) = .0495$. She decided to use $c = 1.65$, so her acceptance region was $(-\infty, 1.65)$. Her rejection region was $[1.65, \infty)$.

For a different significance level, she would choose a different value of c. If she wanted significance level $\alpha = .004$, she would choose $c = 2.65$. Then she would say values of the test statistic less than 2.65 are consistent with the null hypothesis and values greater than or equal to 2.65 are inconsistent with the null hypothesis.

Why do we consider values of the test statistic more extreme than the one actually observed when calculating a p-value? The p-value is the probability of observing a test statistic as extreme as or more extreme (in the direction of the alternative hypothesis) than the one actually observed, if the null hypothesis were really true. The p-value is a measure of our surprise at seeing such a test statistic, if the null hypothesis were true. The "as extreme" part of the definition of p-value makes sense, but why also consider values of the test statistic more extreme than that observed? One way to answer this question is with an example.

Suppose we are in charge of deciding whether the coin that will be used to determine the kick-off in the next Super Bowl appears to be fair. We want to compare the following hypotheses:

Null hypothesis: The coin is fair.
Alternative hypothesis: The coin is not fair.

We toss the coin 100 times and observe 53 heads. Is this result consistent with the null hypothesis or not? Most of us would not be surprised at seeing 53 heads in 100 tosses of a fair coin. How does this lack of surprise relate to a test of hypothesis?

We must formulate a probability model. Let's suppose that the tosses are independent. Under the null hypothesis, the probability of a head on each toss is $\frac{1}{2}$. Then the number of heads observed in 100 tosses has a Binomial$(100, \frac{1}{2})$ distribution.

What is the probability of observing exactly 53 heads if the coin is fair? This is the probability of seeing 53 successes in 100 independent trials, with probability $\frac{1}{2}$ of success on each trial:

$$P(53 \text{ successes when } H_0 \text{ is true}) = \binom{100}{53}\left(\frac{1}{2}\right)^{100} = .067$$

Even though seeing 53 heads in 100 tosses of a fair coin does not surprise us, the probability of its happening is just .067. (In fact, the probability of seeing exactly 50 heads is only slightly larger, .08. If we repeated this 100-toss experiment many times, we would be surprised to see exactly 50 heads and 50 tails in a large percentage of these repetitions. We would expect about 92% of our repetitions to result in something other than an exact 50-50 split between heads and tails.)

Looking at the probability of 53 heads in 100 tosses is not enough. So we ask: If the null hypothesis were true, what results would be more surprising than 53 heads? Fifty-four or more heads would be more surprising than 53 heads. If we add up the probabilities of 53 or more heads, we have

$$P(53 \text{ or more heads when } H_0 \text{ is true}) = \sum_{k=53}^{100} \binom{100}{k} \left(\frac{1}{2}\right)^{100} = .31$$

If the coin were fair, we would expect about 31% of all 100-toss experiments to end in 53 or more heads.

Forty-seven or fewer heads would be just as surprising as 53 or more heads. The chance of 47 or fewer heads under the null hypothesis is also .31. Therefore, the *p*-value associated with our two-sided alternative is equal to .31 + .31 = .62. This large *p*-value corroborates our lack of surprise at seeing 53 heads in 100 tosses, under the fair-coin hypothesis. Exactly 53 heads in 100 tosses has relatively small probability. But 53 is close to what we would expect if the null hypothesis were true.

In general, we consider the probability of seeing a test statistic at least as extreme as the observed value, under the null hypothesis. If this probability is large, we consider the test statistic to be close to what is expected under the null hypothesis; then we say the results are consistent with the null hypothesis. If this probability is small, then we consider the test statistic to be far from what is expected under the null hypothesis; then we say the results are inconsistent with the null hypothesis.

When is a p-value "small" and when is it "large"? Large *p*-values are consistent with the null hypothesis and small *p*-values are inconsistent with the null hypothesis, but what is "large" and what is "small"? This is a difficult question to answer. Our interpretation of experimental results depends on considerations such as sample size and practical importance.

Statistical significance is associated with a small *p*-value. *Practical significance* is associated with a deviation from the null hypothesis that we consider to be important. In Example 9-3, for instance, we obtained a very small *p*-value. The sample average gold thickness was 1.0345 units, compared to the target value of 1.000 unit, and this difference was highly statistically significant. It could be, however, that this difference is not large enough to justify expensive modifications of the electroplating process. Also, such a difference may be necessary to maintain minimum acceptable gold thicknesses. These are practical considerations, distinct from the issue of statistical significance.

In Example 9-2, we had a large *p*-value because the sample size was so small. However, the experimental results suggested that there was a real difference in median distance pedaled under the two experimental conditions (Walkman and no Walkman).

We will consider the issue of statistical significance versus practical importance again in later chapters.

Should we use the significance level or p-value approach to hypothesis testing? This is a matter of personal preference. Many investigators prefer the significance level approach; they interpret their experimental results relative to a specified significance level and decision rule. Other workers prefer the p-value approach; they report a p-value when they write up their results. We can use a combination of the two approaches, as illustrated in Examples 9-1 and 9-3. We can specify a decision rule as outlined in the significance level approach. Then, after collecting our observations, we can calculate a p-value as well.

How do we evaluate the power of a test? The *power* of a test is the likelihood that we will reject the null hypothesis, when a specific alternative is really true. Power depends on how far the particular alternative is from the null hypothesis. The farther the true state of nature is from that specified in the null hypothesis, the greater the likelihood of deciding in favor of the alternative. Power also depends on the sample size.

Let's illustrate how the power of the test can be determined, for a taste test such as the one in Example 9-1. Investigators want to conduct a taste test comparison of Pepsi and Coke. They assume that tasters choose independently of one another and they wish to compare the hypotheses

$$H_0: \; p = \frac{1}{2} \qquad \text{and} \qquad H_a: \; p \neq \frac{1}{2}$$

where p is the probability that a single taster selects Coke. These investigators are making the same simplifying assumptions that the students made in Example 9-1.

The investigators plan to select a random sample of tasters from some larger population. During the process of planning their experiment, they come to us and ask how many tasters they should include. "What sample size do we need?" is a common and extremely sensible question asked by experimenters. Our answer is:

The experimental sample size needed depends on two considerations:
The particular alternative to the null hypothesis that is of interest.
The power of the test desired under this alternative.

We show the investigators what we mean with the following calculations. We will compare the power of the test under three specific alternatives, for sample sizes of 10 and 20.

Consider first a sample size of 10. The test statistic is Y, the number of

Coke selections. Under the no-preference null hypothesis, Y has the Binomial(10, $\frac{1}{2}$) distribution. Suppose we use the following decision rule:

Decision rule for sample size 10
If $2 \leq Y \leq 8$, say the results are consistent with the null hypothesis.
If $Y \leq 1$ or $Y \geq 9$, say the results are inconsistent with the null hypothesis.

Using Table A, we find the significance level α associated with this decision rule to be

$$\text{Significance level} = P(Y \leq 1 \text{ or } Y \geq 9 \text{ when } H_0 \text{ is true}) = .022$$

Now let's calculate the power of the test for three specific alternatives.

Suppose the probability p that a taster selects Coke is really .6. The power is the probability that Y is in the rejection region under this alternative:

$$\text{Power} = P(Y \leq 1 \text{ or } Y \geq 9 \text{ when } p = .6) = .048$$

If the probability p of a taster choosing Coke is really .6, the chance of seeing Y in the rejection region is .048.

If the probability p of a taster choosing Coke is really .8, then the power of the test is

$$\text{Power} = P(Y \leq 1 \text{ or } Y \geq 9 \text{ when } p = .8) = .376$$

With $p = .8$, there is about a 38% chance of saying the observations support the alternative. A similar calculation shows that when $p = .95$ and the sample size is 10, the power of the test is .914.

Let's consider next a sample of 20 tasters. Under the no-preference null hypothesis, the test statistic Y has the Binomial(20, $\frac{1}{2}$) distribution. We would like a decision rule with significance level close to the value ($\alpha = .022$) we used for sample size 10. We have two choices of acceptance and rejection regions with significance levels close to .022, shown in Table 9-6.

We say the first choice in Table 9-6 is more conservative than the second, because we are less likely to decide in favor of the alternative with the first. Let's use this more conservative choice of acceptance and rejection regions. The decision rule then is:

Decision rule for sample size 20
If $5 \leq Y \leq 15$, say the results are consistent with the null hypothesis.
If $Y \leq 4$ or $Y \geq 16$, say the results are inconsistent with the null hypothesis.

The significance level for this decision rule is .012. The probability we say the results are inconsistent with the null hypothesis, when the null hypothesis is really true, is .012.

TABLE 9-6 Two choices of acceptance and rejection regions for sample size 20 in the taste-testing example

Choice	Acceptance region	Rejection region	Significance level, α
1	{5, 6, 7, 8, 9, 10, 11, 12, 13, 14, 15}	{0, 1, 2, 3, 4, 16, 17, 18, 19, 20}	.012
2	{6, 7, 8, 9, 10, 11, 12, 13, 14}	{0, 1, 2, 3, 4, 5, 15, 16, 17, 18, 19, 20}	.041

TABLE 9-7 Power of the test for sample sizes 10 and 20, under three specific alternatives. When the null hypothesis is true, $p = .5$. Here, p is the probability that a taster will select Coke.

Sample size	Rejection region	Significance level	Alternative value of p	Power of the test
10	{0, 1, 9, 10}	.022	.6	.048
			.8	.376
			.95	.914
20	{0, 1, 2, 3, 4, 16, 17, 18, 19, 20}	.012	.6	.051
			.8	.630
			.95	.997

Suppose now the probability p that a taster selects Coke is .6. The power is the probability that Y is in the rejection region:

$$\text{Power} = P(Y \leq 4 \text{ or } Y \geq 16 \text{ when } p = .6) = .051$$

There is about a 5% chance of saying the experimental results are inconsistent with the null hypothesis, when $p = .6$. Similar calculations show that the power of the test is .630 when $p = .8$; the power is .997 when $p = .95$.

The results of our power calculations are summarized in Table 9-7, illustrating the point we want to make for the investigators. The farther the alternative is from the state of nature specified under the null hypothesis, the greater our likelihood of deciding the experimental results are inconsistent with the null hypothesis. Also, for a specific alternative, larger sample sizes increase the chance of seeing results that are inconsistent with the null hypothesis. This is true here, even though for sample size 20 we used a test with smaller significance level than we used for sample size 10. Calculations such as these help us decide on a sample size that balances two opposing considerations: a smaller, less expensive experiment versus a larger, more powerful experiment.

Table 9-7 illustrates another point we addressed earlier. With small sample sizes, we may have little chance of seeing results that are inconsistent with the null hypothesis, even for alternatives of practical interest. This is a serious problem with small experiments.

We have a related concern with larger sample sizes. The likelihood increases that results will appear inconsistent with the null hypothesis, even for alternatives that are only slightly different from the null hypothesis. Then we must make a judgment about whether statistically significant results are really of any practical importance.

9-5 Some Comments on Experimental Design

In formal statistical inference, we make assumptions about the sample observations. We then make probability statements about the population, based on what we observe in the sample.

We should design an investigation so that our assumptions are reasonable. Design of experiments is extremely important, and too often ignored. First, write down the goals of the investigation, stating them in terms of parameters to be estimated and hypotheses to be tested. Then design the investigation to meet these goals.

> **Experimental design** is the area of statistics concerned with designing an investigation to best meet the study goals, as well as the assumptions for statistical inference.

A good experimental design minimizes the effects of factors not of interest. Then the effects of factors that are of interest can be detected more readily. We call the factors not of interest *extraneous factors*.

> An **extraneous factor** is a variable that is not of interest in the experiment, but might affect the outcome.

We design experiments to try to reduce the effects of extraneous factors, so that the effects of factors that are of interest can be detected more readily.

The taste test in Example 9-1 was a double-blind experiment. Neither the experimenter nor the tasters knew the identity of the beverages being tasted. The preconceptions of the experimenter and participants could not influence, whether blatantly or subtly, the results of the study. These preconceptions are extraneous factors.

The student in Example 9-2 tried to reduce the effects of extraneous factors in her bicycle experiment. She wore similar clothing for each trial, went through the same warm-up prior to bicycling, and kept the room temperature constant. The trials were run within a single week, all on weekdays when the student's study and sleeping habits were similar. She covered the odometer so she would not know how far she had pedaled during any trial. She also used a random process to determine the ordering of the experimental conditions. All of these precautions were to reduce the effects of extraneous factors. Because her study was not blinded, she could not eliminate her own preconceptions as possible extraneous factors. How could she have avoided this problem when she designed her experiment?

The manager in Example 9-3 selected a random sample of components from the day's production lot. She knew that random selection of the sample would eliminate her own biases from the selection process. She also hoped that random selection would balance other extraneous factors that might influence the results of the study.

Many investigations provide no information of value because of flaws in the experimental design. A common flaw is that the sample observations are not representative of the population of interest. The results of a taste test, for instance, cannot be generalized to a larger population unless the sample of tasters is representative of that population. One way to get a representative sample is to select the sample at random from the population.

Sometimes a random sample is not possible. Physicians at a cancer research center cannot obtain a random sample of cancer patients for their studies of new treatments. Instead, they rely on volunteers from among the patients referred to the center. These volunteers may not be representative of the population of cancer patients in the country; they may differ with respect to place of residence, racial background, socioeconomic status, and stage of disease. Researchers try to conduct the best experiments possible with the available volunteers. However, their experimental results may not apply to the larger patient population; a treatment that looked promising at the research center may not be successful in the general patient population.

We will see many examples of experimental designs in the chapters that follow. None of these investigations is perfect; few real experiments are. You are encouraged to think of limitations and extraneous factors that could affect the outcome of each investigation.

We begin in Chapter 10 with inferences about a measure of central tendency. Inferences about two measures of central tendency are considered in Chapter 11. In Chapter 12, we discuss comparisons of several means, via single-factor experiments and randomized block experiments. Chapter 13 covers two-factor experimental designs (specifically: balanced, completely randomized, factorial experiments). In Chapter 14 we address the problem of making inferences about one or more population variances. We consider ways to study the relationship between two quantitative variables in Chapter 15, and introduce the linear correlation coefficient as a measure of linear association between two variables. In addition, we discuss the method of least squares as a way of modeling a quantitative variable as a function of other variables. Chapter 16 introduces some simple ways to make inferences about a single qualitative (categorical) variable and about the relationship between two qualitative variables.

The hypothesis testing procedures in Chapters 10–16 are outlined in steps, as the general approach to hypothesis testing is outlined in Section 9-3. The purpose of this approach is to emphasize that the *strategy* of hypothesis testing is *the same,* regardless of the particular technique being used. The procedures are presented this way to help you learn the mechanics of the statistical techniques, not to encourage a mechanical approach to statistical analysis. On the contrary, I *strongly discourage a cookbook approach to statistical analysis.* In any experimental situation, we should consider the experimental design of primary importance. Often, a well-designed experiment needs little formal statistical analysis. On the other hand, a poorly designed experiment often cannot be salvaged, no matter how sophisticated the analysis technique. We should always consider the experimental design when deciding whether a particular procedure is appropriate. In addition, a complete analysis of any experiment requires data analysis, critical thinking, and common sense. The examples in this chapter have illustrated this approach.

After each of Chapters 10–16, there are exercises illustrating use of techniques covered in the chapter. At the end of Part III, there are additional exercises. For all exercises, consider the experimental design, use tools of data analysis, apply techniques of formal statistical inference when appropriate, and use critical thinking and common sense to make a thorough analysis of the problem.

Summary of Chapter 9

We should take care when interpreting the significance level and p-value in hypothesis testing. Some users of statistics *incorrectly* refer to the significance level (or p-value) as the probability that the null hypothesis is true; this is an *incorrect* interpretation! The significance level, as well as the p-value, is calculated under the null hypothesis probability model: It represents the likelihood of certain events occurring *if the null hypothesis were true*. We use the strategy of hypothesis testing to decide whether the experimental results are consistent with the null hypothesis probability model. That is, we assume the null hypothesis is true and then decide whether the experimental results are consistent with this assumption. (We *do not* calculate the probability that the null hypothesis is true!)

We define the power of a test and discuss it in relation to sample size. Power (the likelihood of saying the experimental results are inconsistent with the null hypothesis when a specific alternative is true) depends on the particular alternative of interest and on sample size. When deciding on a sample size for an experiment, we must consider alternatives that are of interest and the power we want under these alternatives.

When we design an experiment, we try to meet the assumptions that will allow us to compare the hypotheses of interest to us. We try to minimize the effects of extraneous factors—variables that could influence our results, but are not the object of our investigation. With careful design and conduct of an experiment, we hope to better assess the effects of factors that we are studying.

A confidence interval is an interval estimate of a population parameter. We will see several uses of confidence intervals in the chapters that follow.

Exercises for Chapter 9

EXERCISE 9-1

When we calculate the p-value based on experimental results, we are finding (select the correct answer):

a. The probability that the null hypothesis is true.

b. The probability that the alternative hypothesis is true.

c. The chance of seeing results at least as extreme in the direction of the alternative hypothesis as the results actually observed, if the null hypothesis were really true.

d. The likelihood of seeing results at least as extreme in the direction of the alternative hypothesis as the results actually observed, if the alternative hypothesis were really true.

EXERCISE 9-2

The significance level associated with a test of hypotheses is (select the correct answer):

a. The probability that the null hypothesis is true.

b. The probability that the alternative hypothesis is true.

c. The probability of seeing experimental results in the acceptance region (consistent with the null hypothesis) if the alternative hypothesis were really true.

d. The probability of seeing experimental results in the rejection region (inconsistent with the null hypothesis) if the alternative hypothesis were really true.

e. The probability of seeing experimental results in the acceptance region (consistent with the null hypothesis) if the null hypothesis were really true.

f. The probability of seeing experimental results in the rejection region (inconsistent with the null hypothesis) if the null hypothesis were really true.

EXERCISE 9-3

The power of the test is (select the correct answer):

a. The probability that the alternative hypothesis is true.

b. The probability that the null hypothesis is true.

c. The probability of seeing experimental results in the acceptance region (consistent with the null hypothesis) if a specific alternative were really true.

d. The probability of seeing experimental results in the rejection region (inconsistent with the null hypothesis) if a specific alternative were really true.

e. The probability of seeing experimental results in the acceptance region (consistent with the null hypothesis) if the null hypothesis were really true.

f. The probability of seeing experimental results in the rejection region (inconsistent with the null hypothesis) if the null hypothesis were really true.

EXERCISE 9-4

We make a Type I error if (select the correct answer):

a. We say the experimental results are consistent with the null hypothesis, when the null hypothesis is really true.

b. We say the experimental results are consistent with the null hypothesis, when the alternative hypothesis is really true.

c. We say the experimental results are inconsistent with the null hypothesis, when the null hypothesis is really true.

d. We say the experimental results are inconsistent with the null hypothesis, when the alternative hypothesis is really true.

EXERCISE 9-5

We make a Type II error if (select the correct answer):

a. We say the experimental results are consistent with the null hypothesis, when the null hypothesis is really true.

b. We say the experimental results are consistent with the null hypothesis, when the alternative hypothesis is really true.

c. We say the experimental results are inconsistent with the null hypothesis, when the null hypothesis is really true.

d. We say the experimental results are inconsistent with the null hypothesis, when the alternative hypothesis is really true.

EXERCISE 9-6

Consider the taste test in Example 9-1. Recall that the decision rule selected by the students was:

If $3 \leq Y \leq 9$, say the results are consistent with the no-preference null hypothesis.
If $Y \leq 2$ or $Y \geq 10$, say the results are inconsistent with the no-preference null hypothesis, suggesting there is a preference.

a. With 12 tasters, the significance level associated with this decision rule is .038, or about .04. What is the correct interpretation of this significance level?

b. Find the power of the test associated with the given decision rule if the probability p that a student selects Coke is: .1, .2, .4, .6, .8, .9.

c. What is the correct interpretation of each power value you found in part (b)? Compare the values of power you calculated under the six separate alternatives in part (b) and discuss your findings.

EXERCISE 9-7

Consider the taste test in Example 9-1, with 12 tasters.

a. Find the significance level associated with this decision rule:

If $2 \leq Y \leq 10$, say the results are consistent with the no-preference null hypothesis.
If $Y \leq 1$ or $Y \geq 11$, say the results are inconsistent with the no-preference null hypothesis, suggesting there is a preference.

What is the correct interpretation of this significance level?

b. Find the power of the test associated with the decision rule in part (a) if the probability p that a student selects Coke is: .1, .2, .4, .6, .8, .9.

c. What is the correct interpretation of each power value you found in part (b)? Compare the values of power that you calculated under the six separate alternatives in part (b) and discuss your findings.

d. Compare your calculations in this exercise with your calculations for the decision rule in Exercise 9-6. Discuss how power depends on the decision rule you select.

EXERCISE 9-8

Consider the taste test in Example 9-1, with 12 tasters.

a. Find the significance level associated with this decision rule:

If $1 \leq Y \leq 11$, say the results are consistent with the no-preference null hypothesis.

If $Y = 0$ or $Y = 12$, say the results are inconsistent with the no-preference null hypothesis, suggesting there is a preference.

What is the correct interpretation of this significance level?

b. Find the power of the test associated with the decision rule in part (a) if the probability p that a student selects Coke is: .1, .2, .4, .6, .8, .9.

c. What is the correct interpretation of each power value you found in part (b)? Compare the values of power that you calculated under the six separate alternatives in part (b) and discuss your findings.

d. Compare your calculations in this exercise with your calculations for the decision rules in Exercises 9-6 and 9-7. Discuss how power depends on the decision rule you select.

EXERCISE 9-9

Consider a taste test such as the one in Example 9-1, only now there are 6 tasters rather than 12.

a. Find the significance level associated with this decision rule:

If $2 \leq Y \leq 4$, say the results are consistent with the no-preference null hypothesis.
If $Y \leq 1$ or $Y \geq 5$, say the results are inconsistent with the no-preference null hypothesis, suggesting there is a preference.

What is the correct interpretation of this significance level?

b. Find the power of the test associated with the decision rule in part (a) if the probability p that a student selects Coke is: .1, .2, .4, .6, .8, .9.

c. What is the correct interpretation of each power value you found in part (b)? Compare the values of power that you calculated under the six separate alternatives in part (b) and discuss your findings.

d. Compare your results in this exercise with what you found in Exercises 9-6, 9-7, and 9-8. Discuss how power depends on sample size.

EXERCISE 9-10

Consider a taste test such as the one in Example 9-1, only now there are 6 tasters rather than 12.

a. Find the significance level associated with this decision rule:

If $1 \leq Y \leq 5$, say the results are consistent with the no-preference null hypothesis.
If $Y = 0$ or $Y = 6$, say the results are inconsistent with the no-preference null hypothesis, suggesting there is a preference.

What is the correct interpretation of this significance level?

b. Find the power of the test associated with the decision rule in part (a) if the probability p that a student selects Coke is: .1, .2, .4, .6, .8, .9.

c. What is the correct interpretation of each power value you found in part (b)? Compare the values of power that you calculated under the six separate alternatives in part (b) and discuss your findings.

d. Compare your calculations in this exercise with your calculations for the decision rule in Exercise 9-9. Discuss how power depends on the decision rule you select.

e. Comparing your results in Exercises 9-9 and 9-10 with your results in Exercises 9-6, 9-7, and 9-8, discuss how power depends on sample size.

EXERCISE 9-11

Consider a stationary bicycle experiment such as the one in Example 9-2.

a. Suppose the student makes four runs with the Walkman and four runs without the Walkman. For each run with the Walkman, she pedals farther than for any run without the Walkman. Using a probability model similar to the one in Example 9-2, calculate the p-value associated with these results. What is the correct interpretation of this p-value?

b. Repeat part (a) if she makes eight runs with the Walkman and eight runs without the Walkman, and she pedals farther in all runs with the Walkman than in any run without the Walkman.

c. Compare the results in parts (a) and (b) with those we found in Example 9-2. Interpret your results. Discuss the sample size considerations.

CHAPTER 10

Inferences About a Measure of Central Tendency

IN THIS CHAPTER

Large-sample inferences about a mean

Large-sample inferences about a proportion

t test and confidence intervals based on a *t* distribution

Wilcoxon signed rank test and associated confidence intervals

Sign test and associated confidence intervals

Is the average birth weight of babies born to exercised goats different from the average birth weight of babies born to unexercised goats? Do bomb bases meet height specifications, on average? Is the median lifetime of a new, cheaper type of electric cord equal to the median lifetime of the older, more expensive type? These questions concern *measures of central tendency,* also called *location parameters.*

A **measure of central tendency,** or **location parameter,** describes the location or center of a distribution or set of numbers.

We would like to use information in a sample to make inferences about central tendency in a population. For instance, we exercise a sample of pregnant goats. When the baby goats are born, we compare their average weight with the average weight of babies born to unexercised goats. We use the results to draw inferences about the effect of exercise on birth weights.

In an industrial setting, we measure a sample of bomb bases. We use the heights of bomb bases in the sample to make inferences about the average height of bomb bases produced. Or, we subject a sample of a new type of electric cord to accelerated life testing. We use the lifetimes in the sample to draw inferences about the median lifetime of this new, cheaper type of cord.

How do we decide on a formal procedure for making inferences about a measure of central tendency? The choice depends on sample size and the assumptions we are willing to make about the observations.

The most stringent assumptions, that we have independent observations from a Gaussian distribution, form the basis for classical statistical analysis. Assuming a specific form of distribution for the observations gives us a *parametric model.* This model leads to the t test and interval estimates for a mean based on a *t distribution.* We will discuss small-sample inferences about a mean based on a t distribution in Section 10-3.

Suppose we drop the assumption that the observations come from a Gaussian distribution. However, we continue to assume that the observations are independent and from a continuous, symmetric distribution. Then we have a *nonparametric model* with inferences based on a *Wilcoxon signed rank distribution.* Small-sample inference about a mean based on a Wilcoxon signed rank distribution is the subject of Section 10-4.

If we assume only that our observations are independent and come from a continuous distribution, then we have a simpler *nonparametric model.* For small samples, we base our inferences on a *binomial distribution,* as we will see in Section 10-5.

Section 10-1 discusses *large-sample inferences* about a population mean. We assume that the observations form a random sample from a population. We also assume that the sample size is large enough to ensure that the sample mean has approximately a Gaussian distribution (see the Central Limit Theorem results in Section 8-3). Then we base tests of hypotheses and interval estimates on the *standard Gaussian distribution.* Large-sample inference about a proportion is a special case, as we see in Section 10-2.

10-1

Large-Sample Inference About a Population Mean Based on the Standard Gaussian Distribution

Recall that in Example 9-3 we considered the thickness of gold applied to computer parts during an electroplating process. Emily was in charge of the process, and she was concerned that too much gold was being used. She took a large random sample from a production lot and used the parts in the sample to make inferences about the average gold thickness on parts in the entire lot. In particular, Emily compared a null hypothesis (mean gold thickness on parts in the lot equals a target value) with an alternative hypothesis (the mean exceeds the target value). Also, she calculated a confidence interval to provide an interval estimate or range of reasonable values for the mean gold thickness on parts in the production lot. Emily used large-sample techniques to make these inferences about the mean gold thickness in the lot.

In this section, we will outline the general approach to making inferences about a population mean based on large samples. Then we will apply this approach to the following example.

EXAMPLE 10-1

When you pick up your 510-gram box of corn flakes or puffed wheat do you ever wonder whether it really weighs 510 grams? You might be surprised if every 510-gram box of cereal weighed exactly 510 grams. After all, slight variations in conditions during processing could affect the amount of cereal deposited in the boxes.

Perhaps the 510-gram label printed on the box refers to an average over many boxes rather than a guaranteed weight in each box. Of course, if the box contains much less than 510 grams, then you, the consumer, feel cheated; and the cereal company does not want to put much more than 510 grams in a box.

A company packaging 510-gram boxes of cereal has an interest in keeping the weight close to 510 grams for each box. Suppose James is in charge of quality control at a company packaging 510-gram boxes of cereal. Out of a large production lot, he randomly selects 81 boxes and weighs the cereal in each box. A dot plot of the 81 weights is shown in Figure 10-1.

Figure 10-1 clearly shows that the weights in the sample do not all equal 510 grams. The weights range from 509.2 grams to 511.2 grams. The distribution appears to be fairly symmetrical about a value somewhat larger than 510 grams.

There are many questions James can ask about the boxes of cereal in the production lot. If government standards specify a range of acceptable weights for 510-gram boxes of cereal, he can ask what proportion of boxes are within this acceptable range (he might address this question with the techniques of Section 10-2). He may want to estimate the variation about the target weight of 510 grams (perhaps using techniques described in Section 14-1). He can also ask if the average weight of cereal packaged equals 510 grams per box.

Since we are considering inferences about a population mean, we will concentrate here on questions about the average weight of cereal packaged

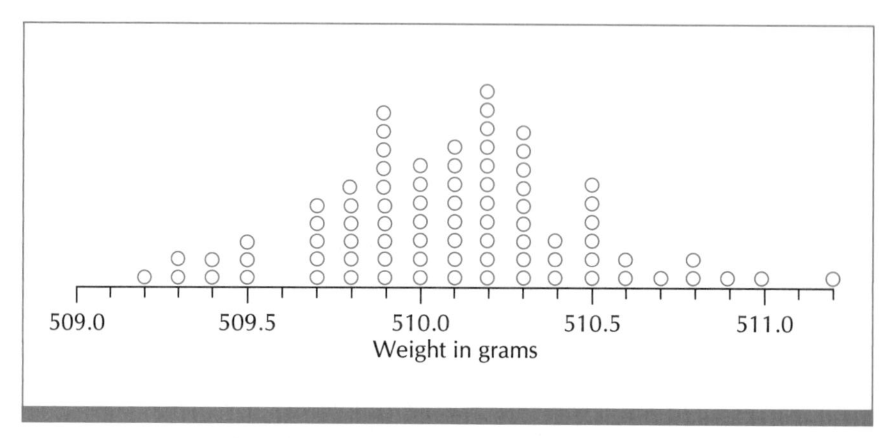

FIGURE 10-1 Dot plot of the weight (in grams) of cereal in 81 boxes, Example 10-1

per box. Let μ denote the mean weight of cereal packaged in boxes in the entire production lot. James wants to compare the hypotheses

$$H_0: \mu = 510 \text{ grams} \qquad \text{and} \qquad H_a: \mu \neq 510 \text{ grams}$$

He also wants to estimate the mean weight of cereal per box packaged in the entire production lot. Let's discuss the general approach to large-sample inferences about a population mean, and then see how James might make inferences about the lot of packaged cereal.

Testing Hypotheses About a Population Mean Based on Large Samples

Suppose we have a large random sample from a population with mean μ, and we want to use the sample to make inferences about μ. Let n represent the sample size, $\bar{X}$ the sample mean, s the sample standard deviation, and $\text{SE} = s/\sqrt{n}$ the standard error of the sample mean. To make large-sample inferences about μ, we assume that the sample size n is large enough that $(\bar{X} - \mu)/\text{SE}$ has approximately the standard Gaussian distribution (see the Central Limit Theorem results in Section 8-3). We now outline the significance level approach to testing hypotheses about a mean based on large samples. Then we will discuss how to find confidence intervals for a population mean.

The significance level approach to testing hypotheses about a population mean μ based on large samples

1. State null and alternative hypotheses, H_0: $\mu = \mu_0$ and H_a: $\mu \neq \mu_0$, for a specific number μ_0.
2. The test statistic is

$$\frac{\bar{X} - \mu_0}{\text{SE}}$$

3. Assume that we have independent observations from a distribution with mean μ. Also assume that the sample size is large enough that $\bar{X}$ has approximately a Gaussian distribution. Then under the null hypothesis, the test statistic has approximately the standard Gaussian distribution.
4. Select a significance level α.
5. The acceptance region is the interval $(-c, c)$ and the rejection region includes the intervals $(-\infty, -c]$ and $[c, \infty)$. The number c is chosen so that $P(Z \leq c) = 1 - \alpha/2$, where Z has the standard Gaussian distribution.
6. The decision rule is:

   If $-c <$ test statistic $< c$, say the results are consistent with the null hypothesis.

   If test statistic $\leq -c$ or test statistic $\geq c$, say the results are inconsistent with the null hypothesis.
7. Collect a large random sample from the population of interest. Calculate the test statistic based on the sample. Use the decision rule in step 6 to decide whether the observations are consistent with the null hypothesis. Draw conclusions based on what we see in the sample.

If we have the one-sided alternative $H_a: \mu > \mu_0$, then in step 5 the acceptance region is the interval $(-\infty, c)$. The rejection region is the interval $[c, \infty)$. The number c is chosen so that $P(Z \leq c) = 1 - \alpha$, where Z has the standard Gaussian distribution. The decision rule in step 6 is:

If test statistic $< c$, say the results are consistent with the null hypothesis.
If test statistic $\geq c$, say the results are inconsistent with the null hypothesis.

If our one-sided alternative is $H_a: \mu < \mu_0$, then in step 5 the acceptance region is the interval $(-c, \infty)$. The rejection region is the interval $(-\infty, -c]$. The number c is chosen so that $P(Z \leq -c) = \alpha$. The decision rule in step 6 is:

If test statistic $> -c$, say the results are consistent with the null hypothesis.
If test statistic $\leq -c$, say the results are inconsistent with the null hypothesis.

Large-Sample Confidence Intervals for a Population Mean

Large-sample confidence intervals for μ are of the form $(\bar{X} - c\text{SE}, \bar{X} + c\text{SE})$. We find the number c from the standard Gaussian distribution to give the desired confidence level. If the area from $-c$ to c under the standard Gaussian curve equals A, then

$$A \doteq P\left(-c \leq \frac{\bar{X} - \mu}{\text{SE}} \leq c\right) = P(\bar{X} - c\text{SE} \leq \mu \leq \bar{X} + c\text{SE})$$

We find c from Table B such that $P(Z \leq c) = (1 + A)/2$, where Z has the standard Gaussian distribution. We say the interval from $\bar{X} - c\text{SE}$ to $\bar{X} + c\text{SE}$ is an approximate $100A\%$ confidence interval for μ. For instance, $(\bar{X} - 1.96\text{SE}, \bar{X} + 1.96\text{SE})$ is an approximate 95% confidence interval for μ and $(\bar{X} - 1.65\text{SE}, \bar{X} + 1.65\text{SE})$ is an approximate 90% confidence interval for μ.

The correct interpretation of a confidence interval is this:

> If the sampling process were repeated over and over, and if a $100A\%$ confidence interval for μ were calculated each time, about $100A\%$ of these confidence intervals would contain μ and about $100(1 - A)\%$ would not. Once a specific confidence interval has been calculated, it either contains μ or it does not. The **confidence level** refers to what we would expect if the sampling process were repeated many times.

Confidence intervals and tests of hypotheses are related. If the value μ_0 specified in the null hypothesis is in the confidence interval, we say the results are consistent with the null hypothesis. If μ_0 is not in the confidence interval, we say the results are inconsistent with the null hypothesis.

EXAMPLE 10-1
(continued)

In Example 10-1, James wants to use his sample to make inferences about the average weight of cereal packaged per box in the production lot. He has a large random sample from a much larger population, so he feels justified in using large-sample inference based on the standard Gaussian distribution. One thing he wants to do is to compare the hypotheses $H_0: \mu = 510$ grams and H_a: $\mu \neq 510$ grams, where μ is the mean weight of cereal packaged in the production lot. He chooses significance level $\alpha = .05$. Since $P(Z \leq 1.96) = .975$, where Z has the standard Gaussian distribution, James chooses the following decision rule:

If $-1.96 <$ test statistic < 1.96, say the results are consistent with the null hypothesis that the mean weight per box in the lot equals 510 grams.
If test statistic ≤ -1.96 or test statistic ≥ 1.96, say the results are inconsistent with the null hypothesis, suggesting that the mean weight per box in the lot does not equal 510 grams.

James calculates the sample mean, standard deviation, and standard error for his observations: $\bar{X} = 510.10$ grams, $s = .39$ gram, SE $= s/9 = .043$ gram. For a test statistic, he calculates

$$\frac{\bar{X} - 510}{\text{SE}} = \frac{510.10 - 510}{.043} = 2.3$$

Since the observed value of his test statistic is 2.3, James decides the results are inconsistent with his null hypothesis that the mean weight per box in the lot equals 510 grams. The approximate p-value is .02, the probability that a standard Gaussian random variable is greater than or equal to 2.3 in absolute value. If the average weight of cereal packaged in the production lot were really 510 grams, there would be about a 2 in 100 chance of seeing a test statistic at least as extreme as the one observed. Since the sample mean is 510.10 grams, these results suggest that the average weight per box in the lot is greater than 510 grams.

James decides to estimate the average weight μ of cereal packaged in the production lot. His point estimate is $\bar{X} = 510.10$ grams. As an interval estimate

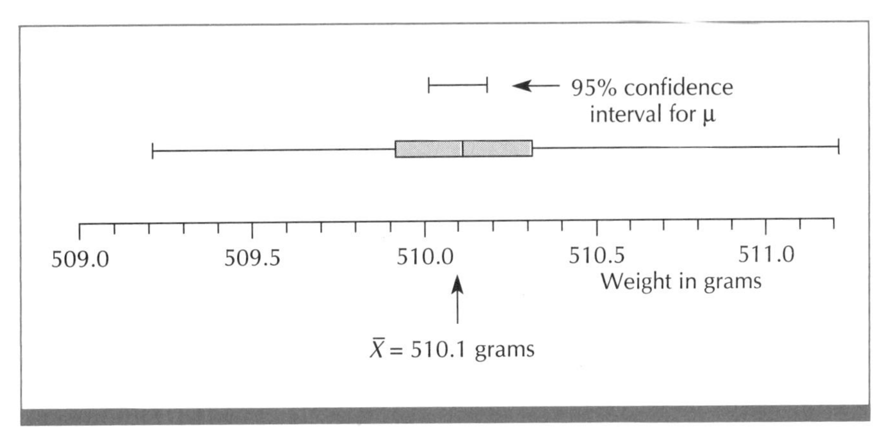

FIGURE 10-2 Box plot of the 81 sample weights in Example 10-1. Also shown is a 95% confidence interval for the population mean weight μ in the production lot.

of μ, he decides to use a 95% confidence interval. He knows that for a standard Gaussian random variable Z, $P(-1.96 \leq Z \leq 1.96) = .95$, so he calculates the confidence interval

$$(\bar{X} - 1.96\text{SE}, \bar{X} + 1.96\text{SE}) = (510.10 - 1.96 \times .043, 510.10 + 1.96 \times .043) = (510.02, 510.18)$$

James estimates that the mean weight of cereal per box in the production lot is from 510.02 grams to 510.18 grams. This confidence interval is illustrated in Figure 10-2, along with a box plot of the observations. Because the sample size is large, the confidence interval for the mean is narrow relative to the range of the observed weights. The null hypothesis mean, $\mu_0 = 510$ grams, is not in the confidence interval, in agreement with what we found using the test of hypotheses.

The sample mean of the 81 weights is 510.1 grams. (The sample median is also 510.1 grams.) From the 95% confidence interval, we estimate that the average packaged weight of cereal is from .02 to .18 gram above the labeled value of 510 grams. These results suggest that the average weight packaged in the cereal boxes during production is greater than the labeled weight of 510 grams. However, such inferences about the mean are not completely satisfactory for this quality control problem. Even though the sample average weight exceeds 510 grams, 29 (almost 36%) of the 81 sample boxes contain less than 510 grams of cereal. Eight (about 10%) of the sample boxes contain 509.5 grams or less. A fairly large proportion of consumers might feel cheated, if they expect at least 510 grams of cereal in a box. On the other hand, the smallest packaged weight in the sample, 509.2 grams, is less than 1 gram short of the labeled weight. Whether this sample represents an acceptable packaging process depends on tolerances or variations in packaged weight that are acceptable to consumers, the government, and the cereal company.

Large-sample inference about a population mean based on the standard Gaussian distribution is *nonparametric* because we assume no specific type of probability distribution for the observations. We do not even assume that the distribution is continuous. We will take advantage of this simplicity in Section 10-2 when we discuss large-sample inference about a proportion.

10-2 Large-Sample Inference About a Proportion

Before we discuss large-sample inference about a proportion in general, let's consider an example.

EXAMPLE 10-2

Tubal infertility refers to infertility resulting from damage to the Fallopian tubes, often caused by infection. Investigators wanted to study the relationship between tubal infertility and method of contraception (Cramer et al., 1987). They interviewed 283 women with tubal infertility and no children. Of these 283 women, 131 had at some time used barrier methods of contraception (for example, use of a diaphragm by the woman or condoms by her partner).

In a large control group of women with children, 51.5% had used barrier methods of contraception at some time. The women in the control group were similar to the women with tubal infertility with respect to age, race, and income status.

The investigators wanted to make inferences about p, the proportion of women with tubal infertility who had sometime used barrier methods of contraception. In particular, they wanted to ask how p compared with .515, the proportion for the control group.

Why do you think these investigators were interested in methods of contraception used by women with tubal infertility? What relationship might method of contraception (in particular, barrier methods versus other methods) have with infection and possible damage to the Fallopian tubes?

We will continue with this example after discussing the general approach to large-sample inference about a proportion.

Large-sample inference about a proportion is really just a special case of large-sample inference about a mean. Let's see why this is true.

Suppose p is the proportion we are interested in. Imagine n independent repetitions of a two-outcome experiment, where p is the probability of success and $1 - p$ is the probability of failure for each repetition. Let the random variable X_i equal 1 if the ith repetition results in success. Let X_i equal 0 if the ith repetition results in failure. Then X_1, X_2, through X_n represent a random sample from a Binomial$(1, p)$ distribution.

A reasonable estimate of the probability of success p is $\hat{p}$, where $\hat{p}$ denotes the proportion of successes observed in the sample. But $\hat{p}$ is the same as $\bar{X}$. Therefore, for large samples, we can base inferences about p on the standard Gaussian distribution.

Suppose we want to test the hypotheses $H_0: p = p_0$ and $H_a: p \neq p_0$ for a specified number p_0. The test statistic has the form

$$\frac{\hat{p} - p_0}{\sqrt{\dfrac{p_0(1 - p_0)}{n}}}$$

where $\sqrt{p_0(1 - p_0)/n}$ is the standard deviation of $\hat{p}$ when $p = p_0$. We carry out the test of hypotheses as outlined for the large-sample case in Section 10-1.

A confidence interval for p has the form

$$\hat{p} \pm c\sqrt{\frac{\hat{p}(1 - \hat{p})}{n}}$$

Here, $\sqrt{\hat{p}(1 - \hat{p})/n}$ is the standard error of $\hat{p}$. We find the number c from the standard Gaussian distribution, as in Section 10-1.

EXAMPLE 10-2 *(continued)*

In Example 10-2, p is the proportion of women with tubal infertility who had sometime used barrier methods of contraception. The investigators wanted to compare p with .515, the proportion of women in the control group who had sometime used barrier methods of contraception. In particular, they wanted to test the hypotheses $H_0: p = .515$ and $H_a: p \neq .515$.

For large-sample inference about p to be appropriate, we must assume that the sample represents 283 independent observations from a much larger population of women with tubal infertility. How likely is it that the sample is a random sample from this population? If these 283 women were all patients at a fertility clinic, how likely is it that the sample is representative of all women with tubal infertility? And, as long as we are posing questions, how reliable would you consider the information on contraception history for the women in the sample and the women in the control group?

A point estimate for p is $\hat{p}$, the proportion of women in the sample who had used barrier methods of contraception: $\hat{p} = 131/283 = .463$. The large-sample test statistic is

$$\text{Test statistic} = \frac{.463 - .515}{\sqrt{\dfrac{(.515)(.485)}{283}}} = -1.75$$

If Z has the standard Gaussian distribution, then $P(Z \geq 1.75) = .0401$. Therefore, the p-value is $P(\text{test statistic} \leq -1.75 \text{ or test statistic} \geq 1.75 \text{ when } H_0 \text{ is true}) \doteq .0802$. If the null hypothesis were true, there would be about an 8% chance of seeing a test statistic at least as far from .515 as the one observed. We might call this a borderline result. We cannot say the sample is strongly consistent with or strongly inconsistent with the null hypothesis.

An approximate 90% confidence interval for p is

$$.463 \pm 1.65\sqrt{\frac{(.463)(.537)}{283}} \quad \text{or} \quad (.414, .512)$$

An approximate 95% confidence interval for p is

$$.463 \pm 1.96 \sqrt{\frac{(.463)(.537)}{283}} \quad \text{or} \quad (.405, .521)$$

The null hypothesis value .515 is in the 95% confidence interval, but not in the 90% confidence interval. This agrees with the borderline significance of the test of hypotheses. There is some evidence that the proportion of women with tubal infertility who had sometime used barrier methods of contraception is smaller than the corresponding proportion for the control group. But the evidence is not overwhelming.

In Section 10-3, we consider the classical approach to inferences about a population mean when we have a small sample.

10-3 Inferences About a Population Mean (or Median) Based on a *t* Distribution

We want to use a small sample to make inferences about the center of a population. In the classical approach, we must assume that the sample represents independent observations from a Gaussian distribution. Then our analysis is based on a t distribution. The corresponding test of hypotheses is called the *t test* or *one-sample t test.* In this section, we will discuss inferences about a population mean based on a t distribution, as applied to the following example.

EXAMPLE 10-3 Does exercise affect birth weights in Pygmy goats? In a study designed to answer this question, investigators trained pregnant Pygmy goats to walk on a motor-driven treadmill (Dhindsa, Metcalfe, and Hummels, 1978). Six of these goats subsequently gave birth to twins. The average weights in grams of the six pairs of twins were:

745.5 1,175.0 1,290.0 1,364.5 1,397.5 1,660.0

In a large control group of similar pregnant Pygmy goats who did not exercise, the average weight of twins was 1,592.0 grams.

A dot plot of the six average weights of twins born to exercised goats is shown in Figure 10-3. We see that five of these six values are less than the average weight (1,592.0 grams) of twins born to unexercised goats. The distribution of the six plotted values is concentrated around an interval from about 1,200 to 1,400 grams. What does the plot suggest about the relative weights of twins born to exercised and unexercised goats?

Let μ denote the mean weight of twins born to exercised goats. We would like to test whether μ equals 1,592.0, the mean weight of twins born to unexercised goats. We would also like to calculate an interval estimate, or range of

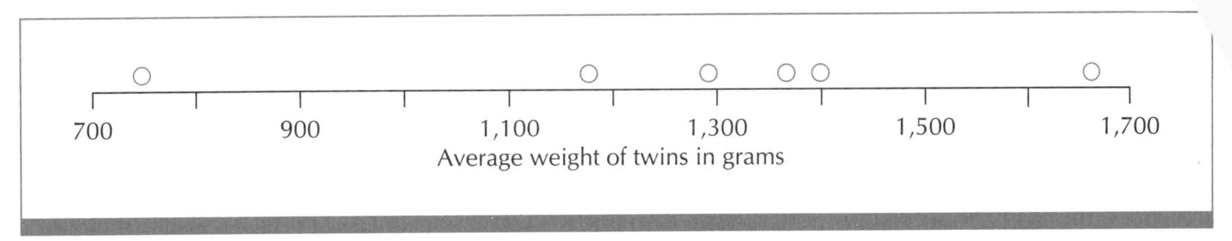

FIGURE 10-3 Dot plot of the average weight of twins born to six exercised goats in Example 10-3

reasonable values, for μ. Each of these analyses will be based on a t distribution. Let's discuss the general approach and then apply it to this example.

First consider the assumptions we must make for analysis based on a t distribution. Suppose we have a sample of size n. We assume that these n observations are independent, with the same Gaussian probability distribution. We want to make inferences about the mean μ of this distribution. Because a Gaussian distribution is symmetric about its mean, μ is also the median of the distribution.

Let $\bar{X}$ denote the sample mean, s the sample standard deviation, and $\text{SE} = s/\sqrt{n}$ the standard error of the mean. Under the assumptions just stated, $(\bar{X} - \mu)/\text{SE}$ has a probability distribution known as the t distribution or Student t distribution with $n - 1$ degrees of freedom.

A t distribution is characterized by a parameter called its *degrees of freedom.* Recall that when we calculate the sample variance, we add up the squared deviations of the observations from the sample mean and then divide by the sample size minus 1. This denominator, $n - 1$, is the degrees of freedom for the t distribution of $(\bar{X} - \mu)/\text{SE}$.

Suppose X_1, X_2, through X_n are independent observations from a Gaussian distribution with mean μ. Let $\bar{X}$ denote the sample mean and SE the standard error of the mean for these observations. Then

$$\frac{\bar{X} - \mu}{\text{SE}}$$

has the t distribution with $n - 1$ degrees of freedom.

A t distribution is a continuous probability distribution that is symmetrical about 0, as illustrated in Figure 10-4a. A graph of a t distribution looks similar to that of the standard Gaussian distribution. However, a random variable with a t distribution is more likely to be far from zero than is a standard Gaussian random variable. We say a t distribution has fatter tails than the standard Gaussian distribution, illustrated in Figure 10-4b. The exact shape of a t distribution depends on its degrees of freedom. As the degrees of freedom (or equivalently, the sample size) increase, the differences between a t distribution and the standard Gaussian distribution diminish. For large sample sizes, probabilities determined from the corresponding t distribution are practically the same as those obtained from the standard Gaussian distribution.

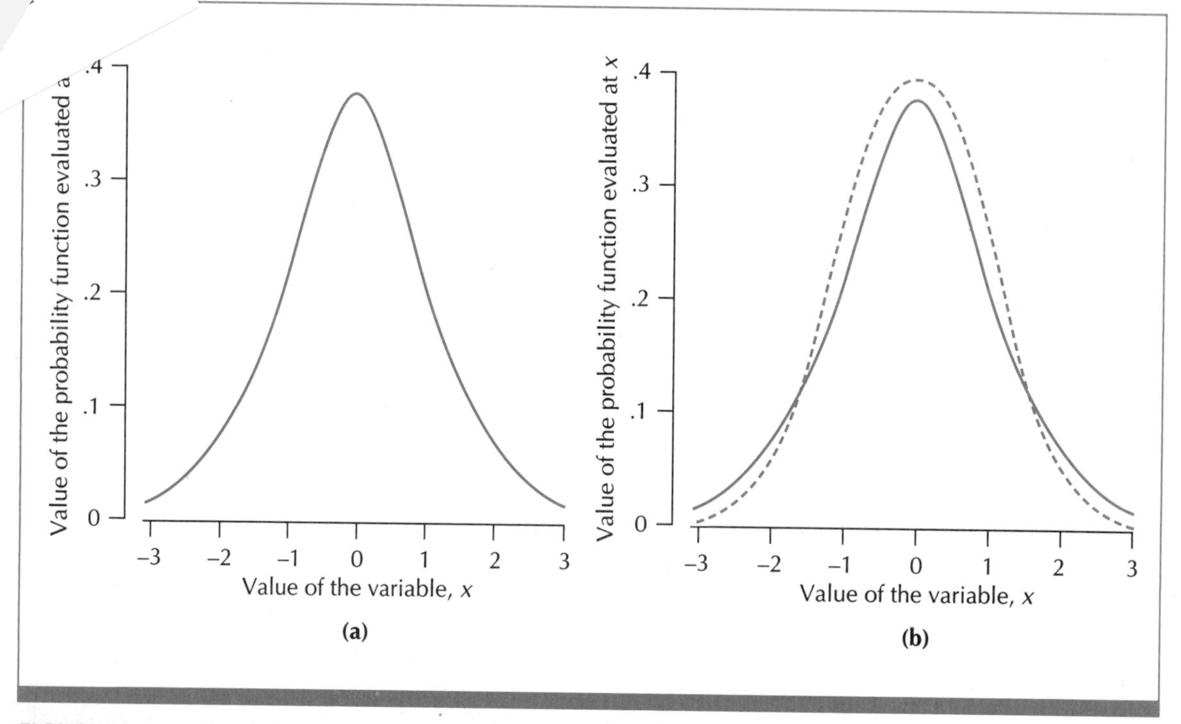

FIGURE 10-4 **a.** Graph of the probability function for the *t* distribution with 5 degrees of freedom. **b.** Comparison of the probability function for the *t* distribution with 5 degrees of freedom (solid line) with the probability function for the standard Gaussian distribution (dashed line). A *t* distribution has fatter tails than does the standard Gaussian distribution.
Adapted from *Mathematical Statistics and Data Analysis,* by J. A. Rice. (Pacific Grove, CA: Brooks/Cole Publishing Co., 1988, p. 170.)

Let T denote a random variable having the t distribution with d degrees of freedom. We are interested in finding probabilities such as $P(T \geq c)$ or $P(-c \leq T \leq c)$, where c is a positive number. These probabilities correspond to areas under the graph of the appropriate probability function. For instance, the area corresponding to the cumulative probability $P(T \leq c)$ is shaded in Figure 10-5. The tail area $P(T \geq c)$ is the unshaded region under the curve in Figure 10-5.

Table C at the back of the book lists numbers c and cumulative probabilities $P(T \leq c)$ for several degrees of freedom d. All the numbers c in Table C are greater than 0. Because a t distribution is symmetrical about 0, we know that

$$P(T \leq -c) = P(T \geq c) = 1 - P(T \leq c)$$

for any positive number c.

Let's see how to use Table C. The line in the table for 5 degrees of freedom is reproduced in Figure 10-5. We see that for a cumulative probability of

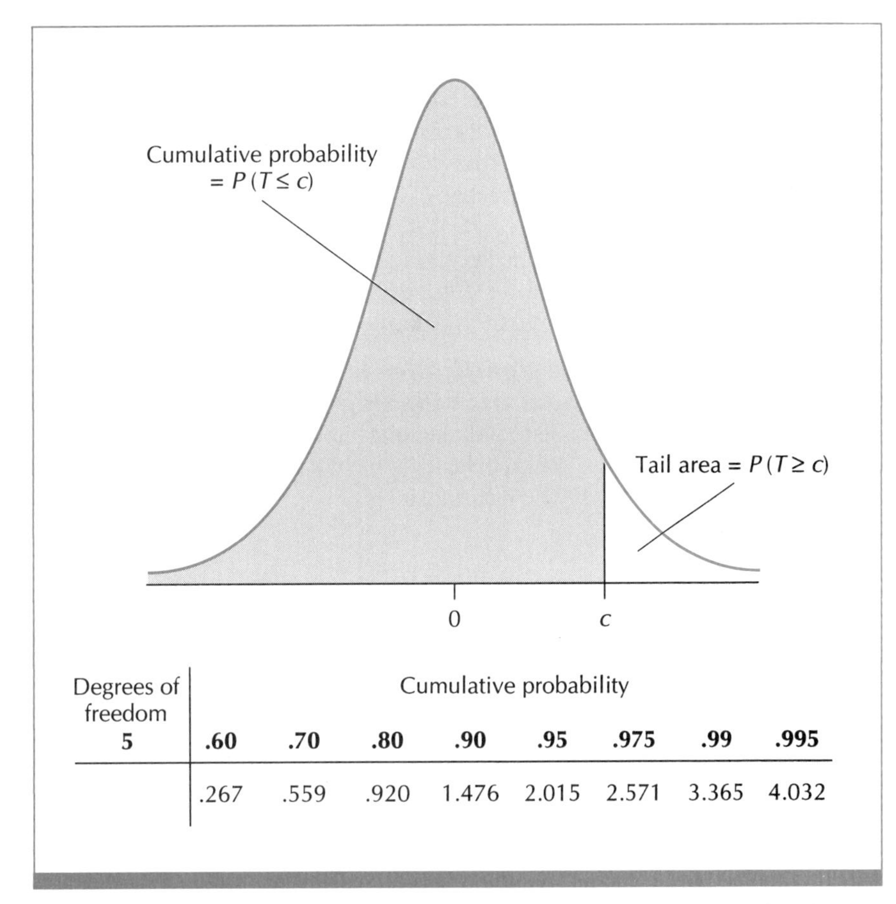

Degrees of freedom	Cumulative probability							
5	**.60**	**.70**	**.80**	**.90**	**.95**	**.975**	**.99**	**.995**
	.267	.559	.920	1.476	2.015	2.571	3.365	4.032

FIGURE 10-5 The line in Table C corresponding to the *t* distribution with 5 degrees of freedom is reproduced here, along with an illustration.

.80, the tabled value of c is .920. This means that if the random variable T has the t distribution with 5 degrees of freedom, then

$$P(T \leq .920) = P(T \geq -.920) = .80$$

From this information, we know that the tail probabilities $P(T > .920)$ and $P(T < -.920)$ both equal .20, and we can find the probability $P(-.920 \leq T \leq .920)$ this way:

$$P(-.920 \leq T \leq .920) = 1 - P(T > .920) - P(T < -.920) = .60$$

Similarly, we can see that

$$P(T \leq 2.571) = P(T \geq -2.571) = .975 \quad \text{and}$$
$$P(-2.571 \leq T \leq 2.571) = .95$$

Table C gives probabilities for t distributions with degrees of freedom up to 30, and three values of d larger than 30. The last line of the table, for $d = \infty$, corresponds to the standard Gaussian distribution. For degrees of freedom greater than 30, we will use the standard Gaussian table (Table B) since $(\bar{X} - \mu)/\text{SE}$ has approximately the standard Gaussian distribution in that case.

A t distribution is a continuous probability distribution, so if a random variable T has a t distribution, then $P(T = c) = 0$ for any number c. Therefore, we know that $P(T \geq c)$ equals $P(T > c)$, for instance, and $P(-c \leq T \leq c) = P(-c < T < c)$. (The argument is the same as for a Gaussian random variable. Recall Figure 8-3d and the discussion in Section 8-1.)

Let's outline the significance level approach to testing hypotheses about a population mean μ based on a t distribution.

The significance level approach to testing hypotheses about a population mean μ based on a t distribution

1. State null and alternative hypotheses, $H_0: \mu = \mu_0$ and $H_a: \mu \neq \mu_0$, for some number μ_0.
2. The test statistic is

$$\frac{\bar{X} - \mu_0}{\text{SE}}$$

3. Assume that we have independent observations from a Gaussian distribution with mean μ. Then the test statistic has the t distribution with $n - 1$ degrees of freedom under the null hypothesis.
4. Specify the significance level α.
5. Find the number c in Table C such that $P(T \leq c) = 1 - \alpha/2$, where T is a random variable having the t distribution with $n - 1$ degrees of freedom. The acceptance region is the interval $(-c, c)$. The rejection region includes $(-\infty, -c]$ and $[c, \infty)$.
6. The decision rule is:

   If $-c <$ test statistic $< c$, say the results are consistent with the null hypothesis.

   If test statistic $\leq -c$ or test statistic $\geq c$, say the results are inconsistent with the null hypothesis.
7. Collect a random sample satisfying the stated assumptions. Calculate the test statistic. Use the decision rule in step 6 to decide whether the observations are consistent or inconsistent with the null hypothesis. Draw conclusions and discuss the experimental results.

If in step 1 we specify a one-sided alternative, then we must alter steps 5 and 6 appropriately. We find the number c in Table C such that $P(T \leq c) = 1 - \alpha$. If we consider the one-sided alternative $H_a: \mu > \mu_0$, then the acceptance region is the interval $(-\infty, c)$ and the rejection region is the interval $[c, \infty)$. We say values of the test statistic less than c are consistent with the null hypothesis, while values greater than or equal to c are inconsistent with the null hypothesis.

If we specify the one-sided alternative $H_a: \mu < \mu_0$, then the acceptance region is the interval $(-c, \infty)$ and the rejection region is the interval $(-\infty, -c]$.

We say values of the test statistic greater than $-c$ are consistent with the null hypothesis, while values less than or equal to $-c$ are inconsistent with the null hypothesis.

EXAMPLE 10-3 *(continued)*

Let's test the hypotheses of interest in Example 10-3. The population is a hypothetical population of Pygmy goats pregnant with twins who might have been subjected to the exercise regimen. If we let μ denote the mean weight of twins born to such goats, then the hypotheses are H_0: $\mu = 1{,}592.0$ grams and H_a: $\mu \neq 1{,}592.0$ grams.

The test statistic is $(\bar{X} - 1{,}592.0)/\text{SE}$, where $\bar{X}$ is the sample mean and SE the standard error of the six observations in the sample. Assume that the observations are independent, from a Gaussian distribution with mean μ. Then the test statistic has the t distribution with 5 degrees of freedom under the null hypothesis.

Independence means that the results for one pregnant goat did not affect in any way the results for another pregnant goat. We cannot judge the appropriateness of this assumption without more details about the experiment.

The dot plot in Figure 10-3 gives us no reason to doubt the assumption that the observations come from a Gaussian distribution. With so few observations, we cannot say we have strong evidence supporting this assumption, either. The t test, however, tends to be *robust* to deviations from the Gaussian assumption. This means that even when the observations do not follow exactly a Gaussian distribution, significance levels (and confidence levels) tend to be close to the values we select.

> We say a statistical procedure is **robust** if the actual significance level (or confidence level) is close to the level we select, even under deviations from assumptions.

Letting the significance level α equal .10, we find $\alpha/2 = .05$. Referring to Table C or Figure 10-5, we see that if T is a random variable having the t distribution with 5 degrees of freedom, then $P(T \leq 2.015) = .95$. The acceptance region is the interval $(-2.015, 2.015)$. The rejection region includes the two intervals $(-\infty, -2.015]$ and $[2.015, \infty)$. The decision rule is:

If $-2.015 <$ test statistic < 2.015, say the results are consistent with the null hypothesis.

If test statistic ≤ -2.015 or test statistic ≥ 2.015, say the results are inconsistent with the null hypothesis.

Using the six sample observations, we find $\bar{X} = 1{,}272.08$ grams and $\text{SE} = 124.07$ grams. The test statistic is

$$\frac{\bar{X} - 1{,}592.0}{\text{SE}} = \frac{1{,}272.08 - 1{,}592.0}{124.07} = -2.6$$

Since -2.6 is less than -2.015, we say the experimental results are inconsistent with the null hypothesis, at the $\alpha = .10$ significance level.

The p-value is the probability of seeing a test statistic as extreme as or more extreme than the one observed, if the null hypothesis were true. Since the test statistic equals -2.6, the p-value equals

$$p\text{-value} = P(T \leq -2.6) + P(T \geq 2.6)$$

where T has the t distribution with 5 degrees of freedom. From Table C (or Figure 10-5), we see the p-value is about .05. If the null hypothesis were true, there would be about a 5% chance of seeing results as extreme as or more extreme (in the direction of the alternative) than those observed.

The experimental results suggest that the population mean μ does not equal 1,592.0 grams. If we want to estimate μ, a reasonable point estimate is $\bar{X} = 1{,}272.08$, or 1,272.1 grams. We know that a different sample would probably result in a different sample mean, so we would like an interval estimate or range of reasonable values for μ. The general procedure for finding a confidence interval for a population mean μ based on a t distribution is outlined below.

Confidence Intervals for a Population Mean Based on a *t* Distribution

A confidence interval for the population mean μ, based on a t distribution, has the form $(\bar{X} - c\text{SE}, \bar{X} + c\text{SE})$, where c is the number that gives the confidence level we want. Suppose A is a number between 0 and 1. If we want a 100A% confidence interval for μ, then we find c such that

$$A = P\left(-c \leq \frac{\bar{X} - \mu}{\text{SE}} \leq c\right) = P(\bar{X} - c\text{SE} \leq \mu \leq \bar{X} + c\text{SE})$$

This is the number c from Table C that satisfies $P(T \leq c) = (1 + A)/2$, where T has the t distribution with $n - 1$ degrees of freedom.

As with large-sample confidence intervals, the correct interpretation is this:

> If the sampling process were repeated over and over, and if a 100A% confidence interval for μ were calculated each time, about 100A% of these confidence intervals would contain μ and about $100(1 - A)$% would not. A specific confidence interval either contains μ or it does not; the **confidence level** refers to what we would expect if the sampling process were repeated many times.

Suppose we want a 90% confidence interval for the population mean in Example 10-3. We can think of μ as the mean weight of twins born to (a hypothetical population of) exercised Pygmy goats. Then $A = .90$ and $(1 + A)/2 = .95$. If T has a t distribution with 5 degrees of freedom, then $P(T \leq 2.015) = .95$. Therefore, a 90% confidence interval for μ is

$$(\bar{X} - 2.015\text{SE}, \bar{X} + 2.015\text{SE}) = (1{,}022.1, 1{,}522.1)$$

This confidence interval is shown graphically in Figure 10-6, along with a dot plot of the six observations. Note that because the sample size is small, the

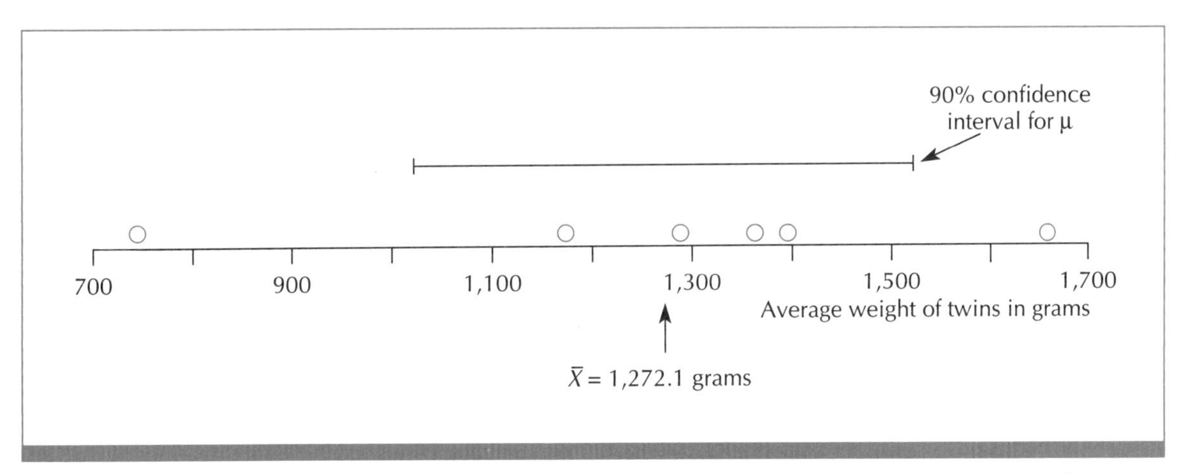

FIGURE 10-6 Dot plot of average weight of twins born to six exercised Pygmy goats in Example 10-3. Also shown is a 90% confidence interval for the mean weight μ of twins born to a hypothetical population of exercised Pygmy goats.

confidence interval is wide compared with the range of data values. We say the interval from 1,022.1 grams to 1,522.1 grams is a range of reasonable values for μ. The null hypothesis value, $\mu_0 = 1{,}592.0$ grams, is not in this interval, in agreement with the test of hypotheses.

What do the results of our analysis suggest about the effect of exercise on average birth weights? If you were asked to develop a theory for why the offspring of exercised goats weighed on average less than the offspring of unexercised goats, what might it be? Would you be willing to extrapolate the results of this experiment on Pygmy goats and hazard a guess about the effect of maternal exercise on birth weights in humans?

Some time after this experiment, the investigators decided a possible extraneous factor might have affected the outcome. A trainer induced a goat to exercise by applying a mild electric shock to a rear leg whenever she stopped exercising! In a subsequent experiment, investigators trained the goats to exercise with positive reinforcement such as petting or food. In this second study, birth weights were on average no different in the exercise and no-exercise groups (Hohimer et al., 1984). How does this new information affect any conclusions we might have drawn based on the first experiment?

In Section 10-4, we consider small-sample inferences about a population mean when we are not willing to assume the observations come from a Gaussian distribution.

10-4 Inferences About a Population Mean (or Median) Based on a Wilcoxon Signed Rank Distribution

Again, we want to use a sample to make inferences about the center of a population. We assume that we have a random sample of observations from a continuous probability distribution that is symmetric about a number μ. Because of the symmetry, μ is both the mean and the median of the population. With these assumptions, we base our inferences about μ on a Wilcoxon signed rank distribution.

For testing hypotheses about μ, we use the Wilcoxon signed rank test. To see how this test works, we will look at an example and then outline the steps for using this test in general.

EXAMPLE 10-4

Workers developed a number of useful techniques in statistics and quality control during World War II, as United States industry prepared for American involvement in the war. Our example here is based on data first reported in 1945 (Kauffman, 1945) and discussed since then in other references on statistics and quality control (Duncan, 1974, page 43; Hollander and Proschan, 1984, page 42).

The height specified for fragmentation bomb bases produced by the American Stove Company was .830 ± .010 inch. A bomb base less than .820 inch or greater than .840 inch in height did not meet specifications because it might cause the bomb to malfunction. (We say that the range from .820 to .840 is a *tolerance range,* the range of tolerated values for height.)

To monitor the production process, quality control workers selected five bomb bases at random from those produced during 15-minute intervals throughout the day. We will consider the heights of the five bomb bases selected in one 15-minute interval:

.826 .829 .831 .836 .840

We will use this sample to make inferences about the average height of all bomb bases produced during that interval.

Whenever possible, we begin an analysis with one or more plots of the observations. Figure 10-7 shows a dot plot of the five sample values. The plot is reasonably symmetric about the interval from .830 to .835 inch. Also, all five heights are within the range of acceptable values, from .820 to .840 inch. As far as tolerances are concerned, this sample seems to represent an acceptable production process.

We want to ask whether the mean of the population sampled equals the target value of .830 inch. That is, we want to compare the hypotheses H_0: $\mu = .830$ inch and H_a: $\mu \neq .830$ inch, where μ is the average height of bomb bases produced during this 15-minute interval.

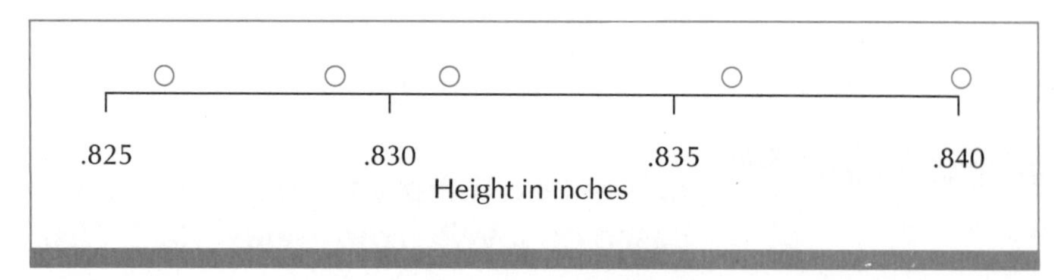

FIGURE 10-7 Dot plot of height in inches for five bomb bases in Example 10-4

TABLE 10-1 Calculating a test statistic for the Wilcoxon signed rank test

Observation: Height (inches)	Deviation: Height − .830	Absolute deviation: |Height − .830|	Rank of absolute deviation	Signed rank
.826	−.004	.004	3	−3
.829	−.001	.001	1.5	−1.5
.831	+.001	.001	1.5	+1.5
.836	+.006	.006	4	+4
.840	+.010	.010	5	+5

$$T+ = \text{Sum of the positive signed ranks} = 1.5 + 4 + 5 = 10.5$$

$$T- = |\text{Sum of negative signed ranks}| = |(-3) + (-1.5)| = 4.5$$

Checking model assumptions, we know we have a random sample of observations from the population of interest, and height is a continuous-type variable. If the size of the production lot is large relative to the size of the sample, then it is reasonable to assume that the observations all have the same probability distribution. The symmetry assumption does not seem unreasonable, as we see in Figure 10-7, but the sample size is too small to provide strong evidence one way or another. We calculate the test statistic for the Wilcoxon signed rank test by following the steps outlined in Table 10-1.

The first column in Table 10-1 lists the observed heights of the five bomb bases in the sample. The second column shows the difference between each height and .830, the null hypothesis mean. The third column gives the absolute value of each difference. We rank these absolute values from smallest (rank 1) to largest (rank 5), with the fourth column recording these ranks.

The smallest absolute deviation is .001, and there are two of them. We say these two values share ranks 1 and 2, and we assign each of them the average of their shared ranks, in this case 1.5. The next largest absolute deviation is .004, which receives the next rank, 3. The absolute deviations .006 and .010 then receive ranks 4 and 5, respectively.

The fifth column contains the same ranks as the fourth column, but with the sign of the deviation in the second column. For instance, the first observation has a signed rank of -3 because the rank in the fourth column is 3 and the sign of the deviation in the second column is negative (meaning the observation is less than the null hypothesis mean).

Now we define two random variables, $T+$ and $T-$. $T+$ is the sum of the positive signed ranks. $T-$ is the absolute value of the sum of the negative signed ranks. For a sample of size 5, the minimum value $T+$ can have is 0, if all the signs are negative. If all the signs are positive, then $T+$ equals the sum of the integers from 1 to 5, or $(5 \times 6)/2 = 15$. Similarly, $T-$ has a minimum of 0 and a maximum of 15. Also, $T+$ and $T-$ add up to 15.

Under the null hypothesis, $T+$ and $T-$ each have a discrete probability distribution, called a Wilcoxon signed rank distribution (see the appendix on Wilcoxon signed rank distributions at the end of this book). Cumulative probabilities for this distribution are given in Table F at the back of the book, for sample sizes from 2 to 15.

Under the null hypothesis, the bomb base heights are symmetrical about .830 inch. Thus, we would expect the positive signed ranks in Table 10-1 to be roughly the same as the negative signed ranks. Therefore, if $T+$ and $T-$ are close to the same value, we say the results are consistent with the null hypothesis. On the other hand, if either $T+$ or $T-$ is very small, we say the results are inconsistent with the null hypothesis.

Let the test statistic equal $T+$ if there are fewer positive signs, $T-$ if there are fewer negative signs. Then we say the results are inconsistent with the null hypothesis if the test statistic is too small.

To use the significance level approach, we choose a significance level α. We find the number c such that $P(W \leq c)$ is close to $\alpha/2$, where W has the Wilcoxon signed rank distribution for sample size 5. Because this probability distribution is discrete, we may not be able to find a value of c for which $P(W \leq c)$ equals $\alpha/2$ exactly. Our decision rule is:

If test statistic $> c$, say the results are consistent with the null hypothesis.
If test statistic $\leq c$, say the results are inconsistent with the null hypothesis.

If we use significance level $\alpha = .05$, then we find c such that $P(W \leq c)$ is close to $\alpha/2 = .025$. Referring to Table F, we see that for sample size 5, $P(W \leq 0) = .031$ and $P(W \leq 1) = .062$. Since .031 is as close to .025 as we can get, we let $c = 0$. Our exact significance level is $2 \times .031 = .062$. The decision rule is:

If test statistic > 0, say the results are consistent with the null hypothesis.
If test statistic $= 0$, say the results are inconsistent with the null hypothesis.

We see in Table 10-1 that we have two negative signs and three positive signs. Therefore, our test statistic is $T-$, with an observed value of 4.5. Since the test statistic is greater than 0, we say the results are consistent with the null hypothesis, at the .062 significance level.

The p-value is the probability of seeing either $T+$ or $T-$ less than or equal to the value of the observed test statistic. Since the test statistic equals 4.5 in Example 10-4, the p-value is

$$p\text{-value} = P(T+ \leq 4.5 \text{ or } T- \leq 4.5 \text{ when } H_0 \text{ is true})$$

From Table F we see that $P(W \leq 4) = .219$ and $P(W \leq 5) = .312$. Therefore, the p-value is between $.219 + .219 = .438$ and $.312 + .312 = .624$:

$$.438 \leq p\text{-value} \leq .624$$

If the null hypothesis were true, there would be a large probability (from .438 to .624) of seeing a test statistic as small as or smaller than the one observed.

Based on the Wilcoxon signed rank test, we say the results are consistent with the null hypothesis that the mean height of bomb bases produced during the 15-minute interval equaled the target value of .830 inch. (This does not imply that the mean height of bomb bases produced in that 15-minute interval really was .830 inch. Why?)

Let's summarize the significance level approach to testing hypotheses about a population mean μ, based on a Wilcoxon signed rank distribution. Suppose for now that n_0 is from 2 to 15, where n_0 is the number of observations different from μ_0.

The significance level approach to testing hypotheses about a population mean based on a Wilcoxon signed rank distribution

1. The hypotheses are H_0: $\mu = \mu_0$ and H_a: $\mu \neq \mu_0$, where μ_0 is a specified number.
2. Subtract μ_0 from each observation and take the absolute value of this difference. Ignore any differences equal to 0. Rank the n_0 nonzero absolute deviations from smallest to largest. If two or more absolute deviations have the same value, assign each of them the average of the ranks they share.

   Give a rank a negative sign if the corresponding observation is less than μ_0 and a positive sign if the observation is greater than μ_0. Let $T+$ be the sum of the positive signed ranks, $T-$ the absolute value of the sum of the negative signed ranks.

   Let the test statistic equal $T+$ if there are fewer positive signs, $T-$ if there are fewer negative signs.
3. Assume that we have a random sample of observations from a continuous symmetric distribution. Then under the null hypothesis, $T+$ and $T-$ each have the Wilcoxon signed rank distribution for sample size n_0.
4. Select a value of α for the significance level of the test.
5. Referring to Table F, find a number c such that $P(W \leq c)$ is close to $\alpha/2$, where W has the Wilcoxon signed rank distribution for sample size n_0. The acceptance region includes numbers greater than c. The rejection region includes numbers less than or equal to c.
6. The decision rule is:

   If test statistic $> c$, say the results are consistent with the null hypothesis.
   If test statistic $\leq c$, say the results are inconsistent with the null hypothesis.
7. Collect a sample that satisfies the assumptions in step 3. Calculate the test statistic based on the sample. Use the decision rule in step 6 to decide whether the observations are consistent with the null hypothesis. Draw conclusions based on the analysis.

If in step 1 we specify a one-sided alternative, we must change steps 2 and 5 appropriately. If we specify the one-sided alternative H_a: $\mu > \mu_0$, then in step 2 we let the test statistic equal $T-$. In step 5, we find the number c such that $P(W \leq c)$ is close to α.

If we specify the one-sided alternative H_a: $\mu < \mu_0$, then in step 2 we let the test statistic equal $T+$. In step 5, we again find the number c such that $P(W \leq c)$ is close to α.

If the number n_0 of observations different from μ_0 is greater than 15, we can use a *large-sample approximation* to make inferences. We use the test statistic

$$\text{Test statistic} = \frac{T+ - \dfrac{n_0(n_0 + 1)}{4}}{\sqrt{\dfrac{n_0(n_0 + 1)(2n_0 + 1)}{24}}}$$

Under the null hypothesis, this test statistic has approximately the standard Gaussian distribution. Values of the test statistic near 0 are consistent with the null hypothesis; values far from 0 are inconsistent with the null hypothesis. We carry out the large-sample test based on the standard Gaussian distribution, as described in Section 10-1.

To illustrate use of this large-sample approximation, let's apply it to Example 10-4. Since the sample size is only 5, we do not necessarily expect results using the large-sample approximation to be close to those based on Wilcoxon signed rank distribution. In the example, $n_0 = 5$ and $T+ = 10.5$. The large-sample test statistic is

$$\text{Test statistic} = \frac{10.5 - \dfrac{5(5 + 1)}{4}}{\sqrt{\dfrac{5(5 + 1)(2 \times 5 + 1)}{24}}} = .81$$

Since the alternative is two-sided, values of the test statistic far from 0 in either the positive or negative direction are in the rejection region. An approximate p-value is $P(Z \geq .81) + P(Z \leq -.81)$, where Z has the standard Gaussian distribution. Referring to Table B and using the symmetry of the standard Gaussian distribution about 0, we see that $P(Z \leq -.81) = P(Z \geq .81) = 1 - P(Z \leq .81) = 1 - .7910 = .209$, so the p-value based on the large-sample approximation is

$$p\text{-value} \doteq .209 + .209 = .418$$

Using the Wilcoxon signed rank distribution for sample size 5, we found a p-value between .438 and .624. Although the p-value is smaller using the large-sample approximation, we would still say the results are consistent with the null hypothesis.

When all the assumptions for the t test are met, the t test is *more powerful* than the Wilcoxon signed rank test.

We say one procedure for testing hypotheses is **more powerful** than another if it has greater probability of rejecting the null hypothesis when the alternative hypothesis is really true.

It makes sense that when either test would be valid, the t test is more powerful than the Wilcoxon signed rank test. The t test uses the actual observations while the Wilcoxon signed rank test uses only the relative magnitudes or ranks of the observations.

We can calculate a *confidence interval for the population mean μ based on a Wilcoxon signed rank distribution* (Lehmann, 1975, pages 181–185). Let O_1, O_2, through O_n represent the observations ordered from smallest to largest. Average each O_i with itself and with each O_j where j is larger than i. There will be $N = n(n + 1)/2$ of these averages. Let A_1 be the smallest of these averages, A_2 the next smallest, and so on.

Confidence intervals for μ have the form (A_k, A_{N-k+1}). We determine k and the confidence level from the relationship

$$P(A_k < \mu < A_{N-k+1}) = 1 - 2P(W \leq k - 1)$$

where W has the Wilcoxon signed rank distribution for sample size n.

EXAMPLE 10-4 *(continued)*

Let's calculate a confidence interval for the mean μ of bomb base heights in the 15-minute interval, for Example 10-4. A convenient way of tabulating the averages of each observation with itself and each higher-ordered observation is shown in Table 10-2. Because the sample size is 5, there are $5(5 + 1)/2 = 15$ of these averages. The ordered averages A_1 through A_{15} are displayed on the right-hand side of Table 10-2.

If $k = 1$, then $(A_1, A_{15}) = (.826, .840)$ is our confidence interval for the mean bomb base height in the 15-minute interval. We find the confidence level from the relationship

$$1 - 2P(W \leq 0) = 1 - 2(.031) = .938$$

TABLE 10-2 Average of each observation with itself and each higher-ordered observation for the sample of size 5 in Example 10-4

Ordered observation	.826	.829	.831	.836	.840	
						A_1: .826
						A_2: .8275
.826	.826	.8275	.8285	.831	.833	A_3: .8285
						A_4: .829
.829	—	.829	.830	.8325	.8345	A_5: .830
						A_6: .831
.831	—	—	.831	.8335	.8355	A_7: .831
						A_8: .8325
.836	—	—	—	.836	.838	A_9: .833
						A_{10}: .8335
.840	—	—	—	—	.840	A_{11}: .8345
						A_{12}: .8355
						A_{13}: .836
						A_{14}: .838
						A_{15}: .840

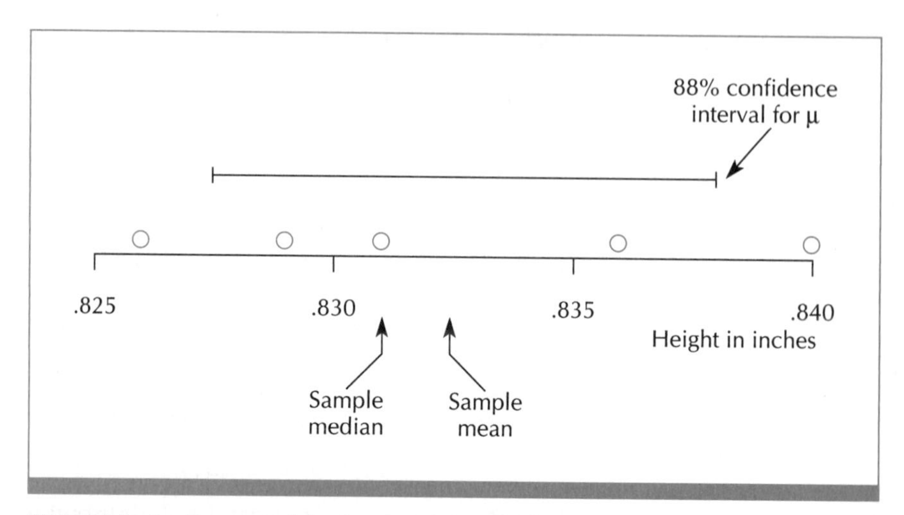

FIGURE 10-8 Dot plot of the five bomb base heights in Example 10-4. Also shown is an 88% confidence interval for the average height μ of bomb bases in the production lot for the 15-minute interval.

We say (.826, .840) is a 93.8% (or 94%) confidence interval for μ. The probability $P(W \leq 0) = .031$ is in Table F for sample size 5.

If $k = 2$, then $(A_2, A_{14}) = (.8275, .838)$ is our confidence interval for μ. The confidence level is determined as $1 - 2P(W \leq 1) = .876$. We say (.8275, .838) is an 87.6% (or 88%) confidence interval for μ. This confidence interval is illustrated in Figure 10-8, along with a dot plot of the observations. Note that because of the small sample size, the confidence interval is wide relative to the range of the data values.

Similarly, we find that $(A_3, A_{13}) = (.8285, .836)$ is an 81% confidence interval and $(A_4, A_{12}) = (.829, .8355)$ is a 69% confidence interval for μ. All of the confidence intervals contain $\mu_0 = .830$, consistent with the null hypothesis. Our analysis gives us no reason to doubt that the average height of bomb bases produced in the 15-minute interval was .830 inch. (Again, this does not imply that the average height of all those bomb bases really is exactly .830 inch.)

We have made inferences about the mean height of bomb bases in Example 10-4. In quality control, we are interested in more than just the mean. We care about variation and about the proportion of acceptable items. In Example 10-4, we would be concerned with the variation in heights of bomb bases produced, and the proportion of bomb bases meeting the tolerance specification. (In this sample, of course, all five bomb bases met the tolerance specification.) We will be discussing inferences about variances and proportions later.

In Section 10-5, we discuss inferences about a population median, based on fewer assumptions than we used in Sections 10-3 and 10-4.

10-5

Inferences About a Population Median Based on a Binomial Distribution

As before, we want to use a sample to make inferences about the center of a population. Now we assume only that our sample consists of independent observations from a continuous probability distribution. Our inferences about the median of this distribution are based on a binomial distribution. We call the test of hypotheses the *sign test.* Let's begin with an example.

EXAMPLE 10-5

How does a new, cheaper type of electric cord compare with an older, more expensive type when used with appliances? An engineer selects a random sample of 12 of the new electric cords. He subjects these 12 cords to *accelerated life testing.* That is, he puts them on a machine that wears the cords out much faster than would occur in actual use. The test runs for 1 week. At the end of the week, the engineer removes the 12 cords from the machine. The 12 times to failure, in hours, are (Nelson, 1986, pages 24–25):

72.4	78.6	81.2	94.0	120.1	126.3
127.2	128.7	141.9	164.1+	164.1+	164.1+

The + after the last three values means that the three cords were still functional when the engineer ended the testing. These are *censored* observations: We know they have values greater than 164.1 hours, but we do not know the exact values. The median time to failure for the more expensive type of electric cord is 114 hours. How does the median time to failure for the new cords compare with this value?

The 12 sample times to failure are plotted in Figure 10-9. A censored observation is indicated with a + sign in the plot. We see that four sample failure times are less than the median failure time of 114 hours for the more expensive type of cord; these four values are in fact all less than 95 hours. The other eight sample failure times are greater than 114 hours, ranging from about 120 hours to more than 164 hours.

The engineer wants to test the hypotheses H_0: $M = 114$ hours and H_a: $M \neq 114$ hours, where M is the median time to failure in hours of the newer,

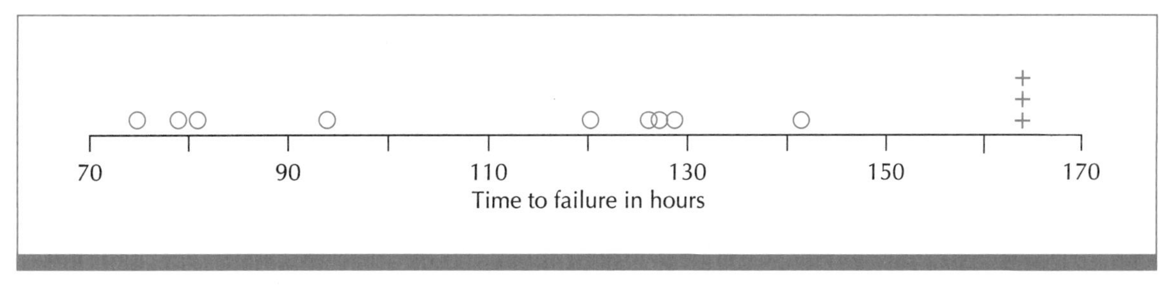

FIGURE 10-9 Dot plot of time to failure for 12 electric cords in Example 10-5. A + indicates a cord that had not failed when the experiment ended at 164.1 hours.

less expensive type of electric cord. He lets his test statistic equal the number of cords that last more than 114 hours.

He assumes that the observations are independent and that each cord has probability p of lasting more than 114 hours. He can then restate his hypotheses as $H_0: p = \frac{1}{2}$ and $H_a: p \neq \frac{1}{2}$. With these assumptions, the test statistic has a Binomial(12, $\frac{1}{2}$) distribution under the null hypothesis.

The engineer wants a significance level α of about .006. He chooses acceptance region = $\{2, 3, 4, 5, 6, 7, 8, 9, 10\}$ and rejection region = $\{0, 1, 11, 12\}$. Using probabilities for a Binomial(12, $\frac{1}{2}$) distribution, which we listed earlier in Table 9-1, he knows that

$$P(\text{test statistic is in the acceptance region when } H_0 \text{ is true}) = .994$$
$$P(\text{test statistic is in the rejection region when } H_0 \text{ is true}) = .006$$

His decision rule is:

If $2 \leq$ test statistic ≤ 10, say the results are consistent with the null hypothesis.
If test statistic ≤ 1 or test statistic ≥ 11, say the results are inconsistent with the null hypothesis.

In the sample of 12 cords, 8 lasted more than 114 hours. The test statistic equals 8, so the engineer says the results are consistent with the null hypothesis. The p-value is P(test statistic ≤ 4 or test statistic ≥ 8 when H_0 is true) = .39. This probability is fairly large, so based on the sign test, we have no reason to say the median time to failure for the new type of electric cord is different from 114 hours. (This does not imply that the median time to failure for the new type of electric cord is exactly 114 hours. Why?) With 8 of 12 new cords lasting more than 120 hours, we might guess that the median time to failure for the new cords is at least as long as for the old cords. Why?

To see why this test of hypotheses is called the *sign test,* subtract the null hypothesis median of 114 hours from each observation in Example 10-5. Use a + sign to indicate a positive result and a − sign to indicate a negative result, as in Table 10-3. The test statistic equals the number of + signs, hence the name *sign test.*

Let's summarize the significance level approach to testing hypotheses about a population median based on a binomial distribution. Let M denote the population median.

The significance level approach to testing hypotheses about a population median M based on a binomial distribution

1. For a number M_0, we want to test the hypotheses H_0: $M = M_0$ and H_a: $M \neq M_0$.
2. The test statistic equals the number of observations greater than M_0.
3. Assume that we have a random sample from a continuous probability distribution with median M. If any observations equal M_0, ignore them for purposes of hypothesis testing. Let n_0 be the number of sample observations that are different from M_0. With our assumptions, the test statistic has a Binomial(n_0, $\frac{1}{2}$) distribution under the null hypothesis.

TABLE 10-3 For each electric cord in Example 10-5, we show the time to failure, time to failure minus 114, and the sign of this difference. The number of plus signs equals the number of observations greater than the null hypothesis median of 114 hours.

Time to failure (hours)	Time − 114	Sign of the difference
72.4	−41.6	−
78.6	−35.4	−
81.2	−32.8	−
94.0	−20.0	−
120.1	6.1	+
126.3	12.3	+
127.2	13.2	+
128.7	14.7	+
141.9	27.9	+
164.1+	50.1+	+
164.1+	50.1+	+
164.1+	50.1+	+

4. Select a value of α for the significance level.
5. The acceptance region contains numbers from $c + 1$ to $n_0 - c - 1$, where c is chosen so that P(test statistic is in the acceptance region when H_0 is true) $= P(c + 1 \leq$ test statistic $\leq n_0 - c - 1$ when H_0 is true) $\doteq 1 - \alpha$.
   The rejection region contains numbers from 0 to c as well as numbers from $n_0 - c$ to n_0, with P(test statistic is in the rejection region when H_0 is true) $= P$(test statistic $\leq c$ or test statistic $\geq n_0 - c$ when H_0 is true) $\doteq \alpha$.
6. The decision rule is:

   If $c + 1 \leq$ test statistic $\leq n_0 - c - 1$, say the results are consistent with the null hypothesis.
   If test statistic $\leq c$ or test statistic $\geq n_0 - c$, say the results are inconsistent with the null hypothesis.
7. Collect a random sample that satisfies the assumptions in step 3. Calculate the test statistic based on the sample. Use the decision rule in step 6 to decide whether the observations are consistent with the null hypothesis. Draw conclusions based on our analysis of the experimental results.

If we have the one-sided alternative H_a: $M > M_0$, then in step 5, the acceptance region contains the numbers from 0 to $c - 1$, where c is chosen so that P(test statistic $\leq c - 1$ when H_0 is true) $\doteq 1 - \alpha$. The rejection region contains numbers from c to n_0 and P(test statistic $\geq c$ when H_0 is true) $\doteq \alpha$. The decision rule in step 6 is:

If test statistic $\leq c - 1$, say the results are consistent with the null hypothesis.
If test statistic $\geq c$, say the results are inconsistent with the null hypothesis.

For the one-sided alternative H_a: $M < M_0$, the acceptance region in step 5 contains numbers from $c + 1$ to n_0, where c is chosen so that P(test statistic $\geq c + 1$ when H_0 is true) $\doteq 1 - \alpha$. The rejection region contains numbers from 0 to c, and $P(\text{test statistic} \leq c \text{ when } H_0 \text{ is true}) \doteq \alpha$. The decision rule in step 6 is:

If test statistic $\geq c + 1$, say the results are consistent with the null hypothesis.
If test statistic $\leq c$, say the results are inconsistent with the null hypothesis.

We find the number c in step 5 from the Binomial(n_0, $\frac{1}{2}$) distribution. Similarly, we calculate exact p-values from the Binomial(n_0, $\frac{1}{2}$) distribution.

For *large samples,* we can use a *large-sample approximation.* If p denotes the probability that an observation is greater than M_0, then we can write the null hypothesis as H_0: $p = \frac{1}{2}$. If the sample size is large enough, we can use the *large-sample test for a proportion* discussed in Section 10-2.

With the sign test, we ignore the actual value of an observation, noting only whether it is greater than or less than M_0. (Because of this, you might guess that the sign test is less powerful than the t test and the Wilcoxon signed rank test when the assumptions for those tests are met. You would be right.) A procedure that ignores actual values is advantageous when we do not know exact values. This was the case for three of the observations in Example 10-5. Since the engineer knew that these three values were all greater than the null hypothesis median, $M_0 = 114$ hours, he was able to use the sign test.

As another illustration, suppose we want to see whether a treatment affects the condition of patients with a chronic condition such as arthritis. We treat a number of patients. At the end of the experiment, we note whether the condition of each patient has improved, deteriorated, or remained unchanged. We assume the change in a patient's condition is a continuous-type variable whose exact value we do not know. If M denotes the median change in patient condition due to treatment, our null hypothesis is

$$H_0: \quad M = 0$$

The test statistic equals the number of patients improved after treatment. For our analysis, we exclude all observations of no change. We then apply the sign test. Of course, when we do not know exact values for our observations, we are not in a position to estimate M, if it seems to be different from the null hypothesis value.

Confidence Intervals for a Population Median Based on a Binomial Distribution

We can obtain *point and interval estimates for the population median M* when exact values for observations are known. The sample median is our point estimate of the population median M. We can calculate interval estimates for M as confidence intervals based on the binomial distribution (Lehmann, 1975, page 182).

Order the observations from smallest to largest. Denote these ordered observations by O_1, O_2, through O_n, where n is the total sample size. Confidence intervals for M are of the form (O_k, O_{n-k+1}), where k is chosen to give the desired confidence level. The confidence level and value of k are determined from the relationship

$$P(O_k < M < O_{n-k+1}) = 1 - 2P(Y \leq k - 1)$$

where Y is a random variable with a Binomial(n, $\frac{1}{2}$) distribution. If $P(O_k < M < O_{n-k+1}) = A$, then we say (O_k, O_{n-k+1}) is a 100A% confidence interval for M. As with all such intervals, the confidence level refers to what we would expect if we repeated the sampling process many times.

EXAMPLE 10-5 *(continued)*

Let's estimate the median time to failure M for the new, cheaper type of electric cord in Example 10-5. Our point estimate is the sample median:

$$\text{Sample median} = \frac{126.3 + 127.2}{2} = 126.75 \text{ hours}$$

We can also find interval estimates for M of the form (O_k, O_{12-k+1}).

Suppose we let $k = 3$. From the ordered list of times to failure in Table 10-3, we see that $O_3 = 81.2$ hours and $O_{12-3+1} = O_{10} = 164.1+$ hours. If Y has a Binomial(12, $\frac{1}{2}$) distribution, then $P(Y \leq k - 1) = P(Y \leq 2) = .019$. The confidence level associated with the interval estimate

$$(O_3, O_{10}) = (81.2, 164.1+)$$

is $1 - 2(.019) \doteq .96$. A 96% confidence interval for the median time to failure for the new type of electric cord is from 81.2 hours to some time greater than 164.1 hours.

If we let $k = 4$, then our interval estimate for M is from O_4 to $O_{12-4+1} = O_9$. In Example 10-5, $O_4 = 94.0$ hours and $O_9 = 141.9$ hours. If Y has a Binomial(12, $\frac{1}{2}$) distribution, then $P(Y \leq k - 1) = P(Y \leq 3) = .073$. The confidence level associated with the interval estimate

$$(O_4, O_9) = (94.0, 141.9)$$

is $1 - 2(.073) \doteq .85$. We say the interval from 94.0 hours to 141.9 hours is an 85% confidence interval for the median time to failure for the new type of electric cord. This confidence interval is illustrated in Figure 10-10, along with the observations. Because the sample size is not large, the confidence interval is fairly wide compared with the range of data values.

The null hypothesis median, $M_0 = 114$ hours, is in both the confidence intervals we calculated here, agreeing with what we found using the sign test.

For the sign test, we assume only that the observations are independent, with a continuous probability distribution. We ignore how far the observations are from the median specified in the null hypothesis. We consider only whether they are above or below that median. This simplicity has an advantage when we do not have exact values, but know only whether an observation is above or below the null hypothesis median.

When we know exact values, we discard a lot of information when we use the sign test. The Wilcoxon signed rank test uses the relative magnitudes,

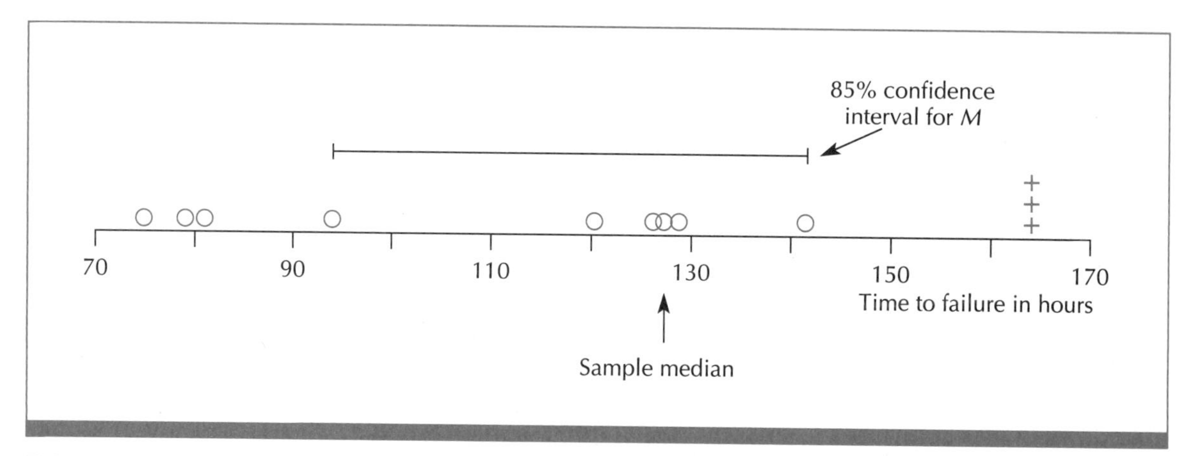

FIGURE 10-10 Dot plot of time to failure for 12 electric cords in Example 10-5. A + indicates a cord that had not failed when the experiment ended at 164.1 hours. The sample median and an 85% confidence interval for the population median M are also shown.

or ranks, of the observations. When the assumptions for it are justified, the Wilcoxon signed rank test is much more powerful than the sign test. That is, using the Wilcoxon signed rank test, we are more likely to say the results are inconsistent with the null hypothesis, when the alternative hypothesis is really true.

The t test uses the actual data values rather than signs or ranks. When the assumptions for the t test are justified, it is more powerful than either the sign test or the Wilcoxon signed rank test.

In Chapter 11, we discuss ways to compare two measures of central tendency.

Summary of Chapter 10

We consider the problem of making inferences about the mean of a population based on a single sample from that population. The procedure we choose to make these inferences depends on the sample size and on the assumptions we are willing to make about the distribution of values in the population.

If we have a random sample from a population and the sample size is large, we can base inferences about the mean of the population on the standard Gaussian distribution. A special case is when the sample size is large and we wish to make inferences about a proportion.

If we have a random sample of observations from a Gaussian distribution, we can base inferences about the population mean (or median) on a t distribution. A t distribution is defined by a parameter called its degrees of freedom. When using a single sample to make inferences about a population mean, the degrees of freedom equal the sample size minus 1.

If we have a random sample of observations from a population having a continuous, symmetric distribution, we can base inferences about the population median (or mean) on a Wilcoxon signed rank distribution. The Wilcoxon signed rank probability distribution for sample size n comes from the probability model for an experiment that randomly assigns + and − signs to ranks 1 through n.

The Wilcoxon signed rank test uses ranks rather than actual values. When the assumptions for the t test are met, the t test is more powerful than the Wilcoxon signed rank test (that is, using the t test we are more likely to say the results are inconsistent with the null hypothesis when the alternative is really true).

If we have a random sample of observations from a continuous distribution, we can base inferences about the population median on a binomial distribution. The corresponding test of hypotheses is known as the sign test. For each observation, we consider only whether it is greater than or less than the null hypothesis median. When the assumptions for the Wilcoxon signed rank test are met, it is much more powerful than the sign test (using the Wilcoxon signed rank test, we are more likely to say the results are inconsistent with the null hypothesis when the alternative is really true).

Procedures based on the t distribution are parametric because we assume the observations follow a Gaussian distribution. Inferences based on the binomial distribution and Wilcoxon signed rank distribution, as well as those based on the standard Gaussian distribution for large samples, are nonparametric because we assume no particular type of probability distribution for the observations.

A confidence interval provides an interval estimate or range of reasonable values for a population measure of central tendency, M. If the confidence level associated with a confidence interval is A, we say the interval is a $100A\%$ confidence interval for M. The interpretation of this confidence level is this: If we repeated the sampling process in the same way many times and calculated a $100A\%$ confidence interval for M each time, we would expect about $100A\%$ of these confidence intervals to contain M and about $100(1 - A)\%$ not to contain M. When we calculate a specific confidence interval, it either contains M or it does not. The confidence level refers to what we would expect if we were able to repeat the sampling process many times.

Minitab Appendix for Chapter 10

Finding Cumulative Probabilities for a *t* Distribution

If we use the CDF command with the T subcommand, we can find cumulative probabilities for a t distribution. We might use the command

```
MTB>   cdf 1.2;
SUBC>  t 5.
 1.2000  0.8581
```

Minitab informs us that if T is a random variable having the t distribution with 5 degrees of freedom, then $P(T \leq 1.2) = .8581$. We can store this cumulative probability as a constant to be used later:

```
MTB>    cdf 1.2 k1;
SUBC>   t 5.
K1      0.858055
```

Minitab prints the cumulative probability after storing it in K1 for later use. Note that Minitab does not always round printed values in the same way. The cumulative probability is printed as 0.8581 after the first command, whereas K1 is printed as 0.858055.

Finding Values of a *t* Probability Function

The PDF command with the T subcommand gives us the value of the appropriate t probability function evaluated at a specified point. For example,

```
MTB>    pdf 1.2;
SUBC>   t 5.
 1.2000  0.1777
```

Minitab tells us that for the t distribution with 5 degrees of freedom, the value of the probability function evaluated at 1.2 is 0.1777. We can save this value as a stored constant if we wish.

Simulating Random Samples from a *t* Distribution

The RANDOM command with the T subcommand generates random values from a specified t distribution. If we use

```
MTB>    random 15 c1;
SUBC>   t 9.
```

Minitab generates 15 random values from the t distribution with 9 degrees of freedom and stores them in column 1. We can use these values in other commands, such as HISTOGRAM. The RANDOM command is useful for seeing what a random sample from a t distribution might look like.

Carrying out a *t* Test

We carry out a one-sample t test in Minitab using the TTEST command. Suppose the six average weights of baby goat twins from Example 10-3 are in column 1 on our worksheet. If we use the command

```
MTB>  ttest 1592 c1
```

```
TEST OF MU = 1592.000 VS MU N.E. 1592.000

                N       MEAN     STDEV   SE MEAN        T    P VALUE
C1              6   1272.083   303.903   124.068    -2.58      0.050
```

FIGURE M10-1 Output from TTEST for Example 10-3

```
TEST OF MU = 1592.000 VS MU G.T. 1592.000

                N       MEAN     STDEV   SE MEAN        T    P VALUE
C1              6   1272.083   303.903   124.068    -2.58       0.98
```

FIGURE M10-2 Output from TTEST with ALTERNATIVE = +1 for Example 10-3

```
TEST OF MU = 1592.000 VS MU L.T. 1592.000

                N       MEAN     STDEV   SE MEAN        T    P VALUE
C1              6   1272.083   303.903   124.068    -2.58      0.025
```

FIGURE M10-3 Output from TTEST with ALTERNATIVE = −1 for Example 10-3

Minitab will calculate the test statistic and two-sided *p*-value for the null hypothesis that $\mu = 1{,}592$ versus the alternative that $\mu \neq 1{,}592$. Here, μ denotes the population mean. The output is in Figure M10-1.

If we want to test the null hypothesis that $\mu = 1{,}592$ versus the one-sided alternative that $\mu > 1{,}592$, we use the ALTERNATIVE subcommand with +1:

```
MTB>   ttest 1592 c1;
SUBC>  alternative=+1.
```

Minitab will print the one-sided *p*-value corresponding to this one-sided alternative, as shown in Figure M10-2.

For the one-sided alternative that $\mu < 1{,}592$, we use the ALTERNATIVE subcommand with −1:

```
MTB>   ttest 1592 c1;
SUBC>  alternative=-1.
```

Minitab prints the one-sided *p*-value for this one-sided alternative, as shown in Figure M10-3.

Calculating a Confidence Interval Based on a *t* Distribution

To get a confidence interval for the population mean μ based on a *t* distribution, we use the TINTERVAL command. In Example 10-3, if we use

```
MTB>  tinterval c1
```

we get the output in Figure M10-4.

```
              N       MEAN     STDEV  SE MEAN     95.0 PERCENT C.I.
C1            6    1272.08    303.90   124.07   (  953.07, 1591.10)
```

FIGURE M10-4 A 95% confidence interval for μ in Example 10-3

```
              N       MEAN     STDEV  SE MEAN     90.0 PERCENT C.I.
C1            6    1272.08    303.90   124.07   ( 1022.01, 1522.15)
```

FIGURE M10-5 A 90% confidence interval for μ in Example 10-3

```
TEST OF MEDIAN = 0.8300 VERSUS MEDIAN N.E. 0.8300

                  N FOR   WILCOXON                ESTIMATED
             N    TEST   STATISTIC   P-VALUE        MEDIAN
C2           5       5        10.5     0.500        0.8325
```

FIGURE M10-6 Wilcoxon signed rank test of the null hypothesis that $\mu = .83$ in Example 10-4

If we do not specify a confidence level, Minitab calculates a 95% confidence interval. If instead we want a 90% confidence interval, we include that in the command:

```
MTB> tinterval 90 c1
```

to get the output in Figure M10-5.

Making Inferences Based on a Wilcoxon Signed Rank Distribution

We can make inferences about a population mean (or median) based on a Wilcoxon signed rank distribution using the WTEST and WINTERVAL commands. Suppose the bomb base heights from Example 10-4 are in column 2 of our worksheet. To test the null hypothesis that the population mean μ equals .83 versus the two-sided alternative, we use the command

```
MTB> wtest 0.83 c2
```

to get the output in Figure M10-6.

Minitab calculates a two-sided *p*-value if no alternative is specified. If we want the upper-tailed alternative, $\mu > .83$, we use the subcommand ALTERNATIVE = +1. For the lower-tailed alternative, $\mu < .83$, we use the subcommand ALTERNATIVE = −1.

To get an approximate 90% confidence interval for μ based on the appropriate Wilcoxon signed rank distribution, we use

```
MTB> winterval 90 c2
```

to get the output in Figure M10-7.

```
                 ESTIMATED    ACHIEVED
            N      MEDIAN   CONFIDENCE   CONFIDENCE INTERVAL
C2          5      0.8325         89.4   (  0.8275,  0.8380)
```

FIGURE M10-7 An approximate 90% confidence interval for μ in Example 10-4, based on a Wilcoxon signed rank distribution

```
SIGN TEST OF MEDIAN = 114.0 VERSUS  N.E.  114.0

              N  BELOW  EQUAL  ABOVE   P-VALUE     MEDIAN
C1           12      4      0      8    0.3877      126.8
```

FIGURE M10-8 Sign test of the null hypothesis that the median time to failure equals 114 hours in Example 10-5

```
SIGN CONFIDENCE INTERVAL FOR MEDIAN

                               ACHIEVED
             N    MEDIAN     CONFIDENCE     CONFIDENCE INTERVAL    POSITION
C1          12     126.8       0.8540       (    94.0,   141.9)           4
                               0.9000       (    91.4,   146.3)         NLI
                               0.9614       (    81.2,   164.1)           3
```

FIGURE M10-9 Confidence intervals for the population median in Example 10-5, based on a binomial distribution

If no confidence level is specified, Minitab calculates an approximate 95% confidence interval.

Making Inferences Based on a Binomial Distribution

For inferences based on a binomial distribution (as in the sign test), we use the STEST and SINTERVAL commands. Suppose the 12 times to failure from Example 10-5 are in column 1 on our worksheet, with the three censored observations each entered as 164.1. To test the null hypothesis that the median time to failure is 114 hours, we use the command

MTB> **stest 114 c1**

to get the output in Figure M10-8.

If we specify no alternative, Minitab calculates a two-sided p-value. For one-sided tests, we use the ALTERNATIVE subcommand, as for the TTEST and WTEST commands.

We get a confidence interval for the population median using the SINTERVAL command:

MTB> **sinterval 90 c1**

We get the output in Figure M10-9.

Making Large-Sample Inferences About a Mean

For large-sample inference about a population mean, we can use the TTEST and TINTERVAL commands, because for large degrees of freedom, a t distribution is close to the standard Gaussian distribution. There is no direct way in Minitab to carry out a large-sample test about a proportion.

Exercises for Chapter 10

In each exercise, describe the population sampled (whether real or hypothetical). Graph the data in any way that seems helpful. For each statistical procedure, state the assumptions that make the analysis appropriate. Do these assumptions seem reasonable? Discuss the results of your analysis.

EXERCISE 10-1

Investigators measured plasma citrate concentrations at 8 A.M., before breakfast, for 10 volunteers (from a contribution by E. B. Jensen to a collection of problems in Andrews and Herzberg, 1985, page 237; from Andersen, Jensen, and Schou, 1981). The measurements are shown below (in μmol per liter).

93 116 125 144 105 109 89 116 151 137

a. Plot the observations.

b. Calculate a confidence interval for the mean plasma citrate concentration, using a t distribution.

c. Calculate a confidence interval for the mean plasma citrate concentration, using a Wilcoxon signed rank distribution.

d. Calculate a confidence interval for the median plasma citrate concentration, using a binomial distribution.

e. Discuss and compare your answers to parts (b), (c), and (d).

EXERCISE 10-2

As part of a patent application for a new cake mix, applicants compared two types of cake mix (Box, Hunter, and Hunter, 1978, page 160; from U. S. Patent 3,505,079, April 7, 1970). They prepared five recipes with each of two types of cake mix, the new mix and an old mix. The difference in volume (units not given) between the two mixes is shown below for each recipe:

Recipe	Difference in volume new − old
1	18
2	8
3	6
4	18
5	8

a. Plot these observations.

b. The patent claims that the new cake mix results in significantly greater volume than the old mix. Is this claim justified by these observations? State and test appropriate hypotheses.

EXERCISE 10-3

In a large set of measurements on nonpregnant women, the average fasting blood glucose level was about 80 milligrams/100 milliliters of blood. Researchers determined fasting blood sugar levels for 52 women during their third trimester of pregnancy (contributed by C. M. Mahan to a collection of problems in Andrews and Herzberg, 1985, pages 211–214; from O'Sullivan and Mahan, 1966). The results (in mg/100 ml) are shown here (sample size = 52, mean = 70.12, sample standard deviation = 9.68).

60	56	80	55	62	74	64	73	68	69	60	70
66	83	68	78	103	77	66	70	75	91	66	75
74	76	74	74	67	78	64	67	78	64	71	63
90	60	48	66	74	60	63	66	77	70	73	78
73	72	65	52								

a. Plot these measurements.

b. How do these blood sugar levels for pregnant women compare with the average level of 80 mg/100 ml for nonpregnant women? State and test appropriate hypotheses.

c. Calculate an interval estimate for the mean fasting blood sugar level of women in their third trimester of pregnancy.

EXERCISE 10-4

Does a heart defect known as patent foramen ovale contribute to the bends (decompression sickness) experienced by some scuba divers? Researchers used echocardiography to examine 30 divers with a history of decompression sickness (*Science News,* March 25, 1989, volume 135, page 188). Eleven of the 30 divers showed evidence of the heart defect. About 5% of the general population has this heart defect.

a. Does the proportion of divers with a history of decompression sickness who have this heart defect seem to differ from that of the general population? State and test appropriate hypotheses.

b. Calculate a confidence interval for the proportion of divers with a history of decompression sickness who have this heart defect. Discuss your findings.

EXERCISE 10-5

An engineer subjected 35 motors to a life test. The engineer set up a machine for recording time to failure, 16 hours after the beginning of the life test. Four motors failed before the recording machine was set up at 16 hours. Ten motors were still working at 30 hours, when the engineer ended the test. The times to failure (in hours) for the other 21 motors are shown below (Shapiro, 1986, page 22; from Brain and Shapiro, 1983).

16.0	16.3	16.7	16.9	17.0	17.1	17.3	17.8	17.9
18.3	18.4	18.6	19.1	19.5	20.6	21.4	22.9	23.0
24.6	25.9	28.6						

a. Plot these times to failure.

b. Test the null hypothesis that the median time to failure is 20 hours versus the alternative that it is not 20 hours.

c. Calculate a confidence interval for the median time to failure for this type of motor.

EXERCISE 10-6

Some health care workers estimate that one-third of people with coronary artery disease may be clinically depressed. In one study, researchers found 9 cases of major depression among 52 patients with newly diagnosed coronary disease (*Science News,* January 7, 1989, volume 135, page 13).

a. Does the health care workers' estimate seem reasonable based on this sample? State and test appropriate hypotheses.

b. Calculate a confidence interval for the proportion of patients with newly diagnosed coronary artery disease suffering from major depression. Discuss your findings.

EXERCISE 10-7

Scientists developed a new method of determining serum iron concentrations. To check the accuracy of the method, they made 20 analyses of control sera, with a concentration of 105 μg serum iron per 100 milliliters (Hollander and Wolfe, 1973, pages 85–86; a portion of the data of Jung and Parekh, 1970). The determinations of serum iron concentration (μg/100 ml) are shown below.

96	98	99	100	103	103	104	104	105	105
106	106	107	108	108	108	110	113	114	114

a. Plot these observations.

b. Test the null hypothesis that the average serum iron determination using the new method is 105 μg/100 ml.

c. Calculate a confidence interval for the average serum iron determination using the new method. Discuss your findings.

EXERCISE 10-8

A major problem in liver transplantation is the short time (at most 10 hours) that donor livers can be preserved. In a large study, researchers treated 185 donor livers with an experimental solution and 180 livers with the traditional solution for organ preservation. Eighty-one of the livers treated with the experimental solution lasted longer than 9.5 hours, while none of the livers treated with the traditional solution lasted that long (numbers calculated from percentages reported in *Science News,* February 4, 1989, volume 135, page 70).

a. Calculate a confidence interval for the proportion of livers treated with the new solution that are still viable after 9.5 hours.

b. Discuss the experimental results.

EXERCISE 10-9

In a tomato processing factory, the drained weight after filling cans of tomatoes in puree averages 21.8 ounces during the morning. One afternoon, a quality control worker selects five cans of tomatoes filled that afternoon and weighs the drained contents in ounces (based on Duncan, 1974, page 569; Grant and Leavenworth, 1972, page 41):

19.0	19.5	19.5	20.5	21.5

a. Plot the observations.

b. Use a t distribution to test the null hypothesis that the average drained weight in cans filled that afternoon equaled 21.8 ounces. Calculate a confidence interval for the average afternoon drained weight.

c. Repeat part (b) using a Wilcoxon signed rank distribution.

d. Use a binomial distribution to test the null hypothesis that the median drained weight in cans filled that afternoon equaled 21.8 ounces. Calculate a confidence interval for the median afternoon drained weight.

e. Compare your results in parts (b), (c), and (d). Discuss your findings.

EXERCISE 10-10

Researchers have worked to identify serum markers for the genetically transmitted disease Duchenne muscular dystrophy. The average serum level of the enzyme creatine kinase in a large group of women who were not carriers of the disease was 39.8 (units not given). Measurements of the serum level of creatine kinase are shown below for 38 women who were genetic carriers of the disease (data contributed by M. Percy of Mount Sinai Hospital in Toronto to a collection of problems in Andrews and Herzberg, 1985, pages 223–228). (Sample size = 38, mean = 175.9, sample standard deviation = 192.8.)

167	104	30	65	440	58	129	265	285	124
53	657	168	286	73	19	113	57	78	69
48	109	925	59	363	37	101	99	560	85
197	154	80	28	57	326	100	115		

a. Plot these observations.

b. Test the null hypothesis that the mean serum creatine kinase level among female carriers equals 39.8.

c. Calculate a confidence interval for the mean serum creatine kinase level among female carriers.

d. Discuss your findings.

EXERCISE 10-11

Refer to the exercise experiment on pregnant Pygmy goats in Example 10-3.

a. Use a Wilcoxon signed rank distribution to test the null hypothesis that the mean weight of twins born to exercised goats equals 1,592.0 grams. Calculate a confidence interval for the mean weight of twins born to exercised goats.

b. Use a binomial distribution to test the null hypothesis that the median weight of twins born to exercised goats equals 1,592.0 grams. Calculate a confidence interval for the median weight of twins born to exercised goats.

c. Compare your results in parts (a) and (b) with our results using a t distribution in Example 10-3.

EXERCISE 10-12 Refer to the quality control problem involving heights of bomb bases in Example 10-4.

a. Use a t distribution to test the null hypothesis that the mean height of bomb bases in the production lot equaled .830 inch. Calculate a confidence interval for the mean height of bomb bases in the production lot.

b. Use a binomial distribution to test the null hypothesis that the median height of bomb bases in the production lot equaled .830 inch. Calculate a confidence interval for the median height of bomb bases in the production lot.

c. Compare your results in parts (a) and (b) with our results using a Wilcoxon signed rank distribution in Example 10-4.

EXERCISE 10-13 In Example 10-5, we asked if the median life of a new type of electric cord equaled 114 hours. Use the large-sample version of the sign test to test this null hypothesis. Compare your results with our results using a binomial distribution in Example 10-5.

EXERCISE 10-14 In the appendix on the Wilcoxon signed rank distributions at the end of the book, we find the Wilcoxon signed rank distribution of $T+$ for sample size 3. Show that $T-$ has the same probability distribution.

EXERCISE 10-15 Find the Wilcoxon signed rank distribution for a sample of size 4.

EXERCISE 10-16 Twelve pairs of siblings were involved in a study to investigate whether children can recognize siblings by the sense of smell (Porter and Moore, 1981). In each pair, the children were full siblings living together with their parents. The younger child was from 36 to 49 months of age and the older child was from 62 to 95 months of age.

Each child was given one of 24 identical new T-shirts. Parents were asked to have each child wear his or her T-shirt to bed three nights in a row. T-shirts were stored during the day in individual sealed plastic bags.

On the morning following the third night of the experiment, each child sniffed each of two T-shirts: one worn during the experiment by his or her sibling and the other worn by another child of about the same age as the sibling. The child was asked to identify the T-shirt worn by his or her sibling. Nineteen of the 24 children made correct selections.

a. Why were T-shirts stored during the day in individual sealed plastic bags?

b. Test the null hypothesis that the performance of the children was no different from what we would expect from chance guessing.

c. What does this experiment suggest about a child's ability to recognize a sibling by the sense of smell?

EXERCISE 10-17 Ten pairs of siblings and 18 of the 20 parents participated in a study to see if parents could distinguish between their two children by sense of smell. (This was a part of the experiment discussed in Exercise 10-16.) Each child wore a

T-shirt for three consecutive nights. Each parent was then asked to sniff each of the T-shirts worn by his or her two children and identify the T-shirt worn by each child. Sixteen of the 18 parents correctly distinguished between the T-shirts worn by their two children (Porter and Moore, 1981).

a. State and test appropriate hypotheses.

b. What does this experiment suggest about a parent's ability to distinguish between his or her children by the sense of smell?

EXERCISE 10-18

Can a growth hormone gene be transferred from one type of fish to another? If so, faster growing fish might be developed, shortening the time for fish farmers to raise full-grown fish. To investigate this idea, experimenters injected a growth hormone gene from rainbow trout into thousands of carp eggs. Of 400 fish that grew from those eggs, 20 incorporated the gene into their DNA (*Science News,* June 11, 1988, volume 133, page 374). Calculate a 95% confidence interval for the proportion of carp that would incorporate this growth hormone gene under the same experimental conditions.

EXERCISE 10-19

The conventional treatment for patients with severe ulcerative colitis, an inflammatory disease of the colon, is surgical removal of the colon. In one study, 11 patients who were candidates for surgery chose to be treated with the experimental drug cyclosporin (known for its ability to prevent organ rejection in organ transplant surgery). At the end of six months, 5 of the 11 patients were in complete remission. A sixth patient had responded to treatment, but still needed drug therapy to control colitis symptoms (*Science News,* May 20, 1989, volume 135, page 310). Calculate a 90% confidence interval for the proportion of patients with severe ulcerative colitis who would respond to drug therapy.

EXERCISE 10-20

Discuss the sampling situations in which the t test, the Wilcoxon signed rank test, and the sign test are appropriate. Which test is preferred in each of these situations?

CHAPTER 11

Inferences About Two Measures of Central Tendency

IN THIS CHAPTER

Two-sample comparisons

Two-sample comparisons based on large samples

Two-sample *t* test

Wilcoxon–Mann–Whitney two-sample test

Median test

Paired-sample comparisons

Paired *t* test

Is the weight of tomatoes canned at a factory the same in the morning and the afternoon? Does a chemical treatment retard tumor growth in mice? Can you squeeze more juice from an orange that has been microwaved for 20 seconds or from an orange that has not been microwaved? Will you run a race faster competing against yourself or someone else?

Each of these questions involves comparison of two groups or experimental conditions: morning versus afternoon, chemical treatment versus no treatment, microwave versus no microwave, competition with self versus competition with another. In this chapter we will be concerned with such comparisons of two measures of central tendency.

As with all statistical inference, the method of analysis we select depends on the assumptions we make about the sampling process. Before considering other assumptions, we first distinguish between *two-sample comparisons* and *paired-sample comparisons:*

> We make a **two-sample comparison** when we compare the means of two independent samples.
>
> If we compare the means of paired samples, we are making a **paired-sample comparison.**

We will look at paired-sample comparisons in Section 11-6. For analysis, we take the differences of the observations within pairs. We then make inferences about the center of these differences using the one-sample methods of Chapter 10.

In Sections 11-1 through 11-5, we discuss two-sample comparisons. When the sample sizes are large, we can base our inferences on the standard Gaussian distribution, as we will see in Section 11-1. Large-sample inference about two proportions is a special case, covered in Section 11-2.

For small samples, we first consider the classical approach, in Section 11-3. This requires the most assumptions about the sample, that we have two independent random samples from Gaussian distributions with equal variances. We compare the means of the two distributions using the *two-sample t test* and obtain interval estimates for the difference between two means based on a *t distribution.*

If we have two independent random samples of continuous-type observations from populations with the same shape and variation, then we can base our analyses on the *ranked observations.* The resulting nonparametric inferences are based on a *Wilcoxon–Mann–Whitney distribution,* covered in Section 11-4.

We might assume only that we have two independent random samples of continuous-type observations. If we consider only whether each observation is above or below the median of the combined observations, we are led to the *median test.* This is a nonparametric method of analysis based on a *hypergeometric distribution,* discussed in Section 11-5.

We begin in Section 11-1 with *two-sample comparisons for large samples.* We assume that we have two independent random samples, one from each of two populations of interest. If the sample sizes are large enough, we can base tests of hypotheses and interval estimates on the *standard Gaussian distribution.*

11-1 Inferences About Two Means When Sample Sizes Are Large

Before we discuss large-sample comparisons of two means in general, let's consider an example.

EXAMPLE 11-1

Wire is wound on plastic spools to make coils for electric motors. When current passes through the wire, the spools heat up. An engineer wants to compare the resulting temperature rise for spools made from two types of plastic. (This example is based on data reported in Nelson, 1986, page 11.) Why do you think the engineer is interested in temperature rise on these plastic spools used in electric motors? Which do you think is preferable: a greater or smaller temperature rise?

The engineer selects a random sample of 30 spools from a large production lot of spools made with an old type of plastic. He selects a separate independent random sample of 30 spools from a large production lot of spools made with a new type of plastic. For each spool, he records the temperature rise after current passes through wire wound around the spool. The results are displayed in stem-and-leaf plots in Figure 11-1.

FIGURE 11-1 Stem-and-leaf plot of temperature rise (in degrees Centigrade) for 30 spools made with an old type of plastic and for 30 spools made with a new type of plastic. The stem shows temperature to the nearest degree. The leaf shows temperature to the nearest tenth of a degree.

New plastic		Old plastic	
Stem	Leaf	Stem	Leaf
44		44	
44		44	7 7
45		45	0 1 3 3 4
45	6 7 9	45	7 8 8 9 9 9
46	0 2 4 4	46	0 0 1 2 2 3 4
46	6 6 7	46	5 5 5 7 7
47	0 0 1 2 2 3 4 4	47	0 2 4
47	6 6 6 7 7 8 9 9	47	
48	0 1 2	48	1
48		48	
49	1	49	
49		49	
50		50	
50		50	6

Note that the scales for the two plots are aligned to make visual comparisons easier. We see there is one large temperature rise for the old plastic. Except for that one extreme value, the spread or variation in observed values is about the same for the two types of plastic. (The interquartile range for both plastics is about 1°C.) In general, however, the distribution of temperature rises for the new plastic is concentrated around higher values than is the distribution for the old plastic. The peak of the distribution for the new plastic is somewhere between 47 and 48 degrees Centigrade, while the peak of the distribution for the old plastic is around 46 degrees Centigrade.

As part of a formal analysis, we would like to test the null hypothesis that the average temperature rise is the same for the two types of plastic. We would also like to estimate the difference in average temperature rise for the two types of plastic. For these inferences, we will use large-sample techniques based on the standard Gaussian distribution. We will outline the approach below and then apply it to this example.

Two-Sample Comparisons of Means Based on Large Samples

Suppose we have two independent random samples, one from each of two populations. Suppose, in addition, that the sample sizes are large. We want to compare the two population means, μ_1 and μ_2.

Let $\bar{X}$ denote the sample mean and SE_1 the standard error of the mean, for the first sample. Let $\bar{Y}$ denote the sample mean and SE_2 the standard error of the mean, for the second sample. A reasonable point estimate for the difference $\mu_1 - \mu_2$ between the two population means is $\bar{X} - \bar{Y}$, the difference between the two sample means. The standard error, or estimated standard deviation, of $\bar{X} - \bar{Y}$ is

$$SE_{\bar{X}-\bar{Y}} = \sqrt{(SE_1)^2 + (SE_2)^2}$$

If the two sample sizes n_1 and n_2 are large enough, the quantity

$$\frac{(\bar{X} - \bar{Y}) - (\mu_1 - \mu_2)}{SE_{\bar{X}-\bar{Y}}}$$

has approximately the standard Gaussian distribution. This result, related to the Central Limit Theorem results in Section 8-3, forms the basis of large-sample inference about two population means. The significance level approach to testing hypotheses about the two population means based on large samples is outlined below.

The significance level approach to testing hypotheses about two population means μ_1 and μ_2 based on large samples

1. The hypotheses are H_0: $\mu_1 = \mu_2$ and H_a: $\mu_1 \neq \mu_2$.
2. The test statistic is

$$\frac{\bar{X} - \bar{Y}}{SE_{\bar{X}-\bar{Y}}}$$

3. Assume that we have two independent random samples, one from a population with mean μ_1 and the other from a population with mean μ_2. Also assume that the two sample sizes are large enough to ensure that the sample means $\bar{X}$ and $\bar{Y}$ have approximate Gaussian distributions. Then under the null hypothesis, the test statistic has approximately the standard Gaussian distribution.
4. Select significance level α.
5. The acceptance region is the interval $(-c, c)$. The rejection region includes the intervals $(-\infty, -c]$ and $[c, \infty)$. The number c is chosen so that $P(Z \leq c) = 1 - \alpha/2$, where Z has the standard Gaussian distribution.
6. The decision rule is:

   If $-c <$ test statistic $< c$, say the results are consistent with the null hypothesis.

   If test statistic $\leq -c$ or test statistic $\geq c$, say the results are inconsistent with the null hypothesis.
7. Collect two large independent random samples, one from each of the two populations of interest. Calculate the test statistic based on the sample. Use the decision rule in step 6 to decide whether the observations are consistent with the null hypothesis. Draw conclusions based on analysis of the experimental results.

If we have a one-sided alternative, then in step 5 we select the number c so that $P(Z \leq c) = 1 - \alpha$, where Z has the standard Gaussian distribution. If the one-sided alternative is $H_a\colon \mu_1 > \mu_2$, then the acceptance region is the interval $(-\infty, c)$; the rejection region is the interval $[c, \infty)$. In step 6 we say values of the test statistic less than c are consistent with the null hypothesis, values greater than or equal to c are inconsistent with the null hypothesis.

If our one-sided alternative is $H_a\colon \mu_1 < \mu_2$, then the acceptance region is the interval $(-c, \infty)$; the rejection region is the interval $(-\infty, -c]$. In step 6 we say values of the test statistic greater than $-c$ are consistent with the null hypothesis, values less than or equal to $-c$ are inconsistent with the null hypothesis.

Large-Sample Confidence Intervals for the Difference Between Two Population Means

Large-sample confidence intervals for $\mu_1 - \mu_2$ are of the form

$$\bar{X} - \bar{Y} \pm c\mathrm{SE}_{\bar{X}-\bar{Y}}$$

We find the number c from the standard Gaussian distribution. If the area from $-c$ to c under the standard Gaussian curve equals A, then we say the interval is an approximate $100A\%$ confidence interval for $\mu_1 - \mu_2$. For instance, if $c = 2.58$, then we have an approximate 99% confidence interval for $\mu_1 - \mu_2$.

EXAMPLE 11-1 *(continued)*

To apply large-sample inference in Example 11-1, the engineer calculates the following statistics from his sample:

New plastic:	$\bar{X} = 47.163$ °C	$s_1 = .824$ °C	$SE_1 = .150$ °C
Old plastic:	$\bar{Y} = 46.230$ °C	$s_2 = 1.132$ °C	$SE_2 = .207$ °C
		$SE_{\bar{X}-\bar{Y}} = .256$ °C	

He wants to know if the average temperature rise in spools is the same for the two types of plastic. So he compares the hypotheses H_0: $\mu_1 = \mu_2$ and H_a: $\mu_1 \neq \mu_2$, where μ_1 and μ_2 represent the mean temperature rise in spools made of the new and old types of plastic, respectively.

The engineer chooses significance level $\alpha = .01$. If Z has the standard Gaussian distribution, then $P(Z \leq 2.58) = .9951$, which is close to $1 - \alpha/2 = .995$. Therefore, he uses the decision rule:

If $-2.58 <$ test statistic < 2.58, say the results are consistent with the null hypothesis.
If test statistic ≤ -2.58 or test statistic ≥ 2.58, say the results are inconsistent with the null hypothesis.

He calculates the test statistic:

$$\frac{\bar{X} - \bar{Y}}{SE_{\bar{X}-\bar{Y}}} = \frac{47.163 - 46.230}{.256} = 3.6$$

Since 3.6 is in the rejection region, he concludes the results are inconsistent with the null hypothesis that mean temperature rise is the same for the two types of plastic.

The engineer notes that his approximate p-value is less than .0004. If the mean temperature rise were really the same for both types of plastic, there would be less than 4 chances in 10,000 of seeing a test statistic at least as extreme as the one observed. The experimental results strongly suggest that mean temperature rise is not the same for the two plastics.

The engineer then decides to estimate the difference $\mu_1 - \mu_2$ in mean temperature rise for the two types of plastic. His point estimate is $\bar{X} - \bar{Y} = 47.163 - 46.230 \doteq .9$ °C. For an interval estimate, he calculates an approximate 99% confidence interval:

$$(47.163 - 46.230 - 2.58 \times .256, 47.163 - 46.230 + 2.58 \times .256) \doteq (.3, 1.6)$$

He estimates that the mean temperature rise for the new type of plastic is from .3 °C to 1.6 °C greater than for the old type of plastic. The null hypothesis value of $\mu_1 - \mu_2$ is 0, not in the confidence interval. This agrees with the test of hypotheses.

From his graphical analysis of the data, the engineer knows that he has similar variation in values of temperature rise for the two types of plastic. The distribution of temperature rises for the new plastic is shifted toward larger values than the distribution for the old plastic. However, the largest temperature rise, which might be considered an outlier because it is so far from the others, was observed for the old plastic. From his formal analysis, the engineer concludes that the difference in average temperature rise is statistically significant, with a p-value less than .0004. Based on a 99% confidence interval for the difference between the two means, he estimates that the mean temperature rise for the new plastic is from .3 °C to 1.6 °C greater than for the old plastic. To evaluate the *practical* significance of these results, the engineer must consider the relative cost of the two types of plastic, the relative lifetimes of spools made from these two plastics, and whether a difference in temperature rise of about 1 °C is of practical concern. Can you think of other issues that should be included in his comparison of the two plastics?

We can use the techniques for large-sample inferences about two population means to compare two proportions when sample sizes are large. We discuss large-sample comparisons of two proportions in Section 11-2.

11-2 Large-Sample Inference About Two Proportions

Suppose we have two independent random samples, one from each of two populations. Each observation has two possible values, say success or failure. We want to compare the proportion of successes in the two populations. Equivalently, we want to compare the probability of success for the two populations.

We can write our null hypothesis as $H_0: p_1 = p_2$, where p_1 represents the proportion of successes for the first population and p_2 the proportion of successes for the second population. If the sample sizes are large, the test statistic is

$$\frac{\hat{p}_1 - \hat{p}_2}{\sqrt{\hat{p}(1 - \hat{p})\left(\frac{1}{n_1} + \frac{1}{n_2}\right)}}$$

where n_1 and n_2 are the two sample sizes, $\hat{p}_1$ and $\hat{p}_2$ are the observed proportions of successes in the two samples, and $\hat{p}$ is the observed proportion of successes in the combined samples. For large sample sizes, this test statistic has approximately the standard Gaussian distribution under the null hypothesis. We test hypotheses using the large-sample techniques discussed in Section 11-1.

Confidence intervals for the difference $p_1 - p_2$ between the two proportions are of the form

$$\hat{p}_1 - \hat{p}_2 \pm c\sqrt{\frac{\hat{p}_1(1 - \hat{p}_1)}{n_1} + \frac{\hat{p}_2(1 - \hat{p}_2)}{n_2}}$$

Here, c comes from the standard Gaussian distribution to give the desired confidence level.

Let's illustrate large-sample inference about two proportions with an example.

EXAMPLE 11-2

Will a vaccine prevent cases of a dread disease? American public health workers in 1954 asked this question about the Salk vaccine for the prevention of polio. Polio is a disease with effects ranging from temporary, mild, flu-like symptoms to permanent disability or death. Parents in the early 1950s lived in fear of polio, often keeping their children away from playgrounds and other public places where polio might be spread. (This dread is perhaps comparable to the fear of AIDS parents today might feel if their child needed a blood transfusion.)

Medical workers must test potential new vaccines for safety and efficacy. A vaccine might not work, or might have unacceptable side effects. A live vaccine, such as the first Salk vaccine, might even cause some cases of the disease it was developed to prevent. To evaluate the effectiveness of the Salk vaccine, American public health workers conducted the largest medical experiment ever. Local health departments chose between two different experimental designs. We will discuss the design that is best from a statistical point of view: the double-blind randomized placebo-controlled experiment. Exercise 11-32 asks you to compare this experimental design with another design, selected by a number of communities.

In one treatment group, children received an injection with the Salk vaccine. In the other treatment group, children received an injection with a biologically inactive (placebo) solution. Public health workers planned to compare the proportions of polio cases for the two groups at the end of the study period. Let p_1 denote the probability of polio in the vaccinated group and p_2 the probability of polio in the placebo control group. Health workers wanted to test the hypotheses

$$H_0:\ p_1 = p_2 \qquad \text{and} \qquad H_a:\ p_1 \neq p_2$$

A two-sided alternative is reasonable here. The vaccine might prevent polio. But it was also possible the vaccine might cause some cases of polio. The two-sided alternative allows for both possibilities.

Health workers asked parents of schoolchildren for their consent to allow their children to participate in the study. They then used a random process to divide the children with parental consent into two groups, a treatment group and a placebo control group. The purpose of the random assignment of children was to balance the groups with respect to extraneous factors that might affect the outcome of the study. Extraneous factors affecting risk of polio included socioeconomic status, age, and geographic location.

The researchers conducted the experiment as a double-blind study. Neither the children, their parents, nor the public health workers treating and examining the children knew how the children had been treated. If parents and children knew the treatment, they might alter their behavior accordingly (as in avoiding or not avoiding swimming pools or playgrounds where polio might be spread). If physicians examining the children knew the treatment, they might be subtly influenced in their diagnosis, since mild cases of polio were sometimes difficult to distinguish from other illnesses such as colds or flu. Use of the double-blind design avoided the influence of such extraneous factors on the experimental outcome.

Parents of about 400,000 schoolchildren gave consent for their children to participate in this experiment. (Parents of about 350,000 children did not give consent.) Results of the study are shown below (Meier, 1989; Francis et al., 1955):

Group	Number of children	Number of polio cases	Proportion of polio cases
Vaccine	200,745	57	$\hat{p}_1 = \dfrac{57}{200,745} \doteq .000284$
Placebo	201,229	142	$\hat{p}_2 = \dfrac{142}{201,229} \doteq .000706$
Total	401,974	199	$\hat{p} = \dfrac{199}{401,974} \doteq .000495$

To test the hypothesis that the probability of polio was the same for vaccinated and unvaccinated children, we use the test statistic

$$\frac{\hat{p}_1 - \hat{p}_2}{\sqrt{\hat{p}(1-\hat{p})\left(\dfrac{1}{n_1} + \dfrac{1}{n_2}\right)}}$$

$$= \frac{.000284 - .000706}{\sqrt{(.000495)(.999505)\left(\dfrac{1}{200,745} + \dfrac{1}{201,229}\right)}} = -6.0$$

Comparing this value of the test statistic with the standard Gaussian distribution, we see that our approximate p-value is less than .0004. This tells us that if the null hypothesis were true and the probability of polio were the same in both groups, there would be less than 4 chances in 10,000 of seeing a test statistic at least this far from 0. The experimental results are inconsistent with the null hypothesis, strongly suggesting that polio incidence is different for vaccinated and unvaccinated children.

A 95% confidence interval for the difference $p_1 - p_2$ between the two probabilities is

$$.000284 - .000706 \pm 1.96\sqrt{\frac{(.000284)(.999716)}{200,745} + \frac{(.000706)(.999294)}{201,229}}$$

$$\doteq (-.00056, -.00028)$$

It is common in public health to report a proportion as the number of cases of disease per 100,000 people. With this in mind, $\hat{p}_1 \doteq .00028$ tells us there were about 28 polio cases per 100,000 children in the vaccine group. Similarly, $\hat{p}_2 \doteq .00071$ tells us there were about 71 polio cases per 100,000 children in the placebo control group. We can interpret the difference $\hat{p}_1 - \hat{p}_2 \doteq -.00042$ this way: We estimate that use of the polio vaccine under the same conditions would result in about 42 fewer cases per 100,000 children than use of the placebo. Our interval estimate is 28 to 56 fewer cases per 100,000 children, using the polio vaccine.

The results of this experiment strongly suggested that the Salk vaccine prevented polio. Wide use of the Salk vaccine after this study indicated that in addition to preventing polio, the vaccine could cause some cases of polio. Researchers subsequently developed better polio vaccines, which are now in use.

In Section 11-3, we discuss the classical approach to comparing two means when the sample sizes are small.

11-3

Inferences About Two Measures of Central Tendency Based on a *t* Distribution

The classical test of hypotheses about two means is called the *two-sample t test*. Let's first consider an example. Then we will describe the classical approach to two-sample comparisons and apply it to this example.

EXAMPLE 11-3

Is the weight of tomatoes canned at a factory the same in the morning and the afternoon? Investigators drained and weighed 10 cans of tomatoes one day. Five cans had been filled in the morning and five in the afternoon. The drained weights in ounces of these ten cans are shown below (Duncan, 1974, page 569; from Grant and Leavenworth, 1972, page 41):

Morning:	22.5	24.5	25.5	20.0	21.0
Afternoon:	22.5	19.5	21.5	20.5	20.0

We will use this information to compare the average canned weights for the two times of day.

Suppose you were the manager of this factory. Why would you be interested in comparing morning and afternoon performance at your factory? Here we are considering average canned weights. What other measures of quality would you be interested in evaluating?

A plot of the observations is shown in Figure 11-2. Three morning weights overlap with the range of afternoon weights; the other two morning weights are at least 2 ounces greater than the largest afternoon weight. Just

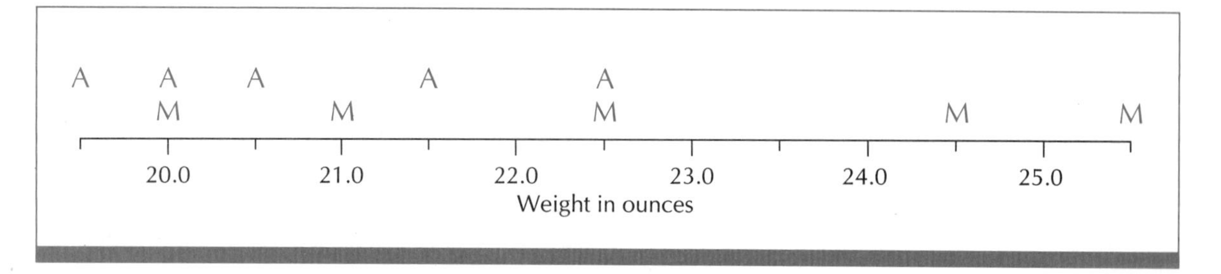

FIGURE 11-2 Plot of drained weights of canned tomatoes in Example 11-3. M represents the weight of a can filled in the morning; A represents afternoon.

from the plot we can see that the average drained weight in the morning sample is larger than the average drained weight in the afternoon sample. The spread or variation in the values also appears somewhat larger in the morning than in the afternoon.

For a formal analysis, we would like to test the null hypothesis that mean drained weight is the same for tomatoes canned in the morning and those canned in the afternoon. We would also like to estimate the difference in mean drained weights for the two times of day.

We will use the two-sample *t* test to decide whether the apparent difference between the morning and afternoon means is statistically significant. We will also use a *t* distribution to calculate an interval estimate for the difference between the morning and afternoon means. Before analyzing the experimental results, let's describe the classical approach to two-sample comparisons in general.

Two-Sample Comparisons of Means Based on a *t* Distribution

Suppose we have two independent random samples, one from each of two populations. Assume that the distribution of values in the two populations is Gaussian and that the two distributions have the same variance, σ^2. Let μ_1 denote the mean of the first population and μ_2 the mean of the second population. We want to test the null hypothesis that μ_1 and μ_2 are equal. We also want to estimate the difference between the means, $\mu_1 - \mu_2$.

The situation is illustrated in Figure 11-3. The two Gaussian distributions have the same variance, but possibly different means, so one distribution is shifted away from the other. The difference $\mu_1 - \mu_2$ between the two means measures the extent of the shift. We want to test whether $\mu_1 - \mu_2 = 0$. If so, the two distributions are identical.

Let $\bar{X}$ denote the mean and s_1 the sample standard deviation of the n_1 observations in the first sample. Similarly, let $\bar{Y}$ and s_2 denote the mean and sample standard deviation of the n_2 observations in the second sample.

The sample variances s_1^2 and s_2^2 are independent estimates (because the samples are independent) of the common variance σ^2 of the two populations.

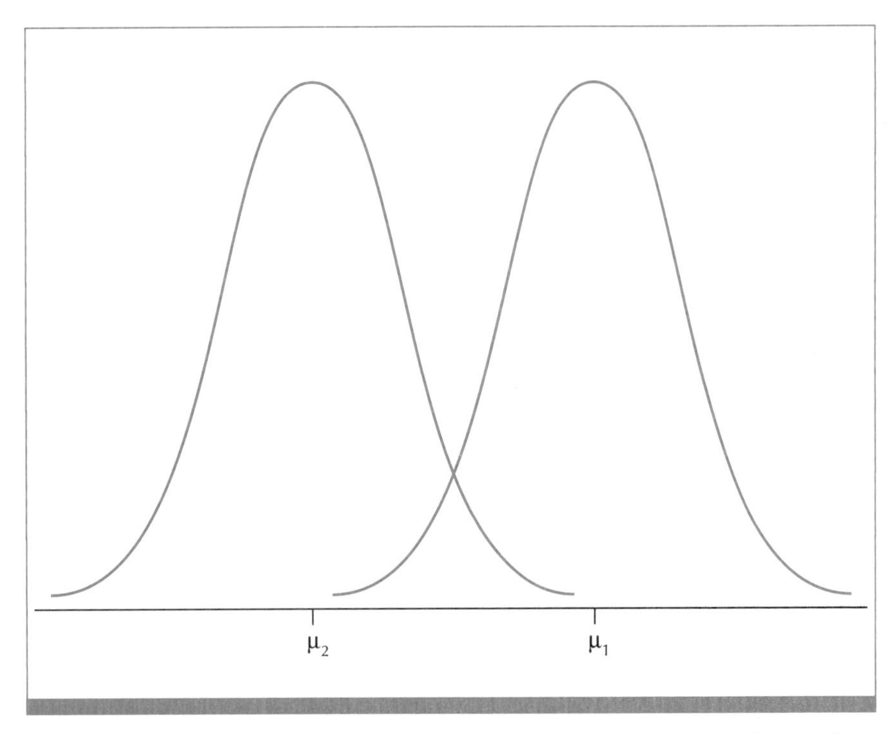

FIGURE 11-3 Probability functions are illustrated for two Gaussian distributions having the same variance.

We get a *pooled* estimate s_p^2 of σ^2 by calculating a weighted average of s_1^2 and s_2^2, with degrees of freedom as weights:

$$s_p^2 = \frac{(n_1 - 1)s_1^2 + (n_2 - 1)s_2^2}{n_1 + n_2 - 2}$$

The standard error $SE_{\bar{X}-\bar{Y}}$ estimates the standard deviation of the difference $\bar{X} - \bar{Y}$ between the two sample means:

$$SE_{\bar{X}-\bar{Y}} = \sqrt{\frac{s_p^2}{n_1} + \frac{s_p^2}{n_2}}$$

Under our model assumptions, the quantity

$$\frac{(\bar{X} - \bar{Y}) - (\mu_1 - \mu_2)}{SE_{\bar{X}-\bar{Y}}}$$

has the t distribution with $n_1 + n_2 - 2$ degrees of freedom. The number $n_1 + n_2 - 2$ defines the t distribution for the quantity of interest here in the two-sample case. This number is the denominator in the definition of the pooled variance estimator s_p^2; it is the sum of the degrees of freedom for the separate sample variances s_1^2 and s_2^2.

Suppose X_1, X_2, through X_{n_1} are independent observations from a Gaussian distribution with mean μ_1 and variance σ^2. Suppose Y_1, Y_2, through Y_{n_2} are independent observations from a Gaussian distribution with mean μ_2 and variance σ^2. The two samples are independent.

Let $\bar{X}$ denote the sample mean of the first sample and $\bar{Y}$ the sample mean of the second sample. Let $SE_{\bar{X}-\bar{Y}}$ denote the estimated standard deviation of $\bar{X} - \bar{Y}$ as defined above. Then the quantity

$$\frac{(\bar{X} - \bar{Y}) - (\mu_1 - \mu_2)}{SE_{\bar{X}-\bar{Y}}}$$

has the t distribution with $n_1 + n_2 - 2$ degrees of freedom.

Now we can outline the significance level approach to testing whether two population means are equal.

The significance level approach to testing for equality of two population means μ_1 and μ_2 based on a t distribution

1. The null and alternative hypotheses are $H_0: \mu_1 = \mu_2$ and $H_a: \mu_1 \neq \mu_2$.
2. The test statistic is

$$\frac{\bar{X} - \bar{Y}}{SE_{\bar{X}-\bar{Y}}}$$

   where $SE_{\bar{X}-\bar{Y}}$ is defined above.
3. Assume that we have two independent random samples from Gaussian distributions with means μ_1 and μ_2. The two distributions have the same variance. Then under the null hypothesis, the test statistic has the t distribution with $n_1 + n_2 - 2$ degrees of freedom.
4. Specify the significance level α.
5. Find the number c in Table C such that $P(T \leq c) = 1 - \alpha/2$, where T is a random variable having the t distribution with $n_1 + n_2 - 2$ degrees of freedom. The acceptance region is the interval $(-c, c)$. The rejection region includes the intervals $(-\infty, -c]$ and $[c, \infty)$.
6. The decision rule is:

   If $-c <$ test statistic $< c$, say the results are consistent with the null hypothesis.

   If test statistic $\leq -c$ or test statistic $\geq c$, say the results are inconsistent with the null hypothesis.
7. Collect a random sample satisfying the given assumptions. Calculate the test statistic. Use the decision rule in step 6 to decide whether the observations are consistent with the null hypothesis. Draw conclusions based on the analysis.

If in step 1 we specify a one-sided alternative, then we must alter steps 5 and 6. In step 5, we find the number c in Table C such that $P(T \leq c) = 1 - \alpha$. If our alternative is $H_a: \mu_1 < \mu_2$, then the acceptance region is the interval $(-c, \infty)$; the rejection region is the interval $(-\infty, -c]$. In step 6, we say values of the test statistic greater than $-c$ are consistent with the null hypothesis; values less than or equal to $-c$ are inconsistent with the null hypothesis.

If our one-sided alternative is $H_a: \mu_1 > \mu_2$, then the acceptance region is the interval $(-\infty, c)$; the rejection region is the interval $[c, \infty)$. In step 6, we say values of the test statistic less than c are consistent with the null hypothesis; values greater than or equal to c are inconsistent with the null hypothesis.

EXAMPLE 11-3 *(continued)*

In Example 11-3, we want to compare the average weight of tomatoes canned in the morning with the average weight of tomatoes canned in the afternoon. In particular, we want to test the hypotheses:

Null hypothesis: There is no difference between the average weight of tomatoes canned in the morning and the average weight of tomatoes canned in the afternoon.

Alternative hypothesis: There is a difference between the average weight of tomatoes canned in the morning and the average weight of tomatoes canned in the afternoon.

Let μ_1 denote the mean weight of cans filled in the morning, μ_2 the mean weight of cans filled in the afternoon. Then we can write our hypotheses as $H_0: \mu_1 = \mu_2$ and $H_a: \mu_1 \neq \mu_2$.

We assume that we have independent random samples from Gaussian distributions with equal variances. Then under the null hypothesis our test statistic has the t distribution with $5 + 5 - 2 = 8$ degrees of freedom.

From the description of the experiment, we do not know whether the independence assumption is reasonable. We would have to know more about how the samples were taken. Looking at Figure 11-2, we see that the variation is somewhat larger among the morning weights. (The two-sample t test is fairly *robust* to small deviations from the equal-variance assumption. This means that actual significance levels and confidence levels are close to the levels we choose, as long as the variances are not too different.) The two sets of observations are roughly symmetrical, so the assumption of Gaussian observations seems reasonable. (The two-sample t test also tends to be robust to deviations from the Gaussian assumption.) We will go ahead with a two-sample t test, aware that the assumptions must be met for the analysis to be valid.

If the significance level α equals .10, then $\alpha/2 = .05$. Referring to Table C, we see that if T has the t distribution with 8 degrees of freedom, then $P(T \leq 1.860) = .95$. The acceptance region is the interval $(-1.860, 1.860)$. The rejection region includes $(-\infty, -1.860]$ and $[1.860, \infty)$. If the test statistic is between -1.860 and 1.860, we will say the experimental results are consistent with the null hypothesis that the mean drained weight of tomatoes is the same for the morning and the afternoon. If the test statistic is less than or equal to -1.860 or else greater than or equal to 1.860, we will say the results are inconsistent with the null hypothesis, suggesting there is a difference between the mean drained weights for tomatoes canned in the morning and the mean for those canned in the afternoon.

To carry out the test, we calculate the following summary statistics for the experimental results:

Morning:	$\overline{X} = 22.7$ ounces	$s_1 = 2.31$ ounces
Afternoon:	$\overline{Y} = 20.8$ ounces	$s_2 = 1.20$ ounces
	$SE_{\overline{X}-\overline{Y}} = 1.16$ ounces	
Test statistic:	$\dfrac{\overline{X} - \overline{Y}}{SE_{\overline{X}-\overline{Y}}} = \dfrac{22.7 - 20.8}{1.16} = 1.6$	

The test statistic is in the acceptance region. We say the results are consistent with the null hypothesis, at the .10 significance level.

The p-value is the probability of seeing a test statistic as extreme as or more extreme than the one observed, if the null hypothesis were true. Since our test statistic equals 1.6, the p-value equals $P(T \leq -1.6) + P(T \geq 1.6)$, where T has the t distribution with 8 degrees of freedom. The p-value is between .1 and .2. Using the two-sample t test to compare the means of the two populations sampled, we say the results are consistent with the "no difference" null hypothesis. Even though the sample morning weights are on average larger than the sample afternoon weights, the difference is not statistically significant.

A reasonable point estimate for the difference $\mu_1 - \mu_2$ between the two population means is $\overline{X} - \overline{Y}$, the difference between the two sample means. Interval estimates for $\mu_1 - \mu_2$ are of the form $\overline{X} - \overline{Y} \pm c\text{SE}_{\overline{X}-\overline{Y}}$. If confidence level A is desired, we obtain c from the t distribution with $n_1 + n_2 - 2$ degrees of freedom to satisfy the relationship

$$A = P\left(-c < \frac{(\overline{X} - \overline{Y}) - (\mu_1 - \mu_2)}{\text{SE}_{\overline{X}-\overline{Y}}} < c\right)$$

Another way to write this is

$$A = P(\overline{X} - \overline{Y} - c\text{SE}_{\overline{X}-\overline{Y}} < \mu_1 - \mu_2 < \overline{X} - \overline{Y} + c\text{SE}_{\overline{X}-\overline{Y}})$$

Then we say $(\overline{X} - \overline{Y} - c\text{SE}_{\overline{X}-\overline{Y}}, \overline{X} - \overline{Y} + c\text{SE}_{\overline{X}-\overline{Y}})$ is a $100A\%$ confidence interval for $\mu_1 - \mu_2$.

In our example, we can estimate $\mu_1 - \mu_2$ with the point estimate $\overline{X} - \overline{Y} = 1.9$ ounces, the difference between the average weight in the morning and the average weight in the afternoon. A 90% confidence interval for $\mu_1 - \mu_2$ is

$$(1.9 - 1.860 \times 1.16, 1.9 + 1.860 \times 1.16) = (-.3, 4.1) \text{ ounces}$$

Zero is in this confidence interval, but near the edge.

In our formal analysis, we say the results are consistent with the null hypothesis that the mean drained weight is the same for morning and afternoon. However, in the sample, the weights of tomatoes canned in the morning are somewhat greater on average than the weights of those canned in the afternoon. This larger average results from the two morning cans that have drained weights at least 2 ounces greater than those recorded for any of the other eight sampled cans. As manager of this factory, you would probably want more information about the canning process, both in the morning and the afternoon. You might decide to look at larger samples of tomatoes canned during these two periods of the day. You would also have to consider such things as the target (labeled) weight of these cans, acceptable variation about this target (both from the company's and the consumer's point of view), and government regulations. How would your evaluation of the canning process differ if the target drained weight for these cans was: 21.0 ounces, 22.0 ounces, 23.0 ounces, 24.0 ounces?

In Section 11-4, we look at two-sample comparisons based on ranks, when we are not willing to assume our observations come from Gaussian distributions.

11-4 Inferences About Two Measures of Central Tendency Based on a Wilcoxon–Mann–Whitney Distribution

Let's assume, as we did in the previous section, that we have two independent random samples of continuous-type observations, one from each of two populations. We assume that the two population distributions have the same shape and variation, but may differ in location. That is, they may be *shifted* away from each other. The difference between the means of the two populations describes the extent of the shift. Such a situation is illustrated in Figure 11-4. A special case is when the distributions are Gaussian with the same variance, as in Figure 11-3. We want to test the null hypothesis that the means of the two populations are equal. This is the same as saying that the two distributions have the same location and are therefore identical. We will test this null hypothesis using the Wilcoxon–Mann–Whitney test based on ranks.

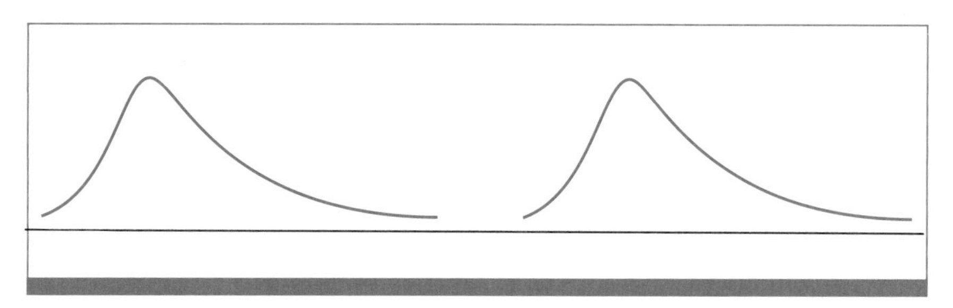

FIGURE 11-4 Illustration of two distributions with the same shape and variation, but different location

Let's look at an example. Then we will describe the Wilcoxon–Mann–Whitney test and apply it to the example.

EXAMPLE 11-4

Does a chemical treatment retard tumor growth in mice? Researchers asked this question in preliminary screening of a potential treatment for cancer in humans (Dunnett, 1972). The experimenters had available nine mice with tumors of the same type and similar size. They randomly divided the mice into two groups, three mice in a treatment group and six mice in a control group. The experimenters administered a chemical treatment to the three mice in the treatment group, for a fixed period of time. They gave no treatment to the six mice in the control group.

The experimenters kept conditions other than treatment the same for all nine mice. With this precaution, along with random assignment of mice to groups, they hoped to control extraneous factors. Then, if they saw differences in tumor size between the two groups of mice, they would be more confident in ascribing those differences to treatment. In particular, they wanted to see if the chemical treatment retarded tumor growth. If so, then the average tumor weight for treated mice would be less than for untreated mice. (The experimenters assigned more mice to the control group because this was one phase in a study of several chemical treatments. They intended to use the same control group for all comparisons. Therefore, they wanted a better estimate of the tumor size in untreated animals.)

At the end of the experiment, the researchers removed the tumors and weighed them. The results are shown below:

Group	Tumor weight in grams
Treatment	.96, 1.14, 1.59
Control	1.29, 1.31, 1.60, 1.88, 2.21, 2.27

A plot of the data is shown in Figure 11-5. Although there is some overlap, the tumor weights tend to be greater in the control group than in the treatment group. The plot suggests that the treatment may be able to retard tumor growth in mice.

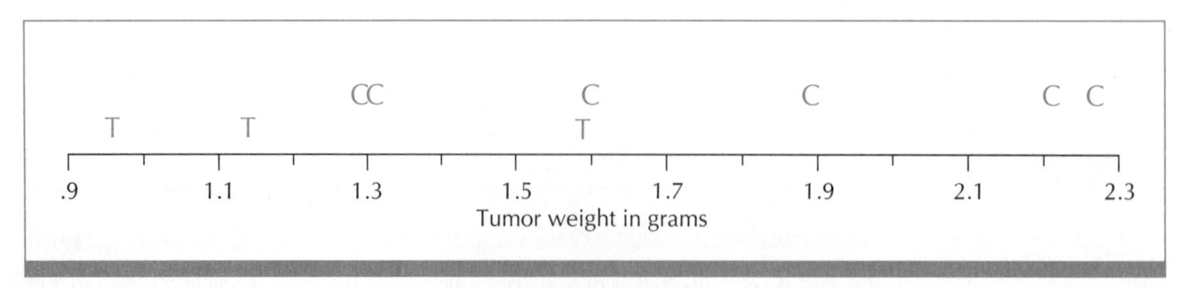

FIGURE 11-5 Plot of the tumor weights after treatment in Example 11-4. T denotes the tumor weight of a treated mouse. C denotes the tumor weight of an untreated control mouse.

For a formal analysis, we would like to test the null hypothesis that the mean tumor size after treatment is the same in the treatment and control groups. The alternative hypothesis of interest is one-sided—that mean tumor size after treatment is less in the treatment group than in the control group.

Let μ_1 denote the mean tumor weight for treated mice, μ_2 the mean tumor weight for control mice. (That is, μ_1 is the mean tumor weight after treatment in a hypothetical population of mice similar to those in this experiment. μ_2 is the mean tumor weight in a hypothetical population of untreated mice similar to those in this experiment.) Then we can state our hypotheses as H_0: $\mu_1 = \mu_2$ and H_a: $\mu_1 < \mu_2$. We will test these hypotheses using the Wilcoxon–Mann–Whitney two-sample test based on ranks.

To use the Wilcoxon–Mann–Whitney test, we assume that we have two independent random samples from distributions with the same shape and variation. Looking at Figure 11-5, we see that there is somewhat greater variation among the tumor weights in the control group than in the treated group (of course, there are also more observations in the control group). There are not really enough observations to compare the shapes of the two distributions. The fact that the mice were randomly divided between the two groups supports the independence assumption. We do not have enough information about how the experiment was conducted to evaluate all of our assumptions; we will keep this in mind when we analyze the experimental results.

Let's outline the steps for the Wilcoxon–Mann–Whitney two-sample test based on ranks. Then we will analyze the results of the experiment in Example 11-4.

The significance level approach to comparing two means using the Wilcoxon–Mann–Whitney two-sample test based on ranks

1. The hypotheses are H_0: $\mu_1 = \mu_2$ and H_a: $\mu_1 \neq \mu_2$, where μ_1 and μ_2 represent the two population means.
2. Rank the observations in the combined samples from smallest to largest. Assign tied observations the average of the ranks they share. Let W_1 be the sum of the ranks in the first sample and W_2 the sum of the ranks in the second sample. Define T_1 and T_2 by

$$T_1 = W_1 - \frac{n_1(n_1 + 1)}{2} \quad \text{and} \quad T_2 = W_2 - \frac{n_2(n_2 + 1)}{2}$$

   where n_1 and n_2 denote the number of observations in the first and second sample, respectively. Let the test statistic equal the minimum of T_1 and T_2.
3. Assume that we have two independent random samples from populations that have the same shape and variation, but possibly different locations. Then under the null hypothesis, T_1 and T_2 have the same Wilcoxon–Mann–Whitney distribution. (See the appendix on the Wilcoxon–Mann–Whitney distributions at the end of the book.)
4. Specify the significance level α.
5. Find the number c in Table G such that $P(W \leq c) = \alpha/2$, where W represents a random variable having the Wilcoxon–Mann–Whitney distribution

for sample sizes n_1 and n_2. The acceptance region includes numbers greater than c. The rejection region includes numbers less than or equal to c.

6. The decision rule is:

   If test statistic $> c$, say the results are consistent with the null hypothesis.
   If test statistic $\leq c$, say the results are inconsistent with the null hypothesis.

7. Collect a random sample satisfying the stated assumptions. Calculate the test statistic. Use the decision rule in step 6 to decide whether the observations are consistent with the null hypothesis. Draw conclusions based on the analysis.

If in step 1 we specify a one-sided alternative, then we must alter steps 2 and 5. In step 5, we find the number c in Table G such that $P(W \leq c) = \alpha$, where W represents a random variable having the Wilcoxon–Mann–Whitney distribution for sample sizes n_1 and n_2. If the one-sided alternative is H_a: $\mu_1 > \mu_2$, then in step 2 we let the test statistic equal T_2. If the one-sided alternative is H_a: $\mu_1 < \mu_2$, then we let the test statistic equal T_1.

EXAMPLE 11-4 *(continued)*

In Example 11-4, our hypotheses are H_0: $\mu_1 = \mu_2$ and H_a: $\mu_1 < \mu_2$. Here, μ_1 is the mean tumor weight for treated mice and μ_2 is the mean tumor weight for untreated mice. With this one-sided alternative, our test statistic is T_1.

Consider significance level $\alpha = .05$. Since the alternative is one-sided, we find the number c in Table G such that $P(W \leq c) = .05$, where W has the Wilcoxon–Mann–Whitney distribution for sample sizes 3 and 6. We will use $c = 2$, since $P(W \leq 2) = .048$, close to .05. The decision rule is:

If $T_1 > 2$, say the results are consistent with the null hypothesis.
If $T_1 \leq 2$, say the results are inconsistent with the null hypothesis.

The calculations for the Wilcoxon–Mann–Whitney test are outlined in Table 11-1.

The observed value of our test statistic, $T_1 = 2$, is in the rejection region. Therefore, we say the results are inconsistent with the null hypothesis, at the .05 significance level.

The *p*-value is the probability under the null hypothesis of observing a test statistic at least as extreme (in the direction of the alternative) as the one observed. Our *p*-value equals $P(T_1 \leq 2) = .048$. We say the results are inconsistent with the null hypothesis, with a *p*-value of less than 5%. These results suggest that the chemical treatment does retard tumor growth in mice. Of course, we would need additional information to do a more thorough evaluation of this experiment.

If the researchers decide that this treatment deserves further investigation, what do you think the next step would be? Describe what you think are the steps in experimentation before the treatment is finally tried on humans. How much time do you think would elapse between this initial experiment and a trial of the treatment involving cancer patients?

TABLE 11-1 Steps in constructing a test statistic for the Wilcoxon–Mann–Whitney two-sample test based on ranks. Tumor weights are in grams.

Treated group		Control group	
Tumor weight	Rank	Tumor weight	Rank
.96	1	1.29	3
1.14	2	1.31	4
1.59	5	1.60	6
		1.88	7
		2.21	8
		2.27	9
$n_1 = 3$		$n_2 = 6$	
$W_1 = 1 + 2 + 5 = 8$		$W_2 = 3 + 4 + 6 + 7 + 8 + 9 = 37$	
$T_1 = 8 - \frac{3(3 + 1)}{2} = 2$		$T_2 = 37 - \frac{6(6 + 1)}{2} = 16$	

A Large-Sample Approximation for the Wilcoxon–Mann–Whitney Test

Table G gives cumulative probabilities for the Wilcoxon–Mann–Whitney distribution in situations where the larger sample size is 12 or less. For larger samples, we can use a *large-sample approximation.* The large-sample test statistic is

$$T^* = \frac{T_1 - \dfrac{n_1 n_2}{2}}{\sqrt{\dfrac{n_1 n_2 (n_1 + n_2 + 1)}{12}}}$$

For large samples, T^* has approximately the standard Gaussian distribution under the null hypothesis. Values of T^* near 0 are consistent with the null hypothesis that the two populations have the same location. We discussed large-sample tests based on the standard Gaussian distribution in Section 10-1.

Although Example 11-4 does not involve large samples, we will calculate the large-sample test statistic for illustration. We have $n_1 = 3$, $n_2 = 6$, and $T_1 = 2$. The large-sample test statistic T^* is

$$T^* = \frac{2 - \dfrac{3 \times 6}{2}}{\sqrt{\dfrac{3 \times 6 \times (3 + 6 + 1)}{12}}} = -1.81$$

Since our alternative hypothesis is H_a: $\mu_1 < \mu_2$, negative values of T^* far from 0 are in the rejection region. Therefore, we calculate an approximate p-value $= P(T^* \leq -1.81) \doteq P(Z \leq -1.81) = .0351$, where Z represents a random variable having the standard Gaussian distribution. This approximate p-value based on the large-sample approximation agrees pretty well with the p-value of .048 we found using the Wilcoxon–Mann–Whitney distribution for sample sizes 3 and 6.

Nonparametric Confidence Intervals for $\mu_1 - \mu_2$

We can obtain nonparametric interval estimates for the difference $\mu_1 - \mu_2$ in location between two populations (Lehmann, 1975, pages 91–95). Let X_1 through X_{n_1} denote the n_1 observations in the first sample. Let Y_1 through Y_{n_2} denote the n_2 observations in the second sample. Find the $N = n_1 \times n_2$ differences of the form $X_i - Y_j$, where i is an integer from 1 to n_1 and j is an integer from 1 to n_2. Let D_1 through D_N denote the ordered values of these differences. Confidence intervals for $\mu_1 - \mu_2$ are of the form (D_k, D_{N-k+1}). The number k and the confidence level are related by

$$P(D_k \leq \mu_1 - \mu_2 \leq D_{N-k+1}) = 1 - 2P(W \leq k - 1)$$

where W has the Wilcoxon–Mann–Whitney distribution for sample sizes n_1 and n_2. A point estimate for $\mu_1 - \mu_2$ is the median of D_1 through D_N.

EXAMPLE 11-4 *(continued)*

Let's use this procedure to find an interval estimate for the difference between the mean tumor weight of treated mice and the mean tumor weight of untreated mice in Example 11-4. We must calculate $3 \times 6 = 18$ differences by subtracting each of the six control group weights from each of the three treatment group weights. The ordered differences D_1 through D_{18} are listed in Table 11-2.

A point estimate for the difference in average tumor weight between the two groups is the median of the 18 differences in Table 11-2. This median is

$$\frac{(-.62) + (-.46)}{2} = -.54 \text{ gram}$$

TABLE 11-2 Listing of the 18 differences obtained by subtracting each of the 6 control group weights from each of the 3 treatment group weights. The differences are ranked from smallest to largest. Sample calculations: $D_1 = .96 - 2.27 = -1.31$ and $D_{18} = 1.59 - 1.29 = .30$.

Rank	Difference	Rank	Difference	Rank	Difference
1	−1.31	7	−.68	13	−.29
2	−1.25	8	−.64	14	−.17
3	−1.13	9	−.62	15	−.15
4	−1.07	10	−.46	16	−.01
5	−.92	11	−.35	17	.28
6	−.74	12	−.33	18	.30

Under the conditions of this experiment, we estimate that the treatment retards tumor growth in mice by about .54 gram on average, compared with no treatment.

Now let's calculate an interval estimate for the difference between groups. Suppose W has the Wilcoxon–Mann–Whitney distribution for sample sizes 3 and 6. From Table G, we see that $P(W \leq 1) = .024$. The confidence level for the interval

$$(D_2, D_{17}) = (-1.25, .28)$$

is $1 - 2(.024) = .952$. We say the interval from -1.25 grams to .28 gram is a 95% confidence interval for the difference in average weights for the two groups. Since $P(W \leq 2) = .048$, the confidence level for the interval

$$(D_3, D_{16}) = (-1.13, -.01)$$

is $1 - 2(.048) = .904$. The interval from -1.13 to $-.01$ gram is a 90% confidence interval for the difference in average weights for the two groups.

The null hypothesis difference of 0 is in the 95% confidence interval and not far outside the 90% confidence interval. We might say our results here are borderline. Although the experiment suggests that the treatment retards tumor growth, the evidence is not overwhelming. On the other hand, with such small sample sizes, the researchers may think the evidence is strong enough to justify continued investigation of the treatment for reduction of tumors.

In Section 11-5, we discuss a test for comparing two population medians when we assume only that we have two independent random samples of continuous-type observations.

11-5 Inferences About Two Medians Based on a Hypergeometric Distribution

Suppose we have two independent random samples of continuous-type observations from populations of interest. To test whether the medians of the two populations are equal, we can use a test of hypotheses called the *median test*. Before outlining the general technique, we illustrate the use of the median test with an example.

EXAMPLE 11-5

Patricia wanted to see if more juice can be squeezed from oranges that have been microwaved for 20 seconds or from oranges that have not been microwaved (Macoul, 1988). She had heard that microwaving for 20 seconds increased yield, and she wanted to check it out for herself as part of a class project. Patricia randomly divided 16 oranges into two equal groups, a microwave group and a no-microwave group. She microwaved each orange in the microwave group for 20 seconds. Two helpers then squeezed as much juice as they could from each of the 16 oranges. The helpers did not know which oranges had been microwaved and which had not. What is the reason for this precaution?

Patricia wanted to compare the hypotheses:

Null hypothesis: The median yield is the same for microwaved and unmicrowaved oranges.

Alternative hypothesis: The median yield is not the same for microwaved oranges and oranges that have not been microwaved.

The yields in ounces squeezed from the oranges are shown below:

Microwave:	1.70	1.90	2.00	2.00
	2.10	2.30	2.45	2.50
No microwave:	1.55	1.60	1.75	1.75
	1.80	1.85	2.10	2.10

We begin our analysis with a plot of the experimental results, shown in Figure 11-6. We see that the microwaved oranges tended to give greater yields than the oranges that were not microwaved, although there is a good deal of overlap in the two distributions. Both distributions seem to be reasonably symmetrical.

To carry out the median test, we need the median of all 16 observations. This overall median is $(1.90 + 2.00)/2 = 1.95$ ounces. We then note how many observations in each of the two samples are less than and greater than the overall median. This information is summarized in Table 11-3.

The test statistic equals the smallest frequency in this table. We assume that the yield squeezed from one orange is unrelated to the yield from any other orange, so the other observations are independent. A reasonable probability model under the null hypothesis says that there is a random distribution across the 16 trials of the 8 results below the overall median, regardless of treatment group. Then under the null hypothesis, each frequency in the table has a hypergeometric distribution determined by the corresponding row and column totals.

For our example, values of the test statistic near 4 are consistent with the null hypothesis that microwaving does not affect median yield. Values of the test statistic near 0 are inconsistent with the null hypothesis. We see from Table 11-3 that the observed value of the test statistic is 2. The probability of seeing a test statistic less than or equal to 2 under the null hypothesis is a sum of hypergeometric probabilities:

$P(\text{test statistic} \leq 2 \text{ when } H_0 \text{ is true})$

$$= \frac{\binom{8}{2}\binom{8}{6}}{\binom{16}{8}} + \frac{\binom{8}{1}\binom{8}{7}}{\binom{16}{8}} + \frac{\binom{8}{0}\binom{8}{8}}{\binom{16}{8}} = .066$$

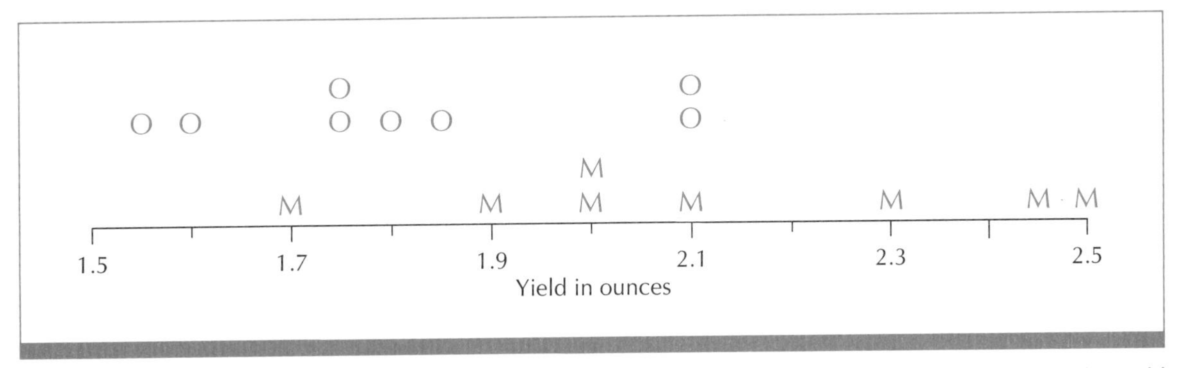

FIGURE 11-6 Yield in ounces of orange juice squeezed from 16 oranges in Example 11-5. M denotes the yield from an orange that had been microwaved. O denotes the yield from an orange that had not been microwaved.

TABLE 11-3 Two-way frequency table for treatment group (microwave, no microwave) and position relative to the overall median of 1.95 ounces.

	Position relative to overall median		
Group	Below	Above	Total
Microwave	2	6	8
No microwave	6	2	8
Total	8	8	16

Since the alternative is two-sided, the *p*-value is twice this value, about .13. (This is the probability under the null hypothesis of seeing 0, 1, 2, 6, 7, or 8 microwave results below the overall median.) If the null hypothesis were true, there would be about 13 chances in 100 of seeing results at least as extreme in the direction of the alternative as those actually observed. This *p*-value is not really small or really large; we might call it borderline. However, the experimental results do suggest that microwaving for 20 seconds may increase the median yield of juice squeezed from oranges by a small amount.

The Median Test

In general, to use the median test we assume that we have two independent random samples of continuous-type observations from populations of interest. We want to test whether the medians of the two populations are equal.

We find the overall median of the combined samples. Then we construct a two-way frequency table, classifying each observation by its grouping and by its position relative to the overall median (above or below). We ignore any observations exactly equal to the combined median.

We let our test statistic equal the smallest frequency in the two-way table. We then calculate the *p*-value, using hypergeometric probabilities based on the row and column totals of the table. When we calculate the *p*-value this way, the median test is the same as *Fisher's exact test* (discussed in Section 16-5). If the sample sizes are large enough, we can carry out the median test as a chi-square test of homogeneity for a two-way frequency table (Section 16-4).

The median test is crude because we consider only whether each observation is above or below the overall median for the two combined samples. We would expect to do better using either the actual data or their ranks, when the assumptions for the corresponding tests are reasonable.

In Section 11-6, we leave the two-sample comparison of means. Instead, we talk about comparing measures of central tendency based on paired samples.

11-6 Inferences About Measures of Central Tendency Based on Paired Samples

Sometimes we have two measures of central tendency to compare, but we have paired rather than independent samples. When observations occur naturally in pairs and we can consider different pairs to be independent, we have a *paired-sample* problem.

Investigators often use paired-sample experiments in medical research. We might make a measurement on each patient at the beginning of a study, then administer a treatment. We make another measurement after treatment. For each patient, we have a pair of observations, a pretreatment measurement and a posttreatment measurement. We design the experiment so that measurements for different patients are independent. However, the two measurements on a given patient are not independent. This is a paired-sample experiment.

Sometimes the two observations in a pair are not from the same individual. Instead they come from similar individuals (people, animals or objects, depending on the experiment) matched according to characteristics that might affect experimental results. Within each pair, one individual is randomly assigned one of two experimental conditions. The other individual is assigned the other experimental condition. Then our pairs of observations are the responses measured on the paired individuals. This is called a *matched-pairs* experiment. We analyze it as a paired-sample problem.

In a matched-pairs experiment, the individuals within a pair are similar with respect to characteristics we think would affect response. Therefore, if there is no difference between the two treatments, we would expect the individuals within a pair to have similar responses. If their responses differ, it may be chance variation or it may be attributable to treatment differences. A test of hypothesis helps us decide which hypothesis seems more reasonable based on the experimental results.

The general setup for a paired-sample analysis is this. We assume that we have a random sample of pairs of continuous-type observations, (X_1, Y_1) through (X_n, Y_n). We let $d_i = X_i - Y_i$ be the difference between the two observations in the ith pair. Then we think of d_1 through d_n as a random sample of continuous-type observations from some population. We want to make inferences about the center of this population of differences. Our analysis is based on the differences d_1 through d_n, using the one-sample techniques discussed in Chapter 10. Let's look at an example.

EXAMPLE 11-6

A high school baseball coach wanted to evaluate the effect of competition on base-running time. (This example is adapted from Loynd, 1985). Forty male high school students participated, although they were not told they were involved in an experiment. The coach matched the 40 students according to running speed. He then randomly selected one student within each matched pair to run a rival-competition trial, the other to run a self-competition trial.

A trial consisted of a student running around a regulation baseball diamond (a square 90 feet on each side), making contact with the base at each corner. The coach used a stopwatch to measure the time it took for the student to complete the run. He conducted such time trials regularly at the end of physical education class one day per week.

The coach told a student running a rival-competition trial that if he beat the time of his rival, he would be excused from half the required wind sprints on Friday. (Rivals were those in the same pair; each student knew who his rival was.) The coach told a student running a self-competition trial that if he beat his own best time, he would be excused from half the required wind sprints on Friday.

The experimental results are shown in Table 11-4. For each pair, the time

TABLE 11-4 Times (to the nearest hundredth of a second) to run a regulation baseball diamond, for 20 pairs of students. Within each pair, one student ran a self-competition trial and one student ran a rival-competition trial.

Pair	Self-competition	Rival-competition	Difference
1	16.20	15.95	.25
2	16.78	16.15	.63
3	17.38	17.05	.33
4	17.59	16.99	.60
5	17.37	17.34	.03
6	17.49	17.53	−.04
7	18.18	17.34	.84
8	18.16	17.51	.65
9	18.36	18.10	.26
10	18.53	18.19	.34
11	15.92	16.04	−.12
12	16.58	16.80	−.22
13	17.57	17.24	.33
14	16.75	16.81	−.06
15	17.28	17.11	.17
16	17.32	17.22	.10
17	17.51	17.33	.18
18	17.58	17.82	−.24
19	18.26	18.19	.07
20	17.87	17.88	−.01

FIGURE 11-7 Stem-and-leaf plots of the 20 self-competition running times and the 20 rival-competition times in Example 11-6. Each stem is in seconds and each leaf is in hundredths of a second. The pairings are ignored in this display.

Self-competition		Rival-competition	
Stem	Leaf	Stem	Leaf
15		15	
15	92	15	95
16	20	16	04 15
16	58 75 78	16	80 81 99
17	28 32 37 38 49	17	05 11 22 24 33 34 34
17	51 57 58 59 87	17	51 53 82 88
18	16 18 26 36	18	10 19 19
18	53	18	

FIGURE 11-8 Stem-and-leaf plot of the difference between the self-competition and rival-competition running times for the 20 pairs of students in Example 11-6. Each stem is in tenths of a second and each leaf is in hundredths of a second.

Difference in running times

Stem	Leaf
−.2	2 4
−.1	2
−.0	1 4 6
.0	3 7
.1	0 7 8
.2	5 6
.3	3 3 4
.4	
.5	
.6	0 3 5
.7	
.8	4

for the self-competition student, the time for the rival-competition student, and the difference between these two times are shown. The times are reported to the nearest hundredth of a second.

First let's look at some graphical displays. Figure 11-7 shows a stem-and-leaf plot of the 20 self-competition running times and a similar plot for the 20 rival-competition times, with pairings ignored. From these plots, we cannot easily compare the running times under the two experimental conditions.

In Figure 11-8, we take into account the pairing of the runners. This is a stem-and-leaf plot of the difference between the self-competition and rival-competition running times for the 20 pairs of students. The distribution of differences is fairly symmetrical, centered about some positive value. Now we can easily see that within the pairs, there is a tendency for the self-competition students to have longer running times than the rival-competition students.

The null hypothesis of interest is that type of competition (self versus rival) does not affect running time, on average. Our analysis will be based on the 20 differences listed in Table 11-4. Let μ_d denote the mean difference in running times under the self-competition and rival-competition conditions, for a hypothetical population of pairs of students similar to those in the study. For our formal analysis, the null hypothesis states that the mean difference in running times under the two competition conditions equals 0. The alternative states that the mean difference is not 0. We can write these hypotheses as H_0: $\mu_d = 0$ and H_a: $\mu_d \neq 0$.

If we want to use a t test (Section 10-3), then we must assume that the 20 differences represent independent observations from a Gaussian distribution. The stem-and-leaf plot of differences in Figure 11-8 does not provide strong evidence either against or in favor of the Gaussian assumption. The reasonableness of the independence assumption depends on how the experiment was conducted. We will proceed with our analysis, aware that we must be cautious in interpreting results.

We calculate the sample average and standard error based on the 20 differences in running times:

$$\bar{d} = .2045 \text{ second} \quad \text{and} \quad \text{SE}_{\bar{d}} = .0672 \text{ second}$$

From Section 10-3, we know that for a t test our test statistic is

$$\frac{\bar{d} - 0}{\text{SE}_{\bar{d}}} = \frac{.2045 - 0}{.0672} = 3.04$$

since $\mu_d = 0$ under the null hypothesis. Our p-value is $P(T \leq -3.04 \text{ or } T \geq 3.04)$ where T is a random variable having the t distribution with 19 degrees of freedom. From Table C, we see that the p-value is less than .01. The experimental results are inconsistent with the null hypothesis, with a p-value less than .01.

Again using the t distribution with 19 degrees of freedom, we find a 95% confidence interval for the mean difference between self-competition and rival-competition running times: $.2045 \pm (2.093)(.0672)$ or (.06, .35) second. We estimate that students run the bases an average of .06 to .35 second faster under rival-competition than under self-competition.

What do these experimental results suggest to you about a runner's performance when trying to beat his or her own best time compared to his or her performance when trying to beat a rival's best time? Would you be willing to guess how a runner would perform when trying to beat his or her own best time, as compared with trying to beat a well-matched rival in an actual race? In general, what might this experiment suggest about human performance under different types of competition?

The Paired *t* Test

When we use a t test to ask whether an average difference is 0, we say we are doing a paired t test.

> A **paired *t* test** is a t test applied to the differences in a paired-sample problem.

If we make fewer assumptions about the sample, then we may choose to do a Wilcoxon signed rank test or a sign test based on the differences (Exercise 11-25).

How do we choose between paired-sample and two-sample experimental designs? When we plotted the 20 self-competition running times separately from the 20 rival-competition times in Figure 11-7, it was difficult to see differences between the two distributions. The variation among running times within a single experimental condition was so large that differences between the experimental conditions were hard to see. Suppose we *incorrectly* ignore the pairing and analyze the experimental results as a two-sample problem. If we assume that we have two independent random samples from Gaussian distributions with equal variances, then we use a two-sample t test. To test the null hypothesis that the self-competition and rival-competition means are equal, our test statistic would be

$$\frac{17.434 - 17.229}{\sqrt{.469\left(\frac{1}{20} + \frac{1}{20}\right)}} = .9$$

Here, 17.434 is the sample average running time for the self-competition condition, 17.229 is the sample average for the rival-competition condition, and .469 is the pooled estimate of the common variance assumed for the two conditions. With our assumptions, the test statistic would have the t distribution with $20 + 20 - 2 = 38$ degrees of freedom under the null hypothesis. The p-value is greater than .30. This is consistent with the null hypothesis that there is no difference in average base-running times under the two experimental conditions. What we saw in examining Figure 11-7 is borne out by the test of hypotheses. The variation within experimental groups is so large that it obscures any differences between experimental groups.

We *correctly* took the pairing into account in Figure 11-8. We saw that within a pair, the self-competition running time tended to be longer than the corresponding rival-competition running time. The results of the paired-sample t test supported this observation.

The experimenter had a choice between a two-sample design and a paired-sample design. If he had chosen a two-sample design, he would have randomly selected 20 students to run under the self-competition condition, the other 20 to run under the rival-competition condition. His total sample size would be 40. For a two-sample t test, he would have 38 degrees of freedom.

Instead of a two-sample design, the experimenter opted for a paired design. He matched students by running speed, to control for differences in running ability, an extraneous variable that could affect the results of the experiment. He hoped that he would then be able to get a clearer picture of any differences between experimental conditions that might exist.

The experimenter pays a price for the increased power he expects from a paired design. The paired-sample analyses were based on 20 differences, effectively half the sample size he would have with a two-sample design. (For the paired-sample t test we had 19 degrees of freedom.)

The paired-sample design is a good idea when variation within groups is large relative to variation between groups. But the two-sample design offers twice the effective sample size for analysis. So, we may prefer the two-sample design when within-group variation is not so large.

There are other times when a two-sample design is preferable. An investigator may not be able to find suitable pairs among the participants in a study, or may not know all the factors that should form the basis for pairing. In these situations, he or she may choose a two-sample design. The investigator hopes that randomization will approximately balance the two experimental groups with respect to extraneous factors.

Summary of Chapter 11

This chapter is about comparing two measures of central tendency. We distinguished between two-sample designs and paired-sample designs. We have a two-sample design if we have two independent random samples, one from each of the two populations of interest; we use the samples to compare the centers of the two populations. In a paired-sample design, we have independent pairs of observations; we make inferences about the mean (or median) difference between observations within pairs. For two-sample and paired-sample designs, the method of analysis we choose depends on sample size and on the assumptions we are willing to make about the sample(s).

With a paired-sample design, we assume that we have a random sample of pairs of continuous-type observations (X_1, Y_1) through (X_n, Y_n). We define the difference between the observations within the ith pair by $d_i = X_i - Y_i$. We want to make inferences about the center of the distribution of these differences. To carry out an analysis of such a paired-sample experiment, we use one-sample techniques on the differences d_1 through d_n.

For a two-sample design, if we have two independent random samples and the sample sizes are large, we base inferences about the population means on the standard Gaussian distribution. A special case is when sample sizes are large and we wish to compare two proportions.

If we have two independent random samples from Gaussian distributions with equal variances, we can compare the two population means using the two-sample t test. Tests of hypotheses and confidence intervals for the difference between the two population means are based on a t distribution.

Suppose we have two independent random samples of continuous-type observations from distributions that have the same shape and variation, but possibly different locations. Then we can test for equality of the two population means using ranked data. We base tests of hypotheses and confidence intervals for the difference between the population means on a Wilcoxon–Mann–Whitney distribution. The Wilcoxon–Mann–Whitney distribution for sample sizes n_1 and n_2 is derived from the probability model for an experiment in which ranks 1 through $n_1 + n_2$ are randomly divided into two groups, one of size n_1 and one of size n_2.

If the assumptions for the two-sample t test are met, it is more powerful (more likely to reject the null hypothesis when the alternative is true) than the Wilcoxon–Mann–Whitney test.

If we have two independent random samples of continuous-type observations, we can test for equality of the two population medians using the median test. We base inferences on a hypergeometric distribution for small samples. For large samples, we can use the chi-square distribution with 1 degree of freedom (see Section 16-4). If the assumptions for the Wilcoxon–Mann–Whitney test are met, it is much more powerful (more likely to reject the null hypothesis when the alternative is true) than the median test.

Procedures based on a t distribution are parametric because we assume that the observations follow Gaussian distributions. Inferences based on a hypergeometric distribution or on a Wilcoxon–Mann–Whitney distribution, as well as those based on the standard Gaussian distribution for large samples, are nonparametric because we assume no particular type of probability distribution for the observations.

For a two-sample design, a confidence interval provides an interval estimate or range of reasonable values for the difference between two population means, $\mu_1 - \mu_2$. If the confidence level is A, we say we have a $100A\%$ confidence interval for the difference between the two population means. The interpretation of such a confidence interval is that if we repeated the sampling process in the same way many times and calculated a $100A\%$ confidence interval for $\mu_1 - \mu_2$ each time, we would expect about $100A\%$ of these confidence intervals to contain $\mu_1 - \mu_2$ and about $100(1 - A)\%$ not to contain $\mu_1 - \mu_2$. When we calculate a specific confidence interval, it either contains the difference between the two population means or it does not. The confidence level refers to what we would expect if we repeated the sampling process many times.

Minitab Appendix for Chapter 11

Carrying Out a Two-Sample *t* Test

Suppose we have the data from Example 11-3 on our worksheet in two columns. Column 1 contains a code for time of day: 1 = morning, 2 = afternoon. Drained weight is in column 2. We can use the TWOT command to carry out the two-sample t test that the two population means are equal and obtain a confidence interval for the difference between the two means. In the TWOT command, we specify the confidence level of the interval, the column for the observations, and the column for the group code. We also use the POOLED

```
TWOSAMPLE T FOR C2
C1  N      MEAN       STDEV    SE MEAN
1   5     22.70        2.31        1.0
2   5     20.80        1.20       0.54

90 PCT CI FOR MU 1 - MU 2: (-0.3, 4.07)

TTEST MU 1 = MU 2 (VS NE): T= 1.63  P=0.14  DF=  8

POOLED STDEV =          1.84
```

FIGURE M11-1 Output from the TWOT command for Example 11-3

subcommand to indicate that we are assuming the two population variances are equal. For Example 11-3, the command

```
MTB>   twot 90 c2 c1;
SUBC>  pooled.
```

produces the output in Figure M11-1.

We can use the ALTERNATIVE subcommand to specify one-sided alternatives. ALTERNATIVES = +1 indicates the alternative $\mu_1 > \mu_2$ and ALTERNATIVE = −1 indicates the alternative $\mu_1 < \mu_2$.

Making Inferences Based on a Wilcoxon–Mann–Whitney Distribution

To use the MANN-WHITNEY command for two-sample inferences based on ranks, the two groups of observations must be in two separate columns. Consider Example 11-4. For the MANN-WHITNEY command, Minitab expects the tumor weights for the treatment group in one column and the tumor weights for the control group in another column. We have not been creating our data files in this way; our files all have one case per row. Suppose we have a worksheet with two columns. Column 1 (named GROUP) contains a code for group: 1 = treatment, 2 = control. Column 2 (named TUMORWT) contains the tumor weights:

ROW	group	tumorwt
1	1	0.96
2	1	1.14
3	1	1.59
4	2	1.29
5	2	1.31
6	2	1.60
7	2	1.88
8	2	2.21
9	2	2.27

To use MANN-WHITNEY, we want to put the tumor weights for the treatment group in column 3 and the tumor weights for the control group in column 4. We can accomplish this using the UNSTACK command:

```
MTB>   unstack 'tumorwt' c3 c4;
SUBC>  subscripts 'group'.
MTB>   name c3 'tumort' c4 'tumorc'
```

This UNSTACK command instructs Minitab to unstack the observations in column 2 ('TUMORWT') according to the codes in column 1 ('GROUP'). Column 3 receives the tumor weights corresponding to the smaller value (1) in 'GROUP'; column 4 receives the tumor weights corresponding to the larger value (2) in 'GROUP'. Our worksheet now looks like this:

ROW	group	tumorwt	tumort	tumorc
1	1	0.96	0.96	1.29
2	1	1.14	1.14	1.31
3	1	1.59	1.59	1.60
4	2	1.29		1.88
5	2	1.31		2.21
6	2	1.60		2.27
7	2	1.88		
8	2	2.21		
9	2	2.27		

In the MANN-WHITNEY command, we specify the type of alternative (0 for two-sided; +1 for one-sided, upper tail; −1 for one-sided, lower tail). We specify the level of the confidence interval for the difference between the two population medians, as well as the columns containing the two groups of observations. For Example 11-4, the command

```
MTB>   mann alternative=-1 90 'tumort' 'tumorc'
```

produces the output in Figure M11-2.

```
Mann-Whitney Confidence Interval and Test

 tumort      N =   3      MEDIAN =        1.1400
 tumorc      N =   6      MEDIAN =        1.7400
 POINT ESTIMATE FOR ETA1-ETA2 IS         -0.5400
 90.7  PCT C.I. FOR ETA1-ETA2 IS (  -1.1301,   -0.0101)
 W =        8.0
 TEST OF ETA1 = ETA2  VS.  ETA1 L.T. ETA2 IS SIGNIFICANT AT  0.0466
```

FIGURE M11-2 Output from the MANN-WHITNEY command in Example 11-4

```
TEST OF MU = 0.0000 VS MU N.E. 0.0000

             N      MEAN     STDEV   SE MEAN        T    P VALUE
diff        20    0.2045    0.3004    0.0672     3.04     0.0067
```

FIGURE M11-3 Paired *t* test for Example 11-6

Carrying Out a Two-Sample Comparison of Means Based on Large Samples

For a two-sample test based on large samples, we can use the TWOT command, without the POOLED subcommand. Minitab does not have a command for the median test. To use Minitab for a comparison of two proportions based on large samples, see the Minitab Appendix for Chapter 16 for the chi-square test of homogeneity.

Carrying Out a Paired-Sample Comparison of Means

We can carry out paired-sample analyses using the appropriate one-sample commands. Suppose we have the data for Example 11-6 in our worksheet. Column 1 contains a code for the pair of runners; column 2, the self-competition time; column 3, the rival-competition time. For a paired-sample *t* test that the mean difference in times under the two conditions is 0, we can use these commands:

```
MTB>   let c4=c2-c3
MTB>   name c4 'diff'
SUBC>  ttest 'diff'
```

If we specify no value for the mean under the null hypothesis in the TTEST command, Minitab uses 0. The output is in Figure M11-3.

We can get a confidence interval for the mean difference in times under the two conditions using the TINTERVAL command on the differences in column 4.

Exercises for Chapter 11

In each exercise, describe the population(s) sampled, whether real or hypothetical. Graph the data in any way that seems helpful. For each statistical procedure, state the assumptions that make the analysis appropriate. Do the assumptions seem reasonable? What additional information about the experiment would you like to have? Discuss the results of your analysis.

EXERCISE 11-1 In 10 trials, the number of female mosquitos collected coming to bite a human and the number killed in an electrocuting device (over a 2-hour period in the same yard) were recorded (Nasci et al., 1983). The results are given below.

Trial	Number of female mosquitos captured: Electrocuting device	Human bait
1	31	94
2	44	146
3	129	194
4	15	54
5	11	39
6	49	90
7	151	172
8	30	219
9	12	60
10	17	21

a. Plot these observations.

b. How does the insect electrocuting device compare with human bait in attracting female mosquitos? State and test appropriate hypotheses.

c. Calculate an interval estimate for the mean difference in number of female mosquitos captured by the two methods. Discuss your findings.

EXERCISE 11-2

Do drugs that suppress abnormal heart rhythms increase the likelihood of sudden cardiac death? A multicenter clinical trial enrolled patients who had had a heart attack and developed cardiac arrhythmias no more than 2 years before start of the study. Seven hundred thirty patients received either encainide or flecainide for cardiac arrhythmia and 730 patients received placebo (*Science News,* April 29, 1989, volume 135, page 260).

The trial began in June 1987. Early review of results by a safety monitoring board found that 33 of the 730 patients receiving encainide or flecainide had experienced either sudden cardiac death or a nonfatal heart attack. Nine of the 730 patients in the placebo group had suffered a nonfatal heart attack or died suddenly from cardiac problems. The trial was stopped based on this early review.

a. Compare the proportion of patients suffering heart attack or sudden cardiac death in the two groups (treatment and placebo). State and test appropriate hypotheses.

b. Calculate a confidence interval for the difference between the proportions for the two groups.

c. Discuss your results.

EXERCISE 11-3

A study was designed to compare the effects of two types of ammunition (.22 Long Rifle versus .22 Magnum) on target-shooting accuracy (Snow, 1986). An experienced shooter used a revolver with interchangeable cylinders to shoot at a circular target 25 yards away. A trial consisted of five shots. The score for a trial was the total score for the five shots. Sixteen trials were run, eight with each type of ammunition. The order in which the two types of ammunition were used was determined by a random process. The results of the experiment are listed here.

Total score of five shots	
.22 Magnum	.22 Long Rifle
42	41
43	43
46	41
47	41
46	40
47	40
39	45
47	47

a. Plot these observations.

b. Is the average score the same for the two types of ammunition? State and test appropriate hypotheses.

c. Calculate an interval estimate for the difference in mean score for the two types of ammunition.

EXERCISE 11-4

Two analysts each made eight independent determinations of the melting point of hydroquinone (Duncan, 1974, pages 575–576; from Wernimont, 1947, page 8).

Analyst	Melting point determination (°C)				
1	174.0	173.5	173.0	173.5	171.5
	172.5	173.5	173.5		
2	173.0	173.0	172.0	173.0	171.0
	172.0	171.0	172.0		

a. Plot the observations.

b. Test the null hypothesis that the mean determination of melting point is the same for the two analysts.

c. Calculate a confidence interval for the difference in mean melting point determinations for the two analysts.

EXERCISE 11-5

Researchers wanted to study effects of regular alcohol consumption (Jerome Hojnacki, 1986, personal communication). The participants in the experiment were 20 adult male squirrel monkeys, of similar age and good health. The researchers randomly divided the monkeys into two equal sized groups. Monkeys in the alcohol group consumed a steady diet of 12% ethyl alcohol (ethyl alcohol constituted 12% of their total calories each meal). Monkeys in the control group did not consume alcohol. At the end of the treatment period, the researchers measured plasma estrogen (in nanograms/deciliter) for each monkey. The results are shown below.

Alcohol:	3.17	2.52	2.59	4.25	3.27	4.92
	5.46	2.83	4.80	2.26		
Control:	6.57	5.81	5.63	5.75	4.54	5.35
	4.16	5.12	4.69	4.52		

a. Plot the observations.

b. Use the median test to test the null hypothesis that the median plasma estrogen level is the same for the two groups of monkeys.

c. Use the Wilcoxon–Mann–Whitney test to test the null hypothesis that the median plasma estrogen level is the same for the two groups of monkeys.

d. Use the two-sample t test to test the null hypothesis that the mean plasma estrogen level is the same for the two groups of monkeys.

e. Compare the results of your analyses in parts (b), (c), and (d). Discuss your findings.

EXERCISE 11-6

In the experiment described in Exercise 11-5, researchers also measured plasma testosterone level (nanograms/deciliter) for each monkey at the end of the treatment period (Jerome Hojnacki, 1986, personal communication). The results are shown below.

Alcohol:	313.99	152.06	145.64	128.86	262.16
	251.29	505.55	94.79	157.49	171.81
Control:	632.92	308.56	1,239.68	440.38	233.02
	142.67	84.91	342.63	1,005.66	735.61

a. Plot the observations.

b. Use the median test to test the null hypothesis that the median testosterone level is the same for the two groups of monkeys.

c. Use the Wilcoxon–Mann–Whitney test to test the null hypothesis that the median testosterone level is the same for the two groups of monkeys.

d. Use the two-sample t test to test the null hypothesis that the mean testosterone level is the same for the two groups.

e. The equal-variance assumption of the two-sample t test is violated for these samples. Take the logarithm base-10 of each observation. Plot these transformed observations, for the two groups. Are the variances for the two groups more similar for these transformed observations? When we transform observations to get more similar variances for different groups, we say we are making a *variance-stabilizing transformation.*

f. Use the two-sample t test to test the null hypothesis that the mean of the logarithm base-10 of testosterone level is the same for the two groups.

g. Compare the results of your tests of hypotheses in parts (b), (c), (d), and (f). Discuss your findings. Note that the test in part (f) applies to the *transformed* observations.

EXERCISE 11-7

Sputum histamine levels (μg/g dry weight sputum) are shown below for 9 allergic and 13 nonallergic people, all smokers (Hollander and Wolfe, 1973, page 74; a subset of data in Thomas and Simmons, 1969).

Allergics:	31.0	39.6	64.7	65.9	67.9	100.0
	102.4	1,112.0	1,651.0			
Nonallergics:	4.7	5.2	6.6	18.9	27.3	29.1
	32.4	34.3	35.4	41.7	45.5	48.0
	48.1					

a. Plot the observations. Because of the wide range for the allergic people, you may wish to take the logarithm base-10 of each value (for both groups) before plotting.

b. Use the median test to test the null hypothesis that the median sputum histamine level is the same for the allergics and nonallergics.

c. Use the Wilcoxon–Mann–Whitney test to test the null hypothesis that the median sputum histamine level is the same for the two groups.

d. The equal-variance assumption of the two-sample t test is violated for these two groups. When you plot the logarithm base-10 of the observations, does the variation seem similar in the two groups? If so, we call this transformation a *variance-stabilizing transformation.* Use the two-sample t test to test the null hypothesis that the mean of the logarithm base-10 of sputum histamine levels is the same for the allergics and nonallergics.

e. Discuss the results of your tests of hypotheses in parts (b), (c), and (d). Note that your test in part (d) applies to the *transformed* observations.

EXERCISE 11-8

This study compared the effectiveness of a bronchodilating aerosol administered by hand and by an automatic inhalation device (Box, Hunter, and Hunter, 1978, page 158; from a larger study reported by F. J. McIlneath and B. M. Cohen in *J. Med.,* 1970, volume 1, page 229). Specific airway resistance 30 minutes after administration is shown below for each of 12 patients using hand administration and 12 patients using an automatic inhalation device.

Hand:	17.00	22.80	21.60	20.40	11.20	14.00	52.25
	7.50	12.20	18.85	6.05	4.05		
Automatic:	11.60	11.60	13.65	17.22	8.25	6.20	41.50
	6.96	8.40	9.00	5.18	3.00		

a. Plot the observations.

b. Test the null hypothesis that mean airway resistance is the same 30 minutes after administration of the aerosol, for the two methods.

c. Calculate a confidence interval for the difference in mean airway resistance after 30 minutes for the two methods.

EXERCISE 11-9

Does diet restriction prolong life? In this experiment, researchers examined the influence of different diets on the aging process in rats (Berger, Boos, and Guess, 1988; from Yu et al., 1982). Lifetimes (in days) of rats on a restricted diet and rats on an unrestricted (*ad libitum*) diet are shown below.

Restricted diet (sample size = 106, mean = 968.7 days, sample standard deviation = 284.6 days)

105	193	211	236	302	363	389	390	391	403	530
604	605	630	716	718	727	731	749	769	770	789
804	810	811	833	868	871	875	893	897	901	906
907	919	923	931	940	957	958	961	962	974	979
982	1,001	1,008	1,010	1,011	1,012	1,014	1,017	1,032	1,039	1,045
1,046	1,047	1,057	1,063	1,070	1,073	1,076	1,085	1,090	1,094	1,099
1,107	1,119	1,120	1,128	1,129	1,131	1,133	1,136	1,138	1,144	1,149
1,160	1,166	1,170	1,173	1,181	1,183	1,188	1,190	1,203	1,206	1,209
1,218	1,220	1,221	1,228	1,230	1,231	1,233	1,239	1,244	1,258	1,268
1,294	1,316	1,327	1,328	1,369	1,393	1,435				

Unrestricted diet (sample size = 90, mean = 682.3 days, sample standard deviation = 134.3 days)

89	104	387	465	479	494	496	514	532	533	536
545	547	548	582	606	609	619	620	621	630	635
639	648	652	653	654	660	665	667	668	670	675
677	678	678	681	684	688	694	695	697	698	702
704	710	711	712	715	716	717	720	721	730	731
732	733	735	736	738	739	741	743	746	749	751
753	764	765	768	770	773	777	779	780	788	791
794	796	799	801	806	807	815	836	838	850	859
894	963									

a. Plot these observations.

b. Test the null hypothesis that mean lifetime is the same for rats on the two diets.

c. Calculate a confidence interval for the difference in mean lifetimes for rats on the two diets.

d. Discuss your findings.

EXERCISE 11-10 Cotton dust in textile factories can lead to a respiratory problem known as byssinosis. A survey of cotton textile workers was carried out to evaluate byssinosis problems (Higgins and Koch, 1977). Of 2,916 men surveyed, 128 had byssinosis complaints. Of 2,503 women surveyed, 37 had byssinosis complaints.

a. Is there a difference between the proportions of men and women with byssinosis complaints? State and test appropriate hypotheses.

b. Calculate a confidence interval for the difference between the proportions of men and women with byssinosis complaints.

c. Calculate a confidence interval for the proportion of men with byssinosis complaints.

d. Calculate a confidence interval for the proportion of women with byssinosis complaints.

EXERCISE 11-11 Cirrhotic patients with bleeding problems were randomly divided into two groups. Patients in one group underwent a standard operation (nonselective shunt). Patients in the other group underwent a new operation (selective shunt). The response is maximal rate of urea synthesis, a measure of liver function; low values correspond to poor liver function. Responses before and after surgery are shown below (Brogan and Kutner, 1980; from *Annals of Surgery,* 1978, volume 188, pages 271–282).

	Maximal rate of urea synthesis (mg urea N/hr/kg BW$^{3/4}$)			Maximal rate of urea synthesis (mg urea N/hr/kg BW$^{3/4}$)	
Patient	Before surgery	After surgery	Patient	Before surgery	After surgery
New operation (selective shunt)			**Standard operation (nonselective shunt)**		
1	51	48	9	34	16
2	35	55	10	40	36
3	66	60	11	34	16
4	40	35	12	36	18
5	39	36	13	38	32
6	46	43	14	32	14
7	52	46	15	44	20
8	42	54	16	50	43
			17	60	45
			18	63	67
			19	50	36
			20	42	34
			21	43	32

a. Plot the observations in any way that seems helpful.

b. Compare average liver function before surgery for the two groups of patients. State and test appropriate hypotheses.

c. Compare before- and after-surgery liver function for patients undergoing the new operation. State and test appropriate hypotheses. Calculate a confidence interval for the mean difference in liver function for this group of patients.

d. Compare before- and after-surgery liver function for patients undergoing the standard operation. State and test appropriate hypotheses. Calculate a confidence interval for the mean difference in liver function for this group of patients.

e. Compare the change in liver function for the two groups of patients. State and test appropriate hypotheses.

f. Discuss your findings.

EXERCISE 11-12

Change in pupil diameter following treatment is shown below for 11 volunteers (Box, Hunter, and Hunter, 1978, page 160; from H. W. Elliott, G. Navarro, and N. Nomof, *J. Med.,* 1970, volume 1, page 77). Six volunteers received several doses of morphine. Five volunteers received several doses of nalbuphine.

Treatment	Change in pupil diameter (millimeters)					
Morphine	.08	.8	1.0	1.9	2.0	2.4
Nalbuphine	−.3	.0	.2	.4	.8	

a. Plot these observations.

b. Test the null hypothesis that the mean change in pupil diameter is the same for the two drugs, using a t distribution.

c. Repeat part (b), using a Wilcoxon–Mann–Whitney distribution.

d. Test the null hypothesis that the median change in pupil diameter is the same for the two drugs, using a hypergeometric distribution.

e. Compare the results of parts (b), (c), and (d).

EXERCISE 11-13

Scientists have developed a new method of determining serum iron concentration that is faster and requires smaller samples than an older method. To compare the accuracy of the two methods, researchers made replicate analyses of control sera containing a concentration of 105 μg serum iron per 100 milliliters. The measurements of serum iron concentration (μg/100 ml) are shown below (Hollander and Wolfe, 1973, pages 85–86; a portion of the data of Jung and Parekh, 1970).

New:	107	108	106	98	105	103	110	105
	104	100	96	108	103	104	114	114
	113	108	106	99				
Old:	111	107	100	99	102	106	109	108
	104	99	101	96	97	102	107	113
	116	113	110	98				

a. Plot the observations.

b. Test the null hypothesis that average serum iron determination is the same using both methods.

c. Calculate a confidence interval for the difference in mean serum iron determination for the two methods.

d. Calculate a confidence interval for the mean serum iron determination using the new method.

e. Calculate a confidence interval for the mean serum iron determination using the old method.

f. Discuss your findings.

EXERCISE 11-14

Investigators tested the effect of a drug with antiarrhythmic properties on patients with frequent premature ventricular contractions (PVCs). For each of 12 patients, the researchers recorded the number of PVCs during a 1-minute electrocardiograph, both before and after treatment with the drug (Berry, 1987).

Patient	Pre-treatment	Post-treatment	Pre − Post (decrease in PVCs)
1	6	5	1
2	9	2	7
3	17	0	17
4	22	0	22
5	7	2	5
6	5	1	4
7	5	0	5
8	14	0	14
9	9	0	9
10	7	0	7
11	9	13	−4
12	51	0	51

a. Plot the observations in any way that seems helpful.

b. Test the null hypothesis of no average change in PVC count before and after treatment, based on a t distribution. Calculate a confidence interval for the average decrease in PVC count.

c. Repeat part (b), based on a Wilcoxon signed rank distribution.

d. Test the null hypothesis of zero median change in PVC count before and after treatment, based on a binomial distribution. Calculate a confidence interval for the median decrease in PVC count.

e. Compare the results of parts (b), (c), and (d).

f. Patient 12 is an outlier in the sense that his pretreatment PVC count is far from the others. Delete the observations for patient 12 and repeat parts (b), (c) and (d). Compare your results.

EXERCISE 11-15 Survival times (units not given) are shown here for skin grafts on 11 burn patients. Each patient received both a closely matched graft and a poorly matched graft (O'Brien and Fleming, 1987; from Woolson and Lachenbruch, 1980; slightly modified from original data in Batchelor and Hackett, 1970). A + indicates a graft still viable at the recorded time.

	Survival time for		
Patient	Closely matched graft	Poorly matched graft	Difference
1	37	29	8
2	19	13	6
3	57+	15	42+
4	93	26	67
5	16	11	5
6	22	17	5
7	20	26	−6
8	18	21	−3
9	63	43	20
10	29	15	14
11	60+	40	20+

a. Plot the observations in any way that seems helpful.

b. Test the null hypothesis of no median difference in survival for the two types of grafts.

c. Calculate a confidence interval for the median difference in survival for the two types of grafts.

d. Calculate a confidence interval for median survival of closely matched grafts.

e. Calculate a confidence interval for median survival of poorly matched grafts.

EXERCISE 11-16 Experimenters wanted to see if practice and training can affect hypnotic susceptibility. Each of six volunteers was evaluated with Stanford profile scales of hypnotic susceptibility by a hypnotist (not one of the experimenters). Then each volunteer went through hypnotic training with one of the experimenters. After this training, each volunteer was evaluated by a different hypnotist (again not one of the experimenters) using equivalent scales of hypnotic susceptibility. Results are shown below (Hollander and Wolfe, 1973, page 45; part of a larger data set of Cooper et al., 1967). A low score indicates low hypnotic susceptibility.

Volunteer	Average score before training	Average score after training
1	10.5	18.5
2	19.5	24.5
3	7.5	11.0
4	4.0	2.5
5	4.5	5.5
6	2.0	3.5

a. Plot these observations.

b. Why was the evaluation of hypnotic susceptibility done by hypnotists other than the experimenters?

c. Does this experiment suggest that training affects hypnotic susceptibility? State and test appropriate hypotheses.

d. Calculate a confidence interval for the average difference in scores before and after training. Discuss your findings.

EXERCISE 11-17 In a study of antibodies associated with malaria, investigators treated adult volunteers for malaria, achieving what is known as a radical cure (Hoffman et al., 1987). By 98 days after radical cure, 60 of the 83 volunteers completing the study were infected with the malaria parasite and 23 were not. One measure of antibody activity is ISI (inhibition of sporozoite invasion of hepatoma cells). Based on ISI, investigators classified the volunteers as having either high antibody levels ($>$ 75% ISI) or low antibody levels ($\leq$ 75% ISI). With this measure, 26 of the 60 infected volunteers and 9 of the 23 uninfected volunteers had high levels of antibodies at the end of the study.

a. Is there any difference between the infected and uninfected volunteers with respect to this measure of antibody levels? State and test appropriate hypotheses.

b. Calculate a confidence interval for the difference in proportions with high antibody levels for the infected and uninfected volunteers. Discuss your findings.

EXERCISE 11-18 Do the two anesthetics enflurane and halothane have the same effects on cardiovascular performance? To answer this question, investigators studied 19 children undergoing elective noncardiac surgery. Ten children received enflurane and nine received halothane. Echocardiographs provided values for shortening fraction and mean blood pressure, two measures of cardiovascular performance. Values of shortening fraction and mean blood pressure before and during low dose of the anesthetic are listed below for each child (Hui and Rosenberg, 1985; from Barash et al., 1979).

	Shortening fraction (percent)		Blood pressure (mm Hg)	
Child	Before	At low dose	Before	At low dose
Enflurane				
1	25	23	79	74
2	39	24	71	76
3	27	30	92	87
4	30	28	90	79
5	31	28	82	69
6	35	37	83	73
7	23	23	77	67
8	27	29	69	63
9	27	21	90	86
10	25	17	87	76
Halothane				
1	34	21	90	89
2	30	29	83	77
3	33	31	68	63
4	23	26	70	69
5	27	26	76	72
6	26	27	64	56
7	28	22	75	67
8	27	29	68	67
9	33	29	79	77

a. Plot these observations in any way that seems helpful.

b. Is there a difference on average between the before and low-dose measurements of shortening fraction for children on enflurane? State and test appropriate hypotheses.

c. Is there a difference on average between the before and low-dose measurements of shortening fraction for children on halothane? State and test appropriate hypotheses.

d. Is the change in shortening fraction for children on enflurane the same as for children on halothane, on average?

e. Is there a difference on average between the before and low-dose measurements of blood pressure for children on enflurane?

f. Is there a difference on average between the before and low-dose measurements of blood pressure for children on halothane?

g. Is the change in blood pressure for children on enflurane the same as for children on halothane, on average?

h. Summarize your findings. Do the two anesthetics appear to have the same effects on cardiovascular performance as measured by shortening fraction and mean blood pressure?

i. In answering parts (b)–(g) you use six tests of hypotheses. Should you be concerned that the overall significance level of the tests taken together is not the same as the significance level(s) you used for the separate tests? We address this question in the next chapter.

EXERCISE 11-19

Eleven healthy young volunteers participated in a study to compare nasal clearance times before and after jogging (Cederlund, Camner, and Svartengren, 1987). Nasal clearance refers to the process by which inhaled particles are transported to the pharynx and then swallowed. People born with impaired nasal clearance tend to be at increased risk of respiratory problems.

This study was designed to investigate the effects of jogging, with associated increased rate of respiration, on nasal clearance time. Nasal clearance was recorded as a transport time: the time between placement of a saccharin particle in the nasal passage and first indication of a sweet taste. Nasal clearance was measured in this way for each volunteer immediately before jogging and approximately 1 hour later after jogging from 8 to 10 kilometers. Personal characteristics and experimental results for the 11 volunteers are shown here.

					Transport time in minutes	
Volunteer	Sex	Age (years)	Smoker	History of allergy	Before jogging	After jogging
1	F	22	No	No	7.5	33.0
2	F	20	No	No	9.5	28.0
3	F	21	No	No	19.5	>45
4	M	30	No	No	4.5	10.5
5	M	19	Yes	Yes	18.0	19.0
6	M	19	Yes	No	13.0	10.0
7	M	19	No	Yes	12.0	17.5
8	M	20	No	No	26.0	29.0
9	M	21	No	No	6.5	17.0
10	F	19	No	No	10.5	40.0
11	F	19	No	No	8.5	13.0

Volunteer 3 reported no sweet taste within 45 minutes of receiving the saccharin particle. The investigators considered this to be an experimental failure and chose to exclude her results from the analysis. Therefore, we are left with the data on the remaining 10 volunteers. In answering parts (a)–(f), ignore the personal information given for the volunteers, considering only the before- and after-jogging transport times.

a. Plot the observations in any way that seems helpful.

b. State null and alternative hypotheses.

c. Test the hypotheses in part (b) and calculate an interval estimate, using a binomial distribution. Discuss the relationship between the test of hypotheses and the confidence interval.

d. Test the hypotheses in part (b) and calculate an interval estimate, using a Wilcoxon signed rank distribution. Discuss the relationship between the test of hypotheses and the confidence interval.

e. Test the hypotheses in part (b) and calculate an interval estimate, using a t distribution. Discuss the relationship between the test of hypotheses and the confidence interval.

f. Compare the results of the tests of hypotheses in parts (c), (d), and (e). Discuss your findings.

g. Discuss the results of the experiment, taking into consideration now the personal characteristics of the volunteers.

h. The investigators chose to exclude the results for volunteer 3 from the analysis because she did not report a sweet taste within 45 minutes. We must always be very cautious when excluding cases. Data values that are different from the others may represent extreme ends of the probability distribution for the observations, or they may represent unusual conditions. Either way, it is unwise to exclude such values without careful thought. In this experiment, the investigators believed that if a volunteer had not detected the sweet taste within 45 minutes, the results should be excluded. (Because of the solubility of saccharin, it would be difficult to interpret responses after 45 minutes.) How does the exclusion of results for this volunteer affect your interpretations of the experimental results?

EXERCISE 11-20

Eight healthy young volunteers participated in a control study of nasal clearance times (Cederlund, Camner, and Svartengren, 1987). The original study was designed to compare nasal transport times before jogging and approximately 1 hour later (see Exercise 11-19). In this control study, nasal transport times were measured while subjects were at rest, the second measurement 1 hour after the first.

	Transport time (minutes)	
Volunteer	First measurement	Second measurement
2	18.5	20.5
4	12.5	20.5
5	25.0	20.0
6	16.0	12.0
7	12.5	10.5
9	15.5	14.0
10	10.0	11.0
11	15.5	15.0

a. Plot the observations in any way that seems helpful.

b. Is there evidence of differences in nasal transport times taken 1 hour apart, on average? State and test appropriate hypotheses.

c. Calculate an interval estimate for the mean difference in transport times taken 1 hour apart. Discuss your findings.

EXERCISE 11-21

Investigators wanted to determine physical characteristics of urine with and without calcium oxalate crystals. They measured calcium concentrations (in millimoles/liter) from urine specimens of 34 men with calcium oxalate crystals and 45 men without crystals (data were obtained from the laboratory of Dr. James S. Elliot and contributed by D. P. Byar to a collection of problems in Andrews and Herzberg, 1985, pages 249–251). The results are shown below.

	Calcium concentrations in urine (millimoles/liter)					
Men with crystals	6.96	13.00	5.54	6.19	7.31	14.34
	4.74	2.50	1.27	4.18	3.10	3.01
	6.81	8.28	2.33	7.18	5.67	12.68
	8.94	3.16	3.30	6.99	.65	4.18
	4.45	.27	7.64	6.63	8.53	9.04
	.58	7.82	12.20	9.39		
	(sample size = 34, mean = 6.143, sample standard deviation = 3.637)					
Men without crystals	2.45	4.49	2.36	2.15	1.16	3.34
	1.40	8.48	1.16	2.21	1.93	1.27
	1.03	1.47	1.53	5.09	1.05	2.03
	7.68	1.45	5.16	.81	1.32	1.55
	1.52	.77	2.17	.17	.83	3.04
	1.06	3.93	5.38	3.53	4.54	3.98
	1.02	3.46	1.19	5.64	2.66	1.22
	2.64	2.31	4.49			
	(sample size = 45, mean = 2.625, sample standard deviation = 1.863)					

a. Graph the calcium concentrations for the two groups of men. Describe and compare the two distributions.

b. State and test appropriate hypotheses to compare mean calcium concentrations for the two groups.

c. Calculate an interval estimate for the difference between the two population means.

d. Discuss your results.

EXERCISE 11-22

In a study comparing a cost-reduced product with a current product, experimenters asked 200 judges their preference on a pair of samples, A and B. Half of the judges tasted sample A first and the other half tasted sample B first. Ignoring the order of tasting, the preference results are shown below (contributed by M. B. Carroll of General Foods Corporation to a collection of problems in Andrews and Herzberg, 1985, pages 189–193):

Preference	Number of judges
Sample A	98
Sample B	88
No preference	14

a. Is there evidence that one sample is preferred over the other? State and test appropriate hypotheses.

b. Calculate an interval estimate for the proportion of judges preferring sample A. Discuss your findings.

EXERCISE 11-23 Investigators studied a method of measuring aflatoxin concentration in contaminated peanuts. (In sampling inspection, inspectors want to protect consumers while not rejecting too many good peanuts.) They ground the peanuts into meal and divided the meal into separate samples. They blended each sample in a chemical solution. For each sample, the experimenters divided the blend equally among 16 centrifuge bottles. The determination of aflatoxin concentration for each bottle (units not given) is shown below (Quesenberry, Whitaker, and Dickens, 1976; from Waltking, Bleffert, and Kiernan, 1968).

Sample 1:	95.33	55.94	72.01	58.96	114.62	41.64
	98.76	53.62	90.23	91.92	66.88	91.81
	100.37	77.26	91.56	66.25		
Sample 2:	20.04	20.30	24.69	22.26	24.92	21.45
	19.44	24.04	24.40	19.85	11.88	24.34
	15.32	14.85	23.10	22.21		

Compare the aflatoxin determinations for the two samples. You may find it useful to take the logarithm of each value for part of your analysis.

EXERCISE 11-24 Recall that in Example 9-2, the student wished to compare the median distance pedaled on an exercise cycle under two experimental conditions: with a Walkman and without a Walkman.

a. Make the comparison using the Wilcoxon–Mann–Whitney test.

b. Make the comparison using the two-sample t test.

c. Compare the results in parts (a) and (b) with what we found using the median test in Example 9-2.

EXERCISE 11-25 Refer to the base-running experiment in Example 11-6.

a. Test the hypotheses and calculate a confidence interval based on ranks.

b. Test the hypotheses using the sign test. Calculate a confidence interval based on a binomial distribution.

c. Compare the tests of hypotheses and confidence intervals in parts (a) and (b) with what we found using a t distribution in Example 11-6.

EXERCISE 11-26 In the appendix on the Wilcoxon–Mann–Whitney distributions at the end of the book, we find the Wilcoxon–Mann–Whitney distribution of T_1 for sample sizes 2 and 4. Show that T_2 has this same probability distribution.

EXERCISE 11-27 Find the Wilcoxon–Mann–Whitney distribution for samples of size 3 and 3.

EXERCISE 11-28 Refer to Example 11-3, comparing weights of tomatoes canned in the morning and afternoon.

a. Test the null hypothesis that the median weight is the same in the morning and the afternoon, using the median test.

b. Repeat part (a) using the Wilcoxon–Mann–Whitney test.

c. Compare the results in parts (a) and (b) with what we found using the two-sample t test in Example 11-3.

EXERCISE 11-29 Refer to Example 11-4, comparing tumor weights in treated and untreated mice.

a. Test the null hypothesis that the median tumor weight is the same for treated and untreated mice, using the median test.

b. Repeat part (a) using the two-sample t test.

c. Compare the results in parts (a) and (b) with what we found using the Wilcoxon–Mann–Whitney test in Example 11-4.

EXERCISE 11-30 Refer to Example 11-5, comparing yields of microwaved and unmicrowaved oranges.

a. Test the null hypothesis that the mean yield is the same for microwaved and unmicrowaved oranges, using the Wilcoxon–Mann–Whitney test.

b. Repeat part (a) using the two-sample t test.

c. Compare the results in parts (a) and (b) with what we found using the median test in Example 11-5.

EXERCISE 11-31 Discuss the sampling situations in which the two-sample t test, the Wilcoxon–Mann–Whitney two-sample test, and the median test are appropriate. Which test is preferred in each of these situations?

EXERCISE 11-32 In Example 11-2 we mentioned that local health departments chose between two different experimental designs for the polio field trials. We discussed the double-blind randomized placebo-controlled design. Under this design, parents of first-, second-, and third-grade children in participating school districts were asked for permission to include their children in the study. Children with parental consent were randomly divided into two groups. The children in the treatment group received a vaccine injection; children in the control group received a placebo (saline solution) injection. The experiment was double-blind, meaning that neither the children nor the immediate caregivers nor the workers making the diagnosis knew the treatment received by any inoculated child.

The other experimental design offered may be termed the NFIP design because it was favored by the National Foundation for Infantile Paralysis. Under this design, parents of second-graders were asked for permission to include their children in the study. All second-grade children with parental consent received a vaccine injection. The second-graders without parental consent, as well as all first- and third-grade children, received no injection, and served as control groups.

a. Discuss the NFIP design. What problems do you see with this experimental design? Under what conditions would this design be likely to bias the results in favor of the polio vaccine? Under what conditions would this design be likely to bias the results against the polio vaccine? What are the advantages of the double-blind randomized placebo-controlled design over the NFIP design? In particular, discuss the control groups to be compared with the vaccinated group in each study.

The results of the two studies can be summarized as shown below, with numbers rounded for this discussion (Freedman, Pisani, and Purves, 1978, page 6; from Francis, 1955). Instead of listing proportion diagnosed with polio in each group, we show the rate, or number of cases of polio per 100,000 children. Rates are often used instead of proportions by workers reporting results in public health.

Group	Number of children	Polio cases per 100,000 children
Double-blind randomized placebo-controlled design		
Vaccinated	200,000	28
Placebo control	200,000	71
No parental consent	350,000	46
NFIP design		
Vaccinated second-graders	225,000	25
No-consent second-graders	125,000	44
All first- and third-graders	725,000	54

b. Discuss the results of these two sets of experiments. Which design do you think gave the clearer comparison of vaccination versus no vaccination? Consider again your discussion of the two experimental designs in light of these results.

c. Show the relationship between the proportion of polio cases in a group and the rate or number of cases per 100,000 children.

Note: In fact, children of parents in higher socioeconomic levels were more likely to receive permission to participate in the study and also were more likely to develop polio. How does this information fit in with your discussion?

CHAPTER 12

Comparing Several Means: Single-Factor and Randomized Block Experiments

IN THIS CHAPTER

Multiple comparisons (Bonferroni method)

One-way analysis of variance

Kruskal–Wallis test

Randomized block experiments

Friedman's test

Are there any differences in average fruit yield among apple trees treated with four different fertilizer supplements? Do four thermometers all give the same average reading of the melting point of a substance? What are the relative biological effects, on average, of three different anesthetics? Each of these questions concerns a comparison of three or more measures of central tendency.

We have already discussed ways to compare two measures of central tendency, considering both two-sample and paired-sample experimental designs. Now we consider inferences about more than two measures of central tendency, in single-factor experiments and randomized block experiments. The single-factor experiment is an extension to several populations of the two-sample experimental design, whereas the randomized block experiment is an extension of the paired-sample experimental design.

In a single-factor experiment, we have several independent random samples, one from each of several populations of interest. We want to use the samples to compare the means of the populations. The single-factor experiment is an extension of the two- (independent) sample experimental design we discussed in Chapter 11. In Section 12-2 we discuss the classical, parametric, way to test whether several means are equal: one-way analysis of variance. We consider a nonparametric analysis, the Kruskal–Wallis test, in Section 12-3.

A randomized block experimental design is an extension of the paired-sample design. In Section 12-4 we discuss the classical, parametric, analysis of a randomized block experiment. We consider a nonparametric analysis, Friedman's test, in Section 12-5.

We begin in Section 12-1 with a fairly crude, but useful, approach to comparing several means: the Bonferroni method of comparing means (or medians) two at a time.

12-1 Comparing Measures of Central Tendency Two at a Time Using the Bonferroni Method

Why do we need special procedures for comparing more than two location parameters? Why can we not just compare them two at a time? There is a problem with that strategy. To illustrate, suppose we have four independent random samples and we want to make inferences about the medians of the populations sampled.

For simplicity, let's consider just two possible comparisons. Suppose we test whether the medians of the first two populations are equal:

$$H_0\colon\ M_1 = M_2 \qquad \text{versus} \qquad H_a\colon\ M_1 \neq M_2$$

and we also test whether the medians of the other two populations are equal:

$$H_0\colon\ M_3 = M_4 \qquad \text{versus} \qquad H_a\colon\ M_3 \neq M_4$$

Since the samples are independent, we can say these two tests are independent.

For each of these two sets of hypotheses, we use a test statistic to measure how far the observed results are from what we would expect under the appropriate null hypothesis. To test the first set of hypotheses, we select a decision rule for deciding whether the two population medians M_1 and M_2 are different—say, with significance level .05. Similarly, we choose a decision rule for deciding whether the two population medians M_3 and M_4 are different, also with significance level .05.

Consider the two tests together. We can think of a combined null hypothesis that the first two population medians are equal *and* the other two population medians are equal:

$$H_0: \quad M_1 = M_2 \text{ and } M_3 = M_4$$

The combined alternative states that at least one of these equalities does not hold:

$$H_a: \quad M_1 \neq M_2 \text{ or } M_3 \neq M_4$$

We reject the combined null hypothesis if either test statistic is in the corresponding rejection region. What is the significance level for this combined criterion?

The probability that the results are consistent with the null hypothesis H_0: $M_1 = M_2$ when these two medians really are equal is $1 - .05 = .95$ (because our significance level is .05). Likewise, the probability that the results are consistent with the null hypothesis H_0: $M_3 = M_4$ when these two medians really are equal is $1 - .05 = .95$.

Suppose the combined null hypothesis is true, so $M_1 = M_2$ and $M_3 = M_4$. The chance the results are consistent with this combined null hypothesis is $.95 \times .95 = .9025$, because the probability of two independent events occurring together is the product of the separate probabilities (Chapter 6). Therefore, the significance level for the combined test is $1 - .9025 = .0975$, almost twice the significance level for either of the separate tests! We have nearly a 10% chance of rejecting the combined null hypothesis when that combined null hypothesis is really true (that is, when $M_1 = M_2$ and $M_3 = M_4$). To test the overall null hypothesis H_0: $M_1 = M_2 = M_3 = M_4$ is even worse, since more pairwise comparisons are necessary and they are not all independent.

One way around this difficulty is to use the *Bonferroni method* for controlling the overall significance level (Rice, 1988, page 384). Suppose m pairwise comparisons are necessary to test a combined null hypothesis. We select a decision rule for each pairwise comparison. Denote the m significance levels associated with these m decision rules by α_1 through α_m. If α is the significance level for the overall test, then α is less than or equal to the sum of α_1 through α_m (Exercise 12-16).

The **Bonferroni method** is a technique for obtaining an upper bound on an overall significance level.

Suppose we have m separate tests of hypotheses. Using the significance level approach, we have a decision rule and associated significance level for each test. For a combined test of hypotheses, we say the results are inconsistent with the combined null hypothesis if any one of the m separate test statistics is in its associated rejection region. The significance level for this combined test is less than or equal to the sum of the significance levels of the separate tests.

By making the pairwise significance levels α_1 through α_m small, we can control the size of the overall significance level α. We will use this idea for the multiple comparisons procedures we discuss in Sections 12-2 and 12-3, when we want to calculate interval estimates for the differences between measures of central tendency for three or more populations.

Multiple comparisons of means refers to the process of comparing several means. For our purposes, multiple comparisons refers to the process of comparing several means, two at a time.

There is another way around the difficulty of comparing several measures of central tendency in a single-factor experiment. We can use one-way analysis of variance (Section 12-2) or the Kruskal–Wallis test (Section 12-3) to test the null hypothesis that the population means (or medians) are all equal. One-way analysis of variance is an extension of the two-sample t test (Section 11-3) to several independent samples. The Kruskal–Wallis test is an extension to several independent samples of the Wilcoxon–Mann–Whitney test for two independent samples (Section 11-4).

12-2 Inferences About Several Means in a Single-Factor Experiment: One-Way Analysis of Variance

In a single-factor experiment, we have several independent random samples and we want to use these observations to make inferences about the populations sampled.

In a **single-factor experiment**, we have several independent random samples and we want to make inferences about the populations sampled.

We are here concerned with making inferences about the means of the populations sampled. Suppose we have k independent random samples, one from each of k populations. For a classical analysis, we assume that the values in each population follow a Gaussian distribution and that all k of these Gaussian distributions have the same variance, σ^2. Let μ_1 through μ_k denote the k population means. We want to test the null hypothesis that these means are all equal:

$$H_0: \quad \mu_1 \text{ through } \mu_k \text{ are all equal}$$

The alternative hypothesis is that these means are not all equal:

$$H_a: \quad \mu_1 \text{ through } \mu_k \text{ are not all equal}$$

We call the procedure for testing these hypotheses *one-way analysis of variance.*

> **One-way analysis of variance** is the classical, parametric, approach to testing the null hypothesis that the population means are all equal in a single-factor experiment.

Let's look at an example; then we will discuss one-way analysis of variance and apply it to this example.

EXAMPLE 12-1

Does a nitrogen supplement improve apple production? To address this question, researchers divided Jonathan apple trees into four treatment groups. They applied no nitrogen to the trees in the control group. They provided a nitrogen supplement to the trees in the other three groups: either urea, potassium nitrate plus calcium, or ammonia plus ammonium sulphate. The researchers stored fruits of 42 trees for 4 months. They then weighed samples of fruit from each tree. Fruit weight, in grams, for each tree is shown below (data contributed by D. A. Ratkowsky to a collection of problems in Andrews and Herzberg, 1985, pages 355–356; from D. A. Ratkowsky and D. Martin, 1974). The researchers wanted to use the sample weights to compare mean fruit production under the four treatments.

Control	Urea	Potassium nitrate and calcium	Ammonia and ammonium sulphate
85.3	117.5	127.1	77.4
113.8	98.9	108.5	91.3
92.9	108.5	99.9	91.3
48.9	104.4	124.8	81.7
99.4	96.8	94.5	89.2
79.1	94.5	99.4	69.6
70.0	90.6	117.5	69.0
86.9	100.8	135.0	73.7
87.7	96.0	85.6	75.1
67.3	99.9	102.5	87.0
	84.6	110.8	

A plot of the observations by treatment group is shown in Figure 12-1. What does this plot suggest about the relative effects of the four treatments upon apple production? What suggestions would you make for carrying out this experiment, in order to reduce the effects of extraneous factors and allow valid comparisons across treatment groups?

We will use one-way analysis of variance to test the null hypothesis that the mean fruit weight is the same for all four treatments in Example 12-1. Before we can discuss this procedure, we need some notation.

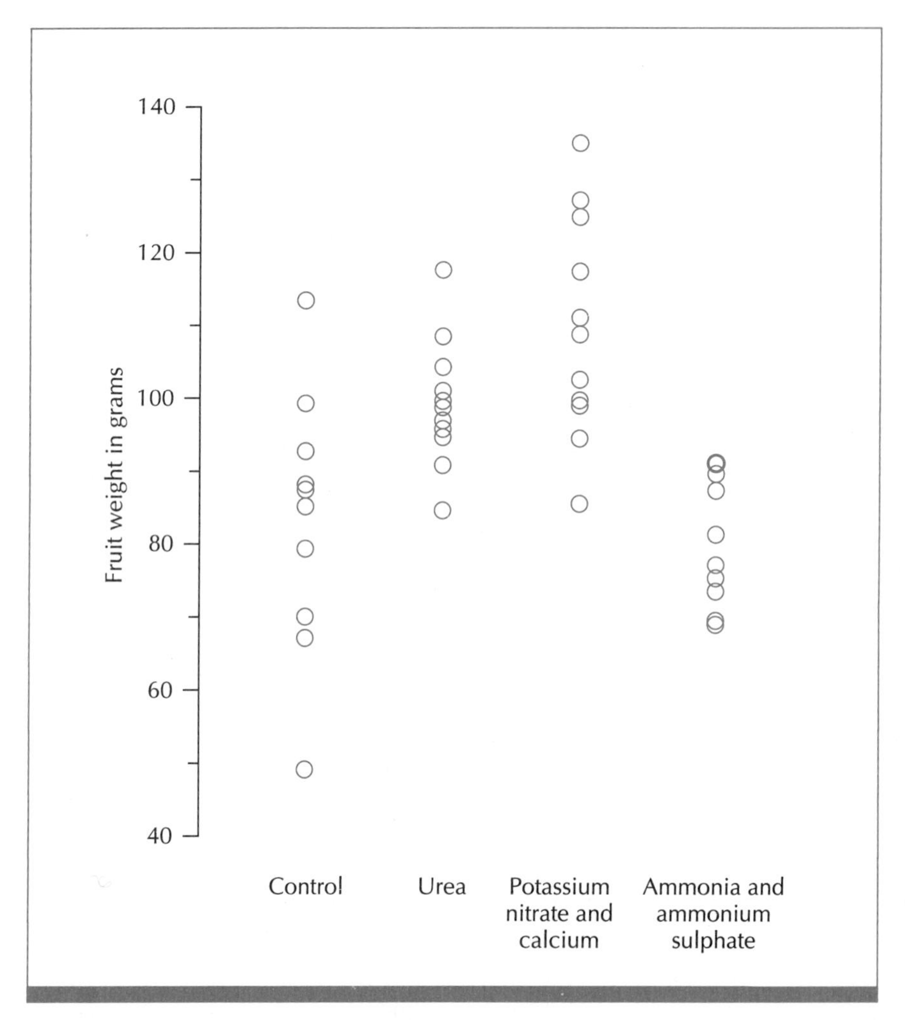

FIGURE 12-1 Fruit weight (in grams) from apple trees in four treatment groups, Example 12-1

One-Way Analysis of Variance for a Single-Factor Experiment

Suppose we have k samples. Let Y_{ij} denote the observation on experimental unit j in sample i. Let $\overline{Y}_i$ denote the sample mean, s_i^2 the sample variance, and n_i the size of sample i. As we did in the two-sample case, we can combine the separate sample variances into a pooled estimate of σ^2. This *pooled estimate of* σ^2 is a weighted average of s_1^2 through s_k^2, calculated as follows:

The pooled estimate of the common population variance σ^2 in a single-factor experiment is

$$s_r^2 = \frac{\sum_{i=1}^{k} (n_i - 1)s_i^2}{\sum_{i=1}^{k}(n_i - 1)} = \frac{\sum_{i=1}^{k}\sum_{j=1}^{n_i} (Y_{ij} - \bar{Y}_i)^2}{N - k}$$

Here, N is the total sample size, the sum of n_1 through n_k. The double summation notation means that for each group i, we sum over the observations in that group, then sum over the groups.

Common names for this variance estimator s_r^2 are **residual mean square, within-groups mean square**, and **mean square within.**

We use the notation s_r^2 because it is an estimate of σ^2 based on residuals. Recall that a residual is the difference between an observation and a summary or estimate for the mean of the observation. In the case of a single-factor experiment, a residual is the difference between an observation and its group mean. (For more on residuals, see Tukey, 1977.)

A **residual** is the difference between an observation and an estimate of its expected value.

In a single-factor experiment, a residual is the difference $Y_{ij} - \bar{Y}_i$ between an observation and the average of all the observations in the same group.

In a single-factor experiment, the group mean is a summary value, estimating the mean of the population sampled. We see from the definition that s_r^2 is an average of the squares of residuals $Y_{ij} - \bar{Y}_i$; hence the name residual mean square.

The *between-groups variance estimate* is another measure of variation we need for our analysis:

The between-groups variance estimate in a single-factor experiment is

$$s_B^2 = \frac{\sum_{i=1}^{k} n_i(\bar{Y}_i - \bar{Y})^2}{k - 1}$$

where $\bar{Y}$ is the average of all N observations in the combined samples.

s_B^2 is sometimes called the **between-groups mean square** or **mean square between.**

We want to test the null hypothesis that the means of the populations sampled are all equal. The test statistic is the ratio of the between-groups variance estimate and the within-groups variance estimate:

The test statistic for testing the null hypothesis that the population means are all equal, in a single-factor experiment, is

$$\text{Test statistic} = \frac{s_B^2}{s_r^2}$$

To find the probability distribution of the test statistic under the null hypothesis, assume that we have k independent random samples from Gaussian populations with the same variance. Then under the null hypothesis the test statistic has the F distribution with $k - 1$ numerator degrees of freedom and $N - k$ denominator degrees of freedom.

The numerator degrees of freedom and denominator degrees of freedom are two constants (or parameters) that define an F distribution. The test statistic defined above has an F distribution under the null hypothesis that all the population means are equal. The numerator degrees of freedom of this F distribution equal $k - 1$, used in calculating the *numerator* of the test statistic, s_B^2. (We call $k - 1$ the degrees of freedom associated with the between-groups mean square s_B^2.) The denominator degrees of freedom of this F distribution equal $N - k$, used in calculating the *denominator* of the test statistic, s_r^2. (We call $N - k$ the degrees of freedom associated with the residual mean square s_r^2.)

In general, an F distribution has numerator degrees of freedom d_1 and denominator degrees of freedom d_2. We denote such an F distribution by $F(d_1, d_2)$. An F distribution is a continuous probability distribution that is skewed to the right. A random variable having an F distribution takes on positive values only. (Our test statistic is a ratio of two variance estimates, and so can have only positive values.) The shape of an F distribution is illustrated in Figure 12-2. Table D at the back of the book lists values of c for which

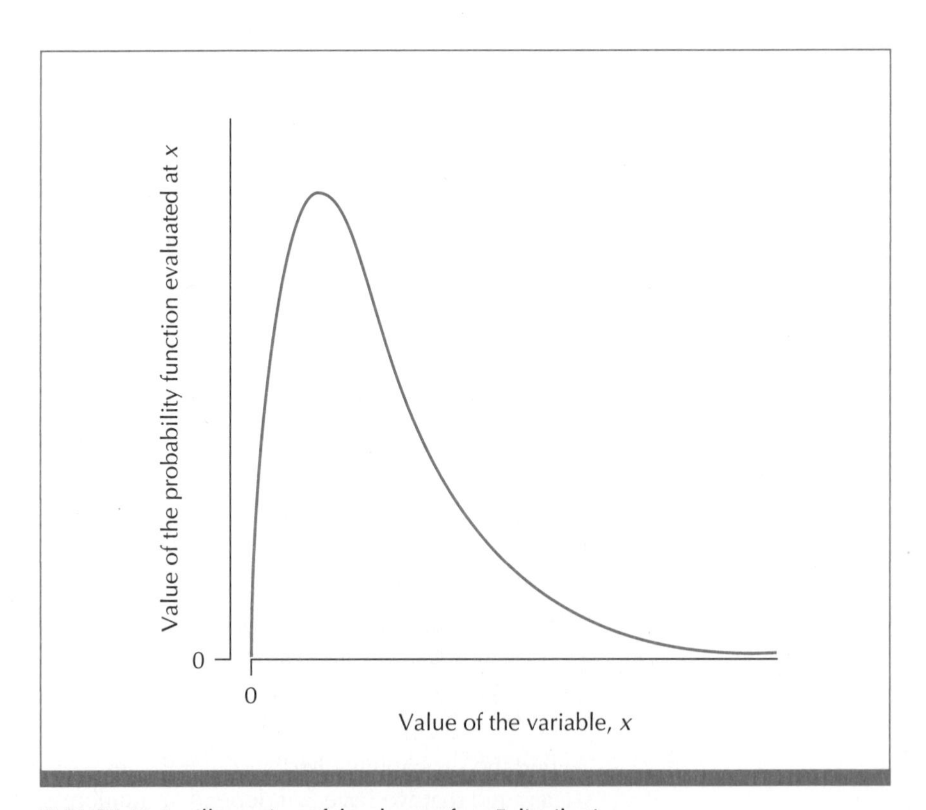

FIGURE 12-2 Illustration of the shape of an F distribution

$P(F \leq c)$ has specified values, where F is a random variable having selected numerator degrees of freedom d_1 and denominator degrees of freedom d_2.

An F distribution is also called a variance ratio distribution. Our test statistic is a ratio of two variance estimates, which has an F distribution if the null hypothesis is true. Even though we are comparing hypotheses about population means, we call the procedure *analysis of variance* because our test statistic is a ratio of two variance estimates.

We know that s_r^2 is an estimate of σ^2, the variance in each population, while s_B^2 is a measure of variation among the sample means. If the null hypothesis is true, the sample means $\overline{Y}_1$ through $\overline{Y}_k$ all estimate the common mean of the k populations. Then s_B^2 is another estimate of σ^2. If the null hypothesis is not true, $\overline{Y}_1$ through $\overline{Y}_k$ do not all estimate the same mean. Then s_B^2 estimates the variation within populations plus the variation between the population means μ_1 through μ_k. Therefore, if the null hypothesis is not true, s_B^2 estimates something larger than σ^2.

With this in mind, we see that values of the test statistic near 1 are consistent with the null hypothesis. Values of the test statistic much larger than 1 are inconsistent with the null hypothesis. Using these ideas, we outline the significance level approach to one-way analysis of variance.

The significance level approach to comparing several means in a single-factor experiment, using one-way analysis of variance

1. The hypotheses are H_0: μ_1 through μ_k are all equal, and H_a: μ_1 through μ_k are not all equal, where μ_1 through μ_k represent the population means.
2. The test statistic is s_B^2/s_r^2, as defined above.
3. Assume that we have independent random samples from k Gaussian distributions with equal variances. Let N denote the sum of the k individual sample sizes. Then under the null hypothesis, the test statistic has the F distribution with $k - 1$ numerator degrees of freedom and $N - k$ denominator degrees of freedom.
4. Select significance level α.
5. Let F denote a random variable having the $F(k - 1, N - k)$ distribution. Find c from Table D at the back of the book such that $P(F \leq c) = 1 - \alpha$. Then the acceptance region is the interval $[0, c)$; the rejection region is the interval $[c, \infty)$.
6. The decision rule is:

   If test statistic $< c$, say the results are consistent with the null hypothesis.
   If test statistic $\geq c$, say the results are inconsistent with the null hypothesis.

7. Carry out an experiment that satisfies the conditions in step 3. Calculate the test statistic in step 2. Use the decision rule in step 6 to decide whether the results are consistent with the null hypothesis.

EXAMPLE 12-1 *(continued)*

In Example 12-1, we want to compare the effects of several nitrogen supplements upon fruit production in Jonathan apple trees. The null hypothesis states that the mean fruit weight is the same for all four treatments. The alternative states that the mean fruit weight is not the same for all four treatments. We can rewrite these hypotheses as

$$H_0: \mu_c = \mu_u = \mu_p = \mu_a$$
$$H_a: \text{The four means are not all equal}$$

where each subscript denotes a treatment group.

Assume that we have four independent random samples from Gaussian distributions with equal variances. Then under the null hypothesis, our test statistic would have the F distribution with $4 - 1 = 3$ numerator degrees of freedom and $42 - 4 = 38$ denominator degrees of freedom (since the total sample size N equals 42 and there are $k = 4$ treatment groups).

To verify the independence assumption, we would have to know more about how the experiment was conducted. What suggestions do you have for ensuring independence?

Figure 12-1 gives us no reason to doubt that each sample comes from a Gaussian distribution, since each of the four sample distributions is fairly symmetric. (One-way analysis of variance tends to be robust to deviations from the Gaussian assumption.)

The variation in fruit weights is somewhat larger in the control group and the potassium nitrate plus calcium group than in the other two groups, the least variation being in the ammonia plus ammonium sulphate group. However, these differences in variation are not extreme enough to make one-way analysis of variance seem inappropriate. (As with the two-sample t test, one-way analysis of variance is fairly robust to deviations from the equal-variance assumption. As long as the variances are not too different, actual significance levels and confidence levels are close to the levels we choose.)

Let's use significance level $\alpha = .01$. Since $1 - \alpha = .99$, we use the last page of Table D. For our test, there are 3 numerator degrees of freedom and 38 denominator degrees of freedom. Table D shows 3 numerator degrees of freedom but not 38 denominator degrees of freedom. We must choose either 30 or 40 for denominator degrees of freedom in the table. To be conservative (less likely to reject the null hypothesis), we will use the smaller value, 30. Then looking in the column for $d_1 = 3$ and the row for $d_2 = 30$, we find $c = 4.51$. The acceptance region is $[0, 4.51)$, the rejection region is $[4.51, \infty)$, and the decision rule is:

If test statistic < 4.51, say the results are consistent with the null hypothesis that there is no difference in mean fruit weight among the four treatments.

If test statistic ≥ 4.51, say the results are inconsistent with the null hypothesis, suggesting there is a difference in mean fruit weight among the four treatments.

The calculations we need for our analysis are outlined in Table 12-1.

Since the test statistic equals 11.28, we say the results are inconsistent with the null hypothesis, at the .01 significance level. The p-value, $P(\text{test statistic} \geq 11.28 \text{ when } H_0 \text{ is true})$, is less than .01. This experiment suggests that mean

TABLE 12-1 Steps in calculating the one-way analysis of variance test statistic

Control	Urea	Potassium nitrate and calcium	Ammonia and ammonium sulphate
$\bar{Y}_c = 83.130$	$\bar{Y}_u = 99.318$	$\bar{Y}_p = 109.600$	$\bar{Y}_a = 80.530$
$s_1^2 = 327.949$	$s_2^2 = 77.662$	$s_3^2 = 230.006$	$s_4^2 = 76.525$
$n_1 = 10$	$n_2 = 11$	$n_3 = 11$	$n_4 = 10$

$N = 42 \quad \bar{Y} = 93.683$

$$s_r^2 = \frac{(10 - 1)(327.949) + (11 - 1)(77.662) + (11 - 1)(230.006) + (10 - 1)(76.525)}{42 - 4} = 176.76$$

$$s_B^2 = \frac{10(83.130 - 93.683)^2 + 11(99.318 - 93.683)^2 + 11(109.600 - 93.683)^2 + 10(80.530 - 93.683)^2}{4 - 1}$$

$$= 1{,}993.27$$

$$\text{Test statistic} = \frac{1{,}993.27}{176.76} = 11.28$$

Numerator degrees of freedom = 4 − 1 = 3
Denominator degrees of freedom = 42 − 4 = 38

TABLE 12-2 Analysis of variance table for a single-factor experiment

Source of variation	Sum of squares	Degrees of freedom	Mean square	Test statistic
Treatments (between groups)	$\sum_{i=1}^{k} n_i(\bar{Y}_i - \bar{Y})^2$	$k - 1$	s_B^2	$\frac{s_B^2}{s_r^2}$
Residual (within groups)	$\sum_{i=1}^{k}\sum_{j=1}^{n_i} (Y_{ij} - \bar{Y}_i)^2$	$N - k$	s_r^2	
Total	$\sum_{i=1}^{k}\sum_{j=1}^{n_i} (Y_{ij} - \bar{Y})^2$	$N - 1$		

fruit weight is not the same for all four treatments. Looking at Figure 12-1, can you make a decision as to which group means are nearly the same and which are very different?

The Analysis of Variance Table for One-Way Analysis of Variance

We often summarize the calculations of one-way analysis of variance in a table called an *analysis of variance table.* Computer output for one-way analysis of variance is displayed in such a table. A general form of analysis of variance table for a single-factor experiment is shown in Table 12-2.

Computer output often has another column at the right of the table, showing the p-value associated with the test statistic. The last row in the table, the total row, is the sum of the previous two rows for sum of squares and degrees of freedom. We include this total row for completeness; we do not

TABLE 12-3 Analysis of variance table for the single-factor experiment in Example 12-1

Source of variation	Sum of squares	Degrees of freedom	Mean square	Test statistic	*p*-value
Between groups	5,979.8	3	1,993.27	11.28	0.0000
Residual (within groups)	6,716.9	38	176.76		
Total	12,696.7	41			

use it in our test of hypotheses. The analysis of variance table for Example 12-1 is shown in Table 12-3. The *p*-value is listed as 0.0000. This means that the *p*-value was less than .0001.

If our results are inconsistent with the null hypothesis, we would like to see where the differences are. Which population means seem to be similar and which seem to be different? To address this question formally, we can use a *multiple comparisons procedure.* There are many ways to make multiple comparisons. We will use a procedure based on the *Bonferroni method.*

We calculate a confidence interval for the difference $\mu_i - \mu_j$ between the means for populations i and j as

$$\bar{Y}_i - \bar{Y}_j \pm c\sqrt{\frac{s_r^2}{n_i} + \frac{s_r^2}{n_j}}$$

The subscripts i and j refer to samples i and j, respectively. The residual mean square s_r^2 is the pooled (within-groups) variance estimate based on the k samples. The number c comes from the t distribution with $N - k$ degrees of freedom, where N is the total sample size and k is the number of samples.

Suppose we calculate m such confidence intervals, making m pairwise comparisons of population means. Denote the confidence levels associated with these intervals by A_1 through A_m. Then the confidence level associated with the m intervals taken together is greater than or equal to $1 - \Sigma_{i=1}^{m}(1 - A_i)$.

The **Bonferroni method** is a technique for obtaining a lower bound on an overall confidence level.

Suppose we make m pairwise comparisons of means, with confidence levels A_1 through A_m. Using the Bonferroni method, we say the confidence level for the m intervals taken together is greater than or equal to

$1 - (1 - A_1) - (1 - A_2) - \cdots - (1 - A_m)$.

EXAMPLE 12-1 *(continued)*

Let's make multiple comparisons of mean fruit weights in Example 12-1, using the Bonferroni method. Because there are four groups, there are $\binom{4}{2} = 6$ possible pairwise comparisons, and we will let $m = 6$. Each separate interval will have confidence level .99. The total sample size is 42 and there are four

TABLE 12-4 Multiple comparisons for Example 12-1

Confidence interval for	Confidence interval
$\mu_c - \mu_u$	$83.1 - 99.3 \pm 2.750\sqrt{176.76\left(\frac{1}{10} + \frac{1}{11}\right)} = (-32.2, -.2)$
$\mu_c - \mu_p$	$83.1 - 109.6 \pm 2.750\sqrt{176.76\left(\frac{1}{10} + \frac{1}{11}\right)} = (-42.5, -10.5)$
$\mu_c - \mu_a$	$83.1 - 80.5 \pm 2.750\sqrt{176.76\left(\frac{1}{10} + \frac{1}{10}\right)} = (-13.8, 19.0)$
$\mu_u - \mu_p$	$99.3 - 109.6 \pm 2.750\sqrt{176.76\left(\frac{1}{11} + \frac{1}{11}\right)} = (-25.9, 5.3)$
$\mu_u - \mu_a$	$99.3 - 80.5 \pm 2.750\sqrt{176.76\left(\frac{1}{11} + \frac{1}{10}\right)} = (2.8, 34.8)$
$\mu_p - \mu_a$	$109.6 - 80.5 \pm 2.750\sqrt{176.76\left(\frac{1}{11} + \frac{1}{10}\right)} = (13.1, 45.1)$

The confidence level for each interval is .99. Therefore, the overall confidence level is greater than or equal to $1 - (.01 + .01 + .01 + .01 + .01 + .01) = .94$.
Note: c = control, u = urea, p = potassium nitrate plus calcium, a = ammonia plus ammonium sulphate.

groups, so we get c from the t distribution with $42 - 4 = 38$ degrees of freedom. In Table C, we have a choice between 30 and 40 degrees of freedom. We will be more conservative (getting wider intervals) and use 30. Then $c = 2.750$. The calculations for the six confidence intervals are outlined in Table 12-4.

Zero is in the confidence interval for $\mu_c - \mu_a$ and the confidence interval for $\mu_u - \mu_p$, but not in the other intervals. Taken together, the intervals suggest that mean fruit weight is the same for the control group and the ammonia plus ammonium sulphate group. Mean fruit weight also seems to be the same for the urea group and the potassium nitrate plus calcium group; these two treatments appear to have mean fruit weights greater than the control and ammonia plus ammonium sulphate groups. This agrees with the visual comparisons we can make by examining the four distributions in Figure 12-1.

In Section 12-3, we discuss nonparametric tests of hypotheses and multiple comparisons for a single-factor experiment.

12-3 Nonparametric Analysis of a Single-Factor Experiment: The Kruskal–Wallis Test

The Kruskal–Wallis test is a nonparametric procedure used to check for equality of several distributions in a single-factor experiment. We will start with an example, then outline the significance level approach to the test of hypotheses, and apply it to the example. Finally, we discuss nonparametric multiple comparisons based on the Bonferroni method.

EXAMPLE 12-2 Maximal oxygen uptake is a measure of physical working capacity or aerobic power. In a survey of aerobic power in world-class athletes, Wilmore lists maximal oxygen uptake for nine young women athletes: three basketball players, four cross-country skiers, and two speed skaters (Wilmore, 1984). The results are shown below:

Sport	Maximal oxygen uptake $\left(\frac{\text{ml}}{\text{kg} \cdot \text{min}}\right)$
Basketball	42.3, 42.9, 49.6
Cross country skiing	56.9, 58.1, 61.5, 68.2
Speed skating	46.1, 52.0

Are there differences in aerobic power among women athletes in these three sports? How would you design an experiment to answer this question? What would you do to reduce the effects of extraneous factors? How should the experiment be conducted to ensure valid comparisons among the three sports?

A plot of the observations is shown in Figure 12-3. What does this plot

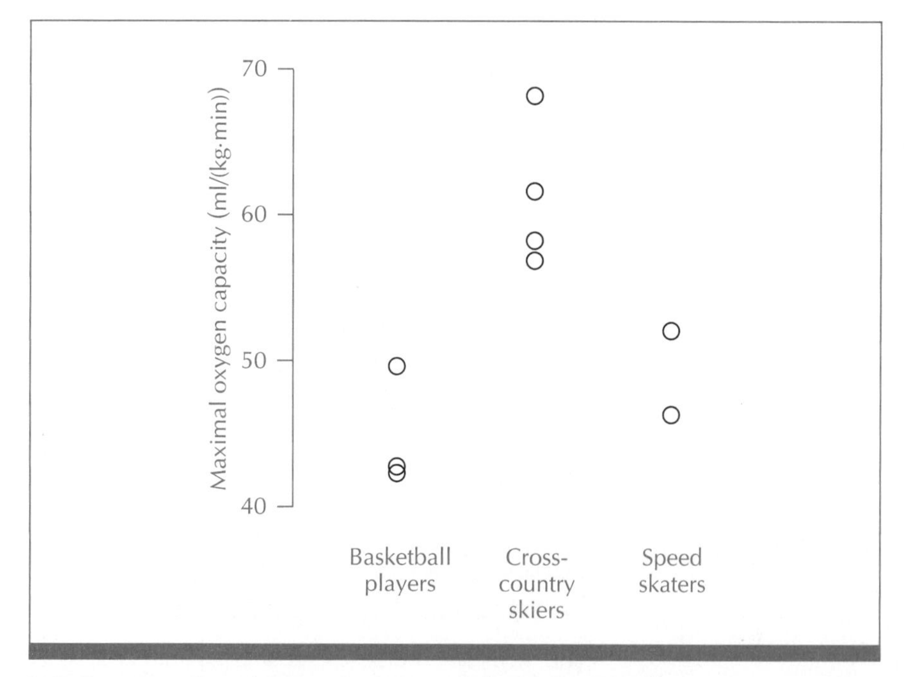

FIGURE 12-3 Plots of maximal oxygen uptake for female basketball players, cross-country skiers, and speed skaters in Example 12-2

suggest about the relative levels of aerobic power in women basketball players, cross-country skiers, and speed skaters? Which groups of athletes seem similar and which seem very different, with respect to aerobic power?

We will use the Kruskal–Wallis test to compare the distributions of maximal oxygen uptake for women in the three sports. Before applying the test to this example, let's first describe it in general.

The Kruskal–Wallis Test for a Single-Factor Experiment

Suppose we have k independent random samples, one from each of k populations. The k distributions are continuous, with the same shape and variation, but possibly different locations (they may be shifted away from each other). Three continuous distributions with the same shape and variation are illustrated in Figure 12-4. A special case is one in which the distributions are all Gaussian with the same variance, the situation discussed in Section 12-2.

If the k distributions have the same shape and variation, then differences or shifts in location are described by differences between the population means (or by differences between the population medians). Our null hypothesis states that the k populations have the same location, and therefore the same distribution. This is the same as saying that the k populations have the same mean (and the same median).

Under the null hypothesis, the exact distribution of the Kruskal–Wallis test statistic described below is the Kruskal–Wallis distribution corresponding to the sample sizes in the experiment. A Kruskal–Wallis probability distribution is derived from the probability model for an experiment in which ranks 1 through n are randomly divided into three or more groups; see the Appendix on the Kruskal–Wallis distributions at the end of the text.

Table H at the back of the book lists probabilities of the form $P(KW \geq c)$, where KW denotes a random variable having a Kruskal–Wallis distribution. Table H covers only three groups ($k = 3$) and sample sizes from 2 to 5. There are many possible values for k and the sample sizes; it is not possible to table probabilities for many Kruskal–Wallis distributions. For situations not covered by Table H, we use an approximation to the distribution of the Kruskal–Wallis test statistic under the null hypothesis, comparing the test statistic with the chi-square distribution for $k - 1$ degrees of freedom. Degrees of freedom

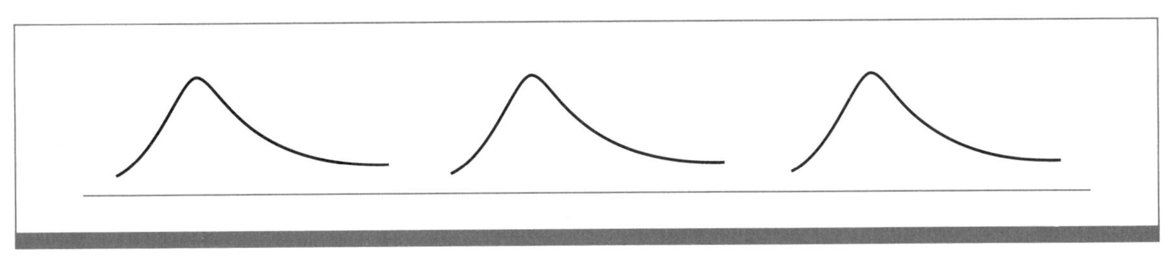

FIGURE 12-4 Illustration of three distributions having the same shape and variation, but different locations. The differences between the means μ_1, μ_2, and μ_3 of the distributions describe the differences in location.

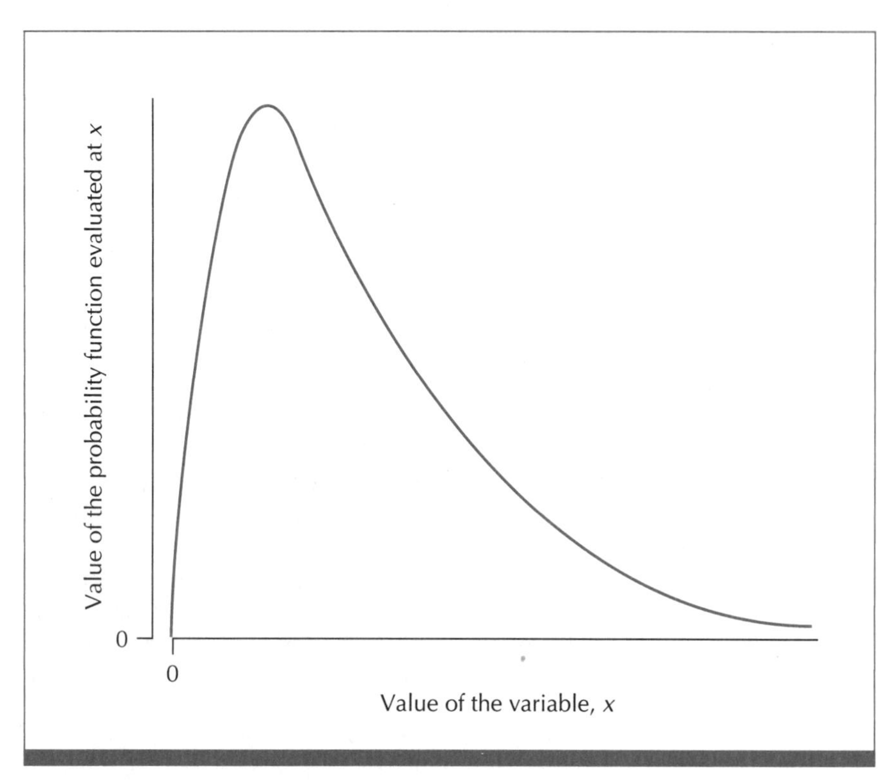

FIGURE 12-5 The shape of a chi-square distribution

here refers to the constant or parameter that defines a particular chi-square distribution.

A chi-square distribution is a continuous probability distribution that is skewed to the right. A random variable with such a distribution takes on only positive values. The general shape of a chi-square distribution is illustrated in Figure 12-5.

A chi-square distribution is characterized by a number called its degrees of freedom. We often denote the chi-square distribution with d degrees of freedom by χ^2_d. Some probabilities associated with several chi-square distributions are listed in Table E at the back of the book.

The Kruskal–Wallis procedure for comparing several distributions is outlined below.

The significance level approach to comparing several distributions in a single-factor experiment, using the Kruskal–Wallis test

1. The null hypothesis states that the k populations all have the same probability distribution. The alternative hypothesis states that the k distributions have the same shape and variation, but are shifted away from each other (they do not all have the same location).

2. Rank the observations in the combined samples from smallest to largest. If two or more observations have the same value, assign each the average of the ranks they share. Let R_i denote the sum of the ranks, n_i the number of observations, and $\bar{R}_i = R_i/n_i$ the average rank in sample i. Calculate the test statistic as

$$\text{Test statistic} = \frac{12}{N(N+1)} \sum_{i=1}^{k} \left(\bar{R}_i - \frac{N+1}{2}\right)^2$$

or, equivalently, as

$$\text{Test statistic} = \frac{12}{N(N+1)} \sum_{i=1}^{k} \frac{(R_i)^2}{n_i} - 3(N+1)$$

where N is the total sample size, the sum of n_1 through n_k. This test statistic measures how far the individual sample rank averages $\bar{R}_i$ differ from the overall average rank, $(N + 1)/2$. If each $\bar{R}_i$ is close to $(N + 1)/2$, then the test statistic is small, consistent with the null hypothesis. If the $\bar{R}_i$'s are not all close to $(N + 1)/2$, then the test statistic is large, inconsistent with the null hypothesis.
3. Assume that the samples are independent random samples of continuous-type observations. Then under the null hypothesis, the test statistic has the Kruskal–Wallis distribution for sample sizes n_1 through n_k. An approximation to the distribution of the test statistic under the null hypothesis is given by the chi-square distribution with $k - 1$ degrees of freedom.
4. Select a significance level α.
5. Using Table H, find the number c such that $P(KW \geq c) \doteq \alpha$, where KW has the Kruskal–Wallis distribution corresponding to the experimental sample sizes. Alternatively, use Table E to find the number c such that $P(X \leq c) = 1 - \alpha$, where X has the chi-square distribution with $k - 1$ degrees of freedom. In either case, the acceptance region is the interval $[0, c)$; the rejection region is the interval $[c, \infty)$.
6. The decision rule is:

   If test statistic $< c$, say the results are consistent with the null hypothesis that the k population distributions are the same.

   If test statistic $\geq c$, say the results are inconsistent with the null hypothesis, suggesting that the k distributions do not all have the same location.
7. Carry out an experiment satisfying the assumptions in step 3. Calculate the test statistic in step 2. Use the decision rule in step 6 to decide whether the observations are consistent with the null hypothesis.

EXAMPLE 12-2 *(continued)*

Let's apply the Kruskal–Wallis test procedure to Example 12-2. We want to test the null hypothesis that the distribution of maximal oxygen uptake is the same for female world-class athletes in the three sports. The alternative hypothesis says these three distributions are not all the same; some are shifted away from each other. We assume that we have independent observations and that the women tested are representative of female world-class athletes in their respective sports. We cannot check these assumptions without additional information about the experiment. We also assume that the three distributions are the same except possibly for differences in location. This assumption does not seem unreasonable from the plots in Figure 12-3.

Implicit assumptions are that we can compare the measurements of maximal oxygen uptake across sports and that these measures truly reflect aerobic power in these women. (A treadmill test may not provide a good measure of aerobic power in swimmers, for example.) We have no way of checking these assumptions from the information provided. We will proceed in our analysis with caution.

We will use significance level .05. Let *KW* denote a random variable having the Kruskal–Wallis distribution for sample sizes 2, 3, and 4. From Table H we see that $P(KW \geq 5.4) = .051$, close to .05. The acceptance region is $[0, 5.4)$, the rejection region is $[5.4, \infty)$, and the decision rule is:

If test statistic < 5.4, say the results are consistent with the null hypothesis that the three distributions are the same.

If test statistic ≥ 5.4, say the results are inconsistent with the null hypothesis, suggesting that the three distributions have different locations (and different medians).

We calculate the test statistic as shown in Table 12-5. We use the second of the two formulas, because it is easier for hand calculations. The test statistic equals 6.444, inconsistent with the null hypothesis, and the *p*-value, $P(KW \geq 6.444$ when H_0 is true), is between .005 and .011. There appear to be differences in maximal oxygen uptake (as measured in this experiment) among female world-class athletes across the three sports.

Suppose we had used the chi-square approximation. Looking in Table E for $3 - 1 = 2$ degrees of freedom, we see that $P(X \leq 5.99) = .95$. Therefore, the cutoff for our acceptance and rejection regions is 5.99 (compared with 5.4

TABLE 12-5 Steps in calculating the Kruskal–Wallis statistic for Example 12-2

Basketball players		Cross-country skiers		Speed skaters	
Value	Rank	Value	Rank	Value	Rank
42.3	1	56.9	6	46.1	3
42.9	2	58.1	7	52.0	5
49.6	4	61.5	8		
		68.2	9		
$n_1 = 3$		$n_2 = 4$		$n_3 = 2$	
$R_1 = 7$		$R_2 = 30$		$R_3 = 8$	
$\bar{R}_1 = 2.333$		$\bar{R}_2 = 7.5$		$\bar{R}_3 = 4$	

$$\text{Test statistic} = \frac{12}{9(9+1)}\left(\frac{7^2}{3} + \frac{30^2}{4} + \frac{8^2}{2}\right) - 3(9+1) = 6.444$$

using the exact Kruskal–Wallis distribution). We still say our results are inconsistent with the null hypothesis; the approximate p-value is between .025 and .05.

To get a feel for the differences among the three sports, we can calculate a confidence interval for the difference between each pair of medians. We will use the Wilcoxon–Mann–Whitney procedure to calculate each confidence interval (see Section 11-4). The Bonferroni method gives a lower bound for the overall confidence level of these intervals taken together. There are $\binom{3}{2} = 3$ ways to compare our three groups two at a time, so we will calculate three separate confidence intervals.

First let's calculate a confidence interval for the difference between medians of maximal oxygen uptake for cross-country skiers and basketball players. We find the $4 \times 3 = 12$ differences between values for skiers and for basketball players. The smallest difference is $56.9 - 49.6 = 7.3$ and the largest difference is $68.2 - 42.3 = 25.9$. From Table G we know that $P(W \leq 0) = .029$, where W is a random variable having the Wilcoxon–Mann–Whitney distribution for sample sizes 3 and 4. The interval (7.3, 25.9) has confidence level $1 - 2(.029) = .942$.

Similarly, (4.9, 22.1) is an interval estimate for the difference between medians of maximal oxygen uptake for female cross-country skiers and speed skaters, with confidence level .866. An interval estimate for the difference between medians of maximal oxygen uptake for female speed skaters and basketball players is $(-3.5, 9.7)$, with confidence level .800.

Using the Bonferroni method, we see that the overall confidence level for these three intervals taken together is greater than or equal to $1 - (1 - .942) - (1 - .866) - (1 - .800) = .608$, or about 61%. This is not very large, but it is the best we can do with such small sample sizes.

The results of our multiple comparisons are summarized in Table 12-6. Zero is not in the first two intervals; median maximal oxygen uptake seems to be greater for the cross-country skiers than for the other two groups of athletes. Zero is in the third interval, so based on these observations we cannot say there is any difference between basketball players and speed skaters with respect to maximal oxygen capacity. This agrees with what we observe in

TABLE 12-6 Nonparametric multiple comparisons for Example 12-2

Confidence interval for	Confidence interval	Individual confidence level
$M_C - M_B$	$(D_1, D_{12}) = (7.3, 25.9)$	.942
$M_C - M_S$	$(D_1, D_8) = (4.9, 22.1)$	.866
$M_S - M_B$	$(D_1, D_6) = (-3.5, 9.7)$	.800

The overall confidence level is greater than or equal to $1 - (.058 + .134 + .200) = .608$.

Note: The subscripts B, C, and S refer to the basketball players, cross-country skiers, and speed skaters, respectively.

Figure 12-3. The distributions for the basketball players and the speed skaters overlap. The distribution for the cross-country skiers is shifted toward larger values, not overlapping the other two sample distributions at all.

In Section 12-4, we discuss the classical, parametric, analysis of randomized block experiments.

12-4 Parametric Analysis of a Randomized Block Experiment

A randomized block design is an extension to several treatments of the paired-sample design we discussed in Section 11-6. We will consider the simplest randomized block design: the number of experimental units in a block equals the number of treatments. Within each block, experimental units are similar with respect to factors that could affect the outcome of the experiment. We randomly assign treatments to experimental units within each block, one unit per treatment. If there are no differences among treatment effects, we expect similar responses from experimental units within a block. If there are differences among treatment effects, we hope the randomized block design will help us see those differences.

> In a **randomized block experiment,** experimental units within a block are similar with respect to factors that could affect the response. In the simplest design, the number of experimental units in a block equals the number of treatments. The treatments are randomly assigned to experimental units within a block.

Let's consider an example.

EXAMPLE 12-3

Are there differences among thermometers in determining melting points? To address this question, three technicians used each of four thermometers to measure the melting point of Hydroquinone (Duncan, 1974, page 632; from Wernimont, 1947, page 8). The recorded melting points (in °C) are shown below.

	Technician			
Thermometer	1	2	3	Thermometer average
1	174.0	173.0	173.5	173.500
2	173.0	172.0	173.0	172.667
3	171.5	171.0	173.0	171.833
4	173.5	171.0	172.5	172.333
Technician average	173.000	171.750	173.000	

The investigators wanted to assess differences among thermometers. However, they were aware that different technicians might obtain different

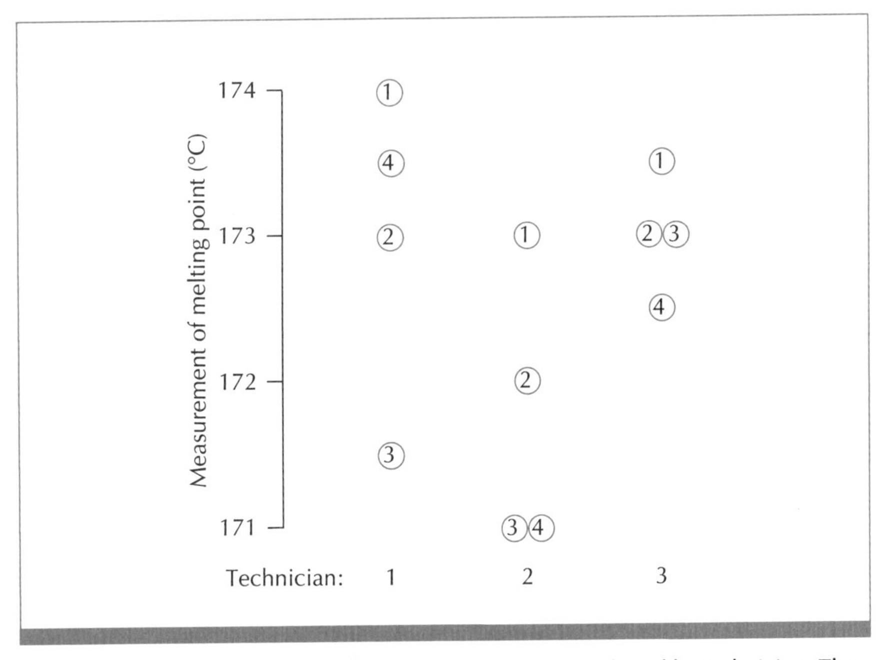

FIGURE 12-6 Recorded melting point of Hydroquinone plotted by technician. Thermometer numbers are shown in the dots.

results, even when using the same thermometer. Therefore, it made sense to have the same technician use each thermometer. Since three technicians were available, the researchers had each of them use each thermometer to determine the melting point of Hydroquinone.

In this experiment, each technician is a block. The idea of blocking in this way is that if there really are no differences among thermometers in measuring melting points, then a single technician should get similar results with each thermometer. The investigators want to control extraneous variation caused by differences among technicians. What suggestions do you have for carrying out this experiment? Should each technician use the thermometers in the same order? Should the first technician make all of his or her measurements, then the second technician, and then the third? Make suggestions regarding these design considerations and any others you can think of, in order to control extraneous sources of variation and make comparisons of thermometers valid.

Plots of the melting point measurements are shown in Figures 12-6 and 12-7. Figure 12-6 shows plots of the readings by technician. The thermometer numbers are shown in the dots. Readings are plotted by thermometer in Figure 12-7. The technician numbers are shown within the dots. What do these plots suggest about differences among thermometers and differences among technicians in determining the melting point of Hydroquinone?

We want to test for differences among thermometers. We will use the classical analysis of a randomized block experiment to assess these differences.

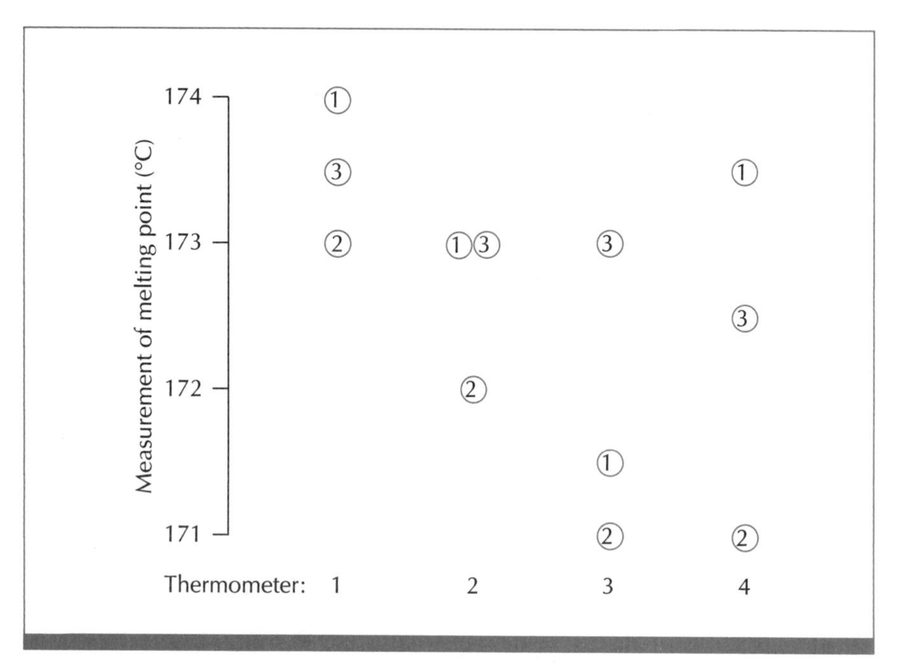

FIGURE 12-7 Recorded melting point of Hydroquinone plotted by thermometer. Technician numbers are shown in the dots.

This classical analysis allows us to test for differences among technicians (blocks) as well. Let's outline the classical, parametric, approach to analyzing a randomized block experiment, and then apply it to this example.

Classical Analysis of a Randomized Block Experiment

To outline the parametric analysis of a randomized block experiment, we need the notation in Table 12-7. Suppose there are b blocks, with k experimental units per block. The number of treatments equals k. Y_{ij} denotes the response of the experimental unit in block j receiving treatment i, $\overline{T}_i$ denotes the average response of the b experimental units receiving treatment i, and $\overline{B}_j$ represents the average response of the k experimental units in block j. The average of all $k \times b$ observations is denoted by $\overline{Y}$. The treatment mean square s_T^2 defined in Table 12-7 is a measure of random variation, plus differences among the means for the k treatments. The block mean square s_B^2 measures random variation, plus differences among the mean responses for the b blocks. The residual mean square s_r^2 is a measure of random variation among observations or responses in the experiment.

To assess differences among treatments on mean response, we compare s_T^2 and s_r^2. If there are really no differences among treatments, then s_T^2 and s_r^2 each estimate random variation among observations in the experiment, so these two variance estimates should be similar in magnitude. If there are dif-

TABLE 12-7 Notation for parametric analysis of a randomized block experiment. Y_{ij} denotes the response of the experimental unit in block j receiving treatment i.

Treatment	Block 1	2	···	b	Treatment average
1	Y_{11}	Y_{12}	···	Y_{1b}	$\bar{T}_1$
2	Y_{21}	Y_{22}	···	Y_{2b}	$\bar{T}_2$
·	·	·	···	·	·
·	·	·	···	·	·
·	·	·	···	·	·
k	Y_{k1}	Y_{k2}	···	Y_{kb}	$\bar{T}_k$
Block average	$\bar{B}_1$	$\bar{B}_2$	···	$\bar{B}_k$	$\bar{Y}$

$$s_T^2 = \text{Treatment mean square} = \frac{b}{k-1}\sum_{i=1}^{k}(\bar{T}_i - \bar{Y})^2$$

$$s_B^2 = \text{Block mean square} = \frac{k}{b-1}\sum_{j=1}^{b}(\bar{B}_j - \bar{Y})^2$$

$$s_r^2 = \text{Residual mean square} = \frac{1}{(k-1)(b-1)}\sum_{i=1}^{k}\sum_{j=1}^{b}(Y_{ij} - \bar{T}_i - \bar{B}_j + \bar{Y})^2$$

ferences among treatments, then s_r^2 still estimates random variation, but s_T^2 estimates random variation plus a measure of differences among the treatment means. Therefore, in the case of treatment differences, we expect s_T^2 to be larger than s_r^2.

Similarly, to assess differences among blocks, we compare s_B^2 and s_r^2. If there really are no differences in average responses among blocks, s_B^2 and s_r^2 should be similar in magnitude, since both then estimate random variation among observations in the experiment. If there are differences among blocks, we expect s_B^2 to be larger than s_r^2, since s_B^2 then estimates random variation plus differences among blocks.

With these ideas in mind, we can outline the significance level approach to the classical analysis of a randomized block experiment.

The significance level approach to classical analysis of a randomized block experiment

1. The hypotheses about treatment differences are:

   H_0: The k treatments all have the same average effect on response.
   H_a: The average effect on response is not the same for all k treatments.

   The hypotheses about block differences can be stated as:

   H_0^*: The average response is the same for all b blocks.
   H_a^*: The average response is not the same for all b blocks.

2. To test the hypotheses about treatment effects, we use the test statistic

$$\text{Test statistic(T)} = \frac{s_T^2}{s_r^2}$$

To test the hypotheses about block effects, we use the test statistic

$$\text{Test statistic(B)} = \frac{s_B^2}{s_r^2}$$

3. Assume that the $k \times b$ observations are all independent, from Gaussian distributions. These distributions have the same variance σ^2. The means may differ, depending on treatment and block. We also assume that the relative treatment effects are the same for each block.

   Under the null hypothesis of no treatment differences, test statistic(T) has the F distribution with $k - 1$ numerator degrees of freedom and $(k - 1)(b - 1)$ denominator degrees of freedom. Small values of test statistic(T), near 1, are consistent with the null hypothesis of no differences among treatments on average. Large values of test statistic(T) are inconsistent with this null hypothesis.

   Under the null hypothesis of no block differences, test statistic(B) has the F distribution with $b - 1$ numerator degrees of freedom and $(k - 1)(b - 1)$ denominator degrees of freedom. Small values of test statistic(B), near 1, are consistent with the null hypothesis of no differences in average response among blocks. Large values of test statistic(B) are inconsistent with this null hypothesis.
4. Select significance level α_1 for the first test of hypotheses, α_2 for the second test.
5. For the test of treatment effects, find the number c_1 from Table D such that $P(F_1 \leq c_1) = 1 - \alpha_1$. Here, F_1 denotes a random variable having the $F(k - 1, (k - 1)(b - 1))$ distribution. The acceptance region is the interval $[0, c_1)$; the rejection region is the interval $[c_1, \infty)$.

   For the test of block effects, find the number c_2 from Table D such that $P(F_2 \leq c_2) = 1 - \alpha_2$. Here, F_2 denotes a random variable having the $F(b - 1, (k - 1)(b - 1))$ distribution. The acceptance region is the interval $[0, c_2)$; the rejection region is the interval $[c_2, \infty)$.
6. To test for treatment differences, the decision rule is:

   If test statistic(T) $< c_1$, say the results are consistent with the null hypothesis of no treatment differences in average response.

   If test statistic(T) $\geq c_1$, say the results are inconsistent with this null hypothesis, suggesting there are treatment differences in average response.

   To test for block differences, the decision rule is:

   If test statistic(B) $< c_2$, say the results are consistent with the null hypothesis of no block differences in average response.

   If test statistic(B) $\geq c_2$, say the results are inconsistent with this null hypothesis, suggesting there are block differences in average response.
7. Carry out an experiment that satisfies the assumptions in step 3. Calculate the test statistics in step 2. Use the decision rules in step 6 to decide whether

there seem to be differences among treatments and differences among blocks. Draw conclusions based on the experimental results.

EXAMPLE 12-3
(continued)

Let's use this parametric approach to analyze the results of the experiment in Example 12-3. The four thermometers represent the treatments in this experiment, while the three technicians represent the blocks. A response is the melting point determination a technician makes with a particular thermometer. The hypotheses about thermometer (treatment) differences are:

H_0: On average, the four thermometers give the same reading for the melting point of Hydroquinone.

H_a: The four thermometers do not give the same reading on average.

The hypotheses about technician (block) differences are:

H_0^*: On average, the three technicians get the same reading for the melting point of Hydroquinone.

H_a^*: The three technicians do not get the same reading on average.

Note that we presented the results of the experiment in Example 12-3 in the format shown in Table 12-7. The only statistic not shown there is the average of all 12 observations, $\bar{Y} = 172.583$.

We assume that the 12 observations are all independent. We cannot check this assumption without more information on how the experiment was conducted. What suggestions would you make about the conduct of the experiment in order to ensure independence of observations?

We also assume that the observations come from Gaussian distributions with the same variance. One way to check this assumption is through plots of *residuals.* Recall that a residual is the difference between an observation and a summary, or predicted value, or estimate of the mean of the observation.

For a randomized block design, the predicted value for observation Y_{ij} (treatment i, block j) is the estimated mean value $\bar{T}_i + \bar{B}_j - \bar{Y}$ based on our model assumptions. The residual for that observation is then $Y_{ij} - \bar{T}_i - \bar{B}_j + \bar{Y}$.

A **residual** is the difference between an observation and an estimate of its expected value. In the simplest randomized block design, a residual has the form $Y_{ij} - \bar{T}_i - \bar{B}_j + \bar{Y}$, where Y_{ij} denotes the observation corresponding to treatment i and block j, $\bar{T}_i$ is the average of all observations for treatment i, $\bar{B}_j$ is the average of observations in block j, and $\bar{Y}$ is the average of all the observations.

The residuals represent what is left over after we fit our randomized block probability model; they are like noise. If all the model assumptions hold, the residuals should (roughly) represent independent observations from the Gaussian distribution with mean 0 and variance σ^2. Note that in our definition of s_r^2 at the bottom of Table 12-7, we add up the squared residuals in order to calculate this variance estimate. This is why we often call s_r^2 the *residual mean square.* The residuals for Example 12-3 are shown in Table 12-8.

TABLE 12-8 Residuals for the randomized block probability model in Example 12-3

	Technician		
Thermometer	1	2	3
1	.08	.33	−.42
2	−.08	.17	−.08
3	−.75	.00	.75
4	.75	−.50	−.25

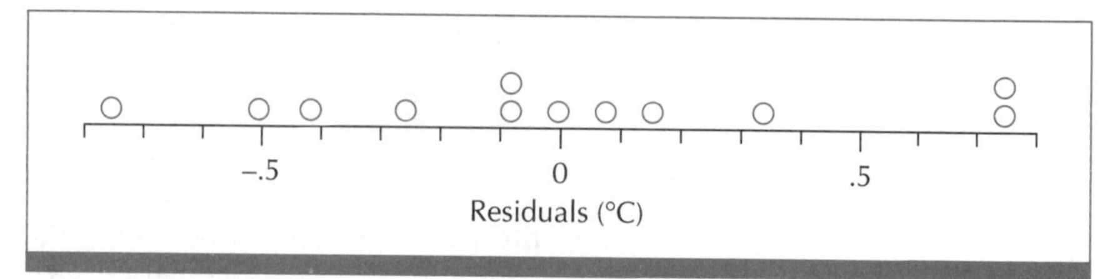

FIGURE 12-8 Dot plot of residuals in Example 12-3

A dot plot of the residuals is shown in Figure 12-8. The plot gives us no reason to doubt the Gaussian assumption. Figure 12-9 shows a plot of residuals by thermometer; we see that the variation among residuals for thermometer 2 is less than for the other three thermometers. Figure 12-10 is a plot of residuals by technician; the variation among residuals is somewhat smaller for technician 2 than for the other two technicians. These differences in variation are not extreme enough to make us avoid the classical analysis of this randomized block experiment. (The analysis is fairly robust to small deviations from the equal-variance assumption, meaning that as long as variances are not too different, significance levels and confidence levels are close to the levels we choose.)

For our analysis, we also assume that the relative thermometer effects are the same for each technician. Consider the dot plots in Figure 12-6. If the assumption held, we would expect the order of the thermometers to be the same for each technician. In fact, all three technicians obtained the highest readings with thermometer 1. But the order varies across technicians for the other three thermometers. We do not have strong evidence for or against the assumption that the relative thermometer effects are the same for all three technicians. We will proceed with our analysis, with our usual caution in interpretations.

If our model assumptions all hold, test statistic(T) has the $F(3, 6)$ distribution under the null hypothesis of no treatment differences. Test statistic(B) has the $F(2, 6)$ distribution under the null hypothesis of no technician differences.

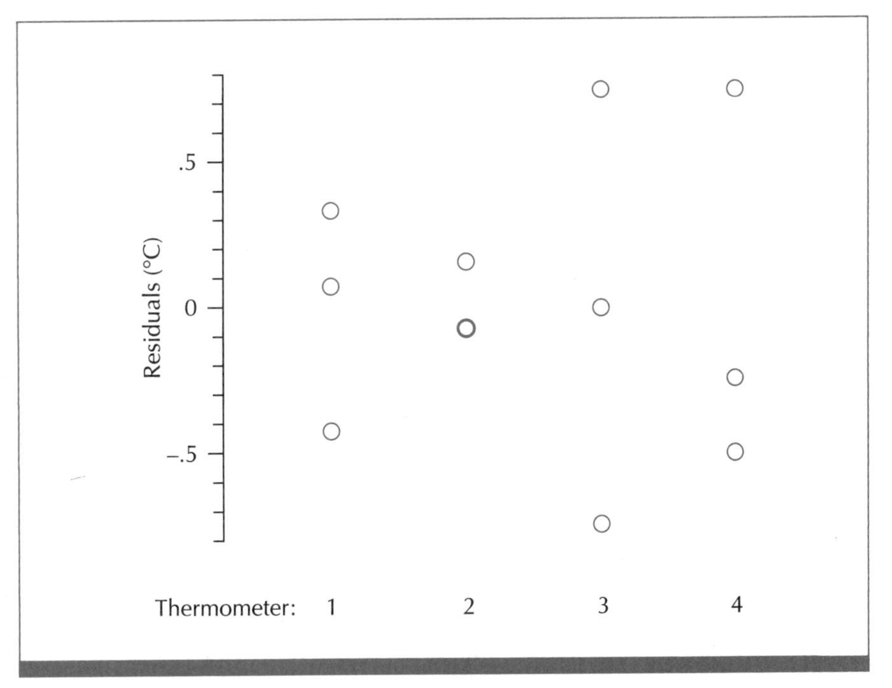

FIGURE 12-9 Plot of residuals by thermometer in Example 12-3

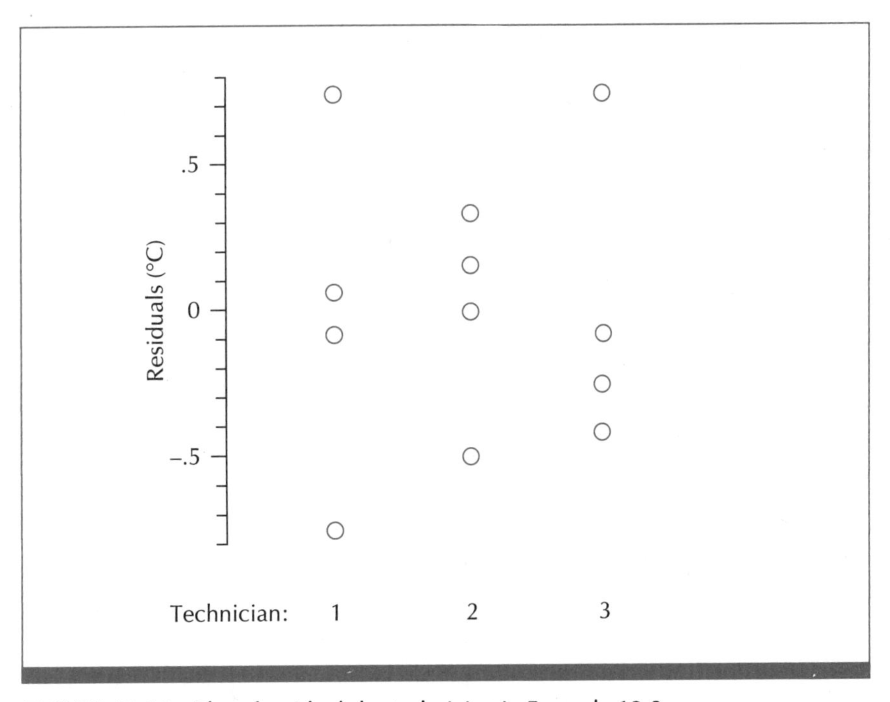

FIGURE 12-10 Plot of residuals by technician in Example 12-3

TABLE 12-9 Calculations for parametric analysis of the randomized block experiment in Example 12-3

k = Number of treatments = 4

b = Number of blocks = 3 $\qquad \bar{Y}$ = Overall sample mean = 172.583

$$s_T^2 = \frac{3}{4-1}[(173.500 - 172.583)^2 + (172.667 - 172.583)^2 + (171.833 - 172.583)^2 + (172.333 - 172.583)^2]$$
$$= 1.47$$

$$s_B^2 = \frac{4}{3-1}[(173.000 - 172.583)^2 + (171.750 - 172.583)^2 + (173.000 - 172.583)^2]$$
$$= 2.08$$

$$s_r^2 = \frac{1}{(4-1)(3-1)}[(.08)^2 + (.33)^2 + (-.42)^2 + (-.08)^2 + (.17)^2 + (-.08)^2 + (-.75)^2 + (0)^2 + (.75)^2 + (.75)^2 + (-.50)^2 + (-.25)^2]$$
$$= .39$$

$$\text{Test statistic(T)} = \frac{1.47}{.39} = 3.8 \qquad \text{Test statistic(B)} = \frac{2.08}{.39} = 5.3$$

$k - 1 = 3 \qquad b - 1 = 2 \qquad (k-1)(b-1) = 6$

We will use significance level .10 for both tests of hypotheses. Looking in Table D, we see that if F_1 has the $F(3, 6)$ distribution, then $P(F_1 \leq 3.29) = .90$. The acceptance region for the test about thermometer effects is [0, 3.29), the rejection region is [3.29, ∞), and the decision rule is:

If test statistic(T) < 3.29, say the results are consistent with the null hypothesis that the thermometers give the same reading on average.

If test statistic(T) ≥ 3.29, say the results are inconsistent with this null hypothesis, suggesting that the thermometers do not give the same reading on average.

If F_2 has the $F(2, 6)$ distribution, then $P(F_2 \leq 3.46) = .90$. The acceptance region for the test about technician effects is [0, 3.46), the rejection region is [3.46, ∞), and the decision rule is:

If test statistic(B) < 3.46, say the results are consistent with the null hypothesis that the technicians get the same reading on average.

If test statistic(B) ≥ 3.46, say the results are inconsistent with this null hypothesis, suggesting that the technicians do not get the same reading on average.

The calculations for our analysis are outlined in Table 12-9.

Test statistic(T) equals 3.8, which is inconsistent with the null hypothesis that there are no differences among thermometers on average, at the .10 significance level. The p-value is between .05 and .10.

Test statistic(B) equals 5.3, which is inconsistent with the null hypothesis that there are no differences among technicians on average, at the .10 significance level. The p-value is a little less than .05.

Our analysis suggests that the thermometers gave somewhat different readings of the melting point of Hydroquinone on average. From Figures 12-6 and 12-7 we see that the highest readings were with thermometer 1. Thermometers 3 and 4 gave lower readings on average and had greater spread in the readings. Thermometer 2 gave intermediate readings.

The analysis also suggests the the technicians got somewhat different results on average. The most striking difference is that technician 2 got lower readings than technicians 1 and 3, for each thermometer (this shows up clearly in Figure 12-7).

The p-value for the test of thermometer differences was between .05 and .10, which we might consider borderline statistical significance. Whether or not we consider the results of this experiment of practical importance depends on the accuracy (closeness to the correct value) and precision (lack of variation, or repeatability) required when melting points are determined in practical situations.

Suppose we had ignored the technicians in our analysis. If we had done a one-way analysis of variance, with three readings for each thermometer, we would have found no significant difference among thermometers, with p-value = .2 (Exercise 12-18). The randomized block design was useful in this experiment. Because there were differences among technicians, blocking helped us see differences among thermometers. Also, from a quality control point of view, it is useful to see that different technicians can get different readings on average.

A general form of analysis of variance table for the simplest randomized block design is shown in Table 12-10. The analysis of variance table for Ex-

TABLE 12-10 Analysis of variance table for a randomized block experiment (number of treatments equals the size of each block)

Source of variation	Sum of squares	Degrees of freedom	Mean square	Test statistic
Treatments	$b\sum_{i=1}^{k}(\bar{T}_i - \bar{Y})^2$	$k - 1$	S_T^2	$\frac{S_T^2}{S_r^2}$
Blocks	$k\sum_{j=1}^{b}(\bar{B}_j - \bar{Y})^2$	$b - 1$	S_B^2	$\frac{S_B^2}{S_r^2}$
Residuals	$\sum_{i=1}^{k}\sum_{j=1}^{b}(Y_{ij} - \bar{T}_i - \bar{B}_j + \bar{Y})^2$	$(k - 1)(b - 1)$	S_r^2	
Total	$\sum_{i=1}^{k}\sum_{j=1}^{b}(Y_{ij} - \bar{Y})^2$	$kb - 1$		

TABLE 12-11 Analysis of variance table for the randomized block experiment in Example 12-3

Source of variation	Sum of squares	Degrees of freedom	Mean square	Test statistic	*p*-value
Thermometers	4.42	3	1.47	3.8	.08
Technicians	4.17	2	2.08	5.3	.05
Residuals	2.33	6	.39		
Total	10.92	11			

ample 12-3 is shown in Table 12-11, with an added column showing the *p*-value for each test statistic.

In Section 12-5, we consider a nonparametric method, called Friedman's test, for analyzing a randomized block experiment.

12-5 Nonparametric Analysis of a Randomized Block Experiment: Friedman's Test

Before considering an example, we will first outline a nonparametric procedure for analysis of a randomized block experiment, *Friedman's test.* This is a test of treatment differences (not block differences). Suppose we have b blocks, with k experimental units per block. There are k treatments, one treatment per experimental unit in each block.

The significance level approach to nonparametric analysis of a randomized block experiment, using Friedman's test

1. The hypotheses are:

   H_0: The treatments have the same average effect on response.
   H_a: The treatments do not all have the same average effect on response.

2. Rank the k observations within each block. The smallest observation gets rank 1 and the largest gets rank k. Tied observations get the average of the ranks they share. Let $\bar{R}_1$ denote the average of the ranks for treatment 1. Let $\bar{R}_2$ denote the average of the ranks for treatment 2, and so on. $\bar{R}_k$ is the average of the ranks for treatment k. The overall average rank is $(k + 1)/2$. The test statistic is

$$\text{Test statistic} = \frac{12b}{k(k+1)} \sum_{i=1}^{k} \left(\bar{R}_i - \frac{k+1}{2} \right)^2$$

3. We assume that the $k \times b$ observations are all independent, from distributions with similar shape and variation. We also assume that the relative treatment effects are the same for each block. Then under the null hypothesis of no treatment differences, the test statistic has approximately the chi-square distribution with $k - 1$ degrees of freedom. Small values of the test statistic

are consistent with the null hypothesis, while large values are inconsistent with the null hypothesis.

4. Select significance level α.
5. Find the number c in Table E such that $P(X \leq c) = 1 - \alpha$, where X has the chi-square distribution with $k - 1$ degrees of freedom. The acceptance region is the interval $[0, c)$. The rejection region is the interval $[c, \infty)$.
6. The decision rule is:

   If test statistic $< c$, say the results are consistent with the null hypothesis of no treatment differences.
   If test statistic $\geq c$, say the results are inconsistent with the null hypothesis, suggesting that there are treatment differences.

7. Carry out an experiment satisfying the assumptions in step 3. Calculate the test statistic in step 2. Use the decision rule in step 6 to decide whether the results are consistent with the null hypothesis. Draw conclusions based on the experimental results.

Let's apply Friedman's test to the following example.

EXAMPLE 12-4

Investigators wanted to compare the effects of three anesthetics upon plasma epinephrine concentration in dogs. They measured plasma epinephrine concentration (in nanograms per milliliter) for ten dogs while under each of these three anesthetics: isofluorane, halothane, and cyclopropane. The measurements are listed below (Rice, 1988, page 431; from Perry, Van Dyke, and Theye, 1974).

	Dog									
Anesthetic	1	2	3	4	5	6	7	8	9	10
Isofluorane	.28	.51	1.00	.39	.29	.36	.32	.69	.17	.33
Halothane	.30	.39	.63	.68	.38	.21	.88	.39	.51	.32
Cyclopropane	1.07	1.35	.69	.28	1.24	1.53	.49	.56	1.02	.30

What suggestions would you make for the design of this experiment? How would you seek to reduce the effects of extraneous factors? Should each dog receive the anesthetics in the same order? Would you worry about carry-over effects of anesthetics from one treatment period to the next? What other concerns would you have and how would your experimental design address those concerns?

Plots of the observations are shown in Figures 12-11 and 12-12. Figure 12-11 shows a plot of the ten measurements of plasma epinephrine concentration, for each of the three anesthetics. Responses under the three anesthetics are plotted for each dog in Figure 12-12. What do these plots suggest about differences among anesthetics and differences among dogs with respect to plasma epinephrine concentration?

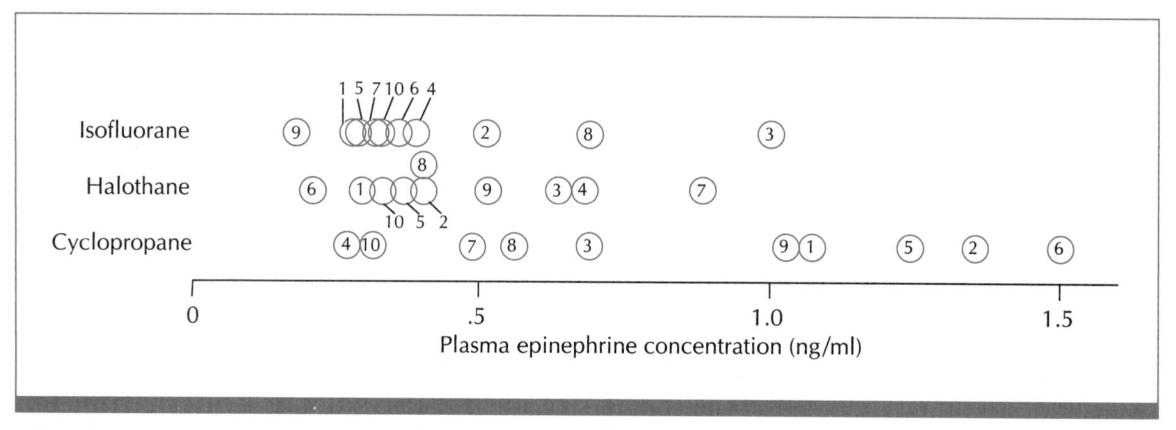

FIGURE 12-11 Plot of plasma epinephrine concentrations by anesthetic, in Example 12-4. An identification number denotes each dog's responses.

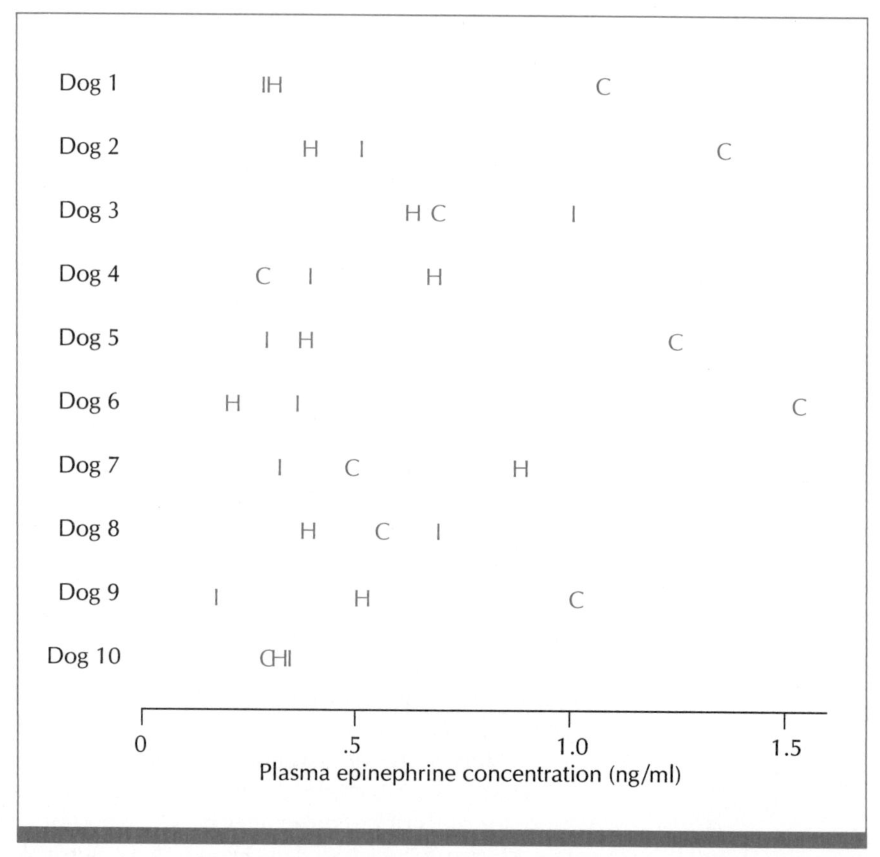

FIGURE 12-12 Plot of plasma epinephrine concentrations for each dog in Example 12-4. The letters I, H, and C denote the response for a dog while under isofluorane, halothane, and cyclopropane, respectively.

In this experiment, the anesthetics are the treatments and the dogs are blocks. We want to test for differences in average effects of the anesthetics on plasma epinephrine concentrations, with hypotheses:

H_0: The three anesthetics have the same average effect on plasma epinephrine concentration.
H_a: The three anesthetics do not have the same average effect on plasma epinephrine concentration.

We must assume that the 30 observations are independent. We cannot check this assumption without more information on how the experiment was carried out. What suggestions would you make in order to ensure independence?

We also assume that the relative effects of the anesthetics are the same for the ten dogs. Looking at Figure 12-12, we see that this assumption is badly violated. For five dogs (dogs 1, 2, 5, 6, and 9), plasma epinephrine concentrations were much higher under cyclopropane than under the other two anesthetics. These are the five largest values plotted for cyclopropane in Figure 12-11. For the other five dogs, there are smaller differences among the anesthetics. Also, cyclopropane did not result in the largest values for these dogs. Clearly, the relative effects of the three anesthetics are not the same for all ten dogs.

For a valid analysis, we must assume that the relative differences among anesthetics (treatments) are the same for each dog (block). Our plots show us that this assumption is not reasonable. For a moment we will ignore this problem and go through the mechanics of the procedure; then we will discuss our results in terms of this violation of assumptions. [This example has appeared in a number of references as a randomized block experiment requiring a standard analysis. In fact, as our plots show, a major assumption of both parametric and nonparametric analysis of a standard (unreplicated) randomized block experiment is badly violated.]

If all assumptions for Friedman's test did hold, then under the null hypothesis the test statistic would have approximately the chi-square distribution with $3 - 1 = 2$ degrees of freedom. We will use significance level .10. From Table E, we find that $P(X \leq 4.61) = .90$, where X has the chi-square distribution with 2 degrees of freedom. The acceptance region is $[0, 4.61)$, the rejection region is $[4.61, \infty)$, and the decision rule is:

If test statistic < 4.61, say the results are consistent with the null hypothesis of no difference in mean plasma epinephrine concentration among the three anesthetics.
If test statistic ≥ 4.61, say the results are inconsistent with the null hypothesis, suggesting that there are differences in mean plasma epinephrine concentrations among the three anesthetics.

The calculations we need for Friedman's test are outlined in Table 12-12.

The test statistic equals 1.4, consistent with the null hypothesis of no difference among anesthetics, at the .10 significance level. The approximate p-value, based on the chi-square distribution with 2 degrees of freedom, is about .5.

TABLE 12-12 Calculations for Friedman's test in Example 12-4. Observations for each dog are ranked from 1 to 3.

	Dog										
Anesthetic	1	2	3	4	5	6	7	8	9	10	Sum of ranks
Isofluorane	1	2	3	2	1	2	1	3	1	3	19
Halothane	2	1	1	3	2	1	3	1	2	2	18
Cyclopropane	3	3	2	1	3	3	2	2	3	1	23

$\bar{R}_1 = \frac{19}{10} = 1.9 \qquad \bar{R}_2 = \frac{18}{10} = 1.8 \qquad \bar{R}_3 = \frac{23}{10} = 2.3$

$\text{Overall average rank} = \frac{3 + 1}{2} = 2$

$k = \text{Number of treatments} = 3 \qquad b = \text{Number of blocks} = 10$

$\text{Test statistic} = \frac{12 \times 10}{3(3 + 1)} [(1.9 - 2)^2 + (1.8 - 2)^2 + (2.3 - 2)^2] = 1.4$

$\text{Degrees of freedom} = k - 1 = 2$

If we ignore the violation of our model assumptions, Friedman's test tells us there do not appear to be differences among the anesthetics. This seems to be true for five of the dogs (dogs 3, 4, 7, 8, and 10). However, as we saw in Figure 12-12, dogs 1, 2, 5, 6, and 9 had plasma epinephrine concentrations much higher under cyclopropane than under isofluorane or halothane. Our plots suggest that there may be differences among anesthetics. Also, the relative effects of the anesthetics vary with dogs. This is called an *interaction* effect of anesthetic and dog upon the response. There is no way to account for this interaction in our analysis of this randomized block experiment. In Example 12-4, the plots are much more useful and informative than the formal analysis, which is misleading because not all the assumptions for the analysis are satisfied.

Exercise 12-19 asks you to use Friedman's test to check for thermometer differences in Example 12-3, where the assumptions for the test seem more reasonable than in Example 12-4.

We say we have an *interaction effect* of treatment and block on response if the relative effects of treatments differ for different blocks. The only way to account for such interaction in our analysis is to have larger blocks. Then we assign each treatment to two or more experimental units within each block. We can analyze this larger experiment using two-way analysis of variance for a replicated randomized block experiment. See, for example, Kirk (1982, Chapter 6).

In Chapter 13 we discuss two-way analysis of variance, for a two-factor experiment.

Summary of Chapter 12

The Bonferroni method provides an upper bound on the overall significance level when we make several tests of hypotheses. The method provides a lower bound on the overall confidence level when several confidence intervals are used for multiple comparisons.

In a single-factor experiment (an extension to several samples of the two-sample design), we assume that we have k independent random samples, one from each of k populations. If we assume that the samples come from Gaussian distributions with the same variance, then we can use one-way analysis of variance to test the null hypothesis that the population means are all equal. If we assume that the samples come from distributions with the same shape and variation, but possibly different locations, we can use the Kruskal–Wallis test to test the null hypothesis that the distributions are equal.

In the simplest randomized block design (an extension of the paired-sample design), experimental units within a block are similar with respect to characteristics that might affect the response. The number of experimental units in each block equals the number of treatments. The treatments are randomly assigned to experimental units within a block.

For a parametric analysis of a randomized block experiment, we assume that the observations are all independent, from Gaussian distributions with the same variance; the means may vary depending on treatment and block. We also assume that the relative treatment effects are the same within each block. We can test the null hypothesis that the average effect is the same for all treatments, as well as the null hypothesis that the average response is the same for all blocks.

For a nonparametric analysis of a randomized block experiment, we assume that the observations are independent, from distributions having similar shape and variation. We also assume that the relative treatment effects are the same for each block. Friedman's test assesses the null hypothesis that the average treatment effect is the same for all treatments.

Residual plots are useful for checking whether model assumptions seem reasonable. A residual is the difference between an observation and a summary, predicted value, or estimate of the mean of the observation.

Minitab Appendix for Chapter 12

Finding Probabilities for *F* and Chi-Square Distributions

We introduced the *F* distributions and the chi-square distributions in Chapter 12. We can use the CDF, PDF, and RANDOM commands with the *F* and CHISQUARE subcommands, as we have discussed for other distributions. With the *F* subcommand, we specify numerator degrees of freedom and then denominator degrees of freedom. With the CHISQUARE subcommand, we specify the degrees of freedom. For instance,

```
MTB>   cdf 12.2;
SUBC>  f 4 6.
 12.2000  0.9952
MTB>   cdf 6.5;
SUBC>  chisquare 2.
 6.5000  0.9612
```

Minitab tells us that if the random variable X has the $F(4, 6)$ distribution, then $P(X \leq 12.2) = .9952$. If the random variable Y has the chi-square distribution with 2 degrees of freedom, then $P(Y \leq 6.5) = .9612$.

Performing One-Way Analysis of Variance with the ONEWAY Command

Suppose the data from Example 12-1 are in two columns on our worksheet. Column 1 (named GROUP) contains a code for group: 1 = control, 2 = urea, 3 = potassium nitrate and calcium, 4 = ammonia and ammonium sulphate. Fruit weights are in column 2 (named FRUITWT). We use the ONEWAY command for a parametric one-way analysis of variance:

```
MTB>  oneway 'fruitwt' 'group'
```

The results are shown in Figure M12-1.

If we specify two additional columns at the end of the ONEWAY command, we can save residuals and estimated (or predicted) values of observations. In Example 12-1, the command

```
MTB>  oneway 'fruitwt' 'group' c3 c4
MTB>  name c3 'resid' c4 'predict'
```

produces the same output as shown in Figure M12-1. In addition, Minitab stores the residuals from the fitted model in C3, which we name RESID. Mini-

```
ANALYSIS OF VARIANCE ON fruitwt
SOURCE      DF         SS         MS          F         p
group        3       5980       1993      11.28     0.000
ERROR       38       6717        177
TOTAL       41      12697
                                       INDIVIDUAL 95 PCT CI'S FOR MEAN
                                       BASED ON POOLED STDEV
LEVEL        N       MEAN      STDEV   --+---------+---------+---------+----
    1       10      83.13      18.11     (----*-----)
    2       11      99.32       8.81                (----*-----)
    3       11     109.60      15.17                          (----*----)
    4       10      80.53       8.75   (-----*----)
                                       --+---------+---------+---------+----
POOLED STDEV =      13.30              75        90        105       120
```

FIGURE M12-1 Output from the ONEWAY command in Example 12-1

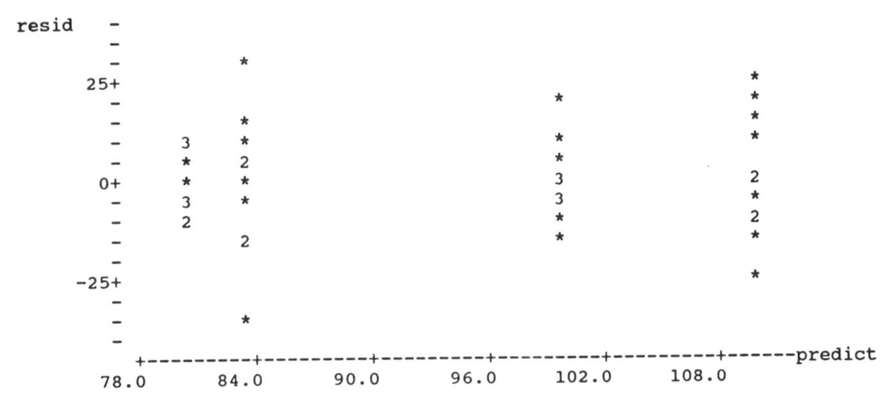

FIGURE M12-2 Scatterplot of residuals versus predicted values in Example 12-1

```
LEVEL      NOBS      MEDIAN  AVE. RANK    Z VALUE
    1         3       42.90        2.3      -2.07
    2         4       59.80        7.5       2.45
    3         2       49.05        4.0      -0.59
OVERALL       9                    5.0

H = 6.444
* NOTE  * ONE OR MORE SMALL SAMPLES
```

FIGURE M12-3 Output from the KRUSKAL-WALLIS command for Example 12-2

tab stores the estimated (or predicted) observations in C4, named PREDICT. We can use these saved values, say in a histogram of residuals or a plot of residuals versus predicted values, to check model assumptions. For instance, the command

```
MTB> plot 'resid' 'predict'
```

produces the scatterplot in Figure M12-2.

We can use the TWOT command and the Bonferroni method to make multiple comparisons, if the results of ONEWAY indicate differences among means.

Carrying Out a Kruskal–Wallis Comparison of Means

The KRUSKAL-WALLIS command carries out the calculations for the Kruskal-Wallis test. Suppose our worksheet contains the data for Example 12-2. Column 1 (named SPORT) contains a code for sport: 1 = basketball, 2 = cross-country skiing, 3 = speed skating. Column 2 (named UPTAKE) contains maximal oxygen uptake. The command

```
MTB> krus 'uptake' 'sport'
```

results in the output in Figure M12-3.

The student edition of Minitab does not print a *p*-value for the Kruskal–Wallis test statistic. We can look it up in a table for the Kruskal–Wallis distribution. If we think the sample sizes are large enough, we can use the large-sample chi-square approximation. In our example, we might want to compare the test statistic with the chi-square distribution with 2 degrees of freedom (since there are three groups). The test statistic equals 6.444 and the approximate *p*-value is $P(X \geq 6.444)$, where X has the chi-square distribution with 2 degrees of freedom. We can find this approximate *p*-value using Minitab as follows:

```
MTB>    cdf 6.444 k1;
SUBC>   chisquare 2.
MTB>    let k2=1-k1
MTB>    print k1 k2
K1      0.960125
K2      0.0398753
```

Minitab prints the cumulative probability K1 = 0.960125 and the approximate *p*-value K2 = 1 − K1 = 0.0398753 corresponding to the observed value 6.444 of the test statistic. (Recall that the *p*-value based on the exact Kruskal–Wallis distribution was between .005 and .011.)

If we want to make multiple comparisons based on Mann–Whitney intervals, we must unstack the observations in 'UPTAKE' based on 'SPORT':

```
MTB>    unstack 'uptake' c3-c5;
SUBC>   subscripts 'sport'.
```

C3 contains uptake values for sport 1; C4, for sport 2; C5, for sport 3. Now we can use the MANN-WHITNEY command two columns at a time on C3 through C5. We apply the Bonferroni method to get the overall confidence level for our multiple comparisons.

Analyzing a Randomized Block Experiment with the TWOWAY Command

To analyze the results of a randomized block experiment, as discussed in Section 12-4, we use the TWOWAY command. Consider the experiment in Example 12-3. Suppose column 1 (named THERM) of our worksheet contains thermometer number, column 2 (named TECH) contains technician number, and column 3 (named MELT) contains measured melting points. We can get descriptive statistics for melting point by thermometer and by technician. The command

```
MTB>    describe 'melt';
SUBC>   by 'therm'.
```

gives the output in Figure M12-4, while the command

```
          therm       N      MEAN    MEDIAN    TRMEAN     STDEV    SEMEAN
melt          1       3    173.50    173.50    173.50      0.50      0.29
              2       3    172.67    173.00    172.67      0.58      0.33
              3       3    171.83    171.50    171.83      1.04      0.60
              4       3    172.33    172.50    172.33      1.26      0.73

          therm     MIN       MAX        Q1        Q3
melt          1  173.00    174.00    173.00    174.00
              2  172.00    173.00    172.00    173.00
              3  171.00    173.00    171.00    173.00
              4  171.00    173.50    171.00    173.50
```

FIGURE M12-4 Descriptive statistics for MELT by thermometer

```
           tech       N      MEAN    MEDIAN    TRMEAN     STDEV    SEMEAN
melt          1       4    173.00    173.25    173.00      1.08      0.54
              2       4    171.75    171.50    171.75      0.96      0.48
              3       4    173.00    173.00    173.00      0.41      0.20

           tech     MIN       MAX        Q1        Q3
melt          1  171.50    174.00    171.87    173.88
              2  171.00    173.00    171.00    172.75
              3  172.50    173.50    172.63    173.37
```

FIGURE M12-5 Descriptive statistics for MELT by technician

```
ANALYSIS OF VARIANCE  melt

SOURCE        DF        SS        MS
therm          3     4.417     1.472
tech           2     4.167     2.083
ERROR          6     2.333     0.389
TOTAL         11    10.917
```

FIGURE M12-6 Analysis of variance table for the randomized block experiment in Example 12-3

```
MTB>   describe 'melt';
SUBC>  by 'tech'.
```

gives the output in Figure M12-5.

The TWOWAY command

```
MTB>   twoway 'melt' 'therm' 'tech'
```

gives the output in Figure M12-6.

Notice that the Minitab output for TWOWAY does not include test statistics or p-values. To test for thermometer differences, we calculate the test statistic from the analysis of variance table and use Minitab to calculate the p-value:

```
MTB>   let k1=1.472/0.389
MTB>   cdf k1 k2;
SUBC>  f 3 6.
MTB>   let k2=1-k2
MTB>   print k1 k2
K1     3.78406
K2     0.0777537
```

Minitab prints the value of the test statistic K1 = 3.78406 and the p-value K2 = 0.0777537. We go through similar steps to test for technician differences:

```
MTB>    let k3=2.083/0.389
MTB>    cdf k3 k4;
SUBC>   f 2 6.
MTB>    let k4=1-k4
MTB>    print k3 k4
K3      5.35476
K4      0.0462980
```

Minitab prints the value of the test statistic K3 = 5.35476 and the p-value K4 = 0.0462980. Because of round-off error, these values do not exactly equal what we found in Example 12-3.

As with ONEWAY, we can specify two extra columns as part of the TWOWAY command to save residuals and predicted values. The command

```
MTB>    twoway 'melt' 'therm' 'tech' c4 c5
```

produces the same output as in Figure M12-6, and saves residuals in C4 and predicted values in C5. We can use these saved values in plots to check model assumptions.

Using Minitab to Carry Out Friedman's Test

The student edition of Minitab does not have a procedure for Friedman's test. We can calculate the test statistic using Minitab, however. Consider the data in Example 12-4. Suppose column 1 (named DOG) of our worksheet contains dog number. Column 2 (named ANES) contains a code for anesthetic: 1 = isofluorane, 2 = halothane, 3 = cyclopropane. Column 3 (named RESPONSE) contains plasma epinephrine concentration. We want to unstack the RESPONSE column into ten columns, one for each dog:

```
MTB>    unstack 'response' c11-c20;
SUBC>   subscript 'dog'.
```

C11 contains the three observations for dog 1, C12 contains the three observations for dog 2, and so on. If we print columns 11–20, we get the output in Figure M12-7.

ROW	C11	C12	C13	C14	C15	C16	C17	C18	C19	C20
1	0.28	0.51	1.00	0.39	0.29	0.36	0.32	0.69	0.17	0.33
2	0.30	0.39	0.63	0.68	0.38	0.21	0.88	0.39	0.51	0.32
3	1.07	1.35	0.69	0.28	1.24	1.53	0.49	0.56	1.02	0.30

FIGURE M12-7 Contents of columns 11–20

ROW	C11	C12	C13	C14	C15	C16	C17	C18	C19	C20
1	1	2	3	2	1	2	1	3	1	3
2	2	1	1	3	2	1	3	1	2	2
3	3	3	2	1	3	3	2	2	3	1

FIGURE M12-8 Columns 11–20, after replacing observations by ranks within columns

Now we want to rank the observations for each dog (block). We will replace the observations with the ranks in columns 11–20:

```
MTB>   rank c11 c11
        .
        .
        .
MTB>   rank c20 c20
```

where the dots indicate that we type this command for all ten columns. If we now print columns 11–20, we get the output in Figure M12-8.

We need the sum of the ranks for each treatment (anesthetic). We will put these sums in column 21:

```
MTB>   rsum c11-c20 c21
```

To find average ranks for the three rows, we divide column 21 by 10:

```
MTB>   let c21=c21/10
```

The overall average rank is (3 + 1)/2, where 3 is the number of treatments. We subtract this value from each element of column 21:

```
MTB>   let c21=c21-(3+1)/2
```

Then we square each element of column 21:

```
MTB>   let c21=c21**2
```

We sum the elements of column 21:

```
MTB>   sum c21 k1
```

Since the number of blocks is 10 and the number of treatments is 3, the test statistic is calculated as

```
MTB>   let k2=k1*12*10/(3*(3+1))
```

We compare this test statistic with the chi-square distribution with 2 degrees of freedom, since there are three treatments:

```
MTB>   cdf k2 k3;
SUBC>  chisquare 2.
MTB>   let k3=1-k3
MTB>   print k2 k3
K2     1.40000
K3     0.496585
```

Minitab prints the value K2 = 1.40000 of Friedman's test statistic and the large-sample approximate p-value K3 = 0.496585.

Exercises for Chapter 12

For each exercise, plot the observations in any ways that seem reasonable. Describe the population(s) sampled, whether real or hypothetical. For each statistical procedure, state appropriate hypotheses. Discuss the assumptions that make the analysis appropriate. Do these assumptions seem reasonable? What additional information would you like to have about the experiment? Discuss the results of each analysis.

EXERCISE 12-1

Does an insect electrocuting device reduce mosquito biting? Researchers equipped suburban yards with either an insect electrocuting device, a standard 6-volt CDC trap, or no device. People serving as bait captured mosquitos coming to bite in each yard. The investigators took steps to allow for differences among yards and differences in attractiveness of the volunteers as mosquito bait. Details of the experiment are given in Nasci, Harris, and Porter (1983). The response for each yard is the percentage of the highest total number of mosquitos collected in any yard that night. The results are shown below.

Device	Percentage of maximum mosquito count						
Electrocuting device	66	57	57	31	87	97	89
	100	85	100	61	58		
CDC trap	100	75	50	77	58	100	62
	82	88	86	100	44		
None	75	84	100	74	40	94	87
	55	91	63	83	87		

a. Plot the observations.

b. Use a parametric analysis to test the null hypothesis that the mean mosquito response is the same for each device. Do the assumptions of the analysis seem reasonable?

c. Go through the steps for a nonparametric analysis. Do the assumptions of this analysis seem reasonable?

d. Compare your results in parts (b) and (c). Discuss your findings.

e. Why did the investigators choose as a response the percentage of highest total number of mosquitos collected in a yard in a night?

EXERCISE 12-2

In a study of a synthetic vaccine for malaria, scientists divided twelve 18–21-year-old male volunteers into four groups. They assigned three volunteers to a saline control group. They divided the other nine men among three different vaccine dose/treatment regimens. After vaccination, the researchers recorded a stimulation index for each volunteer, determined from proliferation assays of peripheral blood mononuclear cells. The results are shown below (Patarroyo et al., 1988).

Group	Stimulation index			
Saline control	1.4	1.0	4.0	
Regimen 1	1.5	5.6	12.4	
Regimen 2	6.6	9.1		
Regimen 3	35.1	13.4	0.8	3.3

a. Plot the observations.

b. Use a parametric analysis to test the null hypothesis that the mean stimulation index is the same for each treatment group. Do the assumptions for this analysis seem reasonable? Use the Bonferroni method to make multiple comparisons.

c. Take the logarithm of each stimulation index. Use a parametric analysis to test the null hypothesis that the mean of the logarithm of stimulation index is the same for each treatment group. Do the assumptions for this analysis seem reasonable? Use the Bonferroni method to make multiple comparisons.

d. Go through the steps for a nonparametric analysis. Do the assumptions for this analysis seem reasonable? Use the Bonferroni method to make multiple comparisons.

e. Compare your results in parts (b), (c), and (d). Discuss your findings.

EXERCISE 12-3

Researchers measured the amount of nitrogen expired by people on four different diets (Devore, 1982, page 600; from "Production of Gaseous Nitrogen in Human Steady-State Conditions," *J. Applied Physiology,* 1972, pages 155–159). The results are shown below.

Diet	Expired nitrogen (liters)					
Fasting	4.079	4.859	3.540	5.047	3.298	4.679
	2.870	4.648	3.847			
23% protein	4.368	5.668	3.752	5.848	3.802	4.844
	3.578	5.393	4.374			
32% protein	4.169	5.709	4.416	5.666	4.123	5.059
	4.403	4.496	4.688			
67% protein	4.928	5.608	4.940	5.291	4.674	5.038
	4.905	5.208	4.806			

a. Plot the observations.

b. Use a parametric analysis to test the null hypothesis that mean expired nitrogen is the same for all four diet groups. Do the assumptions for the analysis seem reasonable? Use the Bonferroni method to make multiple comparisons.

c. Go through the steps for a nonparametric analysis. Do the assumptions for this analysis seem reasonable? Use the Bonferroni method to make multiple comparisons.

d. Compare your answers to parts (b) and (c). Discuss your findings.

EXERCISE 12-4

Researchers measured skin potential (in millivolts) in each of eight volunteers after requesting each of four emotions: fear, happiness, depression, and calmness. The results are shown below (Devore, 1982, page 599; from "Physiological Effects During Hypnotically Requested Emotions," *Psychosomatic Med.*, 1963, pages 334–343).

	Volunteer							
Emotion	1	2	3	4	5	6	7	8
Fear	23.1	57.6	10.5	23.6	11.9	54.6	21.0	20.3
Happiness	22.7	53.2	9.7	19.6	13.8	47.1	13.6	23.6
Depression	22.5	53.7	10.8	21.1	13.7	39.2	13.7	16.3
Calmness	22.6	53.1	8.3	21.6	13.3	37.0	14.8	14.8

a. Plot the observations.

b. Are the relative differences among emotions similar for all eight volunteers?

c. Use a parametric analysis to analyze the results of this experiment. Use residual plots to check model assumptions.

d. Use a nonparametric analysis to test for differences in skin potential under the four emotions. Compare your results with what you found in part (c).

e. Discuss your findings.

EXERCISE 12-5

In a study of the effects of long-term freezing on bread dough, researchers used three types of flour. They made four batches of bread dough using each of the three types of flour. They then froze the dough. After the period of freezing, the researchers removed the dough from the freezer and recorded the volume increase in the bread dough 4 hours later. The results are shown below (from an example in Hocking, 1985, page 7).

Flour type	Volume increase			
1	1.1	1.8	1.0	1.2
2	2.7	2.9	3.3	2.8
3	3.1	3.2	3.3	3.2

a. Plot the observations.

b. Use a parametric analysis to test the null hypothesis that the mean volume increase is the same for the three types of flour. Do the assumptions for the analysis seem reasonable? Use the Bonferroni method to make multiple comparisons.

c. Use a nonparametric analysis to test the null hypothesis that the median volume increase is the same for the three types of flour. Do the assumptions for the analysis seem reasonable? Use the Bonferroni method to make multiple comparisons.

d. Compare your results in parts (b) and (c). Discuss your findings.

EXERCISE 12-6

An investigator wanted to compare the working life of three types of stopwatch (Rice, 1988, page 432; from Natrella, 1963). He tested several of each type, using each stopwatch through repeated cycles (on, off, restart) until it no longer worked. Survival times (thousands of cycles until failure) are listed below.

Type 1	1.7	1.9	6.1	12.5	16.5	25.1	30.5
	42.1	82.5					
Type 2	13.6	19.8	25.2	46.2	46.2	61.1	
Type 3	13.4	20.9	25.1	29.7	46.9		

a. Plot the observations.

b. Use a parametric analysis to test the null hypothesis that mean life is the same for the three types of stopwatch. Do the assumptions for the analysis seem reasonable? Use the Bonferroni method to make multiple comparisons.

c. Use a nonparametric analysis to test the null hypothesis that median life is the same for the three types of stopwatch. Do the assumptions for this analysis seem reasonable? Use the Bonferroni method to make multiple comparisons.

d. Compare your answers to parts (b) and (c). Discuss your findings.

EXERCISE 12-7

W. F. Woodward, a shortstop for the 1970 Cincinnati Reds, compared three methods of rounding first base. Twenty-two volunteers used each method to round first base. Woodward recorded the time it took a volunteer to run from a point between home and first base (35 feet from home plate) to a point between first and second (15 feet short of second base). The response is the average time of two runs (units not given). The results are shown below (Hollander and Wolfe, 1973, pages 140–141; from W. F. Woodward, "A Comparison of Base Running Methods in Baseball," M. Sc. thesis, Florida State University, 1970).

Runner	Round out method	Narrow angle method	Wide angle method
1	5.40	5.50	5.55
2	5.85	5.70	5.75
3	5.20	5.60	5.50
4	5.55	5.50	5.40
5	5.90	5.85	5.70
6	5.45	5.55	5.60
7	5.40	5.40	5.35
8	5.45	5.50	5.35
9	5.25	5.15	5.00
10	5.85	5.80	5.70
11	5.25	5.20	5.10
12	5.65	5.55	5.45
13	5.60	5.35	5.45
14	5.05	5.00	4.95
15	5.50	5.50	5.40
16	5.45	5.55	5.50
17	5.55	5.55	5.35
18	5.45	5.50	5.55
19	5.50	5.45	5.25
20	5.65	5.60	5.40
21	5.70	5.65	5.55
22	6.30	6.30	6.25

a. Plot the observations.

b. Are the relative differences among methods similar for all 22 runners?

c. Use a parametric analysis to analyze the results of this experiment. Use residual plots to check model assumptions.

d. Use a nonparametric analysis to test for differences among methods of rounding first base. Do the assumptions for the analysis seem reasonable? Compare your results with what you found in part (c).

e. Discuss your findings.

EXERCISE 12-8

Researchers scored smoothness of nine types of fabric dried five ways (Devore, 1987, page 447; from "Line-Dried vs. Machine-Dried Fabrics: Comparison of Appearance, Hand, and Consumer Acceptance," *Home Econ. Research J.*, 1984, pages 27–35). The results are listed here.

Fabric	Machine dry	Line dry	Line dry, then 15-minute tumble	Line dry with softener	Line dry with air movement
Crepe	3.3	2.5	2.8	2.5	1.9
Double knit	3.6	2.0	3.6	2.4	2.3
Twill	4.2	3.4	3.8	3.1	3.1
Twill mix	3.4	2.4	2.9	1.6	1.7

Fabric	Machine dry	Line dry	Line dry, then 15-minute tumble	Line dry with softener	Line dry with air movement
Terry	3.8	1.3	2.8	2.0	1.6
Broadcloth	2.2	1.5	2.7	1.5	1.9
Sheeting	3.5	2.1	2.8	2.1	2.2
Corduroy	3.6	1.3	2.8	1.7	1.8
Denim	2.6	1.4	2.4	1.3	1.6

a. Plot the observations.

b. Are the relative differences among drying methods similar for the nine types of fabric?

c. Use a parametric analysis to analyze the results of this experiment. Use residual plots to check model assumptions.

d. Use a nonparametric analysis to test for differences among drying methods. Do the assumptions for the analysis seem reasonable? Compare your results with what you found in part (c).

e. Discuss your findings.

EXERCISE 12-9

Researchers wanted to study the effects of four treatments on earthworm populations. They applied all treatments at concentrations of 1,000 liters/hectare. (A hectare, abbreviated ha, is a metric unit of area equal to 2.471 acres.) The researchers divided a large rectangular field into 40 square plots, separated by buffer areas. They divided the 40 plots into groups of 10. All the plots in a group received one treatment. After treatment, the researchers applied an irritant that caused the earthworms to rise to the surface. They recorded total biomass/m² in equal sized subplots of each of the 40 plots. The results are shown below (part of a data set contributed by R. P. Blackshaw and P. J. Diggle to a collection of problems in Andrews and Herzberg, 1985, pages 301–306).

Treatment	Biomass/meter²					
Water only	17.61	21.19	19.34	33.11	26.63	24.49
	39.12	16.40	53.32	39.26		
.5 kg/ha Benlate	72.61	24.47	9.38	63.90	36.10	28.38
	18.91	36.77	10.65	49.58		
.6 kg/ha Bevistin	57.10	74.06	23.74	28.40	32.31	32.15
	78.15	23.20	21.63	68.21		
1.4 kg/ha Cercobin	32.34	22.17	26.20	59.82	26.90	70.68
	63.01	55.54	49.26	78.62		

a. Plot the observations.

b. Use a parametric analysis to test the null hypothesis that the mean biomass/m² is the same for all four treatments. Do the assumptions of the analysis seem reasonable? Use the Bonferroni method to make multiple comparisons.

c. Use a nonparametric analysis to test the null hypothesis that the median biomass/m² is the same for all four treatments. Do the assumptions for the analysis seem reasonable? Use the Bonferroni method to make multiple comparisons.

d. Compare your results in parts (b) and (c). Discuss your findings.

EXERCISE 12-10

Some researchers wanted to compare yield using four methods of manufacturing penicillin. One important ingredient in producing penicillin is corn steep liquor. Because this ingredient is extremely variable, the researchers decided on a randomized block design. They divided a single blend of corn steep liquor into four parts and randomly assigned the four parts to the four manufacturing methods. These four runs comprised a block. To further reduce the effects of extraneous factors, the researchers used a random process to determine the order of runs within a block. The yields of penicillin (units not given) under the four manufacturing methods are listed below for each of five blends of corn steep liquor. The order of the run within a block is shown in parentheses next to the yield (from an example in *Statistics for Experimenters,* by Box, Hunter, and Hunter, John Wiley and Sons, New York, 1978, page 209).

Manufacturing method	Blend of corn steep liquor				
	1	2	3	4	5
1	89 (1)	84 (4)	81 (2)	87 (1)	79 (3)
2	88 (3)	77 (2)	87 (1)	92 (3)	81 (4)
3	97 (2)	92 (3)	87 (4)	89 (2)	80 (1)
4	94 (4)	79 (1)	85 (3)	84 (4)	88 (2)

a. Plot the observations.

b. Are the relative differences among manufacturers similar within all five blends of corn steep liquor?

c. Use a parametric analysis to analyze the results of this experiment. Use residual plots to check model assumptions.

d. Within each block, plot residuals versus run order. Does there appear to be a trend? If there were a trend, what would it mean?

e. Use a nonparametric analysis to test for differences among manufacturing methods on penicillin yield. Compare your results with what you found using the parametric analysis in part (c).

f. Discuss your findings.

EXERCISE 12-11

Investigators wanted to compare aggressive behavior of three species of mice, labeled I, II, and III. Species III was a cross of species I and II. The experimenters placed a mouse in the center of a box that was 1 meter square. The floor of the box was divided into 49 equal squares. The researchers recorded the number of squares the mouse crossed in 5 minutes. The results are shown below (Rice, 1988, pages 431–432).

Species I	309	229	182	228	326	289	231	225	307
	281	316	290	318	273	328	325	191	219
	216	221	198	181	110	256	240	122	290
	253	164	211	215	211	152	178	194	144
	95	157	240	146	106	252	266	284	274
	285	366	360	237	270	114	176	224	
Species II	37	90	39	104	43	62	17	19	21
	9	16	65	187	17	79	77	60	8
	81	39	133	102	36	19	53	59	29
	47	22	140	41	122	10	41	61	19
	62	86	66	64	53	79	46	89	74
	44	39	59	29	13	11	23	40	
Species III	140	218	215	109	151	154	93	103	90
	184	7	46	9	41	241	118	15	156
	111	120	163	101	170	225	177	72	288
	129								

a. Plot the observations.

b. Use a parametric analysis to test the null hypothesis that the mean number of squares crossed is the same for each species. Do the assumptions for this analysis seem reasonable? Use the Bonferroni method to make multiple comparisons.

c. Use a nonparametric analysis to test the null hypothesis that the median number of squares crossed is the same for each species. Do the assumptions for this analysis seem reasonable?

d. Compare your answers to parts (b) and (c). Discuss your findings.

EXERCISE 12-12 Researchers applied five types of electrode to the arms of 16 volunteers and measured resistance (Berry, 1987). They wanted to see if the different types of electrode gave similar measurements. The results (in k.ohms) are shown below.

	Type of electrode				
Volunteer	1	2	3	4	5
1	500	400	98	200	250
2	660	600	600	75	310
3	250	370	220	250	220
4	72	140	240	33	54
5	135	300	450	430	70
6	27	84	135	190	180
7	100	50	82	73	78
8	105	180	32	58	32
9	90	180	220	34	64
10	200	290	320	280	135
11	15	45	75	88	80
12	160	200	300	300	220
13	250	400	50	50	92
14	170	310	230	20	150
15	66	1,000	1,050	280	220
16	107	48	26	45	51

a. Plot the observations.

b. Are the relative differences among electrode types similar for all 16 volunteers?

c. Use a parametric analysis to analyze the results of this experiment. Use residual plots to check model assumptions.

d. Use a nonparametric analysis to test for differences among electrode types. Compare your results with what you found using the parametric analysis in part (c).

e. There are two very large readings for volunteer 15. The investigators speculated that this may have been due to a large amount of hair on this volunteer's arm. However, we have no information on the amount of arm hair for any of the volunteers. Exclude the observations for volunteer 15 and repeat parts (b), (c), and (d). Compare your results when the observations for volunteer 15 are included and excluded.

f. Discuss your findings.

EXERCISE 12-13 Does knowledge of output improve performance in repetitive work? In this experiment, investigators looked at performance in grinding a piece of metal to meet size and shape specifications (Hollander and Wolfe, 1973, page 121; from Hundal, 1969). They randomly divided 18 men into three groups. The investigators gave the six men in the first group no information on their output. They gave the six men in the second group rough estimates of their output. They gave the six men in the third group accurate and detailed information on their output. The response variable is the number of pieces finished by each worker during a fixed time interval. The results are shown below.

No information						Rough information						Detailed information					
40	35	38	43	44	41	38	40	47	44	40	42	48	40	45	43	46	44

a. Plot the observations.

b. Use a parametric analysis to test the null hypothesis that the mean output is the same under the three conditions. Do the assumptions for the analysis seem reasonable? Use the Bonferroni method to make multiple comparisons.

c. Use a nonparametric analysis to test the null hypothesis that the median output is the same under the three conditions. Do the assumptions for the analysis seem reasonable? Use the Bonferroni method to make multiple comparisons.

d. Compare your answers to parts (b) and (c). Discuss your findings.

EXERCISE 12-14 Researchers treated 12 patients with cardiac arrhythmias with each of three active drugs in a double-blind experiment. The researchers treated a patient with one drug for 1 week. They then made a 24-hour ambulatory electrocardiograph recording. They repeated this regimen for each of the three drugs, with treatment periods widely separated by intervals with no drugs. The response is the number of premature ventricular contractions per hour. The results are shown below (Berry, 1987).

Patient	Drug A	Drug B	Drug C
1	170	7	0
2	19	1.4	6
3	187	205	18
4	10	.3	1
5	216	.2	22
6	49	33	30
7	7	37	3
8	474	9	5
9	.4	.6	0
10	1.4	63	36
11	27	145	26
12	29	0	0

a. Plot the observations.

b. Are the relative differences among drugs similar for all 12 patients?

c. Use a parametric analysis to analyze the results of this experiment. Use residual plots to check model assumptions.

d. Use a nonparametric analysis to test for differences among drugs. Do the assumptions for the analysis seem reasonable? Compare with your results in part (c).

e. Discuss your findings.

EXERCISE 12-15

For a middle-school science project to study possible effects of acid rain, a student planted 12 tomato seeds in loam, in separate containers (Foster, 1986). She randomly divided the 12 containers into three groups. The student watered the four seeds in group 1 every day with water having pH 4.0. She watered the four seeds in group 2 every day with water having pH 5.6. Finally, she watered the four seeds in group 3 every day with distilled water having pH 7.0. When plants came up, she watered the soil (not the leaves). Three of the four plants in group 1 came up; all of the plants in the other two groups came up. After 3 weeks, the student measured the height of each plant. The results are shown below.

Group	Height of tomato plants (centimeters)			
pH 4.0	1.8	1.5	1.9	
pH 5.6	2.1	2.1	2.0	1.8
pH 7.0	2.7	2.6	2.4	2.3

Base your analyses on the 11 plants that came up.

a. Plot the observations.

b. Use a parametric analysis to test the null hypothesis that the mean height is the same for the three levels of pH. Do the assumptions for the analysis seem reasonable? Use the Bonferroni method to make multiple comparisons.

c. Use a nonparametric analysis to test the null hypothesis that the median height is the same for the three levels of pH. Do the assumptions for the analysis seem reasonable? Use the Bonferroni method to make multiple comparisons.

d. Compare your answers to parts (b) and (c). Discuss your findings.

e. How does the fact that one seed in group 1 did not germinate contribute to your discussion of this experiment?

EXERCISE 12-16 Suppose we have m null hypotheses to test. For each test, we state a criterion for deciding whether the data are inconsistent with the null hypothesis. We say the data are inconsistent with the "combined" null hypothesis if at least one of the m criteria is satisfied. Let α denote the significance level associated with this "combined" criterion. Let α_1 through α_m denote the significance levels associated with the m separate criteria. Show that α is less than or equal to the sum of α_1 through α_m. (*Hint:* We know from Chapter 6 that if event E can be written as the union of events E_1 through E_m, then the probability of E is less than or equal to the sum of the probabilities of events E_1 through E_m.)

EXERCISE 12-17 Find the Kruskal–Wallis distribution for samples of size 1, 2, and 2.

EXERCISE 12-18 Consider the experiment in Example 12-3. Ignore the technicians and go through the steps for one-way analysis of variance to test for differences among thermometers. Compare your results with what we found in Example 12-3.

EXERCISE 12-19 Test for differences among thermometers in Example 12-3, using Friedman's test. Compare your results with the results of the parametric analysis in Section 12-3. Do the assumptions for Friedman's test seem reasonable?

EXERCISE 12-20 Discuss the sampling situations in which one-way analysis of variance and the Kruskal–Wallis test are appropriate. Which procedure is preferred in each situation?

EXERCISE 12-21 Discuss the sampling situations in which the classical analysis of a randomized block experiment and Friedman's test are appropriate. Which procedure is preferred in each situation?

CHAPTER 13

Two-Factor Experiments: Balanced, Completely Randomized, Factorial Designs

IN THIS CHAPTER

Balanced design
Completely randomized design
Factorial design
Two-factor experiments
Two-way analysis of variance
Replication

There are many situations in which we are interested in the effects of more than a single factor on a response variable. Imagine, for instance, that you run a weight loss clinic. For any given client, you have a choice of several diets, as well as a choice of several exercise regimens. You realize that different diets might result in different average weight loss among your clients. Also, average weight loss may depend on exercise program. In addition, you suspect that the effectiveness of a given diet may depend on the exercise regimen being followed by a client. If the relative effects of the different diets depend on the particular exercise regimen (or, equivalently, the relative effectiveness of the different exercise programs depends on diet), we say there is an *interaction effect* of diet and exercise on weight loss.

> We say two factors have an **interaction effect** on response if the relative effects of one factor on response depend on the level of the other factor.

Suppose a number of your clients have consented to participate in an experiment that you design. To compare the several diets with respect to weight loss, you could set up a single-factor experiment, assigning a number of clients to each of the diets. You would try to keep extraneous sources of variation to a minimum in order to assess the effects of the diets on weight loss. One possible extraneous factor might be exercise program; you would want all participants in this experiment to maintain the same exercise regimen. What are some other possible extraneous factors?

Similarly, you could carry out a single-factor experiment to compare the several exercise programs with respect to weight loss. A possible extraneous factor in this experiment is diet; you would want all clients participating in this experiment to be on the same diet. Other possible extraneous factors would include those you listed above.

Now you have done two separate single-factor experiments—one to compare diets and one to compare exercise programs. You kept exercise program constant in the diet experiment and you kept diet constant in the exercise experiment. Therefore, you are not able to assess the extent of any interaction effect of diet and exercise on average weight loss.

A way around this problem is to include diet and exercise in the same experiment; we call this a *two-factor experiment.*

> In a **two-factor experiment,** we are interested in the effects of two factors, and possible interaction effects of the two factors, on a response variable.

If you include all combinations of diets and exercise programs in your experiment, you have a *factorial design.*

> In a **factorial design,** all combinations of levels of the factors are included in the experiment.

Your experiment is *balanced* if you assign the same number of clients to each combination of diet and exercise.

> An experimental design is **balanced** if there is the same number of observations for each combination of factors.

When you divide the clients among the diet/exercise combinations, you should use *random assignment* so that no bias, whether subconscious or not, affects the way clients are assigned. We also hope that random assignment will tend to balance any extraneous factors that might affect our experimental results. When we randomly assign experimental units to combinations of the factors (such as diet and exercise), we have a *completely randomized design*. (In the diet/exercise example, the experimental units are the clients participating in the experiment.)

> An experiment is **completely randomized** if experimental units are randomly assigned to combinations of factors.

To look for a possible interaction effect of diet and exercise on average weight loss, you need to assign at least two clients to each combination of diet and exercise. Then you have *replication*.

> We have **replication** in an experiment if we have more than one observation per factor combination.

If you use a balanced, completely randomized, factorial design with replication you will be able to compare diets, compare exercise programs, as well as assess possible interaction effects of diet and exercise on average weight loss. Each of these three comparisons will be independent of the other two. With a well-designed two-factor experiment, you can economically obtain more information than would be available from separate single-factor experiments.

A randomized block experiment (k treatments, b blocks) is a special type of two-factor experiment. One factor is *treatment,* with *k levels,* corresponding to the k treatments. The other factor is *block,* with *b levels,* corresponding to the b blocks. In this chapter we are concerned with two-factor experiments other than randomized block designs.

Suppose the first factor, factor A, has I levels and the second factor, factor B, has J levels. If we include each of the $I \times J$ combinations of factors A and B in the experiment, then we have a *factorial design*. To allow for the possibility of an interaction effect in our analysis, we must have more than one observation per combination of factors. Then we have *replication*. If we have the same number of observations per combination, we have a *balanced* design.

Suppose we have n observations for each of the $I \times J$ factor combinations, with n greater than or equal to 2 (that is, we have n replications). Then our total sample size is $N = I \times J \times n$. If the N experimental units are randomly assigned to the factor combinations, we have a *completely randomized design*. In this chapter, we discuss the classical, parametric, analysis of *balanced, completely randomized, factorial experiments with two factors*.

We will assume that the I levels of factor A are the specific levels of interest to us. We say factor A has *fixed effects*. Similarly, the J levels of factor B are the specific levels of interest to us, so factor B also has fixed effects. (In the diet and exercise example, you are interested in the specific diets and the specific exercise programs included in the experiment.) In contrast, the blocks in a randomized block design are *random effects*. We generally are not interested in the specific levels of a random effect variable, whether people, dogs, or whatever; those included in the experiment are representative of a larger group of subjects of interest to us. We will not consider experiments with random effects (other than the simple randomized block design we have already seen). For discussion of such experiments, see a book on experimental design such as the one by Kirk (1982).

In Section 13-1, we consider the classical, parametric, analysis of two-factor experiments that have balanced, completely randomized, factorial designs. This analysis is called *two-way analysis of variance* or *two-factor analysis of variance.*

Two-way (or **two-factor**) **analysis of variance** is the classical, parametric, approach to analyzing the results of a two-factor experiment that has a balanced, completely randomized, factorial design.

We consider a special case in Section 13-2, when each factor has two levels. Such experiments are especially useful in exploratory studies.

13-1 Two-Factor Analysis of Variance

Let's first look at an example of a balanced, completely randomized, factorial experiment.

EXAMPLE 13-1

How well do different laboratories agree in measuring niacin in bread? Does the extent of agreement depend on the amount of niacin in the bread? Researchers designed a study to answer these questions. They divided samples of bread into three groups. The bread in the first group was not enriched with niacin. The bread in the second group was enriched with 2 milligrams of niacin per 100 grams of bread. The bread in the third group was enriched with 4 milligrams of niacin per 100 grams of bread. The experimenters sent samples to each of six laboratories, where laboratory workers divided a sample into three subsamples. They measured niacin in the sample on three different days (one subsample each day), so there were three replications for each combination of laboratory and niacin level. The measurements of niacin (in milligrams per 100 grams) are shown below (a portion of the data in Rice, 1988, pages 429–430; from Campbell and Pelletier, 1962).

Niacin enrichment	Laboratory					
	a	b	c	d	e	f
0	3.40	3.80	3.66	4.37	4.20	3.76
	3.63	3.80	3.92	3.86	3.60	3.68
	3.52	3.90	4.22	4.46	4.20	3.80
2	5.00	5.30	5.68	6.53	5.80	6.06
	5.27	5.60	5.47	5.85	5.70	5.60
	5.39	5.80	5.84	6.38	5.90	6.05
4	6.54	7.10	7.30	8.32	7.70	7.60
	7.46	7.60	6.40	7.34	7.10	7.50
	6.84	8.00	7.60	8.12	7.20	7.67

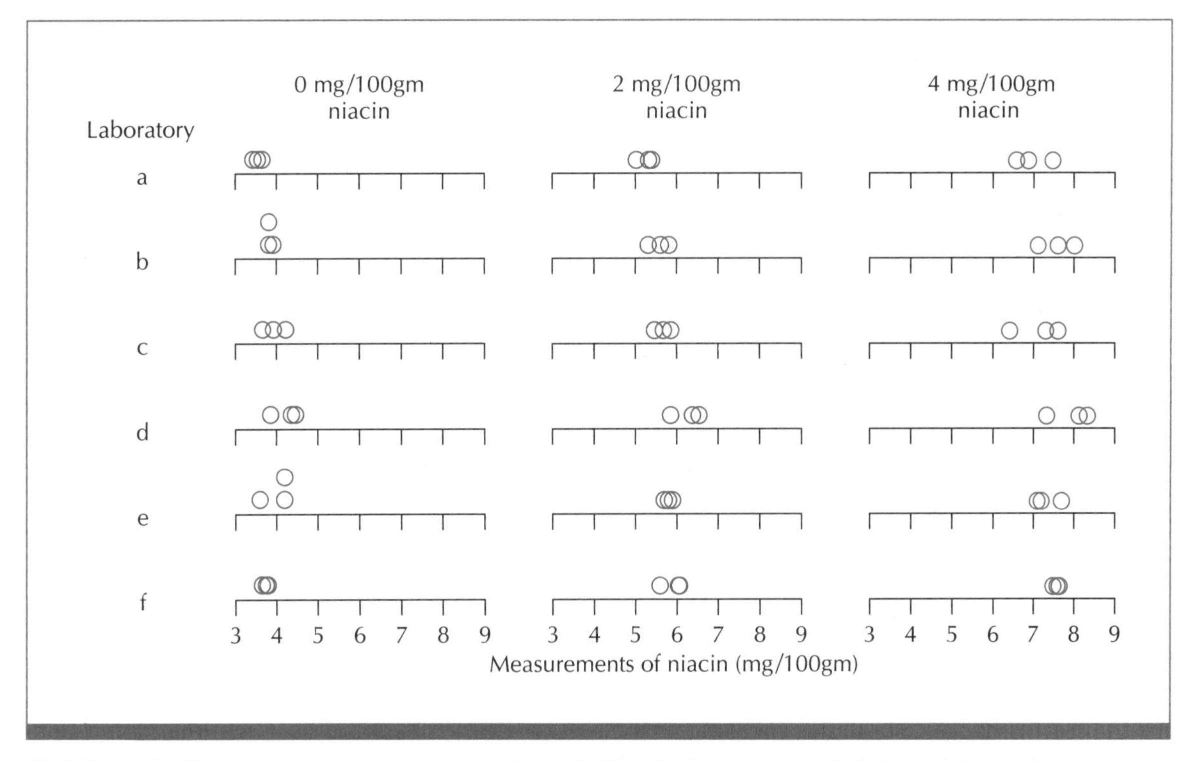

FIGURE 13-1 Plots of niacin measurements in Example 13-1, by laboratory and niacin enrichment level

Do you have suggestions for these investigators? How would you decide which samples go to which laboratories? Should we be concerned about the order in which the samples are analyzed in each laboratory? Would you worry about the time of day, the technician doing the measurements, whether the technician has knowledge of the niacin enrichment level of the sample? Give detailed instructions (called a protocol) for carrying out this experiment.

Plots of the niacin measurements are shown in Figure 13-1 by laboratory and niacin enrichment level. Does it look like the laboratory determinations of niacin depend on the level of niacin enrichment? Do there appear to be differences among laboratories? How does the spread or variation among values compare across the 18 plots?

We are also interested in possible interaction effects on measurements. Do differences among laboratories depend on the level of niacin enrichment? Alternatively, do the relative effects of niacin enrichment level depend on the laboratory making the measurements? One way to answer this question is through a plot such as shown in Figure 13-2. This figure shows a scatterplot of average niacin measurement versus enrichment level for each of the six laboratories.

The points for a single laboratory in Figure 13-2 are connected, creating a *profile* for that laboratory. The slope of each profile clearly illustrates the

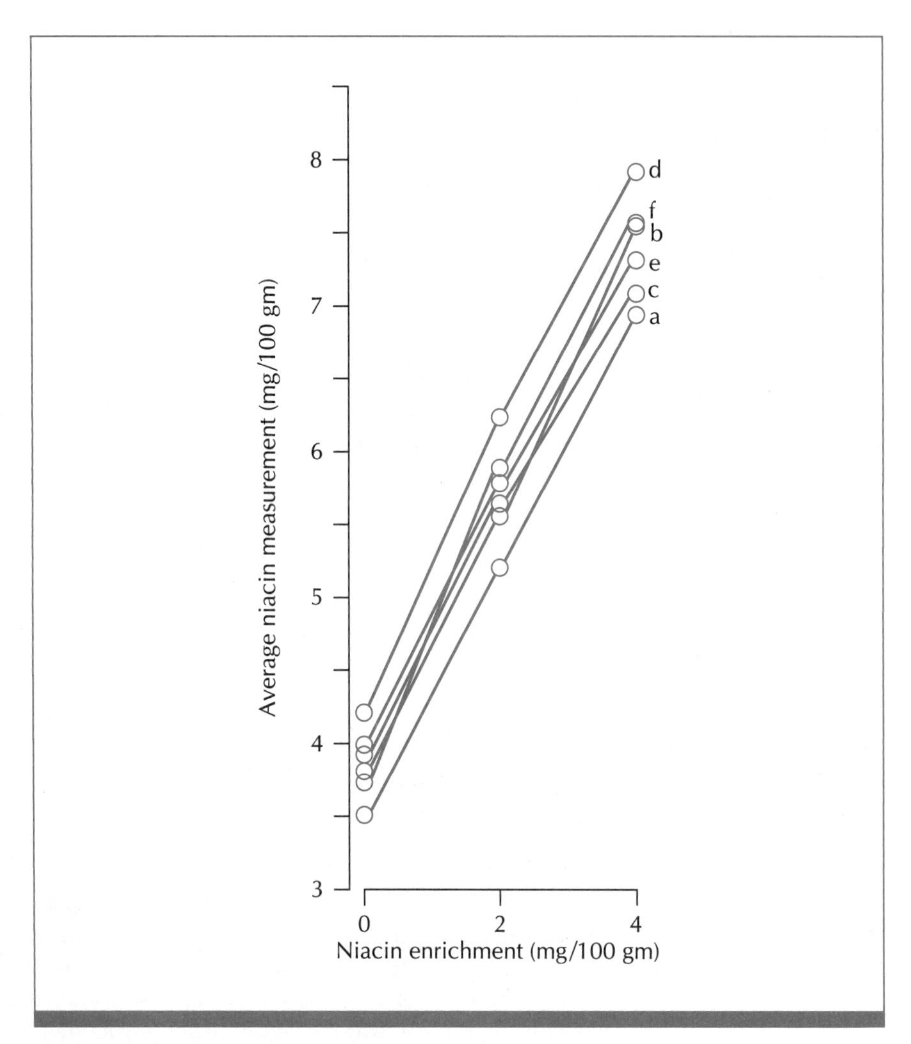

FIGURE 13-2 Scatterplot of average niacin measurement versus niacin enrichment level for each laboratory. The points for a single laboratory are connected.

differences in average measurements for the three niacin enrichment levels. The profiles for the six laboratories do not all coincide (they are not all on top of each other), so there may be differences among laboratories. However, the profiles are parallel: the relative effects of niacin enrichment level seem to be about the same for all six laboratories. That is, there seems to be no interaction effects of laboratory and niacin enrichment level on the niacin measurements.

A formal analysis of this experiment tests hypotheses about differences among laboratories, differences among niacin enrichment levels, and interaction effects of laboratory and niacin enrichment level on average niacin determination. Before outlining the method of analysis, we need some notation.

Two-Factor Analysis of Variance for a Balanced Factorial Design

Suppose factor A has I levels and factor B has J levels. There are n replications per factor combination, with n greater than or equal to 2. Let Y_{ijk} denote the observation for the kth replication at level i of factor A and level j of factor B. $\overline{Y}_{ij}$ represents the average of the n experimental units receiving level i of factor A and level j of factor B. $\overline{A}_i$ denotes the average of the $J \times n$ observations at level i of factor A. Similarly, $\overline{B}_j$ denotes the average of the $I \times n$ observations at level j of factor B. Finally, $\overline{Y}$ represents the average of all $N = I \times J \times n$ observations. The notation for these averages is summarized in Table 13-1, along with four estimates of variation.

The ***residual mean square*** s_r^2 is a measure of random variation among the observations in the experiment. The *mean square for factor A*, s_A^2, is a measure of random variation plus differences in average response at different levels of factor A. The *mean square for factor B*, s_B^2, measures random variation plus differences in average response at different levels of factor B. The *interaction mean square*, s_{AB}^2, estimates random variation plus nonadditive effects of the two factors on average response.

If the I levels of factor A all have the same average effect on response, then s_A^2 and s_r^2 both estimate random variation and should be similar in magnitude. If the levels of factor A do not all have the same average effect on response, then we expect s_A^2 to be larger than s_r^2. This forms the basis for testing hypotheses about differences on average response among the levels of factor A.

TABLE 13-1 Notation for the averages in a two-factor experiment

	Factor B				
Factor A	1	2	...	J	Factor A averages
1	$\overline{Y}_{11}$	$\overline{Y}_{12}$	...	$\overline{Y}_{1J}$	$\overline{A}_1$
2	$\overline{Y}_{21}$	$\overline{Y}_{22}$	...	$\overline{Y}_{2J}$	$\overline{A}_2$
⋮	⋮	⋮		⋮	⋮
I	$\overline{Y}_{I1}$	$\overline{Y}_{I2}$	...	$\overline{Y}_{IJ}$	$\overline{A}_I$
Factor B averages	$\overline{B}_1$	$\overline{B}_2$	...	$\overline{B}_J$	$\overline{Y}$

Variance estimates or mean squares:

$$s_A^2 = \text{Mean square for factor A} = \frac{J \times n}{I - 1} \sum_{i=1}^{I} (\overline{A}_i - \overline{Y})^2$$

$$s_B^2 = \text{Mean square for factor B} = \frac{I \times n}{J - 1} \sum_{j=1}^{J} (\overline{B}_j - \overline{Y})^2$$

$$s_{AB}^2 = \text{Mean square for AB interaction}$$

$$= \frac{n}{(I - 1)(J - 1)} \sum_{i=1}^{I} \sum_{j=1}^{J} (\overline{Y}_{ij} - \overline{A}_i - \overline{B}_j + \overline{Y})^2$$

$$s_r^2 = \text{Residual mean square} = \frac{1}{I \times J \times (n - 1)} \sum_{i=1}^{I} \sum_{j=1}^{J} \sum_{k=1}^{n} (Y_{ijk} - \overline{Y}_{ij})^2$$

Similarly, if the J levels of factor B all have the same average effect on response, then s_B^2 and s_r^2 both estimate random variation, so we expect them to be similar in magnitude. On the other hand, if the levels of factor B do not all have the same average effect on response, then we expect s_B^2 to be larger than s_r^2. We use this fact to test for differences among levels of factor B on average response.

If the relative effects of the levels of factor A are the same for all levels of factor B (or, equivalently, the relative effects of the levels of factor B are the same for all levels of factor A), then s_{AB}^2 and s_r^2 both estimate random variation and should be similar in magnitude. If the relative effects of the levels of factor A depend on the level of factor B (or the relative effects of levels of factor B depend on the level of factor A), then we expect s_{AB}^2 to be larger than s_r^2. This provides the rationale for our test of an interaction effect of factors A and B on average response.

With these ideas in mind, we can outline the significance level approach to the parametric analysis of a two-factor experiment with a balanced, completely randomized, factorial design. After outlining the approach, we will apply it to Example 13-1.

The significance level approach to two-factor analysis of variance for a balanced factorial design

1. We want to test three sets of hypotheses. One test is about the effect of factor A on the response:

   $H_0(A)$: The I levels of factor A have the same average effect on response.
   $H_a(A)$: The average effect on response is not the same for all I levels of factor A.

   Another test is about the effect of factor B on response:

   $H_0(B)$: The J levels of factor B have the same average effect on response.
   $H_a(B)$: The average effect on response is not the same for all J levels of factor B.

   In addition, we want to test the null hypothesis that the levels of factor A have the same relative effects on response within each level of factor B. This is the same as saying that the levels of factor B have the same relative effects on response within each level of factor A. Then we say there is no interaction effect of factors A and B on response.

   $H_0(AB)$: The relative effects of one factor do not depend on the level of the other factor.
   $H_a(AB)$: The relative effects of one factor do depend on the level of the other factor.

2. To test the hypotheses about the factor A effects, we use the test statistic

$$\text{Test statistic(A)} = \frac{s_A^2}{s_r^2}$$

To test the hypotheses about the factor B effects, we use the test statistic

$$\text{Test statistic(B)} = \frac{s_B^2}{s_r^2}$$

To test hypotheses about the interaction effect, we use the test statistic

$$\text{Test statistic(AB)} = \frac{s_{AB}^2}{s_r^2}$$

3. We assume that the $N = I \times J \times n$ observations are all independent, from Gaussian distributions. These distributions all have the same variance σ^2. The means may differ, depending on the combinations of the two factors.

   Under the null hypothesis that the levels of factor A have the same average effect, test statistic(A) has the F distribution with $I - 1$ numerator degrees of freedom and $IJ(n - 1)$ denominator degrees of freedom.

   Under the null hypothesis that the levels of factor B have the same average effect, test statistic(B) has the F distribution with $J - 1$ numerator degrees of freedom and $IJ(n - 1)$ denominator degrees of freedom.

   Under the null hypothesis of no interaction effect on response, test statistic(AB) has the F distribution with $(I - 1)(J - 1)$ numerator degrees of freedom and $IJ(n - 1)$ denominator degrees of freedom.

   For each of the three sets of hypotheses, small values of the test statistic, near 1, are consistent with the corresponding null hypothesis, while large values are inconsistent with that null hypothesis.
4. Select a significance level for each test.
5. Let α denote the significance level for one of the three sets of hypotheses. Find the number c from Table D such that $P(F \leq c) = 1 - \alpha$. Here, F denotes a random variable having the F distribution with degrees of freedom corresponding to the hypotheses being tested. The acceptance region is $[0, c)$; the rejection region is $[c, \infty)$.
6. For each set of hypotheses, the decision rule has the form:

   If test statistic $< c$, say the results are consistent with the null hypothesis.
   If test statistic $\geq c$, say the results are inconsistent with the null hypothesis.
7. Carry out an experiment that satisfies the assumptions in step 3. Calculate the test statistics in step 2. Use appropriate decision rules to decide whether there seem to be effects of the two factors and/or interaction effects on response. Draw conclusions based on the experimental results.

EXAMPLE 13-1 *(continued)*

Let's return now to the experiment described in Example 13-1. Call niacin enrichment level factor A and laboratory factor B. We want to test three sets of hypotheses. The first set is about the effect of niacin enrichment level:

H_0(A): All three niacin enrichment levels have the same average effect on niacin determination.

$H_a(A)$: The three niacin enrichment levels do not all have the same average effect on niacin determination.

The second set of hypotheses is about differences among laboratories:

$H_0(B)$: The six laboratories all have the same average effect on niacin determination.
$H_a(B)$: The six laboratories do not all have the same average effect on niacin determination.

The third set is about interaction effects:

$H_0(AB)$: The relative effects of niacin enrichment on the measurements are the same for all six laboratories. (Equivalently, relative differences among laboratories are the same for all three niacin enrichment levels.)
$H_a(AB)$: The relative effects of niacin enrichment on the measurements are not the same for all six laboratories. (Relative differences among laboratories are not the same for all three niacin enrichment levels.)

For a parametric analysis of this experiment, we assume that the 54 observations are all independent, from Gaussian distributions having the same variance. From the plots of the observations in Figure 13-1, the equal-variance assumption seems reasonable. We cannot assess the independence assumption without more information on how the experiment was conducted. What suggestions would you have for carrying out the experiment, in order to ensure independence? What other suggestions would you make, to control extraneous sources of variation and make tests of hypotheses valid?

To assess the Gaussian assumption, we can look at residuals. Recall that a *residual* is the difference between an observation and a summary, estimate, or predicted value of the observation. To calculate a residual in a balanced two-way factorial design, we subtract from each observation Y_{ijk} the estimated expected value $\bar{Y}_{ij}$ of that observation, based on the two-way analysis of variance model. Therefore, a residual has the form $Y_{ijk} - \bar{Y}_{ij}$.

A **residual** is the difference between an observation and an estimate of its expected value.

In a two-factor analysis of variance model that accounts for possible interaction effects, a residual is the difference $Y_{ijk} - \bar{Y}_{ij}$ between an observation and the average of the observations within the same combination of factors.

The 54 residuals for our example are listed in Table 13-2.

The residuals represent what is left over after we fit a two-way analysis of variance model. If the model is adequate, the residuals should appear to be randomly distributed about a mean of 0. If the Gaussian assumption for the observations is reasonable, the residuals will appear to follow a Gaussian distribution with mean 0. From the histogram of the residuals shown in Figure 13-3, the assumption of Gaussian observations seems reasonable.

TABLE 13-2 Residuals from the two-way analysis of variance model in Example 13-1

Factor A: Niacin enrichment level	Factor B: Laboratory					
	a	b	c	d	e	f
0	−.12	−.03	−.27	.14	.20	.01
	.11	−.03	−.01	−.37	−.40	−.07
	.00	.07	.29	.23	.20	.05
2	−.22	−.27	.02	.28	.00	.16
	.05	.03	−.19	−.40	−.10	−.30
	.17	.23	.18	.13	.10	.15
4	−.41	−.47	.20	.39	.37	.01
	.51	.03	−.70	−.59	−.23	−.09
	−.11	.43	.50	.19	−.13	.08

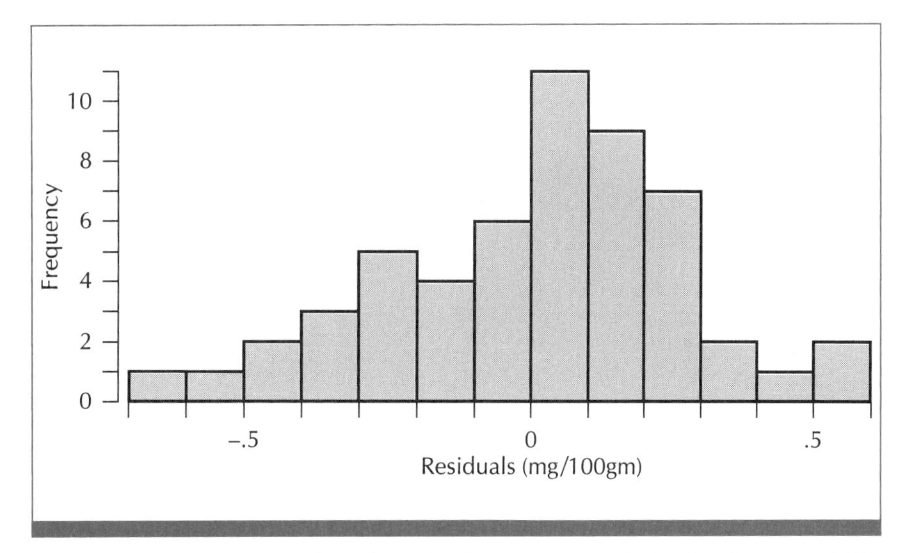

FIGURE 13-3 Histogram of residuals for the two-way analysis of variance model in Example 13-1

Figure 13-4 shows a plot of residuals by niacin enrichment level. The variation in residuals is greater at the highest enrichment level than at the lower levels. The residuals are plotted by laboratory in Figure 13-5. There is a gap in the middle of the plot for laboratory d. Also, the variation in residuals is smaller for laboratory f than for the other laboratories. These plots suggest that not all of our model assumptions may be perfectly met. However, they provide no strong evidence to discourage us from a parametric analysis of the experimental results.

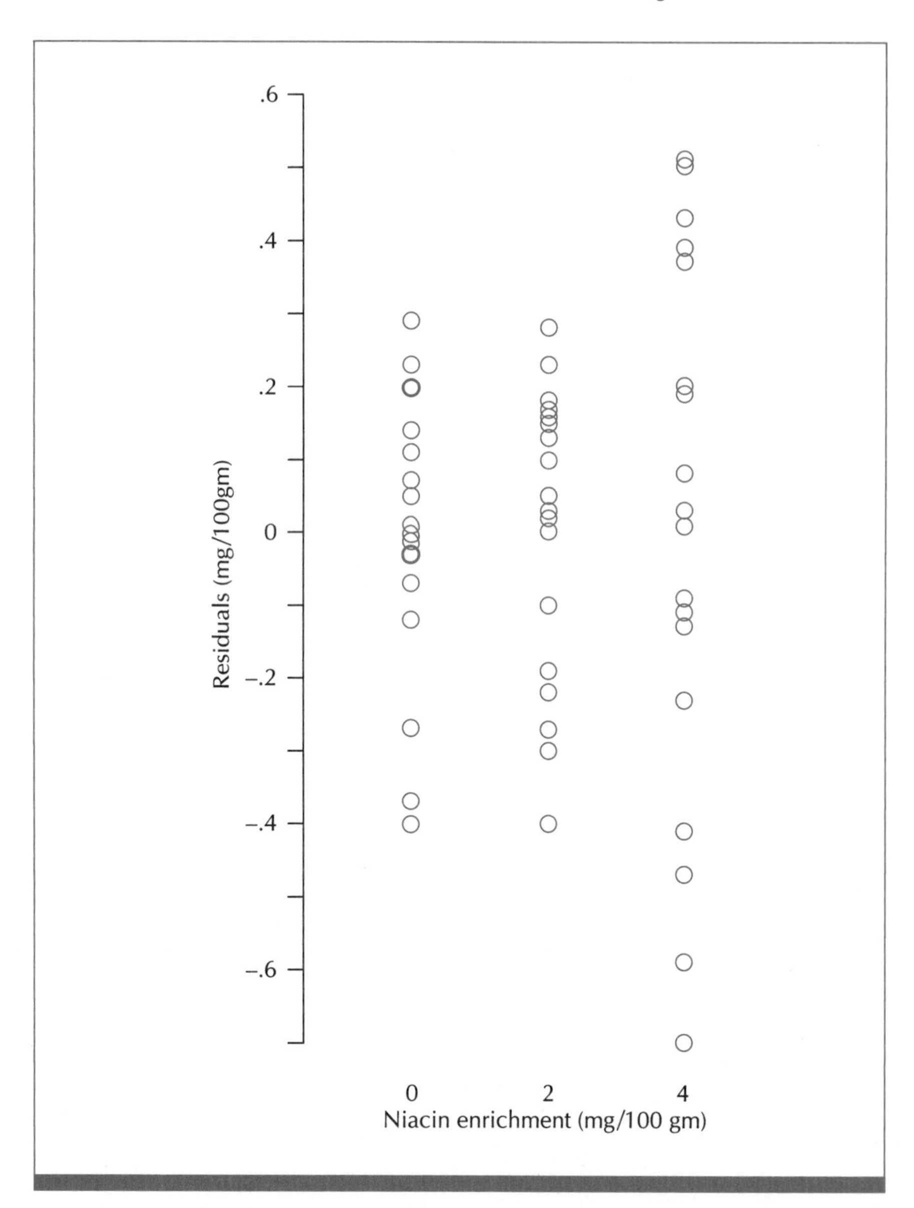

FIGURE 13-4 Plot of residuals by niacin enrichment level in Example 13-1

If all the model assumptions are met, then under the null hypothesis of no difference among niacin enrichment levels with respect to average niacin measurement, test statistic(A) would have the $F(2, 36)$ distribution. Under the null hypothesis of no difference among laboratories in average niacin measurements, test statistic(B) would have the $F(5, 36)$ distribution. Under the null hypothesis of no interaction effects, test statistic(AB) would have the $F(10, 36)$ distribution.

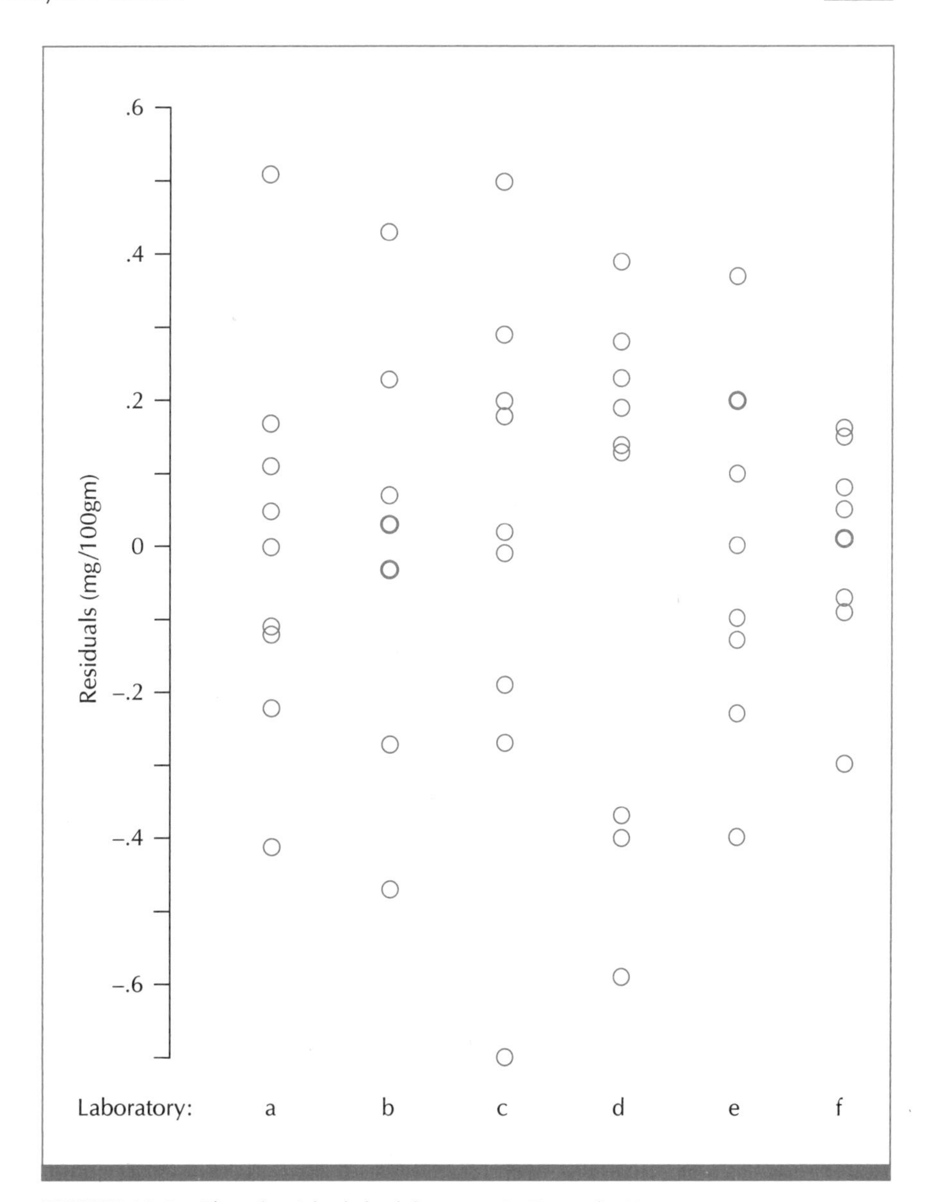

FIGURE 13-5 Plot of residuals by laboratory in Example 13-1

We will use significance level .01 for all three tests of hypotheses. For all three tests of hypotheses, we have 36 denominator degrees of freedom. Since 36 is not a value of d_2 listed in Table D, we will use the value just below 36, $d_2 = 30$. By rounding downward, we get a more conservative test (we are less likely to reject the null hypothesis when it is true).

If F_1 has the $F(2, 30)$ distribution, then $P(F_1 \leq 5.39) = .99$. The acceptance region for the factor A test is $[0, 5.39)$ and the rejection region is

[5.39, ∞). To test for effects of niacin enrichment levels on niacin measurements, the decision rule is:

If test statistic(A) < 5.39, say the results are consistent with the null hypothesis that the levels of niacin enrichment all have the same average effect on niacin determination.

If test statistic(A) ≥ 5.39, say the results are inconsistent with the null hypothesis, suggesting that niacin determinations depend on the level of niacin enrichment.

If F_2 has the $F(5, 30)$ distribution, then $P(F_2 \leq 3.70) = .99$. The acceptance region for the factor B test is [0, 3.70) and the rejection region is [3.70, ∞). To test for laboratory differences, the decision rule is:

If test statistic(B) < 3.70, say the results are consistent with the null hypothesis that the laboratories all have the same average effect on niacin determination.

If test statistic(B) ≥ 3.70, say the results are inconsistent with the null hypothesis, suggesting that the laboratories do not all get the same niacin determinations on average.

If F_3 has the $F(10, 30)$ distribution, then $P(F_3 \leq 2.98) = .99$. The acceptance region for the test of interaction effects is [0, 2.98); the rejection region is [2.98, ∞). To test whether differences among laboratories are the same for all niacin enrichment levels, the decision rule for this interaction effect is:

If test statistic(AB) < 2.98, say the results are consistent with the null hypothesis that the differences among laboratories are the same for all niacin enrichment levels (or the differences among niacin enrichment levels are the same for all laboratories).

If test statistic(AB) ≥ 2.98, say the results are inconsistent with the null hypothesis, suggesting that the differences among laboratories are not the same for all niacin enrichment levels (or the differences among niacin enrichment levels are not the same for all laboratories).

The calculations we need are outlined in Table 13-3.

Test statistic(A) equals 544.7, inconsistent with the null hypothesis that the niacin enrichment levels all have the same average effect on niacin measurements. The p-value is less than .01. We see these differences among niacin enrichment levels in the plots of the observations in Figure 13-1 and in the profile plot of average niacin determination versus niacin enrichment level in Figure 13-2.

Test statistic(B) equals 7.6, inconsistent with the null hypothesis that the laboratories all have the same average effect on niacin measurements. The p-value is less than .01, so that the differences among laboratories that we saw in the plots are statistically significant. From Figures 13-1 and 13-2 we see that laboratory a tended to get the lowest measurements, laboratory d the highest.

Test statistic(AB) equals .7, consistent with the null hypothesis that the differences in measurements among laboratories are the same for all three niacin enrichment levels (or differences in measurements among niacin en-

TABLE 13-3 Calculations for two-way analysis of variance in Example 13-1. The average niacin measurement $\bar{Y}_{ij}$ is shown below for each combination of niacin enrichment level and laboratory. In parentheses are values of $\bar{Y}_{ij} - \bar{A}_i - \bar{B}_j + \bar{Y}$.

Factor A: Niacin enrichment level	Factor B: Laboratory						Factor A averages
	a	b	c	d	e	f	
0	3.52 (.08)	3.83 (−.04)	3.93 (.15)	4.23 (−.12)	4.00 (.08)	3.75 (−.21)	3.88
2	5.22 (−.06)	5.57 (−.15)	5.66 (.03)	6.25 (.05)	5.80 (.03)	5.90 (.09)	5.73
4	6.95 (−.02)	7.57 (.17)	7.10 (−.21)	7.93 (.05)	7.33 (−.12)	7.59 (.10)	7.41
Factor B averages	5.23	5.66	5.57	6.14	5.71	5.75	5.67

$I = 3 \quad J = 6 \quad n = 3 \quad N = 54 \quad I - 1 = 2 \quad J - 1 = 5 \quad (I-1)(J-1) = 10 \quad IJ(n-1) = 36$

$$s_A^2 = \frac{6 \times 3}{3 - 1}[(3.88 - 5.67)^2 + (5.73 - 5.67)^2 + (7.41 - 5.67)^2] = 56.1$$

$$s_B^2 = \frac{3 \times 3}{6 - 1}[(5.23 - 5.67)^2 + (5.66 - 5.67)^2 + (5.57 - 5.67)^2 + (6.14 - 5.67)^2 + (5.71 - 5.67)^2 + (5.75 - 5.67)^2] = .78$$

$$s_{AB}^2 = \frac{3}{(3-1)(6-1)}[\text{Sum of the 18 values of } (\bar{Y}_{ij} - \bar{A}_i - \bar{B}_j + \bar{Y})^2] = .07$$

$$s_r^2 = \frac{1}{3 \times 6 \times (3 - 1)}[\text{Sum of the 54 values of } (Y_{ijk} - \bar{Y}_{ij})^2] = .103$$

$$\text{Test statistic(A)} = \frac{56.1}{.103} = 544.7 \qquad \text{Test statistic(B)} = \frac{.78}{.103} = 7.6 \qquad \text{Test statistic(AB)} = \frac{.07}{.103} = .7$$

richment levels are the same for all laboratories). This agrees with the observations we made about the parallel profiles in Figure 13-2.

The test statistic for niacin enrichment levels equals 544.7; this very large test statistic agrees with the obvious differences in measurements across niacin enrichment levels we observe in Figures 13-1 and 13-2. The test statistic for laboratories, 7.6, is much smaller. While statistically significant, it is more difficult for us to make a visual assessment of laboratory differences in niacin determinations by examining Figures 13-1 and 13-2. The lack of an interaction effect on response is very clear in the profile plot in Figure 13-2.

We can summarize the calculations for two-way analysis of variance in a table. A general form of analysis of variance table for a balanced two-way factorial experiment is shown in Table 13-4.

Computer output often shows an additional column for the p-value, at the right of the table. The analysis of variance table for Example 13-1 is given in Table 13-5. The first p-value is listed as 0.0000, meaning that the actual p-value is less than .0001.

You may notice that the values for the mean squares and test statistics in Table 13-5 are not exactly the same as those listed in Table 13-3. Table 13-5

TABLE 13-4 Analysis of variance table for a balanced two-way factorial experiment

Source of variation	Sum of squares	Degrees of freedom	Mean square	Test statistic
Factor A	$Jn \sum_{i=1}^{I} (\bar{A}_i - \bar{Y})^2$	$I - 1$	s_A^2	$\frac{s_A^2}{s_r^2}$
Factor B	$In \sum_{j=1}^{J} (\bar{B}_j - \bar{Y})^2$	$J - 1$	s_B^2	$\frac{s_B^2}{s_r^2}$
Interaction	$n \sum_{i=1}^{I} \sum_{j=1}^{J} (\bar{Y}_{ij} - \bar{A}_i - \bar{B}_j + \bar{Y})^2$	$(I - 1)(J - 1)$	s_{AB}^2	$\frac{s_{AB}^2}{s_r^2}$
Residuals	$\sum_{i=1}^{I} \sum_{j=1}^{J} \sum_{k=1}^{n} (Y_{ijk} - \bar{Y}_{ij})^2$	$IJ(n - 1)$	s_r^2	
Total	$\sum_{i=1}^{I} \sum_{j=1}^{J} \sum_{k=1}^{n} (Y_{ijk} - \bar{Y})^2$	$IJn - 1$		

TABLE 13-5 Analysis of variance table for the experiment in Example 13-1

Source of variation	Sum of squares	Degrees of freedom	Mean square	Test statistic	*p*-value
Niacin enrichment	112.494344	2	56.2471721	544.30	0.0000
Laboratories	3.88739403	5	0.777478805	7.52	0.0001
Interaction	.708944299	10	0.07089443	0.69	0.7302
Residual	3.72019973	36	0.103338882		
Total	120.810882	53			

summarizes computer output generated by the student version of the personal computer package Stata®, while the values in Table 13-3 were calculated by hand, using fewer decimal places than the computer uses. The resulting *round-off error* accounts for the differences in calculated results.

Many workers consider the test of interaction effects first in a two-way analysis of variance. The reason: If it looks like there is an interaction effect of the two factors on response, then we can say that each factor appears to have an effect on response. Can you explain why this is so?

In Section 13-2, we discuss a two-factor experiment in which both factors have two levels. Graphs are very easy to interpret in this situation.

13-2 Two-Factor Experiments with Each Factor at Two Levels

We are still interested in a two-way, balanced, completely randomized, factorial design. But now we consider experiments in which each factor has two levels.

Simple plots provide a lot of useful information in this case. Let's look at an example.

EXAMPLE 13-2

An engineer was interested in the effect of two factors on strength of sheet castings of a polymer. One factor was time in a polymerization bath with 1% catalyst. The other factor was the temperature of the polymerization bath. The engineer tried two times (20 minutes and 60 minutes) and two temperatures (100° and 120°). He treated samples of polymer under each of the four sets of conditions. Then he measured flexural strength (in pounds per square inch, or psi) on samples of sheet castings of the polymer. The results are shown below (Duncan, 1974, page 685 (temperature scale not given); from Gore, 1947).

Time in bath with 1% catalyst	Temperature of polymerization bath	
	100°	120°
20 minutes	9,500	11,300
	10,650	11,750
	9,700	11,600
	9,950	11,650
	10,100	11,700
60 minutes	11,500	10,900
	11,650	11,500
	11,250	11,850
	11,250	11,700
	11,900	11,650

A plot of the observations is shown in Figure 13-6. What does this plot suggest about the effects of time in bath and bath temperature on strength of these sheet castings?

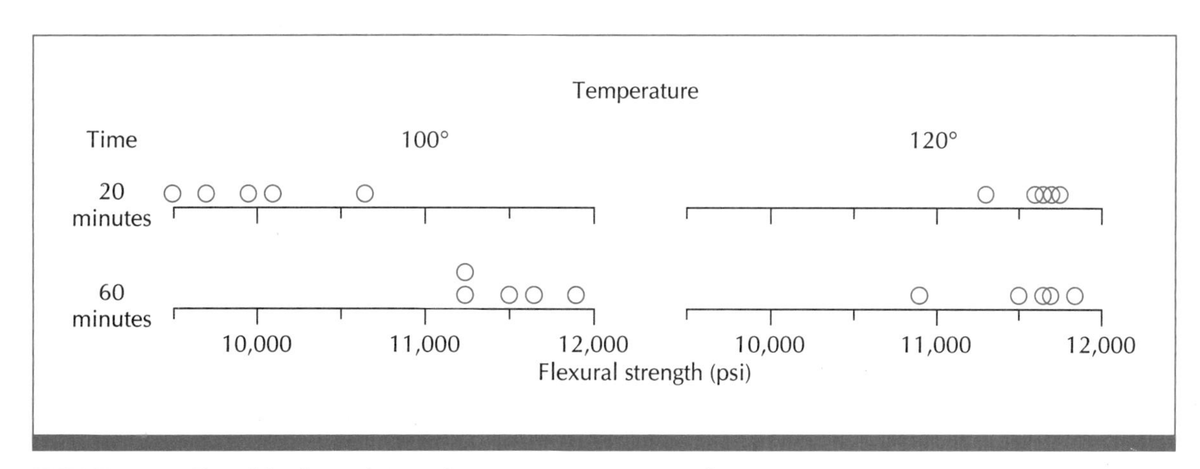

FIGURE 13-6 Plot of the flexural strength measurements in Example 13-2

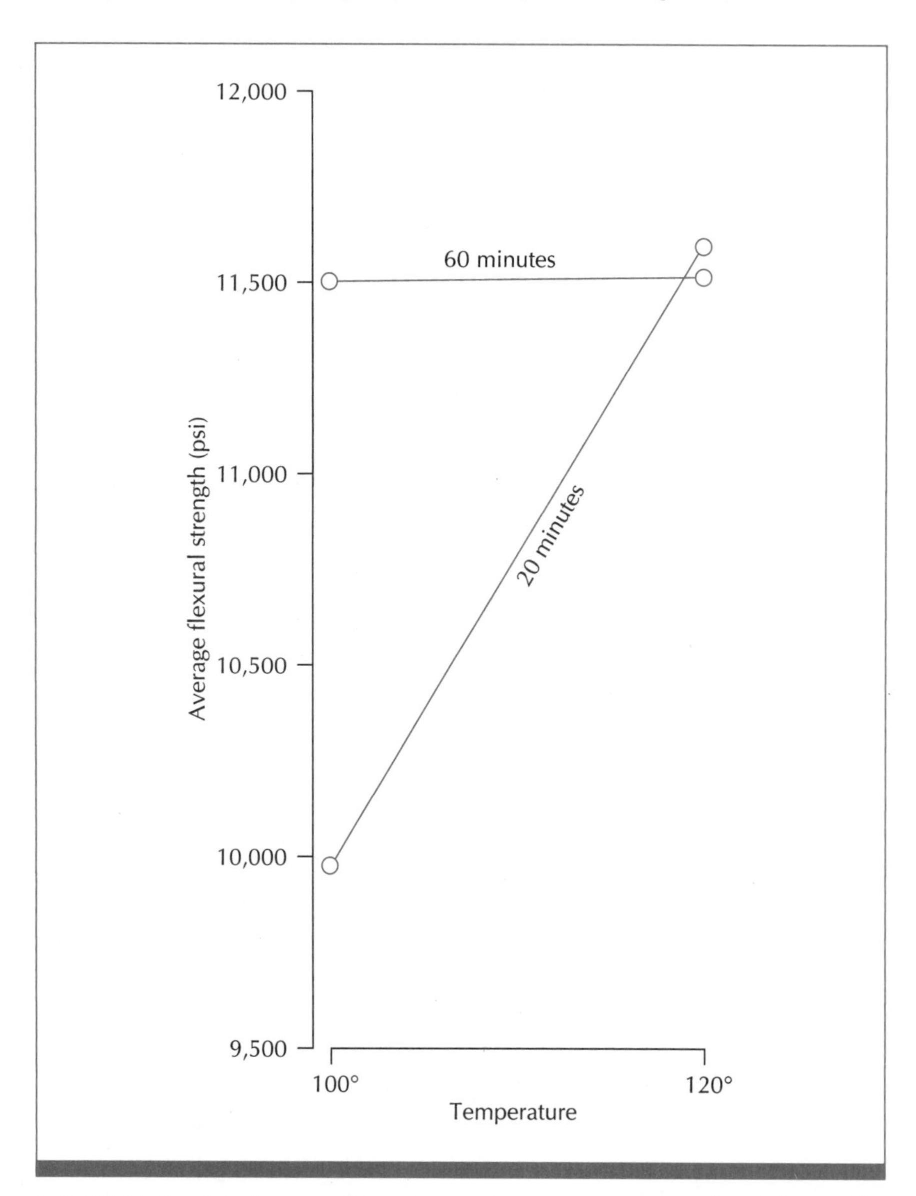

FIGURE 13-7 Plot of average flexural strength versus temperature, for each time. The two points for each time are connected.

Figure 13-7 shows a plot of average flexural strength versus temperature for each of the two bath times. The two points are connected for each time, creating a *profile*. Because the two profiles in Figure 13-7 are not parallel, we can easily see the interaction effect of time and temperature on flexural strength. When time in the bath was 20 minutes, the change in temperature made a big difference in strength of the sheet castings. When time in the bath

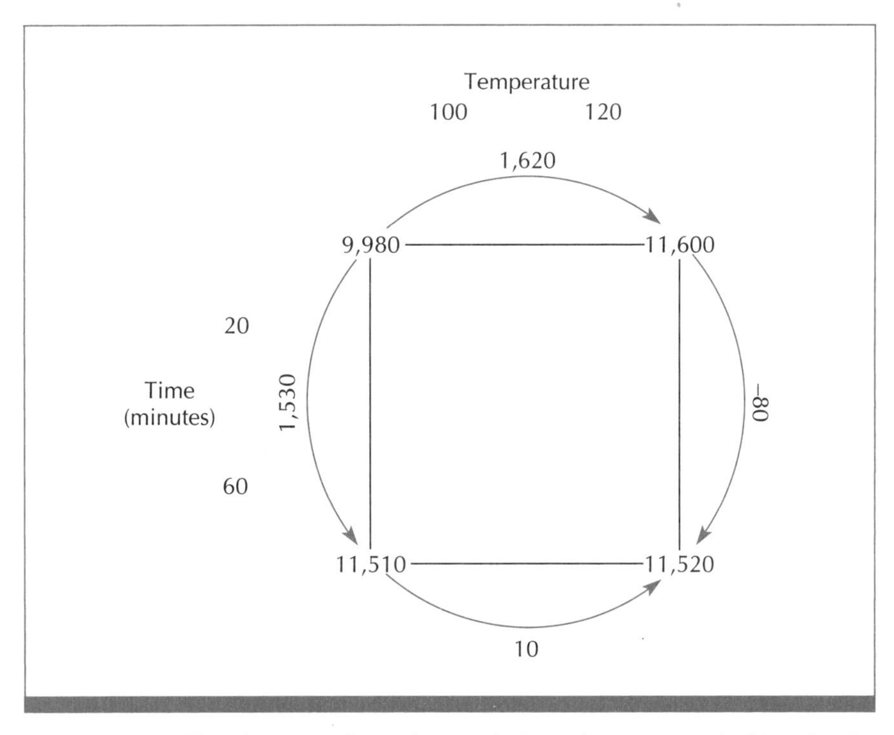

FIGURE 13-8 Plot of average flexural strength (pounds per square inch) under the four sets of conditions in Example 13-2

was 60 minutes, the change in temperature made little difference in strength of the product.

Another useful plot when factors have two levels is shown in Figure 13-8 (Box, Hunter, and Hunter, 1978, Chapter 10). Values of average flexural strength under the four sets of conditions are shown at the corners of the square. We can easily see the direction and extent of the interaction effect of time and temperature on flexural strength. At the low temperature, we see that the longer time in the bath resulted in 1,530 psi greater average flexural strength; at the higher temperature, the longer time in the bath resulted in 80 psi lower average flexural strength. For the shorter time, the higher temperature resulted in 1,620 psi greater average flexural strength; for the longer time, the higher temperature resulted in only 10 psi greater average flexural strength.

For two-way analysis of variance, we assume that we have independent observations from Gaussian distributions with equal variances. From the plot of the observations in Figure 13-6, variation among values does not seem too different across the four samples. We cannot judge the independence assumption without more information on how the experiment was conducted. What suggestions would you make to the experimenter, to ensure independence, reduce the effects of extraneous factors, and allow valid tests of the hypothesis?

To assess the Gaussian assumption, we can look at plots of residuals. A

TABLE 13-6 Residuals for the two-way analysis of variance model in Example 13-2

Time in the bath with 1% catalyst	Temperature of the polymerization bath 100°	120°
20 minutes	−480	−300
	670	150
	−280	0
	−30	50
	120	100
60 minutes	−10	−620
	140	−20
	−260	330
	−260	180
	390	130

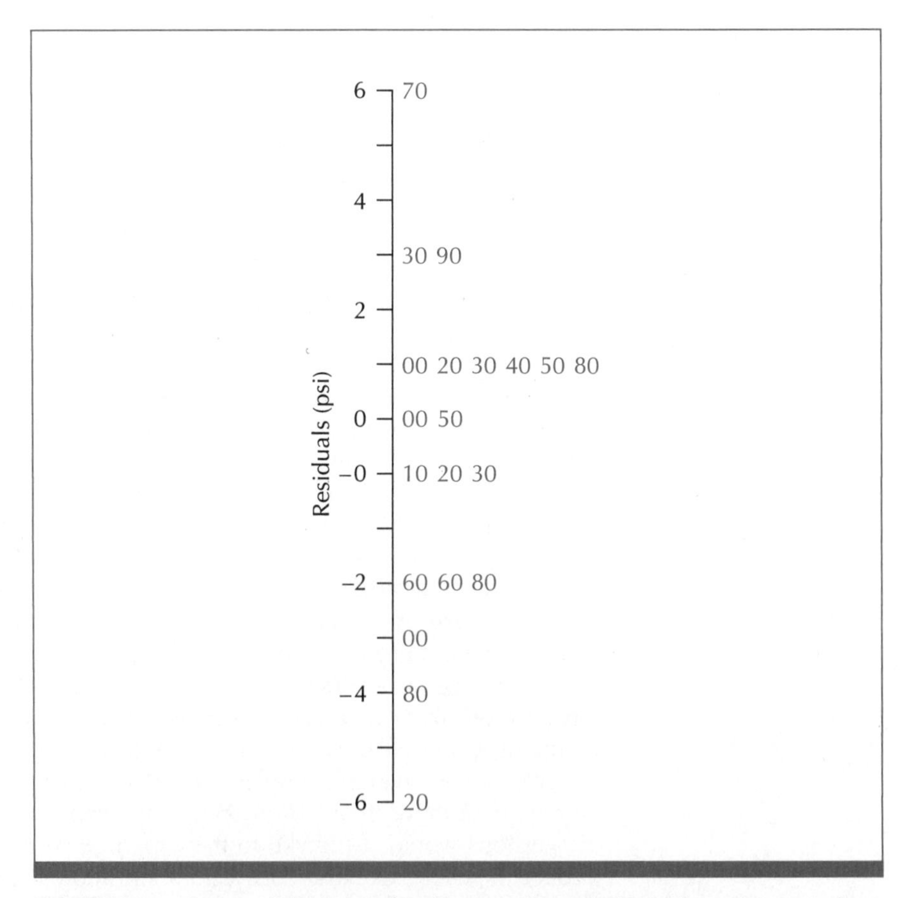

FIGURE 13-9 Stem-and-leaf plot of residuals in Example 13-2. The stems are in hundreds of pounds per square inch. The leaves are in pounds per square inch.

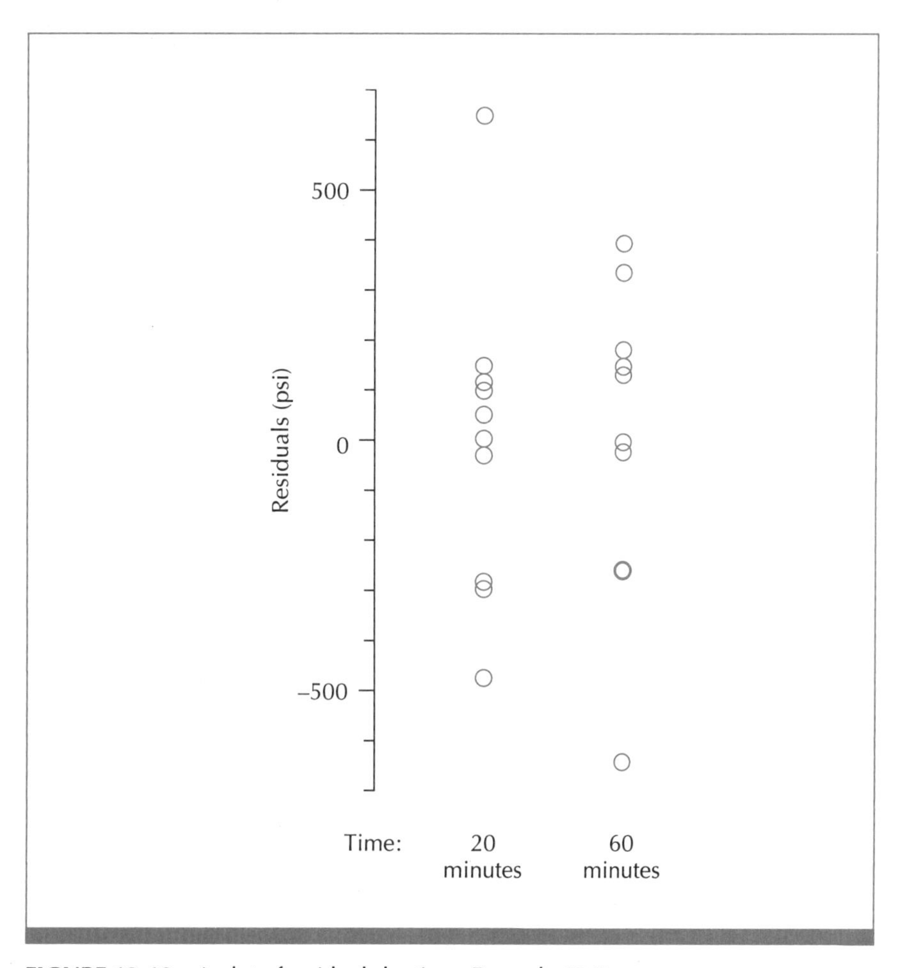

FIGURE 13-10 A plot of residuals by time, Example 13-2

residual for the balanced factorial two-way analysis of variance model is the difference $Y_{ijk} - \overline{Y}_{ij}$ between an observation and the average of all the observations made under the same set of conditions. The residuals for Example 13-2 are shown in Table 13-6.

A stem-and-leaf plot of residuals is shown in Figure 13-9. This plot gives the impression that the assumption of Gaussian observations is reasonable.

Figure 13-10 shows a plot of residuals by time, while Figure 13-11 shows a plot of residuals by temperature. We see that the plot of residuals is shifted up (toward positive values) for the low level of time and temperature. The residuals are shifted down (toward negative values) for the high level of time and temperature. Looking at Table 13-6, we see that the most extreme positive residual is at the low level of time and temperature. The most extreme negative residual is at the high level of time and temperature. Our two-way analysis of variance model cannot account for these anomalies. We will proceed with the analysis, but be cautious in our interpretations. Since all the model assump-

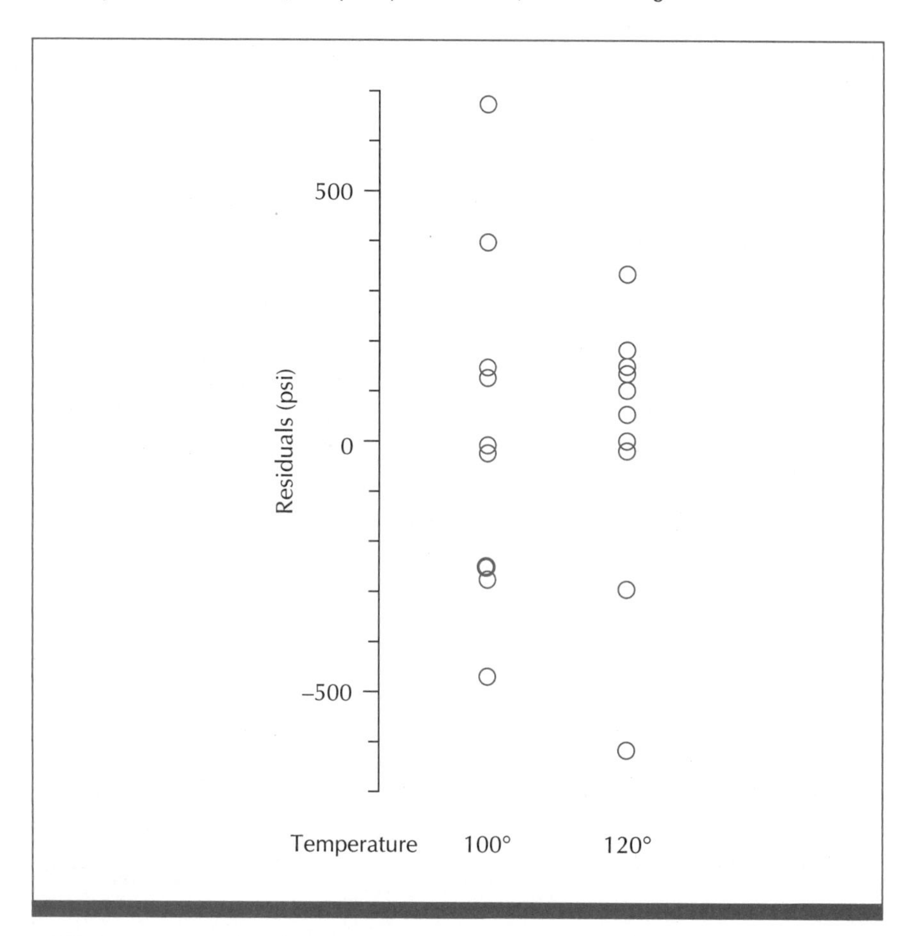

FIGURE 13-11 A plot of residuals by temperature, Example 13-2

TABLE 13-7 Analysis of variance table for the two-factor experiment in Example 13-2

Source of variation	Sum of squares	Degrees of freedom	Mean square	Test statistic	p-value
Time	2,628,125	1	2,628,125	24.06	0.0002
Temperature	3,321,125	1	3,321,125	30.40	0.0000
Interaction	3,240,125	1	3,240,125	29.66	0.0001
Residual	1,748,000	16	109,250		
Total	10,937,375	19			

tions may not be met, we will use the formal analysis to show trends, but not interpret p-values as though they were exact.

The analysis of variance table from the computer package Stata is shown in Table 13-7. The table corroborates what we saw in the plots. There are statistically significant effects of time and temperature, as well as interaction effects of time and temperature on flexural strength of sheet castings of this polymer.

Experiments with factors at two levels are useful for exploratory studies of the effects of many variables (or factors) on a response. Such studies are valuable in product development, for instance. Box, Hunter, and Hunter (1978) have extensive coverage of such experiments. For a general discussion of experiments with three or more factors, see a book on experimental design such as that by Kirk (1982).

In Chapter 14, we talk about comparing variances.

Summary of Chapter 13

Two-way, or two-factor, analysis of variance refers to parametric analysis of a two-factor, balanced, completely randomized, factorial experiment. We can test hypotheses about the effects of each of the factors on the response. With replication, we can also test for interaction effects of the two factors on the response variable.

There is an interaction effect of two variables on the response if the relative effects of levels of one variable depend on the level of the other variable. If there is an interaction effect of the two factors on response, then we say that each factor has an effect on response. (Note that interaction effect does *not* refer to any effect the two factors have on each other, but rather to their mutual effect on the response.)

Simple graphs make experimental results easy to interpret when each factor has just two levels. Such experiments are useful in exploratory studies that proceed in steps. The design of each successive experiment depends on the outcome of the previous experiments.

Minitab Appendix for Chapter 13

Finding Descriptive Statistics Within Levels of Two or More Variables

We can use the TABLE command with the STATS subcommand to print the mean and standard deviation of the responses within each combination of factors in a factorial experiment. Suppose our worksheet contains the data from Example 13-1. Niacin enrichment level is in column 1 (named NIALEVEL) and code for laboratory is in column 2 (named LAB). Niacin measurements are in column 3 (named MEASURE). If we use the command

```
MTB>   table 'nialevel' 'lab';
SUBC>  stats 'measure'.
```

we get the output in Figure M13-1.

```
ROWS: nialevel      COLUMNS: lab

             1         2         3         4         5         6       ALL

 0           3         3         3         3         3         3        18
        3.5167    3.8333    3.9333    4.2300    4.0000    3.7467    3.8767
        0.1150    0.0577    0.2802    0.3236    0.3464    0.0611    0.2992

 2           3         3         3         3         3         3        18
        5.2200    5.5667    5.6633    6.2533    5.8000    5.9033    5.7344
        0.1997    0.2517    0.1856    0.3573    0.1000    0.2628    0.3826

 4           3         3         3         3         3         3        18
        6.9467    7.5667    7.1000    7.9267    7.3333    7.5900    7.4106
        0.4692    0.4509    0.6245    0.5178    0.3214    0.0854    0.5033

ALL          9         9         9         9         9         9        54
        5.2278    5.6556    5.5656    6.1367    5.7111    5.7467    5.6739
        1.5081    1.6387    1.4182    1.6417    1.4650    1.6743    1.5098


 CELL CONTENTS --
         measure:N
                 MEAN
                 STD DEV
```

FIGURE M13-1 Table of means and standard deviations of niacin measurements within combinations of NIALEVEL and LAB

Minitab prints the number of nonmissing values, the mean, and the standard deviation of MEASURE values within each combination of NIALEVEL and LAB. The variables in the main TABLE command must be classification variables, taking integer values between −9999 and +9999 or missing values. We can specify more than two classification variables in the main TABLE command if we have more than two factors to consider. Two or more variables can be listed in the STATS subcommand; the statistics will be printed for each variable listed.

Carrying Out a Two-Factor Analysis of Variance

The TWOWAY command performs some of the calculations for two-way analysis of variance. This command requires the same number of nonmissing observations within each combination of the two factors. The command

```
MTB> twoway 'measure' 'nialevel' 'lab' c4 c5
MTB> name c4 'resid' c5 'predict'
```

produces the two-way analysis of variance table in Figure M13-2. The residuals are stored in column 4, named RESID. The estimated or predicted values of the observations based on the two-way analysis of variance model are stored in column 5, named PREDICT.

We have to calculate the test statistics ourselves. In the following sequence of commands, we calculate and print the test statistic K1 and *p*-value K2 for the hypotheses about effects of the variable NIALEVEL on response, the test statistic K3 and *p*-value K4 for the hypotheses about effects of LAB on

```
ANALYSIS OF VARIANCE  measure

SOURCE         DF        SS        MS
nialevel        2   112.494    56.247
lab             5     3.887     0.777
INTERACTION    10     0.709     0.071
ERROR          36     3.720     0.103
TOTAL          53   120.811
```

FIGURE M13-2 Two-way analysis of variance table for Example 13-1

response, and the test statistic K5 and *p*-value K6 for the hypotheses about interaction effects of NIALEVEL and LAB on response:

```
MTB>   let k1=56.247/0.103
MTB>   cdf k1 k2;
SUBC>  f 2 36.
MTB>   let k2=1-k2
MTB>   let k3=0.777/0.103
MTB>   cdf k3 k4;
SUBC>  f 5 36.
MTB>   let k4=1-k4
MTB>   let k5=0.071/0.103
MTB>   cdf k5 k6;
SUBC>  f 10 36.
MTB>   let k6=1-k6
MTB>   print k1-k6
K1     546.087
K2     0
K3     7.54369
K4     0.000062466
K5     0.689320
K6     0.727334
```

These values for test statistics and *p*-values do not exactly equal those we found in Example 13-1, because we had to use the rounded values printed by Minitab. We can use the residuals stored in column 4 and the predicted values stored in column 5 in plots to check model assumptions. We might like to print the residuals and predicted values within each combination of our two factors. To do this, we use the TABLE command with the DATA subcommand:

```
MTB>   table 'nialevel' 'lab';
SUBC>  data 'resid' 'predict'.
```

Minitab prints all the values of each variable listed in the DATA subcommand, within each combination of levels of the variables listed in the main TABLE command. The results are shown in Figure M13-3.

We can use the TABLE command with the DATA subcommand to print the values of the response variable within combinations of levels of the factors, if we wish.

```
ROWS: nialevel     COLUMNS: lab

           1         2         3         4         5         6

 0  -0.11667  -0.03333  -0.27333   0.14000   0.20000   0.01333
     0.11333  -0.03333  -0.01333  -0.37000  -0.40000  -0.06667
     0.00333   0.06667   0.28667   0.23000   0.20000   0.05333
      3.5167    3.8333    3.9333    4.2300    4.0000    3.7467
      3.5167    3.8333    3.9333    4.2300    4.0000    3.7467
      3.5167    3.8333    3.9333    4.2300    4.0000    3.7467

 2  -0.22000  -0.26667   0.01667   0.27667   0.00000   0.15667
     0.05000   0.03333  -0.19333  -0.40333  -0.10000  -0.30333
     0.17000   0.23333   0.17667   0.12667   0.10000   0.14667
      5.2200    5.5667    5.6633    6.2533    5.8000    5.9033
      5.2200    5.5667    5.6633    6.2533    5.8000    5.9033
      5.2200    5.5667    5.6633    6.2533    5.8000    5.9033

 4  -0.40667  -0.46667   0.20000   0.39333   0.36667   0.01000
     0.51333   0.03333  -0.70000  -0.58667  -0.23333  -0.09000
    -0.10667   0.43333   0.50000   0.19333  -0.13333   0.08000
      6.9467    7.5667    7.1000    7.9267    7.3333    7.5900
      6.9467    7.5667    7.1000    7.9267    7.3333    7.5900
      6.9467    7.5667    7.1000    7.9267    7.3333    7.5900

CELL CONTENTS --
          resid:DATA
        predict:DATA
```

FIGURE M13-3 Residuals and predicted values listed for each combination of NIALEVEL and LAB in Example 13-1

Exercises for Chapter 13

For each exercise, plot the observations in any ways that seem helpful. Describe the population(s) sampled, whether real or hypothetical. For each procedure, describe the assumptions that make the analysis appropriate. Do these assumptions seem reasonable? What additional information would you like to have about the experiment? Discuss the results of your analysis.

EXERCISE 13-1

A softball player wanted to see how far he could hit two brands of softball with two different bats. The two bats were each 34 inches long, and weighed 34 ounces. The balls were all new. A friend placed the balls in a pitching machine. The player did not know which brand of ball he was hitting. He did, of course, know the type of bat he was using for each trial. He hit four balls of each brand with each of the two bats. He and his friend measured the distance (in feet) he hit each ball. The results are shown below (Shaughnessy, 1988). The numbers in parentheses give the order in which the balls were hit.

	Wooden bat	Aluminum bat
Dudley Thunder	242 (9), 230 (11), 250 (12), 242 (8)	270 (4), 282 (5), 265 (6), 277 (3)
Worth Red Dot	258 (15), 264 (10), 265 (7), 275 (14)	290 (1), 318 (16), 302 (13), 310 (2)

a. Plot the observations. Include a plot similar to Figure 13-8.

b. From your plots, does there appear to be an interaction effect of type of ball and bat on distance hit?

c. Go through the steps for two-way analysis of variance.

d. Use residual plots to decide whether the assumptions for the analysis seem reasonable.

e. Plot residuals versus run order. Does there appear to be a time effect on distance hit?

f. Discuss your findings.

EXERCISE 13-2

The supervisor of a resin pilot plant wanted to study the effects of pH and temperature during formulation on the optical density of a polymer latex. Optical density (or absorbance) is a characteristic of a polymer latex that manufacturers generally want to minimize. A skilled technician prepared samples of polymer latex with two levels of pH at each of two temperatures, two samples per factor combination. The technician used standard techniques of emulsion polymerization, keeping all other conditions as constant as possible. The supervisor used a random process to determine the order in which the samples were prepared. A different technician measured optical density for each sample, using a standard electrophotometric instrument. This technician was unaware of the experiment or the conditions under which the samples were prepared. The measured values of optical density (units not given) are shown below (Gasper, 1988; with permission of ICI Resins US, a business unit of ICI Americas, Inc.). The number in parentheses is the order of the run.

	85 °C		95 °C	
pH 9.0	56.6 (6)	38.9 (7)	39.0 (4)	37.5 (5)
pH 9.3	63.0 (1)	96.8 (3)	33.0 (2)	33.3 (8)

a. Plot the observations in any ways that seem helpful. Include a plot similar to Figure 13-8.

b. From your plots, does there appear to be an interaction effect of temperature and pH on optical density?

c. Go through the steps for two-way analysis of variance.

d. Use residual plots to check assumptions. Include a plot of residuals versus run order. Do the assumptions of the analysis seem reasonable?

e. Find the reciprocal of each optical density observation. Plot these values as you did in part (a).

f. Go through the steps for two-way analysis of variance on the reciprocals of the optical density observations.

g. Use residual plots to check assumptions. Include a plot of residuals versus run order. Do the assumptions of the analysis seem reasonable?

h. Discuss your findings. If the supervisor wants to minimize optical density, what does this experiment suggest?

EXERCISE 13-3 In another part of the experiment discussed in Example 13-1, researchers divided samples of bran flakes into three groups. The bran flakes in the first group were not enriched with niacin. The bran flakes in the second group were enriched with 4 milligrams of niacin per 100 grams of flakes. The samples in the third group were enriched with 8 milligrams of niacin per 100 grams of flakes. The experimenters sent samples to the same six laboratories mentioned in Example 13-1. Laboratory workers divided a sample into three subsamples. They measured niacin in the sample on three different days, one subsample each day. The measurements of niacin (in milligrams per 100 grams) are shown below (Rice, 1988, pages 429–430; from Campbell and Pelletier, 1962).

Niacin enrichment	Laboratory					
	a	b	c	d	e	f
0	7.31	8.50	8.20	8.82	8.40	8.32
	7.85	8.50	8.25	8.76	8.60	8.25
	7.92	8.60	8.20	8.52	7.90	8.57
4	11.11	12.00	—	12.90	12.20	12.00
	11.00	13.10	11.68	12.00	11.60	12.40
	11.67	12.60	11.43	13.50	11.60	12.30
8	15.00	17.00	—	17.30	16.10	16.80
	17.00	17.50	16.20	17.60	16.10	16.60
	15.50	17.20	16.60	18.40	15.80	16.30

a. Plot the observations.

b. From the plots, does there appear to be an interaction effect of laboratory and niacin enrichment level on the niacin measurements?

c. Two observations are missing for laboratory c, one for each of two niacin enrichment levels. Therefore, this part of the experiment is not balanced; we do not have the same number of observations per factor combination. Analysis of an unbalanced experiment is more complicated than for a balanced experiment (and beyond the scope of our presentation). When only a few observations are missing, we can avoid a more complicated analysis. Replace a missing observation with the average of the other observations in the same factor combination. Subtract 1 from the residual degrees of freedom for each such substitution. Then proceed with the analysis as we have described. Use this procedure to replace each of the two missing values in this data set.

d. Go through the steps for two-way analysis of variance, accounting for the two missing values in the residual degrees of freedom.

e. Use residual plots to check the assumptions for the analysis.

f. Discuss your findings.

EXERCISE 13-4 This experiment is another phase of the experiment we discussed in Example 13-2. An engineer wanted to study the effect of two factors on strength of sheet castings of a polymer. One factor was time in a polymerization bath with 2% catalyst. The other factor was temperature of the polymerization bath. He tried two times, 20 minutes and 60 minutes. He also tried two temperatures, 100° and 120° (temperature scale not given). The engineer treated samples of polymer under each of the four sets of conditions. Then he measured flexural strength (in pounds per square inch) on samples of sheet castings of the polymer. The results are shown below (Duncan, 1974, page 686; from Gore, 1947).

	100°					120°				
20 minutes	11,800	11,750	11,800	11,950	11,900	10,550	11,000	11,100	11,350	11,200
60 minutes	11,900	11,850	11,850	12,000	12,100	9,900	10,150	9,400	9,800	9,900

a. Plot the observations. Include a plot similar to Figure 13-8.

b. Does there appear to be an interaction effect of temperature and time in the bath on flexural strength?

c. Go through the steps for two-way analysis of variance.

d. Use residual plots to decide whether the assumptions for the analysis seem reasonable.

e. Discuss your findings.

EXERCISE 13-5

Researchers carried out an experiment to study retention of two forms of iron (Fe^{2+} and Fe^{3+}) given to mice at each of three concentrations (10.2, 1.2, and .3 millimolar). They randomly divided 108 mice into six groups, 18 mice per group. Each group of mice received one of the six combinations of type of iron and concentration. The researchers gave radioactively labeled iron to the mice orally. They used a counter to verify the amount of iron administered. They also used the counter to record the amount retained later (time not specified). Percentage of iron retained is shown below for each of the 108 mice (from an example in Rice, 1988, pages 356–357).

10.2 millimolar Fe^{3+}	.71	1.66	2.01	2.16	2.42	2.42	2.56	2.60	3.31
	3.64	3.74	3.74	4.39	4.50	5.07	5.26	8.15	8.24
1.2 millimolar Fe^{3+}	2.20	2.93	3.08	3.49	4.11	4.95	5.16	5.54	5.68
	6.25	7.25	7.90	8.85	11.96	15.54	15.89	18.30	18.59
.3 millimolar Fe^{3+}	2.25	3.93	5.08	5.82	5.84	6.89	8.50	8.56	9.44
	10.52	13.46	13.57	14.76	16.41	16.96	17.56	22.82	29.13
10.2 millimolar Fe^{2+}	2.20	2.69	3.54	3.75	3.83	4.08	4.27	4.53	5.32
	6.18	6.22	6.33	6.97	6.97	7.52	8.36	11.65	12.45
1.2 millimolar Fe^{2+}	4.04	4.16	4.42	4.93	5.49	5.77	5.86	6.28	6.97
	7.06	7.78	9.23	9.34	9.91	13.46	18.40	23.89	26.39
.3 millimolar Fe^{2+}	2.71	5.43	6.38	6.38	8.32	9.04	9.56	10.01	10.08
	10.62	13.80	15.99	17.90	18.25	19.32	19.87	21.60	22.25

a. Plot the observations.

b. From your plots, does there appear to be an interaction effect of form of iron and concentration on percentage retained?

c. Go through the steps for two-way analysis of variance. Use residual plots to check the assumptions for the analysis. Do the assumptions seem reasonable?

d. Plot the logarithm of percentage of iron retained.

e. Go through the steps for two-way analysis of variance, using the logarithms of the observations. Use the residual plots to check the assumptions of the analysis. Do these assumptions seem reasonable?

f. Compare your answers to parts (c) and (e). Discuss your findings.

EXERCISE 13-6

An engineering student carried out a study of the effects of bathing on fecal and total coliform (colon bacillus) bacteria in the bath water. Eight males and eight females participated. Each volunteer took a bath in a 100-gallon polyethylene tub, with dechlorinated 38 °C tap water. Two factors studied were time since the volunteer's last bath and the level of his or her activity in the bath. After a volunteer had been in the tub for 15 minutes, the engineering student recorded two response variables: change in fecal coliform concentration and change in total coliform concentration (organisms per 100 milliliters). The results are shown below (Box, Hunter, and Hunter, 1978, pages 435–436; from Drew, 1971).

Activity	Time since last bath	Females		Males	
Change in fecal coliform concentration after 15 minutes (organisms per 100 ml)					
Lethargic	1 hour	1	2	153	96
Lethargic	24 hours	12	37	129	390
Vigorous	1 hour	16	21	143	300
Vigorous	24 hours	4	2	113	280
Change in total coliform concentration after 15 minutes (organisms per 100 ml)					
Lethargic	1 hour	3	10	426	147
Lethargic	24 hours	57	280	250	1,470
Vigorous	1 hour	323	33	580	665
Vigorous	24 hours	183	10	650	675

a. Plot the values of change in fecal coliform concentration for females. Does there appear to be an interaction effect of time since last bath and bathing activity on change in fecal coliform concentration?

b. Repeat part (a) after taking the logarithm of each observation. Do the assumptions for two-way analysis of variance seem better met by the transformed observations?

c. Analyze the logarithm of change in fecal coliform concentration for females. Discuss your findings.

d. Repeat parts (a) and (b) for males.

e. Repeat part (c) for males.

f. Compare your findings in parts (a), (b), and (c) for the female volunteers with your findings in parts (d) and (e) for the male volunteers.

g. Plot the values of change in total coliform concentration for females. Does there appear to be an interaction effect of time since last bath and bathing activity on the observations?

h. Repeat part (g) after taking the logarithm of each observation. Do the assumptions for two-way analysis of variance seem better met by the transformed observations?

i. Analyze the logarithm of change in total coliform concentration for females. Discuss your findings.

j. Repeat parts (g) and (h) for males.

k. Repeat part (i) for males.

l. Compare your findings in parts (g), (h), and (i) for the female volunteers with your findings in parts (j) and (k) for the males.

EXERCISE 13-7

Researchers wanted to study four treatments in counteracting the effects of three poisons. They randomly assigned 48 animals to treatment/poison combinations, four animals per combination. The researchers measured survival time in hours for each animal. The results are shown below (Rice, 1988, pages 432–433; Box, Hunter, and Hunter, 1978, page 228; from Box and Cox, 1964).

	Poison 1				Poison 2				Poison 3			
Treatment a	3.1	4.5	4.6	4.3	3.6	2.9	4.0	2.3	2.2	2.1	1.8	2.3
Treatment b	8.2	11.0	8.8	7.2	9.2	6.1	4.9	12.4	3.0	3.7	3.8	2.9
Treatment c	4.3	4.5	6.3	7.6	4.4	3.5	3.1	4.0	2.3	2.5	2.4	2.2
Treatment d	4.5	7.1	6.6	6.2	5.6	10.0	7.1	3.8	3.0	3.6	3.1	3.3

a. Plot the observations.

b. From your plots, does there appear to be an interaction effect of poison and treatment on survival?

c. Go through the steps for two-way analysis of variance. Use residual plots to check model assumptions. Do these assumptions seem reasonable?

d. You can interpret the reciprocal of a survival time (time until death) as a death rate (number of deaths per unit time). Take the reciprocal of each of the 48 survival times. Plot these transformed observations.

e. Go through the steps for two-way analysis of variance on the reciprocals of the observations. Use residual plots to check the assumptions for the analysis. Do these assumptions seem reasonable?

f. Compare your answers to parts (c) and (e). Discuss your findings.

EXERCISE 13-8 Managers were interested in the productivity of four technicians on each of five machines. They recorded the number of units produced by a technician on each machine, two different days. The results are shown below. Managers numbered the 84 working days required for the experiment, from 1 to 84. The day the observation was made is shown in parentheses next to the observation (from an exercise on page 285 in *Statistics for Experimenters* by G. E. P. Box, W. G. Hunter, and J. S. Hunter, 1978, John Wiley and Sons, Inc., New York).

	Machine 1	Machine 2	Machine 3	Machine 4	Machine 5
Technician a	18 (9), 17 (76)	17 (1), 13 (71)	16 (3), 17 (77)	15 (2), 17 (72)	17 (17), 18 (84)
Technician b	16 (11), 18 (77)	18 (3), 18 (73)	17 (7), 19 (70)	21 (4), 22 (74)	16 (10), 18 (72)
Technician c	17 (22), 20 (72)	20 (57), 16 (70)	20 (25), 16 (73)	16 (5), 16 (71)	14 (39), 13 (74)
Technician d	27 (3), 27 (73)	28 (2), 23 (78)	31 (33), 30 (72)	31 (6), 24 (75)	28 (7), 22 (82)

a. Plot the observations.

b. Does there appear to be an interaction effect of technician and machine on number of units produced?

c. Go through the steps for two-way analysis of variance. Use residual plots to check model assumptions. Do these assumptions seem reasonable?

d. Plot residuals versus time of observation. Is there a trend? If there were a trend, what would it suggest?

e. Discuss your findings.

CHAPTER 14

Inferences About Variances

IN THIS CHAPTER

Is the variation among drained weights of tomatoes canned in the afternoon the same as the variation among those canned in the morning? Is the variability among plasma estrogen levels the same for monkeys with alcohol in their diets as for monkeys with no alcohol in their diets? Does the variation in rupture times for pieces of stainless steel depend on the level of stress applied to the pieces? Does season affect the variability of earthworm populations in fields?

All of these questions are phrased in terms of variation. Questions about variability are very important in quality control, engineering, and the sciences. In Chapters 10–13 we concentrated on inferences about means (or medians). Now we consider inferences about variances.

In Section 14-1 we consider parametric tests of hypotheses and confidence intervals for a single variance. Then in Section 14-2 we discuss parametric comparisons of two variances, as well as confidence intervals for the ratio of two variances. We cover parametric comparisons of several variances in Section 14-3. Finally, in Section 14-4 we consider inferences about two or more variances that do not require the assumption of Gaussian observations.

14-1 Parametric Inferences About a Variance

Suppose we have a random sample from a Gaussian distribution and we want to make inferences about the variance σ^2 of that distribution. First we will look at an example; then we will outline the significance level approach to a parametric test of hypotheses about a variance, and apply it to the example.

EXAMPLE 14-1

Machines at a factory fill cans with standard-grade tomatoes in puree (based on an example in Duncan, 1974, page 569; from Grant and Leavenworth, 1972, page 41). One responsibility of the quality control manager is to select cans and check the drained weight of the contents. After many checks, the manager has found that for cans filled in the morning, the average drained weight is 21.8 ounces, and the variance is 2.63 ounces^2.

The quality control manager selects a random sample of five cans filled one afternoon. The drained weights (in ounces) are:

22.5 19.5 21.5 20.5 20.0

A plot of these observations is shown in Figure 14-1. We see that the sample values have a fairly symmetrical distribution. The five sample drained weights range from 19.5 ounces to 22.5 ounces.

The manager wants to test the null hypothesis that the variance of drained weights for cans filled during that afternoon equals the morning variance, 2.63 ounces^2. Let's outline the analysis procedure.

The significance level approach to a parametric test of hypotheses about a variance σ^2

1. The hypotheses are H_0: $\sigma^2 = \sigma_0^2$ and H_a: $\sigma^2 \neq \sigma_0^2$, where σ_0^2 is a specified number.

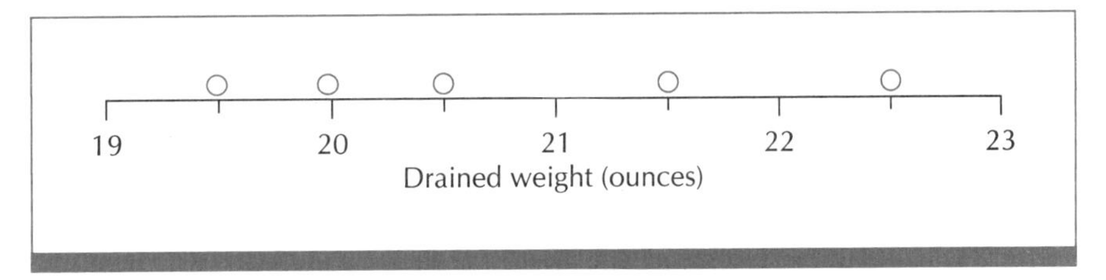

FIGURE 14-1 Drained weights (in ounces) of five cans of tomatoes canned in the afternoon, Example 14-1

2. Let s^2 denote the sample variance and n the sample size. The test statistic is

$$\text{Test statistic} = \frac{(n - 1)\, s^2}{\sigma_0^2}$$

3. Assume that we have a random sample from a Gaussian distribution with variance σ^2. Then under the null hypothesis, the test statistic has the chi-square distribution with $n - 1$ degrees of freedom. Very large or very small values of the test statistic are inconsistent with the null hypothesis.
4. Select significance level α.
5. Let X denote a random variable having the chi-square distribution with $n - 1$ degrees of freedom. Find c_1 and c_2 from Table E such that $P(X \leq c_1) = \alpha/2$ and $P(X \geq c_2) = \alpha/2$. The acceptance region is the interval (c_1, c_2). The rejection region includes the intervals $[0, c_1]$ and $[c_2, \infty)$.
6. The decision rule is:

   If $c_1 <$ test statistic $< c_2$, say the results are consistent with the null hypothesis that the population variance equals σ_0^2.

   If test statistic $\leq c_1$ or test statistic $\geq c_2$, say the results are inconsistent with the null hypothesis, suggesting that the population variance does not equal σ_0^2.

7. Collect a sample that satisfies the assumptions in step 3. Calculate the test statistic in step 2. Use the decision rule in step 6 to decide whether the results are consistent with the null hypothesis. Draw conclusions based on the experimental results.

Suppose that, instead of a two-sided alternative, we have the one-sided alternative H_a: $\sigma^2 < \sigma_0^2$. Then in step 5 we find the number c from Table E such that $P(X \leq c) = \alpha$. Values of the test statistic greater than c are consistent with the null hypothesis; values less than or equal to c are inconsistent with the null hypothesis.

If we have the one-sided alternative H_a: $\sigma^2 > \sigma_0^2$, then in step 5 we find the number c from Table E such that $P(X \geq c) = \alpha$. Values of the test statistic less than c are consistent with the null hypothesis; values greater than or equal to c are inconsistent with the null hypothesis.

EXAMPLE 14-1 *(continued)*

Now we can test the hypotheses of interest in Example 14-1. If σ^2 denotes the variance of drained weights among cans filled that afternoon, then we can state the hypotheses as H_0: $\sigma^2 = 2.63$ ounces^2 and H_a: $\sigma^2 \neq 2.63$ ounces^2.

Assume that the five afternoon observations form a random sample from a large production lot. Assume also that these drained weights follow a Gaussian distribution. Figure 14-1 gives us no reason to doubt the Gaussian assumption, although the sample size is very small. We have no way of checking the other assumptions without more information.

Since the sample size is 5, the test statistic equals $4s^2/2.63$, where s^2 is the sample variance for the five afternoon observations. If the assumptions hold, then under the null hypothesis the test statistic has the chi-square distribution with 4 degrees of freedom.

We will use significance level $\alpha = .05$. From Table E we see that if X has the chi-square distribution with 4 degrees of freedom, then $P(X \leq .484) = .025$ and $P(X \geq 11.14) = .025$. Therefore, the acceptance region is (.484, 11.14), the rejection region consists of [0, .484] and [11.14, ∞), and the decision rule is:

If .484 < test statistic < 11.14, say the results are consistent with the null hypothesis that the afternoon variance equals 2.63 ounces².
If test statistic $\leq$ .484 or test statistic $\geq$ 11.14, say the results are inconsistent with the null hypothesis, suggesting that the afternoon variance does not equal 2.63 ounces².

The sample variance of the five observations is $s^2 = 1.45$ ounces², so the test statistic equals $4 \times 1.45/2.63$, or 2.2. Since 2.2 is in the acceptance region, the results are consistent with the null hypothesis. Based on this test of hypotheses, we have no reason to doubt that the variance for drained weights of tomatoes canned that afternoon equals the morning variance (but, of course, the test does not imply that the variance in the afternoon exactly equals the morning variance of 2.63 ounces²).

Confidence Intervals for a Population Variance

Suppose we want to calculate a 100A% *confidence interval for the population variance* σ^2. Let X denote a random variable having the chi-square distribution with $n - 1$ degrees of freedom. Find c_1 and c_2 such that $P(X \leq c_1) = (1 - A)/2$ and $P(X \geq c_2) = (1 - A)/2$. Then our confidence interval for σ^2 has the form

$$\left(\frac{n-1}{c_2}s^2, \frac{n-1}{c_1}s^2\right)$$

EXAMPLE 14-1 *(continued)*

Let's calculate a 95% confidence interval for the variance of drained weights of tomatoes canned that afternoon in Example 14-1. We have $n = 5$ and $s^2 = 1.45$. Since $A = .95$, we have $(1 - A)/2 = .025$. Referring to Table E, we see that $c_1 = .484$ and $c_2 = 11.14$, the same as we used for our test of hypotheses with significance level .05. So a 95% confidence interval for the variance is

$$\left(\frac{4}{11.14} \times 1.45, \frac{4}{.484} \times 1.45\right) = (.52 \text{ ounces}^2, 11.98 \text{ ounces}^2)$$

Note that our null hypothesis variance, 2.63 ounces², is in this confidence interval, in agreement with our test of hypotheses.

To get a confidence interval for the population standard deviation σ, we can take the square root of the upper and lower limits of the confidence interval for the population variance σ^2. For instance, a 95% confidence interval for the standard deviation of drained weights of tomatoes canned that afternoon in Example 14-1 is ($\sqrt{.52}$ ounce, $\sqrt{11.98}$ ounces) or (.72 ounce, 3.46 ounces).

Our analysis gives us no reason to think that the afternoon variance is different from the morning variance. However, as we have mentioned before, there are other important practical considerations in this type of situation. Is the amount of variation acceptable to the company, to consumers, and to the government? If there are tolerance ranges of acceptable values for drained weights, is the production process adequately meeting these tolerances? Generally, no single formal analysis procedure will address all of the questions relevant to a particular experimental situation.

In Section 14-2, we discuss parametric comparisons of two variances.

14-2 Parametric Inferences About Two Variances

Suppose we have two independent random samples from Gaussian distributions. We want to test the null hypothesis that the two population variances are equal. We will use the *variance ratio test* based on an F distribution.

The **variance ratio test** is a parametric test for equality of two variances.

We also want to calculate a confidence interval for the ratio of the two population variances. Let's begin with an example.

EXAMPLE 14-2

Researchers designed an experiment to study effects of regular alcohol consumption (Jerome Hojnacki, personal communication, 1986). The participants in the study were 20 adult male squirrel monkeys, of similar age and good health. The researchers randomly divided the monkeys into two equal sized groups. Monkeys in the alcohol group consumed a steady diet of 12% ethyl alcohol for approximately 3 months (ethyl alcohol constituted 12% of their total calories each meal). Monkeys in the control group did not consume alcohol. At the end of the treatment period, the researchers measured plasma estrogen (in nanograms/deciliter) for each monkey. The results are shown below.

Group	Plasma estrogen level (ng/dL)							Sample mean	Sample variance
Alcohol	3.17	2.52	2.59	4.25	3.27	4.92	5.46	3.61	1.33
	2.83	4.80	2.26						
Control	6.57	5.81	5.63	5.75	4.54	5.35	4.16	5.21	.556
	5.12	4.69	4.52						

A plot of the observations is shown in Figure 14-2.

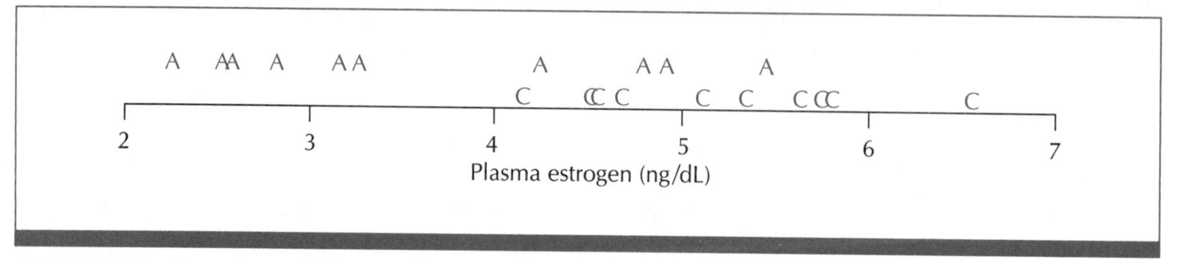

FIGURE 14-2 Plot of plasma estrogen levels in Example 14-2. The symbol A denotes a value for a monkey in the alcohol group; C denotes a value for a monkey in the control group.

The plasma estrogen levels seem to be somewhat lower for the monkeys in the alcohol group (see Exercise 11-5). Is the variation the same for the two groups? Let's outline the significance level approach for comparing the two variances, and then apply it to this example.

The significance level approach to parametric tests of hypotheses about two variances

1. Let σ_1^2 and σ_2^2 denote the variances of the two populations sampled. We want to test the hypotheses H_0: $\sigma_1^2 = \sigma_2^2$ and H_a: $\sigma_1^2 \neq \sigma_2^2$.
2. The test statistic equals the larger sample variance divided by the smaller sample variance:

$$\text{Test statistic} = \frac{\text{Larger sample variance}}{\text{Smaller sample variance}}$$

3. Assume that we have two independent random samples from Gaussian distributions. The first sample is of size n_1 and comes from a Gaussian distribution with variance σ_1^2. The second sample is of size n_2 and comes from a Gaussian distribution with variance σ_2^2. Let s_1^2 denote the first sample variance and s_2^2 the second sample variance. Then under the null hypothesis, the ratio s_1^2/s_2^2 has the $F(n_1 - 1, n_2 - 1)$ distribution. The ratio s_2^2/s_1^2 has the $F(n_2 - 1, n_1 - 1)$ distribution under the null hypothesis.
4. Select significance level α.
5. Suppose s_1^2 is the larger sample variance, so the test statistic is s_1^2/s_2^2. Let F denote a random variable having the $F(n_1 - 1, n_2 - 1)$ distribution. Find the number c from Table D such that $P(F \leq c) = 1 - \alpha/2$. The acceptance region is the interval $[0, c)$; the rejection region is the interval $[c, \infty)$.
6. The decision rule is:

   If test statistic $< c$, say the results are consistent with the null hypothesis that the two population variances are equal.
   If test statistic $\geq c$, say the results are inconsistent with the null hypothesis, suggesting that the two population variances are not equal.
7. Collect two samples that satisfy the assumptions in step 3. Calculate the test statistic in step 2. Use the decision rule in step 6 to decide whether the two population variances seem to be the same or different. Draw conclusions based on the experimental results.

EXAMPLE 14-2
(continued)

In Example 14-2, we want to test the null hypothesis that the variance of plasma estrogen levels is the same for monkeys fed a steady diet of alcohol as for monkeys not fed alcohol. The alternative hypothesis is that the variances are different for the two groups. Since the alcohol group has the larger sample variance, we let the alcohol group be group 1, and our test statistic is s_1^2/s_2^2.

We assume that we have independent random samples from Gaussian distributions. From Figure 14-2, we see that the Gaussian assumption seems reasonable for the control monkeys, but this is not so clear for the alcohol monkeys. We cannot check the independence assumptions without more information about the experiment. What suggestions about experimental design would you make to these researchers, in order to ensure independence and valid inferences?

We will use significance level $\alpha = .05$. Let F denote a random variable having the $F(9, 9)$ distribution. Since $P(F \leq 4.03) = .975$, the acceptance region is $[0, 4.03)$ and the rejection region is $[4.03, \infty)$. The decision rule is:

If test statistic < 4.03, say the results are consistent with the null hypothesis that the two variances are equal.
If test statistic ≥ 4.03, say the results are inconsistent with the null hypothesis, suggesting that the two variances are not equal.

The sample variance for the alcohol monkeys is $s_1^2 = 1.33$. The sample variance for the control monkeys is $s_2^2 = .556$. Therefore, the test statistic equals 1.33/.556, or 2.4, which is in the acceptance region. Using this test of hypothesis, we have no reason to doubt that the variation in plasma estrogen levels is similar for the alcohol and control monkeys.

Confidence Intervals for the Ratio of Two Population Variances

Suppose we would like to calculate a 100A% *confidence interval for the ratio* σ_1^2/σ_2^2 *of the two population variances.* Letting F_1 denote a random variable having an $F(n_1 - 1, n_2 - 1)$ distribution, find the number c_1 such that $P(F_1 \leq c_1) = (1 + A)/2$. Similarly, if F_2 denotes a random variable having the $F(n_2 - 1, n_1 - 1)$ distribution, find the number c_2 such that $P(F_2 \leq c_2) = (1 + A)/2$. A 100$A$% confidence interval for σ_1^2/σ_2^2 has the form

$$\left(\frac{1}{c_1}\frac{s_1^2}{s_2^2},\ c_2\frac{s_1^2}{s_2^2}\right)$$

EXAMPLE 14-2
(continued)

Let's calculate a 95% confidence interval for σ_1^2/σ_2^2 in Example 14-2. Here, σ_1^2 denotes the variance of plasma estrogen values for monkeys consuming alcohol and σ_2^2 denotes the variance for monkeys not consuming alcohol. Since the sample size is 10 for each group, the numerator and denominator degrees of freedom both equal 9. From Table D we see that $c_1 = c_2 = 4.03$. Our 95% confidence interval for σ_1^2/σ_2^2 is

$$\left(\frac{1}{4.03} \times \frac{1.33}{.556},\ 4.03 \times \frac{1.33}{.556}\right) = (.59, 9.6)$$

Note that if the null hypothesis is true, then the ratio of the two population variances is 1. This null hypothesis value is in the confidence interval, agreeing with our test of hypotheses.

Does the ratio σ_1^2/σ_2^2 of the two population variances have units of measurement associated with it? Should the confidence interval (.59, 9.6) for this ratio in Example 14-2 show units? How would you find a confidence interval for the ratio of the two population standard deviations σ_1/σ_2?

The Minitab Appendix for Chapter 14 has an example based on Exercise 14-6, making inferences about the ratio of two variances when the sample sizes are not equal.

Section 14-3 discusses a parametric procedure for comparing more than two variances.

14-3 Parametric Inferences About More Than Two Variances

Suppose we have k independent random samples from Gaussian distributions, where k is greater than or equal to 3. When we test the null hypothesis that the k population variances are equal, we are testing for *homogeneity of variances*. If the null hypothesis of equal variances is true, we say the k population variances are *homogeneous*. If the alternative hypothesis of unequal variances is true, we say the variances are *heterogeneous*.

There are many parametric tests for equality of variances (see, for example, Conover, Johnson, and Johnson, 1981). We will discuss the one most commonly used—*Bartlett's test*.

Bartlett's test is a parametric procedure for testing equality of three or more variances, based on the assumption of independent random samples from Gaussian distributions.

First, we will look at an example.

EXAMPLE 14-3

An engineer subjected uniform pieces of stainless steel to different levels of stress, recording the time to rupture for each piece. He tested six pieces of steel at each of three stress levels. The results are shown below (part of an experiment reported in Schmoyer, 1986; from Garofalo et al., 1961). Stress levels were reported in pounds per square inch (psi).

Stress level (psi)	Rupture time (hours)					
28.84	1,267	1,637	1,658	1,709	1,785	2,437
31.63	170	257	265	570	594	779
34.68	76	87	96	115	122	132

The experimental results are displayed in a scatterplot in Figure 14-3. We see that the time to rupture decreases as stress increases. It also looks like the

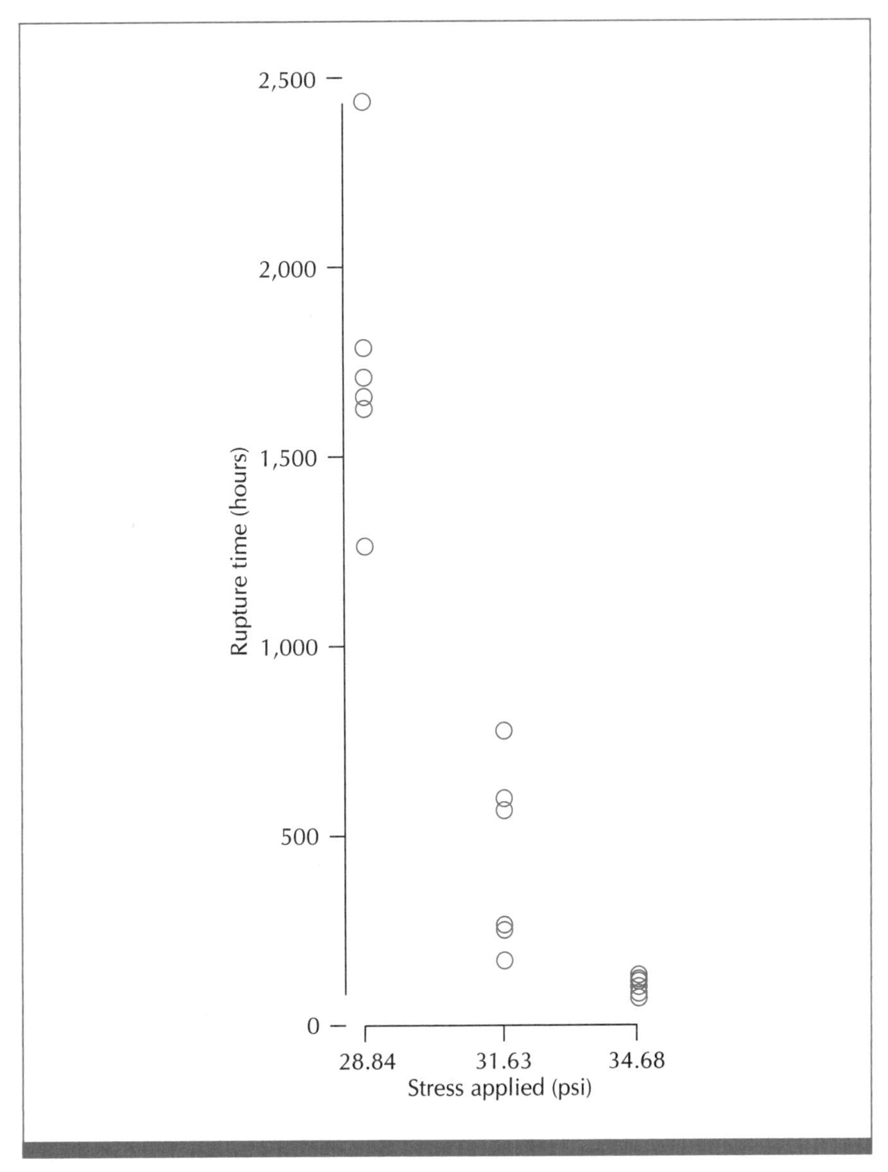

FIGURE 14-3 Scatterplot of rupture time vs. stress applied in Example 14-3

variation in rupture times decreases as stress increases. Let's see how to test the null hypothesis that the three population variances are equal.

The significance level approach to Bartlett's test for equality of several variances

1. Let σ_1^2 through σ_k^2 denote the k population variances. The hypotheses are:

$$H_0\colon\ \sigma_1^2 = \sigma_2^2 = \cdots = \sigma_k^2$$
$$H_a\colon\ \sigma_1^2 \text{ through } \sigma_k^2 \text{ are not all equal}$$

2. Let n_i denote the sample size and s_i^2 the sample variance of sample i. Let N denote the total sample size, the sum of n_1 through n_k. Let s_r^2 denote the pooled variance estimate (the residual mean square we discussed for one-way analysis of variance in Chapter 12). The numerator of the test statistic is

$$\text{Numerator} = 2.3026 \left[(N - k) \log s_r^2 - \sum_{i=1}^{k} (n_i - 1) \log s_i^2 \right]$$

where log denotes logarithm base-10. The denominator of the test statistic is

$$\text{Denominator} = 1 + \frac{1}{3(k - 1)} \left(\sum_{i=1}^{k} \frac{1}{n_i - 1} - \frac{1}{N - k} \right)$$

Then the test statistic is

$$\text{Test statistic} = \frac{\text{Numerator}}{\text{Denominator}}$$

3. Assume that we have k independent random samples, one from each of k Gaussian distributions. Then under the null hypothesis, the test statistic has approximately the chi-square distribution with $k - 1$ degrees of freedom. Large values of the test statistic are inconsistent with the null hypothesis.
4. Select significance level α.
5. Find the number c in Table E such that $P(X \leq c) = 1 - \alpha$, where X has the chi-square distribution with $k - 1$ degrees of freedom. The acceptance region is $[0, c)$; the rejection region is $[c, \infty)$.
6. The decision rule is:

   If test statistic $< c$, say the results are consistent with the null hypothesis that the k population variances are all equal.
   If test statistic $\geq c$, say the results are inconsistent with the null hypothesis, suggesting that the k population variances are not all equal.

7. Carry out an experiment that satisfies the assumptions in step 3. Calculate the test statistic in step 2. Use the decision rule in step 6 to decide whether the population variances seem to be the same or different. Draw conclusions based on the experimental results.

EXAMPLE 14-3 *(continued)*

In Example 14-3, we want to test the null hypothesis that the variance in rupture times of uniform pieces of stainless steel is the same for all three stress levels. The alternative hypothesis is that the three variances are not all equal.

Assume that the three samples represent independent random samples from Gaussian distributions. From the plot in Figure 14-3, the Gaussian assumption does not seem unreasonable because all three sample distributions look fairly symmetric. As always, we cannot assess the independence assumption without more information about the experiment. What suggestions would you make regarding experimental design? How would you try to control extraneous sources of variation and ensure independence of observations? Should

Stress level	Sample size	Sample variance
28.84	6	145,877.8
31.63	6	58,509.37
34.68	6	472.6667

$k = 3 \qquad N = 18 \qquad N - k = 15 \qquad s_r^2 = 68{,}286.61$

$$\text{Numerator} = 2.3026[15\log(68{,}286.61) - 5\log(145{,}877.8) - 5\log(58{,}509.37) - 5\log(472.6667)] = 21.843$$

$$\text{Denominator} = 1 + \frac{1}{3(3-1)}\left(\frac{1}{5} + \frac{1}{5} + \frac{1}{5} - \frac{1}{15}\right) = 1.089$$

$$\text{Test statistic} = \frac{21.843}{1.089} = 20.1 \qquad \text{Degrees of freedom} = 3 - 1 = 2$$

the engineer subject the first six pieces to the first stress level, then reset the equipment and subject the second set of six pieces to the second stress level, then reset the equipment again and subject the third set of pieces to the final stress level? Or would you suggest a different procedure?

If our model assumptions hold, then under the null hypothesis, the test statistic has approximately the chi-square distribution with 2 degrees of freedom. Using significance level $\alpha = .01$, we find $c = 9.21$ from Table E. The acceptance region is $[0, 9.21)$, the rejection region is $[9.21, \infty)$, and the decision rule is:

If test statistic < 9.21, say the results are consistent with the null hypothesis that the variance in rupture times is the same for all three stress levels.
If test statistic ≥ 9.21, say the results are inconsistent with the null hypothesis, suggesting that the variation in rupture times is not the same for all three stress levels.

The calculations we need are outlined in Table 14-1. We see that the test statistic equals 20.1, which is in the rejection region. The results suggest that the variation in rupture times is not the same for all three stress levels, agreeing with what we saw in Figure 14-3.

We can make multiple comparisons to decide which variances seem to be different and which seem to be similar. We use the method of Section 14-2 to calculate confidence intervals for ratios of variances, then use the Bonferroni method (Section 12-1) to get a bound on the overall confidence level for these intervals.

The calculations for our multiple comparisons are outlined in Table 14-2. With three variances, there are three pairwise comparisons. We calculate a 98% confidence interval for each variance ratio. Since each sample size is 6, we use the value $c = 10.97$ from the $F(5, 5)$ distribution for all the intervals. The overall confidence level for the three intervals is at least 94%.

TABLE 14-2 Multiple comparisons of variances in Example 14-3. σ_1^2, σ_2^2, and σ_3^2 denote the variance in rupture times at stress levels 28.84, 31.63, and 34.68 psi, respectively.

Ratio of variances	98% confidence interval
$\frac{\sigma_1^2}{\sigma_2^2}$	$\left(\frac{1}{10.97} \times \frac{145{,}877.8}{58{,}509.37}, 10.97 \times \frac{145{,}877.8}{58{,}509.37}\right) = (.23, 27.35)$
$\frac{\sigma_1^2}{\sigma_3^2}$	$\left(\frac{1}{10.97} \times \frac{145{,}877.8}{472.6667}, 10.97 \times \frac{145{,}877.8}{472.6667}\right) = (28.13, 3{,}385.64)$
$\frac{\sigma_2^2}{\sigma_3^2}$	$\left(\frac{1}{10.97} \times \frac{58{,}509.37}{472.6667}, 10.97 \times \frac{58{,}509.37}{472.6667}\right) = (11.28, 1{,}357.93)$
	Overall confidence level $\geq 1 - (.02 + .02 + .02) = .94$

If 1 is in the confidence interval for a variance ratio, it suggests that the two variances are equal. If 1 is not in the interval, it suggests that the two variances are not equal. From Table 14-2 we see that 1 is in the first confidence interval and not in the other two intervals. These multiple comparisons suggest that the variation in rupture times is similar for stress levels 28.84 and 31.63 psi, while the variation at these two stress levels is much greater than the variation at stress level 34.68 psi. These results agree with our visual evaluation of variation in the three sample distributions illustrated in Figure 14-3.

There is a problem with using the variance ratio test of Section 14-2 for comparing two variances and Bartlett's test of this section for comparing more than two: These procedures are *not robust* to deviations from the Gaussian assumption.

> We say a procedure for testing hypotheses is **robust** if actual significance levels are close to the level we select, even under deviations from assumptions.

When using Bartlett's test and the variance ratio test, if the observations do not exactly follow Gaussian distributions, then the level α we use may be far from the actual significance level of the test. For comparison, t tests and analysis of variance for comparing means *are* quite robust to deviations from the Gaussian assumption, and somewhat robust to small deviations from the equal-variance assumption.

In Section 14-4, we discuss a procedure that does provide a robust test for equality of two or more variances.

14-4 Robust Inferences About Two or More Variances

Suppose we have k independent random samples, one from each of k populations, and we want to test the null hypothesis that the variation in the k

populations is the same. We will discuss a modification of a test proposed by Levene (1960). This modified procedure was recommended by Brown and Forsythe (1974) and shown by Conover, Johnson, and Johnson (1981) to work well in a variety of situations.

> **Levene's (modified) test** is a modified version of a procedure proposed by Levene to test for equality of two or more variances. This test is based on the assumption of independent random samples. The test is robust to deviations from Gaussian observations.

We will outline the p-value approach to Levene's modified test and then apply it to an example.

The p-value approach to Levene's (modified) test for equality of two or more variances

1. The null hypothesis states that the variance is the same in the k populations. The alternative hypothesis is that the variances are not the same in all k populations.
2. Let Y_{ij} denote the jth observation in sample i. Let m_i denote the ith sample median. Define new variables $Z_{ij} = |Y_{ij} - m_i|$. Z_{ij} is the absolute value of the difference between the observation Y_{ij} and the median m_i of sample i. To test our hypotheses, we go through the steps for one-way analysis of variance on the transformed observations Z_{ij}. The test statistic is

$$\text{Test statistic} = \frac{s_B^2}{s_r^2}$$

   where s_B^2 here denotes the between-groups mean square and s_r^2 the residual mean square, based on the Z_{ij}'s.
3. We assume that we have k independent random samples, one from each of k populations. Under the null hypothesis of equality of the k population variances, the test statistic has approximately the $F(k - 1, N - k)$ distribution, where N denotes the total sample size.
4. Carry out an experiment that satisfies the assumptions in step 3. Calculate the test statistic in step 2.
5. Find the p-value $= P(F \geq c_0)$, where c_0 denotes the observed value of the test statistic and F denotes a random variable having the $F(k - 1, N - k)$ distribution.
6. If the p-value is large, say the results are consistent with the null hypothesis that the variances are equal. If the p-value is small, say the results are inconsistent with the null hypothesis, suggesting that the variances are not all equal.

Let's illustrate Levene's modified test for equality of variances with an example.

EXAMPLE 14-4

Does the variation in earthworm populations depend on the time of year? To address this question, researchers divided a field into ten square plots. They watered these plots but did not treat them in any other way. (They added no chemicals, for example.) At three times over a 6-month period, the researchers selected equal sized subplots of the ten plots. (Each subplot was studied just once.) They applied an irritant to the subplots that caused the earthworms to rise to the surface. The researchers recorded total biomass/m² of the earthworms in each subplot. The results are shown below (part of a data set contributed by R. P. Blackshaw and P. J. Diggle to a collection of problems in Andrews and Herzberg, 1985, pages 301–306).

Time	Biomass/m² (values ordered from smallest to largest within the three times)						
1	7.73	8.07	10.61	17.01	17.55	26.98	28.59
	46.42	51.96	81.32				
2	.76	1.82	4.06	4.71	4.73	4.93	5.20
	12.45	37.29	39.57				
3	16.40	17.61	19.34	21.19	24.49	26.63	33.11
	39.12	39.26	53.32				

The observations are plotted in Figure 14-4. The average size of the earthworm populations seems to depend on the time of year. Does the variation

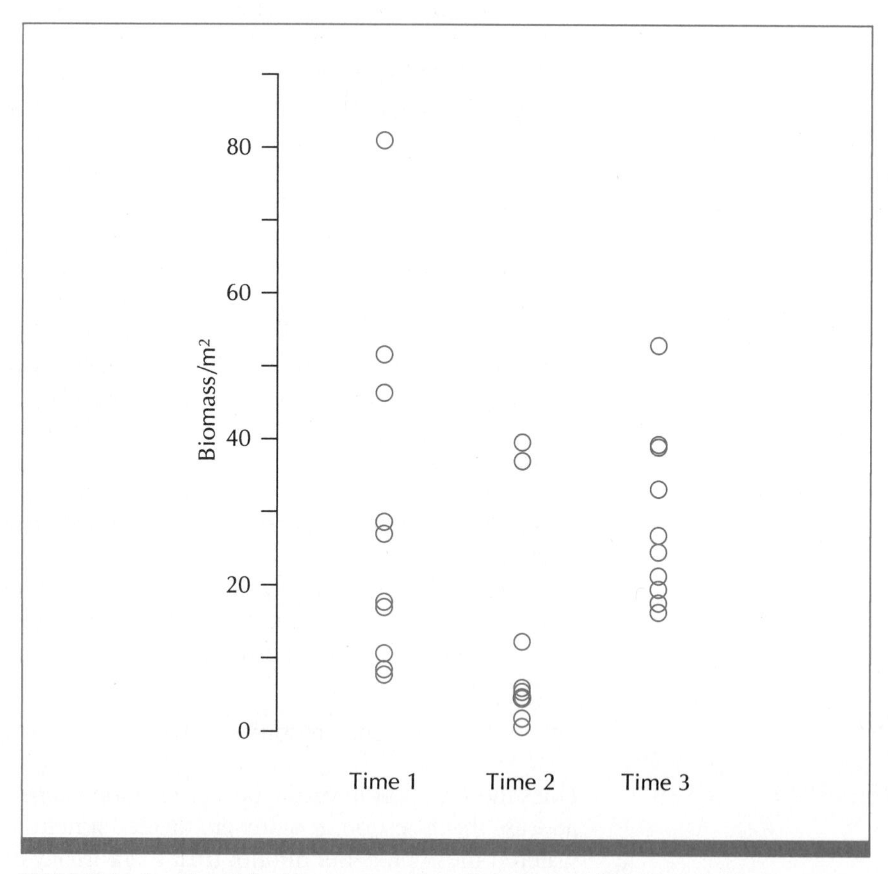

FIGURE 14-4 Plot of the biomass/m² of earthworm populations in Example 14-4

TABLE 14-3 Transformed observations to be used in one-way analysis of variance for Example 14-4

Time	Absolute value of the difference between the observation and the sample median						
1	14.535	14.195	11.655	5.255	4.715	4.715	6.325
	24.155	29.695	59.055				
2	4.07	3.01	.77	.12	.10	.10	.37
	7.62	32.46	34.74				
3	9.16	7.95	6.22	4.37	1.07	1.07	7.55
	13.56	13.70	27.76				

TABLE 14-4 Analysis of variance table for one-way analysis of variance on the values in Table 14-3

Source of variation	Sum of squares	Degrees of freedom	Mean square	Test statistic	p-value
Between groups	502	2	251	1.42	.26
Residual	4,778	27	177		
Total	5,280	29			

also depend on the time of year? We wish to test the null hypothesis that the variance of biomass/m^2 of earthworm populations among equal sized subplots of the field is the same at the three times; the alternative hypothesis is that the three variances are not all the same. We assume that we have three independent random samples, but we cannot judge the validity of this independence assumption without more information about the experiment.

The median of the ten observations at the first time is 22.265, the median at the second time is 4.83, and the median at the third time is 25.56. Table 14-3 shows, for each observation, the absolute value of the difference between the observation and its sample median.

Table 14-4 gives the analysis of variance table resulting from one-way analysis of variance on the values in Table 14-3.

The relatively large p-value of .26 is consistent with the null hypothesis. [If we had found differences, we could have used multiple comparisons (see Section 12-2) on the transformed observations (Table 14-3) to decide which variances seem to be similar and which different.] Looking at the plot of sample values in Figure 14-4, we see that the variation at time 1 was somewhat larger than the variation at the other two times. However, the conclusion that the variation in earthworm populations appears to be similar for the three times does not seem unreasonable.

Summary of Chapter 14

Parametric procedures for making inferences about one or more variances depend on the assumption that the observations are Gaussian distributed. None of these procedures is robust to deviations from the Gaussian assumption. That is, if the observations are not really from Gaussian distributions, the *p*-values for tests of hypotheses and confidence levels for interval estimates may be very wrong (and therefore meaningless).

Levene's modified procedure for testing equality of variances is robust to deviations from the Gaussian assumption. This means that we can feel comfortable interpreting *p*-values even when the observations do not come from Gaussian distributions.

Minitab Appendix for Chapter 14

Making Inferences About a Single Variance

Consider the drained weights of five cans of tomatoes in Example 14-1. Suppose these five values are in column 1 (named DRAINWT) of our worksheet. We want to compare the sample variance with a null hypothesis value of 2.63. We calculate the test statistic and probabilities associated with this test statistic this way:

```
MTB>   stdev 'drainwt' k1
 ST.DEV.   = 1.2042
MTB>   let k2=4*(k1**2)/2.63
MTB>   cdf k2 k3;
SUBC>  chisquare 4.
MTB>   let k4=1-k3
MTB>   print k2 k3 k4
K2     2.20532
K3     0.301945
K4     0.698055
```

Minitab prints the test statistic K2 = 2.20532, the cumulative probability K3 = $P(X \leq \text{K2})$ = 0.301945, and the tail probability K4 = $P(X > \text{K2})$ = 0.698055, where X denotes a random variable having the chi-square distribution with 4 degrees of freedom. The *p*-value is 2 times the smaller of K3 and K4, about .6 in this example.

Let's calculate a 95% confidence interval for the variance of drained weights in the population sampled. Since the sample size is 5, we refer to the chi-square distribution with 4 degrees of freedom, as for the test of hypotheses. Recall that the standard deviation of the drained weights is in K1.

```
MTB>    invcdf 0.025 k7;
SUBC>   chisquare 4.
MTB>    invcdf 0.975 k8;
SUBC>   chisquare 4.
MTB>    let k9=4*(k1**2)/k8
MTB>    let k10=4*(k1**2)/k7
MTB>    print k9 k10
K9      0.520493
K10     11.9731
```

In the calculations above, K7 and K8 are constants such that $P(X \leq \text{K7}) = .025$, $P(X \leq \text{K8}) = .975$, and $P(X > \text{K8}) = .025$, where X has the chi-square distribution with 4 degrees of freedom. K1 is the standard deviation of the five observations. The printed values K9 and K10 are the lower and upper endpoints, respectively, of the 95% confidence interval for the population variance.

Making Inferences About Two Variances

Now let's consider inferences about two variances. Consider the data in Exercise 14-6. Suppose column 1 (named TREATMT) in our worksheet contains a code for treatment: 1 = morphine, 2 = nalbuphine. Column 2 (named CHANGE) contains changes in pupil diameter. We will unstack the changes in column 2 according to treatment in column 1:

```
MTB>    unstack 'change' c3 c4;
SUBC>   subscripts 'treatmt'.
MTB>    name c3 'change1' c4 'change2'
```

Column 3 contains the changes for the six volunteers who received morphine. Column 4 contains the changes for the five volunteers who received nalbuphine. If we DESCRIBE these two columns, we get the display in Figure M14-1.

We will put the sample variance for CHANGE1 in K1 and the sample variance for CHANGE2 in K2:

```
MTB>    stdev 'change1' k1
 ST.DEV.  = 0.87913
MTB>    let k1=k1**2
MTB>    stdev 'change2' k2
 ST.DEV.  = 0.41473
MTB>    let k2=k2**2
```

```
               N      MEAN    MEDIAN    TRMEAN     STDEV    SEMEAN
change1        6     1.363     1.450     1.363     0.879     0.359
change2        5     0.220     0.200     0.220     0.415     0.185

             MIN       MAX        Q1        Q3
change1    0.080     2.400     0.620     2.100
change2   -0.300     0.800    -0.150     0.600
```

FIGURE M14-1 Descriptive statistics for CHANGE1 and CHANGE2 in Exercise 14-6

To test the null hypothesis that the variances under the two treatments are equal, the test statistic is the larger sample variance divided by the smaller one. In this example, K1 is larger than K2. We calculate the test statistic and p-value as follows:

```
MTB>   let k3=k1/k2
MTB>   cdf k3 k4;
SUBC>  f 5 4.
MTB>   let k4=1-k4
MTB>   print k3 k4
K3     4.49341
K4     0.0852627
```

The p-value K4 is about .085, a borderline result. Note that there are 5 degrees of freedom for the larger sample variance (CHANGE1, sample size = 6) and 4 degrees of freedom for the smaller sample variance (CHANGE2, sample size = 5).

We will calculate a 95% confidence interval for σ_1^2/σ_2^2, where σ_1^2 denotes the variance of changes under morphine and σ_2^2 denotes the variance of changes under nalbuphine. Recall that K3 is the ratio of the two sample variances, s_1^2/s_2^2. We calculate the confidence interval this way:

```
MTB>   invcdf 0.975 k5;
SUBC>  f 5 4.
MTB>   invcdf 0.975 k6;
SUBC>  f 4 5.
MTB>   let k7=k3/k5
MTB>   let k8=k3*k6
MTB>   print k7 k8
K7     0.479831
K8     33.1970
```

K7 = 0.479831 is the lower endpoint and K8 = 33.1970 the upper endpoint of the 95% confidence interval.

Carrying Out Levene's Modified Test for Equality of Variances

Minitab does not have a command for Bartlett's test of equality of several variances. Although Minitab commands can be used to calculate the test statistic, we will not do that here. Instead, we will go through the steps for Levene's modified test of equality of several variances, described in Section 14-4. Suppose the data from Example 14-4 are in unstacked form on our worksheet. The observations for times 1, 2, and 3 are in columns 1, 2, and 3, respectively (named TIME1, TIME2, and TIME3, respectively):

```
ROW   time1   time2   time3
  1    7.73    0.76   16.40
  2    8.07    1.82   17.61
  3   10.61    4.06   19.34
  4   17.01    4.71   21.19
  5   17.55    4.73   24.49
  6   26.98    4.93   26.63
  7   28.59    5.20   33.11
  8   46.42   12.45   39.12
  9   51.96   37.29   39.26
 10   81.32   39.57   53.32
```

We find the median of each column of values:

```
MTB> median 'time1' k1
MTB> median 'time2' k2
MTB> median 'time3' k3
```

We subtract each median from the corresponding column and take the absolute value:

```
MTB> let c4=absolute(c1-k1)
MTB> let c5=absolute(c2-k2)
MTB> let c6=absolute(c3-k3)
```

Now we want to carry out a one-way analysis of variance on the data in columns 4, 5, and 6. The data for this single-factor analysis are in unstacked form: the values of the response variable for a single level of the factor (time) are in a single column, one column per level of that factor. Minitab will perform a one-way analysis of variance on such unstacked data if we use the AOVONEWAY command:

```
MTB> aovoneway c4-c6
```

The output is in Figure M14-2. This is the same as we found in Table 14-4.

```
ANALYSIS OF VARIANCE
SOURCE      DF          SS          MS          F         p
FACTOR       2         502         251       1.42     0.260
ERROR       27        4778         177
TOTAL       29        5280
                                       INDIVIDUAL 95 PCT CI'S FOR MEAN
                                       BASED ON POOLED STDEV
LEVEL        N        MEAN       STDEV  --+---------+---------+---------+-----
C4          10       17.43       16.92              (----------*----------)
C5          10        8.34       13.54   (---------*----------)
C6          10        9.24        7.83    (----------*---------)
                                        --+---------+---------+---------+-----
POOLED STDEV =       13.30             0.0       8.0      16.0      24.0
```

FIGURE M14-2 Output for Levene's modified test in Example 14-4

Exercises for Chapter 14

In each exercise, plot the observations in any ways that seem helpful. Describe the population(s) sampled, whether real or hypothetical. For each procedure, state the assumptions that make the analysis valid. Do these assumptions seem reasonable? What additional information would you like to have about the experiment? Discuss the results of your analysis.

EXERCISE 14-1

An engineer studied the time to rupture for pieces of stainless steel at two levels of stress. He tested six uniform pieces of steel at each of the two stress levels. (This is a separate phase of the experiment discussed in Example 14-3.) The results are shown below (Schmoyer, 1986; from Garofalo et al., 1961). Stress levels are in pounds per square inch (psi).

Stress level (psi)	Rupture time (hours)					
41.69	6.6	9.6	11.2	12.3	19.7	20.4
45.71	1.9	3.9	4.3	4.6	5.7	9.0

a. Plot these observations.

b. Test the null hypothesis that the population variances for rupture times are equal at the two stress levels. Do the assumptions for the analysis seem reasonable?

c. Calculate a confidence interval for the ratio of the two population variances.

EXERCISE 14-2

In Example 10-3, we looked at the average weight in grams of six pairs of twins born to exercised Pygmy goats (Dhindsa, Metcalfe, and Hummels, 1978):

745.5 1,175.0 1,290.0 1,364.5 1,397.5 1,660.0

a. Plot the observations.

b. Test the null hypothesis that the variance in average weights of such pairs of twins is 10,000 grams^2.

c. Calculate a 99% confidence interval for the variance in average weights of pairs of twins born to Pygmy goats treated like those in this experiment.

EXERCISE 14-3

In Example 10-4, we considered the height in inches of five bomb bases sampled in a 15-minute interval (Duncan, 1974, page 43; Hollander and Proschan, 1984, page 42; from Kauffman, 1945):

.826 .829 .831 .836 .840

a. Plot the observations.

b. Test the null hypothesis that the variance for heights of bomb bases produced during that 15-minute interval was .0001 inch^2. [Recall that the specifications were .830 $\pm$ .01 inch and $(.01 \text{ inch})^2 = .0001 \text{ inch}^2$.]

c. Calculate a 98% confidence interval for the variance of heights of bomb bases produced during that 15-minute interval.

EXERCISE 14-4

In Exercise 10-7, we considered determinations of serum iron concentrations (μg/100 ml) using a new method (Hollander and Wolfe, 1973, pages 85–86; a portion of the data in Jung and Parekh, 1970):

96	98	99	100	103	103	104	104	105	105
106	106	107	108	108	108	110	113	114	114

a. Plot the observations.

b. Test the null hypothesis that the variance in serum iron concentration determinations using this method is 100 $(\mu\text{g}/100 \text{ ml})^2$.

c. Calculate a 95% confidence interval for the variance in serum iron concentration determinations using this method.

EXERCISE 14-5

In Exercise 10-1, we considered plasma citrate concentrations (μmol/liter) before breakfast for ten volunteers (from a contribution by E. B. Jensen to a collection of problems in Andrews and Herzberg, 1985, page 237; from Andersen, Jensen, and Schou, 1981).

93	116	125	144	105	109	89	116	151	137

a. Plot the observations.

b. Test the null hypothesis that the variance of before-breakfast plasma citrate concentrations is 1,500 $(\mu\text{mol/liter})^2$.

c. Calculate a 90% confidence interval for the variance of before-breakfast plasma citrate concentrations.

EXERCISE 14-6

In Exercise 11-12, we discussed change in pupil diameter (in millimeters) for volunteers after two different treatments (Box, Hunter, and Hunter, 1978, page 160; from H. W. Elliott, G. Navarro, and N. Nomof, *J. Med.,* 1970, volume 1, page 77).

Morphine:	.08	.8	1.0	1.9	2.0	2.4
Nalbuphine:	−.3	.0	.2	.4	.8	

a. Plot the observations.

b. Test the null hypothesis that the variance of change in pupil diameter is the same after both treatments.

c. Calculate a 90% confidence interval for the ratio of the two variances.

EXERCISE 14-7

In Exercise 11-8, we looked at specific airway resistance 30 minutes after administration of a bronchodilating aerosol by patients using either hand administration or an automatic inhalation device (units not given) (Box, Hunter, and Hunter, 1978, page 158; from a larger study reported by F. J. McIlneath and B. M. Cohen in *J. Med.,* 1970, volume 1, page 229):

Hand:	17.00	22.80	21.60	20.40	11.20	14.00
	52.25	7.50	12.20	18.85	6.05	4.05
Automatic:	11.60	11.60	13.65	17.22	8.25	6.20
	41.50	6.96	8.40	9.00	5.18	3.00

a. Plot the observations.

b. Test the null hypothesis that the variance in specific airway resistance is the same for the two methods of administration.

c. Calculate a 98% confidence interval for the ratio of the two variances.

EXERCISE 14-8

In Exercise 11-7, we looked at sputum histamine levels (μg/g dry weight sputum) for 9 allergic people and 13 nonallergic people, all smokers (Hollander and Wolfe, 1973, page 74; a subset of data in Thomas and Simmons, 1969):

Allergics:	31.0	39.6	64.7	65.9	67.9	100.0
	102.4	1,112.0	1,651.0			
Nonallergics:	4.7	5.2	6.6	18.9	27.3	29.1
	32.4	34.3	35.4	41.7	45.5	48.0
	48.1					

a. Plot the observations.

b. Test the null hypothesis that the variance in sputum histamine levels is the same for allergic and nonallergic smokers.

c. Calculate a 98% confidence interval for the ratio of the two variances.

d. Take the logarithm of each observation. Test the null hypothesis that the variance of the logarithm of sputum histamine level is the same for allergic and nonallergic smokers.

e. Calculate a 98% confidence interval for the ratio of the two variances of the logarithm of sputum histamine level.

f. Discuss your findings.

EXERCISE 14-9

In Exercise 11-6, we considered plasma testosterone levels (nanograms/deciliter) of monkeys on two different diets (Jerome Hojnacki, 1986, personal communication):

Alcohol:	313.99	152.06	145.64	128.86	262.16	251.29
	505.55	94.79	157.49	171.81		
Control:	632.92	308.56	1239.68	440.38	233.02	142.67
	84.91	342.63	1005.66	735.61		

a. Plot the observations.

b. Test the null hypothesis that the variance in plasma testosterone levels is the same for monkeys on the two diets.

c. Calculate a 90% confidence interval for the ratio of the two variances.

d. Take the logarithm of each observation. Test the null hypothesis that the variance of the logarithm of plasma testosterone level is the same for monkeys on the two diets.

e. Calculate a 90% confidence interval for the ratio of the two variances for the logarithm of plasma testosterone level.

EXERCISE 14-10

In Exercise 11-4, we looked at eight independent determinations (in °C) of the melting point of hydroquinone by each of two analysts (Duncan, 1974, pages 575–576; from Wernimont, 1947, page 8):

Analyst 1:	174.0	173.5	173.0	173.5	171.5	172.5
	173.5	173.5				
Analyst 2:	173.0	173.0	172.0	173.0	171.0	172.0
	171.0	172.0				

a. Plot the observations.

b. Test the null hypothesis that the variance of determinations is the same for the two analysts.

c. Calculate a 95% confidence interval for the ratio of the variances for the two analysts.

EXERCISE 14-11

In Exercise 11-3, we considered the total score of five shots at a target by an experienced shooter using a revolver. There were eight trials with each of two types of ammunition (Snow, 1986):

.22 Magnum	42	43	46	47	46	47	39	47
.22 Long Rifle	41	43	41	41	40	40	45	47

a. Plot the observations.

b. Test the null hypothesis that the variance of total scores by this shooter is the same for the two types of ammunition.

c. Calculate a 95% confidence interval for the ratio of variances for the two types of ammunition.

EXERCISE 14-12 In Exercise 15-24, we discuss a study of the permeability of concrete (inches per hour) with six different levels of asphalt content (Mendenhall and Sincich, 1988, page 495; from Woelfl et al., 1981):

3% asphalt:	1,189	840	1,020	980	**6% asphalt:**	707	927	1,067	822
4% asphalt:	1,440	1,227	1,022	1,293	**7% asphalt:**	853	900	733	585
5% asphalt:	1,227	1,180	980	1,210	**8% asphalt:**	395	270	310	208

a. Plot the observations.

b. Use a parametric analysis to test for equality of variances of permeability at each of the six asphalt contents.

c. Use a nonparametric analysis to test for equality of variances of permeability at each of the six asphalt contents.

d. Compare your answers to parts (b) and (c).

EXERCISE 14-13 In Exercise 15-17, we consider times to failure (in hours) of samples of insulation for electrical motors in accelerated life testing at four temperatures (Nelson, 1986, pages 20–21):

190 °C:	7,228	7,228	7,228	8,448	9,167	9,167
	9,167	9,167	10,511	10,511		
220 °C:	1,764	2,436	2,436	2,436	2,436	2,436
	3,108	3,108	3,108	3,108		
240 °C:	1,175	1,175	1,521	1,569	1,617	1,665
	1,665	1,713	1,761	1,953		
260 °C:	600	744	744	744	912	1,128
	1,320	1,464	1,608	1,896		

a. Plot the observations.

b. Use a parametric analysis to test for equality of variances in failure times at the four temperatures. Use the Bonferroni method to make multiple comparisons.

c. Use a nonparametric analysis to test for equality of variances in failure times at the four temperatures. Use the Bonferroni method to make multiple comparisons.

d. Compare your answers to parts (b) and (c).

e. Take the logarithm of each failure time. Use a parametric analysis to test for equality of variances of the logarithm of failure time at the four temperatures.

f. Use a nonparametric analysis to test for equality of variances of the logarithm of failure time at the four temperatures.

g. Compare your answers to parts (e) and (f).

h. Discuss your findings.

EXERCISE 14-14

In Exercise 15-16, we discuss instrument response at five concentrations of copper in solution for an experiment in atomic absorption spectroscopy (Carroll, Sacks, and Spiegelman, 1988):

Copper in solution (micrograms/ milliliter)	Instrument response in absorbance units			
.0	.045	.047	.051	.054
.050	.084	.087		
.100	.115	.116		
.200	.183	.191		
.500	.395	.399		

a. Plot the observations.

b. Use a parametric analysis to test for equality of variances of instrument response across copper concentrations.

c. Use a nonparametric analysis to test for equality of variances of instrument response across copper concentrations.

d. Compare your answers to parts (b) and (c).

EXERCISE 14-15

In Exercise 12-6, we compared the working life (thousands of cycles until failure) of three types of stopwatch (Rice, 1988, page 432; from Natrella, 1963):

Type 1:	1.7	1.9	6.1	12.5	16.5	25.1	30.5	42.1
	82.5							
Type 2:	13.6	19.8	25.2	46.2	46.2	61.1		
Type 3:	13.4	20.9	25.1	29.7	46.9			

a. Plot the observations.

b. Use a parametric analysis to test for equality of variances of working life for the three stopwatch types. Use the Bonferroni method to make multiple comparisons.

c. Use a nonparametric analysis to test for equality of variances of working life for the three stopwatch types. Use the Bonferroni method to make multiple comparisons.

d. Compare your results in parts (b) and (c).

EXERCISE 14-16 In Exercise 12-2, we looked at the stimulation index of men treated with one of three different regimens of a synthetic vaccine for malaria or with a saline regimen (Patarroyo et al., 1988):

Saline control:	1.4	1.0	4.0	**Regimen 2:**	6.6	9.1		
Regimen 1:	1.5	5.6	12.4	**Regimen 3:**	35.1	13.4	.8	3.3

a. Plot the observations.

b. Use a parametric analysis to test for equality of variances of stimulation index under the four regimens. Use the Bonferroni method to make multiple comparisons.

c. Use a nonparametric analysis to test for equality of variances of stimulation index under the four regimens. Use the Bonferroni method to make multiple comparisons.

d. Compare your results in parts (b) and (c).

e. Take the logarithm of each observation. Use a parametric analysis to test for equality of variances of the logarithm of stimulation index under the four regimens.

f. Use a nonparametric analysis to test for equality of variances of the logarithm of stimulation index under the four regimens.

g. Compare your results in parts (e) and (f).

EXERCISE 14-17 In Exercise 13-5, we considered an experiment on iron retention in mice. Researchers measured percentage of iron retained for mice under six sets of conditions (from an example in Rice, 1988, pages 356–357):

Fe^{3+}, 10.2 millimolar:	.71	1.66	2.01	2.16	2.42	2.42	2.56	2.60	3.31
	3.64	3.74	3.74	4.39	4.50	5.07	5.26	8.15	8.24
Fe^{3+}, 1.2 millimolar:	2.20	2.93	3.08	3.49	4.11	4.95	5.16	5.54	5.68
	6.25	7.25	7.90	8.85	11.96	15.54	15.89	18.30	18.59
Fe^{3+}, .3 millimolar:	2.25	3.93	5.08	5.82	5.84	6.89	8.50	8.56	9.44
	10.52	13.46	13.57	14.76	16.41	16.96	17.56	22.82	29.13
Fe^{2+}, 10.2 millimolar:	2.20	2.69	3.54	3.75	3.83	4.08	4.27	4.53	5.32
	6.18	6.22	6.33	6.97	6.97	7.52	8.36	11.65	12.45
Fe^{2+}, 1.2 millimolar:	4.04	4.16	4.42	4.93	5.49	5.77	5.86	6.28	6.97
	7.06	7.78	9.23	9.34	9.91	13.46	18.40	23.89	26.39
Fe^{2+}, .3 millimolar:	2.71	5.43	6.38	6.38	8.32	9.04	9.56	10.01	10.08
	10.62	13.80	15.99	17.90	18.25	19.32	19.87	21.60	22.25

a. Plot the observations.

b. Use a parametric analysis to test for equality of variances of percentage of iron retained under the six sets of conditions. Use the Bonferroni method to make multiple comparisons of variances.

c. Use a nonparametric analysis to test for equality of variances of percentage of iron retained under the six sets of conditions. Compare with the parametric test in part (b).

d. Take the logarithm of each observation. Use a parametric analysis to test for equality of variances of the logarithm of percentage of iron retained under the six sets of conditions.

e. Use a nonparametric analysis to test for equality of variances of the logarithm of percentage of iron retained under the six sets of conditions. Compare with the parametric test in part (d).

f. Discuss your findings.

EXERCISE 14-18

In Exercise 13-2, we considered the effects of pH and temperature on optical density (units not given) of a polymer latex (Gasper, 1988; with permission of ICI Resins US):

pH 9.0, 85 °C:	56.6	38.9
pH 9.0, 95 °C:	39.0	37.5
pH 9.3, 85 °C:	63.0	96.8
pH 9.3, 95 °C:	33.0	33.3

a. Plot the observations.

b. Use a parametric analysis to test for equality of variances under the four sets of conditions.

c. Use a nonparametric analysis to test for equality of variances under the four sets of conditions.

d. Compare your answers to parts (b) and (c).

e. Take the reciprocal of each observation. Use a parametric analysis to test for equality of variances of the reciprocal of optical density under the four sets of conditions.

f. Use a nonparametric analysis to test for equality of variances of the reciprocal of optical density under the four sets of conditions.

g. Compare your answers to parts (e) and (f).

h. Discuss your findings.

EXERCISE 14-19

In Exercise 13-1, we looked at the distances (in feet) a player hit a softball under four sets of conditions (Shaughnessy, 1988):

Dudley Thunder, wood bat:	242	230	250	242
Dudley Thunder, aluminum bat:	270	282	265	277
Worth Red Dot, wood bat:	258	264	265	275
Worth Red Dot, aluminum bat:	290	318	302	310

a. Plot the observations.

b. Use a parametric analysis to test for equality of variances under the four sets of conditions.

c. Use a nonparametric analysis to test for equality of variances under the four sets of conditions.

d. Compare your answers to parts (b) and (c).

EXERCISE 14-20 In Example 13-2, we considered flexural strength (in pounds per square inch) of sheet castings of a polymer under four sets of conditions (Duncan, 1974, page 685; from Gore, 1947):

20 minutes, 100°:	9,500	10,650	9,700	9,950	10,100
20 minutes, 120°:	11,300	11,750	11,600	11,650	11,700
60 minutes, 100°:	11,500	11,650	11,250	11,250	11,900
60 minutes, 120°:	10,900	11,500	11,850	11,700	11,650

a. Plot the observations.

b. Use a parametric analysis to test for equality of variances under the four sets of conditions.

c. Use a nonparametric analysis to test for equality of variances under the four sets of conditions.

d. Compare your results in parts (b) and (c).

EXERCISE 14-21 Consider the data on rupture times of pieces of stainless steel at different stress levels, in Example 14-3.

a. Take the logarithm of each rupture time. Use a parametric analysis to test for equality of variances for the logarithm of rupture time at the three stress levels.

b. Use a nonparametric analysis to test for equality of variances for the logarithm of rupture times at the three stress levels. Compare with your results in part (a).

c. Compare your results in part (a) with what we found in Example 14-3.

EXERCISE 14-22 Consider the data on rupture times of pieces of stainless steel at different stress levels, in Example 14-3.

a. Use a nonparametric analysis to test for equal variances of rupture times at the three stress levels. Use the Bonferroni method to make multiple comparisons.

b. Compare your results in part (a) with what we found in Example 14-3.

CHAPTER 15

Correlation, Regression, and the Method of Least Squares

IN THIS CHAPTER

Correlation coefficient
Rank correlation coefficient
Method of least squares
Simple linear regression
Comparing the standard deviation line with the least squares line
Multiple regression

We will now look at ways to study relationships between quantitative variables. For instance: What is the relationship between height and weight in young children? How does income vary with education? Does blood pressure depend on age in adults? What is the relationship between advertising expenditures and sales?

One possible relationship between two quantitative variables is linear: A scatterplot of the two variables looks roughly like a straight line. The correlation coefficient is a measure of the extent of *linear* association between two quantitative variables, as we see in Section 15-1. A parametric test of independence of two quantitative variables is discussed in Section 15-2. In Section 15-3 we consider a correlation coefficient based on ranks, and discuss a nonparametric test of independence of two quantitative variables.

We may want to model one quantitative variable as a straight-line function of another. The method of least squares allows us to fit a straight line to a set of points in a scatterplot. Finding such a straight line and testing hypotheses about the model is called *simple linear regression,* discussed in Section 15-4. The relationship between linear correlation and the least squares line found in simple linear regression is the subject of Section 15-5. Also included are examples of how to interpret the phrase *regression toward the mean.*

Section 15-6 provides a very brief introduction to multiple regression. In multiple regression, we try to model a quantitative variable as a function of other variables.

Let's begin with the linear correlation coefficient. We use the correlation coefficient to measure linear association between two quantitative variables.

15-1 The Linear Correlation Coefficient

The linear correlation coefficient is a descriptive statistic. We use it to measure linear association between two quantitative variables, as a tool in data analysis. In this section we are concerned with the correlation coefficient only as a descriptive statistic; we make no inferences based on it. Therefore, it is appropriate to return to the World Bank data set to provide examples.

Values of four World Bank indicators are shown in Table 15-1 for four high-income, oil-exporting nations (World Bank, 1987). A scatterplot of birth rate versus per capita gross national product is shown in Figure 15-1 for these four countries; the points lie close to a straight line with negative slope. Figure 15-2 is a plot of life expectancy versus per capita gross national product. The association between the variables in this graph is less strongly linear; the points do not appear to lie as close to a straight line as the ones in Figure 15-1. Figure 15-3 shows a scatterplot of calorie supply versus per capita gross national product; there is no linear relationship apparent in this graph.

We use the linear correlation coefficient to measure the extent of *linear* association between two quantitative variables. Let's define the linear correla-

TABLE 15-1 Values of 1985 per capita gross national product, 1985 birth rate per 1,000 population, 1985 life expectancy at birth, and 1985 daily calorie supply per capita are listed for four high-income, oil-exporting nations.

Country	Gross national product	Birth rate	Life expectancy	Calorie supply
Libya	$7,170	45	60	3,612
Saudi Arabia	$8,850	42	62	3,128
Kuwait	$14,480	34	72	3,138
United Arab Emirates	$19,270	30	70	3,625

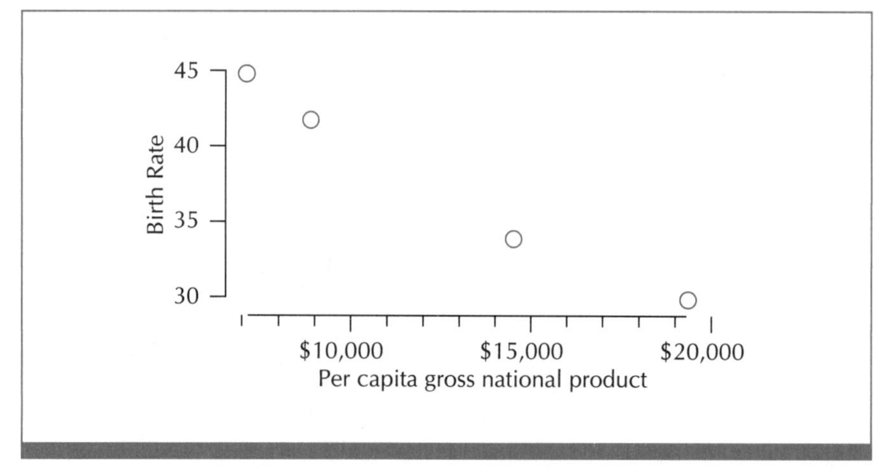

FIGURE 15-1 Scatterplot of birth rate versus per capita gross national product in 1985 for four high-income oil exporters

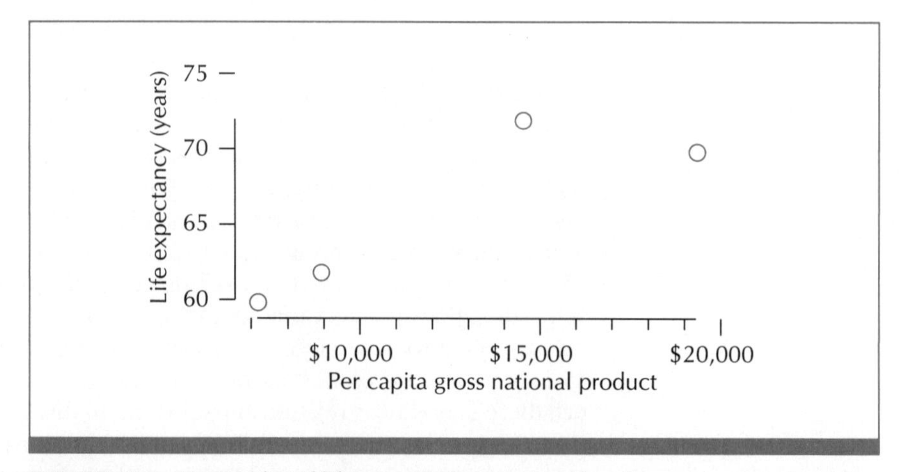

FIGURE 15-2 Scatterplot of life expectancy versus per capita gross national product in 1985 for four high-income oil exporters

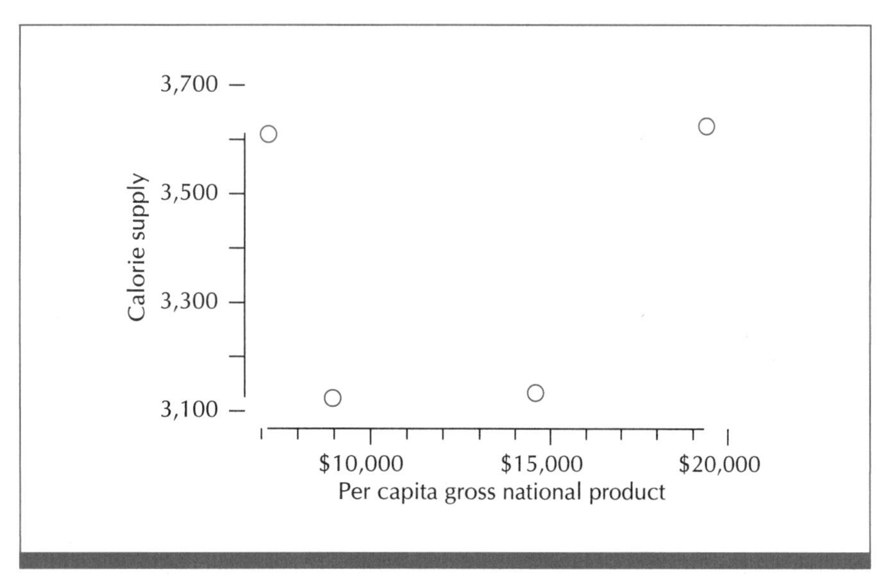

FIGURE 15-3 Scatterplot of daily calorie supply versus gross national product per capita in 1985 for four high-income oil exporters

tion coefficient, then find its value for the three sets of points in Figures 15-1, 15-2, and 15-3.

Suppose we have n pairs of observations (X_i, Y_i), where i goes from 1 to n. We use X_i to denote the ith observation on a variable we call X. Similarly, Y_i denotes the ith observation on a variable we call Y. We want to measure the extent of linear association between the two variables X and Y.

Standardize each value of the first variable by subtracting the sample mean and dividing by the sample standard deviation for that variable. (The mean of a standardized variable equals 0 and the standard deviation equals 1. This is why we call such a variable *standardized.*) Now standardize each value of the second variable. Then for each pair of observations, multiply the standardized values of the two variables. Add up these products, and then divide by the number of pairs minus 1. The result is the ***linear correlation coefficient:***

$$\text{Linear correlation coefficient} = \frac{\text{Sum of the products of the two standardized variables}}{\text{Number of pairs} - 1}$$

Let $\bar{X}$ and SD_x denote the sample mean and sample standard deviation, respectively, for the X variable. Similarly, let $\bar{Y}$ and SD_y denote the sample mean and sample standard deviation for the Y variable. Then we can write the formula for the linear correlation coefficient as

$$\text{Linear correlation coefficient} = \frac{\sum_{i=1}^{n} \left(\frac{X_i - \bar{X}}{SD_x}\right)\left(\frac{Y_i - \bar{Y}}{SD_y}\right)}{n - 1}$$

Another name for the linear correlation coefficient is Pearson's correlation coefficient. We often refer to it simply as the correlation coefficient, and denote it by r. Alternative calculation formulas for the correlation coefficient are shown below.

> The **correlation coefficient,** also called the **linear correlation coefficient** or **Pearson's correlation coefficient,** is a measure of *linear* association between two quantitative variables. If we have n pairs of observations (X_i, Y_i), then we calculate the correlation coefficient r as

$$r = \frac{\sum_{i=1}^{n} (X_i - \bar{X})(Y_i - \bar{Y})}{\sqrt{\left(\sum_{i=1}^{n} (X_i - \bar{X})^2\right)\left(\sum_{i=1}^{n} (Y_i - \bar{Y})^2\right)}}$$

$$= \frac{\sum_{i=1}^{n} X_iY_i - n\bar{X}\bar{Y}}{\sqrt{\left(\sum_{i=1}^{n} X_i^2 - n\bar{X}^2\right)\left(\sum_{i=1}^{n} Y_i^2 - n\bar{Y}^2\right)}}$$

The linear correlation coefficient has no units, and takes values from -1 to 1. A correlation coefficient near 0 suggests there is little or no linear association between the two variables.

A linear correlation coefficient near 1 suggests a strong positive linear association between the two variables. The correlation coefficient equals 1 when and only when all plotted points fall on a straight line with positive slope. The correlation coefficient gives us no information on what this slope is.

A linear correlation coefficient near -1 suggests a strong negative linear association. The correlation coefficient equals -1 when and only when all the points lie on a straight line with negative slope. Again, we cannot determine the slope from the correlation coefficient.

Let's find the correlation coefficient to measure linear association between birth rate and per capita gross national product for the four high-income oil exporters. The calculations are outlined in Table 15-2.

The last two columns in Table 15-2 show the standardized values of gross national product and birth rate. Because of round-off errors in the calculations, the means of our standardized variables may not equal 0 exactly. Similarly, because of rounding errors, these standardized variables may have standard deviations not exactly equal to 1.

At the bottom of Table 15-2, we see that the linear correlation coefficient equals $-.99$. This value indicates a strong negative linear association between birth rate and per capita gross national product for these four countries. As we saw in Figure 15-1, the four plotted points do lie very close to a line with negative slope.

We saw a positive relationship between life expectancy and per capita gross national product in Figure 15-2. The points are not as close to a straight line as the points in Figure 15-1. (United Arab Emirates, with the highest per

TABLE 15-2 Calculating the linear correlation coefficient to measure linear association between per capita gross national product (GNP) and birth rate in 1985 for four high-income, oil-exporting nations

Country	GNP	Birth rate	Standardized GNP	Standardized birth rate
Libya	7,170	45	−.95	1.04
Saudi Arabia	8,850	42	−.65	.61
Kuwait	14,480	34	.37	−.54
United Arab Emirates	19,270	30	1.24	−1.12
Mean	12,442.50	37.75	.0	.0
Standard deviation	5,521.82	6.95	1.0	1.0

Linear correlation coefficient

$$= \frac{(-.95)(1.04) + (-.65)(.61) + (.37)(-.54) + (1.24)(-1.12)}{4 - 1} = -.99$$

capita gross national product, has a life expectancy 2 years shorter than that of Kuwait.) The correlation coefficient equals .88, reflecting the strong positive association between life expectancy and per capita gross national product among these four high-income oil exporters. But .88 is smaller than .99, consistent with our observations that the linear association between life expectancy and per capita gross national product is less than that between birth rate and per capita gross national product among these four countries.

We saw no linear association between daily calorie supply and gross national product per capita in Figure 15-3. The correlation coefficient for these four points is .19. This relatively small value reflects the lack of linear association we saw in the scatterplot for these two variables.

What exactly does the correlation coefficient measure? It measures the extent of clustering of plotted points about a straight line. A correlation coefficient that is large in absolute value suggests strong linear association between the two variables; the variation of points about a line is small relative to the variation in the separate variables. A correlation coefficient near 0 suggests little linear association between the two variables; the variation of points about a line is close to the variation in the separate variables.

Let's discuss these ideas in terms of the scatterplots in Figures 15-4 and 15-5. Figure 15-4 shows a scatterplot of life expectancy versus the logarithm of per capita gross national product for 109 countries. The mean ± 1 standard deviation for life expectancy is graphed near the left vertical axis. The mean ± 1 standard deviation for the logarithm of per capita gross national product is graphed near the top horizontal axis.

A scatterplot of primary school enrollment versus the logarithm of per capita gross national product is shown in Figure 15-5. The mean ± 1 standard deviation is graphed for each variable, as in Figure 15-4.

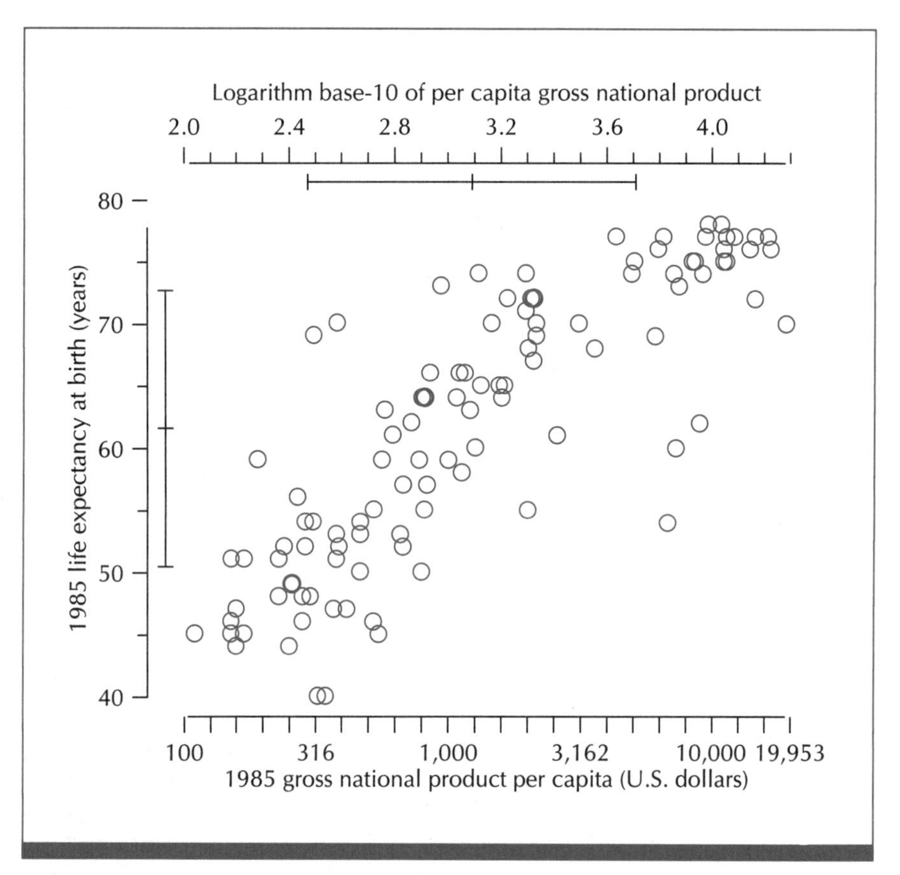

FIGURE 15-4 Scatterplot of life expectancy versus the logarithm of per capita gross national product in 1985 for 109 countries. The mean ± 1 standard deviation for each variable is graphed near the corresponding axis.

We see that there is tighter clustering about a line in Figure 15-4 than in Figure 15-5. The variation about a line drawn through the points in Figure 15-4 is relatively small compared with the variation in the separate variables. In contrast, the variation about a line drawn through the points in Figure 15-5 is close to the variation in the separate variables.

We think there is a stronger linear association illustrated in Figure 15-4 than in Figure 15-5. The correlation coefficients reflect the different impressions we get from these two plots. The correlation coefficient for life expectancy and the logarithm of per capita gross national product is .84. The correlation coefficient for primary school enrollment and the logarithm of per capita gross national product is .49.

Can the correlation coefficient be misleading? Yes, it can. We should always plot two quantitative variables to get a visual feel for their relationship. Then we can use the correlation coefficient to supplement the plot.

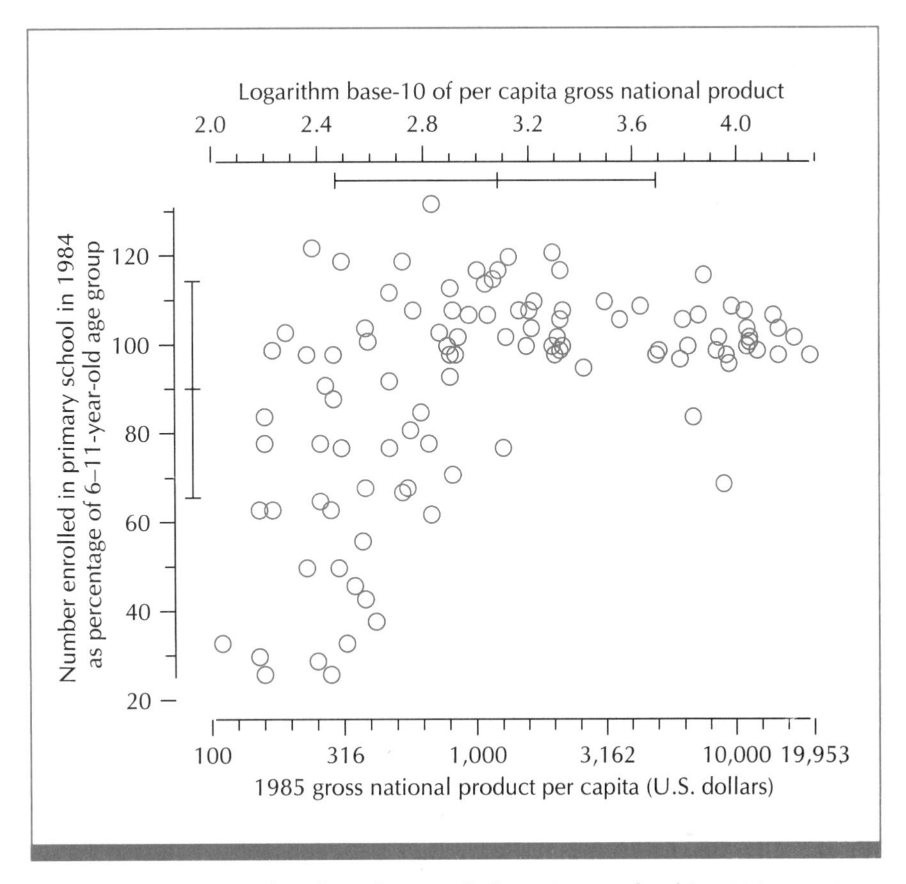

FIGURE 15-5 Scatterplot of number enrolled in primary school in 1984 as percentage of 6–11-year age group and the logarithm of per capita gross national product in 1985 for 104 countries. The mean ± 1 standard deviation for each variable is graphed near the corresponding axis.

Consider the scatterplot of life expectancy versus per capita gross national product in Figure 15-6. The correlation coefficient for the 109 plotted points is .66. By itself, this correlation coefficient might suggest a linear association between these two variables. But we can see in Figure 15-6 a curved relationship. A stronger linear relationship exists between life expectancy and the logarithm of per capita gross national product (Figure 15-4, $r = .84$).

Sometimes a single point or a few points inflate the correlation coefficient (in absolute value) above what it would be if the point(s) were excluded. Consider Figures 15-7 and 15-8, for example. On the vertical axis in each plot is the difference between male and female primary school enrollments in 1985. Overall primary school enrollment is on the horizontal axis.

Figure 15-7 is based on the seven countries in the nonmember economic category with nonmissing information on primary school enrollment. The correlation coefficient for these seven points is .90, a large value. Notice that there

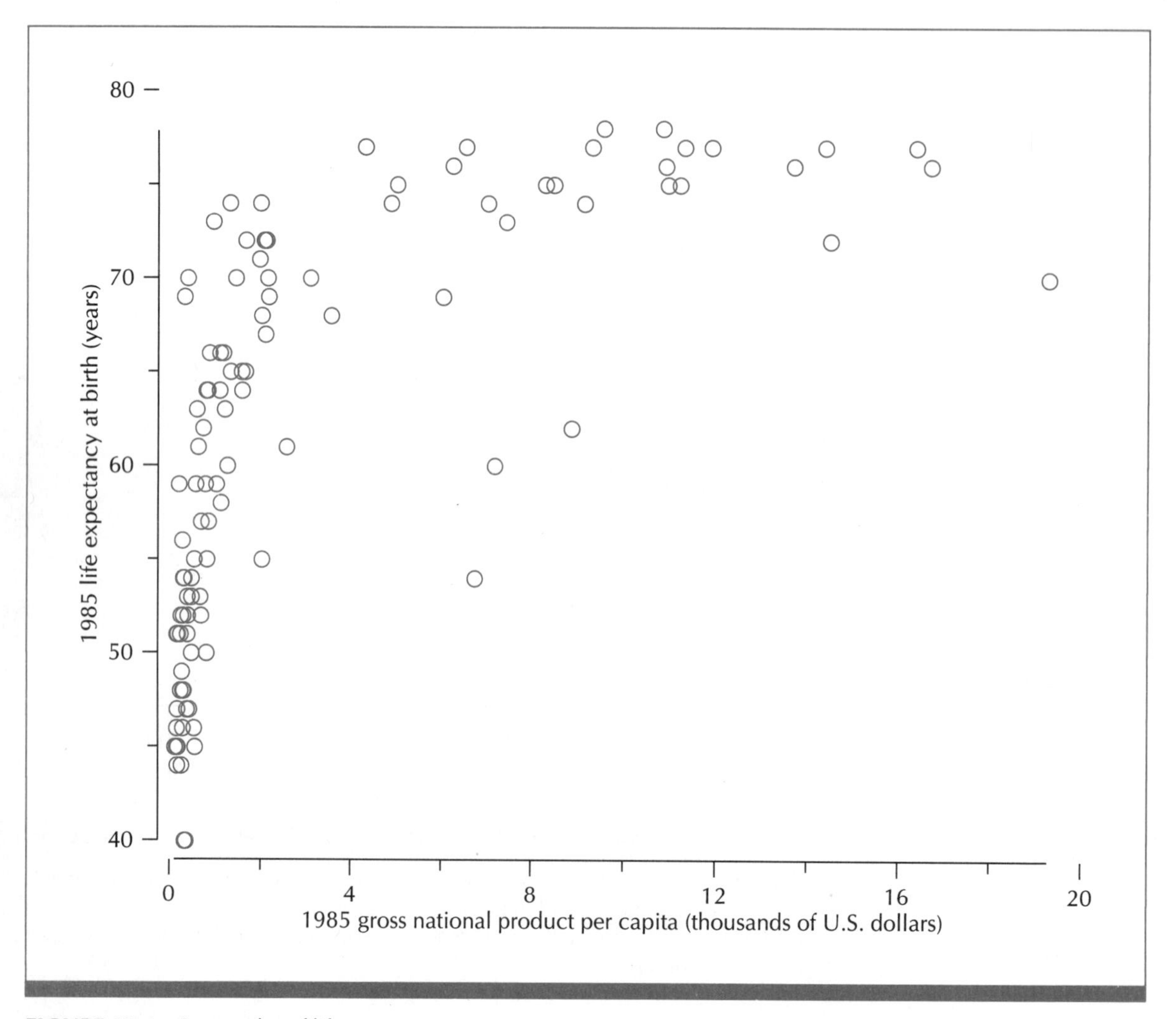

FIGURE 15-6 Scatterplot of life expectancy versus per capita gross national product in 1985 for 109 countries

is a single point by itself in the upper right-hand corner of Figure 15-7. It corresponds to Angola, with a primary school enrollment of 134% and a difference between male and female enrollments of 25%. We might call this point an *outlier:*

An **outlier** is an observation that is far from the other observations.

If we disregard Angola, we get the plot in Figure 15-8. The correlation coefficient for these remaining six points is .40. The large correlation coefficient (.90) for the points in Figure 15-7 results from the relative position of the single point corresponding to Angola. We are unwise to attach much signifi-

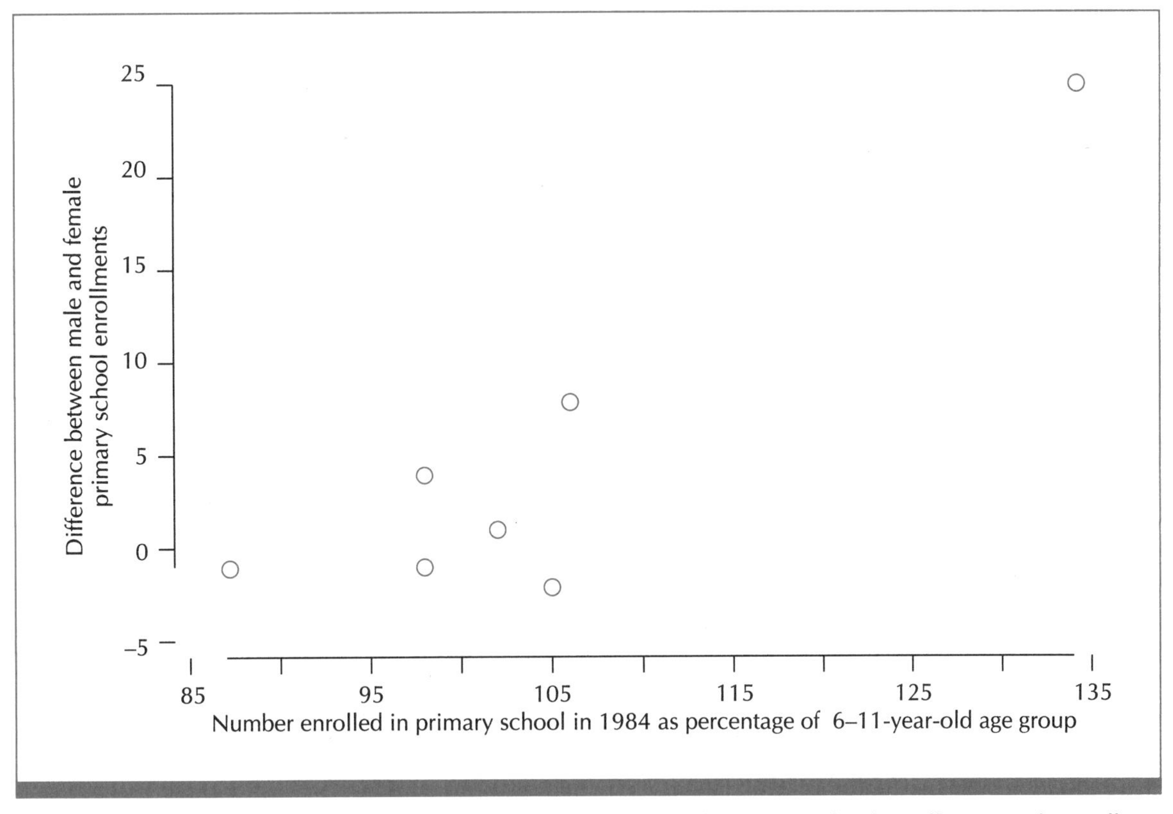

FIGURE 15-7 Scatterplot of the difference between male and female primary school enrollments and overall primary school enrollment in 1984 for seven countries in the nonmember economic category. Two nonmember countries are excluded because of missing values.

cance to a large correlation coefficient that results from the position of a single point.

One point or a few points can also pull a correlation coefficient closer to 0 than it would be if the point(s) were excluded. Figure 15-9 shows a scatterplot of infant mortality rate and per capita gross national product in 1985 for the 20 upper-middle-income countries with nonmissing values for both variables. The correlation coefficient is $-.05$, about as close to 0 as we might expect to see.

Examining Figure 15-9, we see a general trend of decreasing infant mortality rates with increasing per capita gross national product. There is a single striking exception—the point in the upper right-hand corner. This exception is Oman, with per capita gross national product of $6,730 and an infant mortality rate of 109. Removing Oman, we get the plot in Figure 15-10. The correlation coefficient for the remaining 19 points is $-.47$. This is more consistent with our impression from the plot.

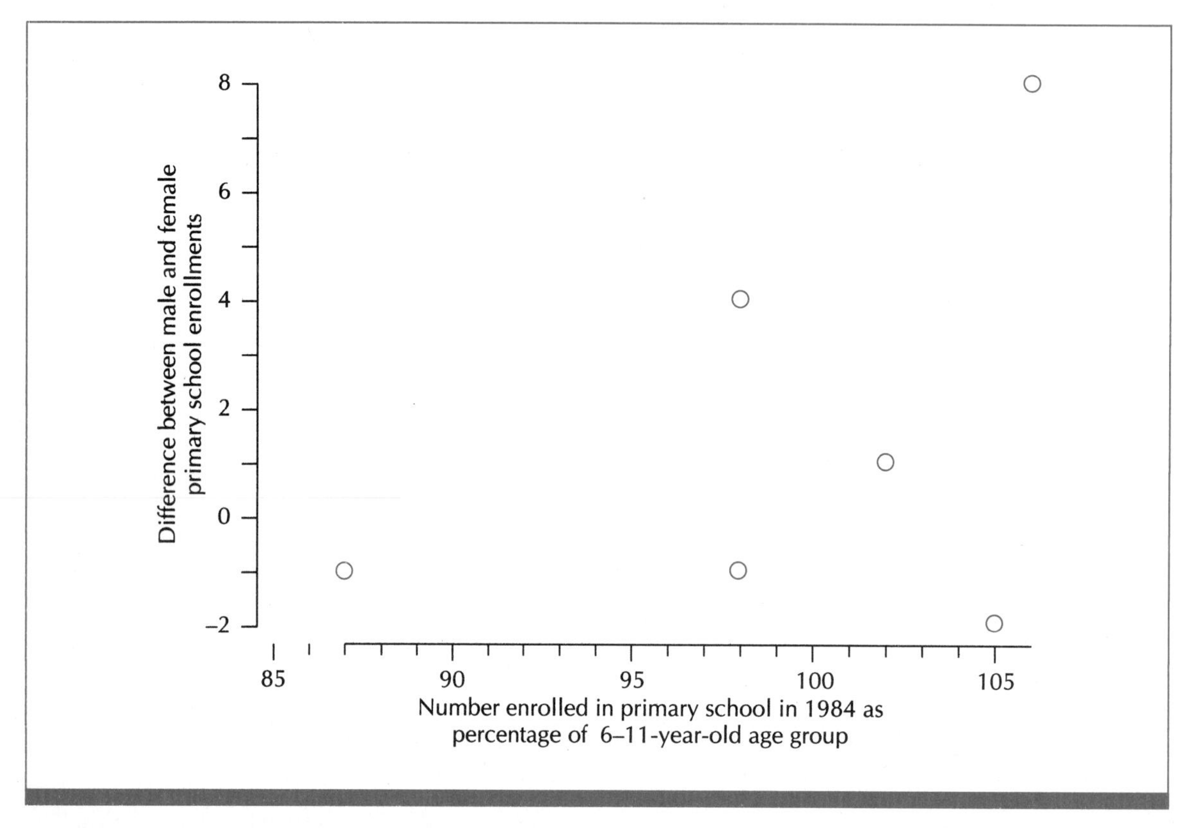

FIGURE 15-8 This scatterplot is the same as the one in Figure 15-7 except that the outlying point in the upper right-hand corner of Figure 15-7 has been excluded.

We can learn a lot in the process of finding out why an outlier is an outlier. Oman has a relatively high per capita gross national product, but a high infant mortality rate typical of the low-income countries. The high infant mortality rate makes Oman different from the other upper-middle-income countries. In fact, in the 1985 *World Development Report,* the World Bank classified Oman as a high-income oil exporter (World Bank, 1985). Recall from Part I that the high-income oil exporters are similar to the low-income countries for some indicators. Therefore, in discussing the relationship between infant mortality and gross national product for upper-middle-income countries, we might want to consider Oman as a special case.

We have to be careful with outliers. We should not exclude a case from an analysis just because it is different from the others. By judicious exclusion of cases, we may "see" characteristics in our data set that are not really there. This is, of course, *not* the purpose of data analysis.

Let's look now at how the difference between female and male life expectancy varies with overall life expectancy. A scatterplot of the difference be-

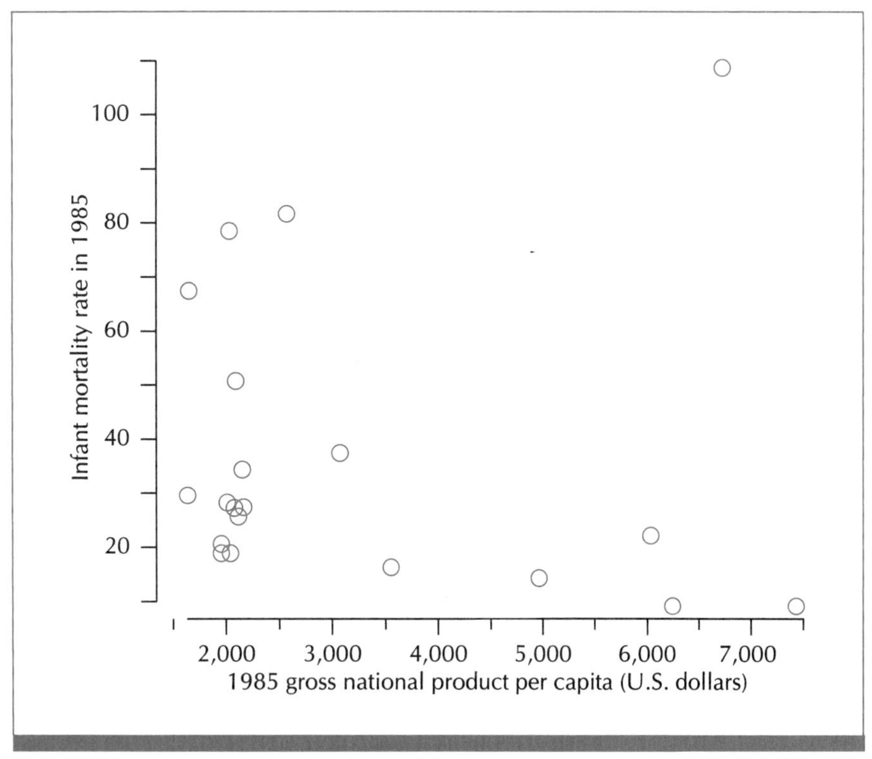

FIGURE 15-9 Scatterplot of infant mortality rate and per capita gross national product in 1985 for 20 upper-middle-income countries. The range of values for each variable is indicated by the line on the corresponding axis.

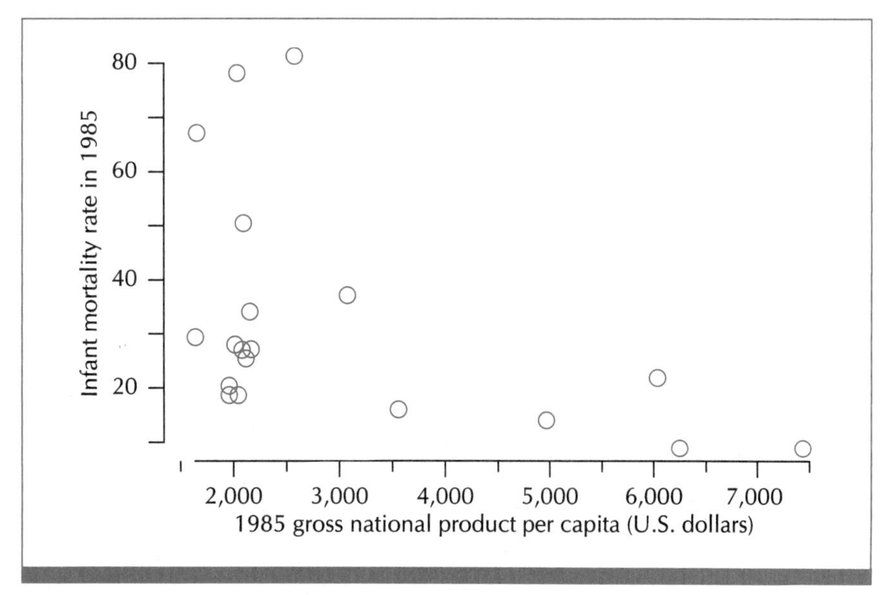

FIGURE 15-10 This scatterplot is the same as the one in Figure 15-9 except that the outlying point in the upper right-hand corner of Figure 15-9 is excluded.

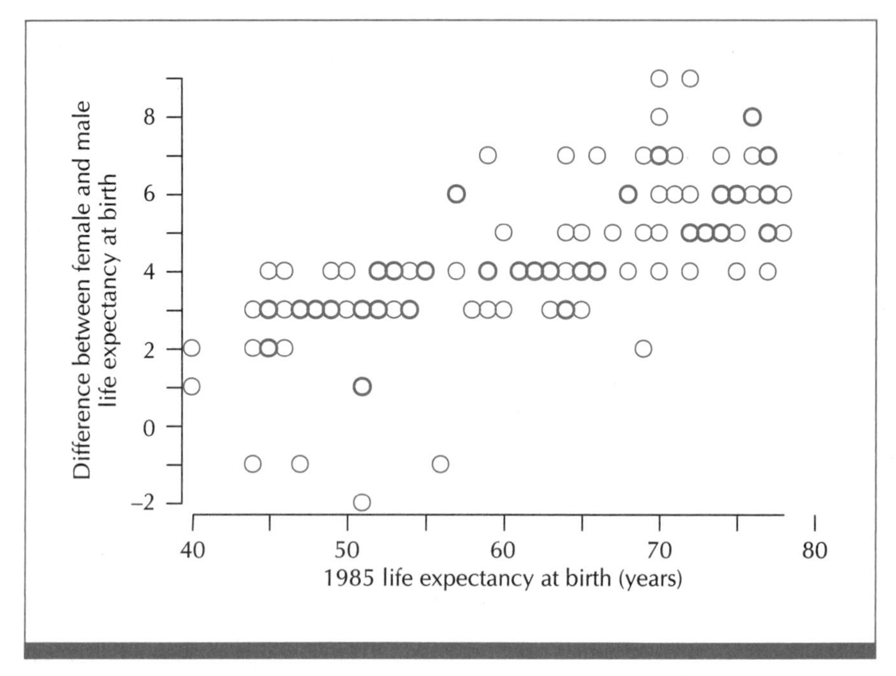

FIGURE 15-11 Scatterplot of the difference between female and male life expectancies and overall life expectancy at birth in 1985 for 125 countries

tween female and male life expectancies and overall life expectancy is shown in Figure 15-11 for 125 countries. There is a fairly strong increasing relationship, with a correlation coefficient of .70.

A scatterplot of the difference between female and male life expectancies and overall life expectancy is shown in Figure 15-12 for the 19 industrial market countries. We no longer see an increasing linear relationship; the correlation coefficient is −.05, very close to 0.

There is a suggestion of another type of relationship between the two variables in Figure 15-12. We see that the differences between female and male life expectancies are largest for countries with overall life expectancy of 76 years (for instance, United States life expectancy was 80 years for females, 72 years for males). The differences are smaller for overall life expectancies shorter than 76 years, as well as for overall life expectancies longer than 76 years. This plot gives us a suggestion of a *quadratic* relationship. Such a relationship is not indicated at all by the correlation coefficient, which measures only *linear* association. (Figure 15-12 might lead us to hope that as life expectancies continue to lengthen, the gap in expected life span between females and males will decrease. This is speculation, of course, because the data plotted in Figure 15-12 are not sufficient for drawing any such conclusion.)

In this section we have discussed the correlation coefficient as a descriptive statistic: a measure of linear association between two quantitative variables.

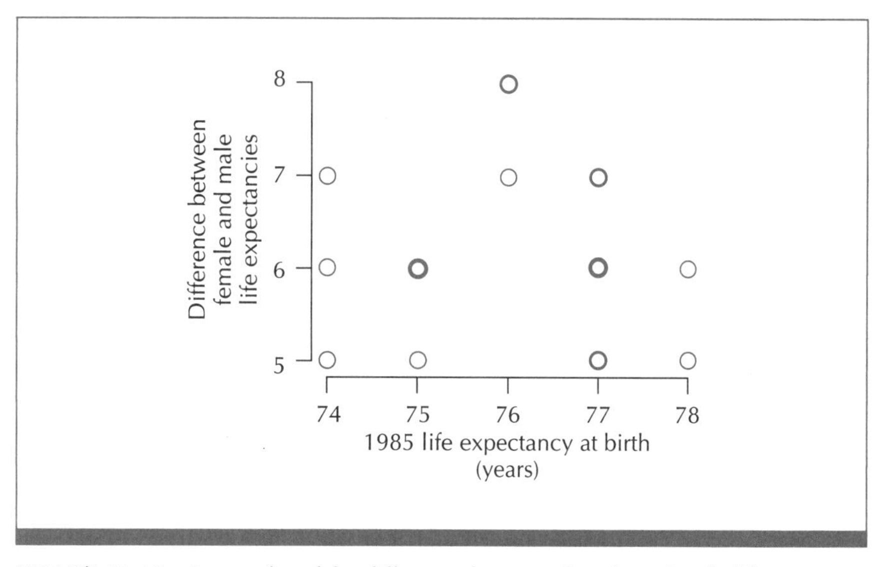

FIGURE 15-12 Scatterplot of the difference between female and male life expectancies and overall life expectancy in 1985 for 19 industrial market countries

When certain assumptions about the variables and the sampling process are met, we can test the null hypothesis that the linear correlation coefficient for two variables is 0. (We used World Bank data in this section to illustrate the correlation coefficient as a descriptive statistic measuring the extent of linear association between two variables. We cannot test hypotheses using the World Bank data set. We have information on the entire population of World Bank countries, rather than on a randomly selected sample of a population.) In Section 15-2, we discuss a parametric test that a linear correlation coefficient equals 0. We can apply this test when our sample meets assumptions described in that section.

15-2 A Parametric Test That a Linear Correlation Coefficient Equals Zero

Suppose we have n independent pairs of observations (X_i, Y_i). We let X_i denote the ith observation on a variable we call X. Y_i denotes the ith observation on a variable Y. We want to test the null hypothesis that the linear correlation coefficient between observations on X and Y is 0.

Let's assume that X_1 through X_n represent a random sample from a Gaussian distribution, and Y_1 through Y_n a random sample from another Gaussian distribution. [We must also assume that the pairs (X_i, Y_i) represent a random sample from what we call a bivariate normal, or bivariate Gaussian, distribution. See, for example, Brownlee (1965, Chapter 12).] With these assumptions, a zero linear correlation coefficient for X and Y is equivalent to independence of the X and Y variables.

Before outlining the test procedure, we'll consider an example.

EXAMPLE 15-1

Researchers designed this experiment to study two methods of measuring breast milk intake by infants. The first method was a deuterium (an isotope of hydrogen) dilution technique; the second method used test weighing. The researchers selected 14 babies, then used both techniques to measure the amount of milk consumed by each baby. The results are shown below (Devore, 1987, page 346, units of measurement not given; from "Evaluation of the Deuterium Dilution Technique Against the Test Weighing Procedure for the Determination of Breast Milk Intake," *Amer. J. Clinical Nutr.*, 1983, pages 996–1003).

Baby	Method 1	Method 2	Baby	Method 1	Method 2
1	1,509	1,498	8	1,198	1,129
2	1,418	1,254	9	1,479	1,342
3	1,561	1,336	10	1,281	1,124
4	1,556	1,565	11	1,414	1,468
5	2,169	2,000	12	1,954	1,604
6	1,760	1,318	13	2,174	1,722
7	1,098	1,410	14	2,058	1,518

A plot of the observations is shown in Figure 15-13. What does the plot suggest about the association between measurements made using the two methods? We will test the null hypothesis that the linear correlation coefficient for measurements using the two methods is 0, but first we will outline the test procedure.

The significance level approach to a parametric test that a linear correlation coefficient equals 0

1. The hypotheses are:

   Null hypothesis: The linear correlation coefficient for the two variables is 0. (The two variables are independent.)

   Alternative hypothesis: The linear correlation coefficient for the two variables is not 0. (The two variables are not independent.)

2. Let r denote the sample linear correlation coefficient calculated from the observations. Then the test statistic is

$$\text{Test statistic} = \sqrt{n-2}\,\frac{r}{\sqrt{1-r^2}}$$

3. Assume that X_1 through X_n represent a random sample from a Gaussian distribution. Assume that Y_1 through Y_n represent a random sample from another Gaussian distribution. [In addition, (X_i, Y_i) represent a random sample from a bivariate normal, or bivariate Gaussian, distribution.] Then

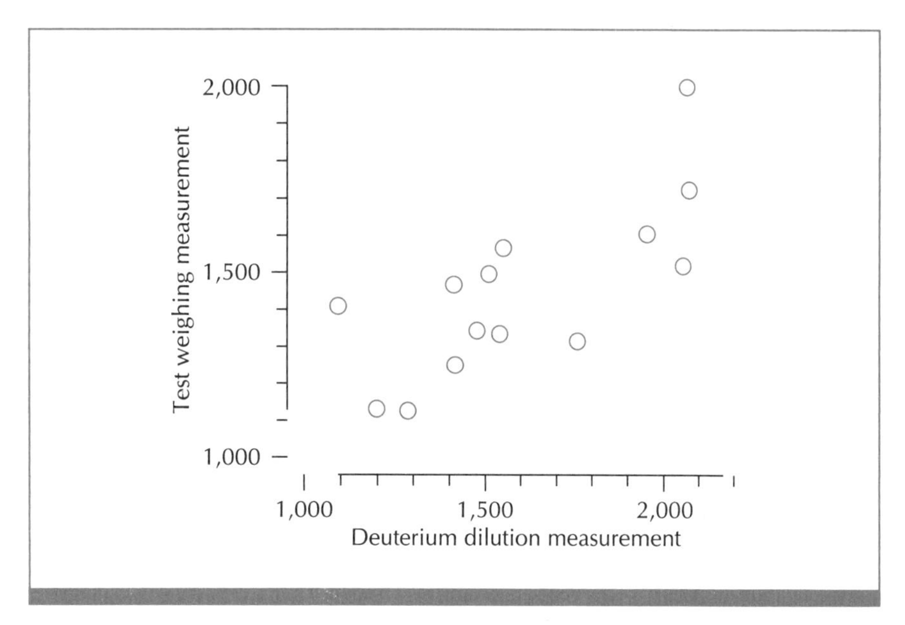

FIGURE 15-13 Scatterplot of measurement of milk consumed using the test weighing method versus measurement using the deuterium dilution technique, in Example 15-1

under the null hypothesis of independence of the X and Y variables, the test statistic has the t distribution with $n - 2$ degrees of freedom.

4. Select a significance level α.
5. Find the number c from Table C such that $P(T \leq c) = 1 - \alpha/2$, where T has the t distribution with $n - 2$ degrees of freedom. The acceptance region is $(-c, c)$. The rejection region includes $(-\infty, -c]$ and $[c, \infty)$.
6. The decision rule is:

   If $-c <$ test statistic $< c$, say the results are consistent with the null hypothesis that the linear correlation between X and Y is 0 (X and Y are independent).

   If test statistic $\leq -c$ or test statistic $\geq c$, say the results are inconsistent with the null hypothesis, suggesting that the linear correlation between X and Y is not 0 (X and Y are not independent).

7. Make observations that satisfy the assumptions in step 3. Calculate the test statistic in step 2. Use the decision rule in step 6 to decide whether the X and Y variables seem to be independent. Draw conclusions based on the experimental results.

EXAMPLE 15-1 *(continued)*

In Example 15-1, we have two measurements of breast milk intake for each of 14 babies. We want to test the hypotheses:

H_0: The linear correlation coefficient for measurements using the two methods is 0.

H_a: The linear correlation coefficient for measurements using the two methods is not 0.

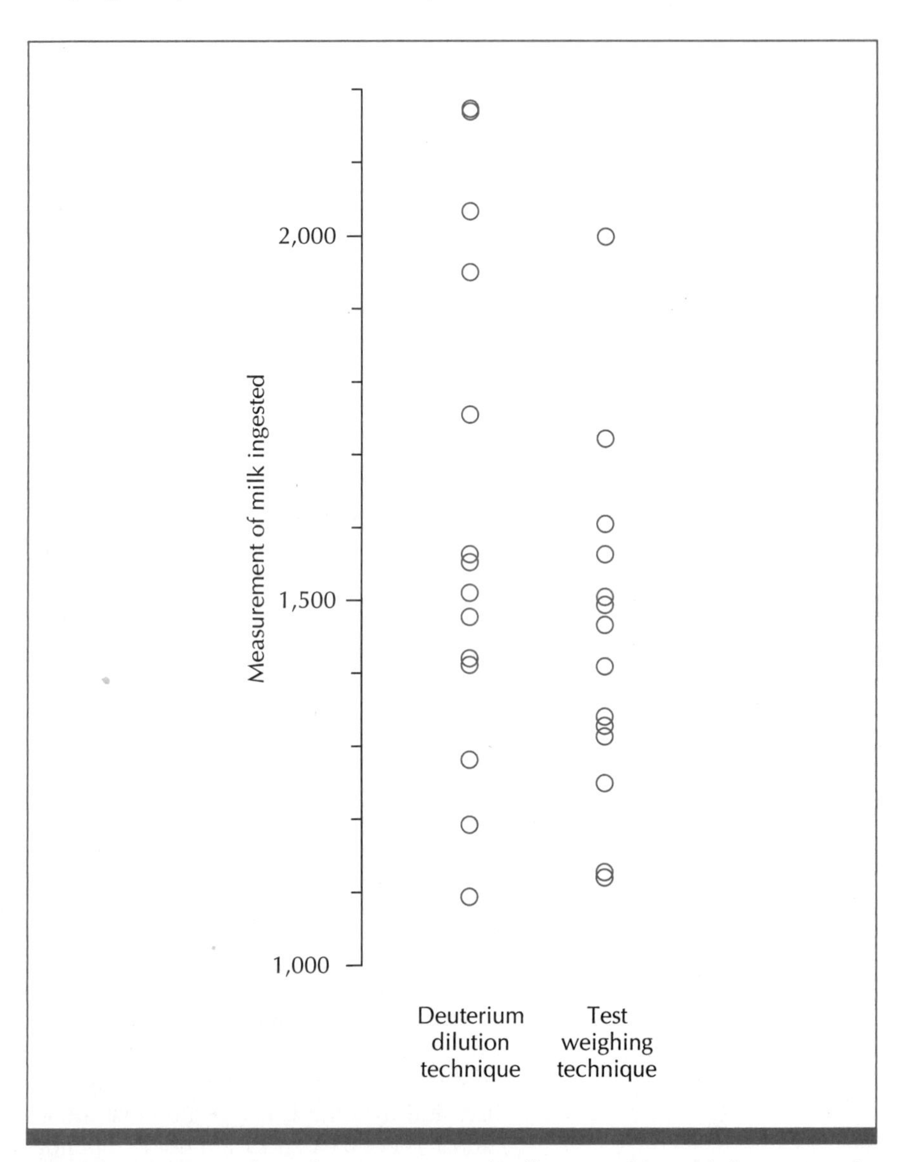

FIGURE 15-14 Dot plots of measurements of milk ingested by 14 babies, using the deuterium dilution technique and using the test weighing technique

We assume that the measurements using the first method represent a random sample from a Gaussian distribution, and the measurements using the second method represent a random sample from another Gaussian distribution. (We will not consider the assumption that the pairs of observations come from a bivariate normal distribution, because a discussion of bivariate normal distributions is beyond the scope of our presentation.) Figure 15-14 shows two dot plots of the measurements, one plot for each of the two methods. Both dot plots appear fairly symmetrical, consistent with the Gaussian assumption. Without further information, we cannot assess the independence assumption. What suggestions would you make about experimental design, in order to ensure independence and reduce the effects of extraneous sources of variation?

We will use significance level $\alpha = .01$. From Table C for the t distribution with $14 - 2 = 12$ degrees of freedom, we find $c = 3.055$. The acceptance region is $(-3.055, 3.055)$; the rejection region includes $(-\infty, -3.055]$ and $[3.055, \infty)$. The decision rule is:

If $-3.055 <$ test statistic < 3.055, say the results are consistent with the null hypothesis that the linear correlation coefficient for the two sets of measurements equals 0.

If test statistic ≤ -3.055 or test statistic ≥ 3.055, say the results are inconsistent with the null hypothesis, suggesting that the linear correlation coefficient for the two sets of measurements does not equal 0.

The sample correlation coefficient for the 14 pairs of measurements is .77. Since $n = 14$ and $r = .77$, the test statistic is

$$\text{Test statistic} = \sqrt{14 - 2}\,\frac{.77}{\sqrt{1 - (.77)^2}} = 4.2$$

The test statistic is in the rejection region, inconsistent with the null hypothesis that the linear correlation coefficient for the two sets of measurements is 0. From Figure 15-13, we see that there is a strong positive linear association between measurements made using the two techniques. Note that the relationship between the two sets of measurements is not exactly linear, because the plotted points do not all fall on a straight line.

In Section 15-3 we discuss a correlation coefficient and test for independence based on ranks.

15-3 Rank Correlation and a Nonparametric Test for Independence of Two Quantitative Variables

Suppose we have n pairs of observations (X_i, Y_i). As before, X_i denotes the ith observation on a variable X, and Y_i the ith observation on a variable Y. We would like a measure of association between the two variables X and Y, based on ranks. We would also like a nonparametric test of independence of the two variables.

We will calculate a measure of association called *Spearman's rank correlation coefficient,* or simply the *rank correlation coefficient.* Rank X_1 through X_n from smallest to largest. Also, rank Y_1 through Y_n from smallest to largest. In each case, assign tied values the average of the ranks they share. Then Spearman's rank correlation coefficient is the linear correlation coefficient based on the pairs of ranks for the X and Y variables. We often denote Spearman's rank correlation coefficient by r_s.

The **rank correlation coefficient,** also called **Spearman's rank correlation coefficient,** is the linear correlation coefficient based on the pairs of ranks for two variables that have been ranked separately. Suppose we have n pairs of observations (X_i, Y_i). Rank the X values from smallest to largest. Rank the Y values from smallest to largest. (Assign tied values the average of the ranks they share.) The rank correlation coefficient, denoted r_s, is the correlation coefficient of the n pairs of ranks.

Let d_i denote the difference between the ranks for variable X and variable Y, for the ith observation. Calculate a statistic D as the sum of the squared values of d_i:

$$D = \sum_{i=1}^{n} d_i^2$$

We can write a calculation formula for the rank correlation coefficient r_s in terms of D:

$$r_s = 1 - \frac{6D}{n^3 - n}$$

Consider the following example.

EXAMPLE 15-2

Investigators measured stiffness and thickness of six samples of a flame-retardant fabric. The results are shown below (Devore, 1987, page 492; from "Sensory and Physical Properties of Inherently Flame-Retardant Fabrics," *Textile Research,* 1984, pages 61–68).

Sample	1	2	3	4	5	6
Thickness (mm)	.28	.65	.32	.27	.81	.57
Stiffness (mg-cm)	7.98	24.52	12.47	6.92	24.11	35.71

A plot of the observations is shown in Figure 15-15. There seems to be a positive association between the two fabric characteristics. The calculations we need for Spearman's rank correlation coefficient are outlined in Table 15-3.

As we can see at the bottom of Table 15-3, Spearman's rank correlation coefficient is $r_s = .77$, indicating a positive association, as we saw in Figure 15-15. We would like to test the null hypothesis that thickness and stiffness are independent in this flame-retardant fabric. Let's outline the general test procedure.

The significance level approach to a nonparametric test that two quantitative variables are independent

1. The null hypothesis states that the two variables are independent. The alternative hypothesis states that the two variables are not independent.

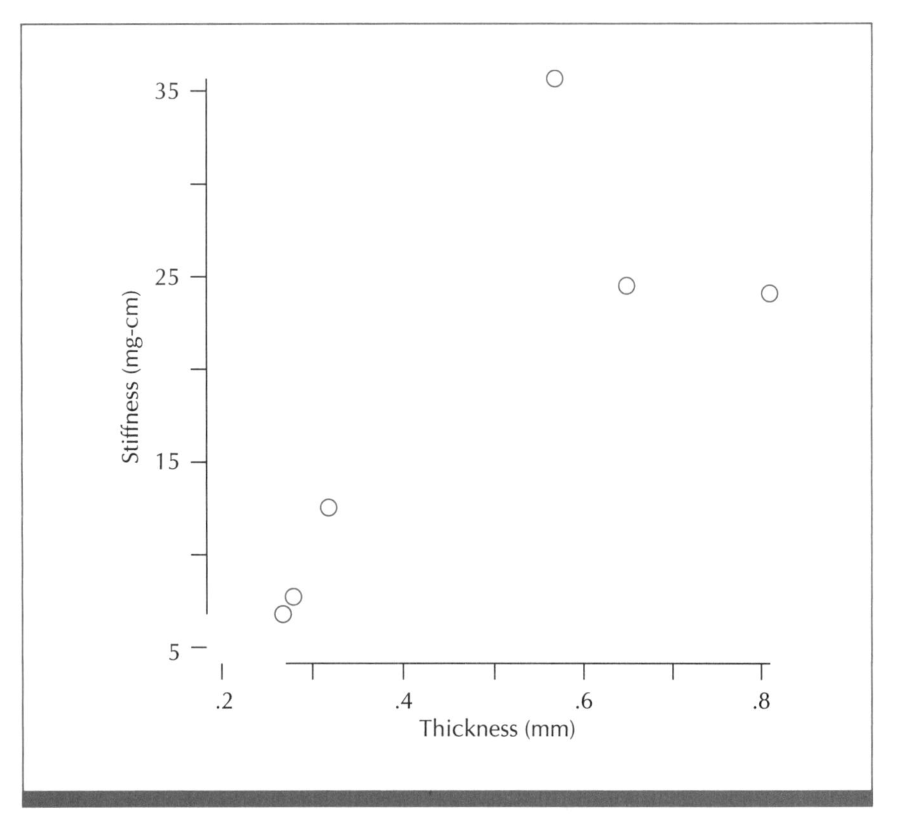

FIGURE 15-15 Scatterplot of stiffness versus thickness of fabric samples in Example 15-2

TABLE 15-3 Calculating the rank correlation coefficient in Example 15-2

Thickness	Rank	Stiffness	Rank	d_i	d_i^2
.28	2	7.98	2	0	0
.65	5	24.52	5	0	0
.32	3	12.47	3	0	0
.27	1	6.92	1	0	0
.81	6	24.11	4	2	4
.57	4	35.71	6	−2	4

$D = 0 + 0 + 0 + 0 + 4 + 4 = 8 \qquad r_s = 1 - \dfrac{6 \times 8}{6^3 - 6} = .77$

2. The test statistic is D, the sum of the squared differences between ranks in the n pairs.
3. We assume that X_1 through X_n denote independent observations of a quantitative variable X, and Y_1 through Y_n denote independent observations of a quantitative variable Y. Under the null hypothesis that the two variables are

independent, the test statistic D has a distribution we will call Spearman's distribution. Probabilities for this distribution are given in Table I for sample sizes from 2 to 11.

4. Select significance level α.
5. Let S denote a random variable having Spearman's distribution for sample size n. Suppose n is an integer from 2 to 11. Find the number c in Table I such that $P(S \leq c) = \alpha/2$. The acceptance region is $(c, (n^3 - n)/3 - c)$. The rejection region includes the intervals $[0, c]$ and $[(n^3 - n)/3 - c, (n^3 - n)/3]$.
6. The decision rule is:

   If $c < D < (n^3 - n)/3 - c$, say the results are consistent with the null hypothesis that the two variables are independent.

   If $D \leq c$ or $D \geq (n^3 - n)/3 - c$, say the results are inconsistent with the null hypothesis, suggesting that the two variables are not independent.
7. Collect pairs of observations that satisfy the assumptions in step 3. Calculate the test statistic D. Use the decision rule in step 6 to decide whether the variables seem to be independent. Draw conclusions based on the experimental results.

Suppose the sample size n is greater than 11. Then use the test statistic (Lehmann, 1975, Chapter 7)

$$\text{Test statistic} = \frac{D - \dfrac{n^3 - n}{6}}{\sqrt{\dfrac{n^2(n+1)^2(n-1)}{36}}}$$

Compare this test statistic with the standard Gaussian distribution. Values of the test statistic near 0 are consistent with the null hypothesis; extreme values (far from 0 in either the positive or negative direction) are inconsistent with the null hypothesis.

EXAMPLE 15-2 *(continued)*

In Example 15-2, the hypotheses are:

Null hypothesis: Stiffness and thickness are independent in this fabric.
Alternative hypothesis: Stiffness and thickness are not independent in this fabric.

We assume that the samples are representative of samples of this flame-retardant fabric, and that the measurements on the six samples are independent. We have no way of checking these assumptions without further information about the experiment.

We will use significance level $\alpha = .05$. Let S denote a random variable having Spearman's distribution for sample size 6. From Table I we see that $P(S \leq 4) = .017$, $P(S \leq 6) = .029$, and $P(S \leq 8) = .051$. Since .029 is closest to $\alpha/2 = .025$, we let $c = 6$. We also calculate $(n^3 - n)/3 = 70$. The acceptance region is (6, 64); the rejection region includes [0, 6] and [64, 70]. The decision rule is:

If $6 < D < 64$, say the results are consistent with the null hypothesis that thickness and stiffness are independent in this fabric.
If $D \leq 6$ or $D \geq 64$, say the results are inconsistent with the null hypothesis, suggesting that thickness and stiffness are not independent in this fabric.

From Table 15-3 we see that $D = 8$, which is in the acceptance region. The p-value $= 2P(D \leq 8) = .102$. Based on this p-value, we might say the results are borderline, especially since the sample size is small. Certainly Figure 15-15 suggests positive association between thickness and stiffness in this flame-retardant fabric.

For the sake of illustration, suppose we try the large-sample test in Example 15-2. We calculate the test statistic:

$$\text{Test statistic} = \frac{8 - \dfrac{6^3 - 6}{6}}{\sqrt{\dfrac{6^2(6 + 1)^2(6 - 1)}{36}}} = -1.72$$

Looking at Table B for the standard Gaussian distribution, we see our approximate p-value is .0854, somewhat smaller than the p-value of .102 we get with the exact distribution of D under the null hypothesis.

In Section 15-4, we discuss the method of least squares for fitting a straight line to a sample of pairs of observations. We also discuss parametric hypothesis testing for the straight-line model.

15-4 Simple Linear Regression and the Method of Least Squares

In many situations we want not only to measure the extent of linear association between two variables, but also to estimate the linear relationship between them. We would like to model one variable as a straight-line function of another, using the *method of least squares.*

Suppose we have n pairs of observations (X_i, Y_i) on two variables. We plot the observations and a linear association seems reasonable. Imagine drawing a straight line $Y = b_0 + b_1X$ through the cloud of plotted points. Here, b_0 denotes the intercept and b_1 the slope of the line. We will use the method of least squares to determine b_0 and b_1.

For any given value X_i of the first variable, we can use the straight-line model to predict the associated Y value. Let $\hat{Y}_i$ denote this predicted or estimated mean Y value. Then $\hat{Y}_i = b_0 + b_1X_i$. The difference $Y_i - \hat{Y}_i$ is a residual, measuring how far the estimated mean Y value is from the actual Y value for the ith observation.

A **residual** is the difference between a Y value and a predicted or estimated mean Y value, when a variable Y is modeled as a function of one or more other variables.

Using the method of least squares, we find the values of b_0 and b_1 that minimize the sum of the squared residual differences $(Y_i - \hat{Y}_i)^2$.

Suppose we have n pairs of observations (X_i, Y_i). Using the **method of least squares** to fit a straight line $Y = b_0 + b_1X$, we find constants b_0 and b_1 to minimize

$$\sum_{i=1}^{n} (Y_i - \hat{Y}_i)^2 = \sum_{i=1}^{n} (Y_i - (b_0 + b_1X_i))^2$$

Provided our observations include at least two distinct values of the X variable, we can find the least squares intercept b_0 and slope b_1. We can calculate these least squares values of b_0 and b_1 using the formulas

$$b_1 = \frac{\sum_{i=1}^{n} (X_i - \bar{X})(Y_i - \bar{Y})}{\sum_{i=1}^{n} (X_i - \bar{X})^2} \quad \text{and} \quad b_0 = \bar{Y} - b_1\bar{X}$$

Let's illustrate the method of least squares with an example.

EXAMPLE 15-3

For each of ten streets with bike lanes, investigators measured the distance between the center line and a cyclist in the bike lane. They used photography to determine the distance between a cyclist and a passing car on those same ten streets, recording all distances in feet. The results are shown below (Devore, 1982, pages 432–433; from "Effects of Bike Lanes on Driver and Bicyclist Behavior," *ASCE Transportation Eng. J.*, 1977, pages 243–256).

Street	1	2	3	4	5	6	7	8	9	10
Center line to cyclist (feet)	12.8	12.9	12.9	13.6	14.5	14.6	15.1	17.5	19.5	20.8
Car to cyclist (feet)	5.5	6.2	6.3	7.0	7.8	8.3	7.1	10.0	10.8	11.0

A plot of the observations is shown in Figure 15-16. Based on a visual inspection of this scatterplot, a linear relationship between the two variables seems reasonable.

Let the X variable be the distance from the center line to the cyclist. Let the Y variable be the distance from the car to the cyclist. We will use the method of least squares to model Y as a straight-line function of X. The necessary calculations are outlined in Table 15-4.

We see from the bottom of Table 15-4 that the least squares line is $Y = -2.18 + .66X$. The positive slope of .66 agrees with the positive association between the two variables that we see in Figure 15-16. The intercept of -2.18 has no physical meaning in this example: We cannot let the distance X from the center line to a cyclist in the bike lane be 0 feet, because the distance Y from the car to the cyclist cannot be -2.18 feet!

The least squares line for our example is plotted in Figure 15-17. The vertical distances from the points (X_i, Y_i) to the least squares line are indicated by dashed lines. These distances are the values of the residuals $Y_i - \hat{Y}_i$. The

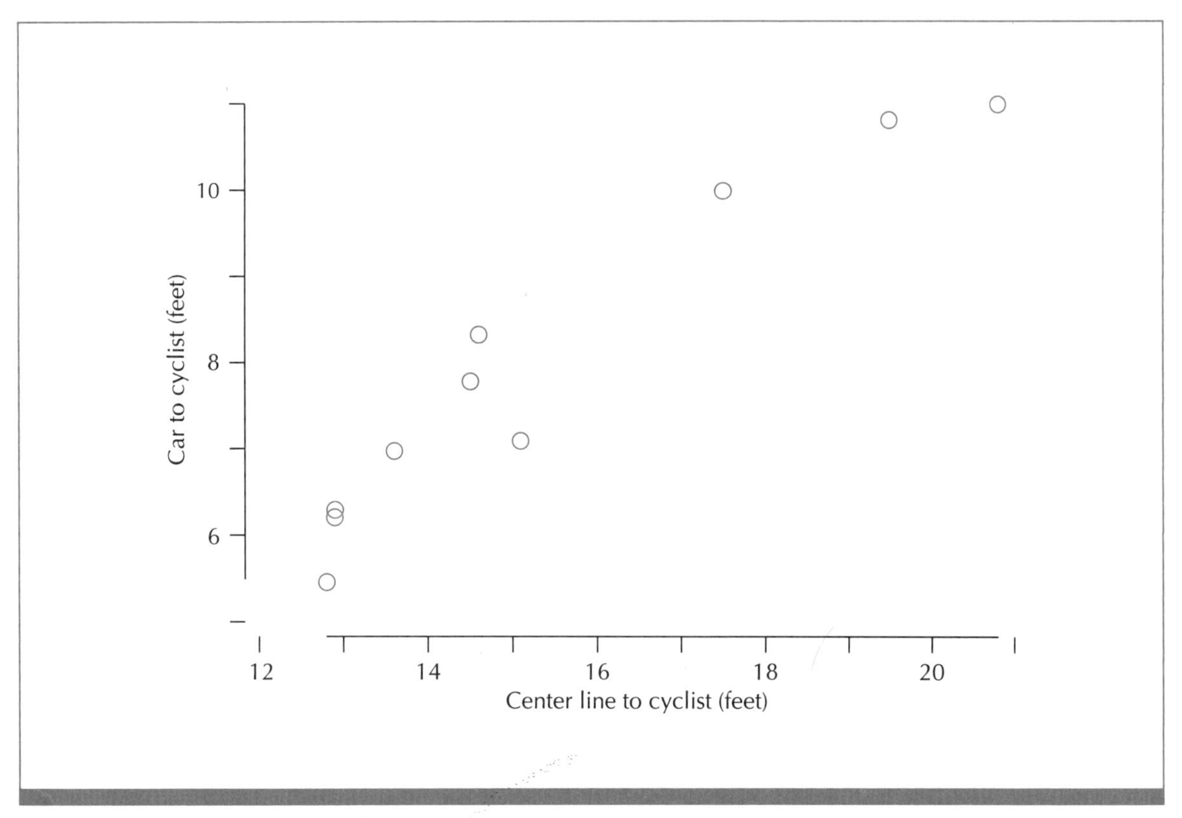

FIGURE 15-16 Scatterplot of the distance from car to cyclist and the distance from center line to cyclist in Example 15-3

TABLE 15-4 Calculations for finding the least squares intercept b_0 and slope b_1 for Example 15-3

X_i	Y_i	$X_i - \bar{X}$	$Y_i - \bar{Y}$	$(X_i - \bar{X})(Y_i - \bar{Y})$	$(X_i - \bar{X})^2$
12.8	5.5	−2.62	−2.5	6.550	6.8644
12.9	6.2	−2.52	−1.8	4.536	6.3504
12.9	6.3	−2.52	−1.7	4.284	6.3504
13.6	7.0	−1.82	−1.0	1.820	3.3124
14.5	7.8	−.92	−.2	.184	.8464
14.6	8.3	−.82	.3	−.246	.6724
15.1	7.1	−.32	−.9	.288	.1024
17.5	10.0	2.08	2.0	4.160	4.3264
19.5	10.8	4.08	2.8	11.424	16.6464
20.8	11.0	5.38	3.0	16.140	28.9444
$\bar{X} = 15.42$	$\bar{Y} = 8$			*Total:* 49.14	74.416

$b_1 = \dfrac{49.14}{74.416} = .66 \qquad b_0 = 8 - (.66)(15.42) = -2.18$

Least squares line: $Y = -2.18 + .66X$

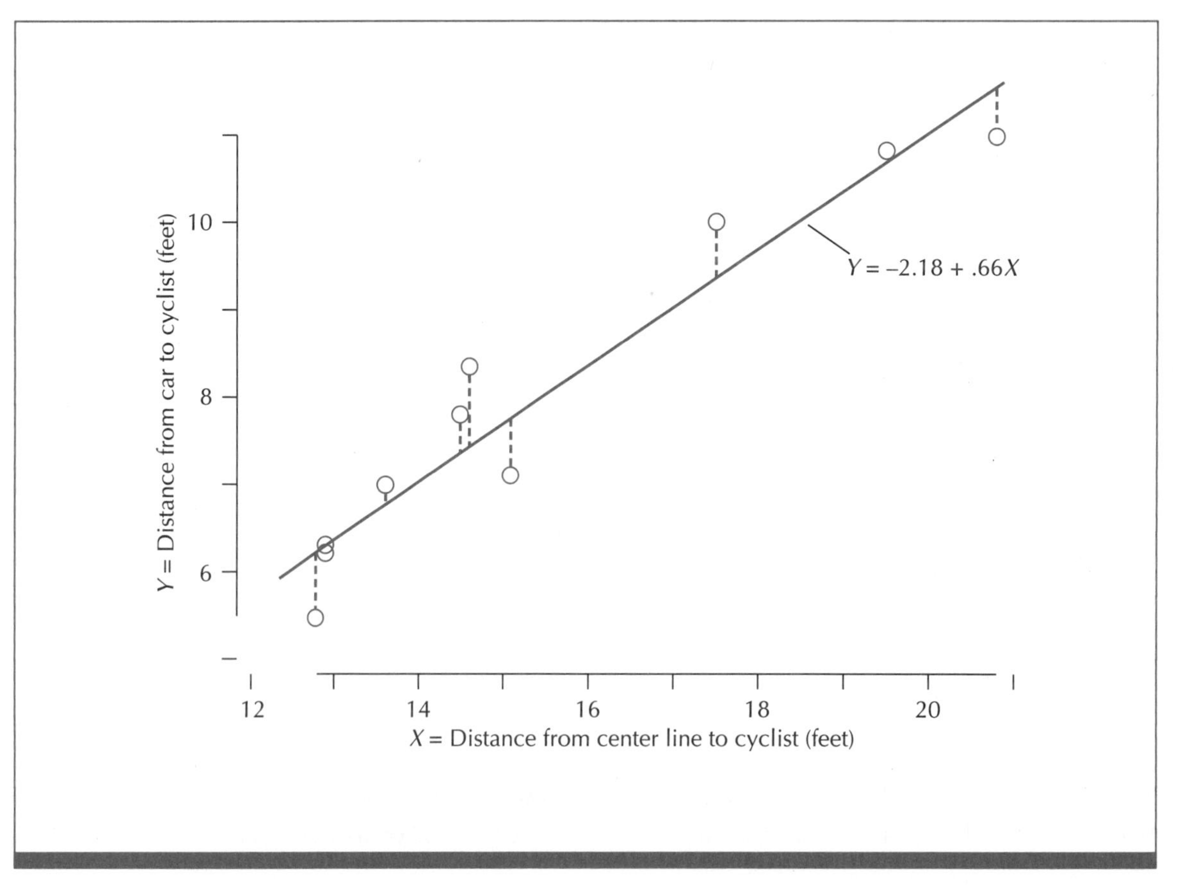

FIGURE 15-17 Scatterplot of distance from car to cyclist versus distance from center line to cyclist in Example 15-3. Also shown is the least squares line.

least squares line is best in the sense of minimizing the sum of the squares of these vertical distances, or residuals.

The slope b_1 in the equation $Y = b_0 + b_1X$ has units equal to the units of the Y variable, divided by the units of the X variable. The slope represents the change in the Y variable for each unit increase in the X variable.

The units of the intercept b_0 are the units of the Y variable. The intercept has a physical interpretation only if there are values of the X variable very close to 0 and if it is possible for the X variable to equal 0.

The method of least squares requires no assumptions. We need to make assumptions about the observations only if we want to test hypotheses about the straight-line model. Suppose we do want to test hypotheses about the straight-line model. We use the term *simple linear regression* to refer to the process of fitting a straight-line model by the method of least squares and testing hypotheses about the model.

Simple linear regression refers to fitting a straight-line model by the method of least squares and then assessing the model.

The *classical assumptions for simple linear regression* are these: Suppose we have n pairs of observations (X_i, Y_i). We observe values of the X variable with no error. Y_1 through Y_n are independent random variables, Y_i coming from a Gaussian distribution with mean equal to $\beta_0 + \beta_1 X_i$ and variance equal to σ^2.

The parameters (or unknown numbers) β_0 and β_1 are the intercept and slope, respectively, of the line describing the relationship between X and Y. The variance σ^2 describes the random variation of the Y values about that line.

We want to estimate the intercept β_0 and the slope β_1, as well as the variance σ^2. We also want to test the null hypothesis that $\beta_1 = 0$ and the null hypothesis that $\beta_0 = 0$.

We estimate β_0 and β_1 using the least squares estimates b_0 and b_1 given previously. We estimate σ^2 with the residual mean square s_r^2:

$$s_r^2 = \frac{1}{n-2}\sum_{i=1}^{n}(Y_i - \hat{Y}_i)^2 = \frac{1}{n-2}\sum_{i=1}^{n} e_i^2$$

where $e_i = Y_i - \hat{Y}_i = Y_i - (b_0 + b_1 X_i)$ is the ith residual, the difference between the observed and estimated Y values.

To test the hypotheses $H_0: \beta_1 = 0$ and $H_a: \beta_1 \neq 0$, we use the test statistic

$$\text{Test statistic(1)} = \frac{b_1}{\sqrt{\dfrac{s_r^2}{\sum_{i=1}^{n}(X_i - \bar{X})^2}}}$$

Under the null hypothesis that $\beta_1 = 0$, test statistic(1) has the t distribution with $n - 2$ degrees of freedom. Values of test statistic(1) far from 0 (in either the positive or negative direction) are inconsistent with the null hypothesis that the slope β_1 equals 0. Note that when we ask whether the slope β_1 equals 0, we implicitly assume that the intercept β_0 is in the model.

To test the hypotheses $H_0: \beta_0 = 0$ and $H_a: \beta_0 \neq 0$, we use the test statistic

$$\text{Test statistic(0)} = \frac{b_0}{\sqrt{s_r^2\left(\dfrac{1}{n} + \dfrac{\bar{X}^2}{\sum_{i=1}^{n}(X_i - \bar{X})^2}\right)}}$$

Under the null hypothesis that $\beta_0 = 0$, test statistic(0) has the t distribution with $n - 2$ degrees of freedom. Extreme values of test statistic(0), far from 0 in either the positive or negative direction, are inconsistent with the null hypothesis that the intercept β_0 equals 0. When we ask whether the intercept β_0 equals 0, we implicitly assume that the slope β_1 is in the model.

EXAMPLE 15-3 *(continued)*

Let's illustrate these ideas by continuing with Example 15-3. Some calculations we need are summarized in Table 15-5.

Before testing hypotheses, we should check our model assumptions. The straight-line relationship between X (distance from center line to cyclist) and

TABLE 15-5 Calculations for simple linear regression in Example 15-3

X	Y	$\hat{Y} = -2.18 + .66X$	$e = Y - \hat{Y}$	$e^2 = (Y - \hat{Y})^2$
12.8	5.5	6.27	−.77	.5929
12.9	6.2	6.33	−.13	.0169
12.9	6.3	6.33	−.03	.0009
13.6	7.0	6.80	.20	.0400
14.5	7.8	7.39	.41	.1681
14.6	8.3	7.46	.84	.7056
15.1	7.1	7.79	−.69	.4761
17.5	10.0	9.37	.63	.3969
19.5	10.8	10.69	.11	.0121
20.8	11.0	11.55	−.55	.3025
			Total:	2.7120

$$s_r^2 = \frac{2.7120}{10 - 2} = .339 \qquad \text{Degrees of freedom} = 10 - 2 = 8$$

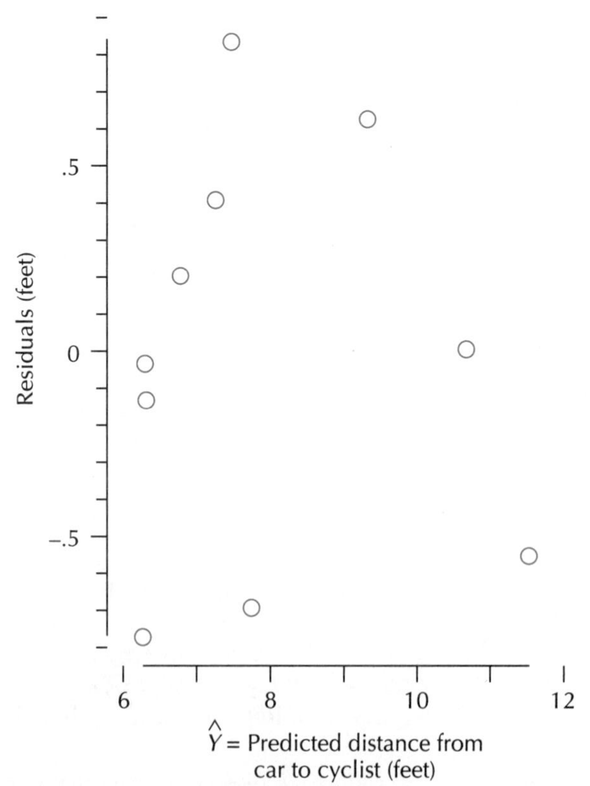

FIGURE 15-18 Scatterplot of residuals versus predicted Y values in Example 15-3

Y (distance from car to cyclist) seems reasonable from Figure 15-17. As another check, we can look at a plot of residuals versus predicted Y values, as in Figure 15-18. If the straight-line model with constant variation holds, the residuals should represent random variation or noise. A residual plot showing a pattern that does not look like random variation or noise suggests that the straight-line model may not be appropriate. We cannot see any particular pattern in Figure 15-18, so this residual plot gives us no reason to doubt the straight-line model.

We must assume that the X variable is measured without error; it seems reasonable that the investigators could measure the distance from the center line to the cyclist with minimal error. Figures 15-17 and 15-18 give us no basis to doubt the assumption that each Y_i has the same variance σ^2.

Also, we assume that the Y_i's are independent; we cannot assess this independence assumption without more information on how the experiment was conducted. What suggestions would you have for carrying out this experiment, in order to ensure independence and reduce the effects of extraneous sources of variation?

We must assume, in addition, that Y_i comes from a Gaussian distribution with mean $\beta_0 + \beta_1 X_i$ and variance σ^2 or, equivalently, that $Y_i - (\beta_0 + \beta_1 X_i)$ comes from a Gaussian distribution with mean 0 and variance σ^2. We use the residual $e_i = Y_i - \hat{Y}_i$ to estimate $Y_i - (\beta_0 + \beta_1 X_i)$. A dot plot of the residuals is shown in Figure 15-19. From this figure, we see no reason to doubt the Gaussian assumption.

Let's test the null hypothesis that the slope β_1 equals 0. Using calculations outlined in Tables 15-4 and 15-5, we see that

$$\text{Test statistic}(1) = \frac{.66}{\sqrt{\dfrac{.339}{74.416}}} = 9.8$$

Referring to Table C for the t distribution with 8 degrees of freedom, we see that our p-value is less than .01. The results are inconsistent with the null hypothesis that the slope β_1 equals 0.

Now let's test the null hypothesis that the intercept β_0 equals 0. We see that

$$\text{Test statistic}(0) = \frac{-2.18}{\sqrt{.339\left(\dfrac{1}{10} + \dfrac{(15.42)^2}{74.416}\right)}} = -2.1$$

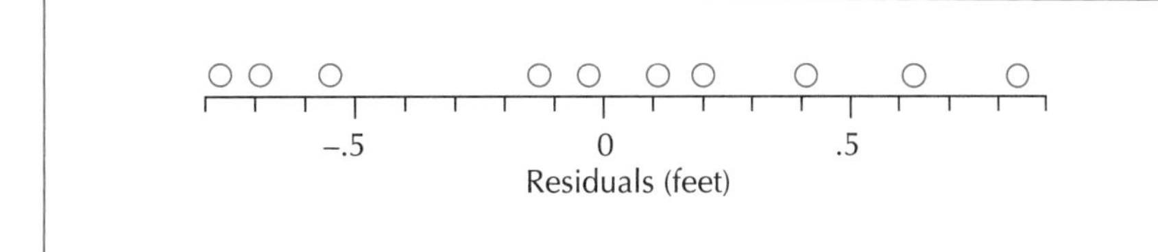

FIGURE 15-19 Dot plot of the residuals in Example 15-3

Again referring to Table C for the t distribution with 8 degrees of freedom, we see that our p-value is between .05 and .10, which is borderline. We will suppose that β_0 is not 0, that we do need a nonzero intercept in our model.

We usually calculate a statistic called R^2 in simple linear regression. R^2 is the square of the simple linear correlation coefficient r for the X and Y variables, so $R^2 = r^2$. R^2 is the proportion of the variation in the Y variable explained by the straight-line model.

> In simple linear regression, the square of the linear correlation coefficient, denoted $\boldsymbol{R^2}$, is the proportion of the variation in the response variable accounted for, or explained, by the straight-line model.

In Example 15-3, the correlation coefficient r for the two variables is .96. Therefore, $R^2 = (.96)^2 = .92$. We say about 92% of the variation in distances between cars and cyclists is explained by the linear relationship between that variable and the distance from center line to cyclist. This is a fairly large value of R^2. From our analysis, including the scatterplot of the data values and the residual plots, it seems that a straight-line model is very reasonable in Example 15-3.

In Section 15-5, we discuss the relation between correlation and simple linear regression.

15-5 Correlation and Simple Linear Regression

Suppose once again that we have n pairs of observations (X_i, Y_i). The linear correlation coefficient r measures the extent of linear association between the X and Y variables. It measures how closely the plotted points cluster about a line. We call this line the ***standard deviation line*** (Freedman, Pisani, and Purves, 1978, page 122).

The standard deviation line is the line most of us would draw freehand through a cloud of plotted points. It passes through the point $(\bar{X}, \bar{Y})$ corresponding to the sample means for the two variables. The slope of the standard deviation line is

$$\text{Slope of standard deviation line} = \begin{cases} \dfrac{\text{SD}_y}{\text{SD}_x} & \text{if the association is positive} \\[2ex] -\dfrac{\text{SD}_y}{\text{SD}_x} & \text{if the association is negative} \end{cases}$$

where SD_y and SD_x denote the sample standard deviations of the Y and X variables, respectively.

Figure 15-20 shows a plot of the observations in Example 15-1. (Recall that investigators used two methods to determine the amount of breast milk ingested by each of 14 babies.) The standard deviation line is also shown. This line goes through the point $(\bar{X}, \bar{Y}) = (1{,}616.4, 1{,}449.1)$. The slope of the stan-

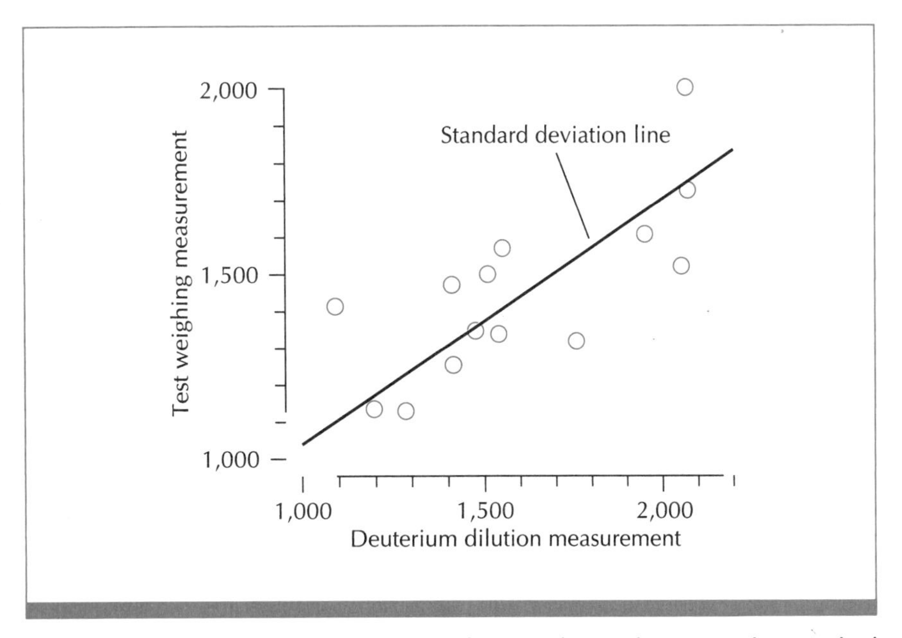

FIGURE 15-20 Plot of measurement of milk ingested using the test weighing method versus measurement using the deuterium dilution technique. The standard deviation line is also shown.

dard deviation line is positive because the association between the two variables is positive:

$$\text{Slope of standard deviation line in Example 15-1: } \frac{SD_y}{SD_x} = \frac{234}{353} = .66$$

How does the standard deviation line compare with the least squares line? We can show that another (equivalent) formula for the slope of the least squares line is

$$\text{Slope of the least squares line} = b_1 = r\frac{SD_y}{SD_x}$$

The slope of the least squares line equals the absolute value of the linear correlation coefficient r times the slope of the standard deviation line. Since the absolute value of r is in the range from 0 to 1, we see that the least squares line is less steep than the standard deviation line.

In Example 15-1, we found the linear correlation coefficient to be $r = .77$. Therefore, the slope of the least squares line is

$$\text{Slope of least squares line in Example 15-1} = (.77)(.66) = .51$$

The least squares line and the standard deviation line are both plotted in Figure 15-21. Note that both lines pass through the point $(\bar{X}, \bar{Y})$. The least squares line is indeed less steep than the standard deviation line. For a large value of X (greater than $\bar{X}$), the corresponding value of Y predicted by the least squares

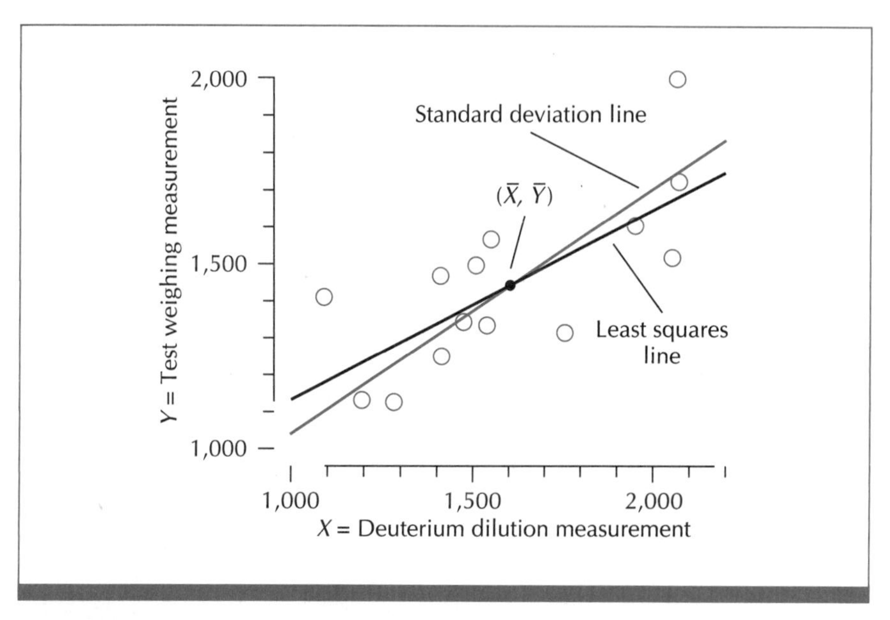

FIGURE 15-21 Scatterplot of the observations in Example 15-1. The least squares line and the standard deviation line are shown.

line is less than that predicted by the standard deviation line. For a small value of X (less than $\bar{X}$), the value of Y predicted by the least squares line is greater than that predicted by the standard deviation line.

Let's look at another example.

EXAMPLE 15-4 Foresters recorded two characteristics of 20 stands of pine trees (Myers, 1986, page 68; from Burkhart et al., 1972). One characteristic was the number of pine trees per acre; the other was the average diameter of pine trees 4.5 feet above the ground (units not given). The values of the two variables are listed in Example 15-5, in Section 15-6. Some descriptive statistics are shown in Table 15-6.

A plot of the observations is given in Figure 15-22. We see that there is a negative relationship between number of pine trees per acre and the average diameter of the trees. The standard deviation line and the least squares line are also shown in the figure, the standard deviation line steeper than the least squares line. For a value of X (number of pine trees per acre) greater than $\bar{X}$, the corresponding value of Y (average diameter) predicted by the least squares line is greater than that predicted by the standard deviation line. For a value of X less than $\bar{X}$, the value of Y predicted by the least squares line is less than that predicted by the standard deviation line.

We use the least squares line to estimate the average value of the Y variable corresponding to a particular value of the X variable. This estimated Y value is generally less extreme than what we might expect by a freehand sketch of a line through the plotted points, because our freehand sketches tend

TABLE 15-6 Descriptive statistics for Example 15-4

X = Number of pine trees per acre Y = Average diameter 4.5 feet above the ground

$\bar{X} = 671.5$ $SD_x = 136.9$ $\bar{Y} = 6.265$ $SD_y = .739$ $r = -.252$

Slope of standard deviation line $= -\dfrac{.739}{136.9} = -.00540$

Slope of least squares line $= (-.252)(.00540) = -.00136$

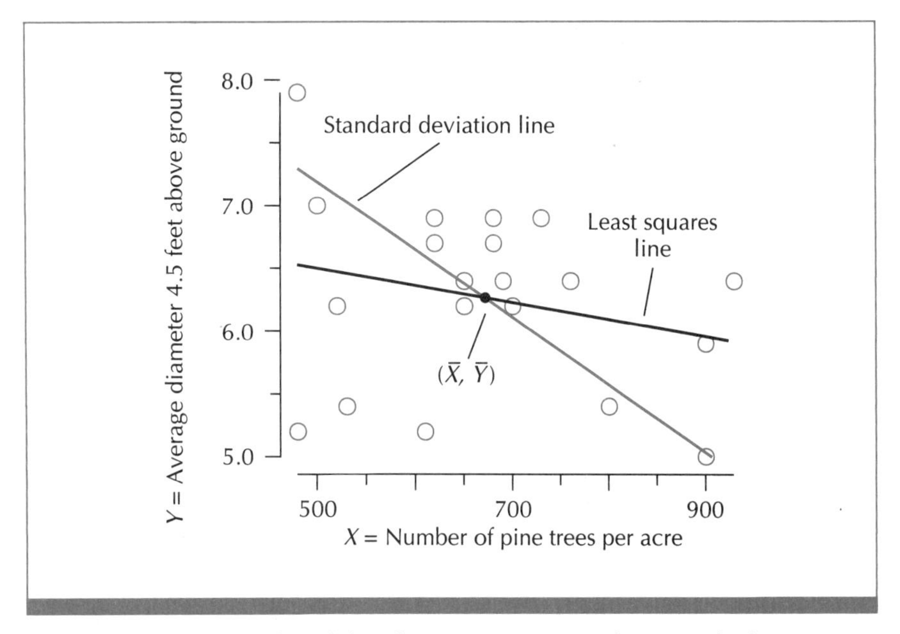

FIGURE 15-22 Scatterplot of the observations in Example 15-4. The least squares line and the standard deviation line are shown.

to be closer to the standard deviation line. The geneticist Sir Francis Galton (1822–1911) noticed this when he studied the sizes of seeds and their offspring and when he studied the heights of fathers and sons. Extremely tall fathers, for instance, had sons who were shorter than they, on average; extremely short fathers had sons who were taller than they, on average. We could turn this around and say extremely tall sons had fathers who were shorter than they, on average; extremely short sons had fathers who were taller than they, on average. Galton called this "regression towards mediocrity" or ***regression toward the mean.*** The term is unfortunately ambiguous; all it means is that the least squares line is less steep than the standard deviation line, as we have noted. It is from Galton that we get the term ***regression,*** as in ***simple linear regression.*** (For a discussion of regression toward the mean or the regression fallacy, see Freedman, Pisani, and Purves, 1978, pages 158–162.)

A final comment on the relationship between correlation and simple linear regression: The test of the null hypothesis that the slope is 0 in simple linear regression (Section 15-4) is equivalent to the test that the linear correlation coefficient is 0 (Section 15-2).

In Section 15-6, we present a very brief introduction to multiple regression.

15-6 A Brief Introduction to Multiple Regression

Suppose we record values of variables X_1 through X_k and Y for each of n observations. We might denote the values for the ith observation by $(X_{1i}, X_{2i}, \ldots, X_{ki}, Y_i)$. We want to model the Y variable as a function of the variables X_1 through X_k, say:

$$Y = b_0 + b_1X_1 + \cdots + b_kX_k$$

We call this a *linear model;* the model is linear in the constants b_0, b_1 through b_k.

A **linear model** is a model that is linear in the parameters, the unknown constants in the model.

We call the process of fitting and assessing such a model *multiple regression.*

By **multiple regression** we mean the process of modeling a quantitative variable as a function of several other variables, and assessing the model. We consider only linear models.

Using such a model, we can estimate the value of the Y variable for any set of values of X_1 through X_k. We use the notation $\hat{Y}_i$ to denote the estimated, or predicted, value of Y for the ith observation:

$$\hat{Y}_i = b_0 + b_1X_{1i} + \cdots + b_kX_{ki}$$

We let $e_i = Y_i - \hat{Y}_i$ denote the ith residual, the difference between Y and the estimated or predicted value of Y for the ith observation.

Using the *method of least squares* we find the values of the constants b_0 through b_k that minimize the sum of the squared residuals:

$$\sum_{i=1}^{n} e_i^2 = \sum_{i=1}^{n} (Y_i - \hat{Y}_i)^2 = \sum_{i=1}^{n} (Y_i - (b_0 + b_1X_{1i} + \cdots + b_kX_{ki}))^2$$

We can find unique values of b_0, b_1 through b_k to minimize the sum of squared residuals, as long as there are at least $k + 1$ distinct sets of values of X_1 through X_k.

We do not need to make any assumptions to fit a linear model using the method of least squares. However, if we want to test hypotheses about the model, we do need to make some assumptions.

The *classical assumptions for multiple regression* are these: The variables X_1 through X_k are measured without error. Y_1 through Y_n represent independent observations from Gaussian distributions, Y_i from a Gaussian distribution with mean $\beta_0 + \beta_1 X_{1i} + \cdots + \beta_k X_{ki}$ and variance σ^2. We use the method of least squares to find the least squares estimates b_0, b_1 through b_k of β_0, β_1 through β_k, respectively.

If these assumptions are met, we can test the hypotheses H_0: $\beta_j = 0$ and H_a: $\beta_j \neq 0$, where β_j is one of the parameters in the model. When we ask whether the parameter β_j equals 0, we implicitly assume that the other parameters are in the model. The test statistic is

$$\text{Test statistic}(j) = \frac{b_j}{\text{SE}(b_j)}$$

where b_j is the least squares estimate of β_j and $\text{SE}(b_j)$ is the standard error, or estimated standard deviation, of b_j. If the null hypothesis is true, test statistic(j) has the t distribution with $n - p$ degrees of freedom, where p is the number of parameters (unknown β_i's) in the model. (We have $p = k + 1$ in the model above.)

The calculations for multiple linear regression are so extensive that we use a computer to perform them.

Let's consider an example.

EXAMPLE 15-5

Foresters studied 20 stands of pine trees. For each stand, they recorded the age of the stand (units not given), the average height in feet of dominant trees, the number of pine trees per acre, and the average diameter 4.5 feet above the ground (units not given). The results are shown below (Myers, 1986, page 68; from Burkhart et al., 1972).

Stand	Age	Height	Number	Average diameter
1	19	51.5	500	7.0
2	14	41.3	900	5.0
3	11	36.7	650	6.2
4	13	32.2	480	5.2
5	13	39.0	520	6.2
6	12	29.8	610	5.2
7	18	51.2	700	6.2
8	14	46.8	760	6.4
9	20	61.8	930	6.4
10	17	55.8	690	6.4
11	13	37.3	800	5.4
12	21	54.2	650	6.4
13	11	32.5	530	5.4
14	19	56.3	680	6.7
15	17	52.8	620	6.7
16	15	47.0	900	5.9
17	16	53.0	620	6.9
18	16	50.3	730	6.9
19	14	50.5	680	6.9
20	22	57.7	480	7.9

The foresters wanted to model Y (average diameter) as a function of one or more of the variables X_1 (age), X_2 (height), X_3 (number), X_4 (age × number), and X_5 (height/number).

Table 15-7 shows the linear correlation coefficient for each pair of the six variables Y, X_1, X_2, X_3, X_4, and X_5. We see that the largest correlation coefficient (in absolute value) for average diameter Y with any other variable is the correlation coefficient of .840 for Y and X_5 = height/number.

We will go through the steps of a ***backward regression analysis*** for this problem. We start with a model that includes all the variables that we think might be important for estimating Y. If any parameter (other than the intercept β_0) seems to be 0, we will drop from the model the one with the largest p-value. Then we will fit a new model excluding the dropped parameter and its corresponding variable. We continue dropping one variable at a time and refitting until all the parameters (other than β_0) appear to be different from 0.

First let's consider the model

$$\text{Expected value of } Y = \beta_0 + \beta_1X_1 + \beta_2X_2 + \beta_3X_3 + \beta_4X_4 + \beta_5X_5$$

Results of a multiple regression analysis using the Student Edition of Minitab are shown in Table 15-8.

TABLE 15-7 The linear correlation coefficient for each pair of variables in Example 15-5

	Y Average diameter	X_1 Age	X_2 Height	X_3 Number	X_4 Age × Number	X_5 Height/ Number
Y	1.000					
X_1	.675	1.000				
X_2	.773	.876	1.000			
X_3	−.252	.016	.229	1.000		
X_4	.244	.678	.755	.732	1.000	
X_5	.840	.735	.634	−.579	.056	1.000

TABLE 15-8 Results of the first multiple regression analysis for Example 15-5

Parameter	Variable	Parameter estimate	Standard error	Test statistic	p-value
β_1	Age	.0526	.1683	.31	.759
β_2	Height	.08246	.04035	2.04	.060
β_3	Number	.003224	.002532	1.27	.224
β_4	Age × Number	−.0002817	.0002300	−1.22	.241
β_5	Height/Number	16.03	27.89	.57	.575
β_0	Intercept	1.233	1.619	.76	.459

p = Number of parameters = 6 $\quad$ n = Sample size = 20

Degrees of freedom for tests of hypotheses = $n - p$ = 14

TABLE 15-9 Results of the second multiple regression analysis for Example 15-5

Parameter	Variable	Parameter estimate	Standard error	Test statistic	p-value
β_2	Height	.07438	.03004	2.48	.026
β_3	Number	.002795	.002065	1.35	.196
β_4	Age × Number	−.0002131	.00006746	−3.16	.006
β_5	Height/Number	22.93	16.53	1.39	.186
β_0	Intercept	1.507	1.321	1.14	.272

p = Number of parameters = 5 $\quad$ n = Sample size = 20

Degrees of freedom for tests of hypotheses = $n - p = 15$

We see that the least squares estimated model is

$$\hat{Y} = 1.23 + .0526X_1 + .0825X_2 + .00322X_3 - .000282X_4 + 16.0X_5$$

where the parameter estimates are shown to three significant figures. The test statistic for a parameter tests the null hypothesis that the parameter is 0, when all the other parameters are in the model. The largest p-value, .759, is consistent with the null hypothesis that β_1 equals 0. That is, if all the other variables are in the model, it looks like we do not need to include X_1 (age).

We drop X_1 and consider the model:

$$\text{Expected value of } Y = \beta_0 + \beta_2X_2 + \beta_3X_3 + \beta_4X_4 + \beta_5X_5$$

Results of the multiple regression analysis are shown in Table 15-9.

The least squares estimated model is

$$\hat{Y} = 1.51 + .0744X_2 + .00280X_3 - .000213X_4 + 22.9X_5$$

The largest p-value (ignoring the intercept) is .196, consistent with the null hypothesis that β_3 equals 0. That is, if the other parameters in Table 15-9 are included in the model, it looks like we need not include X_3 (number of pine trees per acre).

We drop X_3 and consider the model

$$\text{Expected value of } Y = \beta_0 + \beta_2X_2 + \beta_4X_4 + \beta_5X_5$$

Results of the multiple regression analysis are given in Table 15-10.

The least squares estimated model is

$$\hat{Y} = 3.24 + .0974X_2 - .000169X_4 + 3.47X_5$$

The largest p-value, .684, is consistent with the null hypothesis that β_5 equals 0. If the other parameters in Table 15-10 are included in the model, it looks like we do not need to include X_5 (height/number). We drop X_5, even though it had the highest linear correlation coefficient with Y in Table 15-7.

For our final analysis, we drop X_5 and consider the model

$$\text{Expected value of } Y = \beta_0 + \beta_2X_2 + \beta_4X_4$$

Table 15-11 shows the results of the multiple regression analysis.

TABLE 15-10 Results of the third multiple regression analysis for Example 15-5

Parameter	Variable	Parameter estimate	Standard error	Test statistic	*p*-value
β_2	Height	.09741	.02540	3.84	.001
β_4	Age × Number	−.0001689	.00006052	−2.79	.013
β_5	Height/Number	3.467	8.374	.41	.684
β_0	Intercept	3.2357	.3467	9.33	.000

p = Number of parameters = 4 n = Sample size = 20
Degrees of freedom for tests of hypotheses = $n - p = 16$

TABLE 15-11 Results of the final multiple regression analysis for Example 15-5

Parameter	Variable	Parameter estimate	Standard error	Test statistic	*p*-value
β_2	Height	.10691	.01058	10.11	.000
β_4	Age × Number	−.00018975	.00003256	−5.83	.000
β_0	Intercept	3.2605	.3330	9.79	.000

p = Number of parameters = 3 n = Sample size = 20
Degrees of freedom for tests of hypotheses = $n - p = 17$

The least squares estimated model is

$$\hat{Y} = 3.26 + .107X_2 - .000190X_4$$

All the *p*-values in Table 15-11 are less than .001. This suggests that all three parameters—β_0, β_2, and β_4—are necessary to the model.

A descriptive statistic that we often use in multiple regression is the *multiple regression coefficient* or *coefficient of determination,* denoted R^2. The multiple regression coefficient has an interpretation similar to that of R^2 in simple linear regression. The multiple regression coefficient R^2 is the proportion of the variation in the Y values that is explained by the multiple regression model.

> The **multiple regression coefficient** or **coefficient of determination, R^2,** is the proportion of the variation in the response variable that is accounted for, or explained, by the multiple regression model.

For the final model in Table 15-11, we have $R^2 = .87$, a fairly large value. About 87% of the variation in average diameters can be explained by the model that includes height of the dominant trees and age times the number of pine trees per acre.

The predicted Y values and residuals for this final model are shown in Table 15-12. A plot of residuals versus predicted Y values is shown in Figure 15-23. This scatterplot looks like "noise" because we cannot see any pattern or relationship between the residuals and the predicted Y values. This residual

TABLE 15-12 List of the values of Y, the predicted Y values, and the residuals from the model $\hat{Y} = 3.26 + .107X_2 - .000190X_4$ in Example 15-5

Y = Average diameter	$\hat{Y}$ = Predicted value of Y	$e = Y - \hat{Y}$ = Residual
7.0	6.96	.04
5.0	5.29	−.29
6.2	5.83	.37
5.2	5.52	−.32
6.2	6.15	.05
5.2	5.06	.14
6.2	6.34	−.14
6.4	6.25	.15
6.4	6.34	.06
6.4	7.00	−.60
5.4	5.27	.13
6.4	6.47	−.07
5.4	5.63	−.23
6.7	6.83	−.13
6.7	6.91	−.21
5.9	5.72	.18
6.9	7.04	−.14
6.9	6.42	.48
6.9	6.85	.05
7.9	7.43	.47

plot gives us no reason to doubt the equal-variance assumption or the adequacy of the model.

A dot plot of the residuals from this final model is shown in Figure 15-24. The residuals have a fairly symmetrical distribution concentrated around zero. This plot gives us no reason to doubt the Gaussian assumption.

We must assume that height, age, and number of pine trees per acre are observed without error, and that the observed values of average diameter are independent. We would need additional information about the experiment to assess these assumptions. What suggestions would you make about experimental design to help meet model assumptions and reduce extraneous sources of variation?

A scatterplot matrix of the variables height (X_2), age × number (X_4), and average diameter (Y) is shown in Figure 15-25. We can see the positive association between average diameter and height (correlation coefficient = .77). However, there is only a weak association between average diameter and age × number (correlation coefficient = .24). It is common in multiple regression situations that simple bivariate plots do not give us a good impression of which variables are needed in the model.

Notice the strong positive association between height and age × number (correlation coefficient = .75). Sometimes predictor variables are even more

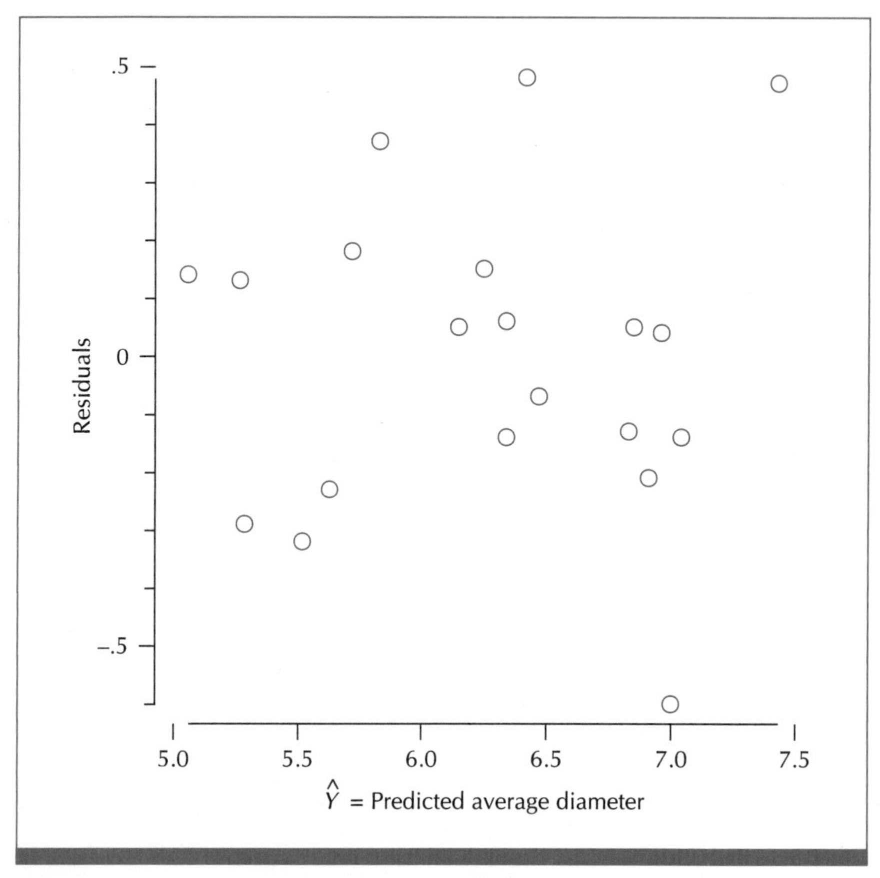

FIGURE 15-23 Plot of residuals versus predicted Y values for the model $\hat{Y} = 3.26 + .107X_2 - .000190X_4$ in Example 15-5

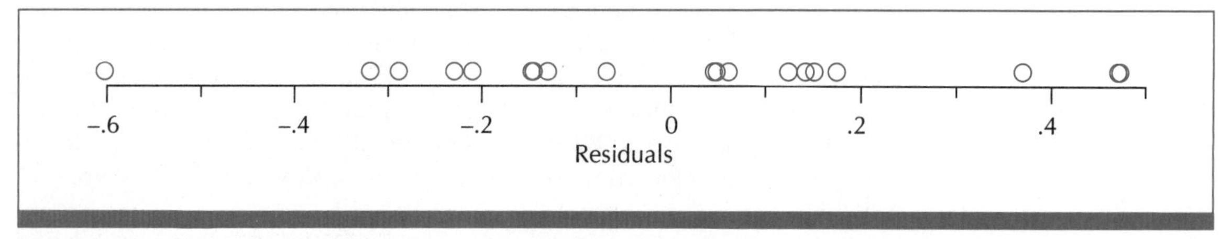

FIGURE 15-24 Dot plot of residuals from the model $\hat{Y} = 3.26 + .107X_2 - .000190X_4$ in Example 15-5

highly correlated. If we try to include in a model two or more predictor variables that are highly correlated with one another, we can run into trouble. It is a problem we call *multicollinearity.*

When two or more predictor variables in a multiple regression analysis are highly correlated with one another, we can get errors in the analysis, a problem called **multicollinearity.**

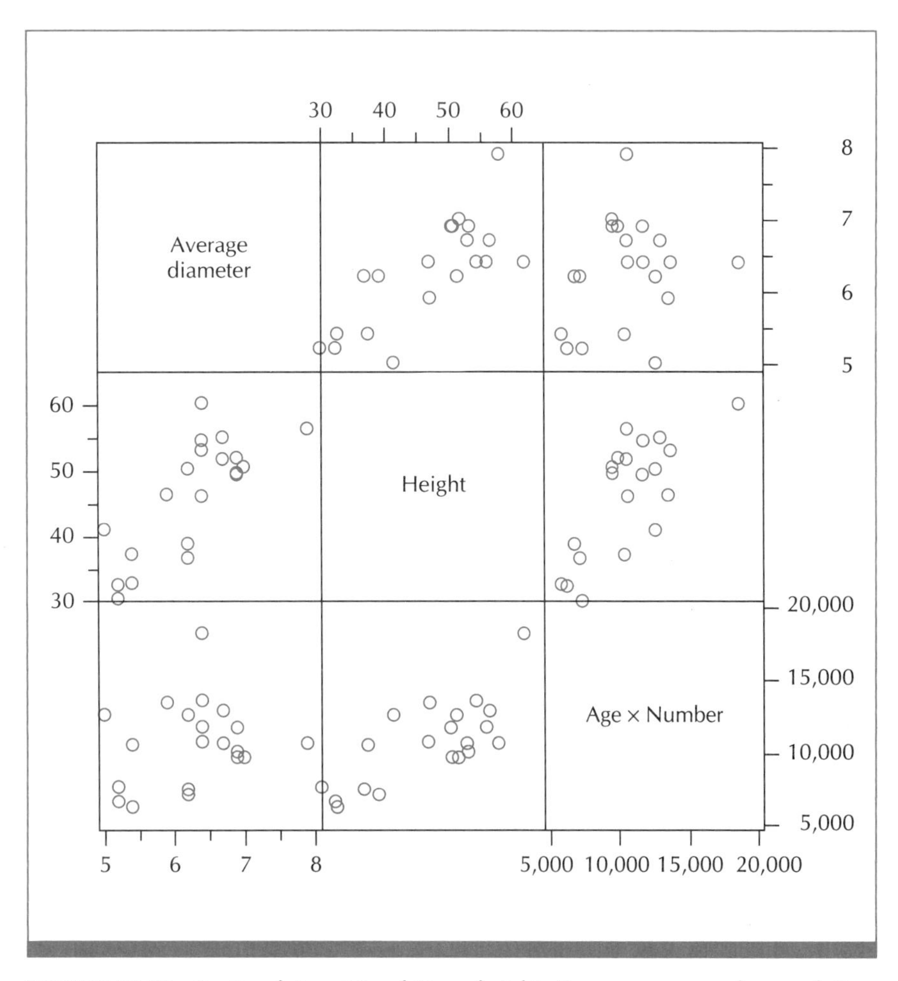

FIGURE 15-25 Scatterplot matrix of X_2 = height, X_4 = age × number, and Y = average diameter in Example 15-5

There are many aspects of multiple regression that we have not addressed, in addition to possible multicollinearity. We have not discussed how to choose predictor variables, how to evaluate observations that especially influence the fitted model, or how to deal with outliers, for instance. Multiple regression requires a course to itself. This section is meant just to give an idea of what it is about. For more information see, for example, Draper and Smith (1981), Daniel and Wood (1980), Myers (1986), or Rawlings (1988).

Summary of Chapter 15

The linear correlation coefficient is a measure of linear association between two quantitative variables. The correlation coefficient is a descriptive statistic.

It may be used to measure linear association between two variables, even in situations when formal statistical inference may not be appropriate.

When certain assumptions about the observations and sampling process are satisfied, we can make inferences about the linear association between two variables. A parametric test that the linear correlation coefficient equals 0 is also a test that the two variables are independent.

The rank correlation coefficient is a measure of association between two quantitative variables, based on ranks. A nonparametric test of independence between two quantitative variables may be based on the rank correlation coefficient.

The method of least squares is one way to model one quantitative variable as a straight-line function of another. We can use the method of least squares to fit a straight line without making any assumptions about the observations. Hypothesis testing in simple linear regression does require that we make assumptions about the observations and sampling process.

Considering the relation between correlation and simple linear regression, we introduce the standard deviation line and compare it with the least squares line. Examples illustrate the idea of regression toward the mean.

In multiple regression, we model a quantitative variable as a function of several other variables. We consider only linear models—that is, models that are linear in the parameters or unknown constants. The method of least squares allows us to calculate a multiple regression model for a set of observations. Hypothesis testing in multiple regression requires that we make assumptions about the observations and sampling process.

Minitab Appendix for Chapter 15

Calculating a Correlation Coefficient

Suppose the data for Example 15-1 are on our worksheet. Column 1 (named BABY) contains an identification code for the babies in the experiment. Breast milk measurements using method 1 are in column 2 (named METHOD1). Measurements using method 2 are in column 3 (named METHOD2). To calculate the linear correlation coefficient between measurements made using the two methods, we use the CORRELATE command:

```
MTB> correlate 'method1' 'method2'
correlation of method 1 and method 2 = 0.770
```

Testing Hypotheses About a Correlation Coefficient

We can calculate the test statistic for the null hypothesis that the population correlation coefficient equals zero. The sample size is 14, so the commands look like this:

```
MTB>    let k1=(sqrt(12))*0.77/sqrt(1-0.77**2)
MTB>    cdf k1 k2;
SUBC>   t 12.
MTB>    let k3=1-k2
MTB>    print k1-k3
K1      4.18052
K2      0.999360
K3      0.000639796
```

Minitab prints the test statistic K1 = 4.18052, the cumulative probability K2 = 0.999360, and the tail probability K3 = .000639796 corresponding to that test statistic. The probabilities are from the t distribution with $14 - 2 = 12$ degrees of freedom. The two-sided p-value is 2 times the smaller of K2 and K3, about .0013 in this example.

The CORRELATE command skips rows in which a missing value occurs for one or both variables.

Calculating a Rank Correlation Coefficient

Suppose we want to calculate the rank correlation coefficient for variables in two columns. We should first delete rows with a missing value for one or both variables. Then rank each column and find the correlation coefficient for the ranks. Consider the data in Example 15-2. Sample number is in column 1 (named SAMPLE), thickness is in column 2 (named THICK), and stiffness is in column 3 (named STIFF). Consider these commands:

```
MTB>  rank 'thick' c4
MTB>  rank 'stiff' c5
MTB>  name c4 'rank1' c5 'rank2'
MTB>  print c1-c5

ROW    sample    thick        stiff      rank1    rank2
 1        1      0.28          7.98        2        2
 2        2      0.65         24.52        5        5
 3        3      0.32         12.47        3        3
 4        4      0.27          6.92        1        1
 5        5      0.81         24.11        6        4
 6        6      0.57         35.71        4        6

MTB>  correlate c4 c5
correlation of rank1 and rank2 = 0.771
```

Testing Hypotheses About a Rank Correlation Coefficient

The following commands calculate the test statistic for a nonparametric test of independence between the two sets of measurements:

```
MTB>  let c6=(c4-c5)**2
MTB>  sum c6 k4
 SUM  = 8.0000
MTB>  print k4
K4    8.00000
```

The stored constant K4 contains D, the sum of squared differences between ranks for the two sets of measurements. We can compare D with a table of Spearman's distribution, for sample size 6. We can also calculate the large-sample test statistic and find the large-sample approximate p-value:

```
MTB>   let k5=k4-(6**3-6)/6
MTB>   let k6=sqrt((6**2)*(7**2)*5/36)
MTB>   let k7=k5/k6
MTB>   cdf k7 k8;
SUBC>  normal.
MTB>   let k9=1-k8
MTB>   print k7-k9
K7     -1.72497
K8     0.0422667
K9     0.957733
```

Minitab prints the large-sample test statistic K7 = −1.72497, the cumulative probability K8 = .0422667, and the upper-tail probability K9 = .957733 from the standard Gaussian distribution. The two-sided p-value is 2 times the smaller of K8 and K9, about .08 in this example.

Carrying Out Simple Linear Regression

We use the REGRESS command for simple linear regression and multiple regression. Consider the data in Example 15-3. Column 1 (named STREET) contains the street code. Distance from center line to cyclist is in column 2 (named CENTER). Distance from car to cyclist is in column 3 (named CAR). The command

```
MTB>  regress 'car' 1 'center'
```

tells Minitab to fit a simple linear regression model of the values in 'CAR' as a function of the one variable 'CENTER'. The output is shown in Figure M15-1.

We can save residuals and estimated (or predicted) values this way:

```
MTB>   regress 'car' 1 'center' c4 c5;
SUBC>  residuals c6.
MTB>   name c4 'stdres' c5 'predict' C6 'resid'
```

Minitab produces the output in Figure M15-1. In addition, standardized residuals (mean 0 and standard deviation 1) are stored in column 4, predicted values

```
The regression equation is
car = - 2.18 + 0.660 center

Predictor       Coef        Stdev    t-ratio       p
Constant      -2.182        1.057      -2.07   0.073
center       0.66034      0.06748       9.79   0.000

s = 0.5821      R-sq = 92.3%      R-sq(adj) = 91.3%

Analysis of Variance

SOURCE        DF          SS          MS         F        p
Regression     1      32.449      32.449     95.76    0.000
Error          8       2.711       0.339
Total          9      35.160
```

FIGURE M15-1 Simple linear regression in Example 15-3

```
            avediam       age   height   number  age*num
age           0.675
height        0.773     0.876
number       -0.252     0.016    0.229
age*num       0.244     0.678    0.755    0.732
ht/num        0.840     0.735    0.634   -0.579    0.056
```

FIGURE M15-2 Linear correlation coefficients for pairs of variables in Example 15-5

in column 5, and residuals in column 6. We can use these stored values in plots to check model assumptions.

Carrying Out Multiple Regression

Let's consider the multiple regression problem in Example 15-5. Suppose our worksheet contains five columns. Column 1 (named STAND) contains stand number. Age, height, number, and average diameter are in columns 2–5, named AGE, HEIGHT, NUMBER, and AVEDIAM, respectively. We calculate two new variables:

```
MTB>  let c6='age'*'number'
MTB>  let c7='height'/'number'
MTB>  name c6 'age*num' c7 'ht/num'
```

We can get the correlation coefficients in Table 15-7 this way:

```
MTB>   correlate 'avediam' 'age' 'height' 'number' &
CONT>  'age*num' 'ht/num'
```

The results are shown in Figure M15-2.

To fit a multiple regression model with AVEDIAM as a function of AGE, HEIGHT, NUMBER, AGE*NUM, and HT/NUM, we use the command

```
MTB>   regress 'avediam' 5 'age' 'height' &
CONT>  'number' 'age*num' 'ht/num'
```

to get the output in Figure M15-3.

```
* NOTE *  age*num is highly correlated with other  predictor variables

The regression equation is
avediam = 1.23 + 0.053 age + 0.0825 height + 0.00322 number -0.000282 age*num
           + 16.0 ht/num

Predictor         Coef         Stdev     t-ratio        p
Constant         1.233         1.619        0.76    0.459
age             0.0526        0.1683        0.31    0.759
height         0.08246       0.04035        2.04    0.060
number        0.003224      0.002532        1.27    0.224
age*num     -0.0002817     0.0002300       -1.22    0.241
ht/num           16.03         27.89        0.57    0.575

s = 0.2952       R-sq = 88.2%      R-sq(adj) = 84.1%

Analysis of Variance

SOURCE        DF          SS          MS          F        p
Regression     5      9.1651      1.8330      21.03    0.000
Error         14      1.2204      0.0872
Total         19     10.3855


CONTINUE? y
SOURCE        DF       SEQ SS
age            1       4.7388
height         1       1.4684
number         1       2.3984
age*num        1       0.5307
ht/num         1       0.0288

Unusual Observations
Obs.      age   avediam        Fit Stdev.Fit  Residual    St.Resid
  2      14.0    5.0000     5.4633    0.1899   -0.4633      -2.05R
 10      17.0    6.4000     6.9457    0.1254   -0.5457      -2.04R

R denotes an obs. with a large st. resid.
```

FIGURE M15-3 Output for the first multiple regression model specified in Example 15-5

For the command

```
MTB>   regress 'avediam' 5 'age' 'height' &
CONT>  'number' 'age*num' 'ht/num' c10 c11;
SUBC>  residuals C12.
```

Minitab will produce the output in Figure M15-3, and save standardized residuals in column 10, predicted values in column 11, and residuals in column 12. We can use these saved values in plots to check model assumptions.

Exercises for Chapter 15

For all exercises, plot the observations in any ways that seem reasonable. Describe the population(s) sampled, whether real or hypothetical. State the assumptions for each test of hypotheses. Do these assumptions seem reasonable? What additional information would you like to have about the experiment? Describe the results of your analysis.

EXERCISE 15-1

In Exercise 4-12, we looked at sodium content and potassium content (no units given) in perspiration of ten healthy women (Oja and Nyblom, 1989; from Johnson and Wichern, 1982, page 182):

Woman	1	2	3	4	5	6	7	8	9	10
Sodium	48.5	65.1	47.2	53.2	55.5	36.1	24.8	33.1	47.4	54.1
Potassium	9.3	8.0	10.9	12.2	9.7	7.9	14.0	7.6	8.5	11.3

a. Plot the observations.

b. Calculate the linear correlation coefficient for the sodium and potassium measurements. Test the null hypothesis that the linear correlation coefficient between sodium and potassium levels in perspiration of healthy women is 0.

c. Calculate the rank correlation coefficient for the sodium and potassium measurements. Carry out a nonparametric test that sodium and potassium levels in perspiration of healthy women are independent.

d. Compare your answers to parts (b) and (c). Discuss your findings.

EXERCISE 15-2

In Exercise 4-11, we considered carbon monoxide concentration (parts per million) and benzo(a)pyrene concentration (μg per 1,000 cubic meters) in 16 different air samples from Herald Square in New York City (Devore, 1982, page 457; from "Carcinogenic Air Pollutants in Relation to Automobile Traffic in New York City," *Environmental Science and Technology,* 1971, pages 145–150). The results are shown below as pairs of readings for each air sample: (carbon monoxide, benzo(a)pyrene).

(2.8, .5)	(15.5, .1)	(19.0, .8)	(6.8, .9)	(5.5, 1.0)
(5.6, 1.1)	(9.6, 3.9)	(13.3, 4.0)	(5.5, 1.3)	(12.0, 5.7)
(5.6, 1.5)	(19.5, 6.0)	(11.0, 7.3)	(12.8, 8.1)	(5.5, 2.2)
(10.5, 9.5)				

a. Plot the observations.

b. Calculate the linear correlation coefficient for the two substances. Test the null hypothesis that the linear correlation coefficient between carbon monoxide readings and benzo(a)pyrene readings at Herald Square under similar conditions is 0.

c. Calculate the rank correlation coefficient for the two substances. Test the null hypothesis that carbon monoxide readings and benzo(a)pyrene readings at Herald Square under similar conditions are independent.

d. Compare your answers to parts (b) and (c). Discuss your findings.

EXERCISE 15-3

Scientists wanted to compare the drop net catch method and the sweep net catch method of collecting grasshoppers (Walpole and Myers, 1989, page 439; from the Department of Entomology, Virginia Polytechnic Institute and State

University). They recorded the average number of grasshoppers caught in each of 17 field quadrants using the two methods. They also recorded the average height of plants in each quadrant. All measurements were made the same day.

Quadrant	Average drop net catch	Average sweep net catch	Average height of plants (centimeters)
1	18.00	4.15	52.7
2	8.88	2.02	42.1
3	2.00	.16	34.8
4	20.00	2.33	27.6
5	2.38	.26	45.9
6	2.75	.57	97.5
7	3.33	.70	102.1
8	1.00	.14	97.8
9	1.33	.12	88.3
10	1.75	.11	58.7
11	4.13	.56	42.4
12	12.88	2.45	31.3
13	5.38	.45	31.8
14	28.00	6.69	35.4
15	4.75	.87	64.5
16	1.75	.15	25.2
17	.13	.02	36.4

a. Construct a scatterplot matrix of these three variables.

b. Calculate the linear correlation coefficient for each pair of variables.

c. Calculate the rank correlation coefficient for each pair of variables.

d. Compare your answers to parts (b) and (c). Discuss your findings.

EXERCISE 15-4 Body weight and heart weight are shown below for each of 19 normal woodchucks (Walpole and Myers, 1989, page 398; from the Department of Veterinary Medicine and the Statistics Consulting Center, Virginia Polytechnic Institute and State University).

Woodchuck	Body (grams)	Heart (grams)	Woodchuck	Body (grams)	Heart (grams)
1	4,050	11.2	11	3,690	10.8
2	2,465	12.4	12	2,800	14.2
3	3,120	10.5	13	2,775	12.2
4	5,700	13.2	14	2,170	10.0
5	2,595	9.8	15	2,370	12.3
6	3,640	11.0	16	2,055	12.5
7	2,050	10.8	17	2,025	11.8
8	4,235	10.4	18	2,645	16.0
9	2,935	12.2	19	2,675	13.8
10	4,975	11.2			

a. Plot the observations.

b. Calculate the linear correlation coefficient for body weight and heart weight. Test the null hypothesis that the linear correlation coefficient for body weight and heart weight in normal woodchucks is 0.

c. Calculate the rank correlation coefficient for body weight and heart weight. Carry out a nonparametric test of the null hypothesis that body weight and heart weight are independent in normal woodchucks.

d. Compare your answers to parts (b) and (c). Discuss your results.

EXERCISE 15-5

Investigators simultaneously measured wind speed (m/s) on the ground and via Seasat satellite at each of 12 times (Milton and Arnold, 1986, pages 325–326; from "Mapping Ocean Winds by Radar," *NASA Tech Briefs,* Fall 1982, page 27).

Ground measurement:	4.46	3.99	3.73	3.29	4.82	6.71	4.61	3.87	3.17	4.42	3.76	3.30
Satellite measurement:	4.08	3.94	5.00	5.20	3.92	6.21	5.95	3.07	4.76	3.25	4.89	4.80

a. Plot the observations.

b. Calculate the linear correlation coefficient for the ground and satellite measurements. Test the null hypothesis that the linear correlation coefficient for ground and satellite measurements of wind speed is 0.

c. Calculate the rank correlation coefficient for the ground and satellite measurements. Test the null hypothesis that ground and satellite measurements of wind speed are independent.

d. Compare your answers to parts (b) and (c). Discuss your findings.

EXERCISE 15-6

Researchers measured inulin clearance (ml/min) of seven living kidney donors and the recipients of their kidneys (Hollander and Wolfe, 1973, page 239; from Shelp et al., 1970).

Recipient:	61.4	63.3	63.7	80.0	77.3	84.0	105.0
Donor:	70.8	89.2	65.8	67.1	87.3	85.1	88.1

a. Plot the observations in any ways that seem helpful.

b. Calculate the linear correlation coefficient for recipient and donor inulin clearance. Test the null hypothesis that the linear correlation coefficient for recipient and donor inulin clearance is 0.

c. Calculate the rank correlation coefficient for recipient and donor inulin clearance. Carry out a nonparametric test that recipient and donor inulin clearance measurements are independent.

d. Compare your answers to parts (b) and (c) and discuss your findings.

EXERCISE 15-7 Investigators wanted to study the effects of illumination on a person's ability to perform a task (Devore, 1982, page 300; from "Performance of Complex Tasks Under Different Levels of Illumination," *J. Illuminating Eng.*, 1976, pages 235–242). A volunteer inserted a fine-tipped probe into the eyehole of a needle, ten times with low light and a black background and ten times with more light and a white background. The average time (units not given) at each light level is shown below for each of nine volunteers.

Volunteer:	1	2	3	4	5	6	7	8	9
Higher light level:	25.85	28.84	32.05	25.74	20.89	41.05	25.01	24.96	27.47
Lower light level:	18.23	20.84	22.96	19.68	19.50	24.98	16.61	16.07	24.59

a. Plot the observations in any ways that seem helpful.

b. Calculate the linear correlation coefficient for the two sets of times. Test the null hypothesis that the linear correlation coefficient of average times under the two light levels is 0.

c. Calculate the rank correlation coefficient for the two sets of times. Carry out a nonparametric test of the null hypothesis that the average times under the two light levels are independent.

d. Compare your answers to parts (b) and (c) and discuss your findings.

EXERCISE 15-8 Consider the measurements of thickness and stiffness on six samples of a flame-retardant fabric in Example 15-2, plotted in Figure 15-15.

a. Calculate the linear correlation coefficient for these two variables. Compare the linear correlation coefficient with the rank correlation coefficient calculated in Example 15-2.

b. Find the least squares line modeling stiffness as a function of thickness.

c. Test the null hypothesis that the slope in the straight-line model is 0.

d. Test the null hypothesis that the intercept in the straight-line model is 0.

e. What percentage of the variation in stiffness is explained by the straight-line model? (That is, what is R^2?)

f. Discuss your findings.

EXERCISE 15-9 In a study of the operation of a factory, investigators recorded 25 observations of amount of steam used per month and average atmospheric temperature (Draper and Smith, 1981, page 9):

Steam used (pounds)	Average temperature (°F)	Steam used (pounds)	Average temperature (°F)
10.98	35.3	9.57	39.1
11.13	29.7	10.94	46.8
12.51	30.8	9.58	48.5
8.40	58.8	10.09	59.3
9.27	61.4	8.11	70.0
8.73	71.3	6.83	70.0
6.36	74.4	8.88	74.5
8.50	76.7	7.68	72.1
7.82	70.7	8.47	58.1
9.14	57.5	8.86	44.6
8.24	46.4	10.36	33.4
12.19	28.9	11.08	28.6
11.88	28.1		

a. Plot steam versus temperature.

b. Calculate the linear correlation coefficient r for steam and temperature.

c. Use the method of least squares to model steam used as a straight-line function of average temperature.

d. Use residual plots to assess the fit of the model.

e. Discuss your findings.

EXERCISE 15-10

Exercise 4-8 described an experiment studying plastic spools used in electric motors. Wire is wound around the spools. When current passes through the wire, the temperature of the spool rises. Investigators made two measurements of temperature rise (°C) on each of 12 such plastic spools. The results are shown below (Nelson, 1986, page 12).

Spool:	1	2	3	4	5	6	7	8	9	10	11	12
First reading:	45.0	45.1	45.4	45.9	45.9	46.0	46.2	46.5	46.5	46.8	47.0	50.6
Second reading:	44.9	44.7	45.8	45.3	45.8	45.2	45.2	45.5	46.0	46.1	45.5	50.0

a. Plot the second reading versus the first reading.

b. Find the least squares line modeling the second reading as a straight-line function of the first reading.

c. Find the least squares line modeling the first reading as a straight-line function of the second reading.

d. Draw the lines you found in parts (b) and (c) on your plot in part (a). Also, draw the standard deviation line. Discuss the meaning of each line.

e. What relationship between the first and second readings would you expect if the two measurements were consistent? Do the two readings appear to be consistent? Discuss your findings.

EXERCISE 15-11 Researchers sampled 22 naval installations to examine man-hours spent monthly on the clerical task of processing items (Myers, 1986, pages 25–26; from *Procedures and Analyses for Staffing Standards Development: Data/Regression Analysis Handbook,* 1979, Navy Manpower and Material Analysis Center, San Diego, California):

Items processed	Man-hours monthly	Items processed	Man-hours monthly
15	85	527	2,158
25	125	533	2,182
57	203	563	2,302
67	293	563	2,202
197	763	932	3,678
166	639	986	3,894
162	673	1,021	4,034
131	499	1,643	6,622
158	657	1,985	7,890
241	939	1,640	6,610
399	1,546	2,143	8,522

a. Plot monthly man-hours versus items processed.

b. Use the method of least squares to model monthly man-hours as a straight-line function of items processed.

c. Test the null hypothesis that the slope is 0.

d. Test the null hypothesis that the intercept is 0.

e. Use residual plots to check model assumptions.

f. What percentage of the variation in monthly man-hours is explained by the straight-line model? (That is, what is R^2?)

g. Discuss your findings.

EXERCISE 15-12 Investigators studied physical characteristics and ability in 13 American football punters. Each volunteer punted a football ten times. The investigators recorded the average distance for the ten punts, in feet. They also recorded the average hang time (time the ball is in the air before the receiver catches it) for the ten punts, in seconds. In addition, the investigators recorded five measures of strength and/or flexibility for each punter: right leg strength (pounds), left leg strength (pounds), right hamstring muscle flexibility (degrees), left hamstring muscle flexibility (degrees), and overall leg strength (foot-pounds). The results are shown below (Walpole and Myers, 1989, pages 444–445, 450; from the study "The Relationship Between Selected Physical Performance Variables and Football Punting Ability" by the Department of Health, Physical Education, and Recreation at the Virginia Polytechnic Institute and State University, 1983).

Punter	Distance	Hang time	Right leg strength	Left leg strength	Right flexibility	Left flexibility	Overall strength
1	162.50	4.75	170	170	106	106	240.57
2	144.00	4.07	140	130	92	93	195.49
3	147.50	4.04	180	170	93	78	152.99
4	163.50	4.18	160	160	103	93	197.09
5	192.00	4.35	170	150	104	93	266.56
6	171.75	4.16	150	150	101	87	260.56
7	162.00	4.43	170	180	108	106	219.25
8	104.93	3.20	110	110	86	92	132.68
9	105.67	3.02	120	110	90	86	130.24
10	117.59	3.64	130	120	85	80	205.88
11	140.25	3.68	120	140	89	83	153.92
12	150.17	3.60	140	130	92	94	154.64
13	165.17	3.85	160	150	95	95	240.57

a. In this exercise, we will consider only the two variables distance and hang time. Plot the observations of distance and hang time in any ways that seem helpful.

b. Find the least squares line modeling distance as a function of hang time.

c. Find the least squares line modeling hang time as a function of distance.

d. In a scatterplot of distance versus hang time, plot the two lines you found in parts (b) and (c). Also, plot the standard deviation line. Label each line. Discuss the meaning and use of each of these three lines.

EXERCISE 15-13 Refer to the experiment described in Exercise 15-12. For this exercise, consider only the two variables distance and hang time.

a. Plot distance versus hang time.

b. Calculate the linear correlation coefficient for distance and hang time. Test the null hypothesis that the linear correlation coefficient between distance and hang time is 0.

c. Calculate the rank correlation coefficient for distance and hang time. Carry out a nonparametric test that distance and hang time are independent.

d. Compare your results in parts (b) and (c). Discuss your findings.

EXERCISE 15-14 Children with congenital heart defects sometimes need a procedure called heart catheterization. Surgeons pass a 3-mm diameter Teflon tube or catheter into a major vein or artery. They push the tube into the heart to get information on the heart's condition. The surgeons have to guess at the appropriate length of the catheter.

In this study, investigators determined the exact length of the catheter needed in 12 children (Rice, 1988, pages 491–492; from Weindling, 1977). The researchers used a fluoroscope to check when the catheter was in place.

Height, weight, and correct catheter length are shown below for each of the 12 children.

Child	Height (inches)	Weight (pounds)	Catheter length (centimeters)
1	42.8	40.0	37.0
2	63.5	93.5	49.5
3	37.5	35.5	34.5
4	39.5	30.0	36.0
5	45.5	52.0	43.0
6	38.5	17.0	28.0
7	43.0	38.5	37.0
8	22.5	8.5	20.0
9	37.0	33.0	33.5
10	23.5	9.5	30.5
11	33.0	21.0	38.5
12	58.0	79.0	47.0

a. Construct a scatterplot matrix of height, weight, and catheter length.

b. Find the linear correlation coefficient for height and weight, for height and catheter length, and for weight and catheter length.

c. Carry out a simple linear regression analysis, modeling catheter length as a straight-line function of height.

d. Carry out a simple linear regression analysis, modeling catheter length as a straight-line function of weight.

e. Carry out a simple linear regression analysis, modeling weight as a straight-line function of height.

f. Discuss your results. Does it look like surgeons could determine the correct catheter length from the child's height or weight?

EXERCISE 15-15 Refer to the experiment described in Exercise 15-12. For this exercise, consider the variables left leg strength, right leg strength, and distance.

a. Construct a scatterplot matrix of left leg strength, right leg strength, and distance.

b. Find the linear correlation coefficient for left and right leg strength, for left leg strength and distance, and for right leg strength and distance.

c. Carry out a simple linear regression analysis modeling distance as a straight-line function of right leg strength.

d. Carry out a simple linear regression analysis modeling distance as a straight-line function of left leg strength.

e. Carry out a simple linear regression analysis modeling left leg strength as a straight-line function of right leg strength.

f. Discuss your results.

EXERCISE 15-16 In a calibration study in atomic absorption spectroscopy, investigators recorded instrument response in absorbance units at each of five concentrations of copper in solution (Carroll, Sacks, and Spiegelman, 1988):

Run	Amount of copper in solution (micrograms/ milliliter)	Instrument response in absorbance units
1	.0	.045
2	.0	.047
3	.0	.051
4	.0	.054
5	.050	.084
6	.050	.087
7	.100	.115
8	.100	.116
9	.200	.183
10	.200	.191
11	.500	.395
12	.500	.399

a. Plot instrument response versus copper concentration.

b. Use the method of least squares to model instrument response as a straight-line function of copper concentration.

c. Test the null hypothesis that the slope is 0.

d. Test the null hypothesis that the intercept is 0.

e. What percentage of the variation in instrument response is explained by the straight-line model? (That is, what is R^2?)

f. Use residual plots to assess the fit.

g. Discuss your results.

EXERCISE 15-17 Investigators studied the time to failure of samples of electrical insulation for motors in accelerated life testing (Nelson, 1986, pages 20–21). The investigators carried out the accelerated life test at four temperatures, with ten samples of insulation at each temperature.

Temperature (°C)	Hours to failure									
190	7,228	7,228	7,228	8,448	9,167	9,167	9,167	9,167	10,511	10,511
220	1,764	2,436	2,436	2,436	2,436	2,436	3,108	3,108	3,108	3,108
240	1,175	1,175	1,521	1,569	1,617	1,665	1,665	1,713	1,761	1,953
260	600	744	744	744	912	1,128	1,320	1,464	1,608	1,896

a. Plot hours to failure versus temperature.

b. Find the linear correlation coefficient for failure time and temperature.

c. Plot the logarithm of failure time versus the reciprocal of temperature.

d. Find the linear correlation coefficient for the logarithm of failure time and the reciprocal of temperature.

e. Carry out a simple linear regression analysis modeling the logarithm of failure time as a straight-line function of the reciprocal of temperature.

f. Use residual plots to assess the fit of the model in part (e).

g. Discuss your findings.

EXERCISE 15-18 Investigators recorded stopping distance of a car on a road, for several velocities (Rice, 1988, page 505; from Brownlee, 1965, pages 371–372):

Velocity (miles per hour)	Stopping distance (feet)
20.5	15.4
20.5	13.3
30.5	33.9
40.5	73.1
48.8	113.0
57.8	142.6

a. Plot stopping distance versus velocity.

b. Use the method of least squares to model stopping distance as a straight-line function of velocity.

c. What percentage of the variation in stopping distance is explained by the model in part (b)? (That is, what is R^2?)

d. Use residual plots to assess the fit of the model in part (b).

e. Plot the square root of stopping distance versus velocity.

f. Use the method of least squares to model the square root of stopping distance as a straight-line function of velocity.

g. What percentage of the variation in the square root of stopping distance is explained by the model in part (f)? (That is, what is R^2?)

h. Use residual plots to assess the fit of the model in part (f).

i. Discuss your findings.

EXERCISE 15-19 An engineer subjected uniform pieces of stainless steel to different levels of stress, and recorded the time to rupture for each piece. He tested six pieces of steel at each of four stress levels. Stress levels are reported in pounds per square inch (psi). The results are shown below (Schmoyer, 1986; from Garofalo et al., 1961):

Stress level (psi)	Rupture time (hours)					
28.84	1,267	1,637	1,658	1,709	1,785	2,437
31.63	170	257	265	570	594	779
34.68	76	87	96	115	122	132
38.02	22	37	39	41	42	43

a. Plot rupture time versus stress level.

b. Plot the logarithm of rupture time versus stress level.

c. Use the method of least squares to model the logarithm of rupture time as a straight-line function of stress level.

d. Test the null hypothesis that the slope is 0.

e. Test the null hypothesis that the intercept is 0.

f. What percentage of the variation in the logarithm of rupture time is explained by the model in part (c)? (That is, what is R^2?)

g. Use residual plots to check model assumptions.

h. Discuss your findings.

i. Schmoyer (1986) plotted the logarithm of rupture time versus the logarithm of stress level. Construct such a plot. Find $R^2 = r^2$ for these two variables. Compare with your answers to parts (b) and (f).

EXERCISE 15-20

As part of an environmental impact study, investigators looked at the relationship between stream depth and rate of flow (Rice, 1988, page 463; from Ryan, Joiner, and Ryan, 1976). The results are shown below (units not given).

Depth:	.34	.29	.28	.42	.29	.41	.76	.73	.46	.40
Flow rate:	.636	.319	.734	1.327	.487	.924	7.350	5.890	1.979	1.124

a. Plot flow rate versus depth.

b. Use the method of least squares to model flow rate as a straight-line function of depth.

c. Plot residuals versus predicted flow rates for the model in part (b). Does the plot look like random scatter, as it would for an adequate model?

d. Plot the logarithm of flow rate versus the logarithm of depth.

e. Use the method of least squares to model the logarithm of flow rate as a straight-line function of the logarithm of depth.

f. Use residual plots to assess the model in part (e).

g. Discuss your findings.

EXERCISE 15-21 Scientists designed this experiment to study a method for extracting crude oil called the carbon dioxide flooding technique (Mendenhall and Sincich, 1988, page 624; from Wang, 1982). In field use, workers flood carbon dioxide into oil pockets. The carbon dioxide replaces the crude oil, making the oil easier to extract. In this experiment, scientists dipped carbon dioxide flow tubes into oil pockets with known amounts of oil. They tried three carbon dioxide flow pressures and three dipping angles for the flow tubes. The response variable is the percentage of oil recovered. Carbon dioxide flow pressure is recorded in pounds per square inch (psi). The dipping angle is recorded in degrees.

Pressure:	1,000	1,000	1,000	1,500	1,500	1,500	2,000	2,000	2,000
Angle:	0	15	30	0	15	30	0	15	30
Recovery:	60.58	72.72	79.99	66.83	80.78	89.78	69.18	80.31	91.99

a. Construct a scatterplot matrix of these three variables.

b. What do you notice about the relationship between pressure and angle? We could think of this as a two-way factorial experimental design. Each factor (pressure and angle) has three levels. The linear correlation coefficient between pressure and angle is 0. This is a feature of a good experimental design.

c. In the plot(s) containing both pressure and recovery, connect the points with the same values for angle. This creates three profiles, one for each angle. Similarly, in the plot(s) containing both angle and recovery, connect the points with the same values for pressure. This creates three more profiles, one for each pressure. In each of these two plots, are the profiles parallel? Parallel profiles suggest no interaction effect of pressure and angle on recovery. Profiles that are not parallel suggest there is an interaction effect. (For a discussion of interaction effects on a response variable, see Chapter 13.)

d. Use multiple regression methods to model recovery as a function of pressure, angle, and the product of pressure and angle. Use residual plots to assess the fit. Discuss your findings.

e. Use multiple regression methods to model recovery as a function of pressure and angle. Use residual plots to assess the fit. Compare these results with those of part (d).

EXERCISE 15-22 Refer to the experiment described in Exercise 15-12.

a. Use multiple regression to model distance as a function of right leg strength, left leg strength, right flexibility, left flexibility, and overall leg strength.

b. Starting with the model in part (a), go through the steps for a backward regression, stopping when the predictor variable(s) have small *p*-values.

c. Use residual plots to assess the final model in part (b).

d. Construct a scatterplot matrix of all the variables in the final model in part (b).

e. Discuss your findings.

EXERCISE 15-23 Refer to the experiment described in Exercise 15-12.

a. Use multiple regression to model hang time as a function of right leg strength, left leg strength, right flexibility, left flexibility, and overall leg strength.

b. Starting with the model in part (a), go through the steps for a backward regression, stopping when the predictor variable(s) have small p-values.

c. Use residual plots to assess the final model in part (b).

d. Construct a scatterplot matrix of all the variables in the final model in part (b).

e. Discuss your findings.

EXERCISE 15-24 Researchers conducted this experiment to evaluate the effect of asphalt content on permeability of a type of concrete (Mendenhall and Sincich, 1988, page 495; from Woelfl et al., 1981). They prepared four samples of concrete with each of six levels of asphalt content. They then measured water permeability as the amount of water lost when de-aired water flowed across a sample. Asphalt content is recorded as percentage by total weight of the concrete mix. Permeability is recorded in inches per hour.

Asphalt content	Perme-ability	Asphalt content	Perme-ability	Asphalt content	Perme-ability
3	1,189	5	1,227	7	853
3	840	5	1,180	7	900
3	1,020	5	980	7	733
3	980	5	1,210	7	585
4	1,440	6	707	8	395
4	1,227	6	927	8	270
4	1,022	6	1,067	8	310
4	1,293	6	822	8	208

a. Plot permeability versus asphalt content.

b. Calculate the linear correlation coefficient for permeability and asphalt content. What percentage of the variation in permeability measurements is explained by a straight-line model of permeability as a function of asphalt content?

c. Does it seem reasonable to model permeability as a straight-line function of asphalt content?

d. Use multiple regression to model permeability as a function of asphalt content and the square of asphalt content.

e. What percentage of the variation in permeability measurements is explained by the quadratic model in part (d)? (That is, what is R^2?)

f. Use residual plots to assess the model in part (d).

g. Discuss your findings.

CHAPTER 16

Inferences About Qualitative (or Categorical) Variables

IN THIS CHAPTER

Chi-square goodness-of-fit test for a single qualitative variable

Small-sample inference about a proportion

Chi-square test of independence of two qualitative variables

Chi-square test of homogeneity of a discrete distribution across populations

Fisher's exact test

When does labor begin for pregnant women? Recording time of onset of labor for many women, investigators consider a day as 24 hour-long periods and ask: Is labor as likely to begin in any of these 24 time periods (Exercise 16-5)? Geneticists studying height (tall, dwarf) and leaf type (cut-leaf, potato-leaf) in a sample of tomato plants ask: Do the numbers of plants across the four combinations of observed height and leaf type suggest that the two characteristics are distributed independently to offspring (Exercise 16-3)? These two questions relate to inferences about a single qualitative or categorical variable, the subject of Sections 16-1 and 16-2. In Section 16-1, we discuss the chi-square goodness-of-fit test for comparing the observed distribution of a qualitative variable with some specified distribution.

> The **chi-square goodness-of-fit test** is a large-sample test for making inferences about a single qualitative variable. We compare the observed distribution of frequencies across categories with a specified distribution.

We consider small-sample inferences about a proportion (or, equivalently, testing for goodness of fit with small samples when a qualitative variable has exactly two categories) in Section 16-2.

Is the incidence of overwhelming infection among children with sickle cell disease different for those children treated with oral penicillin than for those treated with a placebo (Exercise 16-14)? Is there any association between hypertension (yes, no) and obesity (low, average, high) among adults in Western Australia (Exercise 16-9)? Is level of parental encouragement (high, low) related to a male high school senior's plans to attend college (yes, no)? Is the relationship between parental encouragement and college plans different for male and female high school seniors (Exercise 16-10)? These questions deal with inferences about the relationship between two qualitative variables. Sections 16-3 and 16-4 discuss large-sample tests of association between two qualitative variables, called chi-square tests of association.

> A **chi-square test of association** is a large-sample test for making inferences about the relationship between two qualitative variables.

Section 16-3 covers the chi-square test of independence of two qualitative variables.

> The **chi-square test of independence** is a large-sample test of the null hypothesis that two qualitative variables have independent distributions in a population.

We use the chi-square test of homogeneity, discussed in Section 16-4, to compare the distribution of a qualitative variable across populations. Comparing several proportions in a special case.

> The **chi-square test of homogeneity** is a large-sample test of the null hypothesis that a qualitative variable has the same distribution in each of several populations.

Fisher's exact test is a small-sample test of association in a 2×2 frequency table, when each variable has two categories. This test is discussed in Section 16-5.

We use **Fisher's exact test** to make inferences about the relationship between two qualitative variables, each having exactly two categories. Probabilities under the null hypothesis are conditional on the observed row and column totals in the 2 × 2 frequency table summarizing the observed results.

We begin in Section 16-1 with the chi-square goodness-of-fit test.

16-1

The Chi-Square Goodness-of-Fit Test

Suppose a qualitative variable Y has k possible values or categories, denoted by the numbers 1 through k. The probability distribution of Y is determined by the probabilities $P(Y = i)$, where i ranges from 1 to k. We want to test the null hypothesis that the probability distribution of Y is equal to some specified distribution. Equivalently, we want to test the *goodness of fit* of the specified distribution with what we observe in a sample. We call the test the *chi-square goodness-of-fit* test. Consider the following example.

EXAMPLE 16-1

Researchers wanted to study the relationship between tubal infertility and method of contraception (Cramer et al., 1987). (Tubal infertility results from damage to the Fallopian tubes, often caused by infection.) They interviewed 283 women with tubal infertility and no children, classifying the women according to method of contraception longest used. The results are shown below:

Method longest used	Observed frequency
None	36
Intrauterine device	39
Oral contraceptives	144
Barrier methods	64
Total	283

Barrier methods include diaphragms or condoms (or both).

The researchers also interviewed a large control group of women with children. These women were similar to the women with tubal infertility with respect to age, race, and income status. The researchers classified women in this control group according to method of contraception longest used:

Method longest used	Percent of control group
None	13.9
Intrauterine device	6.9
Oral contraceptives	52.7
Barrier methods	26.5

Why would medical workers be interested in looking for a possible relationship between tubal infertility and method of contraception? What would you want to know about the sample of women with tubal infertility and the control group to believe that comparisons are meaningful?

Considering method of contraception longest used, we ask: How does the sample distribution for women with tubal infertility compare with the control distribution? We will use the chi-square goodness-of-fit test to assess the null hypothesis that the distribution for women with tubal infertility is the same as for the control group. First we will outline the steps for carrying out the test.

The significance level approach to the chi-square goodness-of-fit test

1. The qualitative variable Y has k possible categories. The hypotheses are:

   Null hypothesis: $P(Y = i) = p_i$ for all i from 1 to k.
   Alternative hypothesis: $P(Y = i)$ does not equal p_i for all i from 1 to k.

   The specified probabilities p_1 through p_k sum to 1.
2. We have a sample of size n from the population. Let O_i denote the number of observations in category i. Define the expected frequency for category i by $E_i = n \times p_i$. E_i is the frequency we expect on average in category i under the null hypothesis. We want to measure how far the observed frequencies O_1 through O_k are from the null hypothesis expected frequencies E_1 through E_k. The test statistic is

$$\text{Test statistic} = \sum_{i=1}^{k} \frac{(O_i - E_i)^2}{E_i}$$

   If the observed and expected frequencies are close for each category, consistent with the null hypothesis, the test statistic will be small. If the observed and expected frequencies are very different for one or more categories, inconsistent with the null hypothesis, the test statistic will be large.
3. Assume that we have a random sample of size n from the population. Each observation is classified into exactly one of the k categories of variable Y. If n is large enough, the test statistic has approximately the chi-square distribution with $k - 1$ degrees of freedom under the null hypothesis.
4. Choose significance level α.
5. Let X denote a random variable having the chi-square distribution with $k - 1$ degrees of freedom. Find the number c from Table E such that $P(X \leq c) = 1 - \alpha$. The acceptance region is $[0, c)$; the rejection region is $[c, \infty)$.
6. The decision rule is:

   If test statistic $< c$, say the results are consistent with the null hypothesis that the distribution of Y in the population equals the specified distribution.

   If test statistic $\geq c$, say the results are inconsistent with the null hypothesis, suggesting that the distribution of Y in the population does not equal the specified distribution.

7. Collect a random sample of observations on the variable Y. Calculate the test statistic in step 2. Use the decision rule in step 6 to decide whether the distribution of Y in the population appears to equal the specified distribution. Draw conclusions based on the experimental results.

We commonly say that the sample size is large enough to use the chi-square approximation if no expected frequency E_i is less than 1 and no more than 20% of the expected frequencies are less than 5. This is called *Cochran's rule.*

Cochran's rule states that the chi-square approximation (for large-sample tests about one or more qualitative variables) is adequate if no expected frequency is less than 1 and no more than 20% of the expected frequencies are less than 5.

When the sample size is not large enough to justify the chi-square approximation, we may be able to formulate an exact test. For instance, in Section 16-2 we discuss small-sample inference about a qualitative variable that has two categories.

EXAMPLE 16-1 *(continued)*

In Example 16-1, the hypotheses are:

H_0: The distribution across categories for women with tubal infertility is the same as the control distribution.

H_a: The distribution for women with tubal infertility is not the same as the control distribution.

The population includes women with tubal infertility and no children, similar to women in the sample. However, the sample of 283 women is not a random sample from this population. Rather, they were women evaluated for infertility at seven infertility centers over a 30-month period. For our inferences to make sense, we must assume these women are representative of a larger population of interest. We must also assume that the observations for different women are independent. In addition, the control group women must make up a reasonable comparison group. If all these assumptions are met, then under the null hypothesis the test statistic has approximately the chi-square distribution with 3 degrees of freedom.

We will use significance level $\alpha = .005$. Looking in Table E, we find $c = 12.84$. The acceptance region is $[0, 12.84)$ and the rejection region is $[12.84, \infty)$. The decision rule is:

If test statistic < 12.84, say the results are consistent with the null hypothesis.
If test statistic ≥ 12.84, say the results are inconsistent with the null hypothesis.

The calculations for finding the test statistic are outlined in Table 16-1. The test statistic equals 21.5, in the rejection region. This suggests that the distribution for women with tubal infertility is not the same as the distribution for women in the control group.

Comparing the observed and expected frequencies in Table 16-1, we see that the largest differences are in categories 2 and 4. About 20 more women in

TABLE 16-1 Calculating the chi-square goodness-of-fit test statistic for Example 16-1

Category number	Category label	Observed frequency	Category probability under H_0	Expected frequency under H_0
1	None	36	.139	$283 \times .139 = 39.337$
2	Intrauterine device	39	.069	$283 \times .069 = 19.527$
3	Oral contraceptives	144	.527	$283 \times .527 = 149.141$
4	Barrier methods	64	.265	$283 \times .265 = 74.995$
	Totals	283	1.000	283

$$\text{Test statistic} = \frac{(36 - 39.337)^2}{39.337} + \frac{(39 - 19.527)^2}{19.527} + \frac{(144 - 149.141)^2}{149.141} + \frac{(64 - 74.995)^2}{74.995}$$
$$= .283 + 19.419 + .177 + 1.612 \doteq 21.5$$

Degrees of freedom $= 4 - 1 = 3$

the sample were in the intrauterine device category than expected from the control group distribution. About 11 fewer women in the sample were in the barrier methods category than expected from the control group distribution. Compared with the control group, a greater proportion of women with tubal infertility reported intrauterine device as the method of contraception longest used. A smaller proportion of women with tubal infertility reported barrier methods as longest used, compared with the control women.

What do these results suggest to you about a possible relationship between tubal infertility and method of contraception? How might you interpret these results? What suggestions would you have for further research?

When there are just two categories, we can think of the chi-square goodness-of-fit test as a large-sample test about a proportion. We compare the test statistic with the chi-square distribution with 1 degree of freedom. The resulting p-values are exactly the same as we get using the large-sample test for a proportion based on the standard Gaussian distribution (see Section 10-2).

Section 16-2 discusses inferences about a proportion when the sample size is small.

16-2 Small-Sample Inference About a Proportion Based on a Binomial Distribution

Suppose we have a random sample of observations from a population. Each observation has two possible outcomes; call them success and failure. We want to test hypotheses about the probability p of success, or the proportion p of successes in the population. When the sample size is small, we can base inferences on a binomial distribution. (We did this as a special case when we used the sign test in Section 10-5.) Consider the following example.

EXAMPLE 16-2

Researchers wanted to study the relationship between abnormal sex chromosome genotypes and criminal behavior among men in Denmark (Witkin et al., 1976). Since men with abnormal sex chromosome genotypes tend to be taller than average, the researchers included only men at least 184 centimeters in height. They identified 16 men with an extra X chromosome (the XXY genotype). Three of these 16 men had been convicted of at least one crime.

The researchers included more than 4,000 men who were at least 184 centimeters tall and had the normal XY genotype. Of these men, 9.3% had been convicted of one or more crimes.

Define the crime rate of a group to be the proportion in the group who have been convicted of at least one crime. Does this study suggest that the crime rate for men with the XXY genotype is different from the crime rate for men with the XY genotype? We will base our inferences on a binomial distribution. Let's outline the general approach to hypothesis testing and then apply it to this example.

The significance level approach to small-sample inference about a proportion p

1. The hypotheses are H_0: $p = p_0$ and H_a: $p \neq p_0$, where p_0 is a specified number between 0 and 1.
2. The test statistic is the observed number of successes in the sample.
3. We assume that we have a random sample of observations from a population. Each observation has two possible outcomes, success and failure. Then under the null hypothesis, the test statistic has the binomial(n, p_0) distribution.
4. Select significance level α.
5. Let X denote a random variable having the binomial(n, p_0) distribution. Find c_1 and c_2 so that $P(X \leq c_1) = \alpha/2$ and $P(X \geq c_2) \doteq \alpha/2$. The acceptance region is (c_1, c_2). The rejection region includes $[0, c_1]$ and $[c_2, n]$.
6. The decision rule is:

   If $c_1 <$ test statistic $< c_2$, say the results are consistent with the null hypothesis that the proportion p of successes in the population equals p_0.

   If test statistic $\leq c_1$ or test statistic $\geq c_2$, say the results are inconsistent with the null hypothesis, suggesting that the proportion p of successes in the population does not equal p_0.
7. Collect a sample that satisfies the assumptions in step 3. Find the test statistic in step 2. Use the decision rule in step 6 to decide whether the proportion of successes in the population equals p_0. Draw conclusions based on the experimental results.

EXAMPLE 16-2 *(continued)*

In Example 16-2, let p denote the proportion of Danish men with the XXY genotype who have been convicted of at least one crime. We want to test the hypotheses H_0: $p = .093$ and H_a: $p \neq .093$.

The 16 men in the sample were not selected at random from all Danish men with the XXY genotype. Instead, they were identified from a group of over 4,000 Danish men at least 184 centimeters in height. For our purposes, we will assume that the 16 observations are independent and come from the same

TABLE 16-2 Binomial probability distribution for sample size 16 and probability of success .093

Possible value, k	$P(\text{test statistic} = k \text{ when } H_0 \text{ is true}) = \binom{16}{k}(.093)^k(.907)^{16-k}$
0	.2098
1	.3441
2	.2646
3	.1266
4	.0422
5	.0104
6	.0020
7	.0003
8	.0000
9	.0000
10	.0000
11	.0000
12	.0000
13	.0000
14	.0000
15	.0000
16	.0000

probability distribution. The test statistic equals the number in the sample who have been convicted of at least one crime. With our assumptions, the test statistic has the binomial(16, .093) distribution under the null hypothesis. The probabilities for the binomial(16, .093) distribution are shown in Table 16-2.

We will use significance level $\alpha = .05$. From Table 16-2, we see that the probability of 0 successes under the null hypothesis is .2098, much larger than $\alpha/2 = .025$. Therefore, only large values of the test statistic are inconsistent with the null hypothesis. We will let $c = 4$. Under the null hypothesis, the probability of seeing a test statistic greater than or equal to 4 is .0549. The acceptance region is [0, 3], the rejection region is [4, 16], and the decision rule is:

If test statistic ≤ 3, say the results are consistent with the null hypothesis.
If test statistic ≥ 4, say the results are inconsistent with the null hypothesis.

Since three of the 16 men with the XXY genotype had been convicted of at least one crime, the test statistic equals 3, in the acceptance region. The p-value = .1815, the probability of a test statistic at least as extreme as 3 under the null hypothesis. (To get the p-value for a two-sided test about a proportion based on a binomial distribution, we add all the probabilities in the null hypothesis distribution that are less than or equal to the probability for the observed test statistic.) In this study of Danish men at least 184 centimeters tall, the difference in crime rate for men with the XXY genotype and for men with the normal XY genotype is not statistically significant. The sample results are consistent with the null hypothesis that the crime rate for men with the XXY genotype is the same as for men with the normal XY genotype.

In the remainder of this chapter, we consider inferences about the relationship between two qualitative variables.

16-3

The Chi-Square Test of Independence of Two Qualitative Variables

Suppose we have a random sample of observations selected from a population. An observation is classified according to each of two qualitative variables. We want to test the null hypothesis that the two variables are distributed independently in the population. If the variables are independent, we say there is no association between them. We test our hypotheses using a chi-square test of association called the *chi-square test of independence*. Consider the following example.

EXAMPLE 16-3 Investigators wanted to evaluate a no-smoking policy in a health maintenance organization with over 6,000 employees (Rosenstock, Stergachis, and Heaney, 1986). Four months after establishment of a smoking ban, the investigators selected 687 employees at random for an opinion survey. They mailed each of these employees a questionnaire to be answered and returned anonymously. Sixty-three percent, or 434, of these 687 employees returned the questionnaire. Distributions of age, sex, and length of employment for the respondents were similar to those for the entire employee population. Respondents provided information on smoking status and approval of the no-smoking policy, with results shown below.

	Approval of the smoking ban			
Smoking status	Approve	Do not approve	Not sure	Total
Never smoked	237	3	10	250
Ex-smoker	106	4	7	117
Current smoker	24	32	11	67
Total	367	39	28	434

Was there an association between smoking status and approval of the smoking ban in this employee population? Equivalently, was approval of the smoking ban independent of smoking status? We will use the chi-square test of independence to evaluate these questions.

Before outlining the test procedure, we need some notation. Suppose variable I has r categories and variable II has c categories. Let O_{ij} denote the number of observations classified into category i of variable I and category j

of variable II. We might arrange our observations into a two-way frequency table having r rows and c columns (called an $r \times c$ frequency table), with the r categories of variable I forming the rows and the c categories of variable II forming the columns. Then O_{ij} appears in row i and column j of the frequency table:

	Variable II						
Variable I	Category 1	Category 2	...	Category j	...	Category c	Total
Category 1	O_{11}	O_{12}	...	O_{1j}	...	O_{1c}	R_1
Category 2	O_{21}	O_{22}	...	O_{2j}	...	O_{2c}	R_2
⋮							
Category i	O_{i1}	O_{i2}	...	O_{ij}	...	O_{ic}	R_i
⋮							
Category r	O_{r1}	O_{r2}	...	O_{rj}	...	O_{rc}	R_r
Total	C_1	C_2	...	C_j	...	C_c	n

Let R_i denote the total number of observations in category i of variable I, C_j the total number of observations in category j of Variable II, and n the total sample size. With this notation, we can now outline the steps for the chi-square test of independence.

The significance level approach to the chi-square test of independence of two qualitative variables

1. The null hypothesis states that the two variables are independent. The alternative hypothesis states that the two variables are not independent.
2. The expected frequency in category i of variable I and category j of variable II under the null hypothesis is

$$E_{ij} = \frac{R_i \times C_j}{n}$$

   The test statistic is a measure of how far the observed frequencies differ from what we would expect under the null hypothesis:

$$\text{Test statistic} = \sum_{i=1}^{r} \sum_{j=1}^{c} \frac{(O_{ij} - E_{ij})^2}{E_{ij}}$$

   Small values of the test statistic occur when observed and expected frequencies are close to one another, consistent with the null hypothesis. Large values of the test statistic occur when some observed and expected frequencies are not close, inconsistent with the null hypothesis.
3. Assume that we have a random sample of size n from the population. Each observation can be classified into exactly one of the r categories of variable

I and exactly one of the c categories of variable II. Under the null hypothesis that variables I and II are independent, the test statistic has approximately the chi-square distribution with $(r - 1)(c - 1)$ degrees of freedom.

4. Select significance level α.
5. Let X denote a random variable having the chi-square distribution with $(r - 1)(c - 1)$ degrees of freedom. Find c from Table E such that $P(X \leq c) = 1 - \alpha$. The acceptance region is $[0, c)$; the rejection region is $[c, \infty)$.
6. The decision rule is:

   If test statistic $< c$, say the results are consistent with the null hypothesis that variables I and II are independently distributed in the population.

   If test statistic $\geq c$, say the results are inconsistent with the null hypothesis, suggesting that variables I and II are not independently distributed in the population.
7. Collect a sample that satisfies the assumptions in step 3. Calculate the test statistic in step 2. Use the decision rule in step 6 to decide whether variables I and II appear to be independent in the population. Draw conclusions based on the experimental results.

Because the test statistic is based on frequencies, its exact probability distribution is discrete. We say the sample size is large enough to use the chi-square approximation if no expected frequency is less than 1 and no more than 20% of the expected frequencies are less than 5 (Cochran's rule).

EXAMPLE 16-3 *(continued)*

In Example 16-3, we want to test the hypotheses:

H_0: Smoking status and approval of the smoking ban are independent in the employee population.

H_a: Smoking status and approval of the smoking ban are not independent in the employee population.

The investigators selected 687 employees at random to receive questionnaires. However, the 434 actual respondents compose a self-selected subgroup of this random sample. In all survey situations, we must worry about how nonrespondents may differ from respondents. Perhaps in this example, people approving of the smoking ban were more likely to respond to the survey. Or perhaps they were less likely to respond. In any case, we must be careful in interpreting results of the survey.

We will assume that the respondents are representative of a larger group of employees who would have responded to the questionnaire if they had received it. Our inferences apply only to this subgroup of the entire employee population. We also assume that the respondents answered independently. If these assumptions hold, then under the null hypothesis the test statistic has approximately the chi-square distribution with $(3 - 1)(3 - 1) = 4$ degrees of freedom.

We will use significance level $\alpha = .005$. Referring to Table E, we find $c = 14.86$. The acceptance region is $[0, 14.86)$, the rejection region is $[14.86, \infty)$, and the decision rule is:

If test statistic < 14.86, say the results are consistent with the null hypothesis that smoking status and approval of the smoking ban are independent.
If test statistic ≥ 14.86, say the results are inconsistent with the null hypothesis, suggesting that smoking status and approval of the smoking ban are not independent.

If smoking status and approval of the smoking ban were independent in this employee population, we would expect the same distribution for smoking status within each approval category. Equivalently, we would expect the same distribution for approval of the ban within each smoking status category. The expected frequencies under the null hypothesis are shown below:

	Approval of the smoking ban		
Smoking status	Approve	Do not approve	Not sure
Never smoked	$\frac{250 \times 367}{434} = 211.4$	$\frac{250 \times 39}{434} = 22.5$	$\frac{250 \times 28}{434} = 16.1$
Ex-smoker	$\frac{117 \times 367}{434} = 98.9$	$\frac{117 \times 39}{434} = 10.5$	$\frac{117 \times 28}{434} = 7.5$
Current smoker	$\frac{67 \times 367}{434} = 56.7$	$\frac{67 \times 39}{434} = 6.0$	$\frac{67 \times 28}{434} = 4.3$

From these observed and expected frequencies, we calculate the test statistic:

$$\begin{aligned}\text{Test statistic} &= \frac{(237 - 211.4)^2}{211.4} + \frac{(3 - 22.5)^2}{22.5} + \frac{(10 - 16.1)^2}{16.1} \\ &\quad + \frac{(106 - 98.9)^2}{98.9} + \frac{(4 - 10.5)^2}{10.5} + \frac{(7 - 7.5)^2}{7.5} \\ &\quad + \frac{(24 - 56.7)^2}{56.7} + \frac{(32 - 6.0)^2}{6.0} + \frac{(11 - 4.3)^2}{4.3} \\ &= 3.1 + 16.9 + 2.3 + .5 + 4.0 + .0 + 18.9 + 112.7 + 10.4 \\ &= 168.8\end{aligned}$$

This large test statistic is in the rejection region. The survey results strongly suggest that smoking status and approval of the smoking ban are not independent in the population of employees who would have responded to the questionnaire.

Another way to look at the responses is with the two-way frequency table in Table 16-3. Looking at row percentages, we see that 95% of the respondents who had never smoked and 91% of the former smokers approved the smoking ban. Only 36% of the current smokers approved.

The column percentages in Table 16-3 show that among respondents who approved the smoking ban, 65% had never smoked, 29% were former smokers, and 7% were current smokers. In contrast, 82% of respondents who

TABLE 16-3 Frequency table for Example 16-3. Row percentages are shown in parentheses to the right of the observed frequencies. Column percentages are shown in parentheses below the observed frequencies.

	Approval of the smoking ban			
Smoking status	Approve	Do not approve	Not sure	Total
Never smoked	237 (95) (65)	3 (1) (8)	10 (4) (36)	250 (58)
Ex-smoker	106 (91) (29)	4 (3) (10)	7 (6) (25)	117 (27)
Current smoker	24 (36) (7)	32 (48) (82)	11 (16) (39)	67 (15)
Total	367 (85)	39 (9)	28 (6)	434

did not approve the smoking ban were current smokers. This is strong evidence that smoking status and approval of the smoking ban were not independent in the employee population of potential respondents.

In Section 16-4, we discuss the chi-square test of homogeneity for comparing the distribution of a qualitative variable across several populations.

16-4.

Comparing the Distribution of a Qualitative Variable Across Populations

Suppose we have independent random samples, one from each of c populations. We can classify an observation into exactly one of r categories of a qualitative variable we will call variable I. We want to test the null hypothesis that the probability distribution of variable I is the same in each population. The alternative states that the probability distribution of variable I is not the same in all c populations. If the null hypothesis is true, we say the populations are *homogeneous* (alike or the same) with respect to the probability distribution of variable I. We test our hypotheses using the *chi-square test of homogeneity*.

We can think of population as a second qualitative variable in this situation, say variable II. The test statistic and the steps for carrying out the chi-square test of homogeneity are exactly the same as we outlined for the chi-square test of independence, in Section 16-3.

Let's illustrate the chi-square test of homogeneity with an example.

EXAMPLE 16-4

We will compare the distributions of body weight classification for 20–24-year-old women in Britain, Canada, and the United States. We base our inferences on results of surveys carried out during 1976–1981 (Millar and Stephens,

1987). Investigators obtained the samples from the noninstitutionalized populations in the three countries. They used similar techniques in all three surveys.

The investigators defined body weight classification in terms of the Quetelet index. The Quetelet index is weight in kilograms divided by the square of height in meters (so it has units kilograms/meter^2).

A woman is classified as underweight if her Quetelet index is less than or equal to 20 kg/m^2. She has a normal body weight classification if her Quetelet index is greater than 20 kg/m^2 and less than or equal to 25 kg/m^2. She is classified as overweight if her Quetelet index is greater than 25 kg/m^2 and less than or equal to 30 kg/m^2. She has the obese body weight classification if her Quetelet index is greater than 30 kg/m^2.

The researchers studied 547 20–24-year-old women from Britain, 873 from Canada, and 624 from the United States. The women in the United States sample were all white. (Why?) When each woman is classified by country and body weight, we have the following results:

Body weight classification	Britain	Canada	United States	Total
Underweight	126	297	156	579
Normal	306	498	349	1,153
Overweight	88	61	75	224
Obese	27	17	44	88
Total	547	873	624	2,044

We want to test the hypotheses:

H_0: The distribution of 20–24-year-old women across body weight classifications is the same for Britain, Canada, and the United States.

H_a: The distribution of 20–24-year-old women across body weight classifications is not the same for Britain, Canada, and the United States.

We must assume that the sample from each country is a random sample from the population of noninstitutionalized 20–24-year-old (white) women in that country. We also assume that the measurement techniques for determining the Quetelet index were uniform and independent for different women. Then under the null hypothesis, the test statistic has approximately the chi-square distribution with $(4 - 1)(3 - 1) = 6$ degrees of freedom.

We will use significance level $\alpha = .005$. From Table E, we find $c = 18.55$. The acceptance region is $[0, 18.55)$, the rejection region is $[18.55, \infty)$, and the decision rule is:

If test statistic < 18.55, say the results are consistent with the null hypothesis.
If test statistic ≥ 18.55, say the results are inconsistent with the null hypothesis.

The expected frequencies under the null hypothesis are:

Body weight	Britain	Canada	United States
Underweight	$\frac{579 \times 547}{2,044} = 154.9$	$\frac{579 \times 873}{2,044} = 247.3$	$\frac{579 \times 624}{2,044} = 176.8$
Normal	$\frac{1,153 \times 547}{2,044} = 308.6$	$\frac{1,153 \times 873}{2,044} = 492.5$	$\frac{1,153 \times 624}{2,044} = 352.0$
Overweight	$\frac{224 \times 547}{2,044} = 59.9$	$\frac{224 \times 873}{2,044} = 95.7$	$\frac{224 \times 624}{2,044} = 68.4$
Obese	$\frac{88 \times 547}{2,044} = 23.5$	$\frac{88 \times 873}{2,044} = 37.6$	$\frac{88 \times 624}{2,044} = 26.9$

From these observed and expected frequencies, we calculate the test statistic:

$$\begin{aligned}\text{Test statistic} &= \frac{(126 - 154.9)^2}{154.9} + \frac{(297 - 247.3)^2}{247.3} + \frac{(156 - 176.8)^2}{176.8} \\ &\quad + \frac{(306 - 308.6)^2}{308.6} + \frac{(498 - 492.5)^2}{492.5} + \frac{(349 - 352.0)^2}{352.0} \\ &\quad + \frac{(88 - 59.9)^2}{59.9} + \frac{(61 - 95.7)^2}{95.7} + \frac{(75 - 68.4)^2}{68.4} \\ &\quad + \frac{(27 - 23.5)^2}{23.5} + \frac{(17 - 37.6)^2}{37.6} + \frac{(44 - 26.9)^2}{26.9} \\ &= 5.39 + 9.99 + 2.45 + .02 + .06 + .03 \\ &\quad + 13.18 + 12.58 + .64 + .52 + 11.29 + 10.87 = 67.0\end{aligned}$$

The test statistic is in the rejection region. This comparison of survey results strongly suggests that the distribution of body weight classification for 20–24-year-old women is not the same for Britain, Canada, and the United States.

Let's compare the observed and expected frequencies. In the samples from Britain and the United States, there were fewer underweight women and more overweight women than expected under the null hypothesis. In the sample from Canada, there were more underweight women and fewer overweight women than expected under the null hypothesis. We could also look at a frequency table showing row and column percentages, as we did in Table 16-3. What conclusions would you draw based on an analysis of these surveys?

Suppose we have independent random samples from each of several populations. An observation can be classified into exactly one of two categories of a qualitative variable. Call these categories success and failure. We want to compare the proportion in the success category across the populations. Comparing proportions across several populations is a special case of the situation we have discussed in this section, so we can apply the chi-square test of homogeneity.

Now suppose we want to compare two proportions. That is, we want to compare the proportion in the success category for two populations. Then we use the chi-square test of homogeneity, with 1 degree of freedom. Resulting p-values are the same as p-values we get using the large-sample test for two proportions based on the standard Gaussian distribution (Section 11-2).

In Section 16-5, we discuss an exact test for association in a 2 × 2 frequency table, based on a hypergeometric distribution.

16-5

Testing for Association in a 2 × 2 Frequency Table, Using a Hypergeometric Distribution

We will consider two sampling situations. In each situation, we want to test for association in a 2 × 2 frequency table.

In the first situation, we have a random sample from a population. An observation is classified according to each of two qualitative variables—variables I and II. Each variable has two categories. We want to test the hypotheses:

Null hypothesis: The two variables are independent in the population.
Alternative hypothesis: The two variables are not independent in the population.

In the second situation, we have two independent random samples, one from each of two populations. An observation is classified into one of two categories of a qualitative variable—call it variable I. We can think of population as variable II. We want to test the hypotheses:

Null hypothesis: The distribution of variable I is the same in the two populations.
Alternative hypothesis: The distribution of variable I is not the same in the two populations.

For both sampling situations, we can use a test of association called *Fisher's exact test.* The test is based on a hypergeometric distribution, as outlined below.

The p-value approach to Fisher's exact test

1. We state one of the two sets of hypotheses just given, depending on the sampling situation.
2. Suppose we write the 2 × 2 frequency table for our sample as follows:

	Variable II		
Variable I	Category 1	Category 2	Total
Category 1	O_{11}	O_{12}	R_1
Category 2	O_{21}	O_{22}	R_2
Total	C_1	C_2	n

The test statistic equals the smallest of the observed frequencies O_{11}, O_{12}, O_{21}, and O_{22}.

3. Make the assumptions for one of the two sampling situations. Then under the null hypothesis, the conditional probability of the observed results, given the row totals R_1 and R_2 and the column totals C_1 and C_2, is a hypergeometric probability that we can write as

$$P(\text{observed results under } H_0 \text{ given observed row and column totals}) = \frac{R_1!\, R_2!\, C_1!\, C_2!}{O_{11}!\, O_{12}!\, O_{21}!\, O_{22}!\, n!}$$

4. Collect observations that satisfy the appropriate set of assumptions. Let the test statistic equal the smallest of the observed frequencies, O_{11}, O_{12}, O_{21}, and O_{22}.
5. Let m be the minimum of R_1, R_2, C_1, and C_2. Then m is the smallest frequency observed in a category for either variable I or variable II. There are $m + 1$ possible 2×2 frequency tables having row totals R_1 and R_2 and column totals C_1 and C_2. Given these fixed row and column totals, find the probability under the null hypothesis of each of these 2×2 frequency tables, using the formula given in step 3.

   Possible results as extreme as or more extreme (in the direction of the alternative) than those observed are those corresponding to frequency tables with probabilities less than or equal to the probability of the observed table. The *p-value* is the sum of all the probabilities that are less than or equal to the probability for the observed table. If we have a one-sided alternative, the p-value is the sum of only the probabilities in the direction indicated by the alternative.
6. If the p-value is large, say the results are consistent with the null hypothesis. If the p-value is small, say the results are inconsistent with the null hypothesis. Draw conclusions based on the experimental results.

Let's look at an example of the first sampling situation. We have a random sample from a population. Observations are classified according to each of two qualitative variables (each with two categories). We want to test the null hypothesis that the two variables are independent in the population.

EXAMPLE 16-5

Investigators studied the relationship between abnormal sex chromosome genotypes and criminal behavior among men in Denmark (Witkin et al., 1976). As indicated in Example 16-2, they included only men at least 184 centimeters tall. The investigators identified 16 men with an extra X chromosome (genotype XXY) and 12 men with an extra Y chromosome (genotype XYY).

Three of the 16 men with the XXY genotype and 5 of the 12 men with the XYY genotype had been convicted of at least one crime. Does this suggest any difference between men with the two genotypes with respect to conviction of a crime? We will use Fisher's exact test to address this question.

Define a variable called convictions, with two categories: none and at least one. The variable genotype also has two categories: XXY and XYY. We want to test the hypotheses:

H_0: The variables genotype and convictions are independent in the population.
H_a: The variables genotype and convictions are not independent in the population.

As we know from Example 16-2, the investigators did not select these 28 men as a random sample from a larger population of men with XXY or XYY genotypes. Instead, they were identified from among over 4,000 Danish men at least 184 centimeters tall. For our purposes, we will assume that the relationship between genotype and convictions in the sample of 28 men is representative of the relationship between these two variables in the larger population of Danish men at least 184 centimeters tall, with either the XXY or XYY genotype. We also assume that the 28 observations are independent of one another. Then we can calculate conditional probabilities based on a hypergeometric distribution, under the null hypothesis.

We summarize the observations in a 2 × 2 frequency table:

	Genotype		
Convictions	XXY	XYY	Total
None	13	7	20
At least one	3	5	8
Total	16	12	28

The test statistic equals 3, the smallest frequency in the table. The conditional probability for the observed frequency table under the null hypothesis is

$$P(\text{observed results under } H_0, \text{ given } R_1 = 20, R_2 = 8, C_1 = 16, C_2 = 12)$$
$$= \frac{\binom{8}{3}\binom{20}{13}}{\binom{28}{16}} = \frac{20!\ 8!\ 16!\ 12!}{13!\ 7!\ 3!\ 5!\ 28!}$$
$$= .1427$$

This hypergeometric probability equals the number of ways we could select 3 of the 8 men with at least one conviction and 13 of the 20 men with no convictions to have the XXY genotype, divided by the number of ways we could select 16 of the 28 men to have the XXY genotype. We could calculate this same probability as

$$P(\text{observed results under } H_0, \text{ given } R_1 = 20, R_2 = 8, C_1 = 16, C_2 = 12) = \frac{\binom{16}{3}\binom{12}{5}}{\binom{28}{8}} = \frac{20!\ 8!\ 16!\ 12!}{13!\ 7!\ 3!\ 5!\ 28!} = .1427$$

This hypergeometric probability is the number of ways we could select 3 of the 16 XXY men and 5 of the 12 XYY men to have at least one conviction, divided by the number of ways we could select 8 of the 28 men to have at least one conviction.

The smallest row or column total is 8. There are 8 + 1 = 9 possible 2 × 2 tables having the same row and column totals as the observed table. These nine possible frequency tables are listed in Table 16-4, along with the associated conditional probability under the null hypothesis.

TABLE 16-4 For Example 16-5, the conditional probability of each possible frequency table under the null hypothesis, given 16 men with the XXY genotype and 20 men with no convictions

Possible frequency table		Conditional probability under the null hypothesis
16	4	$\frac{20!\ 8!\ 16!\ 12!}{16!\ 4!\ 0!\ 8!\ 28!} = .0002$
0	8	
15	5	$\frac{20!\ 8!\ 16!\ 12!}{15!\ 5!\ 1!\ 7!\ 28!} = .0041$
1	7	
14	6	$\frac{20!\ 8!\ 16!\ 12!}{14!\ 6!\ 2!\ 6!\ 28!} = .0357$
2	6	
13	7	$\frac{20!\ 8!\ 16!\ 12!}{13!\ 7!\ 3!\ 5!\ 28!} = .1427$
3	5	
12	8	$\frac{20!\ 8!\ 16!\ 12!}{12!\ 8!\ 4!\ 4!\ 28!} = .2899$
4	4	
11	9	$\frac{20!\ 8!\ 16!\ 12!}{11!\ 9!\ 5!\ 3!\ 28!} = .3092$
5	3	
10	10	$\frac{20!\ 8!\ 16!\ 12!}{10!\ 10!\ 6!\ 2!\ 28!} = .1700$
6	2	
9	11	$\frac{20!\ 8!\ 16!\ 12!}{9!\ 11!\ 7!\ 1!\ 28!} = .0442$
7	1	
8	12	$\frac{20!\ 8!\ 16!\ 12!}{8!\ 12!\ 8!\ 0!\ 28!} = .0041$
8	0	

Possible frequency tables at least as extreme (in the direction of the alternative) as observed are those with probabilities in Table 16-4 less than or equal to .1427. The p-value is the sum of the corresponding probabilities, so p-value $= .231$. This p-value is fairly large, so our results are consistent with the null hypothesis. From this sample, we cannot see an association between convictions (none or at least one) and genotype (XXY or XYY).

Now let's look at an example of the second sampling situation. We have two independent random samples, one from each of two populations. An observation can be classified into exactly one of two categories of a qualitative variable. We want to test the null hypothesis that the variable has the same distribution in both populations. Equivalently, we want to test whether the proportion in the "success" category is the same in the two populations.

EXAMPLE 16-6

Researchers wanted to study modes of transmission of common cold viruses (Dick et al., 1986). Twenty-four uninfected men and 16 men infected with the common cold participated. Volunteers played poker at tables of five, with two infected and three uninfected men at each table, for 12 hours. Twelve uninfected volunteers were restrained with large collars or arm braces to prevent infection by contact; thus, they could be infected only by airborne viruses. The other 12 uninfected volunteers were not restrained; they could be infected through either airborne viruses or hand contamination. The researchers monitored all 24 uninfected volunteers after the experiment, for subsequent infection with a common cold. The results of the experiment follow.

Developed a common cold	Experimental restraint		
	Restrained	Unrestrained	Total
Yes	6	11	17
No	6	1	7
Total	12	12	24

Does the restraint affect the likelihood of developing a cold? We will test the hypotheses:

H_0: The proportion developing a common cold is the same for restrained and unrestrained volunteers.

H_a: The proportion developing a common cold is smaller for restrained than for unrestrained volunteers.

The two populations sampled are hypothetical. We assume that the volunteers are representative of a larger group of interest. The first experimental population consists of the (hypothetical) responses of this larger group if they were all exposed under the restrained condition. The second experimental population consists of their responses if they were all exposed under the unrestrained condition.

We must assume that the observations are independent. That is, one volunteer developing a cold (or not) does not in any way affect the likelihood of another volunteer developing a cold. We also assume that the volunteers in the restrained and unrestrained experimental groups have similar likelihoods of developing colds if exposed to similar conditions. The best way to ensure this is to select volunteers as similar as possible and then randomly divide them into two experimental groups. We have no way of checking any of these assumptions without more information on how the experiment was conducted.

The test statistic equals 1, the smallest frequency in the table. The conditional probability under the null hypothesis of the observed frequency table is

$$P(\text{observed results under } H_0 \text{, given } R_1 = 17, R_2 = 7, C_1 = 12, C_2 = 12)$$
$$= \frac{\binom{12}{1}\binom{12}{6}}{\binom{24}{7}} = \frac{12!\ 12!\ 7!\ 17!}{1!\ 11!\ 6!\ 6!\ 24!}$$
$$= .032$$

This hypergeometric probability equals the number of ways we could select 1 of the 12 unrestrained volunteers and 6 of the 12 restrained volunteers to remain cold-free, divided by the number of ways we could select 7 of the 24 volunteers to remain cold-free.

What possible results are more extreme (in the direction of the alternative) than those observed? The alternative hypothesis states that restrained volunteers are less likely to develop a cold than unrestrained volunteers. Therefore, seeing no unrestrained volunteers remain free of a cold (test statistic = 0) would be more extreme in the direction of the alternative. The associated frequency table is shown in Table 16-5.

TABLE 16-5 This table displays results that are more extreme in the direction of the alternative hypothesis (that restrained volunteers are less likely to develop colds than unrestrained volunteers) than the observed results in Example 16-6

Developed a common cold	Experimental restraint: Restrained	Unrestrained	Total
Yes	5	12	17
No	7	0	7
Total	12	12	24

Under the null hypothesis, the conditional probability for the frequency table in Table 16-5 is

$$\frac{\binom{12}{0}\binom{12}{7}}{\binom{24}{7}} = \frac{12!\ 12!\ 7!\ 17!}{0!\ 12!\ 7!\ 5!\ 24!} = .002$$

This hypergeometric probability is the number of ways we could select none of the 12 unrestrained volunteers and 7 of the 12 restrained volunteers to remain cold-free, divided by the number of ways we could select 7 of the 24 volunteers to remain cold-free.

The p-value equals $.032 + .002 = .034$. A p-value of .034 is fairly small, so we say the results are inconsistent with the null hypothesis. For the conditions and type of volunteer involved, restrained men seem to be less likely to develop colds than unrestrained men. This experiment suggests that people subject to both hand contamination and airborne transmission are more likely to develop colds than are people exposed only to airborne transmission of viruses.

In most hypothesis testing situations, we greatly simplify the goals of the experiment. The investigators in Example 16-6 were interested in the comparison we made. However, they were also interested in seeing whether the restrained men developed colds at all. (Some previous studies had suggested that airborne viruses were not likely to cause cold infection.) The finding that 6 of 12 restrained volunteers did subsequently develop colds suggested that airborne viruses might be a significant mode of cold transmission.

Summary of Chapter 16

The chi-square goodness-of-fit test is used to make large-sample inferences about the distribution of a qualitative variable. We compare the observed distribution of frequencies across categories with a specified distribution. When there are two categories, we can use the chi-square goodness-of-fit test for large-sample inferences about a proportion. This test is equivalent to the large-sample test based on the standard Gaussian distribution. When the sample size is small, we can base inferences about a proportion on a binomial distribution.

We consider inferences about two qualitative variables for two sampling situations. In the first sampling situation, we have a random sample from a population. An observation is classified according to each of two qualitative variables. We want to test the null hypothesis that the two variables are independent in the population. If the sample size is large enough, we can use the chi-square test of independence.

Statistics

Shelley Rasmussen

good: ① MINITAB Appendix

② more graphics

✗ ③ more about qualitative variables

✗ ④ ch 9, good! concepts related to statistical inference

✗ ⑤ more non-parametric tests

~~ex~~ exercise?

Weekness: ~~① too many chapters, & quite~~

① a few chs we may not want to teach stat 103 students.

e.g. part of ch 11-13 (14?)

② more mathematical than F.P.P.A.
too mathematical for 103!

In the second sampling situation, we have independent random samples, one from each of several populations. Each observation can be classified according to a qualitative variable. (We can think of population as a second qualitative variable.) We want to test the null hypothesis that the variable has the same distribution in each of the populations. With large enough sample sizes, we can use the chi-square test of homogeneity. We can use this test to compare two proportions when sample sizes are large; this test is equivalent to the large-sample test based on the standard Gaussian distribution.

The same large-sample test statistic is used for both sampling situations. If the sample size is large enough, we can compare the test statistic with the chi-square distribution with $(r - 1)(c - 1)$ degrees of freedom. Here, r is the number of rows (categories of the first variable) and c the number of columns (categories of the second variable) in a two-way frequency table summarizing the observed results.

To decide whether the sample size is large enough to use a large-sample test about qualitative variables, we can apply Cochran's rule: No expected frequency should be less than 1 and no more than 20% of the expected frequencies less than 5.

If each of two qualitative variables has two categories, we can test for association using Fisher's exact test. Probabilities are based on a hypergeometric distribution. These null hypothesis probabilities are conditional on the observed row and column totals in the frequency table.

Minitab Appendix for Chapter 16

Carrying Out a Chi-Square Goodness-of-Fit Test

We can use Minitab to calculate the test statistic for the chi-square goodness-of-fit test. Consider Example 16-1. We have observed frequencies in four categories. We want to compare this observed distribution with a specified null hypothesis distribution. We will set the observed frequencies in column 1 of our worksheet and the category probabilities under the null hypothesis in column 2:

```
MTB>   set c1
DATA>  36 39 144 64
DATA>  end
MTB>   set c2
DATA>  0.139 0.069 0.527 0.265
DATA>  end
```

We will put the null hypothesis expected frequencies in column 3, then calculate the test statistic and find the p-value from the chi-square distribution with 3 degrees of freedom:

```
MTB>   sum c1 k1
 SUM  = 283.00
MTB>   let c3=c2*k1
MTB>   let c4=(c1-c3)**2/c3
MTB>   name c1 'observed' c3 'expected' c4 'chisq'
MTB>   sum c4 k2
 SUM  = 21.491
MTB>   cdf k2 k3;
SUBC>  chisquare 3.
MTB>   let k3=1-k3
MTB>   print 'observed' 'expected' 'chisq' k2 k3
K2     21.4914
K3     0.000083506

ROW   observed    expected       chisq
 1       36        39.337        0.2831
 2       39        19.527       19.4191
 3      144       149.141        0.1772
 4       64        74.995        1.6120
```

K2 = 21.4914 is the value of the test statistic and K3 = 0.000083506 is the large-sample approximate p-value, from the chi-square distribution with 3 degrees of freedom.

Carrying Out a Small-Sample Test About a Proportion

Consider now the small-sample test about a proportion, discussed in Section 16-2. In Example 16-2, 3 of 16 men in the sample had been convicted of a crime. We want to test the null hypothesis that the probability of conviction in the population equals .093. We print the probabilities for the binomial(16, .093) distribution, expected under the null hypothesis:

```
MTB>   pdf;
SUBC>  binomial 16 0.093.
```

Minitab will print the probabilities shown in Figure M16-1, as we listed in Table 16-2. Note that Minitab does not print probabilities less than zero to four decimal places. We then carry out the test as described in Section 16-2.

```
BINOMIAL WITH N =  16  P = 0.093000
      K          P( X = K)
      0            0.2098
      1            0.3441
      2            0.2646
      3            0.1266
      4            0.0422
      5            0.0104
      6            0.0020
      7            0.0003
      8            0.0000
```

FIGURE M16-1 Probabilities for the binomial(16, .093) distribution

```
Expected counts are printed below observed counts

              C1         C2         C3    Total
    1        237          3         10      250
          211.41      22.47      16.13

    2        106          4          7      117
           98.94      10.51       7.55

    3         24         32         11       67
           56.66       6.02       4.32

Total        367         39         28      434

ChiSq =   3.099 + 16.866 +  2.329 +
          0.504 +  4.036 +  0.040 +
         18.823 +112.100 + 10.315 = 168.111
df = 4
1 cells with expected counts less than 5.0
```

FIGURE M16-2 CHISQUARE output for Example 16-3

Carrying Out a Chi-Square Test of Association

Suppose we want to use the chi-square test of association in a two-way frequency table, as in Sections 16-3 and 16-4. We put the table into Minitab and use the CHISQUARE command to calculate the test statistic. Consider the data in Example 16-3. We enter the data onto our worksheet and then use the CHISQUARE command:

```
MTB>    read c1-c3
DATA>   237 3 10
DATA>   106 4 7
DATA>   24 32 11
DATA>   END
MTB>    chisquare c1-c3
```

We get the results shown in Figure M16-2.

We find the p-value with the CDF command:

```
MTB>    cdf 168.111 k1;
SUBC>   chisquare 4.
MTB>    let k2=1-k1
MTB>    print k1 k2
K1      1.00000
K2      0
```

Minitab does not have a command for Fisher's exact test (Section 16-5).

Exercises for Chapter 16

For each exercise, describe the population(s) sampled, whether real or hypothetical. For each procedure, state the assumptions that make it valid. Do these assumptions seem reasonable? What additional information would you like to have about the experiment? Describe the results of your analysis.

EXERCISE 16-1

In a genetic study of the relationship between tobacco mosaic virus and the 30-kD protein gene, scientists introduced a gene encoding the 30-kD protein into tobacco plants (Deom, Oliver, and Beachy, 1987). Of 40 seedlings from one such plant, 29 expressed the 30-kD protein and 11 did not. Of 100 seedlings from another plant, 93 expressed the 30-kD protein and 7 did not.

Let p denote the probability that a plant expresses the 30-kD protein. If the 30-kD protein gene is expressed from a single genetic locus, scientists expect a 3:1 ratio (with the 30-kD protein:without the 30-kD protein). That is, they expect $p = \frac{3}{4}$. (See Exercise 6-30.)

If the 30-kD protein gene is expressed at two genetic loci, scientists expect a 15:1 ratio (with the 30-kD protein:without the 30-kD protein). That is, they expect $p = \frac{15}{16}$.

State and test appropriate hypotheses, separately for the two plants. Discuss your findings.

EXERCISE 16-2

In a study of T-DNA insertion mutagenesis in *Arabidopsis thaliana* plants, scientists studied two traits: resistance to kanamycin (resistant or susceptible) and height (dwarf or tall). Scientists recorded the phenotypes (or observed characteristics) of offspring in five experiments with parent plants of the same genotype (or genetic make-up for these two traits). The numbers of offspring with each phenotype are shown below (Feldmann et al., 1989).

Experiment	Category 1 resistant/dwarf	Category 2 resistant/tall	Category 3 susceptible/tall
1	41	81	34
2	54	104	52
3	62	142	54
4	50	125	50
5	46	131	52

In each experiment, the scientists observed no offspring with the susceptible/dwarf phenotype. The researchers wanted to compare the observed results with the 1:2:1 ratio expected under Mendelian genetics. (That is, Mendelian genetics predicts $\frac{1}{4}$ of offspring in category 1, $\frac{1}{2}$ in category 2, and $\frac{1}{4}$ in category 3.)

State and test appropriate hypotheses for each of the five experiments. Discuss your findings.

EXERCISE 16-3

Researchers studied two characteristics of tomatoes: height (tall or dwarf) and leaf (cut-leaf or potato-leaf). The dominant characteristics are tall height and cut-leaf. In a dihybrid cross, both parents have one dominant and one recessive gene for both characteristics. In one such dihybrid cross, the researchers observed the following results (Devore, 1982, page 525; from "Linkage Studies of the Tomato," *Trans. Royal Canadian Institute,* 1931, pages 1–19):

Category:	Tall, cut-leaf	Tall, potato-leaf	Dwarf, cut-leaf	Dwarf, potato-leaf
Observed frequency:	926	288	293	104

Mendelian genetics tells us that if the height and leaf characteristics are distributed independently in such a dihybrid cross, we expect a 9:3:3:1 ratio of the four phenotypes (see Exercise 6-30). That is, we expect $\frac{9}{16}$ of the offspring in the tall, cut-leaf category, $\frac{3}{16}$ of the offspring in the tall, potato-leaf category, $\frac{3}{16}$ in the dwarf, cut-leaf category, and $\frac{1}{16}$ in the dwarf, potato-leaf category.

State and test appropriate hypotheses. Discuss your findings.

EXERCISE 16-4

In a study of the cereal crop sorghum, scientists self-crossed red-seeded plants to produce offspring with red, yellow, and white seeds (Devore, 1982, page 529; from "A Genetic and Biochemical Study on Pericarp Pigments in a Cross Between Two Cultivars of Grain Sorghum, Sorghum Bicolor," *Heredity,* 1976, pages 413–416). Genetic theory predicts red, yellow, and white seeds in a ratio of 9:3:4. That is, the theory predicts that $\frac{9}{16}$ of the seeds will be red, $\frac{3}{16}$ will be yellow, and $\frac{4}{16}$ will be white. The results are shown below.

Category:	Red seeds	Yellow seeds	White seeds
Observed frequency:	195	73	100

State and test appropriate hypotheses. Discuss your findings.

EXERCISE 16-5

Is time of onset of labor in pregnant women uniform across the 24 hours? To address this question, researchers recorded the time of onset of labor for 1,186 pregnant women. They considered 24 1-hour time categories, beginning at midnight. Their observations are shown below (Devore, 1982, pages 527–528; from "The Hour of Birth," *British J. Preventive and Social Medicine,* 1953, pages 43–59).

Hour	Frequency	Hour	Frequency	Hour	Frequency	Hour	Frequency
1	52	7	58	13	21	19	47
2	73	8	47	14	31	20	34
3	89	9	48	15	40	21	36
4	88	10	53	16	24	22	44
5	68	11	47	17	37	23	78
6	47	12	34	18	31	24	59

a. Plot the observations.

b. Test the null hypothesis that labor is just as likely to begin in any of the 24 1-hour periods.

c. Discuss your findings.

EXERCISE 16-6 In Exercise 11-2, we considered a study of two drugs that suppress abnormal heart rhythms, encainide and flecainide (*Science News,* April 29, 1989, volume 135, page 260). In this study, 730 patients received one of these two experimental drugs for cardiac arrhythmia and 730 patients received placebo. In an early review of results, a safety monitoring board found 33 of the 730 patients in the experimental drug group had experienced either sudden cardiac death or a nonfatal heart attack, compared with 9 of the 730 patients in the placebo group.

a. Use a large-sample test based on a chi-square distribution to test the null hypothesis that the probability of suffering heart attack or cardiac death is the same for the two treatments.

b. Compare these results with the results based on the standard Gaussian distribution (Exercise 11-2).

EXERCISE 16-7 Researchers wanted to investigate the response of acute myelogenous leukemia patients to an immunotherapy (Granatek et al., 1981). The researchers measured antibody response on 13 patients before and after immunotherapy treatment. At the end of the study period, they classified patients by the percent change in antibody response following treatment (greater than 40%, less than or equal to 40%) and by survival time (at least 160 weeks, less than 160 weeks).

	Percent change in antibody response	
Survival time	Greater than 40%	Less than or equal to 40%
At least 160 weeks	4	2
Less than 160 weeks	1	6

Use Fisher's exact test to test the null hypothesis that there is no association between these two qualitative variables in patients similar to those in the study.

EXERCISE 16-8 In Exercise 4-27 we considered a study of the response of beetles to an airborne sex pheromone. Scientists exposed 30 beetles at each of four dose rates (units not given) and recorded the number of beetles responding within 60 seconds (Nordheim, Tsiatis, and Shapas, 1983):

Dose rate	Number of beetles responding within 60 seconds
10^{-6}	2
10^{-5}	10
10^{-4}	17
10^{-3}	25

Test the null hypothesis that there is no difference in proportion of beetles responding within 60 seconds across the four dose rates. Discuss your findings.

EXERCISE 16-9 In a study in Western Australia, investigators classified adults by obesity and hypertension (Knuiman and Speed, 1988):

Hyper-tension	Obesity category		
	Low	Average	High
Yes	32	40	59
No	133	121	106

Test the null hypothesis that obesity classification and hypertension are independent in the population sampled. Discuss your findings.

EXERCISE 16-10 In a study of randomly selected Wisconsin high school seniors, investigators classified students by their college plans and by level of parental encouragement to attend college (Bonney, 1987; from Fienberg, 1977, page 101; originally from Sewell and Shah, 1968). Results are shown below separately for male and female students, all with high IQ scores and high socioeconomic status.

Males:	Plans to attend college	Parental encouragement	
		High	Low
	Yes	414	8
	No	54	17

Females:	Plans to attend college	Parental encouragement	
		High	Low
	Yes	360	13
	No	98	49

a. Test the null hypothesis that plans to attend college are independent of level of parental encouragement among high-IQ, high-socioeconomic-status male high school seniors.

b. Test the null hypothesis that plans to attend college are independent of level of parental encouragement among high-IQ, high-socioeconomic-status female high school seniors.

c. Does the relationship between plans to attend college and level of parental encouragement seem to be the same for males and females in the high-IQ, high-socioeconomic-status group?

EXERCISE 16-11

Women under 50 years of age who had breast cancer with minimal inflammation were classified by appearance of the tumor and 3-year survival (Bonney, 1987; from Morrison et al., 1973). The results for a group of patients diagnosed in Boston, Massachusetts, are shown below.

	Appearance	
3-year survival	Malignant	Benign
Yes	11	24
No	6	7

Is 3-year survival independent of tumor appearance among breast cancer patients under 50 years of age? State and test appropriate hypotheses.

EXERCISE 16-12

Researchers studied minor psychiatric disorders among patients receiving primary medical care (Grayson, 1987; from Goldberg et al., 1987). Two symptoms checked for 283 such patients were irritability and poor sleeping. The observed results are shown below.

	Irritability	
Poor sleeping	Yes	No
Yes	77	28
No	60	118

Does there appear to be an association between irritability and poor sleeping in patients receiving primary medical care? State and test appropriate hypotheses. Discuss the experimental results.

EXERCISE 16-13

Researchers painted a tobacco condensate at one of two dose levels onto the backs of mice. They treated 100 mice at each dose level. After 546 days, researchers classified each mouse into one of three categories: developed a skin tumor during the experimental period, died without a tumor before the end of the experiment, alive and no tumor at the end of the experiment. The results are shown below (Gart and Tarone, 1987; from Gart, 1976).

Termination category	Low dose	High dose
Developed tumor before end of experiment	34	53
Died without tumor before end of experiment	17	17
Alive and no tumor at end of experiment	49	30

Is the distribution across the three termination categories the same for the two dose levels? State and test appropriate hypotheses.

EXERCISE 16-14 If a child with sickle cell disease develops an overwhelming infection (called sepsis), he or she has a 30% chance of dying from it. In this study, researchers sought to prevent sepsis in children with sickle cell disease (Kolata, 1987). Of 215 children with sickle cell disease, researchers treated 110 with oral penicillin. They gave the other 105 children a placebo. The researchers followed the children for 15 months, on average. Over that period, 13 children suffered an overwhelming infection in the placebo group and 3 died. In the penicillin group, 2 children suffered an overwhelming infection and none died.

a. Test the null hypothesis that the probability of an overwhelming infection is the same for the two treatments.

b. Use Fisher's exact test to test the null hypothesis that the probability of death is the same for the two treatments.

c. Discuss your findings.

EXERCISE 16-15 In a study of the relationship between depression and coronary artery disease, researchers diagnosed 9 cases of major depression among 52 patients with newly diagnosed coronary artery disease. One year later, 14 of the 43 nondepressed patients had had at least one serious cardiac complication, compared with 7 of the 9 depressed patients (*Science News,* January 7, 1989, volume 135, page 13). Based on this sample, does there appear to be an association between cardiac complications and depression in patients with coronary artery disease?

a. Use a large-sample test to test the null hypothesis that depression and cardiac complications are independent in patients with coronary artery disease.

b. Use Fisher's exact test to test the null hypothesis that depression and cardiac complications are independent in patients with coronary artery disease.

c. Compare the results in parts (a) and (b). Discuss your findings.

EXERCISE 16-16 Early research suggested an association between chronic fatigue syndrome and the Epstein–Barr virus. To study this possibility, some scientists designed an experiment to test the effects of a drug known to stop replication of the Epstein–Barr virus. The experiment included 24 patients with Epstein–Barr antibodies and a history of chronic debilitating fatigue (*Science News,* January 7, 1989, volume 135, page 4). Half the patients received the drug and half a placebo, by intravenous injection every 8 hours for 7 days, followed by 30 days of oral doses. Eleven of the 12 patients on the experimental drug reported improvement at the end of the study, compared with 10 of the 12 placebo patients. How do these experimental findings contribute to an assessment of the link between the Epstein–Barr virus and chronic fatigue syndrome? State and test appropriate hypotheses, using Fisher's exact test. Discuss your findings.

EXERCISE 16-17 Can the nitrous oxide used as an anesthetic by dental workers contribute to infertility? Researchers divided 24 female rats into two groups. They exposed 12 rats to doses of nitrous oxide comparable to what a dentist might breathe, 8 hours a day for 35 days. The other 12 rats served as controls. When mated, 6 of the 12 exposed rats and all 12 of the control rats conceived (*Science News,* March 25, 1989, volume 135, page 182). What does this study suggest about the relationship between nitrous oxide exposure and infertility? State and test appropriate hypotheses, using Fisher's exact test. Discuss your findings.

EXERCISE 16-18 In Exercise 10-4, we considered a study of divers with a history of decompression sickness, or the bends (*Science News,* March 25, 1989, volume 135, page 188). Eleven of 30 such divers showed evidence of a heart defect known as patent foramen ovale. About 5% of the general population has this heart defect.

a. Use a small-sample test to test the null hypothesis that among divers with a history of decompression sickness, the proportion having this heart defect equals .05.

b. Compare the result in part (a) with the result of the large-sample test in Exercise 10-4.

EXERCISE 16-19 In the same investigation discussed in Example 16-4, investigators surveyed 516 20- to 24-year-old men in Britain, 819 in Canada, and 581 in the United States (Millar and Stephens, 1987). The men in the United States survey were all white. (Why?)

Define a variable body weight with two categories. The excessive category corresponds to a Quetelet index greater than 25 kg/m^2, the not excessive category to a Quetelet index less than or equal to 25 kg/m^2. The results of the surveys are summarized below:

Body weight	Britain	Canada	United States
Not excessive	402	614	395
Excessive	114	205	186

Compare the proportion of 20- to 24-year-old men in the excessive body weight category across the three countries. State and test appropriate hypotheses. Discuss your findings.

EXERCISE 16-20 Carry out the chi-square test of independence in Example 16-5. Compare your results with the results of Fisher's exact test in Example 16-5.

EXERCISE 16-21 Carry out the chi-square test of homogeneity in Example 16-6. Compare your results with the results of Fisher's exact test in Example 16-6.

Additional Exercises

For each exercise, plot the observations in any ways that seem helpful. Describe the population (whether real or hypothetical) sampled.

Consider the forms of statistical inference (if any) that are appropriate. For some exercises, no formal statistical analysis may be appropriate. For others, several types of analysis may be reasonable.

For hypothesis testing, state the null and alternative hypotheses. For hypothesis testing and interval estimation, state the assumptions that make the statistical analysis appropriate. Do the assumptions seem reasonable? State conclusions based on your analysis.

When more than one analysis is possible, try each procedure. Use similar confidence levels for confidence intervals calculated under differing assumptions. Compare and discuss results.

Discuss possible extraneous factors that might have influenced results. What additional information would you want about the experiment or how the data were collected? Discuss reasons for caution in interpreting results of the analysis.

Give a complete discussion of the questions that can be answered by the data set. Discuss the information provided by the results of the experiment.

EXERCISE R-1

Experimenters wanted to investigate the effects of four different baking temperatures and five different recipes on the size of cakes. They baked two cakes for each recipe/temperature combination. The recorded response for each cake was the area of a cross-section in square inches. The results are shown below (Johnson, 1976; from Li, 1964). Discuss the results of this experiment.

	Temperature (°C)			
	149	163	190	218
Recipe A:	3.01, 4.08	4.63, 4.63	4.59, 4.45	4.26, 4.49
Recipe B:	3.87, 3.74	4.56, 4.91	4.75, 5.10	5.35, 5.39
Recipe C:	4.13, 4.03	4.80, 4.86	5.30, 5.57	5.67, 5.67
Recipe D:	3.98, 4.11	4.79, 4.88	5.00, 5.02	5.30, 5.67
Recipe E:	4.16, 4.35	4.65, 4.80	5.41, 5.29	5.52, 5.80

EXERCISE R-2 In this dental study, investigators wanted to compare a new treatment with placebo for effects on dental hygiene. The response variable is an index of oral hygiene. The 34 volunteers in group A used the placebo for a fixed period and later used the new treatment for a fixed period. The 30 volunteers in group B used the new treatment for the first fixed period and then used the placebo later during the second fixed period. We call such a study, in which each subject is treated over two or more periods, a *crossover study*. The index of oral hygiene is listed below for each period, for the volunteers in both groups (Brown, 1980; Varma and Chilton, 1974; from Zinner, Duany, and Chilton, 1970).

Group A			Group B		
Volunteer code number	Period 1 Placebo	Period 2 New treatment	Volunteer code number	Period 1 New treatment	Period 2 Placebo
1	.83	1.83	1	1.67	.33
2	1.00	2.17	2	2.50	.50
3	.67	1.67	3	1.00	−.17
4	.50	1.50	4	1.67	.50
5	.50	2.33	5	1.83	.50
6	.83	1.83	6	.50	.33
7	1.00	.50	7	1.33	.67
8	.67	.33	8	1.33	.00
9	.67	.50	9	.50	.17
10	.33	.67	10	2.17	.83
11	.00	.83	11	1.67	.33
12	1.17	1.33	12	1.50	.00
13	.00	.67	13	1.33	.50
14	.50	1.83	14	1.50	.50
15	.33	1.50	15	1.33	.00
16	.33	1.50	16	.67	−.17
17	.50	1.17	17	1.67	.50
18	1.00	1.67	18	2.50	.67
19	.00	1.33	19	1.83	.00
20	.50	1.50	20	.83	.67
21	−.50	2.83	21	2.33	.17
22	.17	2.33	22	1.17	.50
23	1.00	1.33	23	1.33	.00
24	1.00	1.67	24	1.33	.83
25	1.33	.67	25	.33	1.33
26	.33	.83	26	2.17	1.17
27	2.00	1.00	27	1.00	.33
28	4.00	.17	28	.33	1.00
29	.83	1.67	29	1.17	.17
30	.50	1.33	30	.50	.50
31	.50	1.50			
32	.50	1.67			
33	2.17	1.33			
34	.67	1.17			

In any crossover study, we worry that during later treatment periods there may be a residual or carryover effect from earlier treatments. Discuss this concern with respect to this dental study. Discuss the results of this experiment.

EXERCISE R-3

In this experiment, investigators studied a method of determining aflatoxin levels in a lot of contaminated peanuts. (This problem is important in sampling inspection. Inspectors want to protect consumers while not rejecting too many good peanuts.) They ground the peanuts into meal and divided the meal into separate samples. They blended each sample in a chemical solution. For each sample, the investigators divided the blend equally among 16 centrifuge bottles. They measured aflatoxin concentration for each bottle. For three samples, one observation was lost, leaving 15 aflatoxin determinations for each of those samples. The determination of aflatoxin concentration for each bottle (units not given) is shown below (Quesenberry, Whitaker, and Dickens, 1976; from Waltking, Bleffert, and Kiernan, 1968). Discuss your findings as they relate to the problem of quality control and consumer protection.

Sample 1	Sample 2	Sample 3	Sample 4	Sample 5	Sample 6	Sample 7	Sample 8
22.35	30.02	10.84	28.60	34.30	16.71	7.63	52.56
32.54	26.26	19.31	32.59	27.90	18.11	6.37	112.86
23.23	26.48	13.12	37.24	34.87	11.19	6.97	69.89
68.31	36.38	13.20	24.61	32.56	11.98	5.28	100.15
27.51	48.01	12.39	24.71	31.67	24.86	6.74	87.14
27.44	49.50	17.20	35.83	29.06	20.18	6.74	83.42
28.53	15.68	12.61	48.93	32.00	17.49	12.06	82.88
29.49	30.79	18.07	37.99	33.40	9.23	7.53	64.63
51.98	22.03	18.19	29.19	30.55	15.35	8.81	73.98
28.87	26.54	16.65	28.89	31.87	21.12	13.58	112.11
21.80	22.74	15.85	32.46	29.17	17.23	12.64	97.59
29.12	35.36	13.53	40.31	26.05	18.67	10.71	85.36
36.65	51.61	11.06	35.88	36.50	23.17	5.41	82.19
40.76	27.53	15.25	31.04	27.51	17.05	7.46	94.55
23.81	36.71	16.25	32.05	30.67	15.43	4.39	60.24
36.19		12.39		31.76	.16.82	11.85	

EXERCISE R-4

In this experiment, investigators exposed adult flour beetles to several doses of gaseous carbon disulphide. The results are shown below (Prentice, 1976; from Bliss, 1935). Discuss the relationship between dose and percentage of beetles killed.

Dose (logarithm base-10 of CS_2 in mg/liter)	Number of beetles	Number killed
1.6907	59	6
1.7242	60	13
1.7552	62	18
1.7842	56	28
1.8113	63	52
1.8369	59	53
1.8610	62	61
1.8839	60	60

EXERCISE R-5

Mental retardation is associated with some metabolic diseases. In this experiment, investigators studied metabolism of tyrosine among 36 mentally handicapped patients (Geertsema and Reinecke, 1984). The measured response was the total amount of tyrosine catabolites excreted in the urine (in μmoles per 100 ml urine). Ten separate measurements for each patient, plus the average and standard deviation of those ten measurements, are listed below. The patients are listed in order of increasing average tyrosine determination.

	Observation											
ID	1	2	3	4	5	6	7	8	9	10	Patient average	Standard deviation
1	.325	.317	.375	.325	.508	.117	.150	.317	.275	.383	.309	.106
2	.333	.283	.342	.325	.250	.358	.283	.392	.450	.267	.328	.058
3	.208	.483	.317	.300	.217	.217	.433	.392	.575	.408	.355	.118
4	.458	.592	.133	.600	.292	.542	.467	.300	.067	.233	.368	.181
5	.225	.317	.492	.617	.217	.308	.425	.508	.425	.258	.379	.128
6	.100	.317	.675	.625	.133	.542	.150	.642	.183	.442	.381	.219
7	.233	.267	.300	.217	.258	.700	.483	.667	.408	.300	.383	.169
8	.433	.508	.350	.275	.342	.333	.667	.300	.425	.275	.391	.116
9	.133	.800	.575	.617	.283	.433	.275	.308	.283	.225	.393	.198
10	.500	.417	.358	.708	.283	.233	.483	.300	.258	.567	.411	.146
11	.250	.442	.233	.183	.275	.367	.408	.692	.883	.667	.440	.221
12	.133	.200	.417	.750	.083	.833	.183	.842	.767	.233	.444	.301
13	.175	.458	.358	.533	.242	.400	.308	.650	.567	.783	.447	.179
14	.650	.567	.500	.808	.508	.442	.400	.442	.108	.217	.464	.190
15	.217	.183	.667	.767	.900	.183	.608	.700	.233	.592	.505	.259
16	.625	.533	.292	.367	.683	.667	.433	.783	.200	.625	.521	.180
17	.442	.567	.412	.625	.200	.683	.892	.175	.883	.658	.554	.236
18	.642	.408	.500	.733	.317	.700	.617	.667	.683	.667	.593	.131
19	.900	1.067	.950	.442	.542	.317	.183	.483	.942	.300	.613	.305
20	.733	.433	.383	.550	.383	.650	.717	.833	.817	.700	.620	.163
21	.617	.667	.683	.308	.567	.858	.358	.483	.833	.890	.628	.194
22	.842	.808	.408	.492	.683	.500	.350	.833	1.083	.500	.650	.224
23	.683	.700	.633	.850	.775	.750	.067	.867	.950	.467	.674	.240
24	.733	1.025	1.183	.417	.542	.833	.275	.417	.483	1.058	.697	.300
25	.167	.792	.833	.400	1.142	.142	.683	.542	1.375	.867	.699	.378
26	.367	1.075	.700	1.350	.467	.200	.700	.167	.475	1.650	.715	.471
27	1.012	.442	1.000	.583	.633	.783	.383	1.550	.417	.442	.725	.353
28	.983	.600	.417	.450	.667	.800	.550	.750	1.242	1.083	.754	.261
29	.708	2.200	.292	1.633	1.417	.450	.233	.233	.367	.650	.818	.654
30	.583	.575	.533	.292	.408	.783	.625	1.333	1.550	1.667	.835	.469
31	.767	.492	.933	.933	1.358	.683	.692	.700	.992	1.133	.868	.241
32	1.083	.542	.400	.558	.708	1.333	.550	.883	2.100	1.792	.995	.550
33	.542	1.650	.983	.892	.333	1.267	.750	1.142	1.625	1.808	1.099	.468
34	1.550	1.433	.550	.583	.808	1.400	1.133	1.467	1.300	.925	1.115	.355
35	.850	.900	1.242	.917	.092	1.267	.983	1.483	2.367	1.750	1.185	.576
36	2.167	1.600	.717	1.675	1.525	.758	.808	2.283	2.333	3.067	1.693	.743

The investigators selected 1.0 μmole per 100 ml urine as a cutoff for classifying a patient's tyrosine metabolism as clinically negative (less than 1.0, no metabolic problem) or clinically positive (greater than 1.0, a metabolic disorder). As part of your discussion, address the following questions:

a. Is it reasonable to use a single measurement to classify a person as clinically negative or positive for tyrosine metabolism?

b. How consistent are separate determinations in a single patient?

c. How different are the standard deviations for these patients?

d. How different are the tyrosine determinations between patients?

e. Each of these 36 patients had been classified as clinically positive for a tyrosine metabolic disorder after a first screening. Discuss this finding in relation to the later results listed above.

f. Does it appear that the authors use the population or sample formula to calculate their standard deviations? Which would you consider appropriate here? Does this affect your answer to part (c)?

EXERCISE R-6

Investigators estimated specific activity of the enzyme sucrase using samples of intestine from 24 patients who had had intestinal bypass surgery. For each patient, the investigators determined sucrase activity in two ways. The investigators believe the first method, using pellet fractions, is accurate, but it is very time-consuming. The second method, from homogenates, is easier and faster to use. The two recorded levels of sucrase activity (units not given) are listed below for each patient (Carter, 1981; data provided by Dr. Helen Lane). Assess the idea of using the homogenate measurement of sucrase activity as an estimate of the actual (or pellet) level of sucrase activity.

Patient	Homogenate	Pellet	Patient	Homogenate	Pellet
1	18.88	70.00	13	60.78	277.30
2	7.26	55.43	14	77.92	331.50
3	6.50	18.87	15	51.29	133.74
4	9.83	40.41	16	77.91	221.50
5	46.05	57.43	17	36.65	132.93
6	20.10	31.14	18	31.17	85.38
7	35.78	70.10	19	66.09	142.34
8	59.42	137.56	20	115.15	294.63
9	58.43	221.20	21	95.88	262.52
10	62.32	276.43	22	64.61	183.56
11	88.55	316.00	23	37.71	86.12
12	19.50	75.56	24	100.82	226.55

EXERCISE R-7

In this experiment, investigators wanted to compare antagonistic behavior in mice with different brain weights. They bred mice selectively for brain weight (small, medium, large). Under each of two environmental conditions, the experimenters raised seven pairs of mice with each brain weight. After the mice reached maturity, the investigators housed the two mice in a pair together, and noted their fighting behavior. The response variable is a measure of aggressive behavior in mice: a score of seconds of tail rattling per second of fighting. The results are shown below (Scheirer, Ray, and Hare, 1976; from Hahn, Haber, and Fuller, 1973). Discuss the results of this experiment.

	Brain weight		
	Low	Medium	Large
Environmental condition A	9.60	1.98	.77
	8.00	1.81	.66
	5.14	1.37	.37
	3.50	.98	.40
	3.23	.27	.19
	2.66	.00	.13
	1.44	.00	.00
Environmental condition B	4.36	2.61	2.95
	1.49	2.09	2.73
	1.45	1.76	.91
	1.35	1.27	.82
	.33	.84	.69
	.20	.64	.11
	.00	.00	.00

EXERCISE R-8

Investigators carried out a 6-month double-blind randomized study to compare a new treatment with a standard treatment for patients with acute rheumatoid arthritis. At the end of the study, the investigators classified the condition of each patient into one of five categories. The results are summarized in the following frequency table (Mehta, Patel, and Tsiatis, 1984). Does there appear to be a difference in effectiveness between the two treatments or do they appear to be therapeutically equivalent?

	Treatment	
Patient status	New	Standard
Much improved	24	11
Improved	37	51
No change	21	22
Worse	19	21
Much worse	6	7

EXERCISE R-9

At a United States industrial site, public health workers used grab-sample techniques to measure concentration of airborne chlorine. They made 15 chlorine determinations over the course of one working day (Owen and DeRouen, 1980):

Chlorine determination in parts per million (ppm):									
	6	0	6	9	6.5	0	0	0	1
	.5	2	2	0	0	1			

At the time, federal guidelines stated that the maximum allowable exposure for chlorine was an average of 1 ppm over the work day. Was this site in compliance with the federal occupational safety standard for chlorine? You may wish to use the information that the minimum detectable level of chlorine was .25 ppm.

EXERCISE R-10 In this experiment, investigators recorded the length of time for patients with headaches to feel relief. Each patient received a standard treatment and a new treatment, on different occasions. The time until relief from headache (in minutes) is shown below for each patient and treatment (Gross and Lam, 1981; from Gross and Clark, 1975, page 232). Does there appear to be a difference between the two treatments in time until relief?

Patient	New treatment	Standard treatment
1	6.9	8.4
2	6.8	7.7
3	10.3	10.1
4	9.4	9.6
5	8.0	9.3
6	8.8	9.1
7	6.1	9.0
8	7.4	7.7
9	8.0	8.1
10	5.1	5.3

EXERCISE R-11 In a study of cyanotic heart disease in children, investigators recorded the age at first word and the Gesell Adaptive Score for each of 21 children (Ellenberg, 1976; from Mickey, Dunn, and Clark, 1967). How would you describe the relationship between these two variables for this group of children?

Child	Age at first word (months)	Gesell Adaptive Score	Child	Age at first word (months)	Gesell Adaptive Score
1	15	95	12	9	96
2	26	71	13	10	83
3	10	83	14	11	84
4	9	91	15	11	102
5	15	102	16	10	100
6	20	87	17	12	105
7	18	93	18	42	57
8	11	100	19	17	121
9	8	104	20	11	86
10	20	94	21	10	100
11	7	113			

EXERCISE R-12 Investigators carried out a double-blind multiclinic trial to compare a new agent with aspirin in treatment of patients with rheumatoid arthritis. A frequency table summarizing the investigators' assessment of therapeutic effect after 26 weeks of treatment is shown below (Gould, 1980).

Response category	New agent	Aspirin
Excellent	28	23
Satisfactory	44	33
No change	8	10
Worse	2	0
Withdrawn from study	49	72

As we see from the last line of the table, many patients withdrew before the end of the study. Patients may withdraw for reasons related to treatment (feeling cured or feeling worse) or reasons unrelated to treatment (loss of interest, moving away, illness or death unrelated to the condition under study).

a. Compare withdrawal rates for the two treatment groups.

b. Ignoring withdrawals, compare responses for the two treatment groups.

c. Based on your results in parts (a) and (b), how would you assess the relative effectiveness of these two treatments for rheumatoid arthritis?

d. Suppose you learn that all withdrawals were due to treatment intolerance or lack of therapeutic effect. Now how would you assess the relative effectiveness of these two treatments?

EXERCISE R-13

In a study of heart disease in Framingham, Massachusetts, investigators measured serum cholesterol of participants and noted participants who suffered from coronary heart disease 6 years later (Leung and Kupper, 1981; from Walter, 1978):

Coronary heart disease six years later	Initial serum cholesterol (mg %)	
	<220	≥220
Yes	20	72
No	553	684

What can you say about the association between initial serum cholesterol level and coronary heart disease 6 years later for participants of this study?

EXERCISE R-14

Around 1950 medical workers began to administer continuous high concentrations of oxygen to premature babies who showed difficulty breathing. (Premature babies sometimes died because of impaired breathing.) Also around 1950, many premature infants developed retrolental fibroplasia, which led to blindness. Some workers suspected a link between the high doses of oxygen administered and subsequent blindness in these babies. In one study to investigate this idea, experimenters randomly divided 85 premature babies into two treatment groups: higher-dose oxygen and lower-dose oxygen. The researchers noted whether each baby survived to 3 months of age. They classified surviving babies by blindness following treatment. The results are summarized below (Meier, 1979; from Lanman et al., 1954).

Survival:	Treatment group	Number of babies	Number alive at age 3 months
	Higher-dose oxygen	45	36
	Lower-dose oxygen	40	28

Blindness:	Treatment group	Alive at 3 months	Number blind
	Higher-dose oxygen	36	8
	Lower-dose oxygen	28	0

Discuss the results of this study. Because of studies such as this one, medical workers stopped using high doses of oxygen on premature babies and blindness from retrolental fibroplasia became once again a rare condition.

EXERCISE R-15 An investigator carried out this experiment to study the specific retention volume of the organic liquid methylene chloride in the polymer polyethylene terephthalate, at several temperatures. The results are shown below (Gallant, 1977; from P. O. Hsiung, "Study of the Interaction of Organic Liquids with Polymers by Means of Gas Chromatography," unpublished Ph.D. dissertation, North Carolina State University, 1974). Describe the relationship between reciprocal of temperature and the natural logarithm of specific volume seen in this experiment.

Reciprocal $\times 10^3$ of temperature in degrees Kelvin	Natural logarithm of specific volume in cc per gm
2.54323	1.16323
2.60960	1.10458
2.67952	.98832
2.75330	.87471
2.79173	.62060
2.82965	.51175
2.87026	.35371
2.91120	.66954
2.94637	.85555
3.00030	1.07086
3.04228	1.22272
3.09214	1.29113
3.13971	1.38480
3.19081	1.46728

EXERCISE R-16

Listed below are weight/height ratios by age for preschool boys, from a nutritional study of preschool children in the north central United States (Gallant, 1977; Gallant and Fuller, 1973; from Eppright et al., 1972). Age is age in months and W/H is weight divided by height, in pounds per inch. Describe the relationship between weight/height ratio and age for preschool boys suggested by this study.

Age	W/H	Age	W/H	Age	W/H	Age	W/H
.5	.46	18.5	.81	36.5	.87	54.5	.93
1.5	.47	19.5	.78	37.5	.87	55.5	.98
2.5	.56	20.5	.87	38.5	.85	56.5	.95
3.5	.61	21.5	.80	39.5	.90	57.5	.97
4.5	.61	22.5	.83	40.5	.87	58.5	.97
5.5	.67	23.5	.81	41.5	.91	59.5	.96
6.5	.68	24.5	.88	42.5	.90	60.5	.97
7.5	.78	25.5	.81	43.5	.93	61.5	.94
8.5	.69	26.5	.83	44.5	.89	62.5	.96
9.5	.74	27.5	.82	45.5	.89	63.5	1.03
10.5	.77	28.5	.82	46.6	.92	64.5	.99
11.5	.78	29.5	.86	47.5	.89	65.5	1.01
12.5	.75	30.5	.82	48.5	.92	66.5	.99
13.5	.80	31.5	.85	49.5	.96	67.5	.99
14.5	.78	32.5	.88	50.5	.92	68.5	.97
15.5	.82	33.5	.86	51.5	.91	69.5	1.01
16.5	.77	34.5	.91	52.5	.95	70.5	.99
17.5	.80	35.5	.87	53.5	.93	71.5	1.04

EXERCISE R-17

In a study of metabolism, investigators incubated isolated liver cells from starved rats with lactate. They then noted accumulation of pyruvate (a first step in glucose production) over time. The results of duplicate observations at each time are shown below (James and Conyers, 1985). Describe the relationship between pyruvate level and time in this experiment.

Time (minutes)	Pyruvate (μmoles)
.05	.0135, .0115
3.05	.3320, .3500
6.05	.4755, .4885
9.05	.5560, .5430
12.05	.5945, .6025
15.05	.6130, .6195
18.05	.6135, .6225
21.05	.6095, .5640
24.05	.5690, .5505
27.05	.5400, .5250
30.05	.5105, .4965
33.05	.5050, .4630
36.05	.4400, .4305
39.05	.3900, .4190

EXERCISE R-18 In a National Cancer Institute animal carcinogenesis experiment, investigators divided male and female mice and rats into three treatment groups. The control groups received none of the drug tolazamide. Animals in the low-dose groups received tolazamide as .5% of their diet. Animals in the high-dose groups received tolazamide as 1% of their diet. The investigators noted development of leukemia or lymphoma in the animals. The results are shown below as number developing leukemia or lymphoma/number in group (Tarone and Gart, 1980). Discuss the results of this experiment.

	Treatment group		
Sex/Species	Control	Low dose	High dose
Female mice	6/15	2/33	4/34
Male mice	4/14	5/35	1/34
Female rats	4/15	3/33	2/35
Male rats	2/15	1/35	4/35

EXERCISE R-19 In this study of the association between hyperglycemia and relative hyperinsulinemia, investigators administered standard glucose tolerance tests to 13 control and 20 obese patients on the Pediatric Clinical Research Ward, University of Colorado Medical Center. As part of the study, the investigators determined plasma inorganic phosphate levels (mg/dl) from blood samples taken 0, $\frac{1}{2}$, 1, $1\frac{1}{2}$, 2, 3, 4, and 5 hours after a standard-dose oral glucose challenge. The results are listed below (Zerbe, 1979). Discuss the results of this experiment.

	Hours after glucose challenge							
	0	$\frac{1}{2}$	1	$1\frac{1}{2}$	2	3	4	5
Control patients								
1	4.3	3.3	3.0	2.6	2.2	2.5	3.4	4.4
2	3.7	2.6	2.6	1.9	2.9	3.2	3.1	3.9
3	4.0	4.1	3.1	2.3	2.9	3.1	3.9	4.0
4	3.6	3.0	2.2	2.8	2.9	3.9	3.8	4.0
5	4.1	3.8	2.1	3.0	3.6	3.4	3.6	3.7
6	3.8	2.2	2.0	2.6	3.8	3.6	3.0	3.5
7	3.8	3.0	2.4	2.5	3.1	3.4	3.5	3.7
8	4.4	3.9	2.8	2.1	3.6	3.8	4.0	3.9
9	5.0	4.0	3.4	3.4	3.3	3.6	4.0	4.3
10	3.7	3.1	2.9	2.2	1.5	2.3	2.7	2.8
11	3.7	2.6	2.6	2.3	2.9	2.2	3.1	3.9
12	4.4	3.7	3.1	3.2	3.7	4.3	3.9	4.8
13	4.7	3.1	3.2	3.3	3.2	4.2	3.7	4.3

(continued)

(*continued*)

	Hours after glucose challenge							
	0	$\frac{1}{2}$	1	$1\frac{1}{2}$	2	3	4	5
Obese patients								
1	4.3	3.3	3.0	2.6	2.2	2.5	2.4	3.4
2	5.0	4.9	4.1	3.7	3.7	4.1	4.7	4.9
3	4.6	4.4	3.9	3.9	3.7	4.2	4.8	5.0
4	4.3	3.9	3.1	3.1	3.1	3.1	3.6	4.0
5	3.1	3.1	3.3	2.6	2.6	1.9	2.3	2.7
6	4.8	5.0	2.9	2.8	2.2	3.1	3.5	3.6
7	3.7	3.1	3.3	2.8	2.9	3.6	4.3	4.4
8	5.4	4.7	3.9	4.1	2.8	3.7	3.5	3.7
9	3.0	2.5	2.3	2.2	2.1	2.6	3.2	3.5
10	4.9	5.0	4.1	3.7	3.7	4.1	4.7	4.9
11	4.8	4.3	4.7	4.6	4.7	3.7	3.6	3.9
12	4.4	4.2	4.2	3.4	3.5	3.4	3.9	4.0
13	4.9	4.3	4.0	4.0	3.3	4.1	4.2	4.3
14	5.1	4.1	4.6	4.1	3.4	4.2	4.4	4.9
15	4.8	4.6	4.6	4.4	4.1	4.0	3.8	3.8
16	4.2	3.5	3.8	3.6	3.3	3.1	3.5	3.9
17	6.6	6.1	5.2	4.1	4.3	3.8	4.2	4.8
18	3.6	3.4	3.1	2.8	2.1	2.4	2.5	3.5
19	4.5	4.0	3.7	3.3	2.4	2.3	3.1	3.3
20	4.6	4.4	3.8	3.8	3.8	3.6	3.8	3.8

EXERCISE R-20 In a life-testing experiment, investigators recorded the time to breakdown of an insulating fluid under two elevated levels of voltage stress. (They used elevated stress levels because time to breakdown under voltages ordinarily used would be too long to measure.) The results are shown below (Nair, 1984; from Nelson, 1982). Compare the times to breakdown under these two voltage levels.

Voltage level	Time to breakdown (minutes)							
32 Kv	.27	.40	.69	.79	2.75	3.91	9.88	13.95
	15.93	27.80	53.24	82.85	89.29	100.58	215.10	
36 Kv	.35	.59	.96	.99	1.69	1.97	2.07	2.58
	2.71	2.90	3.67	3.99	5.35	13.77	25.50	

EXERCISE R-21 Scientists report using a sensitive genetic test to detect HIV, the AIDS virus, in newborn babies (I. Wickelgren, "Test diagnoses AIDS in newborns," *Science News,* volume 135, June 24, 1989, page 389). Such a test would have great usefulness in diagnosing babies born to HIV-positive mothers; babies needing intensive treatment would be identified earlier and babies not needing the treatment would be spared its toxic effects.

Using their genetic test, the researchers detected HIV DNA in five of seven newborns who later developed AIDS. The researchers detected HIV DNA in one of eight newborns who later showed symptoms of possible HIV infection. The researchers detected no HIV DNA in any of nine newborns who were still healthy after 16 months.

Discuss these experimental results.

EXERCISE R-22

In the 1974 and 1975 General Social Surveys of the National Opinion Research Center, interviewers asked men the following question: "Do you agree with this statement?—Women should take care of running their homes and leave running the country to men." The accompanying table classifies the respondents by years of education and response to the question (Haberman, 1982). What relationship (if any) do you see between years of education and response to this question, for these men?

Years of education	Agree	Disagree
0	4	2
1	2	0
2	4	0
3	6	3
4	5	5
5	13	7
6	25	9
7	27	15
8	75	49
9	29	29
10	32	45
11	36	59
12	115	245
13	31	70
14	28	79
15	9	23
16	15	110
17	3	29
18	1	28
19	2	13
20+	3	20

EXERCISE R-23

Investigators wanted to compare four formulations of the drug verapamil. They randomly divided 26 healthy male volunteers into four groups. They treated each group with one of the drug formulations. The recorded response was area under the plasma time curve. In addition, the investigators recorded the age, height, and weight of each volunteer. The results (no units provided) are listed below (these are the first-period data from a larger experiment reported in Chinchilli, Schwab, and Sen, 1989; data provided by William H. Barr of the

Pharmacy and Pharmaceutics Department, Virginia Commonwealth University). Discuss the results of this experiment.

Volunteer	Treatment	Age	Height	Weight	Area under the plasma time curve
1	A	25	68.0	145.0	224.29
2	B	29	66.5	140.0	231.35
3	C	29	68.5	155.0	253.88
4	D	23	70.0	188.0	327.95
5	A	22	66.5	140.0	326.06
6	B	22	72.0	170.0	259.53
7	D	23	68.0	165.0	347.43
8	A	24	72.0	203.0	270.10
9	B	24	71.0	160.0	618.61
10	C	24	63.0	135.0	476.27
11	D	22	68.5	149.5	337.45
12	A	28	71.0	210.0	483.25
13	B	23	70.0	150.0	223.04
14	C	26	72.0	159.0	399.92
15	D	22	70.0	150.0	117.45
16	B	26	70.5	160.0	183.20
17	C	28	75.0	240.0	344.18
18	D	24	72.0	173.0	181.75
19	A	31	70.0	145.0	94.25
20	B	31	69.0	170.0	195.67
21	C	25	78.0	193.0	458.89
22	D	29	72.0	205.0	383.64
23	A	24	74.0	150.0	413.53
24	B	26	72.0	161.0	132.88
25	C	23	73.0	172.0	245.21
26	D	22	70.0	165.0	298.06

EXERCISE R-24

A frequency table summarizing the sex of each child (by birth order) in completed families of 2, 3, 4, and 5 children in the United States is shown below (Crouchley and Pickles, 1984). Discuss what you find from studying this data set.

Number of children	Sex by birth order	Number of families	Number of children	Sex by birth order	Number of families
2	MM	4,862	3	MMM	2,467
	MF	4,854		MMF	2,504
	FM	5,133		MFM	2,432
	FF	4,432		MFF	2,172
				FMM	2,298
				FMF	2,178
				FFM	2,373
				FFF	1,988

Number of children	Sex by birth order	Number of families	Number of children	Sex by birth order	Number of families
4	MMMM	1,133	5	MMMMM	2,242
	MMMF	1,140		MMMMF	2,187
	MMFM	1,106		MMMFM	2,122
	MMFF	1,046		MMMFF	2,092
	MFMM	1,105		MMFMM	2,062
	MFMF	1,049		MMFMF	1,970
	MFFM	1,094		MMFFM	1,968
	MFFF	982		MMFFF	1,878
	FMMM	1,085		MFMMM	2,187
	FMMF	1,019		MFMMF	2,019
	FMFM	1,071		MFMFM	2,018
	FMFF	935		MFMFF	1,857
	FFMM	1,028		MFFMM	2,011
	FFMF	1,010		MFFMF	1,874
	FFFM	952		MFFFM	1,810
	FFFF	913		MFFFF	1,800
				FMMMM	2,088
				FMMMF	1,948
				FMMFM	1,931
				FMMFF	1,882
				FMFMM	1,957
				FMFMF	1,903
				FMFFM	1,923
				FMFFF	1,770
				FFMMM	1,995
				FFMMF	1,820
				FFMFM	1,859
				FFMFF	1,765
				FFFMM	1,879
				FFFMF	1,760
				FFFFM	1,788
				FFFFF	1,732

EXERCISE R-25 In this study, investigators recorded fill weights of bottles filled by six heads of a multiple-head machine, at five different times. The value recorded is bottle fill weight in grams minus 1,200. The results are shown below (Snee, 1982; from Ott and Snee, 1973). Discuss the results of this experiment.

Time	Head 7	Head 8	Head 9	Head 10	Head 11	Head 12
9:15	68	65	75	57	32	70
12:55	56	52	55	48	65	47
1:55	40	51	52	36	49	45
2:55	84	87	88	73	34	70
3:55	50	52	52	50	45	61

EXERCISE R-26 In this experiment, eight calves from a single herd of Angus cattle arrived at a feedlot. Investigators weighed the calves and divided them among three diet regimens. After a fixed period, the calves were slaughtered. The investigators recorded the slaughter weight for each calf. The slaughter weight and weight on entering feedlot (no units given) are shown below for each calf (Urquhart, 1982).

Energy level in diet	Slaughter weight (Weight on entering feedlot)
Low	1,046 (690)
	1,027 (685)
	1,018 (690)
Medium	874 (665)
	874 (635)
High	1,018 (680)
	874 (605)
	970 (665)

As another part of the experiment, investigators weighed eight calves from a single herd of Brangus cattle and divided them among the same three diet regimens. Weights for these calves are shown below.

Energy level in diet	Slaughter weight (Weight on entering feedlot)
Low	1,133 (765)
	989 (750)
	970 (645)
	1,162 (730)
Medium	1,190 (840)
	1,104 (755)
High	1,104 (755)
	1,114 (755)

Discuss the results of this experiment.

EXERCISE R-27 Fifty-nine women participated in this study to compare an active drug with a placebo for treatment of rheumatoid arthritis. Identification number, age, and response are shown below for the women in each treatment group (Koch et al., 1982). The responses are ranked from best to worst in this order: excellent, good, moderate, fair, poor. Discuss the results of this study.

Active treatment			Placebo treatment		
ID	Age (years)	Response	ID	Age (years)	Response
1	23	Poor	28	23	Fair
2	32	Poor	29	30	Fair
3	37	Moderate	30	30	Fair
4	41	Good	31	31	Moderate
5	41	Fair	32	32	Poor
6	48	Good	33	33	Good
7	48	Poor	34	37	Poor
8	55	Excellent	35	44	Poor
9	55	Good	36	45	Poor
10	56	Good	37	46	Poor
11	57	Good	38	48	Fair
12	57	Good	39	49	Poor
13	57	Good	40	51	Poor
14	58	Poor	41	53	Poor
15	59	Good	42	54	Good
16	59	Excellent	43	54	Poor
17	60	Excellent	44	54	Poor
18	61	Good	45	55	Good
19	62	Good	46	57	Moderate
20	62	Moderate	47	57	Fair
21	66	Excellent	48	58	Moderate
22	67	Good	49	59	Excellent
23	68	Moderate	50	59	Moderate
24	68	Excellent	51	61	Fair
25	69	Moderate	52	63	Moderate
26	69	Poor	53	64	Poor
27	70	Moderate	54	65	Excellent
			55	66	Moderate
			56	66	Fair
			57	66	Poor
			58	68	Moderate
			59	74	Good

EXERCISE R-28

Sixty-nine girls completed this 2-year study to compare three treatments in reducing incidence of dental caries. The three treatments were stannous fluoride, acid phosphate fluoride, and distilled water, all applied topically. The investigators measured the number of decayed, missing, or filled teeth before and after the study. The accompanying table provides this information, in addition to an identification number for each girl, her age in years, and a code for the institution (place) where she was treated (Quade, 1982; from Cartwright, Lindahl, and Bawden, 1968). Note that the recorded result for one girl shows a larger value before than after the study. Discuss the results of this experiment.

ID	Place	Age	Before	After	ID	Place	Age	Before	After
				Distilled water treatment					
1	1	13	7	11	18	2	7	3	4
2	1	17	20	24	19	2	11	4	7
3	1	16	21	25	20	2	15	4	9
4	1	13	1	2	37	3	8	2	4
5	1	10	3	7	38	3	16	13	18
6	1	17	20	23	39	3	14	9	12
7	1	13	9	13	40	3	16	15	18
8	1	9	2	4	41	3	12	13	17
16	2	16	10	14	42	3	8	2	5
17	2	16	13	17	43	3	14	9	12
				Stannous fluoride treatment					
9	1	14	11	13	30	2	10	4	6
10	1	14	15	18	44	3	9	4	6
21	2	14	15	18	45	3	15	10	14
22	2	11	6	8	46	3	14	7	11
23	2	9	4	6	47	3	13	14	15
24	2	17	18	19	48	3	12	7	10
25	2	14	11	12	49	3	12	3	6
26	2	13	9	9	50	3	14	9	12
27	2	9	4	7	51	3	13	8	10
28	2	9	5	7	52	3	14	19	19
29	2	15	11	14	53	3	14	10	13
				Acid phosphate fluoride					
11	1	11	7	10	57	3	9	5	8
12	1	15	17	17	58	3	11	1	3
13	1	11	9	11	59	3	12	8	9
14	1	7	1	5	60	3	14	4	5
15	1	11	3	7	61	3	10	4	7
31	2	10	4	4	62	3	12	14	14
32	2	15	7	7	63	3	11	8	10
33	2	11	0	4	64	3	11	3	5
34	2	9	3	3	65	3	16	11	12
35	2	9	0	1	66	3	15	16	18
36	2	16	8	8	67	3	10	8	8
54	3	14	10	12	68	3	6	0	1
55	3	11	7	11	69	3	7	3	4
56	3	14	13	12					

EXERCISE R-29 In a trial of a vaccine for autoimmune deficiency syndrome (AIDS), 15 of 40 people treated showed an antibody response within 2 months after treatment (*Science News,* 1988). Find an interval estimate for the proportion of the population who would respond to this vaccine.

EXERCISE R-30 Ten volunteers participated in this study of the effects of aerobic exercise on individuals who had amputation of a lower limb (Pitetti et al., 1987). Weight (including prosthesis, trousers, and shoes), resting heart rate, and resting blood pressure were recorded for each volunteer before and after a 15-week aerobic exercise program. The results of the experiment are shown below.

Volunteer	Age (years)	Weight (kilograms)		Resting heart rate (beats/minute)	
		Before	After	Before	After
1	39	81.4	79.1	70	64
2	36	101.2	103.4	114	108
3	33	103.4	105.5	91	80
4	36	91.9	91.2	73	66
5	41	109.1	110.5	70	61
6	61	81.2	81.2	65	62
7	35	94.1	89.1	68	51
8	36	72.3	79.6	75	72
9	35	83.2	83.2	92	61
10	41	101.8	97.7	93	76

Volunteer	Resting blood pressure (mm Hg)	
	Before	After
1	120/75	120/75
2	130/108	130/100
3	133/75	140/85
4	160/87	137/75
5	130/85	125/85
6	140/80	136/80
7	140/85	118/74
8	135/93	135/87
9	125/90	125/85
10	145/75	140/75

a. Is there evidence that the exercise program had an effect on the weight of the volunteers?

b. Is there evidence that the exercise program had an effect on the resting heart rate of the volunteers?

c. Is there evidence that the exercise program had an effect on the resting blood pressure of the volunteers?

d. In parts (a), (b), and (c) we have based several statistical inferences on data from the same experiment. How does this add to our usual need for caution in interpreting results of analyses (p-values and confidence levels in particular)?

Numerical Answers to Selected Exercises

EXERCISE 2-7

minimum = 7 days, first quartile = 440 days, median = 702 days, third quartile = 1,367 days, maximum = 2,509 days

EXERCISE 3-2

mean = 76.05 years, 10% trimmed mean = 76.07 years, weighted mean = 76.31 years

EXERCISE 3-9

mean = 9.65 mg/g, 15% trimmed mean = 9.225 mg/g, median = 8.90 mg/g, range = 11.2 mg/g, interquartile range = 14.0 − 5.1 = 8.9 mg/g, sample standard deviation = 4.91 mg/g

EXERCISE 4-13

b. Listed are, respectively, sample size, mean, median, range, interquartile range, and sample standard deviation for each sport. Units are $ml \cdot kg^{-1} \cdot min^{-1}$. Wrestling: 5, 57.58, 58.30, 13.6, 6.6, 5.36; weightlifting: 6, 45.12, 44.45, 10.6, 8, 4.39; shot/discus: 4, 45.60, 45.15, 6.9, 5.8, 3.45; ice hockey: 3, 56.57, 54.60, 7.9, 7.9, 4.30; cross-country skiing: 4, 72.275, 73.45, 14.4, 7.65, 6.04.

EXERCISE 5-9

a. Listed numbers are, respectively, the minimum, first quartile, median, third quartile, maximum. Units for weights, grams; for wing length, millimeters; for condition index, gm/mm.
Hatch year ducks, weight (n = 31): 940, 1,070, 1,140, 1,180, 1,280
After hatch year, weight (n = 19): 1,050, 1,110, 1,220, 1,280, 1,420
Hatch year ducks, wing length (n = 31): 252, 263, 268, 272, 276
After hatch year, wing length (n = 19): 264, 270, 275, 277, 285
Hatch year ducks, condition index (n = 31): 3.71, 3.99, 4.23, 4.39, 4.74
After hatch year, condition index (n = 19): 3.82, 4.12, 4.50, 4.68, 5.26

EXERCISE 6-14

a. P(lung cancer|smoker) = .015; odds = .01523

b. P(lung cancer|nonsmoker) = .005; odds = .00502

EXERCISE 6-19

a.

Outcome	Probability	Outcome	Probability
SSSS	.00077	FSSF	.01929
SSSF	.00386	FSFS	.01929
SSFS	.00386	SSFF	.01929
SFSS	.00386	SFFF	.09645
FSSS	.00386	FSFF	.09645
SFFS	.01929	FFSF	.09645
FFSS	.01929	FFFS	.09645
SFSF	.01929	FFFF	.48225

b.

k	$P(W = k)$
0	.4823
1	.3858
2	.1157
3	.0154
4	.0008

c. E(W) = .667 correct answers; var(W) = .556 (correct answers)^2; SD(W) = .745 correct answers

d. .1319

EXERCISE 7-1 24

EXERCISE 7-4 84, $\frac{6}{9} = \frac{2}{3}$

EXERCISE 7-11

a. For $N = 10$, $n = 3$, $m_1 = 5$, and $m_2 = 5$, $P(X = k) = \binom{5}{k}\binom{5}{3-k}/\binom{10}{3}$ for $k = 0, 1, 2, 3$

EXERCISE 7-12

c. $P(X = 0) = (1 - p)^3$, $P(X = 1) = 3p(1 - p)^2$, $P(X = 2) = 3p^2(1 - p)$, $P(X = 3) = p^3$

d.

p	$P(X = 0)$	$P(X \geq 1)$	$P(X = 2$ or $3)$	$P(X = 3)$	$E(X)$
.4	.216	.784	.352	.064	1.2
.5	.125	.875	.5	.125	1.5
.6	.064	.936	.648	.216	1.8

EXERCISE 8-2

a. .0475

b. .9992

c. .0471

d. .7486

e. .0471

f. .9544

g. .1587

h. .1587

i. .0038

EXERCISE 8-18

a. $E(X) = 2.5$, $\text{var}(X) = 2.25$

b. (i) .8302, (ii) .2364, (iii) .2712, (iv) .9666, (v) .0334

c. (i) .6826, (ii) .1587, (iii) .1587, (iv) .9050, (v) .0099

EXERCISE 9-7

a. .006

b. power equals .6590, .2749, .0199, .0199, .2749, .6590 when p equals, respectively, .1, .2, .4, .6, .8, .9

EXERCISE 10-1

A 95% confidence interval for the population mean based on the t distribution with 9 degrees of freedom is 103.61 to 133.39 μmol/liter. An approximate 95% confidence interval for the population mean based on the Wilcoxon signed rank distribution for sample size 10 is 102.5 to 133.5 μmol/liter. An approximate 95% confidence interval for the population median based on the binomial(10, .5) distribution is 100.9 to 139.4 μmol/liter.

EXERCISE 10-8

A large sample approximate 95% confidence interval for the proportion of experimentally treated livers expected to last at least 9.5 hours is .3663 to .5093.

EXERCISE 10-9

a. test statistic $= -4.02$, two-sided p-value $= .016$, 90% confidence interval for the population mean is 19.046 to 20.954 ounces

b. test statistic $= 0$, two-sided p-value $= .059$; approximate 90% confidence interval for the population mean is 19.25 to 21.00 ounces

c. test statistic $= 0$, two-sided p-value $= .0625$, approximate 90% confidence interval for the population median is 19.18 to 21.15 ounces

EXERCISE 10-10

A large-sample test statistic $= 4.35$, two-sided p-value $= .0001$, approximate 95% confidence interval for the carrier population mean is 112.5 to 239.3 units.

EXERCISE 11-1

b. For a paired t test, test statistic $= -3.60$, two-sided p-value $= .0058$; for a Wilcoxon signed rank test, test statistic $= 0$, two-sided p-value $= .006$; for a sign test, test statistic $= 0$, two-sided p-value $= .002$.

c. A 95% confidence interval for the mean difference in numbers of mosquitos captured, based on a t distribution is -97.8 to -22.2; approximate 95% confidence interval for the mean difference based on a Wilcoxon signed rank distribution is -105.0 to -28.0; approximate 95% confidence interval for the median difference based on a binomial distribution is -77.67 to -25.6 mosquitos.

EXERCISE 11-2

a. A large-sample test statistic $= 3.76$, two-sided p-value $< .0004$.

b. Approximate 99% confidence interval for the difference between the two proportions is .0104 to .0553.

EXERCISE 11-3

b. For a two-sample t test, test statistic $= 1.72$, two-sided p-value $= .11$; for a Wilcoxon–Mann–Whitney test, test statistic $= 83$, two-sided p-value $= .1$;

for the median test, smallest frequency in 2 × 2 table is 2, two-sided *p*-value = 2 × .143 = .286.

c. A 95% confidence interval for the difference between the two mean scores based on a *t* distribution is −.6 to 5.35; 95–96% confidence interval for the difference between the two mean scores based on a Wilcoxon–Mann–Whitney distribution is −1.001 to 6.000.

EXERCISE 11-9 A large-sample test statistic = 9.2, two-sided *p*-value $< .0001$; approximate 99% confidence interval for the difference between the two mean lifetimes is 206 to 367 days.

EXERCISE 12-5

b. test statistic = 63.45 with 2 and 9 degrees of freedom, *p*-value $< .0001$; separate 99% confidence intervals are (−2.5, −.8) for $\mu_1 - \mu_2$, (−2.6, −1.2) for $\mu_1 - \mu_3$, (−.8, .2) for $\mu_2 - \mu_3$. The mean volume increase for flour type 1 seems to be less than the mean for flour type 2 and the mean for flour type 3; we cannot distinguish between flour types 2 and 3.

c. test statistic = 8.2, *p*-value = .02; separate 97% confidence intervals are (−2.3, −.9) for $\mu_1 - \mu_2$, (−2.3, −1.3) for $\mu_1 - \mu_3$, (−.6, .2) for $\mu_2 - \mu_3$. Conclusions are the same as for part (b).

EXERCISE 12-10

c. test statistic for treatment differences is 1.24 with 3 and 12 degrees of freedom, *p*-value = .3; test statistic for block differences is 3.51 with 4 and 12 degrees of freedom, *p*-value = .04.

e. Friedman's test statistic for treatment difference equals 3.4, *p*-value = .3.

EXERCISE 13-1

c. test statistic for brand differences equals 39.6 with 1 and 12 degrees of freedom, *p*-value $< .0001$; test statistic for material differences equals 65.45 with 1 and 12 degrees of freedom, *p*-value $< .0001$; test statistic for interaction effects equals .62 with 1 and 12 degrees of freedom, *p*-value = .4.

EXERCISE 14-1

b. test statistic = 5.57, *p*-value = .04

c. A 95% confidence interval for the ratio of the two population variances (smaller stress level over greater stress level) is .78 to 39.79.

EXERCISE 14-2

b. test statistic = 46.2, *p*-value $< .0001$

c. A 99% confidence interval for the population variance is 27,570 to 1,121,542 grams^2.

EXERCISE 14-12

b. Bartlett's test statistic = 1.8, *p*-value = .9

c. Levene's test statistic = .41, *p*-value = .8

EXERCISE 15-1

b. linear correlation coefficient = −.22, test statistic = −.64, *p*-value = .5

c. rank correlation coefficient = .055, test statistic = 156, *p*-value = .9

EXERCISE 15-10

b. $Y = 4.68 + .887X$

c. $Y = .71 + .997X$

d. equation of standard deviation line if Y = second reading and X = first reading is $Y = 2.07 + .943\,X$

EXERCISE 15-11

b. $Y = 2.5 + 3.99X$

c. test statistic = 291.37, p-value < .0001

d. test statistic = .20, p-value = 0.8

f. $R^2 = 1.0$

EXERCISE 15-22

Final estimated model is: distance = 12.8 + .556(right leg strength) + .272(overall leg strength) with both p-values less than .03.

EXERCISE 16-3

test statistic = 1.47, p-value = .7

EXERCISE 16-9

test statistic = 11.7, p-value = .003

EXERCISE 16-14

a. test statistic = 9.235, p-value = .002

b. smallest frequency in the 2×2 table is 0, p-value = .2

The Wilcoxon Signed Rank Distributions

The Wilcoxon signed rank probability distribution for sample size n comes from the probability model for an experiment that randomly assigns + and − signs to ranks 1 through n. The Wilcoxon signed rank probability distributions form the basis of the Wilcoxon signed rank test about a population mean (or median). We will describe the probability model that leads to the Wilcoxon signed rank probability distribution for a general sample size, n. First, we will find this probability distribution for sample size 3.

Imagine that we have three objects, labeled 1, 2, and 3. We call these numbers the ranks of the objects, 1 being the smallest rank and 3 the largest. Suppose there is a random process (such as a coin toss) that assigns either a + sign or a − sign to each object. Then there are $2^3 = 8$ possible outcomes, listed in Table 1.

Add up the ranks of the objects that receive a + and call this sum $T+$. Similarly, add up the ranks of the objects that receive a − and call this sum $T-$. Values of $T+$ and $T-$ are listed in Table 1. Notice that for each outcome, $T+$ and $T-$ add up to 6, the sum of the ranks 1, 2, and 3.

To build a probability model for this experiment, suppose that + and − each have probability $\frac{1}{2}$ of being assigned to an object. Also, suppose that assignments to different objects are independent of one another. Then each of the eight outcomes in Table 1 has probability $\frac{1}{2} \times \frac{1}{2} \times \frac{1}{2} = \frac{1}{8}$.

What is the probability distribution of the random variable $T+$? We see from Table 1 that $T+$ has seven possible values: 0, 1, 2, 3, 4, 5, and 6. There are two experimental outcomes for which $T+$ equals 3. Each other possible value of $T+$ is associated with exactly one experimental outcome. Therefore, the probability distribution of $T+$ is the one shown in Table 2. This probability distribution is known as the Wilcoxon signed rank distribution for a sample size of 3. You can easily show that $T-$ has this same probability distribution (Exercise 10-14).

The last column of Table 2 shows cumulative probabilities of the form $P(T+ \leq c)$ for each of the possible values c that $T+$ can take on. These cumulative probabilities are listed in Table F, Appendix 4, for a sample size of 3.

TABLE 1 The eight possible assignments of + and − to objects numbered 1, 2, and 3 are listed. The values of $T+$ and $T-$ for each of the eight outcomes are shown.

Outcome	$T+$	$T-$
1 2 3 + + +	6	0
1 2 3 + + −	3	3
1 2 3 + − +	4	2
1 2 3 − + +	5	1
1 2 3 + − −	1	5
1 2 3 − + −	2	4
1 2 3 − − +	3	3
1 2 3 − − −	0	6

TABLE 2 The Wilcoxon signed rank distribution for sample size of 3

c	$P(T+ = c)$	$P(T+ \leq c)$
0	$\frac{1}{8}$	$\frac{1}{8} = .125$
1	$\frac{1}{8}$	$\frac{2}{8} = .25$
2	$\frac{1}{8}$	$\frac{3}{8} = .375$
3	$\frac{2}{8}$	$\frac{5}{8} = .625$
4	$\frac{1}{8}$	$\frac{6}{8} = .75$
5	$\frac{1}{8}$	$\frac{7}{8} = .875$
6	$\frac{1}{8}$	$\frac{8}{8} = 1$

The Wilcoxon signed rank distribution for general sample size n is developed as follows. Suppose we have n objects with labels 1 through n, called the ranks of the objects. A random process assigns either + or − to each object, with 2^n possible outcomes.

We let $T+$ be the sum of the ranks that receive a + and $T-$ the sum of the ranks that receive a −. $T+$ and $T-$ can take on integer values from 0 to $n(n + 1)/2$. Also, $T+$ and $T-$ add up to $n(n + 1)/2$. [Note that $n(n + 1)/2$ equals the sum of the integers from 1 to n.]

To build a probability model, we assume that + and − each have probability $\frac{1}{2}$ of being assigned to an object. We also assume that assignments to different objects are independent. Then each of the 2^n possible outcomes has probability $1/2^n$.

We can use this probability model to find the probability distribution for $T+$ (and $T-$). This distribution is called the Wilcoxon signed rank distribution for a sample of size n. Table F lists cumulative probabilities of the form $P(T+ \leq c)$ for sample sizes from 2 to 15. Exercise 10-15 asks you to find the Wilcoxon signed rank distribution for a sample of size 4.

The Wilcoxon–Mann–Whitney Distributions

The Wilcoxon–Mann–Whitney distribution for sample sizes n_1 and n_2 is derived from the probability model for an experiment in which ranks 1 through $n_1 + n_2$ are randomly divided into two groups, one of size n_1 and one of size n_2. We use Wilcoxon–Mann–Whitney distributions (or Wilcoxon rank sum distributions) in the Wilcoxon–Mann–Whitney test for comparing two population means. We will describe the probability model that leads to a Wilcoxon–Mann–Whitney distribution in general. First, we will find the Wilcoxon–Mann–Whitney distribution for samples of size 2 and 4.

Imagine that we have six objects, labeled 1, 2, 3, 4, 5, and 6. We call these numbers the ranks of the objects. Suppose there is a random process (such as drawing numbers from a hat) that divides the six objects into two groups, the first group of size 2 and the second group of size 4. There are $\binom{6}{2} = 6!/(2!\,4!) = 15$ possible outcomes, listed in Table 1.

Let W_1 be the sum of the ranks in the first group and W_2 the sum of the ranks in the second group. Values of W_1 and W_2 are listed in Table 1. For each outcome, W_1 and W_2 add up to 21, the sum of the ranks 1, 2, 3, 4, 5, and 6.

If the two smallest ranks are assigned to the first group, then $W_1 = 1 + 2 = 3$. We will define a new random variable T_1 as $T_1 = W_1 - 3$. If the four smallest ranks are placed in the second group, then $W_2 = 1 + 2 + 3 + 4 = 10$. Let the random variable T_2 be defined as $T_2 = W_2 - 10$.

Possible values of T_1 and T_2 are listed in Table 1. Both T_1 and T_2 take integer values from 0 to 8. For each outcome, T_1 and T_2 add up to $21 - 3 - 10 = 8$.

To build a probability model for this experiment, we assume that the process that divides the six objects into two groups is completely random, so each of the 15 outcomes in Table 1 is equally likely. Then each outcome has probability $\frac{1}{15}$ and we can use the information in Table 1 to find the probability distribution of T_1. This probability distribution, shown in Table 2, is known as the Wilcoxon–Mann–Whitney distribution for sample sizes 2 and 4. T_2 has this same probability distribution (Exercise 11-26).

TABLE 1 The 15 possible ways to divide ranks 1 through 6 into two groups, the first group of size 2 and the second of size 4, are listed. Also shown are the values of W_1, T_1, W_2, and T_2 for each outcome. (The order of the ranks within the two groups does not matter.)

Outcome					
Group 1	Group 2	W_1	$T_1 = W_1 - 3$	W_2	$T_2 = W_2 - 10$
1 2	3 4 5 6	3	0	18	8
1 3	2 4 5 6	4	1	17	7
1 4	2 3 5 6	5	2	16	6
1 5	2 3 4 6	6	3	15	5
1 6	2 3 4 5	7	4	14	4
2 3	1 4 5 6	5	2	16	6
2 4	1 3 5 6	6	3	15	5
2 5	1 3 4 6	7	4	14	4
2 6	1 3 4 5	8	5	13	3
3 4	1 2 5 6	7	4	14	4
3 5	1 2 4 6	8	5	13	3
3 6	1 2 4 5	9	6	12	2
4 5	1 2 3 6	9	6	12	2
4 6	1 2 3 5	10	7	11	1
5 6	1 2 3 4	11	8	10	0

TABLE 2 The Wilcoxon–Mann–Whitney distribution for sample sizes 2 and 4

c	$P(T_1 = c)$	$P(T_1 \le c)$
0	$\frac{1}{15}$	$\frac{1}{15}$ = .067
1	$\frac{1}{15}$	$\frac{2}{15}$ = .133
2	$\frac{2}{15}$	$\frac{4}{15}$ = .267
3	$\frac{2}{15}$	$\frac{6}{15}$ = .400
4	$\frac{3}{15}$	$\frac{9}{15}$ = .600
5	$\frac{2}{15}$	$\frac{11}{15}$ = .733
6	$\frac{2}{15}$	$\frac{13}{15}$ = .867
7	$\frac{1}{15}$	$\frac{14}{15}$ = .933
8	$\frac{1}{15}$	$\frac{15}{15}$ = 1

The last column of Table 2 shows cumulative probabilities of the form $P(T_1 \le c)$ for each of the possible values of c that T_1 can take on. These cumulative probabilities are listed in Table G for sample sizes $n_1 = 2$ and $n_2 = 4$.

Now let's develop a general Wilcoxon–Mann–Whitney distribution. Suppose we have n objects with labels 1 through n, called the ranks of the objects. A random process divides the n objects into two groups, the first group of size n_1 and the second group of size n_2, where $n_1 + n_2 = n$. There are $\binom{n}{n_1} = n!/(n_1!\, n_2!)$ possible outcomes of this experiment.

We let W_1 be the sum of the ranks assigned to the first group and W_2 the sum of the ranks in the second group. If the n_1 smallest ranks are all assigned to the first group, then W_1 equals the sum of the ranks from 1 to n_1, $n_1(n_1 + 1)/2$. If the n_2 smallest ranks are all assigned to the second group, then W_2 equals the sum of the ranks from 1 to n_2, $n_2(n_2 + 1)/2$. W_1 and W_2 add up to $n(n + 1)/2$, the sum of the ranks from 1 to n.

Define the random variables T_1 and T_2 by

$$T_1 = W_1 - \frac{n_1(n_1 + 1)}{2} \quad \text{and} \quad T_2 = W_2 - \frac{n_2(n_2 + 1)}{2}$$

T_1 and T_2 each take values from 0 to

$$\frac{n(n + 1)}{2} - \frac{n_1(n_1 + 1)}{2} - \frac{n_2(n_2 + 1)}{2}$$

Also, T_1 and T_2 add up to

$$\frac{n(n + 1)}{2} - \frac{n_1(n_1 + 1)}{2} - \frac{n_2(n_2 + 1)}{2}$$

To build a probability model, we assume that the process dividing the n objects into groups of size n_1 and n_2 is completely random, so each of the $\binom{n}{n_1}$ possible outcomes is equally likely. We can use this probability model to find the probability distribution for T_1 (or T_2). This distribution is called the Wilcoxon–Mann–Whitney distribution for sample sizes n_1 and n_2. Table G lists cumulative probabilities of the form $P(T_1 \leq c)$ for some selected sample sizes. Exercise 11-27 asks you to find the Wilcoxon–Mann–Whitney distribution for sample sizes 3 and 3.

The Kruskal–Wallis Distributions

A Kruskal–Wallis probability distribution is derived from the probability model for an experiment in which ranks 1 through n are randomly divided into three or more groups. We use Kruskal–Wallis distributions in the Kruskal–Wallis test for comparing several medians in a single-factor experiment. We will describe the probability model that leads to a Kruskal–Wallis distribution in general. First, we will find the Kruskal–Wallis distribution when we have samples of size 1, 1, and 3.

Imagine that we have five objects, labeled 1, 2, 3, 4, and 5. These numbers are called the ranks of the objects. Suppose there is a random process (such as drawing numbers from a hat) that divides the five objects into three groups: the first and second groups each with one object and the third group with three objects. By arguments about combinations similar to those we used in Section 7-1, we can show that there are $5!/(1!\,1!\,3!) = 20$ possible outcomes, listed in Table 1.

Let R_1, R_2, and R_3 denote the sum of the ranks in the first, second, and third groups, respectively. R_1, R_2, and R_3 add up to 15, the sum of the ranks 1 through 5. Define a random variable KW by

$$KW = \frac{12}{5 \times 6}\left(\frac{R_1^2}{1} + \frac{R_2^2}{1} + \frac{R_3^2}{3}\right) - 3(5 + 1)$$

Values of R_1, R_2, R_3, and KW are listed in Table 1.

To build a probability model for this experiment, we assume that the process that divides the five objects into three groups is completely random, so each of the 20 possible outcomes is equally likely. Then we can find the probability distribution of KW, shown in Table 2. We call this the Kruskal–Wallis distribution for samples of size 1, 1, and 3. The last column of Table 2 shows tail probabilities of the form $P(KW \geq c)$.

Now we will describe how the Kruskal–Wallis distribution arises in general. Suppose we have N objects labeled with ranks 1 through N. A random process divides the N objects into k groups of size n_1 through n_k, where n_1 through n_k add up to N. The number of possible outcomes of this experiment equals $N!$ divided by the product of $n_1!$ through $n_k!$.

TABLE 1 The 20 possible ways to divide ranks 1 through 5 into groups of size 1, 1, and 3 are listed. Also shown are values of R_1, R_2, R_3, and KW for each outcome.

Outcome						
Group 1	Group 2	Group 3	R_1	R_2	R_3	KW
1	2	3 4 5	1	2	12	3.2
2	1	3 4 5	2	1	12	3.2
1	3	2 4 5	1	3	11	2.1333
3	1	2 4 5	3	1	11	2.1333
1	4	2 3 5	1	4	10	2.1333
4	1	2 3 5	4	1	10	2.1333
1	5	2 3 4	1	5	9	3.2
5	1	2 3 4	5	1	9	3.2
2	3	1 4 5	2	3	10	.5333
3	2	1 4 5	3	2	10	.5333
2	4	1 3 5	2	4	9	.8
4	2	1 3 5	4	2	9	.8
2	5	1 3 4	2	5	8	2.1333
5	2	1 3 4	5	2	8	2.1333
3	4	1 2 5	3	4	8	.5333
4	3	1 2 5	4	3	8	.5333
3	5	1 2 4	3	5	7	2.1333
5	3	1 2 4	5	3	7	2.1333
4	5	1 2 3	4	5	6	3.2
5	4	1 2 3	5	4	6	3.2

TABLE 2 The Kruskal–Wallis distribution for samples of size 1, 1, and 3

c	$P(KW = c)$	$P(KW \geq c)$
.5333	$\frac{4}{20} = .2$	$\frac{20}{20} = 1$
.8	$\frac{2}{20} = .1$	$\frac{16}{20} = .8$
2.1333	$\frac{8}{20} = .4$	$\frac{14}{20} = .7$
3.2	$\frac{6}{20} = .3$	$\frac{6}{20} = .3$

We let R_1 be the sum of the ranks assigned to the first group, R_2 the sum of the ranks in the second group, and so on through R_k. Then we define the random variable KW by

$$KW = \frac{12}{N(N+1)} \sum_{i=1}^{k} \frac{R_i^2}{n_i} - 3(N+1)$$

To get a probability model for this experiment, we assume that the process dividing the N objects into k groups of size n_1 through n_k is completely random, making each possible outcome equally likely. We can use this probability model to find the probability distribution for KW. This distribution is called the Kruskal–Wallis distribution for k samples of size n_1 through n_k.

Table H at the back of the book covers some cases when there are three groups. Probabilities of the form $P(KW \geq c)$ are listed for selected values of n_1, n_2, and n_3. Exercise 12-17 asks you to find the Kruskal–Wallis distribution for three samples of size 1, 2, and 2.

Statistical Tables

TABLE A Binomial distributions: Tabled are probabilities of the form $P(X = k)$, where X is a random variable having the binomial(n, p) distribution for specified sample size n and probability of success p. [This table omits some probabilities that are zero to four decimal places. Because of rounding, listed probabilities for some distributions may not sum exactly to 1. Probabilities were calculated using the Student Edition of Minitab (Schaefer and Anderson, 1989).]

$n = 2$

$p = .05$ k	$P(X = k)$	$p = .95$ k	$p = .1$ k	$P(X = k)$	$p = .9$ k	$p = .2$ k	$P(X = k)$	$p = .8$ k
0	.9025	2	0	.8100	2	0	.6400	2
1	.0950	1	1	.1800	1	1	.3200	1
2	.0025	0	2	.0100	0	2	.0400	0

$p = .3$ k	$P(X = k)$	$p = .7$ k	$p = .4$ k	$P(X = k)$	$p = .6$ k	$p = .5$ k	$P(X = k)$
0	.4900	2	0	.3600	2	0	.2500
1	.4200	1	1	.4800	1	1	.5000
2	.0900	0	2	.1600	0	2	.2500

$n = 3$

$p = .05$ k	$P(X = k)$	$p = .95$ k	$p = .1$ k	$P(X = k)$	$p = .9$ k	$p = .2$ k	$P(X = k)$	$p = .8$ k
0	.8574	3	0	.7290	3	0	.5120	3
1	.1354	2	1	.2430	2	1	.3840	2
2	.0071	1	2	.0270	1	2	.0960	1
3	.0001	0	3	.0010	0	3	.0080	0

$p = .3$ k	$P(X = k)$	$p = .7$ k	$p = .4$ k	$P(X = k)$	$p = .6$ k	$p = .5$ k	$P(X = k)$
0	.3430	3	0	.2160	3	0	.1250
1	.4410	2	1	.4320	2	1	.3750
2	.1890	1	2	.2880	1	2	.3750
3	.0270	0	3	.0640	0	3	.1250

(*continued*)

TABLE A *(continued)*

***n* = 4**

$p = .05$ k	$P(X = k)$	$p = .95$ k	$p = .1$ k	$P(X = k)$	$p = .9$ k	$p = .2$ k	$P(X = k)$	$p = .8$ k
0	.8145	4	0	.6561	4	0	.4096	4
1	.1715	3	1	.2916	3	1	.4096	3
2	.0135	2	2	.0486	2	2	.1536	2
3	.0005	1	3	.0036	1	3	.0256	1
4	.0000	0	4	.0001	0	4	.0016	0

$p = .3$ k	$P(X = k)$	$p = .7$ k	$p = .4$ k	$P(X = k)$	$p = .6$ k	$p = .5$ k	$P(X = k)$
0	.2401	4	0	.1296	4	0	.0625
1	.4116	3	1	.3456	3	1	.2500
2	.2646	2	2	.3456	2	2	.3750
3	.0756	1	3	.1536	1	3	.2500
4	.0081	0	4	.0256	0	4	.0625

***n* = 5**

$p = .05$ k	$P(X = k)$	$p = .95$ k	$p = .1$ k	$P(X = k)$	$p = .9$ k	$p = .2$ k	$P(X = k)$	$p = .8$ k
0	.7738	5	0	.5905	5	0	.3277	5
1	.2036	4	1	.3281	4	1	.4096	4
2	.0214	3	2	.0729	3	2	.2048	3
3	.0011	2	3	.0081	2	3	.0512	2
4	.0000	1	4	.0005	1	4	.0064	1
5	.0000	0	5	.0000	0	5	.0003	0

$p = .3$ k	$P(X = k)$	$p = .7$ k	$p = .4$ k	$P(X = k)$	$p = .6$ k	$p = .5$ k	$P(X = k)$
0	.1681	5	0	.0778	5	0	.0313
1	.3601	4	1	.2592	4	1	.1562
2	.3087	3	2	.3456	3	2	.3125
3	.1323	2	3	.2304	2	3	.3125
4	.0284	1	4	.0768	1	4	.1562
5	.0024	0	5	.0102	0	5	.0313

TABLE A *(continued)*

$n = 6$

$p = .05$ k	P(X = k)	$p = .95$ k	$p = .1$ k	P(X = k)	$p = .9$ k	$p = .2$ k	P(X = k)	$p = .8$ k
0	.7351	6	0	.5314	6	0	.2621	6
1	.2321	5	1	.3543	5	1	.3932	5
2	.0305	4	2	.0984	4	2	.2458	4
3	.0021	3	3	.0146	3	3	.0819	3
4	.0001	2	4	.0012	2	4	.0154	2
5	.0000	1	5	.0001	1	5	.0015	1
6	.0000	0	6	.0000	0	6	.0001	0

$p = .3$ k	P(X = k)	$p = .7$ k	$p = .4$ k	P(X = k)	$p = .6$ k	$p = .5$ k	P(X = k)
0	.1176	6	0	.0467	6	0	.0156
1	.3025	5	1	.1866	5	1	.0937
2	.3241	4	2	.3110	4	2	.2344
3	.1852	3	3	.2765	3	3	.3125
4	.0595	2	4	.1382	2	4	.2344
5	.0102	1	5	.0369	1	5	.0937
6	.0007	0	6	.0041	0	6	.0156

$n = 7$

$p = .05$ k	P(X = k)	$p = .95$ k	$p = .1$ k	P(X = k)	$p = .9$ k	$p = .2$ k	P(X = k)	$p = .8$ k
0	.6983	7	0	.4783	7	0	.2097	7
1	.2573	6	1	.3720	6	1	.3670	6
2	.0406	5	2	.1240	5	2	.2753	5
3	.0036	4	3	.0230	4	3	.1147	4
4	.0002	3	4	.0026	3	4	.0287	3
5	.0000	2	5	.0002	2	5	.0043	2
6	.0000	1	6	.0000	1	6	.0004	1
7	.0000	0	7	.0000	0	7	.0000	0

$p = .3$ k	P(X = k)	$p = .7$ k	$p = .4$ k	P(X = k)	$p = .6$ k	$p = .5$ k	P(X = k)
0	.0824	7	0	.0280	7	0	.0078
1	.2471	6	1	.1306	6	1	.0547
2	.3177	5	2	.2613	5	2	.1641
3	.2269	4	3	.2903	4	3	.2734
4	.0972	3	4	.1935	3	4	.2734
5	.0250	2	5	.0774	2	5	.1641
6	.0036	1	6	.0172	1	6	.0547
7	.0002	0	7	.0016	0	7	.0078

(continued)

TABLE A *(continued)*

$n = 8$

$p = .05$ k	$P(X = k)$	$p = .95$ k	$p = .1$ k	$P(X = k)$	$p = .9$ k	$p = .2$ k	$P(X = k)$	$p = .8$ k
0	.6634	8	0	.4305	8	0	.1678	8
1	.2793	7	1	.3826	7	1	.3355	7
2	.0515	6	2	.1488	6	2	.2936	6
3	.0054	5	3	.0331	5	3	.1468	5
4	.0004	4	4	.0046	4	4	.0459	4
5	.0000	3	5	.0004	3	5	.0092	3
6	.0000	2	6	.0000	2	6	.0011	2
7	.0000	1	7	.0000	1	7	.0001	1
8	.0000	0	8	.0000	0	8	.0000	0

$p = .3$ k	$P(X = k)$	$p = .7$ k	$p = .4$ k	$P(X = k)$	$p = .6$ k	$p = .5$ k	$P(X = k)$
0	.0576	8	0	.0168	8	0	.0039
1	.1977	7	1	.0896	7	1	.0313
2	.2965	6	2	.2090	6	2	.1094
3	.2541	5	3	.2787	5	3	.2187
4	.1361	4	4	.2322	4	4	.2734
5	.0467	3	5	.1239	3	5	.2187
6	.0100	2	6	.0413	2	6	.1094
7	.0012	1	7	.0079	1	7	.0313
8	.0001	0	8	.0007	0	8	.0039

$n = 9$

$p = .05$ k	$P(X = k)$	$p = .95$ k	$p = .1$ k	$P(X = k)$	$p = .9$ k	$p = .2$ k	$P(X = k)$	$p = .8$ k
0	.6302	9	0	.3874	9	0	.1342	9
1	.2985	8	1	.3874	8	1	.3020	8
2	.0629	7	2	.1722	7	2	.3020	7
3	.0077	6	3	.0446	6	3	.1762	6
4	.0006	5	4	.0074	5	4	.0661	5
5	.0000	4	5	.0008	4	5	.0165	4
6	.0000	3	6	.0001	3	6	.0028	3
7	.0000	2	7	.0000	2	7	.0003	2
8	.0000	1	8	.0000	1	8	.0000	1

TABLE A *(continued)*

$p = .3$		$p = .7$	$p = .4$		$p = .6$	$p = .5$	
k	$P(X = k)$	k	k	$P(X = k)$	k	k	$P(X = k)$
0	.0404	9	0	.0101	9	0	.0020
1	.1556	8	1	.0605	8	1	.0176
2	.2668	7	2	.1612	7	2	.0703
3	.2668	6	3	.2508	6	3	.1641
4	.1715	5	4	.2508	5	4	.2461
5	.0735	4	5	.1672	4	5	.2461
6	.0210	3	6	.0743	3	6	.1641
7	.0039	2	7	.0212	2	7	.0703
8	.0004	1	8	.0035	1	8	.0176
9	.0000	0	9	.0003	0	9	.0020

$n = 10$

$p = .05$		$p = .95$	$p = .1$		$p = .9$	$p = .2$		$p = .8$
k	$P(X = k)$	k	k	$P(X = k)$	k	k	$P(X = k)$	k
0	.5987	10	0	.3487	10	0	.1074	10
1	.3151	9	1	.3874	9	1	.2684	9
2	.0746	8	2	.1937	8	2	.3020	8
3	.0105	7	3	.0574	7	3	.2013	7
4	.0010	6	4	.0112	6	4	.0881	6
5	.0001	5	5	.0015	5	5	.0264	5
6	.0000	4	6	.0001	4	6	.0055	4
7	.0000	3	7	.0000	3	7	.0008	3
8	.0000	2	8	.0000	2	8	.0001	2
9	.0000	1	9	.0000	1	9	.0000	1

$p = .3$		$p = .7$	$p = .4$		$p = .6$	$p = .5$	
k	$P(X = k)$	k	k	$P(X = k)$	k	k	$P(X = k)$
0	.0282	10	0	.0060	10	0	.0010
1	.1211	9	1	.0403	9	1	.0098
2	.2335	8	2	.1209	8	2	.0439
3	.2668	7	3	.2150	7	3	.1172
4	.2001	6	4	.2508	6	4	.2051
5	.1029	5	5	.2007	5	5	.2461
6	.0368	4	6	.1115	4	6	.2051
7	.0090	3	7	.0425	3	7	.1172
8	.0014	2	8	.0106	2	8	.0439
9	.0001	1	9	.0016	1	9	.0098
10	.0000	0	10	.0001	0	10	.0010

(continued)

TABLE A *(continued)*
$n = 15$

$p = .05$ k	$P(X = k)$	$p = .95$ k	$p = .1$ k	$P(X = k)$	$p = .9$ k	$p = .2$ k	$P(X = k)$	$p = .8$ k
0	.4633	15	0	.2059	15	0	.0352	15
1	.3658	14	1	.3432	14	1	.1319	14
2	.1348	13	2	.2669	13	2	.2309	13
3	.0307	12	3	.1285	12	3	.2501	12
4	.0049	11	4	.0428	11	4	.1876	11
5	.0006	10	5	.0105	10	5	.1032	10
6	.0000	9	6	.0019	9	6	.0430	9
7	.0000	8	7	.0003	8	7	.0138	8
8	.0000	7	8	.0000	7	8	.0035	7
9	.0000	6	9	.0000	6	9	.0007	6
10	.0000	5	10	.0000	5	10	.0001	5
11	.0000	4	11	.0000	4	11	.0000	4

$p = .3$ k	$P(X = k)$	$p = .7$ k	$p = .4$ k	$P(X = k)$	$p = .6$ k	$p = .5$ k	$P(X = k)$
0	.0047	15	0	.0005	15	0	.0000
1	.0305	14	1	.0047	14	1	.0005
2	.0916	13	2	.0219	13	2	.0032
3	.1700	12	3	.0634	12	3	.0139
4	.2186	11	4	.1268	11	4	.0417
5	.2061	10	5	.1859	10	5	.0916
6	.1472	9	6	.2066	9	6	.1527
7	.0811	8	7	.1771	8	7	.1964
8	.0348	7	8	.1181	7	8	.1964
9	.0116	6	9	.0612	6	9	.1527
10	.0030	5	10	.0245	5	10	.0916
11	.0006	4	11	.0074	4	11	.0417
12	.0001	3	12	.0016	3	12	.0139
13	.0000	2	13	.0003	2	13	.0032
14	.0000	1	14	.0000	1	14	.0005
15	.0000	0	15	.0000	0	15	.0000

TABLE A *(continued)*
***n* = 20**

$p = .05$ k	$P(X = k)$	$p = .95$ k	$p = .1$ k	$P(X = k)$	$p = .9$ k	$p = .2$ k	$P(X = k)$	$p = .8$ k
0	.3585	20	0	.1216	20	0	.0115	20
1	.3774	19	1	.2702	19	1	.0576	19
2	.1887	18	2	.2852	18	2	.1369	18
3	.0596	17	3	.1901	17	3	.2054	17
4	.0133	16	4	.0898	16	4	.2182	16
5	.0022	15	5	.0319	15	5	.1746	15
6	.0003	14	6	.0089	14	6	.1091	14
7	.0000	13	7	.0020	13	7	.0545	13
8	.0000	12	8	.0004	12	8	.0222	12
9	.0000	11	9	.0001	11	9	.0074	11
10	.0000	10	10	.0000	10	10	.0020	10
11	.0000	9	11	.0000	9	11	.0005	9
12	.0000	8	12	.0000	8	12	.0001	8
13	.0000	7	13	.0000	7	13	.0000	7

$p = .3$ k	$P(X = k)$	$p = .7$ k	$p = .4$ k	$P(X = k)$	$p = .6$ k	$p = .5$ k	$P(X = k)$
0	.0008	20	0	.0000	20	0	.0000
1	.0068	19	1	.0005	19	1	.0000
2	.0278	18	2	.0031	18	2	.0002
3	.0716	17	3	.0123	17	3	.0011
4	.1304	16	4	.0350	16	4	.0046
5	.1789	15	5	.0746	15	5	.0148
6	.1916	14	6	.1244	14	6	.0370
7	.1643	13	7	.1659	13	7	.0739
8	.1144	12	8	.1797	12	8	.1201
9	.0654	11	9	.1597	11	9	.1602
10	.0308	10	10	.1171	10	10	.1762
11	.0120	9	11	.0710	9	11	.1602
12	.0039	8	12	.0355	8	12	.1201
13	.0010	7	13	.0146	7	13	.0739
14	.0002	6	14	.0049	6	14	.0370
15	.0000	5	15	.0013	5	15	.0148
16	.0000	4	16	.0003	4	16	.0046
17	.0000	3	17	.0000	3	17	.0011
18	.0000	2	18	.0000	2	18	.0002
19	.0000	1	19	.0000	1	19	.0000

TABLE B Cumulative standard Gaussian (normal) distribution. Tabled are cumulative probabilities of the form $P(Z \leq c)$, where c is a nonnegative number and Z has the standard Gaussian distribution.

$P(Z \leq c)$ $P(Z \leq -c)$ $P(Z \geq c)$

c $-c$ c

c	.00	.01	.02	.03	.04	.05	.06	.07	.08	.09
.0	.5000	.5040	.5080	.5120	.5160	.5199	.5239	.5279	.5319	.5359
.1	.5398	.5438	.5478	.5517	.5557	.5596	.5636	.5675	.5714	.5753
.2	.5793	.5832	.5871	.5910	.5948	.5987	.6026	.6064	.6103	.6141
.3	.6179	.6217	.6255	.6293	.6331	.6368	.6406	.6443	.6480	.6517
.4	.6554	.6591	.6628	.6664	.6700	.6736	.6772	.6808	.6844	.6879
.5	.6915	.6950	.6985	.7019	.7054	.7088	.7123	.7157	.7190	.7224
.6	.7257	.7291	.7324	.7357	.7389	.7422	.7454	.7486	.7517	.7549
.7	.7580	.7611	.7642	.7673	.7704	.7734	.7764	.7794	.7823	.7852
.8	.7881	.7910	.7939	.7967	.7995	.8023	.8051	.8078	.8106	.8133
.9	.8159	.8186	.8212	.8238	.8264	.8289	.8315	.8340	.8365	.8389
1.0	.8413	.8438	.8461	.8485	.8508	.8531	.8554	.8577	.8599	.8621
1.1	.8643	.8665	.8686	.8708	.8729	.8749	.8770	.8790	.8810	.8830
1.2	.8849	.8869	.8888	.8907	.8925	.8944	.8962	.8980	.8997	.9015
1.3	.9032	.9049	.9066	.9082	.9099	.9115	.9131	.9147	.9162	.9177
1.4	.9192	.9207	.9222	.9236	.9251	.9265	.9279	.9292	.9306	.9319
1.5	.9332	.9345	.9357	.9370	.9382	.9394	.9406	.9418	.9429	.9441
1.6	.9452	.9463	.9474	.9484	.9495	.9505	.9515	.9525	.9535	.9545
1.7	.9554	.9564	.9573	.9582	.9591	.9599	.9608	.9616	.9625	.9633
1.8	.9641	.9649	.9656	.9664	.9671	.9678	.9686	.9693	.9699	.9706
1.9	.9713	.9719	.9726	.9732	.9738	.9744	.9750	.9756	.9761	.9767
2.0	.9772	.9778	.9783	.9788	.9793	.9798	.9803	.9808	.9812	.9817
2.1	.9821	.9826	.9830	.9834	.9838	.9842	.9846	.9850	.9854	.9857
2.2	.9861	.9864	.9868	.9871	.9875	.9878	.9881	.9884	.9887	.9890
2.3	.9893	.9896	.9898	.9901	.9904	.9906	.9909	.9911	.9913	.9916
2.4	.9918	.9920	.9922	.9925	.9927	.9929	.9931	.9932	.9934	.9936
2.5	.9938	.9940	.9941	.9943	.9945	.9946	.9948	.9949	.9951	.9952
2.6	.9953	.9955	.9956	.9957	.9959	.9960	.9961	.9962	.9963	.9964
2.7	.9965	.9966	.9967	.9968	.9969	.9970	.9971	.9972	.9973	.9974
2.8	.9974	.9975	.9976	.9977	.9977	.9978	.9979	.9979	.9980	.9981
2.9	.9981	.9982	.9982	.9983	.9984	.9984	.9985	.9985	.9986	.9986
3.0	.9987	.9987	.9987	.9988	.9988	.9989	.9989	.9989	.9990	.9990
3.1	.9990	.9991	.9991	.9991	.9992	.9992	.9992	.9992	.9993	.9993
3.2	.9993	.9993	.9994	.9994	.9994	.9994	.9994	.9995	.9995	.9995
3.3	.9995	.9995	.9995	.9996	.9996	.9996	.9996	.9996	.9996	.9997
3.4	.9997	.9997	.9997	.9997	.9997	.9997	.9997	.9997	.9997	.9998

Note: $P(Z \leq -c) = P(Z \geq c)$
$= 1 - P(Z \leq c)$

Example: $P(Z \leq 1.96) = .9750$
$P(Z \leq -1.96) = P(Z \geq 1.96)$
$= .0250$

From Table 2 of *Mathematical Statistics and Data Analysis* by John A. Rice, Wadsworth & Brooks/Cole, 1988, page 558.

TABLE C Values of c corresponding to cumulative probabilities $P(T \leq c)$ and tail areas $P(T \geq c)$, where T has a t distribution.

Cumulative probability $P(T \leq c)$

Tail area $P(T \geq c)$

c

Cumulative probability	.60	.70	.80	.90	.95	.975	.99	.995
Tail area	.40	.30	.20	.10	.05	.025	.01	.005
Degrees of freedom								
1	.325	.727	1.376	3.078	6.314	12.706	31.821	63.657
2	.289	.617	1.061	1.886	2.920	4.303	6.965	9.925
3	.277	.584	.978	1.638	2.353	3.182	4.541	5.841
4	.271	.569	.941	1.533	2.132	2.776	3.747	4.604
5	.267	.559	.920	1.476	2.015	2.571	3.365	4.032
6	.265	.553	.906	1.440	1.943	2.447	3.143	3.707
7	.263	.549	.896	1.415	1.895	2.365	2.998	3.499
8	.262	.546	.889	1.397	1.860	2.306	2.896	3.355
9	.261	.543	.883	1.383	1.833	2.262	2.821	3.250
10	.260	.542	.879	1.372	1.812	2.228	2.764	3.169
11	.260	.540	.876	1.363	1.796	2.201	2.718	3.106
12	.259	.539	.873	1.356	1.782	2.179	2.681	3.055
13	.259	.538	.870	1.350	1.771	2.160	2.650	3.012
14	.258	.537	.868	1.345	1.761	2.145	2.624	2.977
15	.258	.536	.866	1.341	1.753	2.131	2.602	2.947
16	.258	.535	.865	1.337	1.746	2.120	2.583	2.921
17	.257	.534	.863	1.333	1.740	2.110	2.567	2.898
18	.257	.534	.862	1.330	1.734	2.101	2.552	2.878
19	.257	.533	.861	1.328	1.729	2.093	2.539	2.861
20	.257	.533	.860	1.325	1.725	2.086	2.528	2.845
21	.257	.532	.859	1.323	1.721	2.080	2.518	2.831
22	.256	.532	.858	1.321	1.717	2.074	2.508	2.819
23	.256	.532	.858	1.319	1.714	2.069	2.500	2.807
24	.256	.531	.857	1.318	1.711	2.064	2.492	2.797
25	.256	.531	.856	1.316	1.708	2.060	2.485	2.787
26	.256	.531	.856	1.315	1.706	2.056	2.479	2.779
27	.256	.531	.855	1.314	1.703	2.052	2.473	2.771
28	.256	.530	.855	1.313	1.701	2.048	2.467	2.763
29	.256	.530	.854	1.311	1.699	2.045	2.462	2.756
30	.256	.530	.854	1.310	1.697	2.042	2.457	2.750
40	.255	.529	.851	1.303	1.684	2.021	2.423	2.704
60	.254	.527	.848	1.296	1.671	2.000	2.390	2.660
120	.254	.526	.845	1.289	1.658	1.980	2.358	2.617
∞	.253	.524	.842	1.282	1.645	1.960	2.326	2.576

From Table 4 of *Mathematical Statistics and Data Analysis* by John A. Rice, Wadsworth & Brooks/Cole, 1988, page 560.

TABLE D Values of c corresponding to cumulative probabilities $P(F \leq c)$, where F has an F distribution.

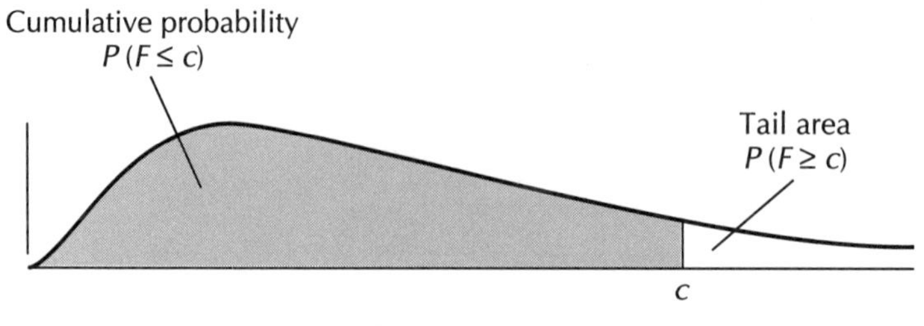

$P(F \leq c) = .90$

d_1 = degrees of freedom for numerator

d_2 = degrees of freedom for denominator

d_2 \ d_1	1	2	3	4	5	6	7	8	9
1	39.86	49.50	53.59	55.83	57.24	58.20	58.91	59.44	59.86
2	8.53	9.00	9.16	9.24	9.29	9.33	9.35	9.37	9.38
3	5.54	5.46	5.39	5.34	5.31	5.28	5.27	5.25	5.24
4	4.54	4.32	4.19	4.11	4.05	4.01	3.98	3.95	3.94
5	4.06	3.78	3.62	3.52	3.45	3.40	3.37	3.34	3.32
6	3.78	3.46	3.29	3.18	3.11	3.05	3.01	2.98	2.96
7	3.59	3.26	3.07	2.96	2.88	2.83	2.78	2.75	2.72
8	3.46	3.11	2.92	2.81	2.73	2.67	2.62	2.59	2.56
9	3.36	3.01	2.81	2.69	2.61	2.55	2.51	2.47	2.44
10	3.29	2.92	2.73	2.61	2.52	2.46	2.41	2.38	2.35
11	3.23	2.86	2.66	2.54	2.45	2.39	2.34	2.30	2.27
12	3.18	2.81	2.61	2.48	2.39	2.33	2.28	2.24	2.21
13	3.14	2.76	2.56	2.43	2.35	2.28	2.23	2.20	2.16
14	3.10	2.73	2.52	2.39	2.31	2.24	2.19	2.15	2.12
15	3.07	2.70	2.49	2.36	2.27	2.21	2.16	2.12	2.09
16	3.05	2.67	2.46	2.33	2.24	2.18	2.13	2.09	2.06
17	3.03	2.64	2.44	2.31	2.22	2.15	2.10	2.06	2.03
18	3.01	2.62	2.42	2.29	2.20	2.13	2.08	2.04	2.00
19	2.99	2.61	2.40	2.27	2.18	2.11	2.06	2.02	1.98
20	2.97	2.59	2.38	2.25	2.16	2.09	2.04	2.00	1.96
21	2.96	2.57	2.36	2.23	2.14	2.08	2.02	1.98	1.95
22	2.95	2.56	2.35	2.22	2.13	2.06	2.01	1.97	1.93
23	2.94	2.55	2.34	2.21	2.11	2.05	1.99	1.95	1.92
24	2.93	2.54	2.33	2.19	2.10	2.04	1.98	1.94	1.91
25	2.92	2.53	2.32	2.18	2.09	2.02	1.97	1.93	1.89
26	2.91	2.52	2.31	2.17	2.08	2.01	1.96	1.92	1.88
27	2.90	2.51	2.30	2.17	2.07	2.00	1.95	1.91	1.87
28	2.89	2.50	2.29	2.16	2.06	2.00	1.94	1.90	1.87
29	2.89	2.50	2.28	2.15	2.06	1.99	1.93	1.89	1.86
30	2.88	2.49	2.28	2.14	2.05	1.98	1.93	1.88	1.85
40	2.84	2.44	2.23	2.09	2.00	1.93	1.87	1.83	1.79
60	2.79	2.39	2.18	2.04	1.95	1.87	1.82	1.77	1.74
120	2.75	2.35	2.13	1.99	1.90	1.82	1.77	1.72	1.68
∞	2.71	2.30	2.08	1.94	1.85	1.77	1.72	1.67	1.63

From Table 5 of *Mathematical Statistics and Data Analysis* by John A. Rice, Wadsworth & Brooks/Cole, 1988, pages 561–564.

TABLE D *(continued)*

10	12	15	20	24	30	40	60	120	∞
60.19	60.71	61.22	61.74	62.00	62.26	62.53	62.79	63.06	63.33
9.39	9.41	9.42	9.44	9.45	9.46	9.47	9.47	9.48	9.49
5.23	5.22	5.20	5.18	5.18	5.17	5.16	5.15	5.14	5.13
3.92	3.90	3.87	3.84	3.83	3.82	3.80	3.79	3.78	3.76
3.30	3.27	3.24	3.21	3.19	3.17	3.16	3.14	3.12	3.10
2.94	2.90	2.87	2.84	2.82	2.80	2.78	2.76	2.74	2.72
2.70	2.67	2.63	2.59	2.58	2.56	2.54	2.51	2.49	2.47
2.50	2.50	2.46	2.42	2.40	2.38	2.36	2.34	2.32	2.29
2.42	2.38	2.34	2.30	2.28	2.25	2.23	2.21	2.18	2.16
2.32	2.28	2.24	2.20	2.18	2.16	2.13	2.11	2.08	2.06
2.25	2.21	2.17	2.12	2.10	2.08	2.05	2.03	2.00	1.97
2.19	2.15	2.10	2.06	2.04	2.01	1.99	1.96	1.93	1.90
2.14	2.10	2.05	2.01	1.98	1.96	1.93	1.90	1.88	1.85
2.10	2.05	2.01	1.96	1.94	1.91	1.89	1.86	1.83	1.80
2.06	2.02	1.97	1.92	1.90	1.87	1.85	1.82	1.79	1.76
2.03	1.99	1.94	1.89	1.87	1.84	1.81	1.78	1.75	1.72
2.00	1.96	1.91	1.86	1.84	1.81	1.78	1.75	1.72	1.69
1.98	1.93	1.89	1.84	1.81	1.78	1.75	1.72	1.69	1.66
1.96	1.91	1.86	1.81	1.79	1.76	1.73	1.70	1.67	1.63
1.94	1.89	1.84	1.79	1.77	1.74	1.71	1.68	1.64	1.61
1.92	1.87	1.83	1.78	1.75	1.72	1.69	1.66	1.62	1.59
1.90	1.86	1.81	1.76	1.73	1.70	1.67	1.64	1.60	1.57
1.89	1.84	1.80	1.74	1.72	1.69	1.66	1.62	1.59	1.55
1.88	1.83	1.78	1.73	1.70	1.67	1.64	1.61	1.57	1.53
1.87	1.82	1.77	1.72	1.69	1.66	1.63	1.59	1.56	1.52
1.86	1.81	1.76	1.71	1.68	1.65	1.61	1.58	1.54	1.50
1.85	1.80	1.75	1.70	1.67	1.64	1.60	1.57	1.53	1.49
1.84	1.79	1.74	1.69	1.66	1.63	1.59	1.56	1.52	1.48
1.83	1.78	1.73	1.68	1.65	1.62	1.58	1.55	1.51	1.47
1.82	1.77	1.72	1.67	1.64	1.61	1.57	1.54	1.50	1.46
1.76	1.71	1.66	1.61	1.57	1.54	1.51	1.47	1.42	1.38
1.71	1.66	1.60	1.54	1.51	1.48	1.44	1.40	1.35	1.29
1.65	1.60	1.55	1.48	1.45	1.41	1.37	1.32	1.26	1.19
1.60	1.55	1.49	1.42	1.38	1.34	1.30	1.24	1.17	1.00

(continued)

TABLE D *(continued)*

$P(F \leq c) = .95$

d_1 = degrees of freedom for numerator

d_2 = degrees of freedom for denominator

d_2 \ d_1	1	2	3	4	5	6	7	8	9
1	161.4	199.5	215.7	224.6	230.2	234.0	236.8	238.9	240.5
2	18.51	19.00	19.16	19.25	19.30	19.33	19.35	19.37	19.38
3	10.13	9.55	9.28	9.12	9.01	8.94	8.89	8.85	8.81
4	7.71	6.94	6.59	6.39	6.26	6.16	6.09	6.04	6.00
5	6.61	5.79	5.41	5.19	5.05	4.95	4.88	4.82	4.77
6	5.99	5.14	4.76	4.53	4.39	4.28	4.21	4.15	4.10
7	5.59	4.74	4.35	4.12	3.97	3.87	3.79	3.73	3.68
8	5.32	4.46	4.07	3.84	3.69	3.58	3.50	3.44	3.39
9	5.12	4.26	3.86	3.63	3.48	3.37	3.29	3.23	3.18
10	4.96	4.10	3.71	3.48	3.83	3.22	3.14	3.07	3.02
11	4.84	3.98	3.59	3.36	3.20	3.09	3.01	2.95	2.90
12	4.75	3.89	3.49	3.26	3.11	3.00	2.91	2.85	2.80
13	4.67	3.81	3.41	3.18	3.03	2.92	2.83	2.77	2.71
14	4.60	3.74	3.34	3.11	2.96	2.85	2.76	2.70	2.65
15	4.54	3.68	3.29	3.06	2.90	2.79	2.71	2.64	2.59
16	4.49	3.63	3.24	3.01	2.85	2.74	2.66	2.59	2.54
17	4.45	3.59	3.20	2.96	2.81	2.70	2.61	2.55	2.49
18	4.41	3.55	3.16	2.93	2.77	2.66	2.58	2.51	2.46
19	4.38	3.52	3.13	2.90	2.74	2.63	2.54	2.48	2.42
20	4.35	3.49	3.10	2.87	2.71	2.60	2.51	2.45	2.39
21	4.32	3.47	3.07	2.84	2.68	2.57	2.49	2.42	2.37
22	4.30	3.44	3.05	2.82	2.66	2.55	2.46	2.40	2.34
23	4.28	3.42	3.03	2.80	2.64	2.53	2.44	2.37	2.32
24	4.26	3.40	3.01	2.78	2.62	2.51	2.42	2.36	2.30
25	4.24	3.39	2.99	2.76	2.60	2.49	2.40	2.34	2.28
26	4.23	3.37	2.98	2.74	2.59	2.47	2.39	2.32	2.27
27	4.21	3.35	2.96	2.73	2.57	2.46	2.37	2.31	2.25
28	4.20	3.34	2.95	2.71	2.56	2.45	2.36	2.29	2.24
29	4.18	3.33	2.93	2.70	2.55	2.43	2.35	2.28	2.22
30	4.17	3.32	2.92	2.69	2.53	2.42	2.33	2.27	2.21
40	4.08	3.23	2.84	2.61	2.45	2.34	2.25	2.18	2.12
60	4.00	3.15	2.76	2.53	2.37	2.25	2.17	2.10	2.04
120	3.92	3.07	2.68	2.45	2.29	2.17	2.09	2.02	1.96
∞	3.84	3.00	2.60	2.37	2.21	2.10	2.01	1.94	1.88

TABLE D *(continued)*

10	12	15	20	24	30	40	60	120	∞
241.9	243.9	245.9	248.0	249.1	250.1	251.1	252.2	253.3	254.3
19.40	19.41	19.43	19.45	19.45	19.46	19.47	19.48	19.49	19.50
8.79	8.74	8.70	8.66	8.64	8.62	8.59	8.57	8.55	8.53
5.96	5.91	5.86	5.80	5.77	5.75	5.72	5.69	5.66	5.63
4.74	4.68	4.62	4.56	4.53	4.50	4.46	4.43	4.40	4.36
4.06	4.00	3.94	3.87	3.84	3.81	3.77	3.74	3.70	3.67
3.64	3.57	3.51	3.44	3.41	3.38	3.34	3.30	3.27	3.23
3.35	3.28	3.22	3.15	3.12	3.08	3.04	3.01	2.97	2.93
3.14	3.07	3.01	2.94	2.90	2.86	2.83	2.79	2.75	2.71
2.98	2.91	2.85	2.77	2.74	2.70	2.66	2.62	2.58	2.54
2.85	2.79	2.72	2.65	2.61	2.57	2.53	2.49	2.45	2.40
2.75	2.69	2.62	2.54	2.51	2.47	2.43	2.38	2.34	2.30
2.67	2.60	2.53	2.46	2.42	2.38	2.34	2.30	2.25	2.21
2.60	2.53	2.46	2.39	2.35	2.31	2.27	2.22	2.18	2.13
2.54	2.48	2.40	2.33	2.29	2.25	2.20	2.16	2.11	2.07
2.49	2.42	2.35	2.28	2.24	2.19	2.15	2.11	2.06	2.01
2.45	2.38	2.31	2.23	2.19	2.15	2.10	2.06	2.01	1.96
2.41	2.34	2.27	2.19	2.15	2.11	2.06	2.02	1.97	1.92
2.38	2.31	2.23	2.16	2.11	2.07	2.03	1.98	1.93	1.88
2.35	2.28	2.20	2.12	2.08	2.04	1.99	1.95	1.90	1.84
2.32	2.25	2.18	2.10	2.05	2.01	1.96	1.92	1.87	1.81
2.30	2.23	2.15	2.07	2.03	1.98	1.94	1.89	1.84	1.78
2.27	2.20	2.13	2.05	2.01	1.96	1.91	1.86	1.81	1.76
2.25	2.18	2.11	2.03	1.98	1.94	1.89	1.84	1.79	1.73
2.24	2.16	2.09	2.01	1.96	1.92	1.87	1.82	1.77	1.71
2.22	2.15	2.07	1.99	1.95	1.90	1.85	1.80	1.75	1.69
2.20	2.13	2.06	1.97	1.93	1.88	1.84	1.79	1.73	1.67
2.19	2.12	2.04	1.96	1.91	1.87	1.82	1.77	1.71	1.65
2.18	2.10	2.03	1.94	1.90	1.85	1.81	1.75	1.70	1.64
2.16	2.09	2.01	1.93	1.89	1.84	1.79	1.74	1.68	1.62
2.08	2.00	1.92	1.84	1.79	1.74	1.69	1.64	1.58	1.51
1.99	1.92	1.84	1.75	1.70	1.65	1.59	1.53	1.47	1.39
1.91	1.83	1.75	1.66	1.61	1.55	1.50	1.43	1.35	1.25
1.83	1.75	1.67	1.57	1.52	1.46	1.39	1.32	1.22	1.00

(continued)

TABLE D *(continued)*

$P(F \leq c) = .975$

d_1 = degrees of freedom for numerator

d_2 = degrees of freedom for denominator

d_2 \ d_1	1	2	3	4	5	6	7	8	9
1	647.8	799.5	864.2	899.6	921.8	937.1	948.2	956.7	963.3
2	38.51	39.00	39.17	39.25	39.30	39.33	39.36	39.37	39.39
3	17.44	16.04	15.44	15.10	14.88	14.73	14.62	14.54	14.47
4	12.22	10.65	9.98	9.60	9.36	9.20	9.07	8.98	8.90
5	10.01	8.43	7.76	7.39	7.15	6.98	6.85	6.76	6.68
6	8.81	7.26	6.60	6.23	5.99	5.82	5.70	5.60	5.52
7	8.07	6.54	5.89	5.52	5.29	5.12	4.99	4.90	4.82
8	7.57	6.06	5.42	5.05	4.82	4.65	4.53	4.43	4.36
9	7.21	5.71	5.08	4.72	4.48	4.32	4.20	4.10	4.03
10	6.94	5.46	4.83	4.47	4.24	4.07	3.95	3.85	3.78
11	6.72	5.26	4.63	4.28	4.04	3.88	3.76	3.66	3.59
12	6.55	5.10	4.47	4.12	3.89	3.73	3.61	3.51	3.44
13	6.41	4.97	4.35	4.00	3.77	3.60	3.48	3.39	3.31
14	6.30	4.86	4.24	3.89	3.66	3.50	3.38	3.29	3.21
15	6.20	4.77	4.15	3.80	3.58	3.41	3.29	3.20	3.12
16	6.12	4.69	4.08	3.73	3.50	3.34	3.22	3.12	3.05
17	6.04	4.62	4.01	3.66	3.44	3.28	3.16	3.06	2.98
18	5.98	4.56	3.95	3.61	3.38	3.22	3.10	3.01	2.93
19	5.92	4.51	3.90	3.56	3.33	3.17	3.05	2.96	2.88
20	5.87	4.46	3.86	3.51	3.29	3.13	3.01	2.91	2.84
21	5.83	4.42	3.82	3.48	3.25	3.09	2.97	2.87	2.80
22	5.79	4.38	3.78	3.44	3.22	3.05	2.93	2.84	2.76
23	5.75	4.35	3.75	3.41	3.18	3.02	2.90	2.81	2.73
24	5.72	4.32	3.72	3.38	3.15	2.99	2.87	2.78	2.70
25	5.69	4.29	3.69	3.35	3.13	2.97	2.85	2.75	2.68
26	5.66	4.27	3.67	3.33	3.10	2.94	2.82	2.73	2.65
27	5.63	4.24	3.65	3.31	3.08	2.92	2.80	2.71	2.63
28	5.61	4.22	3.63	3.29	3.06	2.90	2.78	2.69	2.61
29	5.59	4.20	3.61	3.27	3.04	2.88	2.76	2.67	2.59
30	5.57	4.18	3.59	3.25	3.03	2.87	2.75	2.65	2.57
40	5.42	4.05	3.46	3.13	2.90	2.74	2.62	2.53	2.45
60	5.29	3.93	3.34	3.01	2.79	2.63	2.51	2.41	2.33
120	5.15	3.80	3.23	2.89	2.67	2.52	2.39	2.30	2.22
∞	5.02	3.69	3.12	2.79	2.57	2.41	2.29	2.19	2.11

TABLE D *(continued)*

10	12	15	20	24	30	40	60	120	∞
968.6	976.7	984.9	993.1	997.2	1001	1006	1010	1014	1018
39.40	39.41	39.43	39.45	39.46	39.46	39.47	39.48	39.49	39.50
14.42	14.34	14.25	14.17	14.12	14.08	14.04	13.99	13.95	13.90
8.84	8.75	8.66	8.56	8.51	8.46	8.41	8.36	8.31	8.26
6.62	6.52	6.43	6.33	6.28	6.23	6.18	6.12	6.07	6.02
5.46	5.37	5.27	5.17	5.12	5.07	5.01	4.96	4.90	4.85
4.76	4.67	4.57	4.47	4.42	4.36	4.31	4.25	4.20	4.14
4.30	4.20	4.10	4.00	3.95	3.89	3.84	3.78	3.73	3.67
3.96	3.87	3.77	3.67	3.61	3.56	3.51	3.45	3.39	3.33
3.72	3.62	3.52	3.42	3.37	3.31	3.26	3.20	3.14	3.08
3.53	3.43	3.33	3.23	3.17	3.12	3.06	3.00	2.94	2.88
3.37	3.28	3.18	3.07	3.02	2.96	2.91	2.85	2.79	2.72
3.25	3.15	3.05	2.95	2.89	2.84	2.78	2.72	2.66	2.60
3.15	3.05	2.95	2.84	2.79	2.73	2.67	2.61	2.55	2.49
3.06	2.96	2.86	2.76	2.70	2.64	2.59	2.52	2.46	2.40
2.99	2.89	2.79	2.68	2.63	2.57	2.51	2.45	2.38	2.32
2.92	2.82	2.72	2.62	2.56	2.50	2.44	2.38	2.32	2.25
2.87	2.77	2.67	2.56	2.50	2.44	2.38	2.32	2.26	2.19
2.82	2.72	2.62	2.51	2.45	2.39	2.33	2.27	2.20	2.13
2.77	2.68	2.57	2.46	2.41	2.35	2.29	2.22	2.16	2.09
2.73	2.64	2.53	2.42	2.37	2.31	2.25	2.18	2.11	2.04
2.70	2.60	2.50	2.39	2.33	2.27	2.21	2.14	2.08	2.00
2.67	2.57	2.47	2.36	2.30	2.24	2.18	2.11	2.04	1.97
2.64	2.54	2.44	2.33	2.27	2.21	2.15	2.08	2.01	1.94
2.61	2.51	2.41	2.30	2.24	2.18	2.12	2.05	1.98	1.91
2.59	2.49	2.39	2.28	2.22	2.16	2.09	2.03	1.95	1.88
2.57	2.47	2.36	2.25	2.19	2.13	2.07	2.00	1.93	1.85
2.55	2.45	2.34	2.23	2.17	2.11	2.05	1.98	1.91	1.83
2.53	2.43	2.32	2.21	2.15	2.09	2.03	1.96	1.89	1.81
2.51	2.41	2.31	2.20	2.14	2.07	2.01	1.94	1.87	1.79
2.39	2.29	2.18	2.07	2.01	1.94	1.88	1.80	1.72	1.64
2.27	2.17	2.06	1.94	1.88	1.82	1.74	1.67	1.58	1.48
2.16	2.05	1.94	1.82	1.76	1.69	1.61	1.53	1.43	1.31
2.05	1.94	1.83	1.71	1.64	1.57	1.48	1.39	1.27	1.00

(continued)

TABLE D (*continued*)

$P(F \leq c) = .99$

d_1 = degrees of freedom for numerator

d_2 = degrees of freedom for denominator

d_2 \ d_1	1	2	3	4	5	6	7	8	9
1	4052	4999.5	5403	5625	5764	5859	5928	5982	6022
2	98.50	99.00	99.17	99.25	99.30	99.33	99.36	99.37	99.39
3	34.12	30.82	29.46	28.71	28.24	27.91	27.67	27.49	27.35
4	21.20	18.00	16.69	15.98	15.52	15.21	14.98	14.80	14.66
5	16.26	13.27	12.06	11.39	10.97	10.67	10.46	10.29	10.16
6	13.75	10.92	9.78	9.15	8.75	8.47	8.26	8.10	7.98
7	12.25	9.55	8.45	7.85	7.46	7.19	6.99	6.84	6.72
8	11.26	8.65	7.59	7.01	6.63	6.37	6.18	6.03	5.91
9	10.56	8.02	6.99	6.42	6.06	5.80	5.61	5.47	5.35
10	10.04	7.56	6.55	5.99	5.64	5.39	5.20	5.06	4.94
11	9.65	7.21	6.22	5.67	5.32	5.07	4.89	4.74	4.63
12	9.33	6.93	5.95	5.41	5.06	4.82	4.64	4.50	4.39
13	9.07	6.70	5.74	5.21	4.86	4.62	4.44	4.30	4.19
14	8.86	6.51	5.56	5.04	4.69	4.46	4.28	4.14	4.03
15	8.68	6.36	5.42	4.89	4.56	4.32	4.14	4.00	3.89
16	8.53	6.23	5.29	4.77	4.44	4.20	4.03	3.89	3.78
17	8.40	6.11	5.18	4.67	4.34	4.10	3.93	3.79	3.68
18	8.29	6.01	5.09	4.58	4.25	4.01	3.84	3.71	3.60
19	8.18	5.93	5.01	4.50	4.17	3.94	3.77	3.63	3.52
20	8.10	5.85	4.94	4.43	4.10	3.87	3.70	3.56	3.46
21	8.02	5.78	4.87	4.37	4.04	3.81	3.64	3.51	3.40
22	7.95	5.72	4.82	4.31	3.99	3.76	3.59	3.45	3.35
23	7.88	5.66	4.76	4.26	3.94	3.71	3.54	3.41	3.30
24	7.82	5.61	4.72	4.22	3.90	3.67	3.50	3.36	3.26
25	7.77	5.57	4.68	4.18	3.85	3.63	3.46	3.32	3.22
26	7.72	5.53	4.64	4.14	3.82	3.59	3.42	3.29	3.18
27	7.68	5.49	4.60	4.11	3.78	3.56	3.39	3.26	3.15
28	7.64	5.45	4.57	4.07	3.75	3.53	3.36	3.23	3.12
29	7.60	5.42	4.54	4.04	3.73	3.50	3.33	3.20	3.09
30	7.56	5.39	4.51	4.02	3.70	3.47	3.30	3.17	3.07
40	7.31	5.18	4.31	3.83	3.51	3.29	3.12	2.99	2.89
60	7.08	4.98	4.13	3.65	3.34	3.12	2.95	2.82	2.72
120	6.85	4.79	3.95	3.48	3.17	2.96	2.79	2.66	2.56
∞	6.63	4.61	3.78	3.32	3.02	2.80	2.64	2.51	2.41

TABLE D *(continued)*

10	12	15	20	24	30	40	60	120	∞
6056	6106	6157	6209	6235	6261	6287	6313	6339	6366
99.40	99.42	99.43	99.45	99.46	99.47	99.47	99.48	99.49	99.50
27.23	27.05	26.87	26.69	26.60	26.50	26.41	26.32	26.22	26.13
14.55	14.37	14.20	14.02	13.93	13.84	13.75	13.65	13.56	13.46
10.05	9.89	9.72	9.55	9.47	9.38	9.29	9.20	9.11	9.02
7.87	7.72	7.56	7.40	7.31	7.23	7.14	7.06	6.97	6.88
6.62	6.47	6.31	6.16	6.07	5.99	5.91	5.82	5.74	5.65
5.81	5.67	5.52	5.36	5.28	5.20	5.12	5.03	4.95	4.86
5.26	5.11	4.96	4.81	4.73	4.65	4.57	4.48	4.40	4.31
4.85	4.71	4.56	4.41	4.33	4.25	4.17	4.08	4.00	3.91
4.54	4.40	4.25	4.10	4.02	3.94	3.86	3.78	3.69	3.60
4.30	4.16	4.01	3.86	3.78	3.70	3.62	3.54	3.45	3.36
4.10	3.96	3.82	3.66	3.59	3.51	3.43	3.34	3.25	3.17
3.94	3.80	3.66	3.51	3.43	3.35	3.27	3.18	3.09	3.00
3.80	3.67	3.52	3.37	3.29	3.21	3.13	3.05	2.96	2.87
3.69	3.55	3.41	3.26	3.18	3.10	3.02	2.93	2.84	2.75
3.59	3.46	3.31	3.16	3.08	3.00	2.92	2.83	2.75	2.65
3.51	3.37	3.23	3.08	3.00	2.92	2.84	2.75	2.66	2.57
3.43	3.30	3.15	3.00	2.92	2.84	2.76	2.67	2.58	2.49
3.37	3.23	3.09	2.94	2.86	2.78	2.69	2.61	2.52	2.42
3.31	3.17	3.03	2.88	2.80	2.72	2.64	2.55	2.46	2.36
3.26	3.12	2.98	2.83	2.75	2.67	2.58	2.50	2.40	2.31
3.21	3.07	2.93	2.78	2.70	2.62	2.54	2.45	2.35	2.26
3.17	3.03	2.89	2.74	2.66	2.58	2.49	2.40	2.31	2.21
3.13	2.99	2.85	2.70	2.62	2.54	2.45	2.36	2.27	2.17
3.09	2.96	2.81	2.66	2.58	2.50	2.42	2.33	2.23	2.13
3.06	2.93	2.78	2.63	2.55	2.47	2.38	2.29	2.20	2.10
3.03	2.90	2.75	2.60	2.52	2.44	2.35	2.26	2.17	2.06
3.00	2.87	2.73	2.57	2.49	2.41	2.33	2.23	2.14	2.03
2.98	2.84	2.70	2.55	2.47	2.39	2.30	2.21	2.11	2.01
2.80	2.66	2.52	2.37	2.29	2.20	2.11	2.02	1.92	1.80
2.63	2.50	2.35	2.20	2.12	2.03	1.94	1.84	1.73	1.60
2.47	2.34	2.19	2.03	1.95	1.86	1.76	1.66	1.53	1.38
2.32	2.18	2.04	1.88	1.79	1.70	1.59	1.47	1.32	1.00

(continued)

TABLE E Values of c corresponding to cumulative probabilities $P(X \le c)$ and tail areas $P(X \ge c)$, where X has a chi-square distribution

Cumulative probability $P(X \le c)$

Tail area $P(X \ge c)$

c

Cumulative probability	.005	.01	.025	.05	.10	.90	.95	.975	.99	.995
Tail area	.995	.99	.975	.95	.90	.10	.05	.025	.01	.005
Degrees of freedom										
1	.000039	.00016	.00098	.0039	.0158	2.71	3.84	5.02	6.63	7.88
2	.0100	.0201	.0506	.1026	.2107	4.61	5.99	7.38	9.21	10.60
3	.0717	.115	.216	.352	.584	6.25	7.81	9.35	11.34	12.84
4	.207	.297	.484	.711	1.064	7.78	9.49	11.14	13.28	14.86
5	.412	.554	.831	1.15	1.61	9.24	11.07	12.83	15.09	16.75
6	.676	.872	1.24	1.64	2.20	10.64	12.59	14.45	16.81	18.55
7	.989	1.24	1.69	2.17	2.83	12.02	14.07	16.01	18.48	20.28
8	1.34	1.65	2.18	2.73	3.49	13.36	15.51	17.53	20.09	21.96
9	1.73	2.09	2.70	3.33	4.17	14.68	16.92	19.02	21.67	23.59
10	2.16	2.56	3.25	3.94	4.87	15.99	18.31	20.48	23.21	25.19
11	2.60	3.05	3.82	4.57	5.58	17.28	19.68	21.92	24.73	26.76
12	3.07	3.57	4.40	5.23	6.30	18.55	21.03	23.34	26.22	28.30
13	3.57	4.11	5.01	5.89	7.04	19.81	22.36	24.74	27.69	29.82
14	4.07	4.66	5.63	6.57	7.79	21.06	23.68	26.12	29.14	31.32
15	4.60	5.23	6.26	7.26	8.55	22.31	25.00	27.49	30.58	32.80
16	5.14	5.81	6.91	7.96	9.31	23.54	26.30	28.85	32.00	34.27
18	6.26	7.01	8.23	9.39	10.86	25.99	28.87	31.53	34.81	37.16
20	7.43	8.26	9.59	10.85	12.44	28.41	31.41	34.17	37.57	40.00
24	9.89	10.86	12.40	13.85	15.66	33.20	36.42	39.36	42.98	45.56
30	13.79	14.95	16.79	18.49	20.60	40.26	43.77	46.98	50.89	53.67
40	20.71	22.16	24.43	26.51	29.05	51.81	55.76	59.34	63.69	66.77
60	35.53	37.48	40.48	43.19	46.46	74.40	79.08	83.30	88.38	91.95
120	83.85	86.92	91.58	95.70	100.62	140.23	146.57	152.21	158.95	163.64

From Table 3 of *Mathematical Statistics and Data Analysis* by John A. Rice, Wadsworth & Brooks/Cole, 1988, page 559.

TABLE F Cumulative probabilities $P(W \le c)$, where W has the signed rank distribution for sample size n

n: c	2	3	4	5	6	7	8
0	.250	.125	.062	.031	.016	.008	.004
1	.500	.250	.125	.062	.031	.016	.008
2		.375	.188	.094	.047	.023	.012
3		.625	.312	.156	.078	.039	.020
4			.438	.219	.109	.055	.027
5			.562	.312	.156	.078	.039
6				.406	.219	.109	.055
7				.500	.281	.148	.074
8					.344	.188	.098
9					.422	.234	.125
10					.500	.289	.156
11						.344	.191
12						.406	.230
13						.469	.273
14						.531	.320
15							.371
16							.422
17							.473
18							.527

n: c	9	10	11	12	13	14	15
0	.002	.001	.000	.000	.000	.000	.000
1	.004	.002	.001	.000	.000	.000	.000
2	.006	.003	.001	.001	.000	.000	.000
3	.010	.005	.002	.001	.001	.000	.000
4	.014	.007	.003	.002	.001	.000	.000
5	.020	.010	.005	.002	.001	.001	.000
6	.027	.014	.007	.003	.002	.001	.000
7	.037	.019	.009	.005	.002	.001	.001
8	.049	.024	.012	.006	.003	.002	.001
9	.064	.032	.016	.008	.004	.002	.001
10	.082	.042	.021	.010	.005	.003	.001
11	.102	.053	.027	.013	.007	.003	.002
12	.125	.065	.034	.017	.009	.004	.002
13	.150	.080	.042	.021	.011	.005	.003
14	.180	.097	.051	.026	.013	.007	.003
15	.213	.116	.062	.032	.016	.008	.004
16	.248	.138	.074	.039	.020	.010	.005
17	.285	.161	.087	.046	.024	.012	.006
18	.326	.188	.103	.055	.029	.015	.008
19	.367	.216	.120	.065	.034	.018	.009
20	.410	.246	.139	.076	.040	.021	.011
21	.455	.278	.160	.088	.047	.025	.013
22	.500	.312	.183	.102	.055	.029	.015
23		.348	.207	.117	.064	.034	.018
24		.385	.232	.133	.073	.039	.021
25		.423	.260	.151	.084	.045	.024

(continued)

TABLE F *(continued)*

n: c	10	11	12	13	14	15
26	.461	.289	.170	.095	.052	.028
27	.500	.319	.190	.108	.059	.032
28		.350	.212	.122	.068	.036
29		.382	.235	.137	.077	.042
30		.416	.259	.153	.086	.047
31		.449	.285	.170	.097	.053
32		.483	.311	.188	.108	.060
33		.517	.339	.207	.121	.068
34			.367	.227	.134	.076
35			.396	.249	.148	.084
36			.425	.271	.163	.094
37			.455	.294	.179	.104
38			.485	.318	.196	.115
39			.515	.342	.213	.126
40				.368	.232	.138
41				.393	.251	.151
42				.420	.271	.165
43				.446	.292	.180
44				.473	.313	.195
45				.500	.335	.211
46					.357	.227
47					.380	.244
48					.404	.262
49					.428	.281
50					.452	.300
51					.476	.319
52					.500	.339
53						.360
54						.381
55						.402
56						.423
57						.445
58						.467
59						.489
60						.511

Adapted from Table C of *A Nonparametric Introduction to Statistics* by Charles H. Kraft and Constance van Eeden. Copyright 1968 by the Macmillan Company. Reprinted by permission.

TABLE G Cumulative probabilities $P(T \leq c)$, where T has a Wilcoxon–Mann–Whitney distribution

Smaller sample size	c	Larger sample size 3	4	5	6	7	8	9	10	11	12
2	0	.100	.067	.048	.036	.028	.022	.018	.015	.013	.011
	1	.200	.133	.095	.071	.056	.044	.036	.030	.026	.022
	2	.400	.267	.190	.143	.111	.089	.073	.061	.051	.044
	3	.600	.400	.286	.214	.167	.133	.109	.091	.077	.066
	4		.600	.429	.321	.250	.200	.164	.136	.115	.099
	5			.571	.429	.333	.267	.218	.182	.154	.132
	6				.571	.444	.356	.291	.242	.205	.176
	7					.556	.444	.364	.303	.256	.220
	8						.556	.455	.379	.321	.275
	9							.545	.455	.385	.330
	10								.545	.462	.396
	11									.538	.462
	12										.538
3	0	.050	.029	.018	.012	.008	.006	.005	.003	.003	.002
	1	.100	.057	.036	.024	.017	.012	.009	.007	.005	.004
	2	.200	.114	.071	.048	.033	.024	.018	.014	.011	.009
	3	.350	.200	.125	.083	.058	.042	.032	.024	.019	.015
	4	.500	.314	.196	.131	.092	.067	.050	.038	.030	.024
	5		.429	.286	.190	.133	.097	.073	.056	.044	.035
	6		.571	.393	.274	.192	.139	.105	.080	.063	.051
	7			.500	.357	.258	.188	.141	.108	.085	.068
	8				.452	.333	.248	.186	.143	.113	.090
	9				.548	.417	.315	.241	.185	.146	.116
	10					.500	.388	.300	.234	.184	.147
	11						.461	.364	.287	.228	.182
	12						.539	.432	.346	.277	.224
	13							.500	.406	.330	.268
	14								.469	.385	.316
	15								.531	.442	.367
	16									.500	.420
	17										.473
	18										.527
4	0		.014	.008	.005	.003	.002	.001	.001	.001	.001
	1		.029	.016	.010	.006	.004	.003	.002	.001	.001
	2		.057	.032	.019	.012	.008	.006	.004	.003	.002
	3		.100	.056	.033	.021	.014	.010	.007	.005	.004
	4		.171	.095	.057	.036	.024	.017	.012	.009	.007
	5		.243	.143	.086	.055	.036	.025	.018	.013	.010
	6		.343	.206	.129	.082	.055	.038	.027	.020	.015
	7		.443	.278	.176	.115	.077	.053	.038	.028	.021
	8		.557	.365	.238	.158	.107	.074	.053	.039	.029
	9			.452	.305	.206	.141	.099	.071	.052	.039
	10			.548	.381	.264	.184	.130	.094	.069	.052
	11				.457	.324	.230	.165	.120	.089	.066
	12				.543	.394	.285	.207	.152	.113	.085
	13					.464	.341	.252	.187	.140	.106
	14					.536	.404	.302	.227	.171	.131
	15						.467	.355	.270	.206	.158
	16						.533	.413	.318	.245	.190
	17							.470	.367	.286	.223
	18							.530	.420	.330	.260
	19								.473	.377	.299
	20								.527	.426	.342
	21									.475	.385
	22									.525	.431
	23										.476
	24										.524

Adapted from Table B of *A Nonparametric Introduction to Statistics* by Charles H. Kraft and Constance van Eeden. Copyright 1968 by The Macmillan Company. Reprinted by permission.

(*continued*)

TABLE G (*continued*)

Smaller sample size	c	Larger sample size 5	6	7	8	9	10
5	0	.004	.002	.001	.001	.000	.000
	1	.008	.004	.003	.002	.001	.001
	2	.016	.009	.005	.003	.002	.001
	3	.028	.015	.009	.005	.003	.002
	4	.048	.026	.015	.009	.006	.004
	5	.075	.041	.024	.015	.009	.006
	6	.111	.063	.037	.023	.014	.010
	7	.155	.089	.053	.033	.021	.014
	8	.210	.123	.074	.047	.030	.020
	9	.274	.165	.101	.064	.041	.028
	10	.345	.214	.134	.085	.056	.038
	11	.421	.268	.172	.111	.073	.050
	12	.500	.331	.216	.142	.095	.065
	13		.396	.265	.177	.120	.082
	14		.465	.319	.218	.149	.103
	15		.535	.378	.262	.182	.127
	16			.438	.311	.219	.155
	17			.500	.362	.259	.185
	18				.416	.303	.220
	19				.472	.350	.257
	20				.528	.399	.297
	21					.449	.339
	22					.500	.384
	23						.430
	24						.477
	25						.523
6	0		.001	.001	.000	.000	.000
	1		.002	.001	.001	.000	.000
	2		.004	.002	.001	.001	.000
	3		.008	.004	.002	.001	.001
	4		.013	.007	.004	.002	.001
	5		.021	.011	.006	.004	.002
	6		.032	.017	.010	.006	.004
	7		.047	.026	.015	.009	.005
	8		.066	.037	.021	.013	.008
	9		.090	.051	.030	.018	.011
	10		.120	.069	.041	.025	.016
	11		.155	.090	.054	.033	.021
	12		.197	.117	.071	.044	.028
	13		.242	.147	.091	.057	.036
	14		.294	.183	.114	.072	.047
	15		.350	.223	.141	.091	.059
	16		.409	.267	.172	.112	.074
	17		.469	.314	.207	.136	.090
	18		.531	.365	.245	.164	.110
	19			.418	.286	.194	.132
	20			.473	.331	.228	.157
	21			.527	.377	.264	.184
	22				.426	.303	.214
	23				.475	.344	.246
	24				.525	.388	.281
	25					.432	.318
	26					.477	.356
	27					.523	.396
	28						.437
	29						.479
	30						.521

TABLE G *(continued)*

Smaller sample size	c	Larger sample size 7	8	9	10
7	0	.000	.000	.000	.000
	1	.001	.000	.000	.000
	2	.001	.001	.000	.000
	3	.002	.001	.001	.000
	4	.003	.002	.001	.001
	5	.006	.003	.002	.001
	6	.009	.005	.003	.002
	7	.013	.007	.004	.002
	8	.019	.010	.006	.003
	9	.027	.014	.008	.005
	10	.036	.020	.011	.007
	11	.049	.027	.016	.009
	12	.064	.036	.021	.012
	13	.082	.047	.027	.017
	14	.104	.060	.036	.022
	15	.130	.076	.045	.028
	16	.159	.095	.057	.035
	17	.191	.116	.071	.044
	18	.228	.140	.087	.054
	19	.267	.168	.105	.067
	20	.310	.198	.126	.081
	21	.355	.232	.150	.097
	22	.402	.268	.176	.115
	23	.451	.306	.204	.135
	24	.500	.347	.235	.157
	25		.389	.268	.182
	26		.433	.303	.209
	27		.478	.340	.237
	28		.522	.379	.268
	29			.419	.300
	30			.459	.335
	31			.500	.370
	32				.406
	33				.443
	34				.481
	35				.519

Smaller sample size	c	Larger sample size 8	9	10
8	0	.000	.000	.000
	1	.000	.000	.000
	2	.000	.000	.000
	3	.001	.000	.000
	4	.001	.000	.000
	5	.001	.001	.000
	6	.002	.001	.001
	7	.003	.002	.001
	8	.005	.003	.002
	9	.007	.004	.002
	10	.010	.006	.003
	11	.014	.008	.004
	12	.019	.010	.006
	13	.025	.014	.008
	14	.032	.018	.010
	15	.041	.023	.013
	16	.052	.030	.017
	17	.065	.037	.022
	18	.080	.046	.027
	19	.097	.057	.034
	20	.117	.069	.042
	21	.139	.084	.051
	22	.164	.100	.061
	23	.191	.118	.073
	24	.221	.138	.086
	25	.253	.161	.102
	26	.287	.185	.118
	27	.323	.212	.137
	28	.360	.240	.158
	29	.399	.271	.180
	30	.439	.303	.204
	31	.480	.336	.230
	32	.520	.371	.257
	33		.407	.286
	34		.444	.317
	35		.481	.348
	36		.519	.381
	37			.414
	38			.448
	39			.483
	40			.517

(continued)

TABLE G *(continued)*

Smaller sample size	c	Larger sample size 9	Larger sample size 10
9	0	.000	.000
	1	.000	.000
	2	.000	.000
	3	.000	.000
	4	.000	.000
	5	.000	.000
	6	.001	.000
	7	.001	.000
	8	.001	.001
	9	.002	.001
	10	.003	.001
	11	.004	.002
	12	.005	.003
	13	.007	.004
	14	.009	.005
	15	.012	.007
	16	.016	.009
	17	.020	.011
	18	.025	.014
	19	.031	.017
	20	.039	.022
	21	.047	.027
	22	.057	.033
	23	.068	.039
	24	.081	.047
	25	.095	.056
	26	.111	.067
	27	.129	.078
	28	.149	.091
	29	.170	.106
	30	.193	.121
	31	.218	.139
	32	.245	.158
	33	.273	.178
	34	.302	.200
	35	.333	.223
	36	.365	.248
	37	.398	.274
	38	.432	.302
	39	.466	.330
	40	.500	.360
	41		.390
	42		.421
	43		.452
	44		.484
	45		.516

Smaller sample size	c	Larger sample size 10
10	0	.000
	1	.000
	2	.000
	3	.000
	4	.000
	5	.000
	6	.000
	7	.000
	8	.000
	9	.001
	10	.001
	11	.001
	12	.001
	13	.002
	14	.003
	15	.003
	16	.004
	17	.006
	18	.007
	19	.009
	20	.012
	21	.014
	22	.018
	23	.022
	24	.026
	25	.032
	26	.038
	27	.045
	28	.053
	29	.062
	30	.072
	31	.083
	32	.095
	33	.109
	34	.124
	35	.140
	36	.157
	37	.176
	38	.197
	39	.218
	40	.241
	41	.264
	42	.289
	43	.315
	44	.342
	45	.370
	46	.398
	47	.427
	48	.456
	49	.485
	50	.515

TABLE H Upper-tail probabilities $P(KW \geq c)$ for selected values of c, where KW has the Kruskal–Wallis distribution for three groups (sample sizes in parentheses)

(2, 2, 2)

c	$P(KW \geq c)$
3.714	.200
4.571	.067

(2, 2, 3)

c	$P(KW \geq c)$
4.464	.105
4.500	.067
4.714	.048
5.357	.029

(2, 2, 4)

c	$P(KW \geq c)$
4.167	.105
4.458	.100
4.500	.090
5.125	.052
5.333	.033
5.500	.024
6.000	.014

(2, 2, 5)

c	$P(KW \geq c)$
4.293	.122
4.373	.090
4.573	.085
4.800	.063
5.040	.056
5.160	.034
5.693	.029
6.000	.019
6.133	.013
6.533	.008

(2, 3, 3)

c	$P(KW \geq c)$
4.556	.100
5.000	.075
5.139	.061
5.361	.032
5.556	.025
6.250	.011

(2, 3, 4)

c	$P(KW \geq c)$
4.444	.102
4.711	.079
4.900	.071
4.978	.059
5.144	.054
5.400	.051
5.500	.040
5.800	.030
6.000	.024
6.300	.011
7.000	.005

(2, 3, 5)

c	$P(KW \geq c)$
4.494	.101
4.727	.085
4.814	.071
5.076	.060
5.106	.052
5.251	.049
5.542	.041
5.786	.033
6.004	.025
6.124	.020
6.414	.015
6.822	.010
6.949	.006
7.636	.002

(2, 4, 4)

c	$P(KW \geq c)$
4.446	.103
4.691	.080
4.991	.065
5.236	.052
5.454	.046
5.646	.039
5.946	.028
6.546	.020
6.627	.016
6.873	.011
7.036	.006
7.854	.002

(2, 4, 5)

c	$P(KW \geq c)$
4.518	.101
4.868	.071
5.073	.061
5.268	.051
5.414	.045
5.754	.035
6.041	.025
6.473	.020
6.723	.015
7.118	.010
7.573	.005
8.114	.001

(2, 5, 5)

c	$P(KW \geq c)$
4.508	.100
4.808	.081
5.054	.060
5.246	.051
5.546	.045
5.608	.040
5.915	.030
6.446	.020
7.269	.010
8.131	.005
8.685	.001

(3, 3, 3)

c	$P(KW \geq c)$
4.622	.100
5.422	.071
5.600	.050
5.689	.029
5.956	.025
6.489	.011
7.200	.004

(3, 3, 4)

c	$P(KW \geq c)$
4.700	.101
4.818	.085
5.064	.070
5.500	.056
5.727	.050
5.936	.036
6.154	.025
6.664	.014
6.746	.010
7.000	.006
7.436	.002
8.018	.001

(3, 3, 5)

c	$P(KW \geq c)$
4.412	.109
4.848	.085
5.212	.065
5.515	.051
6.012	.040
6.194	.027
6.376	.020
6.715	.014
7.079	.009
7.515	.005
8.242	.001

(continued)

TABLE H *(continued)*

(3, 4, 4)

c	$P(KW \geq c)$
4.477	.102
5.053	.078
5.303	.061
5.576	.051
5.803	.045
6.000	.040
6.182	.030
6.394	.025
6.659	.020
7.053	.014
7.144	.010
7.598	.004
8.326	.001

(3, 5, 5)

c	$P(KW \geq c)$
4.545	.100
5.064	.070
5.407	.059
5.600	.051
5.802	.045
5.934	.040
6.312	.030
6.488	.025
6.866	.019
6.998	.015
7.543	.010
8.237	.005
9.055	.001

(4, 4, 5)

c	$P(KW \geq c)$
4.619	.100
4.896	.081
5.090	.071
5.410	.060
5.618	.050
6.030	.040
6.343	.030
6.943	.020
7.203	.015
7.760	.009
8.140	.005
8.997	.001

(5, 5, 5)

c	$P(KW \geq c)$
4.560	.100
4.940	.081
5.180	.070
5.460	.060
5.780	.049
6.080	.040
6.540	.030
7.020	.020
7.440	.015
8.000	.009
8.720	.005
9.680	.001

(3, 4, 5)

c	$P(KW \geq c)$
4.523	.103
4.881	.081
5.106	.070
5.342	.061
5.631	.050
5.722	.045
5.814	.040
6.272	.030
6.410	.025
6.635	.020
7.445	.010
7.906	.005
8.503	.001

(4, 4, 4)

c	$P(KW \geq c)$
4.500	.104
4.962	.080
5.346	.063
5.692	.049
6.000	.040
6.500	.030
6.731	.021
7.385	.015
7.538	.011
8.000	.005
8.654	.001

(4, 5, 5)

c	$P(KW \geq c)$
4.520	.101
4.911	.079
5.163	.070
5.400	.061
5.643	.050
6.031	.040
6.440	.030
6.943	.020
7.311	.015
7.766	.010
8.371	.005
9.323	.001

TABLE I Cumulative probabilities $P(S \le c)$, where S has Spearman's distribution for sample size n. Distribution is symmetric about $(n^3 - n)/6$.

c	$P(S \le c)$
$n = 2$	
0	.500
$n = 3$	
0	.167
2	.500
$n = 4$	
0	.042
2	.167
4	.208
6	.375
8	.458
10	.542
$n = 5$	
0	.008
2	.042
4	.067
6	.117
8	.175
10	.225
12	.258
14	.342
16	.392
18	.475
20	.525
$n = 6$	
0	.001
2	.008
4	.017
6	.029
8	.051
10	.068
12	.088
14	.121
16	.149
18	.178
20	.210
22	.249
24	.282
26	.329
28	.357
30	.401
32	.460
34	.500
$n = 7$	
0	.000
2	.001
4	.003
6	.006
8	.012
10	.017
12	.024
14	.033
16	.044
18	.055
20	.069
22	.083
24	.100
26	.118
28	.133
30	.151
32	.177
34	.198
36	.222
38	.249
40	.278
42	.297
44	.331
46	.357
48	.391
50	.420
52	.453
54	.482
56	.518
$n = 8$	
0	.000
2	.000
4	.001
6	.001
8	.002
10	.004
12	.005
14	.008
16	.011
18	.014
20	.018
22	.023
24	.029
26	.035
28	.042
30	.048
32	.057
34	.066
36	.076
38	.085
40	.098
42	.108
44	.122
46	.134
48	.150
50	.163
52	.180
54	.195
56	.214
58	.231
60	.250
62	.268
64	.291
66	.310
68	.332
70	.352
72	.376
74	.397
76	.420
78	.441
80	.467
82	.488
84	.512
$n = 9$	
8	.000
10	.001
12	.001
14	.002
16	.002
18	.003
20	.004
22	.005
24	.007
26	.009
28	.011
30	.013
32	.016
34	.018
36	.022
38	.025
40	.029
42	.033
44	.038
46	.043
48	.048
50	.054
52	.060
54	.066
56	.074
58	.081
60	.089
62	.097
64	.106
66	.115
68	.125
70	.135
72	.146
74	.156
76	.168
78	.179
80	.193
82	.205
84	.218
86	.231
88	.247
90	.260
92	.276
94	.290
96	.307
98	.322
100	.339
102	.354
104	.372
106	.388
108	.405
110	.422
112	.440
114	.456
116	.474
118	.491
120	.509
$n = 10$	
16	.000
18	.001
20	.001
22	.001
24	.001
26	.002
28	.002
30	.003
32	.004
34	.004
36	.005
38	.006
40	.008
42	.009
44	.009
46	.012
48	.013
50	.015
52	.017
54	.019
56	.022
58	.024
60	.027
62	.030
64	.033
66	.037
68	.040
70	.044
72	.048
74	.052
76	.057
78	.062
80	.067
82	.072
84	.077
86	.083
88	.089
90	.096
92	.102
94	.109
96	.116
98	.124
100	.132
102	.139
104	.148
106	.156
108	.165
110	.174
112	.184
114	.193
116	.203
118	.214
120	.224
122	.235
124	.246
126	.257
128	.268
130	.280
132	.292
134	.304
136	.316
138	.328
140	.341
142	.354
144	.367
146	.379
148	.393
150	.406

(continued)

TABLE I (*continued*)

c	$P(S \leq c)$	c	$P(S \leq c)$	c	$P(S \leq c)$	c	$P(S \leq c)$	c	$P(S \leq c)$
152	.419	54	.005	96	.038	138	.130	180	.298
154	.433	56	.006	98	.041	140	.137	182	.307
156	.446	58	.006	100	.044	142	.143	184	.317
158	.459	60	.007	102	.047	144	.150	186	.327
160	.473	62	.008	104	.050	146	.157	188	.337
162	.486	64	.009	106	.054	148	.163	190	.347
164	.500	66	.010	108	.057	150	.171	192	.357
***n* = 11**		68	.011	110	.061	152	.178	194	.367
28	.000	70	.013	112	.065	154	.186	196	.377
30	.001	72	.014	114	.069	156	.193	198	.388
32	.001	74	.015	116	.073	158	.201	200	.398
34	.001	76	.017	118	.077	160	.209	202	.409
36	.001	78	.018	120	.082	162	.217	204	.419
38	.001	80	.020	122	.087	164	.226	206	.430
40	.002	82	.022	124	.091	166	.234	208	.441
42	.002	84	.024	126	.096	168	.243	210	.452
44	.002	86	.026	128	.102	170	.252	212	.462
46	.003	88	.028	130	.107	172	.260	214	.473
48	.003	90	.030	132	.112	174	.270	216	.484
50	.004	92	.033	134	.118	176	.279	218	.495
52	.004	94	.035	136	.124	178	.288	220	.505

Adapted from Table 13.2, Exact Distribution of Spearman's Rank Correlation Coefficient, *Handbook of Statistical Tables* by D. B. Owen.

Glossary of Some Minitab Commands

An introduction to Minitab is provided in the Minitab Appendix for Chapter 1. For all Minitab commands, it is necessary to type only the first four letters of the command name. In the examples that follow, boldface shows how to type the commands. To obtain online help while using Minitab, type the HELP command (**help**).

The term *worksheet* refers to the current set of data, column names, and stored constants we are using. A *classification variable* is a variable that takes integer values from −9999 to +9999, or missing values. (For more details, see Ryan, Joiner, and Ryan, 1985; or Schaefer and Anderson, 1989.)

Command	Description	Minitab Appendix for Chapter
aovoneway	Use AOVONEWAY to perform one-way analysis of variance on unstacked data (that is, the values of the response variable are in several columns, one column per group or level of the single factor) for a single-factor experiment. Suppose values of a response variable for group 1 are in column 1, for group 2 in column 2, and for group 3 in column 3. To test the null hypothesis that the three population means are equal, use the command **aovoneway c1-c3**	14
boxplot	Use BOXPLOT to produce a box plot of values in a column. For example, **boxplot c12** displays a box plot of the values in column 12.	2
	Use the BY subcommand to display separate box plots of a variable within levels of a classification variable. The resulting box plots will all be on the same scale. For example, suppose column 5 contains values for a classification variable. The command **boxplot c12;** **by c5.**	

Command	Description	Minitab Appendix for Chapter
	displays separate box plots of the values in column 12 for each value of column 5, with all plots on the same scale.	2

cdf

Use CDF with an appropriate subcommand to display cumulative probabilities for a specified probability distribution. For example, (7)

```
cdf;
binomial 10 0.25.
```

will display cumulative probabilities for the binomial(10, .25) distribution.

The command (7)

```
cdf 4;
binomial 10 0.25.
```

displays the probability that a random variable having the binomial(10, .25) distribution is less than or equal to 4.

The command (8)

```
cdf 1.2;
normal.
```

displays the probability that a standard Gaussian (normal) random variable is less than or equal to 1.2. To save this value as a stored constant K2, use this command: (8)

```
cdf 1.2 k2;
normal.
```

The command (8)

```
cdf 1.2;
normal 2.5 0.4.
```

displays the probability that a random variable X is less than or equal to 1.2, where X has the Gaussian distribution with mean 2.5 and standard deviation .4. Other distributions can be specified in the CDF command, including T, F, and CHISQUARE. The command (10)

```
cdf 5.1;
t 4.
```

displays the probability that a random variable X is less than or equal to 5.1, where X has the t distribution with 4 degrees of freedom.

If we use (12)

```
cdf 6.1 k4;
F 3 8.
```

the stored constant K4 will contain the probability that a random variable X is less than or equal to 6.1, where X has the F distribution with 3 numerator degrees of freedom and 8 denominator degrees of freedom. The command

Command	Description	Minitab Appendix for Chapter
	cdf 4.0; **chisquare 6.** displays the probability that a random variable X is less than or equal to 4, where X has the chi-square distribution with 6 degrees of freedom.	12
chisquare	Use CHISQUARE to calculate the test statistic for a chi-square test of association. Suppose a table of counts is contained in columns 1–5. Then the command **chisquare c1-c5** displays a table of observed frequencies, plus expected frequencies under the null hypothesis of no association (or independence). The chi-square test statistic and its degrees of freedom are also displayed.	16
code	Use CODE to create a missing value code in Minitab. The missing value code in Minitab is an asterisk, *. Suppose that, in column 14, the value -1 indicates a missing value and we want to change each -1 to *, the Minitab missing value code. If we use the command **code (-1) '*' c14 c14** Minitab changes all values of -1 in column 14 to an asterisk and then puts the altered column of values back in column 14.	1
	Use CODE to recode data values. For example, **code (0:50) 1, (51:100) 2 in c3 store c8** puts a value 1 in column 8 wherever column 3 has values 0 through 50, and puts a value 2 in column 8 wherever column 3 has values 51 through 100.	1
copy	Use COPY to create new columns. For example, **copy c1-c3, c5 to c11-c13, c15** copies the values in columns 1, 2, 3, and 5 into columns 11, 12, 13, and 15, respectively. Columns 1, 2, 3, and 5 are preserved. Another example: **copy c2 c20;** **use only c8=1.** copies rows of column 2 into column 20 only if there is a 1 in the corresponding rows of column 8. A major limitation in creating new columns is that the Student Edition of Minitab allows a worksheet containing at most 2,000 numbers.	1
correlate	Use CORRELATE to calculate the linear correlation coefficient between variables in two columns. For example, **correlate c1 c2**	

Command	Description	Minitab Appendix for Chapter
	displays the linear correlation coefficient for the two variables in columns 1 and 2. The command `corr c1 c2 c3` displays linear correlation coefficients for the pairs of variables in columns 1 and 2, columns 1 and 3, and columns 2 and 3.	15
count	Use COUNT to display the number of values in a column, missing and nonmissing. For example, `count c3` displays the number of entries, both missing and nonmissing, in column 3. To save this count as a stored constant in K9, use the command `count c3 k9`	3
delete	Use DELETE to delete one or more rows of the worksheet. For example, `delete row 8` deletes row 8 from the worksheet.	1
describe	Use DESCRIBE to obtain descriptive statistics for the values in a column. For example, `describe c9` displays descriptive statistics for the values in column 9.	
	Use DESCRIBE with the BY subcommand to describe a variable within levels of a classification variable. For example, suppose column 1 contains values for a classification variable. Then the command `describe c9;` `by c1.` displays descriptive statistics for the variable in column 9 within levels of the variable in column 1.	3
dotplot	Use DOTPLOT to produce a dot plot of values in a column. For example, `dotplot c3` displays a dot plot of the values in column 3.	2
	To produce dot plots for two columns on the same scale, use the SAME subcommand. For example, `dotplot c2 c3;` `same.` will display two dot plots on the same scale, one for the values in column 2 and one for the values in column 3.	2
	To produce separate dot plots of values in one column according to coded values in another column, use the BY subcommand. For example, sup-	

Command	Description	Minitab Appendix for Chapter
	pose column 5 contains values for a classification variable. We could type the command `dotplot c3;` `by c5.` to produce separate dot plots of the values in column 3 within each code (or category) in column 5.	2
end	Use END to indicate the end of data entry.	1
erase	Use ERASE to delete one or more columns of the worksheet. For example, `erase c10` deletes the data in column 10 of the worksheet. Use ERASE to clear columns of the current worksheet before using the READ or RETRIEVE commands to input data files.	1
help	Use HELP to obtain online explanations of commands and subcommands, plus general information about Minitab. For example, `help` provides general information, `help dotplot` provides information on the DOTPLOT command, and `help dotplot by` provides information on the BY subcommand used with the DOTPLOT.	
histogram	Use HISTOGRAM to produce a histogram of values in a column. For example, `histogram c10` displays a histogram of the values in column 10. We can control construction of the histogram with the INCREMENT and START subcommands. INCREMENT specifies interval width and START specifies the midpoint of the first interval. For example, `histogram c10;` `increment=5;` `start=20.` displays a histogram with intervals of width 5, the midpoint of the first interval equal to 20.	2
	To produce several histograms on the same scale, use the BY and SAME subcommands, as for the DOTPLOT command.	2
information	Use INFO to receive a summary of the contents of the current Minitab worksheet.	1

Command	Description	Minitab Appendix for Chapter
invcdf	Use INVCDF with an appropriate subcommand to obtain the number c such that $P(X \leq c)$ equals a specified probability, where X is a random variable having a specified probability distribution. For example,	
	`invcdf 0.05;` `binomial 10 0.25.`	7
	displays values of c for which $P(X \leq c)$ is close to .05, where X is a random variable having the binomial(10, .25) distribution. The command	
	`invcdf 0.95;` `normal.`	8
	displays the value of c for which $P(X \leq c)$ equals .95, where X has the standard Gaussian distribution. The command	
	`invcdf 0.90 k4;` `normal 1 4.`	8
	saves as a stored constant K4 the value of c for which $P(X \leq c)$ equals .90, where X has the Gaussian distribution with mean 1 and standard deviation 4. We can also use T, F, and CHISQUARE subcommands with INVCDF. For example, the commands	
	`invcdf 0.95;` `t 4.`	10
	`invcdf 0.95;` `f 2 6.`	12
	`invcdf 0.95;` `chisquare 3.`	12
	display the value of c for which $P(X \leq c)$ equals .95, where X has, respectively, the t distribution with 4 degrees of freedom, the F distribution with 2 numerator degrees of freedom and 6 denominator degrees of freedom, and the chi-square distribution with 3 degrees of freedom.	
kruskal-wallis	Use KRUSKAL-WALLIS to carry out a Kruskal–Wallis test (nonparametric analysis of a single-factor experiment). Suppose a group code is in column 1 and observations on a variable in column 2. The command	
	`krus c2 c1`	
	carries out the calculations for the Kruskal–Wallis test for equality of the distribution of the variable in column 2 over all the groups specified in column 1.	12
let	Use LET to correct mistakes in the worksheet. For example,	
	`let c6(15)=2.1`	
	will replace whatever value is in row 15 of column 6 with the new value 2.1.	1
	Use LET to calculate functions of columns of data, to create new variables. For example,	
	`let c3=sqrt(c1)`	

Command	Description	Minitab Appendix for Chapter

puts the square root of nonnegative values of column 1 into column 3. If column 1 has a negative or missing value, a missing value is placed in the corresponding position of column 3. Functions we can use with LET include:

- ABSOLUTE (take the absolute value)
- ANTILOG (calculate 10 to the power given)
- EXPO (calculate the mathematical constant e to the power given)
- LOGE (take the natural logarithm)
- LOGTEN (take the logarithm base-10)
- ROUND (round to the nearest integer)
- SIGNS (assign $+1$ to a positive value, -1 to a negative value, 0 to a zero value)
- SQRT (take the square root)

3

Use LET for arithmetic operations on columns of data, with the symbol + for addition, − for subtraction, * for multiplication, / for division, and ** for exponentiation. Minitab uses the usual algebraic order of precedence. Expressions within parentheses are evaluated first, then exponentiation, then multiplication and division, then addition and subtraction. Operations with the same precedence are evaluated from left to right. For example,

```
let c12=sqrt(abs(c1-c2/3))+c4**3
```

causes Minitab to divide each value in column 2 by 3, subtract this from the corresponding value in column 1, take the absolute value of the result, and then take the square root. This result is then added to the cube of the value in column 4. The final result is placed in the corresponding position of column 12.

3

lplot

Use LPLOT to obtain separate scatterplots of two variables within categories of a classification variable, all on the same graph. For example, suppose columns 1 and 2 contain values for quantitative variables and column 5 contains values for a classification variable. Then the command

```
lplot c1 c2 by c5
```

displays separate plots of column 1 (vertical axis) versus column 2 (horizontal axis) within each category of column 5, with all plots on the same graph.

5

mann-whitney

Use MANN-WHITNEY for two-sample inferences (about two population medians) based on ranks and a Wilcoxon–Mann–Whitney distribution. If column 1 contains one sample of values and column 2 contains another sample of values, the command

```
mann c1 c2
```

carries out a test that the two population medians are equal (with the two-sided alternative) and calculates an approximate 95% confidence interval

Command	Description	Minitab Appendix for Chapter

	for the difference between the two population medians. The command
	`mann alternative=+1 90 c1 c2`
	specifies the one-sided alternative that the first population median is greater than the second, and prints an approximate 90% confidence interval for the difference between the two population medians. The command
	`mann alternative=-1 99 c1 c2`
	specifies the one-sided alternative that the first population median is less than the second, and prints an approximate 99% confidence interval for the difference between the two population medians.
		11

maximum	Use MAXIMUM to display the maximum or largest of the nonmissing values in a column. For example,
	`maximum c1`
	displays the largest value in column 1. This maximum can be saved as a stored constant in K4 with the command
	`maximum c1 k4`
		3

mean	Use MEAN to display the average of nonmissing values in a column. For example,
	`mean c8`
	displays the mean of the nonmissing values in column 8, whereas the command
	`mean c8 k7`
	displays that mean and saves it as a stored constant in K7.
		3

median	Use MEDIAN to display the median of nonmissing values in a column. For example,
	`median c6`
	displays the median of the nonmissing values in column 6, whereas the command
	`median c6 k14`
	displays that median and saves it as a stored constant in K14.
		3

minimum	Use MINIMUM to display the minimum or smallest of the nonmissing values in a column. For example,
	`minimum c1`
	displays the smallest value in column 1. This minimum can be saved as a stored constant in K5 with the command
	`minimum c1 k5`
		3

Command	Description	Minitab Appendix for Chapter
mplot	Use MPLOT to superimpose two or more scatterplots on the same graph. For example, `mplot c1 c2 c11 c12` plots column 1 (vertical axis) versus column 2 (horizontal axis) and also plots column 11 (vertical axis) versus column 12 (horizontal axis) on the same graph, using different plotting symbols for the two plots.	5

n	Use N to display the number of nonmissing values in a column. For example,
	`n c2`
	displays the number of nonmissing values in column 2. To save this count as a stored constant in K6, use the command
	`n c2 k6`
		3

name	The NAME command names columns for use in commands and output. For example, `name c1 'height' c2 'weight'` gives column 1 the name 'height' and column 2 the name 'weight'. The designation 'height' may then be used in commands instead of c1 and 'weight' may be used instead of c2. The single quotes around the column names are required.	1

nmiss	Use NMISS to display the number of missing values in a column. For example,
	`nmiss c4`
	displays the number of missing values in column 4. The command
	`nmiss c4 k2`
	displays the number of missing values in column 4 and saves that number as a stored constant in K2.
		3

nooutfile	The NOOUTFILE command stops the recording of commands and Minitab responses initiated by the OUTFILE command.	1
nopaper	The NOPAPER command stops routing output to the printer (which was initiated by the PAPER command).	1
oneway	Use ONEWAY to perform one-way analysis of variance. Suppose a group code is in column 1 and observations on a variable in column 2. The command `oneway c2 c1` carries out one-way analysis of variance, testing the null hypothesis that	

Command	Description	Minitab Appendix for Chapter
	the population means (of the variable in column 2) are equal for all the groups (specified in column 1). To save residuals in column 3 and estimated or predicted values in column 4, we use the command **oneway c2 c1 c3 c4**	12
outfile	Use OUTFILE to save all commands and Minitab's responses on a disk file that can be accessed later for editing or printing. For example, **outfile 'sample'** saves output in a file called SAMPLE.LIS on a mainframe computer. If we type OUTFILE when the command PAPER is in effect, we stop the PAPER command.	1
paper	The PAPER command sends the Minitab session directly to the printer. If the command OUTFILE is in effect and we type PAPER, we stop the OUTFILE command.	1
pdf	Use PDF with an appropriate subcommand to display probabilities (for a discrete probability distribution) or values of the probability density function (for a continuous probability distribution). For example, **pdf;** **binomial 10 0.25.**	7
	displays the probabilities for the binomial(10,.25) distribution. The command **pdf 6;** **binomial 10 0.25.**	7
	displays the probability that a random variable having the binomial(10, .25) distribution will equal 6. The command **pdf 1.1;** **normal.**	8
	displays the value of the standard Gaussian probability function evaluated at 1.1. The command **pdf 5.4 k6;** **normal 4 2.1.**	8
	saves as a stored constant K6 the value of the probability function for the Gaussian distribution with mean 4 and standard deviation 2.1, evaluated at 5.4. Other distributions can be specified in the PDF command, including T, F, and CHISQUARE. The command **pdf −1.1;** **t 10.**	10
	displays the value of the probability function for the *t* distribution with 10 degrees of freedom, evaluated at −1.1. The command	

Command	Description	Minitab Appendix for Chapter

```
pdf 3.3;
f 15 12.
```

(12)

displays the value of the probability function for the F distribution with 15 numerator degrees of freedom and 12 denominator degrees of freedom, evaluated at 3.3. The command

```
pdf 2.0;
chisquare 8.
```

(12)

displays the value of the probability function for the chi-square distribution with 8 degrees of freedom, evaluated at 2.0.

plot

Use PLOT to create scatterplots. For example,

```
plot c4 c2
```

displays a scatterplot with the variable in column 4 on the vertical axis and the variable in column 2 on the horizontal axis. Numerals (2, 3, 4, etc.) indicate the number of cases with the same plotted points. To control the scales of the axes, use the YSTART and XSTART subcommands. For example,

```
plot c4 c2;
ystart 10 50;
xstart 90 120.
```

displays a plot with the vertical axis extending from 10 to 50 and the horizontal axis from 90 to 120. (4)

print

Use PRINT to display all or part of the data in the Minitab worksheet. For example,

```
print c5-c10
```

causes Minitab to display the data in columns 5–10. (1)

random

Use RANDOM to generate simulated random samples of numbers from a specified distribution. For example,

```
random 50 c1;
normal 10.1 2.2.
```

(8)

stores in column 1 a simulated random sample of 50 values from the Gaussian distribution with mean 10.1 and standard deviation 2.2. The command

```
random 50 c11 c12;
normal 10.1 2.2.
```

(8)

stores two such samples, 50 values in column 11 and 50 values in column 12. Other distributions can be specified in the RANDOM command, including BINOMIAL, BERNOULLI, T, F, and CHISQUARE. For example,

```
random 20 c1;
binomial 6 0.5.
```

(8)

Command	Description	Minitab Appendix for Chapter

	puts 20 simulated random values from the binomial(6, .5) distribution in column 1. The command
	`random 10 c2;`
	`bernoulli 0.5.`
		8

	simulates ten random values from the Bernoulli(.5) = binomial(1, .5) distribution and places them in column 2. To place 30 simulated random values from the *t* distribution with 9 degrees of freedom in column 4, use the command
	`random 30 c4;`
	`t 9.`
		10

	If we want column 5 to contain 40 simulated random values from the *F* distribution with 6 numerator degrees of freedom and 8 denominator degrees of freedom, we use
	`random 40 c5;`
	`F 6 8.`
		12

	To place 25 simulated random values from the chi-square distribution with 14 degrees of freedom in column 6, we can use
	`random 25 c6;`
	`chisquare 14.`
		12

rank	The RANK command assigns ranks based on values in a column, without rearranging the worksheet. For example, `rank c4 c10` puts the ranks for values in column 4 into column 10. The smallest value is assigned rank 1, the next smallest is assigned rank 2, and so on. Tied values are assigned the average of the ranks they share.	1
rcount	Use RCOUNT to count the entries, missing and nonmissing, across rows of several columns and store the results in another column. For example, `rcount c1-c10 c12` counts the entries in columns 1–10 for each row and stores the results in corresponding positions in column 12.	3
read	Use READ to enter data row-wise. For example, `read c1 c2` indicates that we will enter two values per row, the first to be saved in column 1 and the second in column 2 of the worksheet. Minitab accepts rows of data until we type END.	1
	Use READ to input data from an ASCII file, perhaps saved using the WRITE command or created using a text editor. For example, `read 'c:exercise' c1, c3-c10`	

Command	Description	Minitab Appendix for Chapter
	enters the data from file EXERCISE.DAT on hard disk drive C of a personal computer into columns 1 and 3–10 of the worksheet.	1

regress	Use REGRESS for simple and multiple linear regression. For example, to model the response variable in column 1 as a straight-line function of the predictor variable in column 2, use the command
	`regress c1 1 c2`
	where the number 1 indicates that there is one predictor variable in the model. To save standardized residuals in column 3, predicted or estimated values of the response variable in column 4, and residuals in column 5, use
	`regress c1 1 c2 c3 c4;`
	`residuals c5.`
	To regress the response variable in column 1 on five predictor variables in columns 11–15, use
	`regress c1 5 c11-c15`
	Residuals and predicted values of the response variable can be saved as indicated above.
		15

retrieve	Use RETRIEVE to access a worksheet previously saved using the SAVE command. For example,
	`retrieve 'b:example'`
	enters into the current worksheet the file EXAMPLE.MTW from drive B on a personal computer. The command
	`retrieve 'example'`
	retrieves the file EXAMPLE.MTW on a mainframe computer (the extension may vary with mainframe installations).
		1

rmaximum	Use RMAXIMUM to determine the maximum of nonmissing values across rows of several columns and store the results in another column. For example, `rmaximum c1-c10 c22` finds the maximum of the nonmissing values in columns 1–10 for each row, and places the results in corresponding positions in column 22.	3
rmean	Use RMEAN to average nonmissing values across rows of several columns and store the results in another column. For example, `rmean c1-c10 c15` averages the nonmissing values in columns 1–10 for each row, and places the results in corresponding positions in column 15.	3

Command	Description	Minitab Appendix for Chapter
rmedian	Use RMEDIAN to determine the median of nonmissing values across rows of several columns and store the results in another column. For example, **rmedian c1-c10 c21** finds the median of the nonmissing values in columns 1–10 for each row, and places the results in corresponding positions in column 21.	3
rminimum	Use RMINIMUM to determine the minimum of nonmissing values across rows of several columns and store the results in another column. For example, **rminimum c1-c10 c22** finds the minimum of the nonmissing values in columns 1–10 for each row, and places the results in corresponding positions in column 22.	3
rn	Use RN to count the nonmissing entries across rows of several columns and store the results in another column. For example, **rn c1-c10 c13** counts the nonmissing entries in columns 1–10 for each row and stores the results in corresponding positions in column 13.	3
rnmiss	Use RNMISS to count the missing entries across rows of several columns and store the results in another column. For example, **rnmiss c1-c10 c14** counts the missing entries in columns 1–10 for each row and stores the results in corresponding positions in column 14.	3
rssq	Use RSSQ to calculate the sum of the squared (nonmissing) values across rows of several columns and store the results in another column. For example, **rssq c1-c10 c24** calculates the sum of the squared (nonmissing) values in columns 1–10 for each row, and places the results in corresponding positions in column 24.	3
rstdev	Use RSTDEV to calculate the standard deviation of nonmissing values across rows of several columns and store the results in another column. For example, **rstdev c1-c10 c20** calculates the standard deviation of the nonmissing values in columns 1–10 for each row, and places the results in corresponding positions in column 20.	3
rsum	Use RSUM to add values across rows of several columns or to add one or	

Command	Description	Minitab Appendix for Chapter

	more constants to each row of a column, storing the results in another column. For example,
	`rsum c1-c3 c5`
	adds the values in columns 1, 2, and 3 for each row and places the results in corresponding positions of column 5. The command
	`rsum 5.2 c10 c11`
	adds 5.2 to each value in column 10 and places the results in column 11. The command
	`rsum k5 c6 c7`
	adds the value in the stored constant K5 to each value in column 6 and places the results in corresponding positions of column 7.
		3

save	Use SAVE to save the contents of the current Minitab worksheet. For example,
	`save 'b:example'`
	saves the worksheet in a file called EXAMPLE.MTW on drive B, if we are using a personal computer. Another example:
	`save 'example'`
	saves the file EXAMPLE.MTW on a mainframe computer (the extension after the file name may vary with the mainframe installation).
		1

set	Use SET to enter data column-wise. For example, `set c5` indicates that we will enter values to be stored in column 5 of the worksheet. Minitab accepts values until we type END.	1
sinterval	Use SINTERVAL to obtain a confidence interval for a population median based on a binomial distribution. For example, `sinterval 90 c4` displays an approximate 90% confidence interval based on the values in column 4. If no confidence level is specified, Minitab displays an approximate 95% confidence interval.	10

sort	Use SORT to rearrange a data file in increasing order of a single variable. For example,
	`sort c3 c1 c2 c13 c11 c12`
	rearranges the values in columns 3, 1, and 2 according to increasing order of values in column 3 and puts the rearranged rows into columns 13, 11, and 12, respectively. Another example:
	`sort c3 c1 c2 c3 c1 c2`	

Command	Description	Minitab Appendix for Chapter
	rearranges the values in columns 3, 1, and 2 the same way and then puts the rearranged rows back into columns 3, 1, and 2, respectively.	1
sum	Use SUM to display the sum of the nonmissing values in a column. For example, `sum c1` displays the sum of the values in column 1. To save this sum as a stored constant in K2, use the command `sum c1 k2`	3
ssq	Use SSQ to display the sum of the squares of the nonmissing values in a column. For example, `ssq c2` displays the sum of the squares of the values in column 2. To save this result as a stored constant in K10, use the command `ssq c2 k10`	3
stdev	Use STDEV to display the standard deviation of nonmissing values in a column. For example, `stdev c8` displays the standard deviation of the values in column 8, whereas the command `stdev c8 k2` displays that standard deviation and saves it as a stored constant in K2.	3
stem-and-leaf	Use STEM-AND-LEAF to produce a stem-and-leaf plot of values in a column. For example, `stem c3` displays a stem-and-leaf plot of the values in column 3.	2
	If we use the command `stem c3;` `by c5.` where column 5 contains values for a classification variable, we get a separate stem-and-leaf plot of the values in column 3 for each value in column 5.	2
stest	Use STEST to test hypotheses about a population median based on a binomial distribution. To use the values in column 4 to test the null hypothesis that the population median equals 50, against the two-sided alternative, the command is `stest 50 c4`	

Command	Description	Minitab Appendix for Chapter
	To specify one-sided alternatives, we can use the ALTERNATIVE subcommand as described for the TTEST command.	10
table	Use TABLE to create frequency tables for classification variables. For example, `table c1 c2` displays a two-way frequency table for the two classification variables in columns 1 and 2. The subcommand ROWPERCENTS provides row percentages and the subcommand COLPERCENTS provides column percentages. For example, `table c1 c2;` `rowpercents;` `colpercents.` displays the same two-way frequency table as above, but with row percentages and column percentages shown as well. If columns 1, 2, and 3 all contain classification variables, the command `table c1 c2 c3` displays a three-dimensional frequency table.	4
	We can use the TABLE command with the STATS subcommand to print descriptive statistics of a variable within levels of two or more classification variables. Suppose columns 1 and 2 contain levels of two classification variables and column 5 contains values of another variable. The command `table c1 c2;` `stats c5.` displays the number of nonmissing values, and the mean and standard deviation of the variable in column 5 within each combination of values in columns 1 and 2. Two or more classification variables can be specified. If two or more variables are specified in the STATS subcommand, Minitab displays descriptive statistics for each one.	13
tally	Use the TALLY command to obtain a frequency table for a classification variable. For example, `tally c6` displays a frequency table for the classification variable in column 6. The ALL subcommand produces output that includes counts, cumulative counts, percentages, and cumulative percentages. For example, `tally c6;` `all.` displays such an enhanced frequency table for the variable in column 6.	2

Command	Description	Minitab Appendix for Chapter

	Use TINTERVAL to obtain a confidence interval for the population mean based on a t distribution. If we use
	`tinterval c1`
	we obtain a 95% confidence interval for the population mean. If we want a different confidence level, we can specify it. For instance, to obtain a 90% confidence interval, we can type
	`tinterval 90 c1`
		10

ttest	Use TTEST to carry out a one-sample t test. The command
	`ttest 10.1 c1`
	uses the values in column 1 to test the null hypothesis that the population mean equals 10.1, versus the two-sided alternative that the population mean does not equal 10.1. To consider only the one-sided alternative that the population mean is greater than 10.1, use the command
	`ttest 10.1 c1;`
	`alternative=+1.`
	To consider only the one-sided alternative that the population mean is less than 10.1, use the command
	`ttest 10.1 c1;`
	`alternative=-1.`
		10

twot	Use TWOT to carry out a two-sample t test or a large-sample comparison of two means. Suppose column 1 contains a code for the two groups and column 2 contains values of a variable. If we assume equal variances for a two-sample t test comparing the means of the variable across the two groups, we use the command
	`twot c2 c1;`
	`pooled.`
	If the two sample sizes are large, the command
	`twot c2 c1`
	gives us a large-sample comparison of the two population means.
		11

twoway	Use TWOWAY to carry out a two-way analysis of variance for a balanced, factorial, two-factor experiment. Suppose levels of one factor are in column 1, levels of another factor in column 2, and values of the response variable in column 3. The command
	`twoway c3 c1 c2`
	carries out calculations for two-way analysis of variance. If we want to save residuals in column 7 and estimated or predicted values in column 8, we use the command
	`twoway c3 c1 c2 c7 c8`
		12

Command	Description	Minitab Appendix for Chapter

unstack	Use UNSTACK to place values in a column into other columns, according to codes of a classification variable. Suppose column 1 contains codes for three categories or groups and column 2 contains values for another variable. Suppose we want to place in columns 4, 5, and 6 the values in column 2 corresponding to the smallest value, middle value, and largest value, respectively, in column 1. Then we use the command
	`unstack c2 c4-c6;`
	`subscripts c1.`
		11

winterval	Use WINTERVAL to obtain a confidence interval for the population median based on a Wilcoxon signed rank distribution. For example, `winterval 99 c3` displays an approximate 99% confidence interval based on the values in column 3. If no confidence level is specified, Minitab displays an approximate 95% confidence interval.	10
write	Use WRITE to save data in an ASCII file. For example, `write 'c:exercise' c1, c3-c10` saves these nine columns of data (column 1 and columns 3–10) in file EXERCISE.DAT on hard disk drive C of a personal computer.	1
wtest	Use WTEST to test hypotheses about a population median based on a Wilcoxon signed rank distribution. For example, to use the values in column 3 to test the null hypothesis that the population mean equals 2.5, against the two-sided alternative, the command is `wtest 2.5 c3` We can specify one-sided alternatives with the ALTERNATIVE subcommand, as described for the TTEST command.	10

BIBLIOGRAPHY

ALLMAN, WILLIAM F. "Staying alive in the 20th century." *Science* 85, October 1985, pages 31–41.

AMERICAN COUNCIL ON EDUCATION *Minorities in Higher Education, Sixth Annual Status Report* 1987. American Council on Education Office of Minority Concerns, Washington, D.C.

ANDERSEN, A. H., E. B. JENSEN, and G. SCHOU "Two-way analysis of variance with correlated errors." *International Statistical Review,* 1981, volume 49, pages 153–167.

ANDREWS, D. F. and A. M. HERZBERG *Data: A Collection of Problems from Many Fields for the Student and Research Worker.* Springer-Verlag, New York, 1985.

BAIN, L. J. *Statistical Analysis of Reliability and Life-Testing Models.* Marcel Dekker, New York, 1978.

BAIRD, D. D. and A. J. WILCOX "Cigarette smoking associated with delayed conception." *Journal of the American Medical Association,* 1985, volume 253, pages 2979–2983.

BARASH, P. G., J. D. KATZ, S. FIRESTONE, S. L. HUI, K. TAUNT, C. JAFFE, N. S. TALNER, and C. S. KLEINMAN "Cardiovascular performance in children during induction: An echocardiographic comparison of enflurane and halothane." *Anesthesiology,* 1979, volume 51, page 5315.

BARNES, DEBORAH M. "New questions about AIDS test accuracy." *Science,* 1987, volume 238, pages 884–885.

BASHAM, T. Y., M. S. KAMINSKY, K. KITAMURA, R. LEVY, and T. C. MERIGAN "Synergistic antitumor effect of interferon and anti-idiotype monoclonal antibody in murine lymphoma." *Journal of Immunology,* 1986, volume 137, pages 3019–3024.

BATCHELOR, J. R. and M. HACKETT "HL-A matching in treatment of burned patients with skin allografts." *Lancet,* 1970, volume 2, pages 581–583.

BECK, ALAN M., SUSANNE R. FELSER, and LAWRENCE T. GLICKMAN "An epizootic of rabies in Maryland, 1982–84." *American Journal of Public Health,* 1987, volume 77, pages 42–44.

BERGER, ROGER L., DENNIS D. BOOS, and FRANK M. GUESS "Tests and confidence sets for comparing two mean residual life functions." *Biometrics,* 1988, volume 44, pages 103–115.

BERRY, D. A. "Logarithmic transformations in ANOVA." *Biometrics,* 1987, volume 43, pages 439–456.

BERRY, DONALD A. and TIMOTHY D. BERRY "The probability of a field goal: Rating kickers." *The American Statistician,* 1985, volume 39, pages 152–155.

BIOMEDICAL INFORMATION CORPORATION "Patient information on cholesterol and triglyceride disorders." *Compendium on Patient Information,* 1987.

BISHOP, Y. M. M. "Examples of graphical methods." In *Statistics by Example, Volume* 1, edited by F. Mosteller, W. H. Kruskal, R. S. Pieters, G. R. Rising, and R. F. Link. Addison-Wesley, Reading, Massachusetts, 1973, pages 33–48.

BLISS, C. I. "The calculation of the dosage–mortality curve." *Annals of Applied Biology,* 1935, volume 22, pages 134–167.

BONNEY, GEORGE EBOW "Logistic regression for dependent binary observations." *Biometrics,* 1987, volume 43, pages 951–973.

BOX, G. E. P. and D. R. COX "An analysis of transformations (with Discussion)." *Journal of the Royal Statistical Society, Series B,* 1964, volume 26, pages 211–246.

BOX, GEORGE E. P., WILLIAM G. HUNTER, and J. STUART HUNTER *Statistics for Experimenters: An Introduction to Design, Data Analysis and Model Building.* Wiley, New York, 1978.

BRAIN, C. W. and S. S. SHAPIRO "A regression test for exponentiality: Censored and complete samples." *Technometrics,* 1983, volume 25, pages 69–76.

BRESLOW, NORMAN "Comment: Risk assessment: Science or policy." *Statistical Science,* 1988, volume 3, pages 28–33.

BROGAN, DONNA R. and MICHAEL H. KUTNER "Comparative analyses of pretest-posttest research designs." *The American Statistician,* 1980, volume 34, pages 229–232.

BROOK, RICHARD J. "How much does a kilogram of milk powder weigh?" In *The Fascination of Statistics,* edited by Richard J. Brook, Gregory C. Arnold, Thomas H. Hassard, and Robert M. Pringle. Marcel Dekker, New York, 1986.

BROWN, BYRON W., JR. "The crossover experiment for clinical trials." *Biometrics,* 1980, volume 36, pages 69–79.

BROWN, M. B. and A. B. FORSYTHE "Robust tests for the equality of variances." *Journal of the American Statistical Association,* 1974, volume 69, pages 364–367.

BROWNLEE, K. A. *Statistical Theory and Methodology in Science and Engineering,* 2nd edition. Wiley, New York, 1965.

BUECHNER, JAY S., DONALD K. PERRY, H. DENMAN SCOTT, BEVERLY EHRICH FREEDMAN, JOHN T. TIERNEY, and WILLIAM J. WATERS "Cigarette smoking behavior among Rhode Island physicians, 1963–83." *American Journal of Public Health,* 1986, volume 76, pages 285–286.

BURKHART, HAROLD E., ROBERT C. PARKER, MIKE R. STRUB, and RICHARD ODERWALD "Yields of old-field loblolly pine plantations." Division of Forestry and Wildlife Resources Publication FWS-3-72, Virginia Polytechnic Institute and State University, Blacksburg, Virginia, 1972.

CAMPBELL, J. A. and O. PELLETIER "Determination of niacin (niacinamide) in cereal products." *Journal of the Association of Official Analytical Chemists,* 1962, volume 45, pages 449–453.

CARLBORG, F. "Cancer risk assessment for saccharin." *Food and Chemical Toxicology,* 1985, volume 23, pages 499–506.

CARROLL, R. J., J. SACKS, and C. H. SPIEGELMAN "A quick and easy multiple-use calibration-curve procedure." *Technometrics,* 1988, volume 30, pages 137–141.

CARTER, RANDY L. "Restricted maximum likelihood estimation of bias and reliability in the comparison of several measuring methods." *Biometrics,* 1981, volume 37, pages 733–741.

CARTER, RANDY L. and BARRY J. N. BLIGHT "A Bayesian change-point problem with an application to the prediction and detection of ovulation in women." *Biometrics,* 1981, volume 37, pages 743–751.

CARTWRIGHT, H. V., R. L. LINDAHL, and J. W. BAWDEN "Clinical findings on the effectiveness of stannous fluoride and acid phosphate fluoride as caries-reducing agents in children." *Journal of Dentistry for Children,* 1968, volume 35, pages 36–40.

CEDERLUND, ANNA, PER CAMNER, and MAGNUS SVARTENGREN "Nasal mucociliary transport before and after jogging." *The Physician and Sportsmedicine,* 1987, volume 15, pages 93–95.

CHAMBERS, JOHN M., WILLIAM S. CLEVELAND, BEAT KLEINER, and PAUL A. TUKEY *Graphical Methods for Data Analysis.* Wadsworth Advanced Books and Software, Pacific Grove, California, 1983.

Chance "Supreme Court ruling on death penalty." 1988, volume 1, pages 7–8.

Chance "Random testing for AIDS?" 1988, volume 1, page 9.

CHENG, R. C. H. "Confidence bands for two-stage design problems." *Technometrics,* 1987, volume 29, pages 301–309.

CHINCHILLI, VERNON M., BARRY H. SCHWAB, and PRANAB K. SEN "Inference based on ranks for the multiple-design multivariate linear model." *Journal of the American Statistical Association,* 1989, volume 84, pages 517–524.

CLEVELAND, WILLIAM S. *The Elements of Graphing Data.* Wadsworth Advanced Books and Software, Pacific Grove, California, 1985.

CLONINGER, C. ROBERT "Neurogenetic adaptive mechanisms in alcoholism." *Science,* 1987, volume 236, pages 410–416.

COLE, A. F. W. and M. KATZ "Summer ozone concentrations in southern Ontario in relation to photochemical aspects and vegetation damage." *Journal of the Air Pollution Control Association,* 1966, volume 16, pages 201–206.

CONOVER, W. J., MARK E. JOHNSON, and MYRLE M. JOHNSON "A comparative study of tests for homogeneity of variances, with applications to the outer continental shelf bidding data." *Technometrics,* 1981, volume 23, pages 351–361.

COOPER, L. M., E. SCHUBOT, S. A. BANFORD, and C. T. TART "A further attempt to modify hypnotic susceptibility through repeated individualized experience." *International Journal of Clinical and Experimental Hypnosis,* 1967, volume 15, pages 118–124.

COOPER, N. R. "A statistical theory of failure for brittle materials." Memorandum 232, U. K. Ministry of Defence, Procurement Executive, Royal Armament Research and Development Establishment, 1984.

COSGROVE, RICHARD "Thirty thousand years of human colonization in Tasmania." *Science,* 1989, volume 243, pages 1706–1708.

CRAMER, DANIEL W., MARLENE B. GOLDMAN, ISAAC SCHIFF, SERGE BELISLE, BRUCE ALBRECHT, BRUCE STADEL, MARK GIBSON, EMERY WILSON, ROBERT STILLMAN, and IRWIN THOMPSON "The relationship of tubal infertility to barrier method and oral contraceptive use." *Journal of the American Medical Association,* 1987, volume 257, pages 2446–2450.

CROUCHLEY, R. and A. R. PICKLES "Methods for the identification of Lexian, Poisson, and Markovian variations in the secondary sex ratio." *Biometrics,* 1984, volume 40, pages 165–175.

DANIEL, CUTHBERT and FRED S. WOOD *Fitting Equations to Data, Computer Analysis of Multifactor Data,* 2nd edition. Wiley, New York, 1980.

DEHAVEN, KENNETH E., and DAVID M. LINTNER "Athletic injuries: Comparison by age, sport and gender." *The American Journal of Sports Medicine,* 1986, volume 14, pages 218–221.

DEOM, CARL M., MELVIN J. OLIVER, and ROGER N. BEACHY "The 30-kilodalton gene product of tobacco mosaic virus potentiates virus movement." *Science,* 1987, volume 237, pages 389–394.

DEVORE, JAY L. *Probability and Statistics for Engineering and the Sciences.* Brooks/Cole, Pacific Grove, California, 1982.

DEVORE, JAY L. *Probability and Statistics for Engineering and the Sciences,* 2nd edition. Brooks/Cole, Pacific Grove, California, 1987.

DHINDSA, DHARAM S., JAMES METCALFE, and DEAN H. HUMMELS "Responses to exercise in the pregnant Pygmy goat." *Respiration Physiology,* 1978, volume 32, pages 299–311.

DICK, E. C., K. A. MINK, T. A. DEMKE, and S. C. INHORN "Aerosol transmission of rhinovirus infection." Abstract of a presentation to the annual meeting of the American Association for Microbiology, Washington, D.C., March 23–28, 1986.

DIENSTFREY, STEPHEN J. "Women veterans' exposure to combat." *Proceedings of the*

Social Statistics Section of the American Statistical Association Annual Meeting in Chicago, August 18–21, 1986, pages 177–180.

DIETZ, E. JACQUELIN "Teaching regression in a nonparametric statistics course." *The American Statistician,* 1989, volume 43, pages 35–40.

DRAPER, N. R. and H. SMITH *Applied Regression Analysis,* 2nd edition. Wiley, New York, 1981.

DREW, GLEN D. "The effects of bathers on the fecal coliform, total coliform and total bacteria density of water." Master's thesis, Department of Civil Engineering, Tufts University, 1971.

DUNCAN, ACHESON J. *Quality Control and Industrial Statistics,* 4th edition, Richard D. Irwin, Homewood, Illinois, 1974.

DUNNETT, CHARLES W. "Drug screening: The never-ending search for new and better drugs." In *Statistics: A Guide to the Unknown,* edited by Judith M. Tanur. Holden-Day, San Francisco, 1972, page 25.

EFRON, B. and R. TIBSHIRANI "Bootstrap methods for standard errors, confidence intervals, and other measures of statistical accuracy." *Statistical Science,* 1986, volume 1, pages 54–77.

ELLENBERG, JONAS H. "Testing for a single outlier from a general linear regression." *Biometrics,* 1976, volume 32, pages 637–645.

EPPRIGHT, E. S., H. M. FOX, B. A. FRYER, G. H. LAMKIN, V. M. VIVIAN, and E. S. FULLER "Nutrition of infants and preschool children in the north central region of the United States of America." *World Review of Nutrition and Dietetics,* 1972, volume 14, pages 269–332.

FELDMANN, KENNETH A., M. DAVID MARKS, MICHAEL L. CHRISTIANSON, and RALPH S. QUATRANO "A dwarf mutant of *Arabidopsis* generated by T-DNA insertion mutagenesis." *Science,* 1989, volume 243, pages 1351–1354.

FIENBERG, S. E. *The Analysis of Cross-Classified Categorical Data.* MIT Press, Cambridge, Massachusetts, 1977.

FOSTER, STEPHEN C. "The effect of acid rain on the growth of vegetable seeds." Report for a class project in experimental design, 1986.

FOX, J. R. and J. E. RANDALL "Relationship between forearm tremor and the biceps electromyogram." *Journal of Applied Physiology,* 1970, volume 29, pages 103–108.

FRANCIS, THOMAS, JR. et al. "An evaluation of the 1954 poliomyelitis vaccine trials—Summary report." *American Journal of Public Health,* 1955, volume 45(5), pages 1–63.

FREEDMAN, DAVID, ROBERT PISANI, and ROGER PURVES *Statistics.* W. W. Norton, New York, 1978.

FRIEDMAN, MEYER, CARL E. THORESEN, JAMES J. GILL, LYNDA H. POWELL, DIANE ULMER, LEONTI THOMPSON, VIRGINIA A. PRICE, DAVID D. RABIN, WILLIAM S. BREALL, THEODORE DIXON, RICHARD LEVY, and EDWARD BOURG "Alteration of type A behavior and reduction in cardiac recurrences in post myocardial infarction patients." *American Heart Journal,* 1984, volume 108, pages 237–248.

GALLANT, A. RONALD "Testing a nonlinear regression specification: A nonregular case." *Journal of the American Statistical Association,* 1977, volume 72, pages 523–530.

GALLANT, A. RONALD and W. A. FULLER "Fitting segmented polynomial regression models whose join points have to be estimated." *Journal of the American Statistical Association,* 1973, volume 68, pages 144–147.

GAROFALO, F., R. W. WHITMORE, W. F. DOMIS, and F. VON GEMMINGEN "Creep and creep-rupture relationships in an austenitic stainless steel." *Transactions of the Metallurgical Society of AIME,* 1961, volume 221, pages 310–319.

GART, J. J. "Statistical analysis of the first mouse skin-painting study." In *Toward Less Hazardous Cigarettes,* edited by G. B. Gori. Department of Health, Education and Welfare, Washington, D.C., 1976, pages 109–121.

GART, JOHN J. and ROBERT E. TARONE "On the efficiency of age-adjusted tests in animal carcinogenicity experiments." *Biometrics,* 1987, volume 43, pages 235–244.

GASPER, JAMES D. "Factors affecting optical density of a polymer latex." Class project in experimental design, 1988.

GASTWIRTH, JOSEPH P. "The statistical precision of medical screening procedures: Application to polygraph and AIDS antibodies test data." *Statistical Science,* 1987, volume 2, pages 213–238.

GEERTSEMA, J. C. and C. J. REINECKE "The application of a sequential procedure with elimination as a method for the screening for metabolic diseases." *Biometrics,* 1984, volume 40, pages 663–673.

GLADUE, BRIAN A., RICHARD GREEN, and RONALD E. HELLMAN "Neuroendocrine response to estrogen and sexual orientation." *Science,* 1984, volume 225, pages 1496–1499.

GOLDBERG, D. P., K. BRIDGES, P. DUNCAN-JONES, and D. GRAYSON "Dimensions of neurosis seen in primary care settings." *Psychological Medicine,* 1987, volume 17, pages 461–470.

GORE, W. L. "Statistical techniques for research and development." *Conference Papers, First Annual Convention, American Society for Quality Control and Second Midwest Quality Control Conference,* June 5–6, 1947. John L. Swift Company, Chicago, Illinois, 1947.

GOULD, A. LAWRENCE "A new approach to the analysis of clinical drug trials with withdrawals." *Biometrics,* 1980, volume 36, pages 721–727.

GRANATEK, CHRISTINE H., KOHJI EZAKI, EVAN M. HERSH, MICHAEL J. KEATING, and SHELLEY RASMUSSEN "Antibody responses of remission leukemia patients receiving active specific and nonspecific immunotherapy." *Cancer,* 1981, volume 47, pages 272–279.

GRANT, E. L. and R. S. LEAVENWORTH *Statistical Quality Control,* 4th edition. McGraw-Hill, New York, 1972.

GRANT, JAMES P. *The State of the World's Children.* Published for the United Nations Children's Fund by Oxford University Press, New York, 1987.

GRAYSON, D. A. "Statistical diagnosis and the influence of diagnostic error." *Biometrics,* 1987, volume 43, pages 975–984.

GROSS, A. J. and V. A. CLARK *Survival Distributions: Reliability Applications in the Biomedical Sciences.* Wiley, New York, 1975.

GROSS, ALAN J. and CHAN F. LAM "Paired observations from a survival distribution." *Biometrics,* 1981, volume 37, pages 505–511.

HABERMAN, S. J. *Analysis of Qualitative Data, Volume 2: New Developments.* Academic Press, New York, 1979.

HABERMAN, SHELBY J. "Analysis of dispersion of multinomial responses." *Journal of the American Statistical Association,* 1982, volume 77, pages 568–580.

HAHN, M. E., S. B. HABER, and J. L. FULLER "Differential antagonistic behavior in mice selected for brain weight." *Physiology and Behavior,* 1973, volume 10, pages 759–762.

HARTL, DANIEL L. *Human Genetics.* Harper & Row, New York, 1983.

HASTINGS, N. A. J. and J. B. PEACOCK *Statistical Distributions.* Butterworth and Company, London, 1975.

HATTON, W. M. and M. D. CLARKE-HUNDLEY *Cancer in Western Australia 1982, An Analysis of Age and Sex-Specific Rates.* Epidemiology Branch, Health Department of Western Australia, Perth, W. A., 1984.

HAUGH, LARRY D., BRIAN S. FLYNN, PETER W. CALLAS, ROGER H. SECKER-WALKER, and MUN SON "Factors influencing transitions in cigarette smoking status in children: Logistic regression analysis." *Proceedings of the Social Statistics Section of the American Statistical Association Annual Meeting in Chicago,* August 18–21, 1986, pages 421–426.

HEGLAND, D. "Robotics growth—No end in sight." *Production Engineering,* April 1983, pages 46–51.

HIGGINS, JAMES E. and GARY G. KOCH "Variable selection and generalized chi-square analysis of categorical data applied to a large cross-sectional occupational health survey." *International Statistical Review,* 1977, volume 45, pages 51–62.

HILL, N. J. and A. R. PADMANABHAN "Robust comparison of two regression lines and biomedical applications." *Biometrics,* 1984, volume 40, pages 985–994.

HOAGLIN, DAVID C., FREDERICK MOSTELLER, and JOHN W. TUKEY (editors) *Understanding Robust and Exploratory Data Analysis.* Wiley, New York, 1983.

HOCKING, R. R. *The Analysis of Linear Models.* Brooks/Cole, Pacific Grove, California, 1985.

HOFFMAN, STEPHEN L., CHARLES N. OSTER, CHRISTOPHER V. PLOWE, GILLIAN R. WOOLLETT, JOHN C. BEIER, JEFFREY D. CHULAY, ROBERT A. WIRTZ, MICHAEL R. HOLLINGDALE, and MUTUMA MUGAMBI "Naturally acquired antibodies to sporozoites do not prevent malaria: Vaccine development implications." *Science,* 1987, volume 237, pages 639–642.

HOHIMER, A. ROGER, JOHN M. BISSONNETTE, JAMES METCALFE, and THOMAS A. McKEAN "Effect of exercise on uterine blood flow in the pregnant Pygmy goat." *American Journal of Physiology,* 1984, volume 246, pages H207–H212.

HOLLANDER, MYLES and FRANK PROSCHAN *The Statistical Exorcist, Dispelling Statistics Anxiety.* Marcel Dekker, New York, 1984.

HOLLANDER, MYLES and DOUGLAS A. WOLFE *Nonparametric Statistical Methods.* Wiley, New York, 1973.

HSIEH, H. K. "An exact test for comparing location parameters of k exponential distributions with unequal scales based on Type II censored data." *Technometrics,* 1986, volume 28, pages 157–164.

HUI, SIU L. and SAUL H. ROSENBERG "Multivariate slope ratio assay with repeated measurements." *Biometrics,* 1985, volume 41, pages 11–18.

HUNDAL, P. S. "Knowledge of performance as an incentive in repetitive industrial work." *Journal of Applied Psychology,* 1969, volume 53, pages 224–226.

INGEBRITSEN, S. E., D. R. SHERROD, and R. H. MARINER "Heat flow and hydrothermal circulation in the Cascade Range, north-central Oregon." *Science,* 1989, volume 243, pages 1458–1462.

ISMAIL, M. M., J. I. BRUCE, S. L. RASMUSSEN, M. ATTIA, and M. SALAMA "Schistosomiasis and other helminthic infections in Kafr Soliman Village, Sharkia Governorate, Egypt." *Journal of the Egyptian Society of Parasitology,* 1988, volume 18, pages 47–61.

JAMES, ALAN T. and ROBERT A. J. CONYERS "Estimation of a derivative by a difference quotient: Its application to hepatocyte lactate metabolism." *Biometrics,* 1985, volume 41, pages 467–476.

JAMES, I. R. and M. W. KNUIMAN "An application of Bayes methodology to the analysis of diary records from a water use study." *Journal of the American Statistical Association,* 1987, volume 82, pages 705–711.

JOHNSON, DALLAS E. "Some new multiple comparison procedures for the two-way AOV model with interaction." *Biometrics,* 1976, volume 32, pages 929–934.

JOHNSON, R. A. and D. W. WICHERN *Applied Multivariate Statistical Analysis.* Prentice-Hall, Englewood Cliffs, New Jersey, 1982.

JUNG, D. H. and A. C. PAREKH "A semi-micromethod for the determination of serum iron and iron-binding capacity without deproteinization." *American Journal of Clinical Pathology,* 1970, volume 54, pages 813–817.

KAHNEMAN, DANIEL and AMOS TVERSKY "The psychology of preferences." *Scientific American,* January 1982, pages 160–173.

KANKI, PHYLLIS J., SOULEYMANE M'BOUP, DOMINIQUE RICARD, FRANCIS BARIN, FRANÇOIS DENIS, CHIEKH BOYE, LASANA SANGARE, KARIN TRAVERS, MICHAEL ALBAUM, RICHARD MARLINK, JEAN-LOUP ROMET-LEMONNE, and MYRON ESSEX

"Human T-lymphotropic virus type 4 and the human immunodeficiency virus in West Africa." *Science,* 1987, volume 236, pages 827–831.

KARELITZ, S., V. R. FISICHELLI, J. COSTA, R. KARELITZ, and L. ROSENFELD "Relation of crying in early infancy to speech and intellectual development at age three years." *Child Development,* 1964, volume 35, pages 769–777.

KAUFFMAN, LESTER A. *Statistical Quality Control at the St. Louis Division of American Stove Company.* War Production Board, Office of Production, Research and Development. *Quality Control Reports,* No. 3, August 1945.

KIRK, ROGER E. *Experimental Design,* 2nd edition. Brooks/Cole, Pacific Grove, California, 1982.

KISS, KLARA "Quantitative electron probe analysis of low atomic number samples with irregular surfaces." *Applied Spectroscopy,* February 1983, pages 19–24.

KNUIMAN, M. W. and T. P. SPEED "Incorporating prior information into the analysis of contingency tables." *Biometrics,* 1988, volume 44, pages 1061–1071.

KOCH, GARY G., INGRID A. AMARA, GORDON W. DAVIS, and DENNIS B. GILLINGS "A review of some statistical methods for covariance analysis of categorical data." *Biometrics,* 1982, volume 38, pages 563–595.

KOLATA, GINA "Panel urges newborn sickle cell screening." *Science,* 1987, volume 236, pages 259–260.

LANMAN, J. et al. "Retrolental fibroplasia and oxygen therapy." *Journal of the American Medical Association,* 1954, volume 155, pages 223–226.

LARSEN, RICHARD J. and MORRIS L. MARX *An Introduction to Mathematical Statistics and Its Applications,* 2nd edition. Prentice-Hall, Englewood Cliffs, New Jersey, 1986.

LEE, LARRY and R. G. KRUTCHKOFF "Mean and variance of partially-truncated distributions." *Biometrics,* 1980, volume 36, pages 531–536.

LEHMANN, E. L. *Nonparametrics: Statistical Methods Based on Ranks.* Holden-Day, San Francisco, California, 1975.

LEUNG, HOI M. and LAWRENCE L. KUPPER "Comparisons of confidence intervals for attributable risk." *Biometrics,* 1981, volume 37, pages 293–302.

LEVENE, H. "Robust tests for equality of variances." In *Contributions to Probability and Statistics,* edited by I. Olkin. Stanford University Press, Palo Alto, California, 1960, pages 278–292.

LI, J. C. R. *Statistical Inference* I, Edwards Brothers, Ann Arbor, Michigan, 1964.

LOYND, JACK "Competition and base-running time trials." Class project in experimental design, 1985.

MACOUL, PATRICIA "Orange juice experiment." Class project in experimental design, 1988.

MARTY, PHILLIP J., ROBERT J. McDERMOTT, and TOM WILLIAMS "Patterns of smokeless tobacco use in a population of high school students." *American Journal of Public Health,* 1986, volume 76, pages 190–192.

MATHER, MONICA H. and BERNARD D. ROITBERG "A sheep in wolf's clothing: Tephritid flies mimic spider predators." *Science,* 1987, volume 236, pages 308–310.

McENTEE, W. J. and R. G. MAIR "Memory impairment in Korsakoff's psychosis: A correlation with brain noradrenergic activity." *Science,* 1978, volume 202, pages 905–907.

McMAHAN, C. ALEX "An index of tracking." *Biometrics,* 1981, volume 37, pages 447–455.

MEHTA, CYRUS R., NITIN R. PATEL, and ANASTASIOS A. TSIATIS "Exact significance testing to establish treatment equivalence with ordered categorical data." *Biometrics,* 1984, volume 40, pages 819–825.

MEIER, PAUL "Terminating a trial—the ethical problem." *Clinical Pharmacology and Therapeutics,* 1979, volume 25, pages 633–640.

MEIER, PAUL "The biggest public health experiment ever: The 1954 field trial of the Salk poliomyelitis vaccine." In *Statistics: A Guide to the Unknown,* edited by

Judith M. Tanur et al. Wadsworth and Brooks/Cole Advanced Books & Software, Pacific Grove, California, 1989, pages 3–14.

MENDENHALL, WILLIAM and TERRY SINCICH *Statistics for the Engineering and Computer Sciences,* 2nd edition. Dellen, San Francisco, California, 1988.

MICKEY, M. R., O. J. DUNN, and V. CLARK "Note on the use of stepwise regression in detection outliers." *Computers and Biomedical Research,* 1967, volume 1, pages 105–111.

MILLAR, WAYNE J. and THOMAS STEPHENS "The prevalence of overweight and obesity in Britain, Canada, and United States." *American Journal of Public Health,* 1987, volume 77, pages 38–41.

MILTON, J. S. and JESSE C. ARNOLD *Probability and Statistics in the Engineering and Computing Sciences.* McGraw-Hill, New York, 1986.

MORRISON, A. S., M. M. BLACK, C. R. LOWE, B. MACMAHON, and S. YUASA "Some international differences in histology and survival in breast cancer." *International Journal of Cancer,* 1973, volume 11, pages 216–267.

MOSTELLER, FREDERICK and JOHN W. TUKEY *Data Analysis and Regression.* Addison-Wesley, Reading, Massachusetts, 1977.

MYERS, RAYMOND H. *Classical and Modern Regression with Applications.* Duxbury Press, Boston, Massachusetts, 1986.

NAIR, VIJAYAN N. "On the behavior of some estimators from probability plots." *Journal of the American Statistical Association,* 1984, volume 79, pages 823–831.

NASCI, ROGER S., CEDRIC W. HARRIS, and CYRESA K. PORTER "Failure of an insect electrocuting device to reduce mosquito biting." *Mosquito News,* 1983, volume 43, pages 180–184.

NATIONAL WEATHER SERVICE *Operations Manual* U.S. Department of Commerce, National Oceanic and Atmospheric Administration, WSOM Issuance 84-11, July 11, 1984, pages 21–25.

NATRELLA, M. *Experimental Statistics.* National Bureau of Standards Handbook 91, Washington, D.C., 1963.

NELSON, WAYNE *How to Analyze Data with Simple Plots.* Volume 1 of Basic References in Quality Control: Statistical Techniques. American Society for Quality Control, Madison, Wisconsin, 1986.

NELSON, W. *Applied Life Data Analysis.* Wiley, New York, 1982.

NORDHEIM, E. V., A. TSIATIS, and T. J. SHAPAS "Incorporating extra information in experimental design for bioassay." *Biometrics,* 1983, volume 39, pages 87–96.

O'BRIEN, PETER C. and THOMAS R. FLEMING "A paired Prentice–Wilcoxon test for censored paired data." *Biometrics,* 1987, volume 43, pages 169–180.

OJA, HANNU and JUKKA NYBLOM "Bivariate sign tests." *Journal of the American Statistical Association,* 1989, volume 84, pages 249–252.

OLSEN, ALEXANDER Personal communication, 1986.

O'SULLIVAN, J. B. and C. M. MAHAN "Glucose tolerance test: Variability in pregnant and non-pregnant women." *American Journal of Clinical Nutrition,* 1966, volume 19, pages 345–351.

OTT, E. R. and R. D. SNEE "Identifying useful differences in a multiple-head machine." *Journal of Quality Technology,* 1973, volume 5, pages 45–57.

OWEN, W. J. and T. A. DEROUEN "Estimation of the mean for lognormal data containing zeroes and left-censored values, with applications to the measurement of worker exposures to air contaminants." *Biometrics,* 1980, volume 36, pages 707–719.

PARTRIDGE, C. J., M. JOHNSTON, and S. EDWARDS "Recovery from physical disability after stroke: Normal patterns as a basis for evaluation." *The Lancet,* February 14, 1987, pages 373–374.

PATARROYO, MANUEL E., ROBERTO AMADOR, PEDRO CLAVIJO, ALBERTO MORENO, FANNY GUZMAN, PEDRO ROMERO, RICARDO TASCON, ANTONIO FRANCO,

LUIS A. MURILLO, GABRIEL PONTON, and GUSTAVO TRUJILLO "A synthetic vaccine protects humans against challenge with asexual blood stages of *Plasmodium falciparum* malaria." *Nature,* 1988, volume 332, pages 158–161.

PERRY, L., R. VAN DYKE, and R. THEYE "Sympathoadrenal and hemodynamic effects of isoflurane, halothane, and cyclopropane in dogs." *Anesthesiology,* 1974, volume 40, pages 465–470.

PIEGORSCH, WALTER W., CLARICE R. WEINBERG, and BARRY H. MARGOLIN "Exploring simple independent action in multifactor tables of proportions." *Biometrics,* 1988, volume 44, pages 595–603.

PITETTI, KENNETH H., PETER G. SNELL, JAMES STRAY-GUNDERSEN, and FRANK A. GOTTSCHALK "Aerobic training exercises for individuals who had amputation of the lower limb." *The Journal of Bone and Joint Surgery,* 1987, volume 69-A, pages 914–915.

POLLOCK, KENNETH H., SCOTT R. WINTERSTEIN, and MICHAEL J. CONROY "Estimation and analysis of survival distributions for radio-tagged animals." *Biometrics,* 1989, volume 45, pages 99–109.

PORTER, RICHARD H. and JOHN D. MOORE "Human kin recognition by olfactory cues." *Physiology & Behavior,* 1981, volume 27, pages 493–495.

PRENTICE, ROSS L. "A generalization of the probit and logit methods for dose response curves." *Biometrics,* 1976, volume 32, pages 761–768.

PROSCHAN, F. "Theoretical explanation of observed decreasing failure rate." *Technometrics,* 1963, volume 5, pages 375–383.

QUADE, DANA "Nonparametric analysis of covariance by matching." *Biometrics,* 1982, volume 38, pages 597–611.

QUESENBERRY, C. P., T. B. WHITAKER, and J. W. DICKENS "On testing normality using several samples: An analysis of peanut aflatoxin data." *Biometrics,* 1976, volume 32, pages 753–759.

RAMSAY, J. O. "Monotone regression splines in action." *Statistical Science,* 1988, volume 3, pages 425–461.

RATKOWSKY, D. A. and D. MARTIN "The use of multivariate analysis in identifying relationships among disorder and mineral element content in apples." *Australian Journal of Agricultural Research,* 1974, volume 25, pages 783–790.

RAWLINGS, JOHN O. *Applied Regression Analysis: A Research Tool.* Wadsworth & Brooks/Cole Advanced Books & Software, Pacific Grove, California, 1988.

RENNEKER, MARK "Surfing: The sport and the life-style." *The Physician and Sportsmedicine,* 1987, volume 15, pages 156–158.

RICE, JOHN A. *Mathematical Statistics and Data Analysis.* Wadsworth and Brooks/Cole Advanced Books & Software, Pacific Grove, California, 1988.

ROBERTS, LESLIE "Is acid deposition killing West German forests?" *Bioscience,* 1983, volume 33, pages 302–305.

ROSENSTOCK, IRWIN M., ANDY STERGACHIS, and CATHERINE HEANEY "Evaluation of smoking prohibition policy in a health maintenance organization." *American Journal of Public Health,* 1986, volume 76, pages 1014–1015.

ROTHWELL, NORMAN V. *Human Genetics.* Prentice-Hall, Englewood Cliffs, New Jersey, 1977.

RYAN, BARBARA F., BRIAN L. JOINER, and THOMAS A. RYAN JR. *Minitab Handbook.* 2nd edition. Duxbury Press, Boston, Massachusetts, 1985.

RYAN, THOMAS P. *Statistical Methods for Quality Improvement.* Wiley, New York, 1989.

RYAN, THOMAS A., JR., BRIAN L. JOINER, and BARBARA F. RYAN *Minitab Student Handbook.* Duxbury Press, Boston, Massachusetts, 1976.

SCHAEFER, ROBERT L. and RICHARD B. ANDERSON *The Student Edition of Minitab, Statistical Software Adapted for Education.* Addison-Wesley, Reading, Massachusetts, 1989.

SCHEIRER, C. JAMES, WILLIAM S. RAY, and NATHAN HARE "The analysis of ranked data derived from completely randomized factorial designs." *Biometrics,* 1976, volume 32, pages 429–434.

SCHILLING, EDWARD G. *Acceptance Sampling in Quality Control.* Marcel Dekker, New York, 1982.

SCHMOYER, RICHARD L. "An exact distribution-free analysis for accelerated life testing at several levels of a single stress." *Technometrics,* 1986, volume 28, pages 165–175.

Science News "AIDS virus in bone, vaccine on trial." May 7, 1988, volume 133, number 19, page 292.

SEWELL, W. H. and V. P. SHAH "Social class, parental encouragement, and educational aspirations." *American Journal of Sociology,* 1968, volume 73, pages 559–572.

SHAPIRO, SAMUEL S. *The ASQC Basic References for Quality Control: Statistical Techniques, Volume 3, How to Test Normality and Other Distribution Assumptions.* American Society for Quality Control, Milwaukee, Wisconsin, 1986.

SHAUGHNESSY, JAMES "Softball superstitions: Are they imagined or do these notions have substance?" Class project in experimental design, 1988.

SHELP, W. D., F. H. BACH, W. A. KISKEN, M. NEWTON, R. E. RIESELBACH, and A. B. WEINSTEIN "Long-term integrity of renal function in cadaver allografts." *Journal of the American Medical Association,* 1970, volume 213, pages 1443–1447.

SIDDIQUI, M. M. and E. A. GEHAN *Statistical Methodology for Survival Time Studies.* Communication of National Cancer Institute, 1966.

SNEDECOR, GEORGE W. and WILLIAM G. COCHRAN *Statistical Methods,* 6th edition. The Iowa State University Press, Ames, Iowa, 1967.

SNEE, RONALD D. "Nonadditivity in a two-way classification: Is it interaction or nonhomogeneous variance?" *Journal of the American Statistical Association,* 1982, volume 77, pages 515–519.

SNOW, STEVEN R. "The effect of ammunition type on target shooting scores." Class project in experimental design, 1986.

TANNER, MARTIN A. and WING HUNG WONG "The calculation of posterior distributions by data augmentation." *Journal of the American Statistical Association,* 1987, volume 82, pages 528–540.

TARONE, ROBERT E. and JOHN J. GART "On the robustness of combined tests for trends in proportions." *Journal of the American Statistical Association,* 1980, volume 75, pages 110–116.

The 1987 Information Please Almanac Houghton Mifflin, New York, 1986.

THOMAS, H. V. and E. SIMMONS "Histamine content in sputum from allergic and nonallergic individuals." *Journal of Applied Physiology,* 1969, volume 26, pages 793–797.

TUFTE, EDWARD R. *The Visual Display of Quantitative Information.* Graphics Press, Cheshire, Connecticut, 1983.

TUFTE, EDWARD R. *Envisioning Information.* Graphics Press, Cheshire, Connecticut, 1990.

TUKEY, JOHN W. *Exploratory Data Analysis.* Addison-Wesley, Reading, Massachusetts, 1977.

TURSZ, ANNE and MONIQUE CROST "Sports-related injuries in children." *The American Journal of Sports Medicine,* 1986, volume 14, pages 294–299.

U. S. COMMISSION ON CIVIL RIGHTS *Insurance Redlining: Fact Not Fiction.* A report prepared by the Illinois, Indiana, Michigan, Minnesota, Ohio, and Wisconsin Advisory Committees to the U.S. Commission on Civil Rights, Washington, D.C., 1979.

U. S. DEPARTMENT OF AGRICULTURE, WAR FOOD ADMINISTRATION, OFFICE OF MARKETING SERVICES "Results of fiber and spinning tests of some varieties of upland cotton grown in the United States, crop of 1944." (Processed April 1945)

URQUHART, N. SCOTT "Adjustment in covariance when one factor affects the covariate." *Biometrics,* 1982, volume 38, pages 651–660.

VAN RYZIN, JOHN and KAMTA RAI "A dose–response model incorporating nonlinear kinetics." *Biometrics,* 1987, volume 43, pages 95–105.

VARMA, A. O. and N. W. CHILTON "Crossover designs involving two treatments." *Journal of Periodontal Research,* 1974, volume 9, supplement 14, pages 160–170.

VELLEMAN, PAUL F. and DAVID C. HOAGLIN *Applications, Basics and Computing of Exploratory Data Analysis.* Duxbury Press, Boston, 1981.

WALLACH, D. and B. GOFFINET "Mean squared error of prediction in models for studying ecological and agronomic systems." *Biometrics,* 1987, volume 43, pages 561–573.

WALPOLE, RONALD E. and RAYMOND H. MYERS *Probability and Statistics for Engineers and Scientists,* 4th edition. Macmillan, New York, 1989.

WALSH, MICHELE "The bicycle experiment." Class project in experimental design, 1988.

WALTER, S. D. "Calculation of attributable risks from epidemiologic data." *International Journal of Epidemiology,* 1978, volume 7, pages 175–182.

WALTKING, A. E., G. BLEFFERT, and M. KIERNAN "An improved rapid physiochemical assay method for aflatoxin in peanuts and peanut products." *Journal of the American Oil Chemists' Society,* 1968, volume 49, pages 590–593.

WANG, G. C. "Microscopic investigation of CO_2 flooding process." *Journal of Petroleum Technology,* 1982, volume 34, pages 1789–1797.

WEINBERG, CLARICE RING and BETH C. GLADEN "The beta–geometric distribution applied to comparative fecundability studies." *Biometrics,* 1986, volume 42, pages 547–560.

WEINDLING, S. Statistics report: Math 80B, 1977.

WEISS, S. H., J. J. GOEDERT, M. G. SARNGADHARAN, and A. J. BODNER, THE AIDS SEREOEPIDEMIOLOGY WORKING GROUP, R. C. GALLO, and A. BLATTNER "Screening test for HTLF-III (AIDS agent) antibodies." *Journal of the American Medical Association,* 1985, volume 253, pages 221–225.

WERNIMONT, GRANT "Quality control in the chemical industry, II. Statistical quality control in the chemical laboratory." *Industrial Quality Control,* May 1947.

WILMORE, JACK H. "The assessment of and variation in aerobic power in world class athletes as related to specific sports." *The American Journal of Sports Medicine,* 1984, volume 12, pages 120–127.

WITKIN, HERMAN A., SARNOFF A. MEDNICK, FINI SCHULSINGER, ESKILD BAKKESTROM, KARL O. CHRISTIANSEN, DONALD R. GOODENOUGH, KURT HIRSCHHORN, CLAES LUNDSTEEN, DAVID R. OWEN, JOHN PHILIP, DONALD B. RUBIN, and MARTHA STOCKING "Criminality in XYY and XXY men." *Science,* 1976, volume 193, pages 547–555.

WOELFL, G., I. WEI, C. FAULSTICH, and H. LITWACK "Laboratory testing of asphalt concrete for porous pavements." *Journal of Testing and Evaluation,* 1981, volume 9, pages 175–181.

WOODWARD, W. F. "A comparison of base running methods in baseball." Master's Thesis, Florida State University, 1970.

WOOLSON, R. F. and P. A. LACHENBRUCH "Rank tests for censored matched pairs." *Biometrika,* 1980, volume 67, pages 597–606.

WORLD BANK *World Development Report 1987.* Oxford University Press, New York, 1987.

WORLD BANK *World Development Report 1985.* Oxford University Press, New York, 1985.

YAMAMOTO, JANET K., HERMAN HANSEN, ESTHER W. HO, TERESA Y. MORISHITA, TAKIKO OKUDA, THOMAS R. SAWA, ROBERT M. NAKAMURA, and NIELS C. PEDERSEN "Epidemiologic and clinical aspects of feline immunodeficiency virus infection in cats from the continental United States and Canada and possible mode of transmission." *Journal of the American Veterinary Medical Association,* 1989, volume 194, pages 213–218.

YOUNG, B. A. and J. L. CORBETT "Maintenance energy requirement of grazing sheep in relation to herbal availability." *Australian Journal of Agricultural Research,* 1972, volume 23, pages 57–76.

YU, B. P., E. J. MASORO, I. MURATA, H. A. BERTRAND, and F. T. LYND "Lifespan study of SPF Fisher 344 male rats fed *ad libitum* or restricted diets: Longevity, growth, lean body mass and disease." *Journal of Gerontology,* 1982, volume 37, pages 130–141.

ZERBE, GARY O. "Randomization analysis of the completely randomized design extended to growth and response curves." *Journal of the American Statistical Association,* 1979, volume 74, pages 215–221.

ZINNER, D. D., L. F. DUANY, and N. W. CHILTON "Controlled study of the clinical effectiveness of a new oxygen gel on plaque, oral debris and gingival inflammation." *Pharmacology and Therapeutics in Dentistry,* 1970, volume 1, pages 7–15.

Subject Index

Source Index to Referenced Examples and Exercises

A

B

C

N

O

P

Q

R

S